AF167627

Advances in Fungal Biotechnology for Industry, Agriculture, and Medicine

Advances in Fungal Biotechnology for Industry, Agriculture, and Medicine

Edited by

JAN S. TKACZ
Department of Biologics Research
Merck Research Laboratories
Rahway, New Jersey

and

LENE LANGE
Department of Molecular Biotechnology
Novozymes A/S
Bagsvaerd, Denmark

Springer Science+Business Media, LLC

Library of Congress Cataloging-in-Publication Data

Advances in fungal biotechnology for industry, agriculture, and medicine / edited by
Jan S. Tkacz and Lene Lange.
p. cm.
Includes bibliographical references and index.
ISBN 978-1-4613-4694-4 ISBN 978-1-4419-8859-1 (eBook)
DOI 10.1007/978-1-4419-8859-1
1. Fungi—Biotechnology. I. Tkacz, Jan S. II. Lange, Lene.

TP248.27.F86A36 2004
660.6′2—dc22

2003060293

ISBN 978-1-4613-4694-4

©2004 Springer Science+Business Media New York
Originally published by Kluwer Academic / Plenum Publishers, New York in 2004
Softcover reprint of the hardcover 1st edition 2004

http://www.wkap.nl/

10 9 8 7 6 5 4 3 2 1

A C.I.P. record for this book is available from the Library of Congress

All rights reserved

No part of this book may be reproduced, stored in a retrieval system, or transmitted in any form or by any means, electronic, mechanical, photocopying, microfilming, recording, or otherwise, without written permission from the Publisher, with the exception of any material supplied specifically for the purpose of being entered and executed on a computer system, for exclusive use by the purchaser of the work.

Permissions for books published in Europe: *permissions@wkap.nl*
Permissions for books published in the United States of America: *permissions@wkap.com*

Contributors

Concepcion Azcon-Aguilar
CSIC, Departamento de Microbiologia del Suelo y Sistemas Simbioticos
Estacion Experimental del Zaidin
Profesor Albareda 1
18008 Granada
Spain

Guillaume Bécard
Université Paul Sabatier/UMR 5546 CNRS
Equipe de Mycologie Végétale
Pôle de Biotechnologie Végétale
24 Chemin de Borde-Rouge
BP 17 Auzeville
31326 Castanet-Tolosan
France

Ralf G. Berger
Zentrum Angewandte Chemie
Universität Hannover
Institut für Lebensmittelchemie
Wunstorfer Strasse 14
D-30453 Hannover
Germany

Paola Bonfante
Università degli Studi di Torino
Dipartimento di Biologia Vegetale
Mycology IPP/CNR and Molecular Biology
Viale Mattioli 25
10125 Torino
Italy

Rainer Borriss
Institute of Biology
Humboldt-University Berlin
Chausseestrasse 117
D-10055 Berlin
Germany

Russell J. Cox
School of Chemistry
University of Bristol
Cantock's Close, Clifton
Bristol
BS8 1TS
United Kingdom

Yutaka Ebizuka
Graduate School of Pharmaceutical Sciences
The University of Tokyo
7-3-1 Hongo, Bunkyo-ku
Tokyo 113-0033
Japan

John E. Edwards, Jr.
Division of Infectious Diseases
UCLA School of Medicine
Harbor-UCLA Medical Center
1000W Carson St.
Torrance CA 90502

Nuria Ferrol
CSIC, Departamento de Microbiologia del Suelo y Sistemas Simbioticos
Estacion Experimental del Zaidin
Profesor Albareda 1
18008 Granada
Spain

Philipp Franken
Institute for Vegetable and Ornamental Plants
Theodor-Echtermeyer-Weg 1
D-14979 Grossbeeren
Germany

Isao Fujii
Graduate School of Pharmaceutical Sciences
The University of Tokyo
7-3-1 Hongo, Bunkyo-ku
Tokyo 113-0033
Japan

David M. Geiser
Department of Plant Pathology
The Pennsylvania State University
121 Buckhout Laboratory
University Park PA 16802

Vivienne Gianinazzi-Pearson
UMR 1088 INRA/Université de Bourgogne BBCE-IPM
INRA-CMSE, BP 86510
21065 Dijon cedex
France

Frank Glod
School of Chemistry
University of Bristol
Cantock's Close, Clifton

Bristol, BS8 1TS
United Kingdom

Armelle Gollotte
UMR 1088 INRA/Université de Bourgogne BBCE-IPM
INRA-CMSE, BP 86510
21065 Dijon cedex
France

Heather E. Hallen
Department of Energy Plant Research Laboratory
Michigan State University
East Lansing MI 48824

John E. Hamer
Paradigm Genetics Inc.
P.O. Box 14528
Research Triangle Park NC 27709

Lucy Alexandra Harrier
The Scottish Agricultural College
Biotechnology Department
Plant Science Division
West Mains Road
EH9 3JG Edinburgh
United Kingdom

Paul J. J. Hooykaas
Institute of Molecular Plant Sciences
Clusius Laboratory
Leiden University
Wassenaarseweg 64
2333 AL Leiden
The Netherlands

Ashraf S. Ibrahim
Division of Infectious Diseases
UCLA School of Medicine
1000W Carson St.
Torrance CA 90502

Geoffrey B. Jameson
Institute of Fundamental Sciences
College of Sciences
Massey University
Private Bag 11 222
Palmerston North
New Zealand

Joanna M. Jenkinson
School of Biological Sciences
University of Exeter
Washington Singer Laboratories
Perry Road
Exeter, EX4 4QG
United Kingdom

Luisa Lanfranco
Università degli Studi di Torino
Dipartimento di Biologia Vegetale
Mycology CSMT/CNR and Molecular Biology
Viale Mattioli 25
10125 Torino
Italy

Linda L. Lasure
Pacific Northwest National Laboratory
902 Battelle Blvd.
P.O. Box 999, MSIN: K2-12
Richland WA 99352

Jan Lehmbeck
Novozymes A/S, Krogshoejvej 36
DK-2880, Bagsvaerd
Denmark

Jon K. Magnuson
Pacific Northwest National Laboratory
902 Battelle Blvd.
P.O. Box 999, MSIN: K2-12
Richland WA 99352

Daniel G. Panaccione
Division of Plant & Soil Sciences
West Virginia University
401 Brooks Hall
P.O. Box 6058
Morgantown WV 26506-6058

Emily J. Parker
Institute of Fundamental Sciences
College of Sciences
Massey University
Private Bag 11 222
Palmerston North
New Zealand

Frieder Schauer
Institute of Microbiology
E.-M.-Arndt-University Greifswald
F.-L.-Jahn-Strasse 15
D-17487 Greifswald
Germany

Ulrich Schulte
Institut für Biochemie
Heinrich-Heine-Universität
D-40225 Düsseldorf
Germany

D. Barry Scott
Institute of Molecular BioSciences
College of Sciences,
Massey University
Private Bag 11 222

Palmerston North
New Zealand

Donald C. Sheppard
Division of Infectious Diseases
Harbor-UCLA Research and Education Institute
1000W Carson St.
Torrance CA 90502

Nicholas J. Talbot
School of Biological Sciences
University of Exeter
Washington Singer Laboratories
Perry Road
Exeter, EX4 4QG
United Kingdom

Diederik van Tuinen
UMR 1088 INRA/Université de Bourgogne BBCE-IPM
INRA-CMSE, BP 86510
21065 Dijon cedex
France

Jesper Vind
Novozymes
NOVO Alle 1, 1U.1.20
DK-2880 Bagsvaerd
Denmark

Jonathan D. Walton
Department of Energy Plant Research Laboratory
Michigan State University
East Lansing MI 48824

Akira Watanabe
Structural Biology Laboratory
The Salk Institute for Biological Studies
San Diego CA 92186-5800

Wendy Thompson Yoder
Novozymes Biotech Inc.
1445 Drew Ave.
Davis CA 95616

Holger Zorn
Zentrum Angewandte Chemie
Universität Hannover
Institut für Lebensmittelchemie
Wunstorfer Strasse 14
D-30453 Hannover
Germany

Contents

4. Transformation Mediated by *Agrobacterium tumefaciens*

Paul J. J. Hooykaas

Part II. Special (Secondary) Metabolism

5. Fungal Polyketide Synthases in the Information Age

Russell J. Cox and Frank Glod

Part III. Enzymes and Green Chemistry

9. Heterologous Expression and Protein Secretion in Filamentous Fungi

Wendy Thompson Yoder and Jan Lehmbeck

10. Artificial Evolution of Fungal Proteins

Jesper Vind

11. Biocatalysis and Biotransformation

Frieder Schauer and Rainer Borriss

12. Organic Acid Production by Filamentous Fungi

Jon K. Magnuson and Linda L. Lasure

13. Flavors and Fragrances

Ralf G. Berger and Holger Zorn

Part IV. Host–Fungal Interactions

14. Human Mycoses: The Role of Molecular Biology

Donald C. Sheppard, Ashraf S. Ibrahim, and John E. Edwards Jr.

15. Molecular Interactions of Phytopathogens and Hosts

Joanna M. Jenkinson and Nicholas J. Talbot

16. Structural and Functional Genomics of Symbiotic Arbuscular Mycorrhizal Fungi

V. Gianinazzi-Pearson, C. Azcon-Aguilar, G. Bécard, P. Bonfante, N. Ferrol, P. Franken, A. Gollotte, L.A. Harrier, L. Lanfranco, and D. van Tuinen

Preface

Over the past 15 years, the tools of molecular biology have been successfully adapted for the study of filamentous fungi. Their application has elevated the status of fungal genetics from a fascinating and, at times, truly insightful field of study, but one focused upon only a handful of model organisms, to a discipline spanning the broad spectrum of organisms and topics that engage the interest of mycologists as a whole and biotechnologists in general. Molecular genetics has provided a toolbox of immensely powerful experimental approaches and, additionally, has become a magnet attracting young investigators to formerly intractable or unstudied organisms. It has embraced virtually all groups of economically, medically, and environmentally important fungi and is having a significant impact on commercial bioprocesses.

Fungi, by virtue of their metabolic versatility, ecological diversity, complex life cycles, and essential role in Nature, have attracted the attention of naturalists, biologists, geneticists, ecologists, chemists, and biochemists in myriad ways over many years. Understandably, knowledge of the fungal world developed as areas of specialization that interdigitated with one another loosely, if at all. Molecular biology has put more of the fungal world within experimental reach, but more importantly, it is permitting us to make connections where the parochial views of individual sub-disciplines had prevailed. It is allowing us rapidly to gain at least a rudimentary appreciation of some universal cellular mechanisms at work within the fungal realm. Our aim with this book project was to gather expert contributing authors whose chapters collectively would be more than a snapshot of the biotechnology of filamentous fungi today; we hoped that their chapters would vividly illustrate the unifying force that molecular genetics has become in fungal biology and technology. As the book was readied for publication, we were pleased to see that the authors have, to a large extent, achieved this goal and have gone on to provide glimpses of where our new techniques and insights may allow us to venture in the near future.

As we view fungi in relation to the biological world in its entirety, another contribution of molecular genetics becomes apparent in the compelling evidence that the fungal kingdom is related more closely to animals than to plants. Recently we have been reminded that man's relationship with fungi extends over a very long period of history: the 5000-year-old "Ice Man" found in a receding Tyrolean glacier in 1991 carried mycelia of *Fomes fomentarius* in a leather pouch and tissue of the bracket fungus, *Piptoporus betulinus*, purposefully strung on a leather strap. Yet the modern view of phylogeny points to a biological relatedness between man and fungi extending back over many more millennia. For evidence of this shared heritage we need only consider the magnitude of the challenge inherent in developing non-toxic and selective agents to fight the life-threatening human mycoses that are increasing in prevalence as populations of immunosuppressed patients expand.

For mycology, the analysis of divergence in selected DNA sequences has generally affirmed the conventional, morphologically-based taxonomical classification. However, it has also revealed strikingly surprising relationships and exposed artificial assemblages of anamorphs, as well. Discussions and disputes about various species concepts (taxonomical, biological, phylogenetic, or ecological) are likely to continue for years to come. Very importantly for the applied aspects of mycology, a phylogeny based on DNA divergence provides the framework for the facile identification of new isolates, especially strains known only in their anamorphic state.

Genomic sequencing has been completed for a few filamentous fungi, and some general conclusions are emerging. However, a significant portion of the sequence data has not been widely available, having been generated by commercial entities intent upon utilizing the information in the development of agricultural fungicides or therapeutic agents for human and animal health. Soon, ongoing international academic sequencing initiatives will increase significantly the number of sequenced fungal genomes in the public domain. Gleaning information from this mass of new data will likely require an effort in the next decade approximating that expended acquiring the data over the last 10 years. It will call for the development or improvement of bioinformatics tools and user-friendly platforms that will allow convenient and reliable annotation of the fungal open reading frames (ORFs) and offer the possibility of conducting comparative genomic inquiries more expediently. To verify provisional annotations and deepen the understanding of given ORFs, techniques for functional genomic analyses that can be applied globally, or at least in a high throughput mode, will need to be implemented. The point of departure for this book is a series of chapters on modern fungal systematics, genomics, and genetic tools for functional analysis. Grouped under the heading of Genetic Technology, they set the stage for the subsequent chapters that are organized into three additional groups.

The two sections at the heart of the book are devoted to the economic impacts of fungi. (Though yeasts are, of course, fungi and have a long history of use in biotechnology, we asked the chapter authors to focus specifically upon the filamentous fungi, rather than their well-studied cousins.) Today a wide range of industries make use of filamentous fungi or their products. Fungal organisms are also increasingly employed for waste treatment and bioremediation. Biotechnology can trace its historical roots to the use of microorganisms in the production and processing of beverages and foods, and the value of filamentous fungi in some of these processes was undoubtedly recognized in antiquity. Modern-day palates are quite familiar with the flavors, textures, and metabolites provided by filamentous fungi in cheeses, oriental sauces, and other food products. The submerged fermentation processes developed near the middle of the past century for citric acid and penicillin production by filamentous fungi are still paradigms for the production aspects of modern biotechnology. In recent decades the commercial value of filamentous fungi has grown as they have become important sources of pharmaceutical agents for the treatment of infectious and metabolic diseases and of a broad range of enzymes used to process foods and animal feeds, fortify detergents, treat textiles and wood pulp by-products, and perform highly specific biotransformations. Today the commercial goods produced by industries that make use of fungi represent a segment of the world economy valued in billions of dollars annually.

Despite the variety of ways that fungi provide economic benefits to mankind, their most significant economic impact is measured in terms of losses, especially agricultural losses. Molds are widespread pathogens of crop plants, they are agents of food spoilage and deterioration, and they are responsible for the contamination of grains with toxic and, in some instances, carcinogenic compounds. Although we humans tend to view their activities as favorable or damaging in economic terms, for the fungus there is no dichotomy. The ability to make a special metabolite that is not part of general metabolism, whether the metabolite is a potent carcinogen or an inhibitor of sterol synthesis that can be beneficial in the management of serum cholesterol levels, is for the fungus simply an evolutionary adaptation, perhaps of utility in exploiting and inhabiting a specific natural niche. Similarly enzymes that enable fungi to play their crucial role as Nature's recyclers are also the ones that degrade our wooden structures or that can be used to improve the nutritional value of animal feed. In the discussion of genomic structure in Chapter 2, one is struck with the realization that genes found uniquely in fungi are, for the most part, the very ones that mediate processes with economic consequences.

Thus, the chapters related to special metabolites (traditionally known as secondary metabolites) that comprise the second section of the book, treat their subjects in biosynthetic terms rather than from the perspective of positive or negative consequences in human affairs. The three major classes of fungal special metabolites are considered: polyketides derived from acetate equivalents, non-ribosomal peptides made from amino and hydroxy acids, and isoprenoids, also known as terpenes, that arise from isoprenyl units. The enzymes necessary to produce members of each class are encoded by genes that are carried in the genome principally as clusters, and at least for polyketides, it is clear that one organism may retain several clusters for different products. Considering the significant proportion of genomic DNA devoted to these clusters as well as the conservation of functional clusters in wild strains and in unrelated members of the fungal kingdom, one can clearly appreciate just how misleading it is to consider special metabolites as *secondary* or *dispensable*.

The theme of evolutionary adaptation continues in the third section of the book that is devoted to enzymes produced by fungi. Here the metabolic versatility of fungi is in the limelight as the chapters explore the wide-ranging uses that have been found for fungal enzymes. Through molecular technology, the selection of proteins produced commercially in fungi has been broadened to include enzymes of plants, animals, and humans, and strategies have been developed to tailor the characteristics of an enzyme to meet the needs of a specific industrial or agricultural process. Also to be found in this section are examples of fungal cells exploited as biocatalysts to transform synthetic organic substrates selectively and stereospecifically or to convert a common or inexpensive precursor into a product with commercial value. Interest in biocatalysis is high in the fine chemicals industry that is faced with the need to minimize production and energy costs and any negative environmental consequences of its large-scale processes. Biocatalysis is also receiving increasing attention for its potential in environmental remediation. The chapters in this third section are grouped under the heading of Enzymes and Green Chemistry.

The last section of the book, with three chapters under the heading Host–Fungal Interactions, returns the perspective to the level of the organism to illustrate how the unique capabilities of fungi, discussed in genomic, enzymatic, and chemical terms in earlier portions of the book, translate into the success of the whole organism, as a pathogen or symbiont.

At the conclusion the chapters leave us standing on the threshold of an era of "functional genomics" which promises to provide an appreciation of the complexity of cellular processes that could not have been attained previously. The limitations of linearity in conceptualization and experimental approaches, exemplified by the 'one gene, one protein, one function' mind-set, are giving way as networks of metabolic and regulatory functions are revealed. Here biotechnology may have an opportunity to repay a portion of its debt to basic science through its well-developed mathematics and algorithms for principal component analysis and multivariable analyses that are widely used in the development of industrial fermentation processes. We may also expect the next decade to bring a wider appreciation of the regulatory implications and metabolic consequences (metabolite channeling) of protein/protein interactions. Multiple functions for a single gene product are likely to come to light. Already today a variety of new fluorescent probes and imaging techniques make it possible to determine the subcellular location of a specific protein throughout the division cycle, and these methods should become a standard part of the characterization of a gene product in the future.

For fungal biotechnology, new doors are opening allowing consideration of strategies not previously feasible. Gene homologs can be sought in a collection of organisms to identify species capable of more robust production of the gene product or to provide DNA sequence information for the design of a 'consensus' protein (see Chapter 10). Segments of gene homologs from several organisms can be shuffled to identify products with improved characteristics. Although information of this kind will increasingly be available by comparative genomics, it can also be obtained with DNA extracted from a natural habitat, without ever isolating the organisms present in that habitat. It may become possible to circumvent the bottleneck that the random mutation approach has represented in strain improvement and process development by utilizing rational strategies. For example, a cluster of genes or, perhaps, a key gene from a cluster, may be moved into an organism that already has favorable characteristics for growth in large-scale fermentor tanks or has already proven amenable to the efficient production or secretion of the type of product desired. As an alternative or adjunct, genes derived from global signaling cascades and regulatory circuits bearing mutations making their products functionally dominant can be introduced into the producing organism to deregulate metabolite synthesis. Deregulation via this approach could also prove useful in assessing the full potential of a new isolate to produce bioactive compounds. For fungal strains that have already been engineered to improve the efficiency of protein folding and secretion, it may soon prove feasible to alter their glycosylation pathways to produce mammalian-type oligosaccharides, a development which could make fungal fermentation a realistic alternative to mammalian cell culture for the production of therapeutic glycoproteins such as antibodies. With new possibilities such as these, fungal biotechnology will surely continue to flourish.

Thirty-nine authors from laboratories around the world prepared chapters for this book, and we conclude this introduction by expressing our gratitude to them for the time and effort they generously contributed. Our hope would be that their texts will help attract, train, and inspire a new generation of scientists who will advance the field of mycology that we find a continual source of fascination.

Jan S. Tkacz
Lene Lange

I

Genetic Technology

1

Practical Molecular Taxonomy of Fungi

David M. Geiser

1. Introduction

With fungi increasingly used in industry as sources of bioactive compounds and as agents to produce such compounds, it is becoming more important that fungal taxonomy be a practical discipline able to convey as much biological meaning as possible. An evolutionary approach to define fungal taxa is preferred because it is objective and provides outstanding predictive value. Traditional morphological approaches to fungal systematics are problematic owing to a lack of characters useful for grouping, and they frequently fail to provide a solid evolutionary framework, particularly at the species level. However, in the past decade new molecular and analytical tools have been developed that allow robust evolutionary inferences that can be used to define and identify fungal taxa. One or two gene sequences can be used to identify unknown fungi, often to the species level, with increasing accuracy. At the species level, a multigene phylogenetic approach is advocated to define and recognize fungal species. With this approach, species boundaries are perceived where different genes share evolutionary partitions, which serve as evidence for a lack of genetic exchange among reproductively isolated taxa. In this chapter, I will outline molecular phylogenetic approaches for addressing specific questions that arise in fungal biotechnology, from the species level upward.

2. Identifying a Fungus to Species—What Does It Mean?

2.1. Useful Species Definitions

The desired outcome of a fungal identification is usually a species name. Currently, approximately 80,000 species of true Fungi are formally recognized (Kirk et al., 2001), approximately 60% of which are members of the Phylum Ascomycota and their asexual relatives, a group that includes the most important industrial fungi. There is little doubt that the vast majority of fungal species remain undiscovered, with estimates that only 0.7–15% of actual species have been formally recognized (Hawksworth, 2001). Of course, there are

---

David M. Geiser • Department of Plant Pathology, The Pennsylvania State University, 121 Buckhout Laboratory, University Park, Pennsylvania.

Advances in Fungal Biotechnology for Industry, Agriculture, and Medicine. Edited by Jan S. Tkacz and Lene Lange, Kluwer Academic/Plenum Publishers, 2004.

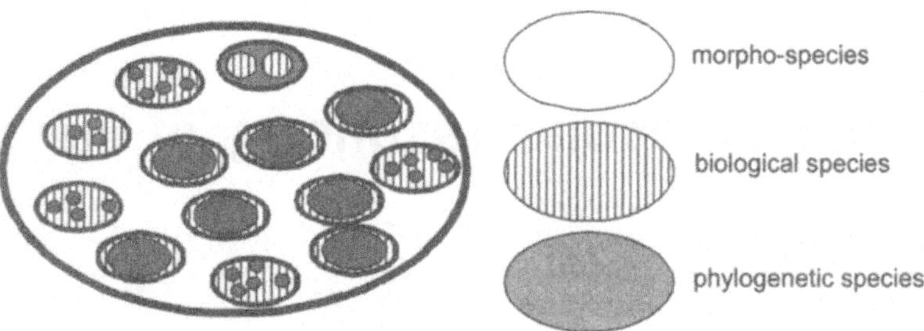

Figure 1.1. A schematic representation of expected relationships between fungal species defined on the basis of morphology (morpho-species), mating (biological species) and phylogenetics (phylogenetic species). Fungal morpho-species tend to be broadly defined and encompass multiple biological and phylogenetic species. There is often a very good one-to-one correspondence between biological species and phylogenetic species, but multiple phylogenetic species may be defined within a biological species, and multiple biological species could theoretically exist within a phylogenetic species.

different ways of defining fungal species, and different definitions provide different answers (Figure 1.1.). Virtually all fungal species have been defined through a morphological species concept, where a species is recognized as a group of isolates that share a set of morphological characters distinguishing it from other so-defined groups ("morpho-species"). Other approaches include the biological species concept, which uses mating compatibility as the criterion for defining species (Mayr, 1940), and the phylogenetic species concept, which defines species as the smallest diagnosable phylogenetic unit (Cracraft, 1983); these definitions tend to recognize multiple fungal species ("cryptic species") within species defined morphologically. The biological and phylogenetic species concepts tend to be correlated, although not perfectly. The disjunction among different concepts strongly suggests that we currently lack a good species-level framework in fungal taxonomy and that careful approaches need to be undertaken to provide a system that is truly useful.

Species concepts have been a subject of debate for a long time, and it has been argued that such debates are bound to remain unresolved (Hey, 2001). However, in practice, some concept of species is necessary so that we can go about the business of working with fungi. For example, patent claims often cover the use of a particular isolate or group of isolates of a particular fungal species. Obviously, any patent claim covering a poorly defined species is vulnerable to challenge. Similarly, "Generally Regarded As Safe" (GRAS) status may be placed on a particular fungal strain based on its presumed species identity. The Latin binomial applied to a fungus is expected to convey a tremendous amount of information about the organism's biology, including the toxins it may produce, the useful compounds it may produce, and the diseases it may cause, attributes that have important implications regarding safety, patent rights, and quarantine. This is, of course, too much to ask from two Latin words, but it is clear that new phylogenetic approaches to defining and recognizing species will help.

2.2. Genealogical Concordance as a Means to Recognize Fungal Species

Most taxonomists advocate an evolutionary perspective for recognizing fungal species because of its predictive value. By identifying species as clearly sharing a common ancestry and a history of reproductive exchange, we can predict that individuals within such species will share many important biological characteristics, from ecology to metabolism. Both mating and phylogenetic criteria for species recognition fall under this evolutionary heading, and the two criteria are, in some ways, two sides of the same coin. Taylor and his colleagues (Taylor *et al.*, 2000) advocated a multigene phylogenetic approach for recognizing fungal species called Genealogical Concordance Phylogenetic Species Recognition (GCPSR). Within a species, different unlinked genes are expected to have different evolutionary histories because recombination and chromosomal re-assortment occur in sexual reproduction. At the same time, there is a lack of genetic exchange between species, so the genealogies of the same genes are expected to share a common history at this level. Species boundaries are recognized at the transition point between concordance and non-concordance of gene genealogies (Figure 1.2.). Of course, the application of the genealogical concordance criterion to truly asexual species is trivial; in these organisms, all genes act as a single linkage group because the entire genome is inherited intact. Therefore, in the case of a truly asexual fungus, GCPSR would identify any two clones as separate species. Approximately 40% of described Fungi are known only as asexual stages, including almost all industrially important fungi. However, where the null hypothesis of recombination has actually been tested in these fungi, evidence for some sort of historical recombination has been identified (Taylor *et al.*, 1999). Therefore, it is probably best to assume that any apparently asexual fungus undergoes recombination on occasion, until proven otherwise. At the same time, it is possible that GCSPR may identify highly differentiated populations: groups of isolates that have a long history of not interbreeding, perhaps due to geographic or ecological separation, but retain the potential to interbreed (Leslie *et al.*, 2001). In these theoretical cases, some may argue that GCPSR splits to too great an extent. Whether the approach is identifying lineages that deserve official taxonomic recognition as a species is not very important in a practical sense because GCPSR recognizes an actual lack of interbreeding. Even if they are not formally described as species, phylogenetic species recognized

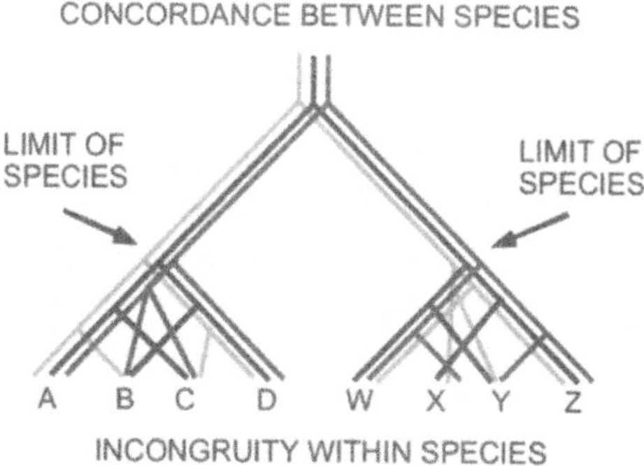

Figure 1.2. A schematic representation of how multiple gene genealogies are used in genealogical concordance phylogenetic species recognition (GCPSR). The genealogies of three genes (three different shades) are inferred in eight individuals (A–D, W–Z). All three genes are concordant in showing the separation of individuals A–D from W–Z, reflecting the lack of reproductive exchange between these two inferred species. At the same time, sexual reproduction within species scrambles the genes among individuals, such that the genealogies are non-concordant. A species boundary is inferred at the boundary between concordance and non-concordance. Reprinted from Taylor *et al.* (2000) with permission from Elsevier Science.

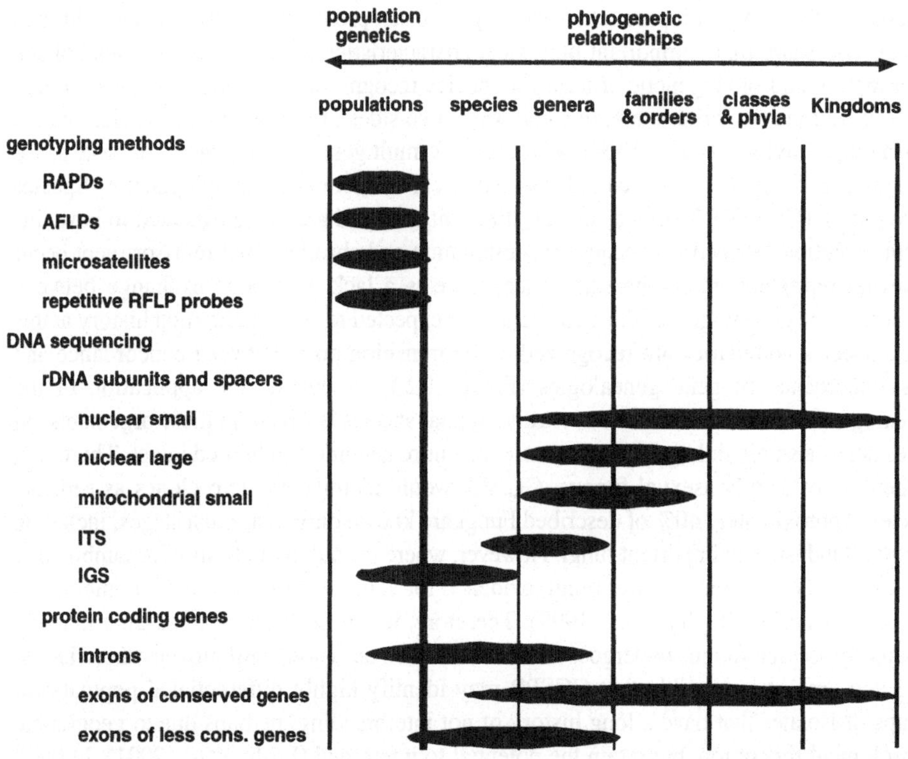

Figure 1.3. Predicted utility of various molecular tools for characterizing and identifying fungi at different taxonomic levels. This figure is adapted from Figure 6 of Bruns *et al.* (1991).

by GCPSR are useful categorical tools because they may reflect important attributes such as geographic origin, biochemical characters, and pathogenicity.

2.3. Molecular Taxonomy in Practice

Different molecular tools provide different levels of resolution for exploring taxonomic questions, and no single tool will serve perfectly for any given set of taxa. The taxonomic utility of various molecular tools is summarized in Figure 1.3. GCPSR requires the use of multiple (two or more) gene genealogies to define species, but a single gene sequence is often useful to provide a "quick and dirty" species identification. An important consideration is that GCPSR is not based solely on divergence for the diagnosis of species, and one cannot interpret a certain number of nucleotide differences as corresponding to a species difference.

3. How do I Identify an Unknown Fungus?

In asking this question, researchers are usually coming from one of three directions: (1) they have a vague idea of what type of fungus they have (e.g., an asexual fungus, probably a member of the ascomycotan lineage) and want to know its identity at the ordinal, familial, generic or species level; (2) they are fairly confident that they know what genus they have and want an exact species identification; or (3) they know what species they have but would like to know what genetic individual they have within that species. There is no set of methods that will always work perfectly for answering any of these questions, so some thought and creativity is generally required to get the answer one really wants. Overall it is just as important to consider what biological question one has in mind (e.g., what toxins might this fungus produce or what diseases might it cause), as it is to consider the taxonomic question.

In many cases, a cursory morphological analysis may provide a suitable identification. However, as molecular tools are becoming easier to use, less expensive, and more useful, they are also becoming far more attractive options. In the following sections, I will outline some of the general molecular approaches one might use to answer these different questions from different perspectives.

3.1. The Molecular Toolbox

3.1.1. DNA Sequence Tools

Tools for Higher Taxonomic Levels. The advent of molecular phylogenetic approaches based upon the polymerase chain reaction (PCR) in the late 1980s marked the beginning of a revolution in the way we understand the relationships among fungi and in our ability to identify unknown isolates. The genes encoding nuclear and mitochondrial ribosomal RNA genes and associated spacer regions were quickly identified as ideal candidates for analysis. Based on the universal and conserved nature of these genes, primers could be designed that were applicable to a broad range of fungi and that, in many cases, were useful in plants and animals as well as in fungi (White *et al.*, 1990). The fact that the nuclear ribosomal genes are arranged in a long tandem repeat turned out to be another useful characteristic, as any region of interest is already largely amplified in the genome (Figure 1.4A). An early review (Bruns *et al.*, 1991) summarized the known and predicted

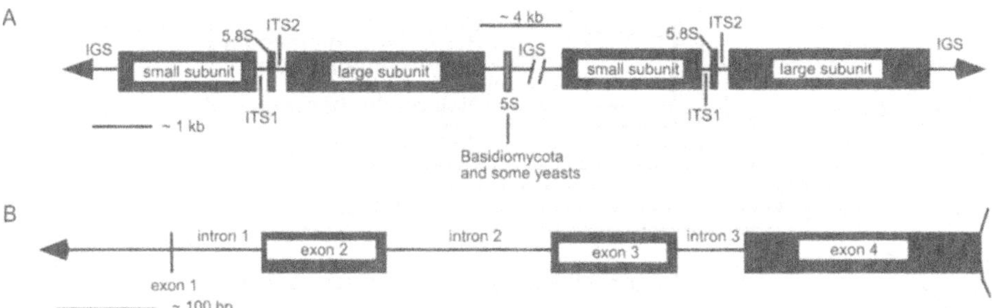

Figure 1.4. A. Map of the nuclear rRNA gene repeat in fungi. A 5S subunit gene is inserted in the intergenic spacer (IGS) region of the Basiciomycota and some Ascomycotan yeasts. The cluster of ribosomal genes is tandemly repeated in the genome. B. Map of the intron-rich 5′ end of the translation elongation factor 1-alpha gene from *Neurospora crassa*, a typical intron-rich, evolutionarily conserved protein-coding gene useful for species-level phylogenetics. Intron number and position varies among fungal groups. See the Deep Hypha web site (*http://ocid.nacse.org/research/deephyphae/index.php?id=research_tools*) for links to primer locations and sequences.

utility of various molecular approaches to characterize fungi at different taxonomic levels. In particular, the authors noted that the nuclear small ribosomal RNA subunit gene, often referred to simply as the nuclear small subunit or "18S" gene, was a particularly useful molecule for characterizing a wide variety of taxonomic levels, from the Kingdom level down to genera. They also pointed to the internal transcribed spacer ("ITS") regions of the nuclear ribosomal repeat as potentially useful regions for looking at closely related species. These two spacer regions are transcribed, but they do not encode a gene product, so they evolve at a faster rate than the ribosomal subunit genes. Other genes, such as the nuclear large subunit ("28S" or sometimes "25S"), and the mitochondrial small and large subunit genes, have proven utility at intermediate taxonomic levels.

The nuclear small ribosomal subunit gene is currently our keystone tool for understanding broad fungal relationships. A phylogenetic tree of the major fungal groups based on the sequences of this gene is provided in Figure 1.5. Many of these relationships have been corroborated by data from other genes, including protein-coding genes such as RNA polymerase II subunit *rpb2* (Liu *et al.*, 1999), mitochondrial *atp-6* gene (Kretzer and

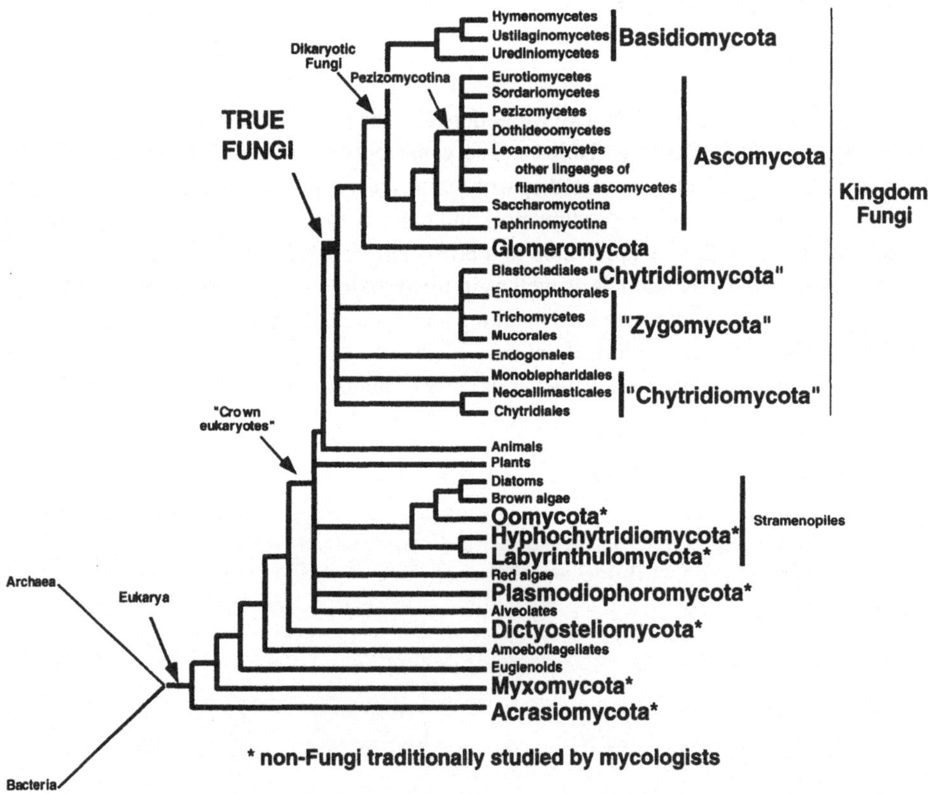

Figure 1.5. Known phylogenetic relationships among the major groups of true Fungi and other lineages of organisms traditionally studied by mycologists (*) (Alexopoulos, Blackwell, & Mims, 1996; Eriksson *et al.*, 2001; James, Porter, Leander, Vilgalys,& Longcore, 2000). Currently, five phyla are recognized within the Kingdom Fungi: Ascomycota, Basidiomycota, Glomeromycota, Zygomycota and Chytridiomycota. Glomeromycota was recently separated from Zygomycota (Schussler, Schwarzott, & Walker, 2001). As commonly defined, neither Zygomycota nor Chytridiomycota comprise natural evolutionary groups, and thus await formal taxonomic revision.

Bruns, 1999), and translation elongation factor 1-alpha (Stephen Rehner, unpublished), and the production of extensive databases of these gene regions is in progress. These new tools are expected to help resolve major portions of the fungal tree where the ribosomal small subunit gene has not been completely informative, particularly the major lineages of the filamentous Ascomycota, or Pezizomycotina (Eriksson *et al.*, 2001), the group that includes most fungi of industrial and biotechnological interest.

Tools for the Species Level. In addition to the *rpb2, atp-6*, and elongation factor sequences which are most useful for characterizing fungi at higher taxonomic levels, a wide variety of other protein-coding genes have proven useful at the species level. Since the mid-1990s, work with a number of fungi (Tsai *et al.*, 1994; Koufopanou *et al.*, 1997; O'Donnell and Cigelnik, 1997) has led to a number of very important findings: (1) intron-rich protein-coding genes, particularly highly conserved genes such as translation elongation factor 1-alpha, beta-tubulin, and histone genes, provide a more robust and useful phylogenetic signal at the species level than the ITS regions of the nuclear ribosomal repeat; (2) cryptic phylogenetic species are frequently identified within morphologically defined species; and (3) the most popular tool for characterizing fungi at the species level, the ITS regions, fails to resolve very closely related phylogenetic species (Koufopanou *et al.*, 1997; Geiser *et al.*, 1998; O'Donnell *et al.*, 1998a; O'Donnell *et al.*, 1998b; O'Donnell, 2000; O'Donnell *et al.*, 2000). The higher resolution of these protein-coding gene regions is due to the presence of introns, which evolve at a rate even higher than that of the ITS regions. PCR is performed from conserved exon sites flanking intron-rich regions, producing a small (usually 500–800 bp) region that may contain up to three introns comprising more than half of the amplicon. Such an amplicon may provide a phylogenetic signal with 3–6 times more information than that of ITS in a given study (Figure 1.4.B). The failure of ITS to resolve completely the most closely related species, however, should not discourage its use as an identification tool. Because of its universality, ease of use, and the breadth of data available, ITS remains our best and most reliable tool for identifying any given fungus to the species level (Bruns, 2001). Other genes will be more useful for basic identification only in genera where a great deal of groundwork has already been established, such as in the genus *Fusarium*, where extensive databases of elongation factor 1-alpha sequences make it the tool of choice (see later section). The exon portions of these genes are informative at higher taxonomic levels, based on either DNA or amino acid sequences.

Below the Species Level. The intron-rich coding regions of genes may also provide extensive levels of intraspecific variation that are useful for analyses at the intraspecific level, including making indirect inferences as to whether a fungus is sexual or asexual in nature (Koufopanou *et al.*, 1997; Geiser *et al.*, 1998; Carbone and Kohn, 2001). Carbone and Kohn (2001) utilized the intergenic spacer (IGS) region of the nuclear ribosomal gene repeat, which shows a great deal of sequence variation over its 3–4 kb length, to analyze population dynamics in the plant pathogenic fungus *Sclerotinia sclerotiorum*. This gene region should be used with caution, however, as intra-molecular recombination could make interpretation difficult in some cases.

3.1.2. Genotyping Methods: Comparing and Identifying Isolates within a Species

A variety of genotyping methods are now available that provide unique band patterns useful for identifying and comparing genetic individuals within a species. These methods involve the electrophoretic separation of DNA bands generated via PCR or restriction digestion, and comparing patterns based on mobility (i.e., size) of the DNA fragments. In terms of isolate identification, these patterns can be thought of as simple "bar codes" that can be the basis for a match. Prior to the availability of PCR, most genotyping studies in fungi relied either on electrophoretic isoenzyme analysis, or hybridization of restriction-digested genomic DNA with probes corresponding to single-copy or repeated elements, producing highly variable restriction fragment length polymorphism (RFLP) patterns. RFLP analyses provide excellent data, but also require considerable groundwork to identify useful probes, and the analyses themselves are rather labor-intensive (Boeger *et al.*, 1993; Geiser *et al.*, 1994; Kohli *et al.*, 1995).

Soon after the advent of PCR, randomly amplified polymorphic DNA (RAPD analysis) became available as the first PCR-based genotyping method. This method is universally applicable, requires virtually no groundwork, and needs only a single PCR amplification step, and in theory the same PCR primers can be useful for any species. The reliability and robustness of this method is often questioned because the presence or absence of bands is based on unknown hybridization and amplification properties of particular primers. However, with conservative treatment and care, RAPD analysis can be a very powerful tool because of its ease of use: one can generate RAPD patterns on a set of fungi in a single day.

Following the development of RAPD analysis, a technique emerged that combined the reliability of restriction enzyme analysis and the power of PCR. It was known as amplification fragment length polymorphism (AFLP) analysis (Vos *et al.*, 1995). This method is a bit more labor intensive than its predecessor, requiring a restriction digestion/linker ligation step in addition to one or two rounds of PCR amplification, but it is less cumbersome than RFLP analysis. Like RAPD, AFLP relies on universally applicable primers and requires relatively little groundwork. A single set of AFLP primers may provide dozens of bands that are polymorphic within a species, and the technique is highly repeatable.

Finally, microsatellite markers provide an extremely powerful population genetics tool, revealing a high level of allelic variation within a known genetic locus (Rassmann *et al.*, 1991). A profile of an individual isolate's genotype at several microsatellite markers provides identification at very high-resolution, similar to that employed in human DNA typing based on analysis of variable nucleotide tandem repeats (VNTR). Microsatellites are similar to RFLPs in that considerable work is generally required to identify useful microsatellite loci in a particular fungus. However, this work can be greatly reduced when a full genome sequence is available. While the isolation of a microsatellite locus usually requires the isolation of clones containing a particular di-, tri- or tetranucleotide microsatellite motif from a genomic DNA library, a search feature in a word-processing program can be used to identify microsatellites from a text file corresponding to a full genome sequence (J. Dettman, J. Platt, and J. Taylor, personal communication). PCR primers can be designed based on sequences flanking the microsatellite repeat, and the characterization of many such loci could provide the highest level of identification resolution possible.

One problem with genotyping methods is that different laboratories use different tools, and we generally lack comprehensive databases of genotypic data for fungi. A notable exception is the rice blast fungus *Magnaporthe grisea*, hundreds of isolates of which have been analyzed using the highly variable MGR-fingerprint probes identified in that species (Hamer, 1991). While identifying a fingerprint match is not as simple as submitting a sequence for a BLAST search, it is possible to identify a given isolate by comparing its MGR RFLP fingerprint with those of known isolates. One of the major challenges facing mycologists in the coming years will be the development of useful databases of such genotypic data from a wide variety of fungi (Kang *et al.*, 2002).

While genotyping methods are successful in distinguishing individuals within a species, they are generally not as useful for inter-specific comparisons. Within a genus, there may be little or no basis for comparing the genotypes of anything but the most closely related species, because few bands are actually shared. As the distance of the relationship between two isolates increases, the chances that two bands of equal size not actually representing the same DNA fragment also increases.

3.2. Using the Toolbox

3.2.1. Tools for Identifying any Fungus

Context is key in identifying an unknown fungus, and the context for performing a molecular identification is a comprehensive database of sequences from all major groups of fungi. Since the advent of fungal molecular systematics, the GenBank database has accumulated fungal ribosomal sequences at an accelerating rate, particularly for the 18S and ITS regions. Now it is often possible to identify a fungus reasonably well simply by amplifying and sequencing one or both of these regions and performing a BLAST search of the GenBank database. Primers and protocols for using these tools are available in the links provided at the Deep Hypha web site (*http://ocid.nacse.org/research/deephyphae/*). An exact 18S match strongly suggests a match at the generic level, while an exact ITS match strongly suggests a match at the species level. While one cannot always draw firm conclusions from such an analysis, this is often enough information to guide further exploration. For example, an exact ITS match provides a reasonably strong inference of species identity, but members of different species diagnosed by GCPSR may have identical ITS sequences. Likewise, some degree of ITS variation may exist within a well-defined species, so one or a few nucleotide differences are impossible to interpret precisely without more data from independent genes.

A wealth of data regarding the relationships of fungi has been assembled and is available to all researchers, thanks to the efforts of many individual research groups combined with the coordinating power that the GenBank database has provided. Additionally, the TREEBASE database (*www.treebase.org*) has served as a depository for DNA sequence alignments from published phylogenetic studies. More recently, coordinated research networks and collaborative groups such as the US Deep Hypha and Assembling the Fungal Tree of Life initiatives, The Fungal Web initiative, and other international efforts, have coordinated research with the goals of providing comprehensive databases and complete analyses of fungal phylogeny. It is expected that within 3–5 years sequences from as many as seven different gene regions will be available for as many as 1500 species of fungi representing all of the major groups, with universal internet access to the data.

3.2.2. Tools for Identifying Fungi in a Particular Taxonomic Group of Intensive Study

What species is this? The most commonly used tool for identifying fungi to species is the ITS region of the nuclear ribosomal RNA gene repeat. Fungal genomes harbor hundreds of copies of ribosomal RNA genes in a tandem repeat, and separating these genes are two spacer regions, two ITS, with an intervening 5.8S ribosomal subunit. The ITS region is generally approximately 500–1000 bp in length, and usually can be amplified as a single PCR amplicon and sequenced completely from its two ends. The PCR-primers generally used for ITS analysis are highly conserved and will work for practically any fungus (White *et al.*, 1990). Consequently they provide a particularly attractive option for characterizing unknown fungi. Also, ITS sequences are very well-represented in sequence databases, so one can often identify a fungus quite accurately simply by sequencing the ITS region, performing a BLAST search of public databases, and finding an exact match. However, it is known that species recognized clearly using GCPSR may have identical ITS sequences, and there is sometimes ITS variation within species recognized using GCPSR. Restriction digestion of the ITS amplicon using enzymes that recognize four-base sites followed by comparisons to databases of known restriction profiles provides an easy and less expensive means of screening large numbers of unknowns (Horton and Bruns, 2001).

While ITS is the tool of choice for performing most routine fungal molecular identifications, some genera of intensive interest have superior tools available. Communities of researchers interested in particularly important genera such as *Fusarium* and *Trichoderma* have developed robust multi-gene phylogenies to identify species lineages (see next section). Because ribosomal genes, including the ITS regions, do not resolve phylogenetic species very well, large databases of multiple intron-rich protein coding genes are being developed that provide a very high level of resolution. In the genus *Fusarium*, elongation factor sequences are currently available for most of the major species, and these collectively form the tool of choice for performing molecular identifications. For example, not only can elongation factor sequences be used to identify the species *Fusarium graminearum*, they can also be used to identify the affiliation of an isolate with one of the eight known phylogenetic lineages that exist within this morpho-species, lineages that correlate with biogeographic origins (O'Donnell *et al.*, 2000). Similar structure is revealed within the biocontrol fungus, *Trichoderma harzianum*, which shows similar biogeographic phylogenetic structure within a morpho-species (Chaverri *et al.*, 2003).

4. What is Next?

At this time, fungal researchers have a variety of useful tools at their disposal for identifying and characterizing fungi and are well on their way to having comprehensive databases of molecular information on all groups of fungi. What is lacking now is established high-throughput infrastructure for characterizing large numbers of isolates quickly and reliably. Gene chip technology combined with real-time molecular methods should provide that, and one can expect the development of such systems in the near future.

References

Alexopoulos, C.J., Blackwell, M., and Mims, C.W. (1996). *Introductory Mycology*. New York: Wiley & Sons.

Boeger, J.M., Chen, R.S., and McDonald, B.A. (1993). Gene flow between geographic populations of *Mycosphaerella graminicola* (Anamorph *Septoria tritici*)—Detected with Restriction-Fragment-Length-Polymorphism Markers. *Phytopathology* **83**, 1148–1154.

Bruns, T.D. (2001). ITS reality. *Inoculum* **52**, 2–3.

Bruns, T.D., White, T.J., and Taylor, J.W. (1991). Fungal molecular systematics. *Annu. Rev. Ecol. Syst.* **22**, 525–564.

Carbone, I., and Kohn, L.M. (2001). A microbial population-species interface: Nested cladistic and coalescent inference with multilocus data. *Molecular Ecology* **10**, 947–964.

Chaverri, P.C., Castlebury, L.A., Samuels, G.J., and Geiser, D.M. (2003). Multilocus phylogenetic structure within the *Trichoderma harzianum/Hypocrea lixii* complex. *Mol. Phyl. Evol.* **27**, 302–313.

Cracraft, J. (1983). Species concepts and speciation analysis. *Current Ornithology* **1**, 159–187.

Eriksson, O.E., Baral, H.-O., Currah, R.S., Hansen, K., Kurtzmann, C.P., Rambold, G., and Laessøe, T. (2001). Outline of Ascomycota—2001. *Myconet* **7**, 1–88.

Geiser, D.M., Arnold, M.L., and Timberlake, W.E. (1994). Sexual origins of British *Aspergillus nidulans* isolates. *Proc. Natl. Acad. Sci. U.S.A.* **91**, 2349–2352.

Geiser, D.M., Pitt, J.I., and Taylor, J.W. (1998). Cryptic speciation and recombination in the aflatoxin-producing fungus *Aspergillus flavus. Proc. Natl. Acad. Sci. U.S.A.* **95**, 388–393.

Hamer, J.E. (1991). Molecular probes for rice blast disease. *Science* **252**, 632–633.

Hawksworth, D.L. (2001). The magnitude of fungal diversity: The 1.5 million species estimate revisited. *Mycol. Res.* **105**, 1422–1432.

Hey, J. (2001). *Genes, categories and species: The evolutionary and cognitive causes of the species problem.* Oxford, UK: Oxford University Press.

Horton, T.R., and Bruns, T.D. (2001). The molecular revolution in ectomycorrhizal ecology: Peeking into the black-box. *Molec. Ecol.* **10**, 1855–1871.

James, T.Y., Porter, D., Leander, C.A., Vilgalys, R., and Longcore, J.E. (2000). Molecular phylogenetics of the Chytridiomycota supports the utility of ultrastructural data in chytrid systematics. *Can. J. Bot. -Revue Canadienne De Botanique* **78**, 336–350.

Kang, S.C., Ayers, J.E., DeWolf, E.D., Geiser, D.M., Kuldau, G., Moorman, G.W., Mullins, E., Uddin, W., Correll, J.C., Deckert, G., Lee, Y.H., Lee, Y.W., Martin, F.N., and Subbarao, K. (2002). The internet-based fungal pathogen database: A proposed model. *Phytopathology* **92**, 232–236.

Kirk, P.M., David, J.C., and Stalpers, J.A. (2001). *Ainsworth & Bisby's Dictionary of the Fungi* (9th ed.). Wallingford, UK: CAB International.

Kohli, Y., Brunner, L.J., Yoell, H., Milgroom, M.G., Anderson, J.B., Morrall, R.A.A., and Kohn, L.M. (1995). Clonal dispersal and spatial mixing in populations of the plant-pathogenic fungus, *Sclerotinia sclerotiorum. Molec. Ecol.* **4**, 69–77.

Koufopanou, V., Burt, A., and Taylor, J.W. (1997). Concordance of gene genealogies reveals reproductive isolation in the pathogenic fungus *Coccidioides immitis. Proc. Natl. Acad. Sci. U.S.A.* **94**, 5478–5482.

Kretzer, A.M., and Bruns, T.D. (1999). Use of *atp6* in fungal phylogenetics: An example from the Boletales. *Mol. Phylogenet. Evol.* **13**, 483–492.

Leslie, J.F., Zeller, K.A., and Summerell, B.A. (2001). Icebergs and species in populations of Fusarium. *Phys. Mo. Plant Path.* **59**, 107–117.

Liu, Y.J.J., Whelen, S., and Benjamin, D.H. (1999). Phylogenetic relationships among ascomycetes: Evidence from an RNA polymerase II subunit. *Mol. Biol. Evolution* **16**, 1799–1808.

Mayr, E. (1940). Speciation phenomena in birds. *Amer. Naturalist* **74**, 249–278.

O'Donnell, K. (2000). Molecular phylogeny of the *Nectria haematococca-Fusarium solani* species complex. *Mycologia* **92**, 919–938.

O'Donnell, K., and Cigelnik, E. (1997). Two divergent intragenomic rDNA ITS2 types within a monophyletic lineage of the fungus *Fusarium* are nonorthologous. *Mol. Phylogenet. Evol.* **7**, 103–116.

O'Donnell, K., Cigelnik, E., and Nirenberg, H.I. (1998a). Molecular systematics and phylogeography of the *Gibberella fujikuroi* species complex. *Mycologia* **90**, 465–493.

O'Donnell, K., Kistler, H.C., Cigelnik, E., and Ploetz, R.C. (1998b). Multiple evolutionary origins of the fungus causing Panama disease of banana: Concordant evidence from nuclear and mitochondrial gene genealogies. *Proc. Nat. Acad. Sci. U.S.A.* **95**, 2044–2049.

O'Donnell, K., Kistler, H.C., Tacke, B.K., and Casper, H.H. (2000). Gene genealogies reveal global phylogeographic structure and reproductive isolation among lineages of *Fusarium graminearum*, the fungus causing wheat scab. *Proc. Nat. Acad. Sci. U.S.A.* **97**, 7905–7910.

Rassmann, K., Schlotterer, C., and Tautz, D. (1991). Isolation of simple-sequence loci for use in polymerase chain reaction-based DNA fingerprinting. *Electrophoresis* **12**, 113–118.

Schussler, A., Schwarzott, D., and Walker, C. (2001). A new fungal phylum, the Glomeromycota: Phylogeny and evolution. *Mycol. Res.* **105**, 1413–1421.

Taylor, J.W., Jacobson, D.J., and Fisher, M.C. (1999). The evolution of asexual fungi: Reproduction, speciation and classification. *Annu. Rev. Phytopath.* **37**, 197–246.

Taylor, J.W., Jacobson, D.J., Kroken, S., Kasuga, T., Geiser, D.M., Hibbett, D.S., and Fisher, M.C. (2000). Phylogenetic species recognition and species concepts in fungi. *Fungal Genet. Biol.* **31**, 21–32.

Tsai, H.F., Liu, J.S., Staben, C., Christensen, M.J., Latch, G.C.M., Siegel, M.R., and Schardl, C.L. (1994). Evolutionary diversification of fungal endophytes of tall fescue grass by hybridization with *Epichlöe* species. *Proc. Nat. Acad. Sci. U.S.A.* **91**, 2542–2546.

Vos, P., Hogers, R., Bleeker, M., Reijans, M., Vandelee, T., Hornes, M., Frijters, A., Pot, J., Peleman, J., Kuiper, M., and Zabeau, M. (1995). AFLP—a new technique for DNA-fingerprinting. *Nucl. Acids Res.* **23**, 4407–4414.

White, T.J., Bruns, T.D., Lee, S., and Taylor, J.W. (1990). Amplification and direct sequencing of fungal ribosomal RNA genes for phylogenetics. In M. Innis, D. Gelfand, J. Sninsky, and T.J. White (eds) *PCR Protocols: A guide to methods and applications* (Chapter 38). Orlando, FL: Academic Press.

2

Genomics of Filamentous Fungi

Ulrich Schulte

1. Introduction

Large-scale sequence analysis of fungal genomes has been focused on yeast for a long time. The publication of the complete genome of budding yeast (*Saccharomyces cerevisiae*) heralded the era of eukaryotic genome analysis (Goffeau *et al.*, 1996). Six years later the complete analysis of the fission yeast (*Schizosaccharomyces pombe*) genome was presented (Wood *et al.*, 2002). The genomes of these yeast proved similar in size and gene content: the *S. cerevisiae* genome codes for approximately 6,500 proteins in a total of 12.6 Mb, while the 13.8-Mb genome of *S. pombe* includes about 5,000 genes. These yeasts diverged 400 million years ago, making them as distant from each other as from ascomycetous filamentous fungi like *Aspergillus nidulans* and *Neurospora crassa*, and yet about 80% of their genes reveal significant sequence similarity. In fact, the gene complement of these organisms may represent the minimal set of genes needed by a free-living eukaryote (Wood *et al.*, 2002). Genomic sequencing is now being extended to the pathogenic yeast, *Candida albicans*, a hemiascomycete which can grow in unicellular form as well as filamentous form, and *Cryptococcus neoformans*, a basidiomycete (Heitman *et al.*, 1999).

From a mycologist's point of view, yeast, despite their widespread use in biological research, represent only a small section of the tremendous variety in fungal biology. Many filamentous fungi differ considerably from yeast regarding habitat, nutrient requirements, and physiology. Properties and peculiarities of filamentous fungi are therefore not adequately reflected in yeast genomes. This became the impetus in the late 1990s for an initiative to expand genome sequencing efforts to filamentous fungi (Bennett, 1997). From the data available to date, it is apparent, that the genomes of most filamentous fungi comprise a significantly extended set of genes compared to known yeast genomes (Schulte *et al.*, 2002). This chapter gives an overview of current projects and summarizes publicly accessible data. The main focus will be on the *N. crassa* genome, for which the most advanced databases are currently available.

Ulrich Schulte • Institut für Biochemie, Heinrich-Heine-Universität, D-40225 Düsseldorf, Germany.

Advances in Fungal Biotechnology for Industry, Agriculture, and Medicine. Edited by Jan S. Tkacz and Lene Lange, Kluwer Academic/Plenum Publishers, 2004.

2. Genomic Projects Focusing on Fungi

The main emphasis in the genomics of filamentous fungi has been on the genus *Aspergillus* (Table 2.1). No less than four species of this genus have been sequenced. *A. nidulans*, one of the model genetic systems for filamentous fungi, was sequenced by Cereon (USA). Of immediate commercial interest are *A. niger* and *A. oryzae* sequenced for DSM (Dutch State Mines) and the Japanese National Institute of Technology and Evaluation, respectively. Finally, the pathogenic *A. fumigatus* is sequenced in a joint effort of the Sanger Centre (Great Britain) and TIGR (USA) (Denning *et al.*, 2002). Genomic sequences of *N. crassa* were assembled by the German *Neurospora* Genome Project as well as the Whitehead Genome Center (USA). The plant pathogen *Magnaporthe grisea*, causing rice blast disease, is also sequenced at the Whitehead Genome Center. The filamentous fungus *Ashbya gossypii* was sequenced at the Biozentrum (Switzerland). Despite its filamentous growth habit, *A. gossypii* is a close relative of *S. cerevisiae* resembling its genome structure as well as its gene complement (Altmann-Jöhl and Philippsen, 1996). *Pneumocystis carinii*, a parasitic fungus causing pneumonia, is sequenced at the Children's Hospital Medical Center of Cincinnati (USA) (Cushion and Arnold, 1998). *C. albicans*, mentioned previously, completes the list of genome projects involving ascomycetes. Basidiomycetes for which genome projects are completed or underway include *Ustilago maydis*, a plant pathogen, that was sequenced for Bayer (Germany), *Phanaerochaete chrysosporium*, a wood-degrading, white-rot fungus, sequenced at the DOE Joint Genome Institute (USA), and *C. neoformans*. Genomes of zygomycetes have not yet been tackled by large-scale sequencing. A first project has, however, been proposed by the Whitehead Genome Center, recently (discussed later).

Unfortunately, the data of many fungal genome projects are not publicly accessible. Just one *Aspergillus* project provides free online access to its data (Table 2.1). Projects financed by private funds have restricted use of their databases, most being locked completely to date. Public access to these databases has been announced in some cases but has not yet been granted.

Recently the pace in the genomics of fungi was increased considerably by the Whitehead Genome Center. Supported by the National Human Genome Research Institute (USA), a program to sequence the genomes of up to 15 filamentous fungi was proposed (*www-genome.wi.mit.edu/seq/fgi/FGI_whitepaper_Feb8.pdf*). A list of seven primary candidates includes four organisms, whose genomes have already or are currently being sequenced. The genomic sequences of *A. nidulans* and *U. maydis* will thus be put in the public domain. Also included is *C. neoformans*, serotype A, which will complement the data for serotype D sequenced at TIGR. *P. carinii* falls into divergent populations, and the Whitehead proposal aims at two distinct forms. One is infectious for humans (*P. carinii* f. sp. *hominis*) the other is a mouse pathogen (*P. carinii* f. sp. *muris*). The ongoing project mentioned earlier focuses on the population specific for rats (*P. carinii* f. sp. *carinii*). The other three primary candidates of the Whitehead proposal for which no sequencing projects are under way, are *Coccidioides immitis*, an euascomycetous, human pathogen, *Rhizopus arrhizus*, which would be the first zygomycete to be sequenced, and *Coprinus cinereus*, a basidiomycetous mushroom.

Initially eukaryotic genomes were sequenced based on mapped, large-insert clones, for example, cosmids and BACs. This approach was also applied to most fungal projects initiated in the last millennium, for example, *S. cerevisiae* (Anonymous, 1997), *S. pombe*

Table 2.1. Fungal Genomes

Organism	Phylum	Size (Mb)	Number of chromosomes	Putative genes	Database link
Aspergillus fumigatus	Euascomycete	32			*www.tigr.org/tdb/e2k1/afu1/*
Ashbya gossypii	Archaeascomycete	9	7	4800	
Aspergillus nidulans	Euascomycete	29	8		
Aspergillus niger	Euascomycete	30	8	14000*	
Aspergillus oryzae	Euascomycete				
Candida albicans	Hemiascomycete	16	8	6500*	*sequence-www.stanford.edu/group/candida/index.html*
Cryptococcus neoformans	Basidiomycete	21			*www.tigr.org/tdb/e2k1/cna1/*
Magnaporthe grisea	Euascomycete	40	7		*www-genome.wi.mit.edu/annotation/fungi/magnaporthe/*
Neurospora crassa	Euascomycete	40	7	11000*	*mips.gsf.de/proj/neurospora/* *www-genome.wi.mit.edu/annotation/fungi/neurospora/*
Pneumocystis carinii	Archaeascomycete	8			*pneumocystis.chmcc.org*
Penicillium chrysosporium	Basidiomycete	30	10		*www.jgi.doe.gov/programs/whiterot.htm*
Saccharomyces cerevisiae	Hemiascomycete	13	16	6500	*mips.gsf.de/proj/yeast/CYGD* *genome-www.stanford.edu/Saccharomyces*
Schizosaccharomyces pombe	Archaeascomycete	14	3	5000	*www.sanger.ac.uk/Projects/S_pombe/*
Ustilage maydis	Basidiomycete	20			

* Estimated.

(Wood *et al.*, 2002), *A. niger, N. crassa* (Aign *et al.*, 2001), and *U. maydis*. In view of the large sequencing capacities available today, whole-genome shotgun sequencing, regularly used for smaller genomes, has proven to be an efficient approach for fungal genomes as well. The *N. crassa* genome, for which a chromosome oriented project was conceived originally and started by sequencing two single chromosomes (Schulte *et al.*, 2002), was ultimately tackled by a whole-genome shotgun approach. Although the bulk of the sequence is obtained randomly from plasmids, efficient assembly of shotgun sequences relies on data obtained from large-insert libraries. At least for the *N. crassa* genome, these libraries proved to be the bottleneck for the ordered clone approach as well as the shotgun approach. A significant amount of sequence is not covered even by large cosmid or BAC libraries (Aign *et al.*, 2001; Mannhaupt *et al.*, 2003). A draft sequence assembled from shotgun data at the Whitehead Genome Center reveals 169 clone gaps despite both a 20-fold sequence coverage and a 100-fold physical coverage of the genome by analyzed clones (Galagan *et al.*, 2003). As an advantage of sequencing small insert clones, the accessibility of sequences close to centromeres is considerably improved compared to cosmids and BACs.

3. Genome Structure

The smallest of all characterized fungal genomes is found in *P. carinii*. Less than 8 Mb are sufficient for this parasitic fungus. Most yeast have genomes about twice that size, while the ascomycetous filamentous fungi contain genomes of 30–40 Mb. Though similar in size, the genome structures of yeast differ widely. The number of chromosomes varies between 3 in *S. pombe* and 16 in *S. cerevisiae*. Most fungi, however, have seven or eight chromosomes.

A conserved genome structure has been found between closely related fungi. Synteny among hemiascomycetous yeast has been studied extensively and various degrees have been revealed (Llorente *et al.*, 2000). A conserved genome structure has been proposed as a way to analyze genomes by low coverage sequencing for *C. albicans* as well as *A. gossypii*, both of which show synteny to *S. cerevisiae* (Altmann-Jöhl and Philippsen, 1996; Hartung *et al.*, 1998; Seoighe *et al.*, 2000). Studies on synteny among filamentous fungi are emerging as well. Synteny with the genomic sequence of *N. crassa* was reported for *Blumeria graminis* (Pedersen *et al.*, 2002) and for *M. grisea* (Hamer *et al.*, 2001).

A major conclusion drawn from the complete sequence of the *S. cerevisiae* genome was that the entire genome duplicated itself 100 million years ago (Wolfe and Shields, 1997). Large stretches of duplicated sequence are still apparent by a conserved order of paralogous genes. As a consequence of the duplication event, the number of paralogous genes retained in today's *S. cerevisiae* genome is very high. In contrast, no indications of large scale duplications have been detected in the genome of *S. pombe* (Wood *et al.*, 2002). Less is known about the genomes of filamentous fungi. In *N. crassa* and closely related ascomycetes, duplicated sequences are affected by repeat-induced point mutations (RIP) (Cambareri *et al.*, 1989; Graia *et al.*, 2001; Ikeda *et al.*, 2002). The RIP effect leads to numerous mutations in both copies of a duplicated sequence during meiosis and thus selects against the duplication of essential sequences (Selker, 1997). The genome of *N. crassa* has been shaped significantly by the RIP effect. Duplicated sequences are rare. Screening two chromosomes of *N. crassa* for duplicated sequences yielded just a few

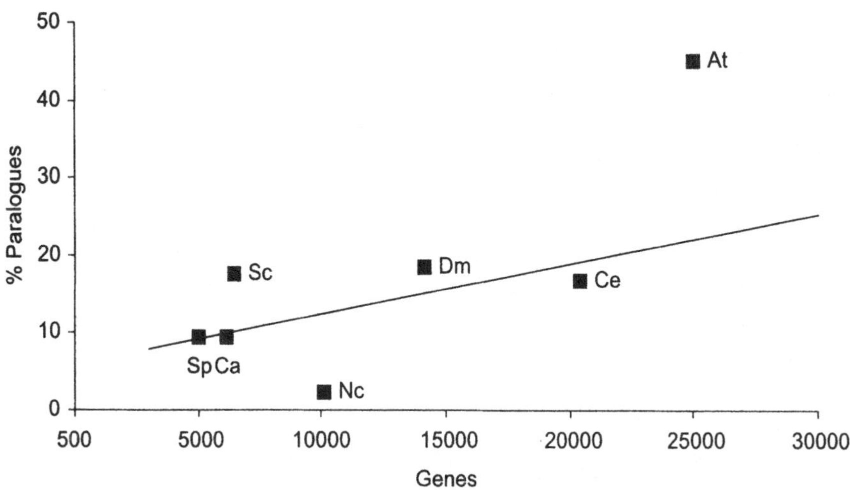

Figure 2.1. Correlation of gene number and frequency of paralogues in eukaryotes. The number of paralogues, defined as ORFs sharing at least 80% sequence identity and differing in their length by no more than 20%, was calculated for *S. pombe* (Sp), *C. albicans* (Ca), *S. cerevisiae* (Sc), *N. crassa* (Nc), *D. melanogaster* (Dm), *Caenorhabditis elegans* (Ce), and *A. thaliana* (At). A best fit line is given for data points Sp, Ca, Dm, and Ce.

small stretches about 10 Kb in length duplicated or triplicated in one or both of the two chromosomes (Mannhaupt *et al.*, 2003). Remarkably these sequences are devoid of genes.

RIP also interferes with the activity of mobile elements (Kinsey *et al.*, 1994). In contrast to yeast, most strains of *Neurospora* and *Aspergillus* species lack active transposons (Li Destri Nicosia *et al.*, 2001). Though remnants of transposon related sequences are present in the genome of *N. crassa*, they have been victims of RIP (Galagan *et al.*, 2003; Mannhaupt *et al.*, 2003). Active transposons like Tad are confined to special strains (Kinsey and Helber, 1989).

Probably the most dramatic impact of RIP is reflected in the gene complement of *N. crassa*. Genome evolution through duplication of genes has been virtually abolished in the organism since emergence of RIP. The size of multigene families is extremely small, and sequence similarities among family members are rarely above 80% (Galagan *et al.*, 2003). In general, the number of paralogous genes does not fit the size of the *N. crassa* genome (Mannhaupt *et al.*, 2003). While a linear relationship between genome size and frequency of paralogues is found for most genomes, an unusually high proportion of paralogues is present in genomes known to have undergone entire duplications, for example, *S. cerevisiae* and *Arabidopsis thaliana* (Figure 2.1). In contrast, the number of paralogues in the genome of *N. crassa* is extremely low. On the bright side, RIP suppresses selfish, foreign DNA, but on its dark side, it limits expansion of a gene set.

4. Gene Identification and Annotation

Fungal genomes are generally compact, showing a high gene density. In *S. cerevisiae* an average of one gene per 2.1 kb has been calculated (Goffeau *et al.*, 1996). For *S. pombe* this number is about 2.5 kb per gene (Wood *et al.*, 2002). Gene numbers and genome sizes of filamentous fungi are not yet known exactly. One gene every 3.6 kb has been estimated for *N. crassa* (Mannhaupt *et al.*, 2003). The identification of 14,000 genes in *A. niger* was recently reported (European Conference on Fungal Genetics, Pisa, 2002) which would result in a gene density of 2.2 kb per gene.

Introns are usually small in fungal genes, ranging from 50 to 300 bp in the majority of cases. In *S. cerevisiae* introns are rare; only 4% of genes are interrupted by introns (Goffeau *et al.*, 1996). The frequency of introns in other fungi is considerably higher. Among *S. pombe* genes, 43% have introns (Wood *et al.*, 2002). An average number of 2.3 introns per gene was reported for *N. crassa* (Schulte *et al.*, 2002). Splice sites resemble general eukaryotic consensus sequences (Bruchez *et al.*, 1993; Edelmann and Staben, 1994; Spingola *et al.*, 1999).

While the determination of gene structure is straightforward in *S. cerevisiae* due to its high gene density and low frequency of introns, identification of genes within the genomic sequences of other fungi has proven to be much more complicated. Several software tools have been used for gene prediction in fungi. The genome of *S. pombe* was scanned for genes by GENEFINDER (P. Green and L. Hillier, unpublished) trained on a set of experimentally validated genes. Gene prediction programs trained on filamentous fungi are not yet widespread. Fgenesh (Salamov and Solovyev, 2000) has been applied to analyze the *N. crassa* and *A. niger* genome sequences. Gene prediction programs, even when trained for a specific organism, provide gene models of limited accuracy. To improve the predictions, combinations of different algorithms are employed, results of database searches are considered, and models are manually edited. The MIPS *N. crassa* database (MNCDB) provides for part of the *N. crassa* genome a set of genes based on predictions by Fgenesh subjected to a manual correction process (Mannhaupt *et al.*, 2003). A set of 10,000 genes was identified in the assembled draft sequence of the *N. crassa* genome by a combination of Fgenesh and GENEWISE (Galagan *et al.*, 2003). Generally more than half of the genes do not require any changes of the exon–intron structure predicted by Fgenesh. Quite often the boundaries of 5′ or 3′ exons need to be corrected based on available EST sequences. A peculiarity of *N. crassa* genes are very small 5′ exons. Some are less than 3 bp and thus do not even encode a single amino acid. These small exons are especially difficult to predict. While Fgenesh rarely misses a gene completely, two closely linked genes are occasionally fused artificially. Of uncertain significance are the numerous very small genes. In the manually reviewed database, MNCDB, genes with less than 100 codons are retained only if supported by significant matches with ESTs or known genes. Small genes are less likely to find a significant match in searches through sequence databases (Schulte *et al.*, 2002). Nevertheless many small genes have been shown to be transcribed in *S. cerevisiae* (Velculescu *et al.*, 1997), and several are already known in *N. crassa* as well. Due to the uncertainties in gene prediction and the gaps in the available sequence data, the exact gene number in *N. crassa* is still unknown. It is expected to be about twice as high as the number in *S. cerevisiae*.

Annotated databases provide data for the proteins predicted from genes deduced through sequence analysis. Included are simple calculations of molecular weight and isoelectric point as well as predictions of secondary structure elements and membrane spanning helices. Functional properties of proteins are concluded from a variety of database searches. Thus, the sequence similarity to known or predicted proteins is determined.

The most reliable prediction of protein function is based on a high sequence identity to known proteins from different organisms covering the entire sequence length. Searches in special databases like COGs and BLOCKS point to related, well-characterized groups of proteins. Sequence similarities restricted to certain parts of a protein are less significant. Nevertheless, the identification of protein domains and characteristic sequence motifs can provide valuable hints concerning the function of proteins.

The genome of *S. cerevisiae* has been extensively subjected to annotation procedures. Several comprehensive databases concerning this genome have been established and provide a benchmark for all large-scale sequencing projects of eukaryotic organisms (Dwight *et al.*, 2002; Mewes *et al.*, 2002). A complete annotation is also available for the genome of *S. pombe* (Wood and Bahler, 2002). The most advanced database of a filamentous fungus has been built for *N. crassa* in a collaboration of the Whitehead Genome Center and MIPS (Munich Information Center for Protein Sequences). The entire genome is subjected to a comprehensive manual annotation. This provides a major advancement for research related to filamentous fungi as the gene complement of filamentous fungi comprises numerous genes not found in yeast or other organisms as we shall now consider.

5. Gene Complement of a Filamentous Fungus

Knowledge about the gene complement of yeast, especially *S. cerevisiae*, is far advanced. A substantial amount of data regarding different hemiascomycetes enables efficient comparative genomics. The "genolevure" program extends the knowledge basis analyzing the genomes of many close relatives of *S. cerevisiae* (Souciet *et al.*, 2000). Recently the results of a comprehensive knock out program in *S. cerevisiae* were reported (Giaever *et al.*, 2002). For filamentous fungi the knowledge about the function and cellular location of proteins is far less advanced. Annotation of the *N. crassa* genes deduced from the genomic sequence provides functional classifications for almost one half of recognized genes (Mannhaupt *et al.*, 2003). This leaves the other half without reliable indications concerning their function. Roughly one fourth of annotated genes are so-called conserved hypothetical genes. The amino acid sequences deduced from these genes are related to open reading frames (ORFs) identified in the genomes of other organisms. However, no experimentally characterized protein yields sufficient sequence similarity (FASTA score > 150) when compared to the conserved hypothetical genes. Sequence conservation among different organisms in this case supports identification of bona fide genes but does not provide clues permitting a functional assessment. One third of genes identified in the genomic sequence of *N. crassa* represent novel genes lacking readily detectable relatives in the accessible databases. This rather large portion of orphan genes is an indication for the small sampling of known genes from filamentous fungi. While a complete gene set of numerous bacteria and archaea, several yeasts and metazoa, and one plant are present in the public databases, no genome of a filamentous fungus has been completely annotated and put in the databases, yet. In addition to data for *N. crassa*, sequences covering most of the genomic sequence of other filamentous fungi, for example, *A. fumigatus* and *M. grisea*, are available but not yet annotated. A TBLASTN search in the genomic sequence of *A. fumigatus* reveals significant ($P < 10^{-8}$) hits for one third of the orphan genes of *N. crassa* (Mannhaupt *et al.*, 2003). Compared to the genomic sequence of *M. grisea*, half

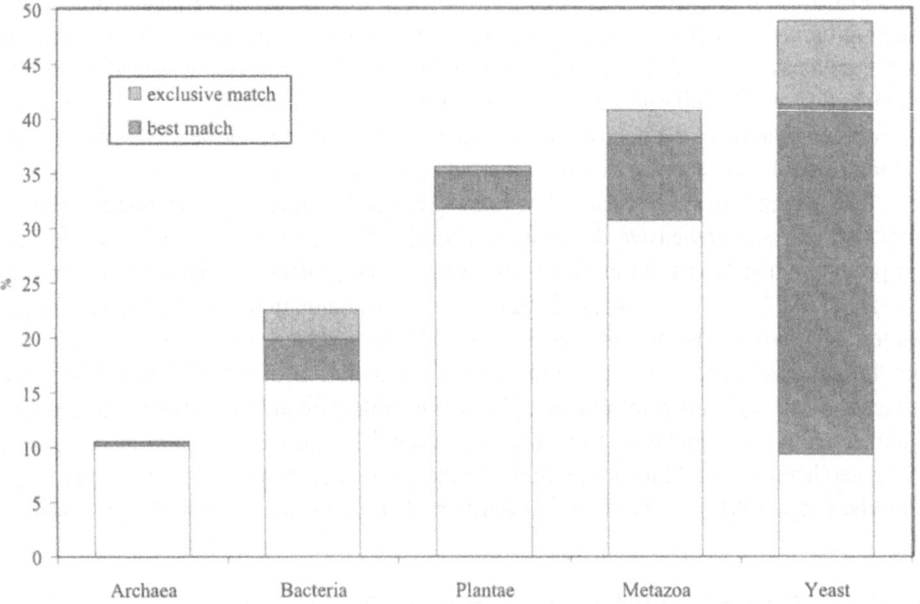

Figure 2.2. Distribution of homologues to *N. crassa* ORFs among kingdoms. Percentages are given for 3218 *N. crassa* ORFs taken from MNCDB (manually edited) matching an ORF from an organism within the group indicated (FASTA score > 150). Dark gray areas represent *N. crassa* ORFs that yielded the highest FASTA score with an ORF from that group of organisms. Light gray areas give the percentage of *N. crassa* ORFs exclusively matching an ORF from that group of organisms.

of the current orphan genes of *N. crassa* yield a significant hit. Once genes have been identified within the genomic sequence of these and other filamentous fungi, the number of orphan genes in *N. crassa* will drop below 15%.

The impact of many genes on fungal biology remains enigmatic and much experimental work remains to be done to reach a better knowledge base for filamentous fungi. Since currently more than 3,000 genes of *N. crassa* lack a sequence relative in other organisms, characterization of these genes and the proteins they encode will rely on experimental work with *N. crassa* as well as other filamentous fungi.

Nevertheless, with today's knowledge interesting insights into the peculiarities of the protein complement of a filamentous fungus like *N. crassa* are possible (Mannhaupt *et al.*, 2003). To understand what makes a fungus filamentous, a comparison is first made with the well-characterized ascomycetous relatives *S. cerevisiae* and *S. pombe*. A match to a gene in *S. cerevisiae* or *S. pombe* is found for 48% of all *N. crassa* genes. For 81% of those, a yeast gene yields the best match (Figure 2.2). Only 8% of genes with a known yeast relative lack a match in non-fungal organisms and thus appear to be specific for fungi. Table 2.2 summarizes the prevalence of fungal-specific genes in the *N. crassa* genome according to functional categories in which the deduced proteins can be placed. Though enzymes involved in metabolism are especially numerous in *N. crassa*, few proteins in the category metabolism appear to be fungal specific. Among the few fungal-specific metabolic enzymes, there are mostly specific hydrolases, for example, glucanases and lipases. Remarkably, no proteins

Table 2.2. Functional categories of fungal-specific genes

Functional category	*N. crassa* genes specific for Fungi incl. yeast (%)	*N. crassa* genes specific for Filamentous fungi (%)	All annotated *N. crassa* genes(%)
Metabolism	10	66	23
Energy	3	0	5
Cell cycle and DNA processing	13	0	8
Transcription	23	0	12
Protein synthesis	5	0	6
Protein fate (folding, modification, destination)	11	6	12
Cellular transport and transport mechanism	7	6	7
Cellular communication / Signal transduction	0	0	4
Cell rescue, defence, and virulence	3	11	5
Interaction with cellular environment	2	0	1
Cell fate	11	11	7
Transposable elements, viral and plasmid proteins	2	0	1
Control of cellular organization	3	0	3
Transport facilitation	7	0	6
	100	100	100

As described for Figure 2.2., *N. crassa* genes exclusively matching with fungal genes (yeasts and filamentous fungi) or only with filamentous fungi were identified. The percentage of these genes attributed to a given functional category are listed. For comparison the percentage of functional categories among all annotated *N. crassa* genes are given (data taken from MNCDB)

involved in signal transduction appear to be fungal-specific. Especially frequent are fungal-specific genes involved in transcriptional processes. Many transcription factors are shared between yeast and filamentous fungi but lack a detectable non-fungal homologue. A relatively high abundance of fungal-specific proteins is also apparent for two categories: Cell cycle and DNA processing and Cell fate. Thus several regulatory mechanisms are specifically conserved among fungi, while metabolic pathways as well as signal transduction are either shared with non-fungal organisms or divergent in yeast and filamentous fungi.

For a large number of *N. crassa* genes no significant sequence similarity with yeast genes is detectable. Only about half of the genes currently identified in the genome of *N. crassa* yield a significant BLAST match ($P<10^{-8}$) when compared to genes of *S. cerevisiae* or *S. pombe*. As mentioned, one third of *N. crassa* genes lack any relative at all. Of those genes related to known or predicted genes in other organisms, 30% have no relative among yeast genes. A number of genes appear to be missing specifically in yeast, i.e., relatives are present in filamentous fungi, animals, and plants but not in yeast. Table 2.3 lists several examples. It is apparent, that yeasts have lost a number of specific metabolic enzymes. The mitochondrial respiratory chain of fermentative yeasts lacks a complex I and a transhydrogenase (Nosek and Fukuhara, 1994). Oxygenases, especially FAD-monooxygenase and cytochrome P450 proteins, are frequent in filamentous fungi, but rare in yeasts. These enzymes are usually involved in detoxification or secondary metabolism. Due to their specialized natural niche, yeasts can dispense with these and other genes to streamline their genomes.

Table 2.3. Common eukaryotic genes missing in yeast

Code	Description	Functional sub-category
b23g1_170	Probable 4-hydroxyphenylpyruvate dioxygenase	Amino acid metabolism
b14a6_230	L-amino-acid oxidase	Amino acid metabolism
b23e9_040	Probable xanthine dehydrogenase	Purine nucleotide metabolism
b23l4_170	Related to glyoxal oxidase	Carbohydrate metabolism
b13h18_180	Related to beta-galactosidase	Carbohydrate metabolism
b2f7_180	Related to beta-1,3 exoglucanase precursor	Degradation of exogenous compounds
b7f21_020	Related to hydroxymethylglutaryl-CoA lyase	Lipid and isoprenoid metabolism
b13h18_150	Related to lipoxygenase 1	Lipid and isoprenoid metabolism
b8j22_180	Related to calcium-independent phospholipase A2	Lipid and isoprenoid metabolism
93g11_210	Related to C-8,7 sterol isomerase	Lipid and isoprenoid metabolism
7c14_070	Related to molybdopterin synthase large subunit	Biosynthesis of cofactors
90c4_150	Tyrosinase	Secondary metabolism
123a4_360	NADH:ubiquinone oxidoreductase subunit b17.2	Respiration
b1o14_280	NADH:ubiquinone oxidoreductase subunit 22K	Respiration
b13d15_210	Complex I intermediate-associated protein CIA30	Respiration
13e11_040	Related to nicotinamide nucleotide transhydrogenase	Respiration
20h10_190	Related to DNA-damage repair protein DRT111	DNA repair
b9b11_030	Related to pirin	mRNA transcription
5e6_250	Related to NAD^+-ADP-ribosyltransferase	Protein modification
b12k8_010	Related to UDP-N-acetylglucosamin-peptide N-acetylglucosaminyltransferase	Protein modification
80a10_160	Related to endopeptidase Clp	Proteolytic degradation
b8g12_220	Related to metalloprotease MEP1	Proteolytic degradation
93g11_240	Related to calmodulin	Intracellular signalling
b24n4_170	Related to dimethylaniline monooxygenase	Detoxification
b24n4_150	Related to monooxigenase	Detoxification
b2o8_350	Related to cytochrome P450	Detoxification
65e11_080	Related to membrane protein	Located in plasma membrane
7f4_330	Related to cleft lip transmembrane protein 1 (CLPTM1)	Located in plasma membrane
93g11_060	Related to Werner syndrome helicase	Located in nucleus
5e6_200	Probable aldehyde reductase 6	Located in peroxisome
b24p7_290	Related to hxB protein	
65e11_140	Regulator of conidiation rca-1	
12f11_010	Related to microsomal glutathione S-transferase 3	
71b5_140	Related to P element somatic inhibitor	
b24p11_050	Related to DRPLA protein	

N. crassa genes matching genes (FASTA score > 150) in animals as well as plants but finding no match in *S. cerevisiae* or *S. pombe* are listed (data taken from MNCDB).

The number of genes *N. crassa* shares exclusively with other filamentous fungi is rather small. Currently just 2% of *N. crassa* genes belong to this group (Mannhaupt *et al.*, 2003). Table 2.2 shows the distribution of functional categories among these genes. A clear bias for metabolic enzymes is noticeable. The majority of proteins in this category are enzymes that degrade complex carbohydrates. It is evident from the data presented in Table 2.2, that genes specific for fungi in general and those specific for filamentous fungi in particular differ significantly in function. Regulation of transcription, development and differentiation, common among fungal-specific genes are rather rare among genes specific

Table 2.4. Functional Categories of *N. crassa* genes with homologues restricted to one kingdom

	N. crassa genes with homologues restricted to		
Functional category	Metazoa	Plantae	Bacteria
Metabolism	10	22	79
Energy	5	0	0
Cell cycle and DNA processing	10	0	0
Transcription	0	0	0
Protein synthesis	0	0	0
Protein fate (folding, modification, destination)	21	10	6
Cellular transport and transport mechanism	16	0	0
Cellular communication / Signal transduction	20	7	0
Cell rescue, defense, and virulence	4	18	12
Interaction with cellular environment	0	0	0
Cell fate	5	9	0
Transposable elements, viral and plasmid proteins	0	0	0
Control of cellular organization	0	0	0
Transport facilitation	4	23	3
	~100	~100	~100

As described for Figure 2.2, *N. crassa* genes exclusively matching with genes from metazoa, plants or bacteria, respectively, were identified. The percentage of these genes attributed to a given functional category are listed (data taken from MNCDB).

for filamentous fungi. On the other hand, genes involved in signal transduction and especially in metabolism are rare among fungal-specific genes but abundant among filamentous fungal-specific genes.

As mentioned, the number of known genes from filamentous fungi is about to increase tremendously and so will the number of genes identified as specific for filamentous fungi. The small sampling of filamentous fungal genes is reflected not only in the number of specific genes identified but also in the distribution of functions attributed to these genes. Hydrolytic, metabolic enzymes are particularly well studied in filamentous fungi (see Chapter 11), while many genes involved in specific aspects of cell differentiation, development, and communication are still unknown. Table 2.2, therefore, gives a necessarily preliminary and biased collection of genes specific for filamentous fungi.

Approximately 40% of *N. crassa* genes have relatives in metazoa. For 10% of *N. crassa* genes, animal genes yield the best FASTA score (Figure 2.2). However, rather few genes (2.5%) appear to be present exclusively in metazoa and filamentous fungi. Table 2.4 lists the functional categories of those with a known relative. A bias is apparent for gene products involved in protein degradation (protein fate) and signal transduction. Few genes code for metabolic functions. Though frequently found among *N. crassa* genes in general (Table 2.2), a function in transcription or protein synthesis is not attributed to any *N. crassa* gene exclusively shared with animals.

Similar to what has been found for metazoa, about 36% of *N. crassa* genes have relatives in plants. Most of these are common to both, plants and animals. A best score to a plant gene is obtained for 4% of *N. crassa* genes. Very few (0.5%) *N. crassa* genes have relatives exclusively in plants. Those with known relatives code predominantly for transporters and proteins involved in the degradation of toxins (Table 2.4).

Finally, one third of genes find relatives in prokaryotes. Archaeal genes do not play a prominent role among the relatives of *N. crassa* genes. Just 0.4% of genes yield a best score to an archaeal gene, and no *N. crassa* gene is exclusively related to genes of archaea. A different picture is found for bacterial genes. A best score to a bacterial gene is obtained for 6% of *N. crassa* genes, and 3% of *N. crassa* genes are exclusively related to bacterial genes, which is more than the portion of *N. crassa* genes exclusively related to plants or animals, respectively (Figure 2.2). The proteins coded for by these genes are predominantly related to specific metabolic pathways (Table 2.4). Many degrade complex carbohydrates and thus complement the set of genes specific for filamentous fungi (Table 2.2). In addition to these, several enzymes are attributed to secondary metabolism. Genes shared exclusively between filamentous fungi and bacteria can be considered to be specific for microorganisms, prokaryotic as well as eukaryotic. These genes were apparently abandoned by higher eukaryotes as they adapted to specialized living conditions, became multicellular, and adopted a strategy of mutual interactions of specialized cells to satisfy their needs. As mentioned, yeasts have been restricted to specific ecological niches and do not need many genes.

6. Novel Aspects of Fungal Biology

In addition to providing an overview of the gene complement and immediate access to the sequence and location of any gene, analysis of the *N. crassa* genome sequence highlights so far unknown aspects of the biology of this fungus. Though different regulatory pathways, light sensing, and circadian rhythms have been studied extensively in *N. crassa*, an impressive number of new and unexpected genes were identified, complementing the existing set of genes attributed to signalling pathways (Galagan *et al.*, 2003). Among the highlights are: two genes coding for putative phytochromes, known in other organisms to be involved in red light sensing (a phenomenon which has not yet been observed in *N. crassa*); nine new histidine kinases, expanding a set that previously was limited to only two; ten seven-transmembrane helix, G protein-coupled receptors (GPCRs) including members of cAMP GPCRs until now unknown in fungi; 25 genes coding for proteins involved in Ca^{++} signaling, none of which was known before.

N. crassa also encodes a significant variety of proteins that could plausibly participate in secondary metabolism. The formation of secondary metabolites has been well characterized in parasitic and saprophytic fungi (see Chapters 5–8). Except for carotenoid synthesis, the presence of these pathways in *N. crassa* remained unclear. Key enzymes in the formation of pigments and toxins are polyketide synthases (PKS) and non-ribosomal peptide synthetases (NRPS). *N. crassa* encodes seven PKSs and three NRPSs which is fewer than found in typical pathogenic fungi but significantly more than in the yeasts *S. cerevisiae* and *S. pombe* (Galagan *et al.*, 2003; Yoder and Turgeon, 2001). Other enzymes typically involved in the synthesis of secondary metabolites are various monooxygenases and transporters. Homologues of these, previously unknown in *N. crassa*, are abundant in the *N. crassa* genome (Mannhaupt *et al.*, 2003). While the secondary metabolites formed by *N. crassa* remain elusive, comparing the gene complement of parasitic and saprophytic fungi will become an important tool to define the determinants of pathogenicity.

Finaly, among the proteins encoded by genes conserved between filamentous fungi and bacteria but absent in other eukaryotes (Table 2.3) are many which appear to be involved in metabolic pathways novel for fungi. These include putatively degrading enzymes, like various oxygenases, as well as homologues of enzymes involved in the formation of glycolipids or polyhydroxyalkonates in bacteria (Mannhaupt *et al.*, 2003).

7. Summary

Genome analysis of fungi has been shifted from a focus on few yeast models to a much broader approach including filamentous fungi. The number of genes in filamentous fungi is roughly twice that reported for yeasts and close to the number of genes in *Drosophila*. Currently, analysis of the *N. crassa* genome is most advanced. Since more than half of all *N. crassa* genes cannot yet be functionally evaluated, any explanation as to why a filamentous fungus needs so many genes is necessarily preliminary. Evaluation of genes amenable to functional description based on sequence analysis reveals a high abundance of genes attributed to metabolism. According to the protein domains identified in *N. crassa* genes, monooxygenases and, to a lesser extent, short chain dehydrogenases and AMP binding proteins are exceptionally abundant (Mannhaupt *et al.*, 2003). A large portion of *N. crassa* genes with short chain dehydrogenase and monooxygenase domains are most closely related to bacterial genes and lack yeast relatives. Metabolic enzymes are particularly common among proteins specific for either filamentous fungi alone or filamentous fungi as well as bacteria (Tables 2.2 and 2.4). While yeasts have streamlined their protein complement, optimizing themselves for specific ecological niches, other more versatile microorganisms including filamentous fungi adjust to various environmental conditions and use various nutrients. Thus filamentous fungi have, on one hand, retained metabolic enzymes no longer needed by yeasts and other eukaryotes and, on the other hand, have developed specific enzymes especially useful in the degradation of recalcitrant, complex substrates.

Analysis of the *N. crassa* genome sequence demonstrates that filamentous fungi hold many more secrets to be uncovered. The complete, annotated genomes of filamentous fungi will become treasure chests opening new and exciting ways to explore their cellular functions and to apply the knowledge so that we may deal more effectively with them, either in fighting them where detrimental or using them where beneficial.

References

Aign, V., Schulte, U., and Hoheisel, J.D. (2001). Hybridization-based mapping of *Neurospora crassa* linkage groups II and V. *Genetics*. **157**, 1015–1020.

Altmann-Jöhl, R., and Philippsen, P. (1996). Ag*THR4*, a new selection marker for transformation of the filamentous fungus *Ashbya gossypii*, maps in a four-gene cluster that is conserved between *A. gossypii* and *Saccharomyces cerevisiae*. *Mol Gen. Genet.* **250**, 69–80.

Anonymous, (1997). The yeast genome directory. *Nature* **387** (suppl.). 1–105.

Bennett, J.W. (1997). White paper: Genomics for filamentous fungi. *Fungal Genet. Biol.* **21**, 3–7.

Bruchez, J.J.P., Eberle, J., and Russo, V.E.A. (1993). Regulatory sequences in the transcription of *Neurospora crassa* genes: CAAT box, TATA box, introns, poly(A) tail formation sequences. *Fung. Genet. Newslett.* **40**, 89–96.

Cambareri, E.B., Jensen, B.C., Schabacht, E., and Selker, E.U. (1989). Repeat-induced G-C to A-T mutations in *Neurospora. Science.* **244**, 1571–1575.

Cushion, M.T., and Arnold, J.A. (1998). Proposal for a *Pneumocystis* genome project. *J. Euk. Microbiol.* **44**, 7S.

Denning, D.W., Anderson, M.J., Turner, G., Latgé, J.P., and Bennett, J.W. (2002). Sequencing the *Aspergillus fumigatus* genome. *Lancet Infect. Dis.* **2**, 251–253.

Dwight, S.S., Harris, M.A., Dolinski, K., Ball, C.A., Binkley, G., Christie, K.R., Fisk, D.G., Issel-Tarver, L. *et al.* (2002). *Saccharomyces* genome database (SGD) provides secondary gene annotation using the gene ontology (GO). *Nucleic Acids Res.* **30**, 69–72.

Edelmann, S.E., and Staben, C. (1994). A statistical analysis of sequence features within genes from *Neurospora crassa. Exp. Mycol.* **18**, 70–81.

Galagan, J.E., Calvo, S.E., Borkovich, K.A., Selker, E.U., Read, N.D., FitzHugh, W., Ma L.-J., Smirnov, S. *et al.* (2003). The genome sequence of the filamentous fungus *Neurospora crassa. Nature.* **422**, 859–868.

Giaever, G., Chu, A.M., Ni, L., Connelly, C., Riles, L., Veronneau, S., Dow, S., Lucau-Danila, A. *et al.* (2002). Functional profiling of the *Saccharomyces cerevisiae* genome. *Nature.* **418**, 387–391.

Goffeau, A., Barrell, B.G., Bussey, H., Davis, R.W., Dujon, B., Feldmann, H., Galibert, F., Hoheisel, J.D. *et al.* (1996). Life with 6000 genes. *Science.* **274**, 546–567.

Graia, F., Lespinet, O., Rimbault, B., Dequard-Chablat, M., Coppin, E., and Picard, M. (2001). Genome quality control: RIP (repeat-induced point mutation) comes to *Podospora. Mol. Microbiol.* **40**, 586–595.

Hamer, L., Pan, H., Adachi, K., Orbach, M.J., Page, A., Ramamurthy, L., and Woessner, J.P. (2001). Regions of microsynteny in *Magnaporthe grisea* and *Neurospora crassa. Fungal Genet. Biol.* **33**, 137–143.

Hartung, K., Frishman, D., Hinnen, A., and Wölfl, S. (1998). Single-read sequence tags of a limited number of genomic DNA fragments provide an inexpensive tool for comparative genome analysis. *Yeast.* **14**, 1327–1332.

Heitman, J., Casadevall, A., Lodge, J.K., and Perfect, J.R. (1999). The *Cryptococcus neoformans* genome sequencing project. *Mycopathologia.* **148**, 1–7.

Ikeda, K., Nakayashiki, H., Kataoka, T., Tamba, H., Hashimoto, Y., Tosa, Y., and Mayama, S. (2002). Repeat-induced point mutation (RIP) in *Magnaporthe grisea*: Implications for its sexual cycle in the natural field context. *Mol. Microbiol.* **45**, 1355–1364.

Kinsey, J.A., and Helber, J. (1989). Isolation of a transposable element from *Neurospora crassa. Proc. Natl. Acad. Sci. U.S.A.* **86**, 1929–1933.

Kinsey, J.A., Garrett-Engele, P.W., Cambareri, E.B., and Selker, E.U. (1994). The *Neurospora* transposon Tad is sensitive to repeat-induced point mutation (RIP). *Genetics.* **138**, 657–664.

Li Destri Nicosia, M.G., Brocard-Masson, C., Demais, S., Hua Van, A., Daboussi, M.J., and Scazzocchio, C. (2001). Heterologous transposition in *Aspergillus nidulans. Mol. Microbiol.* **39**, 1330–1344.

Llorente, B., Malpertuy, A., Neuveglise, C., de Montigny, J., Aigle, M., Artiguenave, F., Blandin, G., Bolotin-Fukuhara, M. *et al.* (2000). Genomic exploration of the hemiascomycetous yeasts: 18. Comparative analysis of chromosome maps and synteny with *Saccharomyces cerevisiae. FEBS Lett.* **487**, 101–112.

Mannhaupt, G., Montrone, C., Haase, D., Mewes, H.W., Aign, V., Hoheisel, J.D., Fartmann, B., Nyakatura, G. *et al.* (2003). What's in the genome of a filamentous fungus? Analysis of the *Neurospora* genome sequence. *Nucleic Acids Res.* **31**, 1944–1954.

Mewes, H.W., Frishman, D., Guldener, U., Mannhaupt, G., Mayer, K., Mokrejs, M., Morgenstern, B., Munsterkotter, M. *et al.* (2002). MIPS: A database for genomes and protein sequences. *Nucleic Acids Res.* **30**, 31–34.

Nosek, J., and Fukuhara, H., (1994). NADH dehydrogenase subunit genes in the mitochondrial DNA of yeast. *J. Bact.* **17**, 5622–5630.

Pedersen, C., Rasmussen, S.W., and Giese, H. (2002). A genetic map of *Blumeria graminis* based on functional genes, avirulence genes, and molecular markers. *Fungal Genet. Biol.* **35**, 235–246.

Salamov, A.A., and Solovyev, V.V. (2000). *Ab initio* gene finding in *Drosophila* genomic DNA. *Genome Res.* **10**, 516–522.

Schulte, U., Becker, I., Mewes, H.W., and Mannhaupt, G. (2002). Large scale analysis of sequences from *Neurospora crassa. J. Biotechnol.* **94**, 3–13.

Selker, E.U. (1997). Epigenetic phenomena in filamentous fungi: Useful paradigms or repeat-induced confusion? *Trends Genet.* **13**, 296–301.

Seoighe, C., Federspiel, N., Jones, T., Hansen, N., Bivolarovic, V., Surzycki, R., Tamse, R., Komp, C. *et al.* (2000). Prevalence of small inversions in yeast gene order evolution. *Proc Natl Acad Sci U.S.A.* **97**, 14433–14437.

Spingola, M., Grate, L., Haussler, D., and Ares, M.Jr. (1999). Genome-wide bioinformatic and molecular analysis of introns in *Saccharomyces cerevisiae*. *RNA*. **5**, 221–234.

Souciet, J., Aigle, M., Artiguenave, F., Blandin, G., Bolotin-Fukuhara, M., Bon, E., Brottier, P., Casaregola, S. *et al.* (2000). Genomic exploration of the hemiascomycetous yeasts: 1. A set of yeast species for molecular evolution studies. *FEBS Lett*. **487**, 3–12.

Velculescu, V.E., Zhang, L., Zhou, W., Vogelstein, J., Basrai, M.A., Bassett, D.E., Hieter, P., Vogelstein, B. *et al.* (1997). Characterization of the yeast transcriptome. *Cell*. **88**, 243–251.

Wolfe, K.H., and Shields, D.C. (1997). Molecular evidence for an ancient duplication of the entire yeast genome. *Nature*. **387**, 708–713.

Wood, V., and Bahler, J. (2002). Website review: How to get the best from fission yeast genome data. *Comp. Funct. Genomics*. **3**, 282–288.

Wood, V., Gwilliam, R., Rajandream, M.A., Lyne, M., Lyne, R., Stewart, A., Sgouros, J., Peat, N. *et al.* (2002). The genome sequence of *Schizosaccharomyces pombe*. *Nature*. **414**, 871–880.

Yoder, O.C. and Turgeon, B.G. (2001). Fungal genomics and pathogenicity. *Curr. Opin. Plant Biol*. **4**, 315–321.

3

A Molecular Tool Kit for Fungal Biotechnology

John E. Hamer

1. Introduction

Fungal biotechnology increasingly relies on a set of tools to facilitate genetic manipulations for better economic outcomes (Table 3.1). In this chapter I will review some of the newer tools and technologies that can be used to advance biotechnology programs with fungi. For the purposes of this chapter I have focussed on tools applicable to heterotrophic filamentous ascomycetes, by far the most important fungi for biotechnology, but with some creativeness, these techniques can likely be adapted for a broader range of fungi. It is important to point out that the success or efficiency of any particular technology is generally fungal-specific. For example, rates of protoplast regeneration, transformation frequency, and degree of gene targeting are all variable depending on the fungal species and even the genetic background of particular strains.

Various tools of molecular genetics that are applicable to fungi have been reviewed over the past 8–10 years, many covered in an excellent volume edited by Talbot (2001), including transformation technology, and a variety of cytological, biochemical, and genetic techniques. Chapters from this volume cover recent developments in transformation, gene expression, and genomics. A constant source of information on techniques and tools for filamentous fungi is the Fungal Genetics Newsletter, available on-line (*http://www.fgsc.net*); fungal transformation vectors and modifications to techniques are routinely posted.

A wide variety of genomic technologies can be applied to any organism including fungi (e.g., yeast two-hybrid libraries and DNA micro-arrays), and these have been well described elsewhere (e.g., Phizicky and Fields, 1995; Fields and Bartel, 2001; Yang and Speed, 2002). Similarly, genomic technologies applied to studies in yeast are developing rapidly, and there have been excellent reviews of these developments (e.g., Kumar and Snyder, 2001). For the purposes of this chapter, I will review technologies for filamentous fungi that can take advantage of sequenced genomes or large expressed-sequence-tagged (EST) libraries. In particular, I will focus on technologies that can be used to evaluate large

John E. Hamer • Paradigm Genetics Inc., Research Triangle Park, North Carolina.

Advances in Fungal Biotechnology for Industry, Agriculture, and Medicine. Edited by Jan S. Tkacz and Lene Lange, Kluwer Academic/Plenum Publishers, 2004.

Table 3.1. Generic fungal biotechnology tool kit

Starter kit	Advanced kit
Selectable markers (drug resistance, auxotrophic)	Excisable selectable markers
Transformation system	Recombination-based vectors
Cosmid and plasmid vectors for transformation	Annotated genome sequence
Controllable promoters	DNA microarray
Reporter genes (*lacZ*, GFP, GUS)	Deep EST sequence (>8,000 unigenes)
	Transposon-arrayed library
	Arrayed physical map as cosmid or BAC clones

numbers of genes, rather than single genes. Although genome approaches are underway for fungi (see *http://www.genome.wi.mit.edu/annotation/fungi*), it will take some time to successfully complete the annotation of a fungal genome. This effort will certainly leave a large number of genes poorly annotated for any specific biochemical or cellular function, and tools for mutagenesis and analysis are greatly needed. Given the small size of the fungal research community, tool development and improvement remains a constant challenge.

2. Vectors and Transformation

Since the first publication of *Gene Manipulations in Fungi* (Bennett and Lasure, 1985) and *More Gene Manipulations in Fungi* (Bennett and Lasure, 1991), fungal vectors and transformation technology have remained largely unchanged for the majority of fungi. For most fungi, protoplast or spheroplast transformation and regeneration remains the stalwart of transformation technology. Biolistic transformation has seen widespread use in *Cryptococcus* (Davidson *et al.*, 2000) and *Agrobacterium*-based transformation is proving valuable in certain situations (see Chapter 4). Transformation systems have now been reported for the obligate plant pathogen, *Blumeria graminis* (Chaure *et al.*, 2000), although follow-up studies will be needed to judge the utility of this particular tool for studying obligate pathogens.

Vector technology also remains relatively unchanged with integrative vectors predominating for filamentous fungi. Replicating vectors appear to have only been useful in *Aspergillus nidulans* thus far (Aleksenko and Clutterbuck, 1997), despite reports of replicating vectors for fungi that are now at least 8 years old (e.g., see Javerzat *et al.*, 1993). New vectors are constructed almost daily in modern fungal biotechnology laboratories, and the polymerase chain reaction has facilitated the design and production process of new vectors. Recent modifications have included constructing vectors that can be used to sequentially disrupt genes (Royer *et al.*, 1999).

3. Gene Cloning Tools for Genomic Approaches

As fungal genome sequence information and gene sequence collection grows, it is imperative to develop efficient methods for rapidly manipulating cloned genes or gene segments from fungi. The best approaches today use site-specific recombinases to move cloned gene segments between a variety of expression vectors for bacterial and fungal expression. For example, several sets of vectors have recently been described for *Saccharomyces cerevisiae* (Liu *et al.*, 1998), allowing a cloned gene segment to be rapidly moved between yeast and bacterial expression vectors that contain in-frame epitope tags for later purification and localization. The vectors use the site-specific recombination specificity of the *Cre-lox* system (Sauer, 1996). *Cre* recombinase, supplied as pure enzyme, can be used to recombine a DNA segment cloned between adjacent *lox* sites. The *lox* sites are arranged to link DNA segments, in-frame, to a variety of epitope tags, promoters, and fusion proteins.

A commercially available system called Gateway™ is available from Invitrogen (*http://www.invitrogen.com*) that uses the lambda recombination system (*att*B × *att*P → *att*L × *att*R). In this system, genes are cloned once into an "Entry" vector, which contains flanking *att*L sites. The *att*L flanked entry clones can then be efficiently moved between a series of *att*B "Destination" vectors in parallel. Current Gateway™ destination vectors contain expression modules for bacteria, yeast, and mammalian hosts, but could also be adapted for fungi.

While it may be inefficient to generate such vectors for a large number of fungi, vectors could be generated for generalized interrogation of over-expression effects or subcellular targeting. The advantage is that libraries of cloned gene segments can be rapidly and accurately moved into a number of expression vectors. Such approaches could facilitate the identification of genes affecting complex and important biotechnological processes.

One ideal use of such systems is to generate a cloned proteome consisting of a set of full-length cDNAs for a subset or all of the genome. The cloned proteome can be used to provide targeted protein arrays for small molecule screening or a starting point for expression screening. Approaches such as these are being used in systematic proteome analysis at the Harvard Institute for Proteomics in the Flexgene Consortium (*http://www.hip.harvard.edu*). *In vivo* recombinational methods have been efficiently used to create rapidly gene-targeting vectors for *Aspergillus*. The techniques make use of the phage lambda Redαβγ functions that promote homologous recombination in *E. coli* with linear PCR cassettes (Chaveroche *et al.*, 2000).

4. Fungal Transposons as Tools

Although transposons have been identified for sometime and studied in yeast, their utility has only recently been exploited in filamentous fungi (Nicosia *et al.*, 2001). The transposable elements employed belong to the *Mariner* superclass of TC-1 like transposons. All these transpose by a "cut and paste" mechanism employing an element encoded transposase that inserts the elements at ubiquitous TA positions throughout the genome. Elegant work in *Fusarium* and *Magnaporthe* has described the use of these elements in classical transposon-tagging schemes (Migheli *et al.*, 2000; Villalba *et al.*, 2001). The next iteration of this technology would employ pooling strategies used to screen mutagenized strain collections with primers designed to identify a transposon end and a specific gene. An alternative, albeit longer, procedure is to collect end clones by PCR and generate an easily searchable sequence library. A variety of these approaches have been used recently for *Arabidopsis* (Speulman *et al.*, 1999) and may be useful for some fungi.

Identifying such IR-transposons in filamentous fungi is relatively straightforward as they generally are well distributed in the genome, are small in size (<2 kb), and are identified by the short inverted repeats at their ends (see Daboussi, 1997). Steps must be taken to reconstitute the transposase coding region, as this region will often contain inactivating mutations in the majority of genomic copies.

5. Tools for Identifying Essential Genes

Identifying novel, essential genes to serve as targets for developing antimycotic leads for agrochemical or pharmaceutical research remains a challenge. First, although the set of essential *S. cerevisiae* genes is largely known, it is not clear how effectively this list translates to other fungi. There are well documented examples of essential *Saccharomyces* genes that proved dispensable when analyzed in a related fungal pathogen (Kelly *et al.*, 2000). For predominantly haploid strains with poorly developed parasexual or sexual stages, the technical challenges increase. Several approaches are described below.

5.1. Generating Conditional Lethal Mutants

A gene disruption or knockout that results in a conditional phenotype can be easily identified and selected for during transformation. Conditions can include nutrient composition of the growth medium, salt concentration, environmental conditions such as temperature, or developmental state (e.g., aconidial strains). However, while such conditional mutations are easy to identify, they may not always represent ideal targets for triggering fungal cell death, and fungi are uniquely adept at enduring harsh environmental conditions.

5.2. Inference

In a haploid organism, attempts to create a mutation in a gene through homologous recombination will almost always be futile if the gene is essential. By screening sufficient numbers of transformants, one may infer that a mutation in that gene is most likely lethal. While unlikely to constitute sufficient proof for publication, the approach can be supplemented by deleting the resident gene in a strain carrying an additional copy at an unlinked location elsewhere in the genome. Essentiality can be inferred by failing to recover meiotic segregants carrying only the deleted copy. Other inferential approaches include detecting the mutated copy by PCR in a mixture of unregenerated protoplasts, but not from regenerated transformants. Similarly, where parasexual genetics and heterokaryons are available, essential gene mutations can be identified in living heterokaryons. Essentiality is again inferred by the failure to recover haploid segregants containing the mutation either from heterokaryons or from heterozygous diploids (e.g., see, Harris and Hamer, 1995). Variations of these inferential approaches include using transposons or insertional mutations to generate mutations in diploid strains.

5.3. Using Controllable Promoters

Perhaps the most widely used tool in any molecular geneticist tool kit is the controllable promoter. Because of the diversity of fungal organisms it is difficult to develop a single controllable promoter that would have wide utility. However, by cDNA screening it is reasonably straightforward to identify genes, and by inference promoters, that are controlled by simple nutritional shifts. Alternatively, there are combinations of promoter/transcription factor systems that are inducible by addition of a low molecular weight compound. Well understood promoters are generally genus/species specific and require use of a specific transcription factor when used in heterologous systems (e.g., Waring *et al.*, 1989; Caddick *et al.*, 1998). Controllable promoters can be used to induce a conditional state on any gene allowing its role in growth to be accurately assessed. Examples include the generation of so-called "alcoholic" strains of *Aspergillus nidulans* (Xiang *et al.*, 1994).

Recently, such an approach was used systematically to place a large number of putative essential genes of *Candida albicans* under the control of a repressible promoter (*http://www.elitra.com*). The approach, called GRACE for Gene-Replacement And Conditional–Expression, replaces one allele with a gene-disruption cassette and replaces the second allele with a tetracycline repressible promoter. The strains developed can be screened in animal infection studies supplying tetracycline to turn-off genes *in vivo* and thereby investigate whether they play an essential role (Haselbeck *et al.*, 2002). The small genome and efficient gene targeting in *C. albicans* makes this a particularly attractive approach.

5.4. Post-Transcriptional Gene Silencing (PTGS)

Fungi were among the first organisms to be used to uncover the basis of PTGS (Romano and Macino, 1992). However, systematic use of this phenomenon has not been widely adopted in fungal research. There are reports of anti-sense (likely a trigger for PTGS) working in *Candida* (de Backer *et al.*, 2001; de Backer *et al.*, 2002) and some filamentous fungi (Gorlach *et al.*, 2002). With an understanding of the mechanisms of PTGS now at hand, it may be possible to build vector systems that can be used to direct gene silencing more efficiently. However, in many eukaryotic cells, PTGS results in a mosaic phenotype where not all cells or nuclei experience the same degree of silencing. The heterokaryotic growth capability of many filamentous fungi means that PTGS effects could be missed. Efficient mechanisms for inducing PTGS must be found for this to see widespread use for biotechnological processes with fungi.

6. Genome-Based Tools

The best genome-based tool is a well-annotated and meticulously curated genome sequence. While there are still possibilities that many fungi will eventually be sequenced, it is unlikely that many will find enough scientists willing to engage in hand curated annotation of the data. For many fungi, other tools or resources that can be assembled by a handful of researchers will have to suffice, such as EST collections, physical maps, and micro-arrayed DNA or cDNA collections. Other chapters in this volume deal with assembling this type of information. Below I focus on some genome-wide tools that can be used, or modified for use, with filamentous fungi.

6.1. Genome-Wide Insertional Mutagenesis

Restriction enzyme mediated insertions (REMI) have been used to create small (a few thousand) collections of insertion mutants that can be screened for desired phenotypes (for reviews see Brown *et al.*, 1998; Sweigard and Ebbole, 2001). Insertions of plasmid DNA and flanking regions can be recovered by plasmid shuffling. While useful for finding a few genes (<10), the technique suffers the drawback that more than 50% of mutations are unlinked with the insertion loci, and larger mutant searches have resulted in finding recurrent alleles rather than more unique genes, suggesting that insertion sites are biased. Modifications of REMI have included the use of promoters (Shuster and Connelley, 1999) or expression constructs to modify and discover new types of genes through mis-expression or localization (Marhoul and Adams, 1995). Overall the approach allows phenotype driven gene identification, but cannot be used to test larger panels of genes in a systematic way.

Recently, REMI has been modified and used in Signature Tag Mutagenesis (STM) approaches (Chiang *et al.*, 1999) in *Aspergillus fumigatus* (Brown *et al.*, 2000), *Cryptococcus neoformans* (Nelson *et al.*, 2001), and *Candida glabrata* (Cormack *et al.*, 1999). While the approach allowed identification of some virulence genes, it was inefficient, particularly in filamentous fungi with large genomes.

We have recently used a systematic approach that can simultaneously collect gene sequence information and create libraries of constructs suitable for generating mutations at will (Hamer *et al.*, 2001). Transposon-Arrayed-Gene-Knock-Outs (TAG-KO) uses transposition to create insertional mutations in large-insert DNA libraries (lambda, cosmids, BACs) *in vitro*. The transposon is engineered to carry a hygromycin phosphotransferase cassette that expresses in *E. coli* and fungi. Library DNA is mutagenized by *in vitro* transposition and transformed into *E. coli* cells that are subsequently plated on hygromycin containing media. Insertion events are collected and sequencing is performed from the transposon ends. The end sequences are used to sort and discover genes, and gene knockouts can be created at will by transforming selected cosmids into fungal hosts. Because the cosmid vector can be linearized and carries a large DNA insert, gene targeting is very efficient. The approach has an advantage in that new genomes can be surveyed as rapidly as sequencing can be performed, and an array of mutagenesis vectors is created simultaneously.

This technique can be easily modified for sequenced genomes (see Figure 3.1). A minimal cosmid tile can be created from a sequenced genome and aligned by cosmid-end sequencing. By this approach, any gene or gene cluster can be selected and rapidly mutagenized. TAG-KO can be modified in a number of ways. We have found that bacterial artificial chromosomes and lambda and plasmid vectors can be used with high efficiency. Also, transposons may be modified to carry promoters, terminators, or reporter genes.

6.2. Genome-Shuffling

For many biotechnological applications it is advantageous to enhance the ability of fungi to perform various biochemical activities. There are extensive reviews on pathway engineering and on using promoters to increase expression of desired gene products (Punt *et al.*, 2002). However, traditional mutagenesis and screening are still used for strain improvement. Recently, a modification of this traditional method has been described for bacteria in which recurrent cycles of protoplast fusions and selection are used to "shuffle" the genome and yield strain improvements (Zhang *et al.*, 2002). The rationale behind the

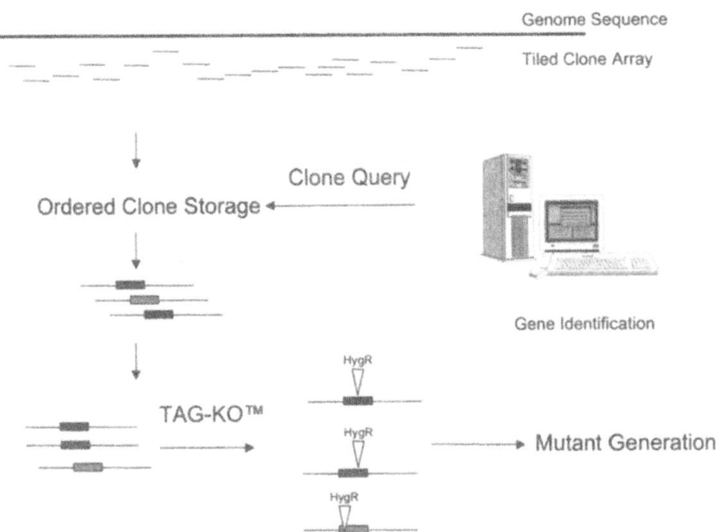

Figure 3.1. Schematic outline for employing TAG-KO™ for screening a sequenced fungal genome. From a completed fungal genome sequence, a Tiled Clone Array consisting of cosmid or BAC clones is generated by end-sequencing a cosmid or BAC library and aligning a minimal set of clones with the genome. The desired genes or gene clusters to be analyzed can be queried and selected clones can be picked and mutagenized to a desired degree of saturation using the TAG-KO™ approach (see text). The mutagenized clones can then be used to generate a desired set of mutants.

approach is that during protoplast fusion, nuclear exchanges take place at diverse genomic sites allowing a form of sexual recombination to occur. Since protoplast fusion has been a traditional approach in fungal strain improvements, it seems likely that some form of genome shuffling may prove useful to fungal biotechnology.

7. Summary

A molecular tool kit for fungi has been developing over the last two decades. Recent additions could include recombinational tools and vectors that can be more efficient than traditional molecular biology techniques of restriction and ligation. A set of fungal vectors that incorporates these tools will be a powerful asset for any laboratory. A variety of approaches are available for identifying fungal genes important for biotechnological processes. Among these are approaches that incorporate the availability of genome sequences and physical maps. A focussed effort on securing some of these tools will advance fungal biotechnology programs in industry and academia.

References

Aleksenko, A., and Clutterbuck, A.J. (1997). Autonomous plasmid replication in *Aspergillus nidulans: AMA1* and *MATE* elements. *Fungal Genet. Biol.* **21**, 373–387.

Bennett, J.W., and Lasure, L.L. (1985). *Gene Manipulations in Fungi.* Academic Press, San Diego.

Bennett, J.W., and Lasure, L.L. (1991). *More Gene Manipulations in Fungi.* Academic Press, San Diego.

Brown, J.S., Aufauvre-Brown, A., Brown, J., Jennings, J.M., Arst, H.Jr., and Holden, D.W. (2000). Signature-tagged and directed mutagenesis identify PABA synthetase as essential for *Aspergillus fumigatus* pathogenicity. *Mol. Microbiol.* **36**, 1371–1380.

Brown, J.S., Aufauvre-Brown, A., and Holden, D.W. (1998). Insertional mutagenesis in *Aspergillus fumigatus. Mol. Gen. Genet.* **259**, 327–335.

Caddick, M.X., Greenland, A.J., Jepson, I., Krausę, K.P, Qu, N., Riddell, K.V., Salter, M.G., Schuch, W., *et al.* (1998). An ethanol inducible gene switch for plants used to manipulate carbon metabolism. *Nature Biotechnol.* **16**, 177–180.

Chaure, P., Gurr, S.J., and Spanu, P. (2000). Stable transformation of *Erysiphe graminis*, an obligate biotrophic pathogen of barley. *Nature Biotechnol.* **18**, 205–207.

Chaveroche, M.K., Ghigo, J.M., and d'Enfert, C. (2000). A rapid method for efficient gene replacement in the filamentous fungus *Aspergillus nidulans. Nucleic Acids Res.* **28**, E97.

Chiang, S.L., Mekalanos, J.J., and Holden, D.W. (1999). *In vivo* genetic analysis of bacterial virulence. *Annu. Rev. Microbiol.* **53**, 129–154.

Cormack, B.P., Ghori, N., and Falkow, S. (1999). An adhesin of the yeast pathogen *Candida glabrata* mediating adherence to human epithelial cells. *Science* **285**, 578–582.

Daboussi, M.J. (1997). Fungal tranponsable elements and genome evolution. *Genetica* **100**, 253–260.

Davidson, R.C., Cruz, M.C., Sia, R.A., Allen, B., Alspaugh, J.A., and Heitman, J. (2000). Gene disruption by biolistic transformation in serotype D strains of *Cryptococcus neoformans. Fungal Genet. Biol.* **29**, 38–48.

de Backer, M.D., Nelissen, B., Logghe, M., Viaene, J., Loonen, I., Vandoninck, S., de Hoogt, R., Dewaele, S. *et al.* (2001). An antisense-based functional genomics approach for identification of genes critical for growth of *Candida albicans. Nature Biotechnol.* **19**, 235–241.

de Backer, M.D., Raponi, M., and Arndt, G.M. (2002). RNA-mediated gene silencing in non-pathogenic and pathogenic fungi. *Curr. Opin. Microbiol.* **5**, 323–329.

Fields, S., and Bartel, P.L. (2001). The two-hybrid system. A personal view. *Methods Mol. Biol.* **177**, 3–8.

Gorlach, J.M., McDade, H.C., Perfect, J.R., and Cox, G.M. (2002). Anti-sense repression in *Cryptococcus neoformans* as a laboratory tool and potential antifungal strategy. *Microbiology* **148**, 213–219.

Hamer, L., Adachi, K., Montenegro-Chamorro, M.V., Tanzer, M.M., Mahanty, S.K., Lo, C., Tarpey, R.W., Skalchunes, A.R. *et al.* (2001). Gene discovery and gene function assignment in filamentous fungi. *Proc. Natl. Acad. U.S.A.* **98**, 5110–5115.

Harris, S.D., and Hamer, J.E. (1995). *sepB*: An *Aspergillus nidulans* gene involved in chromosome segregation and the initiation of cytokinesis. *EMBO J.* **14**, 5244–5257.

Haselbeck, R., Wall, D., Jiang, B., Ketela, T., Zyskind, J., Bussey, H., Foulkes, J.G., and Roemer, T. (2002). Comprehensive essential gene identification as a platform for novel anti-infective drug discovery. *Curr. Pharm. Design.* **8**, 1155–1172.

Javerzat, J.P., Bhattacherjee, V., and Barreau, C. (1993). Isolation of telomeric DNA from the filamentous fungus *Podospora anserina* and construction of a self-replicating linear plasmid showing high transformation frequency. *Nucleic Acids Res.* **21**, 497–504.

Kelly, R., Card, D., Register, E., Mazur, P., Kelly, T., Tanaka, K.I., Onishi, J., Williamson, J.M. *et al.* (2000). Geranylgeranyltransferase I of *Candida albicans*: Null mutants or enzyme inhibitors produce unexpected phenotypes. *J. Bacteriol.* **182**, 704–713.

Kumar, A., and Snyder, M. (2001). Emerging technologies in yeast genomics. *Nat. Rev. Genet.* **2**, 302–312.

Liu, Q., Li, M.Z., Leibham, D., Cortez, D., and Elledge, S.J. (1998). The univector plasmid-fusion system, a method for rapid construction of recombinant DNA without restriction enzymes. *Curr. Biol.* **8**, 1300–1309.

Marhoul, J.F., and Adams, T.H. (1995). Identification of developmental regulatory genes in *Aspergillus nidulans* by overexpression. *Genetics* **139**, 537–547.

Migheli, Q., Steinberg, C., Daviere, J.M., Olivain, C., Gerlinger, C., Gautheron, N., Alabouvette, C., and Daboussi, M.J. (2000). Recovery of mutants impaired in pathogenicity after transposition of *Impala* in *Fusarium oxysporum* f. sp *melonis. Phytopathol.* **90**, 1279–1284.

Nelson, R.T., Hua, J., Pryor, B., and Lodge, J.K. (2001). Identification of virulence mutants of the fungal pathogen *Cryptococcus neoformans* using signature-tagged mutagenesis. *Genetics* **157**, 935–947.

Nicosia, M.G.L., Brocard-Masson, C., Demais, S., Van, H., Daboussi, M.J., and Scazzocchio, C. (2001). Heterologous transposition in *Aspergillus nidulans. Mol. Microbiol.* **39**, 1330–1344.

Phizicky, E.M., and Fields, S. (1995). Protein-protein interactions: Methods for detection and analysis. *Microbiol. Rev.* **59**, 94–123.

Punt, P.J., van Biezen, N., Conesa, A., Albers, A., Mangnus, J., and van den Hondel, C.A.M.J.J. (2002). Filamentous fungi as cell factories for heterologous protein production. *Trends Biotechnol.* **20**, 200–206.

Romano, N., and Macino, G. (1992). Quelling: Transient inactivation of gene expression in *Neurospora crassa* by transformation with homologous sequences. *Mol. Microbiol.* **6**, 3343–3353.

Royer, J.C., Christianson, L.M., Yoder, W.T., Gambetta, G.A., Klotz, A.V., Morris, C.L., Brody, H., and Otani, S. (1999). Deletion of the trichodiene synthase gene of *Fusarium venenatum*: Two systems for repeated gene deletions. *Fungal Genet. Biol.* **28**, 68–78.

Sauer, B. (1996). Multiplex *Cre/lox* recombination permits selective site-specific DNA targeting to both a natural and an engineered site in the yeast genome. *Nucl. Acids Res.* **24**, 4608–4613.

Shuster, J.R., and Connelley, M.B. (1999). Promoter-tagged restriction enzyme-mediated insertion (PT-REMI) mutagenesis in *Aspergillus niger. Mol. Gen. Gent.* **262**, 27–34.

Speulman, E., Metz, P., Arkel, G., Hekkert, B., Stiekema, W., and Perira, A. (1999). A two component enhancer-inhibitor transposon mutagenesis system for functional analysis of the *Arabidopsis* genome. *Plant Cell* **11**, 1853–1866.

Sweigard, J.A., and Ebbole, D.J. (2001). Functional analysis of pathogenicity genes in a genomics world. *Curr. Opin. Microbiol.* **4**, 387–392.

Talbot, N. (2001). *Molecular and Cellular Biology of Filamentous Fungi. A Practical Approach.* Oxford University Press, Oxford.

Villalba, F., Lebrun, M.H., Hua-Van, A., Daboussi, M.J., and Grosjean-Cournoyer, M.C. (2001). Transposon *Impala*, a novel tool for gene tagging in the rice blast fungus *Magnaporthe grisea. Mol. Plant-Microbe Interact.* **14**, 308–315.

Waring, R.B., May, G.S., and Morris, N.R. (1989). Characterization of an inducible expression system in *Aspergillus nidulans* using *alcA* and tubulin-coding genes. *Gene* **79**, 119–130.

Xiang, X., Beckwith, S.M., and Morris, N.R. (1994). Cytoplasmic dynein is involved in nuclear migration in *Aspergillus nidulans. Proc. Natl. Acad. Sci. U.S.A.* **91**, 2100–2104.

Yang, Y.H., and Speed, T. (2002). Design issues for cDNA microarray experiments. *Nature Rev. Genet.* **3**, 579–588.

Zhang, Y.-Z., Perry, K., Vinci, V.A., Powell, K., Stemmer, W.P.C., and del Cardayre, S.B. (2002). Genome shuffling leads to rapid phenotypic improvement in bacteria. *Nature* **415**, 644–646.

4

Transformation Mediated by *Agrobacterium tumefaciens*

Paul J. J. Hooykaas

1. Introduction

The soil bacterium *Agrobacterium tumefaciens* can be used nowadays as a vector for the genetic transformation of organisms as diverse as plants, yeasts, and filamentous fungi. For a century this bacterium has been recognized as the etiological agent of the plant disease crown gall, a disease characterized by tumorous overgrowths that occur mostly on the root crown of the plant. In the laboratory crown gall can be induced at any plant part by wounding and infecting with the bacterium. Fifty years ago it was discovered that crown gall cells differ from normal plant cells by their ability to grow in the absence of plant growth regulators (Braun, 1958). They can do so even in the absence of the inducing Agrobacteria, and it was concluded, therefore, that crown gall cells are tumor cells. How the bacterium transforms normal plant cells into tumor cells was elucidated several years later. The bacterium introduces a segment of oncogenic DNA, called T(transferred)-DNA into the plant cells at the infection sites (Chilton *et al.*, 1977), and this converts the normal cells into tumor cells because the DNA directs the synthesis of plant growth regulators. By the receipt of the T-DNA, the crown gall cells concomitantly acquire the ability to produce tumor-specific metabolites called opines (Figure 4.1). Different *Agrobacterium* strains may provoke tumors that produce different types of opines. Thus the strains are classified as octopine strains, nopaline strains, leucinopine strains, etc. The opines that are produced in (and secreted by) the tumor cells are the dividend gained by the bacterium for having provoked the tumor, as only the inducing agrobacteria can catabolize the opines formed by the tumor (Petit *et al.*, 1970). For example, octopine strains, which induce tumors producing octopine, can utilize octopine (but not, for instance, nopaline) for growth, whereas conversely nopaline strains induce tumors making nopaline and can degrade nopaline (but not octopine). The process by which *Agrobacterium* induces plant tumors, is often referred to as genetic colonization, because genetic modification of the host creates a favorable niche for the inducing bacterium. The horizontal transfer of T-DNA from *Agrobacterium* to

---

Paul J. J. Hooykaas • Institute of Biology, Clusius Laboratory, Leiden University, Wassenaarseweg 64, 2333 Al Leiden, The Netherlands.

Advances in Fungal Biotechnology for Industry, Agriculture, and Medicine. Edited by Jan S. Tkacz and Lene Lange, Kluwer Academic/Plenum Publishers, 2004.

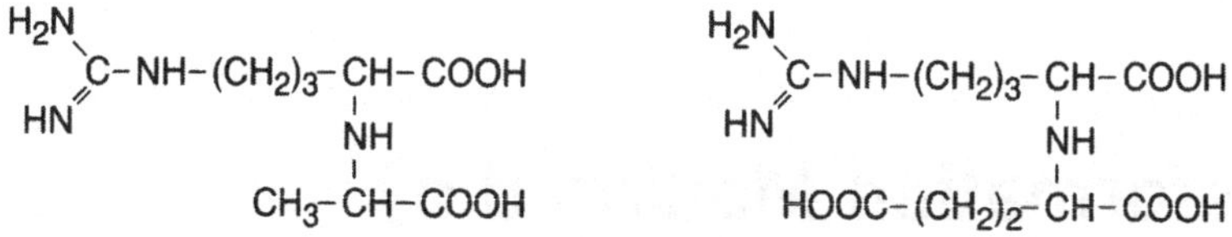

H_2N $C-NH-(CH_2)_3-CH-COOH$ HN NH $CH_3-CH-COOH$

Octopine

H_2N $C-NH-(CH_2)_3-CH-COOH$ HN NH $HOOC-(CH_2)_2-CH-COOH$

Nopaline

Figure 4.1. The structure of opines.

plants may have played a role in evolution, as can be seen in the plant genus *Nicotiana*, where segments of T-DNA are naturally present in the genomes of several species (Aoki and Syono, 1999).

For further information on tumor formation by *Agrobacterium* the reader is referred to books edited by Kahl and Schell (1982) and by Spaink *et al.* (1998). Before specifically considering the use of *Agrobacterium* for the transformation of filamentous fungi, this chapter will provide a detailed summary of the biological mechanisms through which the bacterium can mediate trans-kingdom genetic transfer.

2. *Agrobacterium*

Agrobacteria are Gram-negative organisms that belong to the bacterial family, *Rhizobiaceae*, and thus are closely related to the rhizobia, which induce nitrogen-fixing root nodules on leguminous plants. Besides the species *Agrobacterium tumefaciens*, the inducer of crown gall, the species *Agrobacterium radiobacter*, which is non-pathogenic, and *Agrobacterium rhizogenes*, which is responsible for the plant disease hairy root, are known. Hairy root is characterized by a massive proliferation of roots from the infection sites on plants and is due to the genetic transformation of plant cells by the T-DNA from *A. rhizogenes*. The latter T-DNA contains oncogenes differing from those present on the T-DNA of *A. tumefaciens*, but opines are produced by the hairy roots just as they are by crown gall tumor cells. The classification of *Agrobacterium* into the species *A. tumefaciens, A. radiobacter*, and *A. rhizogenes* was questioned, when it became clear that the phytopathogenic properties of the bacteria were largely determined by genes on a plasmid of about 200–300 kb (van Larebeke *et al.*, 1974). Such a virulence plasmid was absent from the non-pathogenic species *A. radiobacter*. The plasmid in *A. tumefaciens* was called the Tumor inducing (Ti)-plasmid, and that in *A. rhizogenes* the Root inducing (Ri)-plasmid. The T-region forms a small part (about 20 kb or less) of the Ti and Ri plasmids, and in addition to this region, the Ti and Ri plasmids contain an area with virulence genes, which are not transferred to plant cells (see next paragraph), and a segment with genes that confer upon the bacterium the ability to utilize specific opines as carbon, nitrogen, and energy source (Figure 4.2). Ti and Ri plasmids are conjugative plasmids, but their conjugation genes are not involved in virulence. It is interesting that the conjugation genes are under opine control (Petit *et al.*, 1978); their expression depends on the presence of a specific opine: octopine in the case of octopine Ti plasmids. Thus their transfer to recipient bacteria occurs preferentially in crown gall tumors, which explains early findings of transfer of virulence from virulent to avirulent agrobacteria in the tumors (Kerr, 1969).

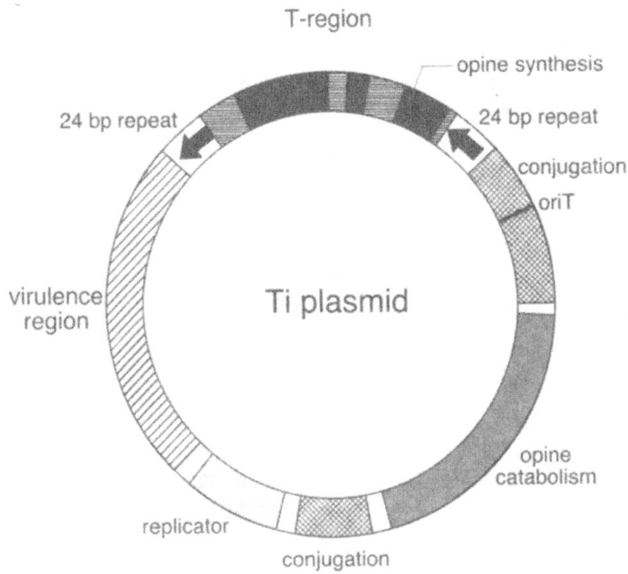

Figure 4.2. Map of the Ti plasmid.

On the basis of a large number of characteristics, three so-called biotypes can be distinguished in the agrobacteria (Kerr and Panagopoulos, 1977). Biotype 1 contains most of the *A. tumefaciens* and *A. radiobacter* strains, which have a maximum growth temperature of 37°C and produce ketolactose from lactose. Biotype 2 does not grow at temperatures above 29°C, does not produce ketolactose, and embraces most of the *A. rhizogenes* strains. Biotype 3 contains intermediate strains that have a maximum growth temperature of 35°C and that are often virulent specifically on grapevines. Hence the latter strains are considered a separate species called *A. vitis* (Ophel and Kerr, 1990). That rhizobia are closely related to the agrobacteria follows not only from a large number of shared phenotypic characteristics but also the corroborating finding that transfer of a Ti plasmid to *Rhizobium trifolii* renders this bacterium tumorigenic on plants (Hooykaas *et al.*, 1977). Conversely introducing the large Sym (Symbiosis) plasmid of this *Rhizobium* into *A. tumefaciens* allows the latter bacterium to induce root nodules in clover (Hooykaas *et al.*, 1981). However, the tumorigenicity of rhizobia with the Ti plasmid is always less compared to that of *Agrobacterium* with the same Ti plasmid. *Sinorhizobium meliloti* strains with a Ti plasmid remain completely avirulent (van Veen *et al.*, 1989). Similarly, nodulation is always delayed and is usually not accompanied by nitrogen fixation after inoculation with an *Agrobacterium* strain harboring a Sym plasmid. This suggests that co-evolution must have occurred between the large plasmid and the chromosome, leading to an optimal interplay between the genes on these two different genetic elements.

Several chromosomal genes that are needed for tumorigenesis by *Agrobacterium* have been identified. Some of these are involved in the capacity of the bacterium to attach to plant cells, one of the essential and first steps in tumorigenesis (Matthysse, 1987). Attachment is a two step process: after initial association, the bacteria become firmly bound to the plant cells by the production of cellulose fibrils (Matthysse *et al.*, 1981). Other chromosomal virulence genes, such as that encoding the global pH sensor ChvG, regulate the expression of genes that are necessary in the acidic plant environment (Li *et al.*, 2002). One of these genes encodes for a catalase that may protect *Agrobacterium* against peroxide produced by plant cells. Absence of this catalase leads to reduced virulence (Xu and Pan, 2000). Recently, the genomic sequence of *Agrobacterium tumefaciens* strain C58 was determined by two consortia (Goodner *et al.*, 2001; Wood *et al.*, 2001). This showed that *Agrobacterium* has a 5.7 Mb genome with four replicons, a circular chromosome (2.8 Mb), a linear chromosome (2.1 Mb), and two plasmids: a Ti plasmid (214 kb) and a cryptic plasmid (543 kb).

3. Host Range

Agrobacterium tumefaciens has long been known as a phytopathogen for a large number of dicotyledonous plants (de Cleene and de Ley, 1976). As crown gall did not occur on monocots (lilies, cereals), it was believed that the bacterial host range was restricted to the dicots. Several explanations were given for this, including that the walls of the cells of monocots were different from those of dicots and would not allow attachment of the bacteria. An alternate explanation was that if T-DNA transfer occurs in monocots, it might not lead to crown gall formation because monocot cell division is not responsive to the growth regulators produced under the direction of the T-DNA. In this case then, T-DNA transfer might be detected by the local production of opines at the infection sites. When opines and opine synthase activity were indeed found in infection sites on monocots, this hypothesis was validated (Hooykaas-van Slogteren *et al.*, 1984). Since these initial experiments, T-DNA transfer has been demonstrated to almost all plant species analyzed including important cereals such as rice (Hiei *et al.*, 1997) and maize (Ishida *et al.*, 1996). The range of organisms into which *Agrobacterium* can deliver its T-DNA has, however, expanded much more. When co-cultivations were done on (nitrocellulose) filters (Figure 4.3), the yeast *Saccharomyces cerevisiae*, was found to be an effective recipient of the *Agrobacterium* T-DNA (Bundock *et al.*, 1995). Similar observations were subsequently made with other yeasts (Bundock *et al.*, 1999) and with filamentous fungi (de Groot *et al.*, 1998). Recently, DNA transfer from *Agrobacterium* even to mammalian cells has been demonstrated (Kunik *et al.*, 2001), but in as much as transfer occurred in the absence of *vir*-inducers (see below) and at 37°C, it is possible that mammalian cell transformation occurs by a molecular mechanism that is different than that for the transformation of plants, yeasts, and filamentous fungi.

4. T-DNA Transfer Resembles Bacterial Conjugation

The method by which *Agrobacterium* delivers the T-DNA into recipient cells resembles bacterial conjugation. The genes involved in T-DNA transfer were initially identified by mutations that led to avirulence in the bacterium. Hence they were named: *vir(ulence)* genes. They are located outside the T-DNA, in a region of the Ti plasmid that is adjacent to the T-DNA, a segment known as the Virulence region (Hooykaas and Beijersbergen

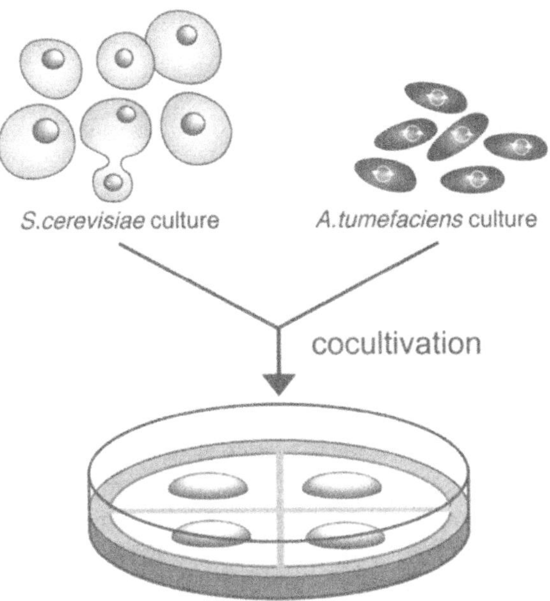

Figure 4.3. T-DNA transfer to yeast.

1994; Zhu *et al.*, 2000). This region is distinct from the area with the genes that mediate conjugative transfer of the Ti plasmid between bacteria. The Vir-region occupies about 40 kb with 20–30 genes encoding proteins that either are directly involved in the transfer process or provide accessory functions needed for successful interaction with (particular) plants. The *vir*-regions of different types of Ti plasmids vary somewhat in gene composition. The *vir*-genes are organized in a number of operons (Figure 4.4) that together form a regulon controlled by a two-component regulatory system consisting of the chemoreceptor VirA and the transcriptional activator VirG (Winans, 1992). The genes of the *vir*-regulon are silent, unless they are induced by so-called *vir*-inducers, compounds that activate the chemoreceptor VirA. Naturally *vir*-inducers are present in plant exudates. Purification of the active factors from tobacco resulted in the identification of (hydroxy)acetosyringone (Figure 4.5) as a *vir*-inducer (Stachel *et al.*, 1985). Further work showed that many plant phenolic compounds, including well-known lignin precursors such as coniferyl alcohol (Figure 4.5), can act as *vir*-inducers (Spencer and Towers, 1988; Melchers *et al.*, 1989b). It is noteworthy that induction requires not only a phenolic inducer but also the appropriate physical conditions, that is a low pH (5–6) and a temperature below 28 °C (Stachel *et al.*, 1985). These conditions resemble the natural circumstances encountered by *Agrobacterium* in plants. Upon stimulation the chemoreceptor VirA, which forms a dimer in the inner membrane of the bacterium, becomes autophosphorylated on a histidine residue. It can then phosphorylate the transcriptional activator VirG on a specific aspartate residue (Figure 4.6). The phosphorylated VirG then induces transcription from the *vir*-promoters after binding to a specific DNA sequence in these *vir*-promoters, known as the *vir*-box, with the consensus sequence 5′-CAATTGAAA-3′ (Winans, 1992). Certain monosaccharides enhance the expression of the virulence genes by a synergistic action with the phenolic signal molecules

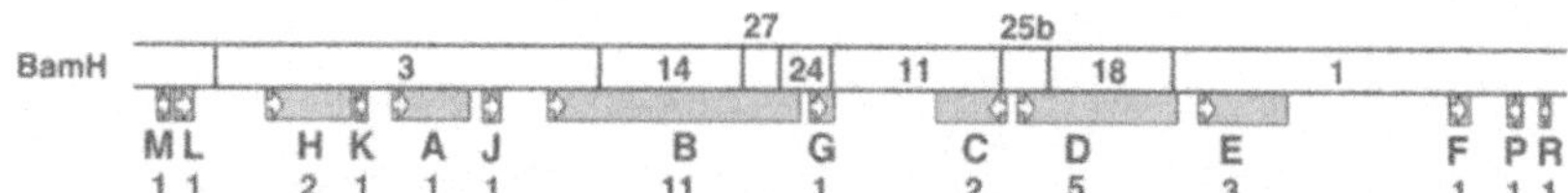

Figure 4.4. Map of the virulence region of the octopine Ti plasmid. The different operons are indicated and the numbers of genes they contain.

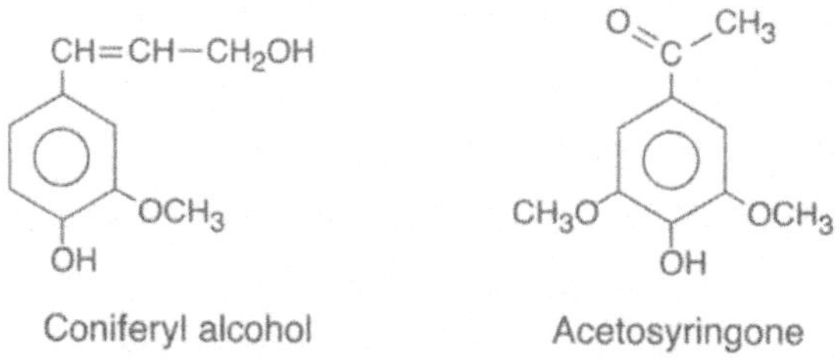

CH=CH−CH$_2$OH
OCH$_3$
OH
Coniferyl alcohol

O=C−CH$_3$
CH$_3$O
OCH$_3$
OH
Acetosyringone

Figure 4.5. Structures of two *vir*-inducers.

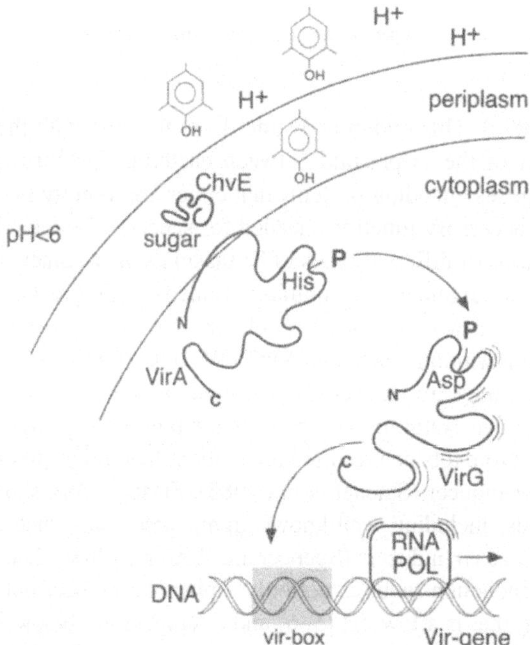

Figure 4.6. Activation of the *vir*-genes by a phosphorylation cascade.

(Shimoda *et al.*, 1990). Active monosaccharides include glucose, xylose, arabinose, glucuronic acid, and galacturonic acid (Ankenbauer and Nester, 1990). These sugars mediate their effect by binding to a periplasmic binding protein called ChvE (Cangelosi *et al.*, 1990), which can interact with the periplasmic domain of the chemoreceptor VirA.

Different *Agrobacterium* strains may differ somewhat in virulence. For instance, octopine strains are more virulent than nopaline strains on certain solanaceous plants. That is so because their Ti plasmid carries a virulence gene called *virF*, which is absent from the nopaline Ti plasmid (Melchers *et al.*, 1990). The succinamopine strain A281 is hypervirulent on several (but not all) plant species. This is due to the presence of a *virG* ortholog on the succinamopine Ti plasmid that is activated more easily than its counterparts on other Ti plasmids (Jin *et al.*, 1987). The physical requirements (pH, temperature) for *vir*-gene induction also differ somewhat among different *Agrobacterium* strains (Turk *et al.*, 1991). Mutants of octopine strains with elevated *vir*-gene expression have been isolated. These have mutations in *virG* leading to the production of VirG protein with an N54D, an I106L, or an I77V substitution (Pazour and Das, 1992; Scheeren-Groot *et al.*, 1994). The first two mutations lead to *vir*-expression in the absence of *vir*-inducer, and the latter leads to maximal *vir*-expression in the presence of low levels of inducer. Similar supersensitivity for *vir*-inducer is seen after substitution of the periplasmic domain of VirA by that of the *E. coli* chemoreceptor Tar (Melchers *et al.*, 1989a; Turk *et al.*, 1993). The presence of such substitutions generates *Agrobacterium* strains that have an extended host range for transformation and are more proficient in the transformation of recalcitrant plants (Turk *et al.*, 1993; Hansen *et al.*, 1994). The presence of mutant or wild-type *virG* on a multicopy plasmid may also lead to an increased T-DNA donor activity (van der Fits *et al.*, 2000; Ke *et al.*, 2001).

Besides the *virA* and *virG* genes, the *virB, virC*, and *virD* operons are conserved in all the Ti and Ri plasmids. Transcription and translation of the *virB* operon results in the production of 11 different VirB proteins that form a type IV secretion channel, through which the T-DNA can move into the host cells. The *virC* and *virD* operons generate proteins that are involved in the formation of the T-DNA molecule that migrates to the host. Sequence analysis has revealed that the genes in these two *vir*-operons resemble genes involved in DNA transfer and replication (*dtr* genes) of conjugative plasmids and in mobilization (*mob* genes) of non-conjugative plasmids (Lessl and Lanka, 1994). The T-region in the Ti plasmid is surrounded by a 24-bp direct repeat. The right repeat is essential for T-DNA transfer, but the left repeat is much less important. It is now known that these 24-bp sequences functionally resemble the origins of transfer (*ori*T) of conjugative plasmids and are nicked by the VirD2 strand-transferase with assistance of the VirD1 protein (Lessl *et al.*, 1992; Lessl and Lanka, 1994). The VirD2 protein becomes covalently linked to the 5′end of the nicked strand. It is thought that the free 3′OH end of the nicked strand in the right repeat forms the starting point for rolling circle replication, liberating the DNA strand with the VirD2 protein attached (Figure 4.7). Upon arrival of the replication complex at the nick site in the left repeat, a single stranded copy of the T-region will have been released. Such ssDNA molecules can indeed be found in extracts of induced agrobacteria and are called T-strands (Stachel *et al.*, 1986). A variety of experiments have shown that these T-strands, with VirD2 still attached at the 5′end, are introduced into recipient cells by *Agrobacterium* (Chaudhury *et al.*, 1994; Tinland *et al.*, 1994; Yusibov *et al.*, 1994). The role of the VirC proteins in T-DNA processing is not clear, but the VirC1 protein can bind to a sequence called overdrive or T-DNA transfer enhancer that is located in the vicinity of the right 24-bp border sequence (Toro *et al.*, 1989). In the absence of the *virC* genes or the overdrive sequence tumorigenicity is attenuated, and T-strand production is diminished (van Haaren *et al.*, 1987). The VirC1/overdrive complex facilitates nicking of the right

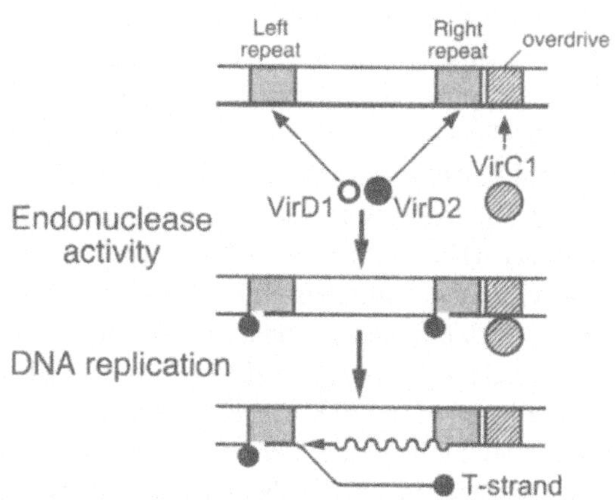

Figure 4.7. Model for the generation of T-strands in *Agrobacterium*.

border repeat by VirD2 and thereby probably enhances T-strand formation (Toro *et al.*, 1988). Although the 24-bp border repeats and adjacent "overdrive" areas differ somewhat in different Ti and Ri plasmids, these are functionally identical as seen in experiments in which "octopine" border areas were replaced by those from other Ti plasmids (van Haaren *et al.*, 1988). Generation of the T-strand resembles the formation of the ssDNA molecules that are introduced into recipient cells during conjugation; in fact, mobilizable *incQ* plasmids are successfully mobilized by the *Agrobacterium* virulence system into plant cells (Buchanan-Wollaston *et al.*, 1987). Such transfer, however, does not require the VirD2 protein or the 24-bp repeat, suggesting that the MobA strand transferase encoded by the *incQ* plasmid and the plasmid *oriT* can functionally replace these Ti plasmid functions. It has subsequently been shown that the *Agrobacterium* Vir-system can mediate bacterial conjugation: in the presence of a phenolic *vir*-inducer, *Agrobacterium* strains with the Vir-system mediate mobilization of *incQ* plasmids to other recipient bacteria, including other *Agrobacterium* strains, and *E. coli* (Beijersbergen *et al.*, 1992). Recently *vir*-mediated DNA transfer from *Agrobacterium* to *Streptomyces lividans* was described (Kelly and Kado, 2002).

A few years ago the similarity between the VirB transport system and the structure that bridges donor and recipient during bacterial conjugation became apparent. More specifically the genes of the *virB* operon resemble the genes that are present in the *tra*-operons of certain broad host range conjugative plasmids such as the *incP* plasmid RP4 (Lessl *et al.*, 1992) and the *incN* plasmid pKM101 (Pohlman *et al.*, 1994). These *tra*-operons encode a transport complex that is associated with structures on the surface of bacterial cells known as the sex-pili. Similar pili called T-pili can be detected on the surface of agrobacteria following induction of the *vir*-regulon (Lai and Kado, 2000). Formation of T-pili is a thermosensitive process, occurring much more readily at 20°C than at 28°C (Baron *et al.*, 2001). The T-pili are built from processed and circularized VirB2 monomers (Lai and Kado, 2002). The other VirB proteins are necessary for the formation of these T-pili and form a structure that spans the inner and outer membrane and that functions as a transport-pore (Lai *et al.*, 2000). The VirB1 protein is the only one among the 11 VirB proteins which is important, but not absolutely essential, for T-DNA transport. VirB1 plays an accessory role as a lysozyme-like transglycosylase which can locally open the rigid bacterial cell wall to allow formation of the VirB transport system (Bayer *et al.*, 1995; Mushegian *et al.*, 1996). The complex structure of the transport system is now beginning to be unravelled (Kumar *et al.*, 2000; Krall *et al.*, 2002; Ward *et al.*, 2002). The VirD4 protein is an inner membrane protein, which is thought to act as a so-called coupling protein, forming the interface between the nicked T-DNA or T-strand and the secretion channel (Cabezón *et al.*, 1997; Schröder *et al.*, 2002). The VirD4 protein is essential for T-DNA transfer, but not for elaboration of the T-pili. Structural analysis has shown that coupling proteins (such as VirD4) form a hexameric ring in the inner membrane with an opening of 20 Å (Gomis-Rüth *et al.*, 2002). VirD4 has a cytoplasmic ATP-binding domain, necessary for the localization of VirD4 at the cell poles (Kumar and Das, 2002).

5. Accessory Functions Enabling Trans-Kingdom DNA Transfer

The finding that the *virB* and *virCD* operons of the Ti plasmid are evolutionary related to the mating pair formation (*mpf*) and (*dtr*) operons of broad host range conjugative plasmids and thus that T-DNA transfer is a distinctive form of conjugation led to the question of what is special in the *Agrobacterium* system that enables "trans-kingdom conjugation" to plants. Some insight into this has now been obtained. First, *Agrobacterium* has chromosomal and (Ti) plasmid functions that allow it to proliferate around, on, and in plants. What these functions are is largely unknown, but one protein, called VirH2, apparently acts as a cytochrome P450-like enzyme and demethylates certain plant phenolic compounds, thereby converting them to less toxic molecules (Kalogeraki *et al.*, 1999). These compounds include *vir*-inducers such as ferulic acid, and thus VirH2 may play a role in turning off the Vir-system. Continuous exposure to acetosyringone leads to the loss of virulence by the selection of mutants with loss-of-function mutations in the *vir*-regulators *virA, virG*, or *chvE* (Fortin *et al.*, 1993). Secondly, it is now clear that *Agrobacterium* delivers into host cells not only the T-strand but also a number of virulence proteins. One of these is VirD2, which, in addition to being attached to the 5′end of the T-strand, plays an essential role in delivering the T-strand to the nucleus. Its C-terminal domain has a nuclear localization sequence (NLS) that is essential for efficient transformation (Gelvin, 2000; Tzfira *et al.*, 2000). A protein in the host, called importin α, binds to NLS and mediates the import of NLS proteins into the nucleus. Other proteins which play a role in the host cell during transformation are VirE2 and VirF. Mutations have been found in the *virE2* and *virF* genes that diminish transformation frequency and that can be complemented by expressing their wild-type alleles under the control of plant transcriptional activators in (transgenic) plant recipient cells (Citovsky *et al.*, 1992; Regensburg-Tuïnk and Hooykaas, 1993). The VirE2 protein is a ssDNA binding protein which binds co-operatively to the T-strand and thus can completely coat the T-strand. This is important to protect the T-strand against host nucleases (Rossi *et al.*, 1996) and also to unknot the T-strand so that a long thin thread is formed that can pass through the nuclear pore (Tzfira *et al.*, 2000). The VirE2 protein also interacts with a host protein called Vip1 (Tzfira *et al.*, 2001), and this interaction is thought to facilitate nuclear import of the T-complex (the complex of T-strand with VirD2 and VirE2). The susceptibility of plants to *Agrobacterium* infection can indeed be increased by overexpression of the Vip1 protein (Tzfira *et al.*, 2002).

As mentioned, the *virF* gene is present on octopine, but not on nopaline Ti plasmids. This generates a difference between octopine and nopaline strains in their host range for tumor formation (Hooykaas *et al.*, 1984; Melchers *et al.*, 1990). The VirF protein is a 22 kD hydrophilic protein that interacts in the host cell with orthologs of the yeast Skp1 protein (Schrammeijer *et al.*, 2001). To date, the VirF protein is the only prokaryotic protein known to contain an F-box domain by which it can interact with Skp1 proteins. Mutations in the F-box of VirF lead to a loss of the capacity to interact with Skp1 proteins and to diminished virulence of the *Agrobacterium* host (Schrammeijer *et al.*, 2001). This indicates that the interaction of VirF with Skp1 proteins is part of the virulence function of VirF. It is known that both F-box proteins and Skp1 proteins are subunits of a class of E3 ubiquitin ligases referred to as SCF complexes (Skp1/Cdc53/F-box protein). These complexes target specific proteins for proteolysis by the proteasome and play an important role in processes such as the cell cycle and signal transduction (Patton *et al.*, 1998). Thus, VirF may be involved in the targeted proteolysis of specific proteins in the host cell in early stages of the transformation process. Further work will be necessary to determine whether this hypothesis is correct and, if so, which proteins need to be removed from the cell to enable efficient transformation. There is independent evidence that interaction with *Agrobacterium* leads to immediate and significant changes in the expression of a large number of genes, including defense and stress response genes, in the plant host cell (Ditt *et al.*, 2001). Whether this is due to the mere presence of bacterial factors, to the transport of the T-complex, or to specific proteins is not known at the moment.

6. Protein Translocation from *Agrobacterium* into Host Cells

The VirD2 protein is introduced into host cells because it is covalently linked to the T-strand. Initially it was also thought that the ssDNA binding protein VirE2 was bound to the T-strand in the bacterium and that a T-complex consisting of T-strand, one copy of VirD2, and numerous copies of VirE2 was transferred from the bacterium into the host cells (discussed in the previous Section). This model did not explain how VirF, which does not bind to DNA, is transferred, and it left open the question of how oncogenic *virE2* mutants could be complemented by co-inoculation with a non-oncogenic helper strain, that is a strain with a Ti plasmid from which the T-region has been deleted (Otten *et al.*, 1984). These observations of "extracellular complementation" led to a new model, in which the VirD2-T-strand and the other transferred proteins (VirE2, VirF) move independently to the recipient cell. Evidence has accumulated over the last few years that this model is correct. In fact, binding of VirE2 to the T-strand in the bacterium is prevented by the co-expression of the small VirE1 protein together with VirE2. The VirE1 protein binds to VirE2 domains that are involved in ssDNA binding and self-association, thus preventing both the premature binding of VirE2 to ssDNA and VirE2 aggregation (Sundberg and Ream, 1999). On the other hand, VirE1–VirE2 complex formation may stabilize VirE2 and assist in VirE2 translocation to host cells (Zhao *et al.*, 2001). The carboxy-termini of VirE2 and VirF are required for their translocation to host cells by the VirB system (Vergunst *et al.*, 2000; Simone *et al.*, 2001). These parts of the proteins can be used to move heterologous proteins into host cells (Vergunst *et al.*, 2000). Fusions between VirE2 or VirF and the Cre recombinase (as a reporter) were constructed to enable monitoring of protein transport in the absence of any T-DNA transport. Targeted plant cells contained a selectable marker (the neomycin phosphotransferase II gene) that was transcriptionally inactive due to the insertion of a DNA segment between the promoter and the *nptII* coding sequence. This "blocking DNA" was flanked by 34 bp *lox* sites, and its excision by the Cre recombinase could restore activity of the *nptII* gene. Plant cells into which the Vir–Cre fusion protein had entered could be selected on medium with kanamycin. This experimental system gave direct evidence that proteins are translocated into plant cells from *Agrobacterium* even in the absence of any DNA-transport (Vergunst *et al.*, 2000). Similar experiments with yeast cells harboring an *URA3* gene flanked by *lox* sites (loss of which can be positively selected by growth on medium with 5-fluoroorotate) revealed that the translocation of these virulence proteins by *Agrobacterium* is not specific to plant cells (Schrammeijer *et al.*, 2003). The VirE2 protein can insert itself into artificial membranes and form anion-selective channels (Dumas *et al.*, 2001). It has been proposed that such VirE2 channels might facilitate the translocation of T-DNA and Vir-proteins into host cells. However, we have found that VirF protein translocation is independent of the presence of VirE2, and that virulence protein translocation occurs only if the donor *Agrobacterium* contains an intact VirB/VirD4 system. Apparently, the VirB transport system and the coupling factor together mediate not only DNA transfer but also protein translocation into host cells.

Recently, genome sequencing programs have identified VirB-like systems in a variety of human pathogens, and these appear to play important roles in the virulence of these pathogens (Baron *et al.*, 2002). *Helicobacter pylori*, the bacterium responsible for peptic ulcer disease and gastric adenocarcinoma, uses its VirB-system to deliver a 145-kDa protein called CagA into human cells; the protein interferes in signal transduction processes (Odenbreit *et al.*, 2000). The etiological agent of whooping cough, *Bordetella pertussis*, secretes the pertussis toxin by a VirB-like system (Farizo *et al.*, 2002). Other, intracellular pathogens, including *Bartonella henselae, Brucella suis*, and *Legionella pneumoniae*, which are responsible for cat scratch disease, brucellosis, and Legionnaire's disease, respectively, translocate effector proteins into macrophages. These effector proteins protect the bacteria from the degradative enzymes of the lysosomes by preventing fusion of these organelles with the vacuoles in which the bacteria reside (O'Callaghan *et al.*, 1999; Segal and Shuman, 1999; Schmiederer and Anderson, 2000). The VirB-like systems are now classified as type IV secretion systems (Christie and Vogel, 2000).

7. T-DNA Integration

The T-DNA integrates randomly into the plant cell genome by nonhomologous recombination (Tinland, 1996). During integration the ends of the T-DNA are usually fairly well preserved. Often the two nucleotides of the right border repeat that were the linkage points for VirD2 during transport into the plant cell nucleus are still present in the integrated T-DNA. The left end is less well preserved: several nucleotides of the left border repeat are usually absent; sometimes the deletions are larger and include small internal parts of the T-DNA. In as much as microhomology of 2–5 nucleotides is often present between this left end of the T-DNA and the integration site in the host, it has been postulated that the integration reaction starts by annealing of the left end of the T-strand and host genomic sequences at the site of microhomology. The right end is possibly directly ligated to the host genomic sequence in a later step of the integration process. This end is better preserved and may be prevented from initial annealing/recombination reactions by the presence of the VirD2 protein. Integration of the T-DNA almost invariably results in the formation of small (less than 100 bp) deletions in the host genome, and is regularly accompanied by the co-integration of "filler" DNA, sequences originating from elsewhere in the host genome or from an unknown origin. In rare cases, large parts from another area of the genome may be copied to the integration site during the integration resulting in genomic duplication (Risseeuw *et al.*, 1995; Tax and Vernon, 2001). Infrequently T-DNA integration may be accompanied by a large deletion or a genomic translocation (Kaya *et al.*, 2000; Tax and Vernon, 2001). Transformed plant cells may have a single copy of the T-DNA in their genome or multiple copies, which can be present at one integration site or at multiple loci. When present at one locus, the T-DNAs may occur as direct repeats or as inverted repeats or a mixture of both. Precise end-to-end fusions can be found between two right border ends, but imprecise fusions and filler DNA are seen with fusions involving the left border end (de Buck *et al.*, 1999; Krizkova and Hrouda, 1998). The T-DNA can also be targeted to a specific locus in the plant genome by homologous recombination (Offringa *et al.*, 1990), but this is a rare event occurring only in 1 out of 100,000 T-DNA integrations. It is the plant host which determines this bias for non-homologous integration. With the yeast *Saccharomyces cerevisiae*, as a T-DNA recipient, integration into the genome occurs preferentially by homologous recombination (Bundock *et al.*, 1995). However, the T-DNA integrates by non-homologous recombination in the yeast genome, if the T-DNA lacks homology with the yeast genome (Bundock and Hooykaas, 1996). This enabled a search for the recombination proteins that mediate T-DNA integration in yeast which led to the identification of *YKU70, YKU80, RAD50, XRS2, MRE11*, and *LIG4* genes as necessary participants (van Attikum *et al.*, 2001). The products of these genes are known to be involved in a process called nonhomologous end-joining (NHEJ), which mediates repair of genomic DNA double strand breaks (DSBs) by ligation of the broken ends. These repair events are often accompanied by the integration of extrachromosomal DNA molecules such as transposons or mitochondrial DNA molecules (Moore and Haber, 1996; Ricchetti *et al.*, 1999). Further work has shown that the T-DNA can indeed be captured at sites where DSBs are deliberately introduced in the genomes of plants (Salomon and Puchta, 1998) or yeast (van Attikum and Hooykaas, unpublished). In *YKU70* mutants no T-DNA integrations were observed, but in the *RAD50, XRS2*, and *MRE11* mutants some residual T-DNA integrations were seen. Determination of the integration positions revealed that these invariably were present at the (sub)telomeric repeats (van Attikum *et al.*, 2001). Apparently in the absence of the Rad50-Xrs2-Mre11 complex, the T-DNA can still be infrequently integrated by a protein complex which includes Yku70 at the chromosomal telomeres. It is interesting to note that the proteins identified in yeast, *viz.*, the Yku70–Yku80 heterodimer, the Rad50–Xrs2–Mre11 complex, and Ligase 4, are conserved in other eukaryotic organisms including animals and plants. This can explain why T-DNA integration occurs by a similar mechanism in all the eukaryotic organisms where this has been studied.

8. *Agrobacterium*-Based Vector Systems

Transfer of the T-DNA to host cells is mediated by the products of the genes that are located in the Virulence region of the Ti plasmid. None of the genes that are naturally located within the T-region is involved in transfer. Rather, these genes encode enzymes for the production of phytohormones (indole acetic acid, isopentenyl-AMP) and opines, genes that are expressed in the plant cells because they possess plant transcription regulation signals. This, together with the finding that transposons that land in the T-region in the bacterium are co-transferred to plant cells with the rest of the T-region, suggested that it would be possible to use the Ti plasmid as a vector for plant transformation by adding the genes of interest to the T-region. At the same time, the oncogenes responsible for tumorigenesis could be removed. As the Ti plasmid is very large, direct cloning of new genes into the T-region is not readily accomplished; two vector strategies were developed to solve this problem: intermediate vectors, and binary vectors. Intermediate vectors are standard *E. coli* vectors which contain a small part from the T-region of the Ti plasmid. As they cannot replicate in *Agrobacterium*, they can only survive in *Agrobacterium* by homologous recombination with the Ti plasmid, and this leads to a Campbell-type integration of the intermediate vector in the T-region of the Ti plasmid. *Agrobacterium* then delivers the T-region with the entire intermediate vector into host cells. An acceptor Ti plasmid frequently used is pGV3850, a Ti plasmid in which the internal part of the T-region including the oncogenes has been replaced by the *E. coli* vector pBR322 (Zambryski *et al.*, 1983). This allows the utilization of standard *E. coli* vectors as intermediate vectors. Binary vectors, on the other hand, are based on the principle that a T-region physically separated from the rest of the Ti plasmid and present on another replicon (a plasmid compatible with the Ti plasmid or the chromosome) is still efficiently transferred to host cells by the virulence system (Hoekema *et al.*, 1983; Hoekema *et al.*, 1984). Binary vectors contain an artificial T-region with restriction enzyme sites into which genes of interest can be directly cloned (Bevan, 1984). The T-regions of such binary vectors can be delivered into host cells by *Agrobacterium* strains harboring so-called helper plasmids, that is Ti plasmids with an intact virulence system but lacking the T-region. Oncogenic strains may also be used as donors, but then there is the risk that the oncogenic T-DNA is cotransferred. Well known helper strains include the octopine strains LBA4404 and MOG101, the nopaline strain MOG301 and the succinamopine strains EHA105 and AGL0 (Hoekema *et al.*, 1983; Lazo *et al.*, 1991; Hood *et al.*, 1993). The octopine strain GV2260 can be similarly used as a helper strain; in this strain the entire T-region has been replaced by pBR322, which allows maintenance of introduced pBR322 derivatives by homologous recombination (Deblaere *et al.*, 1985). If such a pBR322 derivative carries a T-region, this can be re-introduced into the helper plasmid by homologous recombination after genes of interest have been cloned into it. Helper strains may also contain extra copies of wild-type or constitutive *virG* on a separate, compatible plasmid to increase T-DNA transfer (Hansen *et al.*, 1994; van der Fits and Memelink, 2000). The size of the T-region does not seem to have a maximum. For transfer of large segments of DNA, derivatives of BAC-vectors called BIBAC-vectors have been developed that allow large segments of DNA to be cloned between 24-bp border repeats. With these vectors large 150-kb segments of DNA were transferred and integrated into the plant genome without alteration (Hamilton *et al.*, 1996). Transfer of such large T-DNAs occurs with a frequency much lower than that of the standard sized T-DNAs.

Plant cells that have been transformed by the T-DNA can be recognized by the presence of a selectable marker or reporter gene that was inserted on the T-DNA. Selectable markers include antibiotic resistance genes from bacterial R plasmids, tailored for expression in plants by the addition of plant transcription signals at both their 5′ and 3′ ends.

Plant cells are sensitive to antibiotics due to the prokaryotic heritage of their chloroplasts and mitochondria. Frequently used resistance genes include those encoding the neomycin phosphotransferase II (nptII) giving resistance to kanamycin and geneticin (G418) and the hygromycin phosphotransferase (hpt). Otherwise herbicide resistance genes can used, including mutant forms of acetolactate synthase conferring resistance to sulfonylurea herbicides (Haughn *et al.*, 1988). Reporter genes have been similarly adapted for use in plants. They include those encoding β-glucuronidase (Jefferson *et al.*, 1987), luciferase (Ow *et al.*, 1986), and green fluorescent protein (Haseloff *et al.*, 1997). Inappropriate expression of these plant adapted genes in *Agrobacterium* can be silenced by the addition of an intron to the gene construct (Vancanneyt *et al.*, 1990). With such reporters, T-DNA transfer can be detected within 48 hours of co-cultivation (Janssen and Gardner, 1989).

In time, reporter gene activity fades away in many cells, suggesting that the incoming T-DNA is expressed while still in an extrachromosomal state and that, in many cells receiving T-DNA, it does not integrate into the genome and is lost (Janssen and Gardner, 1989). Cells that maintain stable expression have an integrated copy of the T-DNA. However, these results can also be explained by silencing of the introduced T-DNA in many cells (Weld *et al.*, 2001). Nevertheless, experiments with T-DNA constructs expressing the Cre-recombinase provide unequivocal evidence for the notion that introduced T-DNAs can be expressed in an extrachromosomal state and then lost (Vergunst and Hooykaas, 1998). As mentioned earlier, plant cells transformed by *Agrobacterium* sometimes carry a single integrated copy of the T-DNA in their genome. More often multiple copies are present, either in one locus as a direct or inverted repeat, or in two or more loci. Co-transformation experiments have shown that plant cells may be transformed by multiple *Agrobacterium* strains at the same time (de Picker *et al.*, 1985; Prosen and Simpson, 1987). Sometimes the DNA that is introduced into the host cell is not restricted to the T-DNA segment of the binary vector. This can be due to left border skipping resulting in the transfer of DNA molecules that are larger than the expected T-DNA and include "vector" sequences (Martineau *et al.*, 1994; Ramanathan and Veluthambi, 1995; Kononov *et al.*, 1997). It has also been observed that T-DNA processing may start at the left border repeat instead of the right border repeat resulting in transfer of the "vector part" instead of the "T-DNA part." As the selection marker is located in the T-DNA part such events are detected only if they are accompanied by (right) border skipping resulting again in transfer of the vector part together with the T-DNA part (van der Graaff *et al.*, 1996). Transfer of "vector" sequences has also been observed in experiments with yeast (Bundock *et al.*, 1995) and fungi (Covert *et al.*, 2001).

9. Transformation of Yeasts and Filamentous Fungi

The yeast *S. cerevisiae* can be most efficiently transformed by plasmids that contain either a yeast replicator, such as an autonomous replicating sequence (ARS) from one of the chromosomes or the replicator of the $2\,\mu$ plasmid, or a segment of DNA homologous with the yeast genome enabling integration by homologous recombination. If this is not the case, transformation is inefficient due to the low frequency of integration by non-homologous recombination. Using *Agrobacterium* for transformation similar results are obtained. Binary vectors with the $2\,\mu$ replicator (Bundock *et al.*, 1995), an ARS (Piers *et al.*, 1996), or a segment of yeast DNA (Bundock *et al.*, 1995; Risseeuw *et al.*, 1996) between the 24-bp border repeats gave transformation at a frequency 100-fold higher than that of binary vectors lacking any of these elements (Bundock and Hooykaas, 1996). The T-DNA with an independent 2 μ plasmid replicator was maintained as a circular plasmid by the transformants. Circularization appeared to occur by exact fusion of the transferred parts of the left and right 24-bp repeats, resulting in one complete (mixed) border repeat. The T-DNAs with a segment of yeast DNA integrated into the genome by homologous recombination. Integration of non-homologous T-DNA occurred by non-homologous recombination and was mediated by the non-homologous end-joining enzymes (van Attikum *et al.*, 2001). Almost invariably, single copy T-DNA integrations were obtained in yeast (Bundock and Hooykaas, 1996; Bundock *et al.*, 2002). Markers that were used to select for T-DNA transfer to yeast included genes such as *URA3* and *TRP1* by which auxotrophic mutations were complemented and the KanMX gene conferring resistance to the antibiotic G418. T-DNA transfer to yeast by *Agrobacterium* was dependent upon the proper conditions of *vir*-induction (presence of a phenolic inducer such as acetosyringone, low pH, and room temperature). A detailed protocol for yeast transformation by *Agrobacterium* can be found at the BioMed Net website of Technical Tips Online (van Attikum *et al.*, 2002). The essential virulence operons (*virA, virB, virD, virG*) are also essential for T-DNA transfer to yeast. In the absence of the *virE* operon the frequency of transfer is up to 10-fold lower, but the lack of *virC, virH,* or *virF* mutations did not reduce the efficiency of transfer to yeast (Bundock *et al.*, 1995). It might be questioned whether the T-DNA is a good substrate for the homologous recombination machinery of the cell, since it is complexed by both VirD2 and VirE2 when it enters the nucleus. However, experiments with yeast have shown that T-DNA is an excellent substrate for homologous recombination. Both replacement and integration (after circularization of the T-DNA) events were obtained, and the T-DNA could be used as an insertion vector for gap repair in the yeast genome (Risseeuw *et al.*, 1996). Homologous T-DNA integration was dependent on the Rad52p, and less so on the Rad51p protein of yeast (van Attikum and Hooykaas, 2003). The yeast *Kluyveromyces lactis* has machinery for non-homologous recombination that is more efficient than that of *S. cerevisiae*, and gene disruption by gene targeting is therefore more difficult in *K. lactis*. Interestingly, gene replacement in this organism was more effective with homologous DNA presented as T-DNA rather than as a plasmid construct. The proportion of disruptants among the transformants was 71% with the T-DNA as compared to 1% for the plasmid construct (Bundock *et al.*, 1999). This indicates that the T-complex may be a substrate which can interact with the homologous recombination apparatus of the cell more efficiently than plasmid DNA.

In 1998 de Groot *et al.* published the finding that filamentous fungi can also be transformed by *Agrobacterium.* Species included ascomycetes, but also the basidiomycete *Agaricus bisporus* (the cultivated white button mushroom). Since then the list of transformed fungal species has become larger every year (Table 4.1). Fungal transformants have been selected by the use of the hygromycin phosphotransferase gene as a selection marker. Fungal protoplasts, mycelia, and conidia have been used as targets for transformation. For *Agaricus bisporus* fruiting body gill tissue turned out to be an excellent target tissue for transformation after vacuum infiltration with the bacterial suspension (Chen *et al.*, 2000). As with other species transformation of fungi by *Agrobacterium* requires the presence of a phenolic inducer in the cocultivation medium to induce the virulence system, and low pH, and room temperature. Fungal transformation also requires the virulence genes essential in

Table 4.1. Fungal species transformed by *Agrobacterium*

Fungus	Reference
Agaricus bisporus	*de Groot et al.*, 1998; Chen *et al.*, 2000; Mikosch *et al.*, 2001
Aspergillus awamori	de Groot *et al.*, 1998; Gouka *et al.*, 1999
Aspergillus niger	de Groot *et al.*, 1998
Calonectria morganii	Malonek and Meinhardt, 2001
Coccidioides immitis	Abuodeh *et al.*, 2000
Colletotrichum gloeosporioides	de Groot *et al.*, 1998
Fusarium circinatum	Covert *et al.*, 2001
Fusarium oxysporum	Mullins *et al.*, 2001
Fusarium venenatum	de Groot *et al.*, 1998
Glarea lozoyensis	Zhang *et al.*, 2003
Kluyveromyces lactis	Bundock *et al.*, 1999
Mycosphaerella graminicola	Zwiers and de Waard, 2001
Neurospora crassa	de Groot *et al.*, 1998
Saccharomyces cerevisiae	Bundock *et al.*, 1995; 1996; Piers *et al.*, 1996; Risseeuw *et al.*, 1996
Suillus bovinus	Hanif *et al.*, 2002
Trichoderma reesei	de Groot *et al.*, 1998
Verticillium dahliae	Mullins *et al.*, 2001

other species, that is besides the *virA* and *virG* regulatory genes, the *virB* and *virD* genes are obligatory. However, the absence of *virE* leads only to a 2-fold lower transformation frequency, while the absence of *virC* reduces transformation 10-fold (Michielse, Ram, van den Hondel, and Hooykaas, unpublished). As in yeast, the number of (independent) transformants increases with the duration of the cocultivation time (Mullins, Chen, Romaine, Raina, Geiser, and Kang, 2001; our unpublished results), but after an incubation time longer than 2 days fungal growth may be so prominent that selection of transformants becomes difficult. Increasing the *Agrobacterium*-fungal ratio from 1 : 1 to 500 : 1 may result in 10-fold higher transformation frequencies (Abuodeh *et al.*, 2000; Mullins *et al.*, 2001). Usually, cocultivation is conducted at room temperature; however, small differences in incubation temperature may have a significant, effect on the efficiency of transformation (Michielse, Ram, van den Hondel, and Hooykaas, unpublished results). In fungi both single copy T-DNA integrations as well as multiple integrations have been seen. Whether this is dependent on the species or on the co-cultivation conditions is not clear at the moment. T-DNA integration in the fungal genome seems to occur by a similar mechanism of nonhomologous recombination as in yeasts and plants, whereby the right end of the T-DNA is in general better preserved than the left end (de Groot *et al.*, 1998). As far as this has been studied, the T-DNA shows both mitotic and meiotic stability as is the case in plants. The T-DNA can be targeted to a predetermined site in the genome by adding a segment of homology with the fungal genome to the T-region. By selection for restoration of a 3′ deleted *pyrG* locus, in *Aspergillus awamori*, the only transformants obtained were those with the *pyrG* locus restored by homologous recombination (Gouka *et al.*, 1999). Transgenes that were present in the T-DNA between the areas at which homologous recombination occurred were integrated at the same time at the *pyrG* locus. Gene disruptions were made in the genome of the phytopathogenic fungus *Mycosphaerella graminicola* by transforming the organism with a T-DNA containing a gene targeting construct. Among the transformants obtained, about

44% were disruptants (Zwiers and de Waard, 2001). Relatively long flanking sequences of 2.6 and 4.1 kb of homologous DNA were required to obtain homologous recombinants. The use of constructs containing homologous flanking sequences of up to 2.2 kb did not lead to successful gene disruption. Similar results were recently obtained with the fungus *Glarea lozoyensis* (Zhang *et al.*, 2003). It seems that the *Agrobacterium* vector system forms an ideal tool for the transformation of fungi.

10. Concluding Remarks

The plant research community has adopted the *Agrobacterium* vector system as the standard tool for the genetic modification of plants, and this system can offer many advances to the fungal research community as well. For species such as the phytopathogenic ascomycete *Calonectria morganii* and the mushroom *Agaricus bisporus*, for which no stable transformants can be obtained by a variety of other transformation protocols, the *Agrobacterium* system offers a unique tool for research and genetic engineering (Malonek and Meinhardt, 2001; Mikosch *et al.*, 2001). Even for species that can be transformed by other means, the *Agrobacterium* system may have certain advantages and novel assets. The use of *Agrobacterium* as a vector is easy and cheap, and with some optimization, delivers more transformants than other methodologies. Moreover, the T-DNA integrates often in a single copy and in the absence of vector backbone sequences and remains stable in mitosis and meiosis. The types of genome rearrangements seen to accompany T-DNA integration in plants have so far not been observed in yeasts and fungi. In contrast, significant plasmid and genomic DNA rearrangements have been observed in the genome of *Aspergillus fumigatus* after the integration of electroporated plasmid constructs (Firon *et al.*, 2002), and the use of electroporation to build libraries of fungal insertional mutants was therefore considered to have relatively limited value. *Agrobacterium tumefaciens* as an insertional mutagen is well established. Libraries of T-DNA insertion mutants of *Arabidopsis thaliana* are available to the *Arabidopsis* research community from which specific knock out mutants can now easily be obtained (Krysan *et al.*, 1999). Also a small survey has already shown that T-DNA can be used as an effective insertional mutagen in the yeast *Saccharomyces cerevisiae* (Bundock *et al.*, 2002). T-DNA vectors meant to be used for insertional mutagenesis have been modified for simultaneous use as promoter/enhancer traps or for activation tagging (Figure 4.8). In promoter/enhancer trap T-DNAs a promoterless/enhancerless reporter gene coding for GUS or GFP is located adjacent to one of the T-DNA ends. When the T-DNA lands in a transcriptionally active area, the reporter gene will be turned on and the reporter (GUS, GFP) can be used to determine the expression regulation of the integration locus. Activation tagging T-DNA vectors have a strong promoter/enhancer located next to one of the T-DNA ends. Upon insertion the presence of this promoter/enhancer will increase the transcription of genes around the integration site. This can lead to dominant mutations that become apparent by phenotypic alterations (e.g., van der Graaff *et al.*, 2002). Proper selection schemes in combination with activation tagging has led to the identification of key regulators of plant secondary metabolism (Borevitz *et al.*, 2000; van der Fits and Memelink, 2000). Because of gene redundancy the role of such genes is often difficult to determine by classic loss-of-function mutation. Similar use of activation tagging for the analysis of the secondary metabolism in fungi can be foreseen.

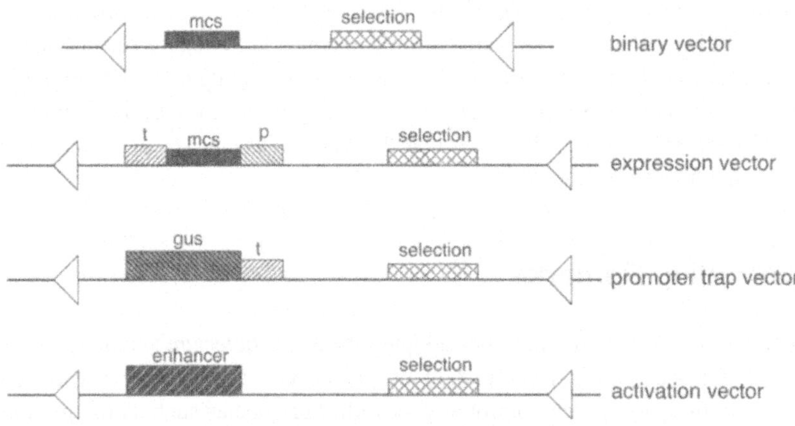

Figure 4.8. Special purpose T-DNA vectors.

The T-DNA can also be used to make gene disruptions by gene targeting, that is targeting the T-DNA to a specific locus by the presence of a segment of DNA homology. The host cell determines the efficiency with which this occurs. In plants this is a rare event occurring only in 1 out of 10,000–100,000 transformants (Offringa *et al.*, 1990). In the yeast *S. cerevisiae* this is the preferred mode of integration. In virtually 100% of the cells the T-DNA integrates by homologous recombination. In other yeasts such as *K. lactis* and the fungus *M. graminicola* there is less preference for integration by homologous recombination (Bundock *et al.*, 1999; Zwiers and de Waard, 2001). However, it has become clear that the T-complex which enters the cell nucleus seems to be a better substrate for the homologous recombination machinery than plasmid DNA (Bundock *et al.*, 1999). By increasing the length of the homologous flanking sequences in the T-DNA the frequency of integration by homologous recombination can be strongly increased in fungi (Zwiers and de Waard, 2001; Zhang *et al.*, 2003). Alternatively, it has been found that the T-DNA can be targeted to a specific site in the genome by the incorporation of a site specific recombination system such as the bacterial Cre/*lox*-system in the T-DNA so that Cre recombinase targeting of the *lox*-site containing T-DNA to a *lox*-site previously introduced in the genome becomes possible (Vergunst *et al.*, 1998). Reversely, such site-specific recombination systems have been used to remove unwanted sequences such as antibiotic resistance markers from the genome (Dale and Ow, 1991). The *Agrobacterium* Vir-system can be employed to deliver such site-specific recombinases directly in the plant or yeast cell to mediate such recombination reactions (Vergunst *et al.*, 2000).

The analysis of the conditions that are required for optimal transformation of a new species by *Agrobacterium* may be time consuming and cumbersome, but the rewards may be extremely high due to the versatility of this natural vector system.

References

Abuodeh, R.O., Orbach, M.J., Mandel, M.A., Das, A., and Galgiani, J.N. (2000). Genetic transformation of *Coccidioides immitis* facilitated by *Agrobacterium tumefaciens*. *J. Infect. Dis.* **181**, 2106–2110.

Ankenbauer, R.G., and Nester, E.W. (1990). Sugar-mediated induction of *Agrobacterium tumefaciens* virulence genes: Structural specificity and activities of monosaccharides. *J. Bacteriol.* **172**, 6442–6446.

Aoki, S., and Syono, K. (1999). Function of Ng*rol* genes in the evolution of *Nicotiana glauca*: Conservation of the function of NgORF13 and NgORF14 after ancient infection by an *Agrobacterium rhizogenes*-like ancestor. *Plant Cell Physiol.* **40**, 222–230.

Baron, C., Domke, N., Beinhofer, M., and Hapfelmeier, S. (2001). Elevated temperature differentially affects virulence, VirB protein accumulation, and T-pilus formation in different *Agrobacterium tumefaciens* and *Agrobacterium vitis* strains. *J. Bacteriol.* **183**, 6852–6861.

Baron, C., O'Callaghan, D., and Lanka, E. (2002). Bacterial secrets of secretion: EuroConference on the biology of type IV secretion processes. *Mol. Microbiol.* **43**, 1359–1365.

Bayer, M., Eferl, R., Zellnig, G., Teferle, K., Dijkstra, A., and Koraimann, G. (1995). Gene 19 of plasmid R1 is required for both efficient conjugative DNA transfer and bacteriophage R17 infection. *J. Bacteriol.* **177**, 4279–4288.

Beijersbergen, A., den Dulk-Ras, A., Schilperoort, R.A., and Hooykaas, P.J.J. (1992). Conjugative transfer by the virulence system of *Agrobacterium tumefaciens. Science.* **256**, 1324–1327.

Bevan, M. (1984). Binary *Agrobacterium* vectors for plant transformation. *Nucl. Acids Res.* **12**, 8711–8721.

Borevitz, J.O., Xia, Y., Blount, J., Dixon, R.A., and Lamb, C. (2000). Activation tagging identifies a conserved MYB regulator of phenylpropanoid biosynthesis. *Plant Cell.* **12**, 2383–2393.

Braun, A.C. (1958). A physiological basis for autonomous growth of crown gall tumor cell. *Proc. Natl. Acad. Sci. U.S.A.* **44**, 344–349.

Buchanan-Wollaston, V., Passiatore, J.E., and Cannon, F. (1987). The *mob* and *oriT* mobilization functions of a bacterial plasmid promote its transfer to plants. *Nature.* **328**, 172–175.

Bundock, P., den Dulk-Ras, A., Beijersbergen, A., and Hooykaas, P.J.J. (1995). Trans-kingdom T-DNA transfer from *Agrobacterium tumefaciens* to *Saccharomyces cerevisiae. EMBO J.* **14**, 3206–3214.

Bundock, P., and Hooykaas, P.J.J. (1996). Integration of *Agrobacterium tumefaciens* T-DNA in the *Saccharomyces cerevisiae* genome by illegitimate recombination. *Proc. Natl. Acad. Sci. U.S.A.* **93**, 15272–15275.

Bundock, P., Mroczek, K., Winkler, A.A., Steensma, H.Y., and Hooykaas, P.J.J. (1999). T-DNA from *Agrobacterium tumefaciens* as an efficient tool for gene targeting in *Kluyveromyces lactis. Mol. Gen. Genet.* **261**, 115–121.

Bundock, P., van Attikum, H., den Dulk-Ras, A., and Hooykaas, P.J.J. (2002). Insertional mutagenesis in yeasts using T-DNA from *Agrobacterium tumefaciens. Yeast.* **19**, 529–536.

Cabezón, E., Sastre, J.I., and de la Cruz, F. (1997). Genetic evidence of a coupling role for the TraG protein family in bacterial conjugation. *Mol. Gen. Genet.* **254**, 400–406.

Cangelosi, G.A., Ankenbauer, R.G., and Nester, E.W. (1990). Sugars induce the *Agrobacterium* virulence genes through a periplasmic binding protein and a transmembrane signal protein. *Proc. Natl. Acad. Sci. U.S.A.* **87**, 6708–6712.

Chaudhury, A.M., Dennis, E.S., and Brettell, R.I.S. (1994). Gene-expression following T-DNA transfer into plant cells is aphidicolin-sensitive. *Aust. J. Plant Physiol.* **21**, 125–131.

Chen, X., Stone, M., Schlagnhaufer, C., and Romaine, C.P. (2000). A fruiting body tissue method for efficient *Agrobacterium*-mediated transformation of *Agaricus bisporus. Appl. Env. Microbiol.* **66**, 4510–4513.

Chilton, M.-D., Drummond, M.H., Merlo, D.J., Sciaky, D., Montoya, A.L., Gordon, M.P., and Nester, E.W. (1977). Stable incorporation of plasmid DNA into higher plant cells: The molecular basis of crown gall tumorigenesis. *Cell.* **11**, 263–271.

Christie, P.J., and Vogel, J.P. (2000). Bacterial type IV secretion: Conjugation systems adapted to deliver effector molecules to host cells. *Trends Microbiol.* **8**, 354–360.

Citovsky, V., Zupan, J., Warnick, D., and Zambryski, P. (1992). Nuclear localization of *Agrobacterium* VirE2 protein in plant cells. *Science.* **256**, 1802–1805.

Covert, S.F., Kapoor, P., Lee, M.-H., Briley, A., and Nairn, C.J. (2001). *Agrobacterium tumefaciens*-mediated transformation of *Fusarium circinatum. Mycol. Res.* **105**, 259–264.

Dale, E.C., and Ow, D.W. (1991). Gene transfer with subsequent removal of the selection gene from the host genome. *Proc. Natl. Acad. Sci. U.S.A.* **88**, 10558–10562.

de Buck, S., Jacobs, A., van Montagu, M., and de Picker, A. (1999). The DNA sequences of T-DNA junctions suggest that complex T-DNA loci are formed by a recombination process resembling T-DNA integration. *Plant Journal.* **20**, 295–304.

de Cleene, M., and de Ley, J. (1976). The host range of crown gall. *Botan. Rev.* **42**, 389–466.

de Groot, M.J.A., Bundock, P., Hooykaas, P.J.J., and Beijersbergen, A.G.M. (1998). *Agrobacterium tumefaciens*-mediated transformation of filamentous fungi. *Nature Biotechnol.* **16**, 839–842.

Deblaere, R., Bytebier, B., de Greve, H., Deboeck, F., Schell, J., van Montagu, M., and Leemans, J. (1985). Efficient octopine Ti plasmid-derived vectors for *Agrobacterium*-mediated gene transfer to plants. *Nucl. Acids Res.* **13**, 4777–4788.

de Picker, A., Herman, L., Jacobs, A., Schell, J., and van Montagu, M. (1985). Frequencies of simultaneous transformation with different T-DNAs and their relevance to the *Agrobacterium* plant cell interaction. *Mol. Gen. Genet.* **201**, 477–484.

Ditt, R.F., Nester, E.W., and Comai, L. (2001). Plant gene expression response to *Agrobacterium tumefaciens. Proc. Natl. Acad. Sci. U.S.A.* **98**, 10954–10959.

Dumas, F., Duckely, M., Pelczar, P., van Gelder, P., and Hohn, B. (2001). An *Agrobacterium* VirE2 channel for transferred-DNA transport into plant cells. *Proc. Natl. Acad. Sci. U.S.A.* **98**, 485–490.

Farizo, K.M., Fiddner, S., Cheung, A.M., and Burns, D.L. (2002). Membrane localization of the S1 subunit of pertussis toxin in *Bordetella pertussis* and implications for pertussis toxin secretion. *Infect. Immun.* **70**, 1193–1201.

Firon, A., Beauvais, A., Latgé, J.-P., Couvé, E., Grosjean-Cournoyer, M.-C., and d' Enfert, C. (2002). Characterization of essential genes by parasexual genetics in the human fungal pathogen *Aspergillus fumigatus*: Impact of genomic rearrangements associated with electroporation of DNA. *Genetics* **161**, 1077–1087.

Fortin, C., Marquis, C., Nester, E.W., and Dion, P. (1993). Dynamic structure of *Agrobacterium tumefaciens* Ti plasmids. *J. Bacteriol.* **175**, 4790–4799.

Gelvin, S.B. (2000). *Agrobacterium* and plant genes involved in T-DNA transfer and integration. *Ann. Rev. Plant Physiol. Plant Mol. Biol.* **51**, 223–256.

Gomis-Ruth, F.X., Moncalián, G., de la Cruz, F., and Coll, M. (2002). Conjugative plasmid protein TrwB, an integral membrane type IV secretion system coupling protein. Detailed structural features and mapping of the active site cleft. *J. Biol. Chem.* **277**, 7556–7566.

Goodner, B., Hinkle, G., Gattung, S., Miller, N., Blanchard, M., Qurollo, B., Goldman, B.S., Cao, Y. *et al.* (2001). Genome sequence of the plant pathogen and biotechnology agent *Agrobacterium tumefaciens* C58. *Science.* **294**, 2323–2328.

Gouka, R.J., Gerk, C., Hooykaas, P.J.J., Bundock, P., Musters, W., Verrips, C.T., and de Groot, M.J.A. (1999). Transformation of *Aspergillus awamori* by *Agrobacterium tumefaciens*-mediated homologous recombination. *Nature Biotechnol.* **17**, 598–601.

Hamilton, C.M., Frary, A., Lewis, C., and Tanksley, S.D. (1996). Stable transfer of intact high molecular weight DNA into plant chromosomes. *Proc. Natl. Acad. Sci. U.S.A.* **93**, 9975–9979.

Hanif, M., Pardo, A.G., Gorfer, M., and Raudaskiski, M. (2002). T-DNA transfer and integration in the ectomycorrihizal fungus *Suillus bovinus* using hygromycin B as a selectable marker. *Curr. Genet.* **41**, 183–188.

Hansen, G., Das, A., and Chilton, M.-D. (1994). Constitutive expression of the virulence genes improves the efficiency of plant transformation by *Agrobacterium. Proc. Natl. Acad. Sci. U.S.A.* **91**, 7603–7606.

Haseloff, J., Siemering, K.R., Prasher, D.C., and Hodge, S. (1997). Removal of a cryptic intron and subcellular localization of green fluorescent protein are required to mark transgenic *Arabidopsis* plants brightly. *Proc. Natl. Acad. Sci. U.S.A.* **94**, 2122–2127.

Haughn, G.W., Smith, J., Mazur, B., and Somerville, C. (1988). Transformation with a mutant *Arabidopsis* acetolactate synthase gene renders tobacco resistant to sulfonylurea herbicides. *Mol. Gen. Genet.* **211**, 266–271.

Hiei, Y., Komari, T., and Kubo, T. (1997). Transformation of rice mediated by *Agrobacterium tumefaciens. Plant Mol. Biol.* **35**, 205–218.

Hoekema, A., Hirsch, P.R., Hooykaas, P.J.J., and Schilperoort, R.A. (1983). A binary plant vector strategy based on separation of *vir* and T- region of the *Agrobacterium tumefaciens* Ti-plasmid. *Nature.* **303**, 179–180.

Hoekema, A., Roelvink, P.W., Hooykaas, P.J.J., and Schilperoort, R.A. (1984). Delivery of T-DNA from the *Agrobacterium tumefaciens* chromosome into plant cells, *EMBO J.* **3**, 2485–2490.

Hood, E.E., Gelvin, S.B., Melchers, L.S., and Hoekema, A. (1993). New *Agrobacterium* helper plasmids for gene transfer to plants. *Transgenic Res.* **2**, 208–218.

Hooykaas-van Slogteren, G.M., Hooykaas, P.J.J., and Schilperoort, R.A. (1984). Expression of Ti plasmid genes in monocotyledonous plants infected with *Agrobacterium tumefaciens. Nature.* **311**, 763–764.

Hooykaas, P.J.J., and Beijersbergen, A.G.M. (1994). The virulence system of *Agrobacterium tumefaciens. Ann. Rev. Phytopathol.* **32**, 157–179.

Hooykaas, P.J.J., Hofker, M., den Dulk-Ras, H., and Schilperoort, R.A. (1984). A comparison of virulence determinants in an octopine Ti plasmid, a nopaline Ti plasmid, and an Ri plasmid by complementation analysis of *Agrobacterium tumefaciens* mutants. *Plasmid.* **11**, 195–205.

Hooykaas, P.J.J., Klapwijk, P.M., Nuti, M.P., Schilperoort, R.A., and Rörsch, A. (1977). Transfer of the *Agrobacterium tumefaciens* Ti plasmid to avirulent agrobacteria and to *Rhizobium ex planta. J. Gen. Microbiol.* **98**, 477–484.

Hooykaas, P.J.J., van Brussel, A.A.N., den Dulk-Ras, H., van Slogteren, G.M.S., and Schilperoort, R.A. (1981). Sym plasmid of *Rhizobium trifolii* expressed in different rhizobial species and *Agrobacterium tumefaciens. Nature.* **291**, 351–353.

Ishida, Y., Saito, H., Ohta, S., Hiei, Y., Komari, T., and Kumashiro, T. (1996). High efficiency transformation of maize (*Zea mays* L.) mediated by *Agrobacterium tumefaciens. Nature Biotechnol.* **14**, 745–750.

Janssen, B.J., and Gardner, R.C. (1989). Localized transient expression of GUS in leaf discs following co-cultivation with *Agrobacterium. Plant Mol. Biol.* **14**, 61–72.

Jefferson, R.A., Kavanagh, T.A., and Bevan, M.W. (1987). GUS fusions: β-glucuronidase as a sensitive and versatile gene fusion marker in higher plants. *EMBO J.* **6**, 3901–3907.

Jin, S., Komari, T., Gordon, M.P., and Nester, E.W. (1987). Genes responsible for the supervirulence phenotype of *Agrobacterium tumefaciens* A281. *J. Bacteriol.* **169**, 4417–4425.

Kahl, G., and Schell, J. (1982). *Molecular Biology of Plant Tumors.* Academic Press, New York.

Kalogeraki, V.S., Zhu, J., Eberhard, A., Madsen, E.L., and Winans, S.C. (1999). The phenolic *vir* gene inducer ferulic acid is O-demethylated by the VirH2 protein of an *Agrobacterium tumefaciens* Ti plasmid. *Mol. Microbiol.* **34**, 512–522.

Kaya, H., Sato, S., Tabata, S., Kobayashi, Y., Iwabuchi, M., and Araki, T. (2000). *Hosoba toge toge*, a syndrome caused by a large chromosomal deletion associated with a T-DNA insertion in *Arabidopsis. Plant Cell Physiol.* **41**, 1055–1066.

Ke, J., Khan, R., Johnson, T., Somers, D.A., and Das, A. (2001). High-efficiency gene transfer to recalcitrant plants by *Agrobacterium tumefaciens. Plant Cell Reports.* **20**, 150–156.

Kelly, B.A., and Kado, C.I. (2002). *Agrobacterium*-mediated T-DNA transfer and integration into the chromosome of *Streptomyces lividans. Mol. Plant Pathol.* **3**, 125–134.

Kerr, A. (1969). Transfer of virulence between isolates of *Agrobacterium. Nature.* **223**, 1175–1176.

Kerr, A., and Panagopoulos, C.G. (1977). Biotypes of *Agrobacterium radiobacter var. tumefaciens* and their biological control. *Phytopath. Z.* **90**, 172–179.

Kononov, M.E., Bassuner, B., and Gelvin, S.B. (1997). Integration of T-DNA binary vector 'backbone' sequences into the tobacco genome: Evidence for multiple complex patterns of integration. *Plant Journal.* **11**, 945–957.

Krall, L., Wiedemann, U., Unsin, G., Weiss, S., Domke, N., and Baron, C. (2002). Detergent extraction identifies different VirB protein subassemblies of the type IV secretion machinery in the membranes of *Agrobacterium tumefaciens. Proc. Natl. Acad. Sci. U.S.A.* **99**, 11405–11410.

Krizkova, L., and Hrouda, M. (1998). Direct repeats of T-DNA integrated in tobacco chromosome: Characterization of junction regions. *Plant Journal.* **16**, 673–680.

Krysan, P.J., Young, J.C., and Sussman, M.R. (1999). T-DNA as an insertional mutagen in *Arabidopsis. Plant Cell.* **11**, 2283–2290.

Kumar, R.B., and Das, A. (2002). Polar location and functional domains of the *Agrobacterium tumefaciens* DNA transfer protein VirD4. *Mol. Microbiol.* **43**, 1523–1532.

Kumar, R.B., Xie, Y.-H., and Das, A. (2000). Subcellular localization of the *Agrobacterium tumefaciens* T-DNA transport pore proteins: VirB8 is essential for the assembly of the transport pore. *Mol. Microbiol.* **36**, 608–617.

Kunik, T., Tzfira, T., Kapulnik, Y., Gafni, Y., Dingwall, C., and Citovsky, V. (2001). Genetic transformation of HeLa cells by *Agrobacterium. Proc. Natl. Acad. Sci. U.S.A.* **98**, 1871–1876.

Lai, E.-M., and Kado, C.I. (2000). The T-pilus of *Agrobacterium tumefaciens. Trends Microbiol.* **8**, 361–369.

Lai, E.-M., Chesnokova, O., Banta, L.M., and Kado, C.I. (2000). Genetic and environmental factors affecting T-pilin export and T-pilus biogenesis in relation to flagellation of *Agrobacterium tumefaciens. J. Bacteriol.* **182**, 3705–3716.

Lai, E.-M., Eisenbrandt, R., Kalkum, M., Lanka, E., and Kado, C.I. (2002). Biogenesis of T pili in *Agrobacterium tumefaciens* requires precise VirB2 propilin cleavage and cyclization. *J. Bacteriol.* **184**, 327–330.

Lazo, G.R., Stein, P.A., and Ludwig, R.A. (1991). A DNA transformation-competent *Arabidopsis* genomic library in *Agrobacterium. Bio/Technology.* **9**, 963–967.

Lessl, M., Balzer, D., Pansegrau, W., and Lanka, E. (1992). Sequence similarities between the RP4 Tra2 and the Ti VirB region strongly support the conjugation model for T-DNA transfer. *J. Biol. Chem.* **267**, 20471–20480.

Lessl, M., and Lanka, E. (1994). Common mechanisms in bacterial conjugation and Ti-mediated T-DNA transfer to plant cells. *Cell.* **77**, 321–324.

Lessl, M., Pansegrau, W., and Lanka, E. (1992). Relationship of DNA-transfer-systems: Essential transfer factors of plasmids RP4, Ti and F share common sequences. *Nucl. Acids Res.* **20**, 6099–6100.

Li, L., Jia, Y., Hou, Q., Charles, T.C., Nester, E.W., and Pan, S.Q. (2002). A global pH sensor: *Agrobacterium* sensor protein ChvG regulates acid-inducible genes on its two chromosomes and Ti plasmid. *Proc. Natl. Acad. Sci. U.S.A.* **99**, 12369–12374.

Malonek, S., and Meinhardt, F. (2001). *Agrobacterium tumefaciens*-mediated genetic transformation of the phytopathogenic ascomycete *Calonectria morganii. Nat. Rev. Mol. Cell Biol.* **40**, 152–155.

Martineau, B., Voelker, T.A., and Sanders, R.A. (1994). On defining T-DNA. *Plant Cell.* **6**, 1032–1033.

Matthysse, A.G. (1987). Characterization of nonattaching mutants of *Agrobacterium tumefaciens. J. Bacteriol.* **169**, 313–323.

Matthysse, A.G., Holmes, K.V., and Gurlitz, R.H.G. (1981). Elaboration of cellulose fibrils by *Agrobacterium tumefaciens* during attachment to carrot cells. *J. Bacteriol.* **145**, 583–595.

Melchers, L.S., Maroney, M.J., den Dulk-Ras, A., Thompson, D.V., van Vuuren, H.A.J., Schilperoort, R.A., and Hooykaas, P.J.J. (1990). Octopine and nopaline strains of *Agrobacterium tumefaciens* differ in virulence; molecular characterization of the *virF*-locus. *Plant Mol. Biol.* **14**, 249–259.

Melchers, L.S., Regensburg-Tuïnk, A.J.G., Bourret, R.B., Sedee, N.J.A., Schilperoort, R.A., and Hooykaas, P.J.J. (1989a). Membrane topology and functional analysis of the sensory protein VirA of *Agrobacterium tumefaciens. EMBO J.* **8**, 1919–1925.

Melchers, L.S., Regensburg-Tuïnk, A.J.G., Schilperoort, R.A., and Hooykaas, P.J.J. (1989b). Specificity of signal molecules on the activation of *Agrobacterium* virulence gene expression. *Mol. Microbiol.* **3**, 969–977.

Mikosch, T.S.P., Lavrijssen, B., Sonnenberg, A.S.M., and van Griensven, L.J.L.D. (2001). Transformation of the cultivated mushroom *Agaricus bisporus* (Lange) using T-DNA from *Agrobacterium tumefaciens. Curr. Genet.* **39**, 35–39.

Moore, J.K., and Haber, J.E. (1996). Capture of retrotransposon DNA at the sites of chromosomal couble-strand breaks. *Nature.* **383**, 644–646.

Mullins, E.D., Chen, X., Romaine, P., Raina, R., Geiser, D.M., and Kang, S. (2001). *Agrobacterium*-mediated transformation of *Fusarium oxysporum*: An efficient tool for insertional mutagenesis and gene transfer. *Phytopathology.* **91**, 173–180.

Mushegian, A.R., Fullner, K.J., Koonin, E.V., and Nester, E.W. (1996). A family of lysozyme-like virulence factor in bacterial pathogens of plants and animals. *Proc. Natl. Acad. Sci. U.S.A.* **93**, 7321–7326.

O'Callaghan, D., Cazevieille, C., Allardet-Servent, A., Boschiroli, M.L., Bourg, G., Foulongne, V., Frutos, P., Kulakov, Y. *et al.* (1999). A homologue of the *Agrobacterium tumefaciens* VirB and *Bordetella pertussis* Ptl type IV secretion systems is essential for intracellular survival of *Brucella suis. Mol. Microbiol.* **33**, 1210–1220.

Odenbreit, S., Püls, J., Sedlmaier, B., Gerland, E., Fischer, W., and Haas, R. (2000). Translocation of *Helicobacter pylori* CagA into gastric ephithelial cells by type IV secretion. *Science.* **287**, 1497–1499.

Offringa, R., de Groot, M.J.A., Haagsman, H.J., Does, M.P., van den Elzen, P.J., and Hooykaas, P.J.J. (1990). Extrachromosomal homologous recombination and gene targeting in plant cells after *Agrobacterium*-mediated transformation. *EMBO J.* **9**, 3077–3084.

Ophel, K., and Kerr, A. (1990). *Agrobacterium vitis* sp. nov. for strains of *Agrobacterium* biovar 3 from grapevines. *Int. J. Syst. Bact.* **40**, 236–241.

Otten, L.A.B.M., deGreve, H., Leemans, J., Hain, R., Hooykaas, P.J.J., and Schell, J. (1984). Restoration of virulence of *vir* region mutants of *Agrobacterium tumefaciens* strain B6S3 by coinfection with normal and mutant *Agrobacterium* strains. *Mol. Gen. Genet.* **195**, 159–163.

Ow, D.W., Wood, K.V., de Luca, M., de Wet, J.R., Helinski, D.R., and Howell, S.H. (1986). Transient and stable expression of the firefly luciferase gene in plant cells and transgenic plants. *Science.* **234**, 856–859.

Patton, E.E., Willems, A.R., and Tyers, M. (1998). Combinatorial control in ubiquitin-dependent proteolysis: Don't Skp the F-box hypothesis. *Trends Genetics* **14**, 236–243.

Petit, A., Delhaye, S., Tempé, J., and Morel, G. (1970). Recherches sur les guanidines des tissus de crown-gall. Mise en evidence d'une relation biochimique specifique entre les souches d'*Agrobacterium tumefaciens* et les tumeurs qu'elles induisent. *Physiol. Veget.* **8**, 205–213.

Petit, A., Tempé, J., Kerr, A., Holsters, M., van Montagu, M., and Schell, J. (1978). Substrate induction of conjugative activity of *Agrobacterium tumefaciens* Ti plasmids. *Nature.* **271**, 570–572.

Piers, K.L., Heath, J.D., Liang, X., Stephens, K.M., and Nester, E.W. (1996). *Agrobacterium tumefaciens*-mediated transformation of yeast. *Proc. Natl. Acad. Sci. U.S.A.* **93**, 1613–1618.

Pohlman, R.F., Genetti, H.D., and Winans, S.C. (1994). Common ancestry between incN conjucal transfer genes and macromolecular export systems of plant and animal pathogens. *Mol. Microbiol.* **14**, 655–668.

Prosen, D.E., and Simpson, R.B. (1987). Transfer of a ten-member genomic library to plants using *Agrobacterium tumefaciens. Bio/Technology* **5**, 966–971.

Ramanathan, V., and Veluthambi, K. (1995). Transfer of non-T-DNA portions of the *Agrobacterium tumefaciens* Ti plasmid pTiA6 from the left terminus of TL-DNA. *Plant Mol. Biol.* **28**, 1149–1154.

Regensburg-Tuïnk, A.J.G., and Hooykaas, P.J.J. (1993). Transgenic *N. glauca* plants expressing bacterial virulence gene *virF* are converted into hosts for nopaline strains of *A. tumefaciens. Nature.* **363**, 69–71.

Ricchetti, M., Fairhead, C., and Dujon, B. (1999). Mitochondrial DNA repairs double-strand breaks in yeast chromosomes. *Nature.* **402**, 96–100.

Risseeuw, E., Franke-van Dijk, M.E.I., and Hooykaas, P.J.J. (1996). Integration of an insertion-type transferred DNA vector from *Agrobacterium tumefaciens* into the *Saccharomyces cerevisiae* genome by gap repair. *Mol. Cell. Biol.* **16**, 5924–5932.

Risseeuw, E., Offringa, R., Franke-van Dijk, M.E.I., and Hooykaas, P.J.J. (1995). Targeted recombination in plants using *Agrobacterium* coincides with additional rearrangements at the target locus. *Plant Journal.* **7**, 109–119.

Rossi, L., Hohn, B., and Tinland, B. (1996). Integration of complete transferred DNA units is dependent on the activity of virulence E2 protein of *Agrobacterium tumefaciens. Proc. Natl. Acad. Sci. U.S.A.* **93**, 126–130.

Salomon, S., and Puchta, H. (1998). Capture of genomic and T-DNA sequences during double-strand break repair in somatic plant cells. *EMBO J.* **17**, 6086–6095.

Scheeren-Groot, E.P., Rodenburg, K.W., den Dulk-Ras, A., Turk, S.C.H.J., and Hooykaas, P.J.J. (1994). Mutational analysis of the transcriptional activator VirG of *Agrobacterium tumefaciens. J. Bacteriol.* **176**, 6418–6426.

Schmiederer, M., and Anderson, B. (2000). Cloning, sequencing, and expression of three *Bartonella henselae* genes homologous to the *Agrobacterium tumefaciens VirB* region. *DNA and Cell Biology.* **19**, 141–147.

Schrammeijer, B., den Dulk-Ras, A., Vergunst, A.C., Jurado-Jacome, E., and Hooykaas, P.J.J. (2003). Analysis of Vir protein translocation from *Agrobacterium tumefaciens* using *Saccharomyces cerevisiae* as a model: Evidence for transport of a novel effector protein VirE3. *Nucl. Acids Res.* (in press).

Schrammeijer, B., Risseeuw, E., Pansegrau, W., Regensburg-Tuïnk, A.J.G., Crosby, W.L., and Hooykaas, P.J.J. (2001). Interaction of the virulence protein VirF of *Agrobacterium tumefaciens* with plant homologs of the yeast Skp1 protein. *Curr. Biology.* **11**, 258–262.

Schröder, G., Krause, S., Zechner, E.L., Traxler, B., Yeo, H.J., Lurz, R., Waksman, G., and Lanka, E. (2002). TraG-Like proteins of DNA transfer systems and of the *Helicobacter pylori* type IV secretion system: Inner membrane gate for exported substrates? *J. Bacteriol.* **184**, 2767–2779.

Segal, G., and Shuman, H.A. (1999). Possible origin of the *Legionella pneumophila* virulence genes and their relation to *Coxiella burnetii. Mol. Microbiol.* **33**, 669–670.

Shimoda, N., Toyoda-Yamamoto, A., Nagamine, J., Usami, S., Katayama, M., Sakagami, Y., and Machida, Y. (1990). Control of expression of *Agrobacterium vir* genes by synergistic actions of phenolic signal molecules and monosaccharides. *Proc. Natl. Acad. Sci. U.S.A.* **87**, 6684–6688.

Simone, M., McCullen, C.A., Stahl, L.E., and Binns, A.N. (2001). The carboxy-terminus of VirE2 from *Agrobacterium tumefaciens* is required for its transport to host cells by the *virB*-encoded type IV transport system. *Mol. Microbiol.* **41**, 1283–1293.

Spaink, H.P., Kondorosi, A., and Hooykaas, P.J.J. (1998). *The Rhizobiaceae.* Kluwer Academic Publications, Dordrecht, The Netherlands.

Spencer, P.A., and Towers, G.H.N. (1988). Specificity of signal compounds detected by *Agrobacterium tumefaciens. Phytochemistry.* **27**, 2781–2785.

Stachel, S.E., Messens, E., van Montagu, M., and Zambryski, P. (1985). Identification of the signal molecules produced by wounded plant cells that activate T-DNA transfer in *Agrobacterium tumefaciens. Nature.* **318**, 624–629.

Stachel, S.E., Timmerman, B., and Zambryski, P. (1986). Generation of single-stranded T-DNA molecules during the initial stages of T-DNA transfer from *Agrobacterium tumefaciens* to plant cells. *Nature.* **322**, 706–712.

Sundberg, C.D. and Ream, W. (1999). The *Agrobacterium tumefaciens* chaperone-like protein, VirE1, interacts with VirE2 at domains required for single-stranded DNA binding and cooperative interaction. *J. Bacteriol.* **181**, 6850–6855.

Tax, F.E., and Vernon, D.M. (2001). T-DNA-associated duplication/translocations in *Arabidopsis*. Implications for mutant analysis and functional genomics. *Plant Physiology.* **126**, 1527–1538.

Tinland, B. (1996). The integration of T-DNA into plant genomes. *Trends Plant Science.* **1**, 178–184.

Tinland, B., Hohn, B., and Puchta, H. (1994). *Agrobacterium tumefaciens* transfers single-stranded transferred DNA (T-DNA) into the plant cell nucleus. *Proc. Natl. Acad. Sci. U.S.A.* **91**, 8000–8004.

Toro, N., Datta, A., Carmi, O.A., Young, C., Prusti, R.K., and Nester, E.W. (1989). The *Agrobacterium tumefaciens virC1* gene product binds to overdrive, a T-DNA transfer enhancer. *J. Bacteriol.* **171**, 6845–6849.

Toro, N., Datta, A., Yanofsky, M., and Nester, E.W. (1988). Role of the overdrive sequence in T-DNA border cleavage in *Agrobacterium. Proc. Natl. Acad. Sci. U.S.A.* **85**, 8558–8562.

Turk, S.C.H.J., Melchers, L.S., den Dulk-Ras, H., Regensburg-Tuïnk, A.J.G., and Hooykaas, P.J.J. (1991). Environmental conditions differentially affect *vir* gene induction in different *Agrobacterium* strains. Role of the VirA sensor protein. *Plant Mol. Biol.* **16**, 1051–1059.

Turk, S.C.H.J., van Lange, R.P., Sonneveld, E., and Hooykaas, P.J.J. (1993). The chimeric VirA-Tar receptor protein is locked into a highly responsive state. *J. Bacteriol.* **175**, 5706–5709.

Tzfira, T., Rhee, Y., Chen, M.-H., Kunik, T., and Citovsky, V. (2000). Nucleic acid transport in plant-microbe interactions: the molecules that walk through the walls. *Annu. Rev. Microbiol.* **54**, 187–219.

Tzfira, T., Vaidya, M., and Citovsky, V. (2001). VIP1, an *Arabidopsis* protein that interacts with *Agrobacterium* VirE2, is involved in VirE2 nuclear import and *Agrobacterium* infectivity. *EMBO J.* **20**, 3596–3607.

Tzfira, T., Vaidya, M., and Citovsky, V. (2002). Increasing plant susceptibility to *Agrobacterium* infection by overexpression of the *Arabidopsis* nuclear protein VIP1. *Proc. Natl. Acad. Sci. U.S.A.* **99**, 10435–10440.

van Attikum, H., Bundock, P., den Dulk-Ras, A., Vergunst, A.C., and Hooykaas, P.J.J. (2002). *Agrobacterium tumefaciens*-mediated transformation of *Saccharomyces cerevisiae. BioMedNet Technical Tips Online.* **1**, T02732.

van Attikum, H., Bundock, P., and Hooykaas, P.J.J. (2001). Non-homologous end-joining proteins are required for *Agrobacterium* T-DNA integration. *EMBO J.* **20**, 6550–6558.

van Attikum, H., and Hooykaas, P.J.J. (2003). Genetic requirements for the targeted integration of *Agrobacterium* T-DNA in *Saccharomyces cerevisiae. Nucl. Acids Res.* (in press).

van der Fits, L., Deakin, E.A., Hoge, J.H.C., and Memelink, J. (2000). The ternary transformation system: Constitutive *virG* on a compatible plasmid dramatically increases *Agrobacterium*-mediated plant transformation. *Plant Mol. Biol.* **43**, 495–502.

van der Fits, L., and Memelink, J. (2000). ORCA3, a jasmonate-responsive transcriptional regulator of plant primary and secondary metabolism. *Science* **289**, 295–297.

van der Graaff, E., den Dulk-Ras, A., and Hooykaas, P.J.J. (1996). Deviating T-DNA transfer from *Agrobacterium tumefaciens* to plants. *Plant Mol. Biol.* **31**, 677–681.

van der Graaff, E., Hooykaas, P.J.J., and Keller, B. (2002). Activation tagging of the two closely linked genes *LEP* and *VAS* independently affects vascular cell number. *Plant Journal* **32**, 819–830.

van Haaren, M.J.J., Sedee, N.J.A., Krul, M., Schilperoort, R.A., and Hooykaas, P.J.J. (1988). Function of heterologous and pseudo border repeats in T region transfer via the octopine virulence system of *Agrobacterium tumefaciens. Plant Mol. Biol.* **11**, 773–781.

van Haaren, M.J.J., Sedee, N.J.A., Schilperoort, R.A., and Hooykaas, P.J.J. (1987). Overdrive is a T-region transfer enhancer which stimulates T- strand production in *Agrobacterium tumefaciens. Nucl. Acids Res.* **15**, 8983–8997.

van Larebeke, N., Engler, G., Holsters, M., van den Elsacker, S., Zaenen, I., Schilperoort, R.A., and Schell, J. (1974). Large plasmid in *Agrobacterium tumefaciens* essential for crown gall-inducing ability. *Nature* **252**, 169–170.

van Veen, R.J.M., den Dulk-Ras, H., Schilperoort, R.A., and Hooykaas, P.J.J. (1989). Ti plasmid containing *Rhizobium meliloti* are non-tumorigenic on plants, despite proper virulence gene induction and T-strand formation. *Arch. Microbiol.* **153**, 85–89.

Vancanneyt, G., Schmidt, R., O'Connor-Sanchez, A., Willmitzer, L., and Rocha-Sosa, M. (1990). Construction of an intron-containing marker gene: Splicing of the intron in transgenic plants and its use in monitoring early events in *Agrobacterium*-mediated plant transformation. *Mol. Gen. Genet.* **220**, 245–250.

Vergunst, A.C., and Hooykaas, P.J.J. (1998). Cre/*lox*-mediated site-specific integration of *Agrobacterium* T-DNA in *Arabidopsis thaliana* by transient expression of *cre. Plant Mol. Biol.* **38**, 393–406.

Vergunst, A.C., Jansen, L.E.T., and Hooykaas, P.J.J. (1998). Site-specific integration of *Agrobacterium* T-DNA in *Arabidopsis thaliana* mediated by Cre recombinase. *Nucl. Acids Res.* **26**, 2729–2734.

Vergunst, A.C., Schrammeijer, B., den Dulk-Ras, A., de Vlaam, C.M.T., Regensburg-Tuïnk, A.J.G., and Hooykaas, P.J.J. (2000). VirB/D4-dependent protein translocation from *Agrobacterium* into plant cells. *Science* **290**, 979–982.

Ward, D.V., Draper, O., Zupan, J.R., and Zambryski, P.C. (2002). Inaugural article: Peptide linkage mapping of the *Agrobacterium tumefaciens vir*-encoded type IV secretion system reveals protein subassemblies. *Proc. Natl. Acad. Sci. U.S.A.* **99**, 11493–11500.

Weld, R., Heinemann, J., and Eady, C. (2001). Transient GFP expression in *Nicotiana plumbaginifolia* suspension cells: The role of gene silencing, cell death and T-DNA loss. *Plant Mol. Biol.* **45**, 377–385.

Winans, S.C. (1992). Two-way chemical signaling in *Agrobacterium*-plant interactions. *Microbiol. Rev.* **56**, 12–31.

Wood, D.W., Setubal, J.C., Kaul, R., Monks, D.E., Kitajima, J.P., Okura, V.K., Zhou, Y., Chen, L. *et al.* (2001). The genome of the natural genetic engineer *Agrobacterium tumefaciens* C58. *Science* **294**, 2317–2323.

Xu, X.Q., and Pan, S.Q. (2000). An *Agrobacterium* catalase is a virulence factor involved in tumorigenesis. *Mol. Microbiol.* **35**, 407–414.

Yusibov, V.M., Steck, T.R., Gupta, V., and Gelvin, S.B. (1994). Association of single-stranded transferred DNA from *Agrobacterium tumefaciens* with tobacco cells. *Proc. Natl. Acad. Sci. U.S.A.* **91**, 2994–2998.

Zambryski, P., Joos, H., Genetello, C., Leemans, J., van Montagu, M., and Schell, J. (1983). Ti plasmid vector for the introduction of DNA into plant cells without alteration of their normal regeneration capacity. *EMBO J.* **2**, 2143–2150.

Zhang, A., Lu, P., Dahl-Roshak, A.M., Paress, P.S., Kennedy, S., Tkacz, J.S., and An, Z. (2003). Efficient disruption of a polyketide synthase gene (*pks*1) required for melanin synthesis through *Agrobacterium*-mediated transformation of *Glarea lozoyensis. Mol. Genet. Genomics* **268**, 645–655.

Zhao, Z., Sagulenko, E., Ding, Z., and Christie, P.J. (2001). Activities of *virE1* and the VirE1 secretion chaperone in export of the multifunctional VirE2 effector via an *Agrobacterium* type IV secretion pathway. *J. Bacteriol.* **183**, 3855–3865.

Zhu, J., Oger, P.M., Schrammeijer, B., Hooykaas, P.J.J., Farrand, S.K., and Winans, S.C. (2000). The bases of crown gall tumorigenesis. *J. Bacteriol.* **182**, 3885–3895.

Zwiers, L.-H., and de Waard, M.A. (2001). Efficient *Agrobacterium tumefaciens*-mediated gene disruption in the phytopathogen *Mycosphaerella graminicola. Curr. Genet.* **39**, 388–393.

II

Special (Secondary) Metabolism

5

Fungal Polyketide Synthases in the Information Age

Russell J. Cox and Frank Glod

1. Introduction

1.1. Secondary Metabolites

In contrast to primary metabolism, that is, all processes for cell growth and function such as protein synthesis, DNA replication, and fatty acid biosynthesis, secondary metabolism encompasses the biochemical pathways that are *not essential* to growth and reproduction of the organism. Secondary metabolites constitute a huge family of chemically diverse compounds produced in the idiophase after the growth phase (tropophase) of the organism (Demain, 1986; Bennett, 1995). The true function of these compounds in the producer remains largely obscure, and many theories have been put forward as to the role of secondary metabolites in nature.

Firn and Jones (2000) propose a model that unifies conflicting views arising from the observations that most secondary metabolites show no (known!) biological activity and that only few natural products contribute to the fitness of the producer. They suggest that retaining the ability to generate chemical diversity is a prerequisite to producing a biologically active compound that benefits the organism. Their model accounts for the vast array of structurally different compounds ranging from small molecules such as 6-methylsalicylic acid (6MSA **1**) to macrolides such as rapamycin **2** (Figure 5.1). Although the role of these compounds within the organism is often unknown, man has harnessed the bioactivity of these products to his benefit since prehistoric times. Active chemicals contained in crude plant or fungal extracts formed the basis of the first herbal remedies, and they are responsible for the well-known poisonous and hallucinogenic properties associated with macrofungi.

The isolation of penicillin from a *Penicillium* mold saw the first development of a microbial secondary metabolite as a drug and marked the beginning of the modern antibiotic era. Since then more than 100,000 secondary metabolites of low molecular weight (less than 2,500 Da) have been isolated from various plant and microbial species

---

Russell J. Cox and Frank Glod • School of Chemistry, University of Bristol, Cantock's Close, Clifton, Bristol, UK.

Advances in Fungal Biotechnology for Industry, Agriculture, and Medicine. Edited by Jan S. Tkacz and Lene Lange, Kluwer Academic/Plenum Publishers, 2004.

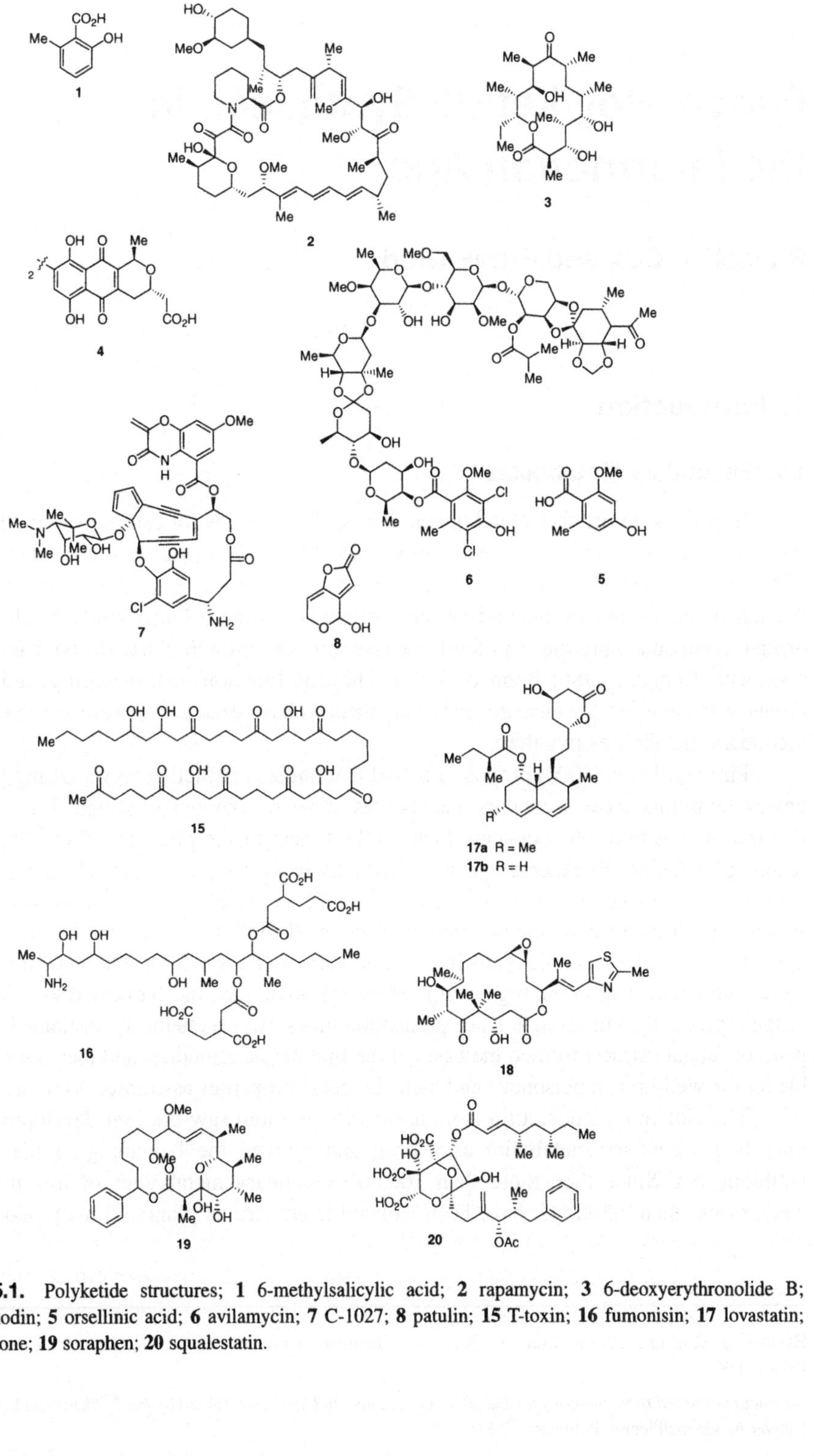

Figure 5.1. Polyketide structures; **1** 6-methylsalicylic acid; **2** rapamycin; **3** 6-deoxyerythronolide B; **4** actinorhodin; **5** orsellinic acid; **6** avilamycin; **7** C-1027; **8** patulin; **15** T-toxin; **16** fumonisin; **17** lovastatin; **18** epothilone; **19** soraphen; **20** squalestatin.

(Demain, 1999). The classes of chemical compounds isolated to date include terpenes (also known as isoprenoids), alkaloids, nonribosomally derived peptides, polyketides, and shikimate metabolites, and their pharmaceutical activities are wide-ranging. Most prolific among producers of secondary metabolites are the soil-dwelling actinomycete bacteria and filamentous fungi. They exemplify the observation that secondary metabolite production is related to occupancy of particular ecological niches rather than with phylogeny (Hunter, 1992).

The chemical study of these organisms has led to the identification of the biosynthetic pathways of these natural products starting from small precursor molecules derived from primary metabolism. With the advent of recombinant DNA technology the biosynthesis of the metabolites can now be studied on a genetic level, and this led to the isolation of numerous genes associated with secondary metabolite production, some of which will be discussed here and also in Chapters 6–8.

1.2. Polyketides

Polyketides form the largest family of structurally diverse secondary metabolites synthesized in both prokaryotic and eukaryotic organisms. Biological activities associated with polyketide metabolites encompass antibacterial, antiviral, antitumor, antihypertensive, antiinsect, and immunosuppressant compounds as well as mycotoxins and the highly carcinogenic aflatoxins (Simpson, 1995).

The biosynthesis of these compounds and others such as 6-deoxyerythronolide B (*ery*, **3**), a macrolide polyketide and antibiotic precursor produced by the actinomycete *Saccharopolyspora erythrae*, have been studied in depth. It is clear that polyketides, independent of their structural diversity, have a common biosynthetic origin. Polyketides are derived from highly functionalized carbon chains, whose assembly mechanism bears close resemblance to the fatty acid synthetic pathway (O'Hagan, 1991).

Early chemical and biochemical approaches established the relationship between polyketide and fatty acid biosynthesis, in which the carbon backbones of the molecules are assembled by the successive condensation of small acyl units (O'Hagan, 1991). The assembly process (Scheme 5.1) is controlled by multifunctional enzyme complexes called polyketide synthases (PKS) (O'Hagan, 1991; Simpson, 1995). The biosynthesis of fatty acids and polyketides are highly analogous. During fatty acid biosynthesis a *starter* acyl unit, normally acetate, is transfered from coenzyme A (CoA) onto the active site thiol of the β-ketoacyl synthase (KS). A thiolester linkage is also formed between a malonate extender unit from CoA and the pantetheine thiol of the *holo* form of the acyl carrier protein (ACP). These loading reactions are generally catalyzed by specific acyl transferase (AT) enzymes. The KS catalyzes the decarboxylative Claisen condensation between the acyl and malonyl units to generate acetoacetyl ACP. This process is repeated several times. Each condensation is followed by a cycle of modifying reactions that involves the enzymes ketoreductase (KR), dehydratase (DH), and enoylreductase (ER) in the subsequent reduction steps. During fatty acid biosynthesis the fully reduced intermediate is then subjected to further extension.

At this stage however, a major difference between fatty acid and polyketide biosynthesis becomes apparent. Fatty acid synthases (FAS) catalyze the full reduction of each β-keto moiety prior to further chain extension in every cycle. Polyketide biosynthesis, however, shows a higher degree of complexity as the reduction steps following condensation

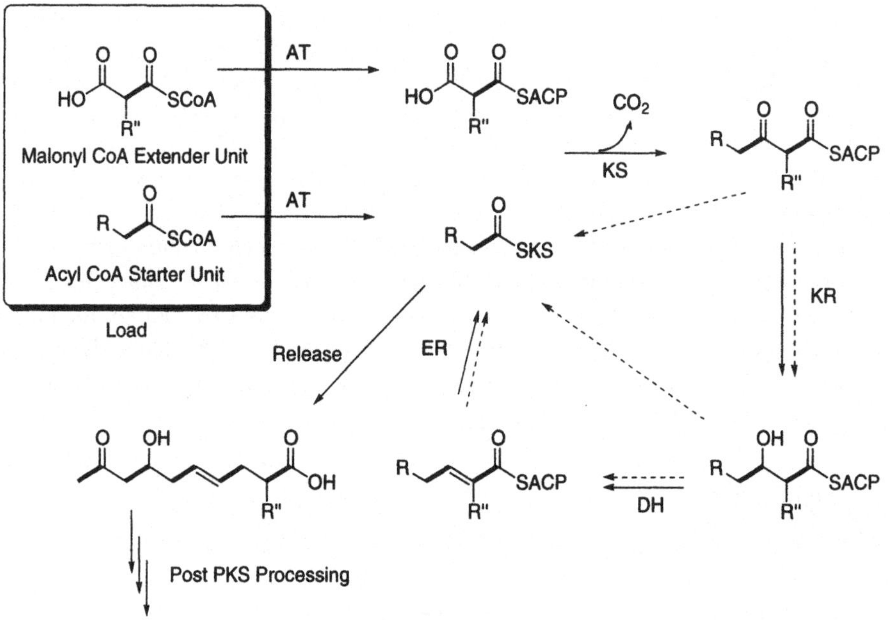

Scheme 5.1. Reactions catalyzed by FAS and PKS complexes. AT, acyl transferase; ACP, acyl carrier protein; KS, β-ketoacyl synthase; KR, β-ketoacyl reductase; DH, β-hydroxyacyl dehydratase; ER, enoyl reductase. Solid arrows show obligatory reactions catalyzed by FAS, dashed arrows show optional reactions catalyzed by PKS.

can be fully, or partially, omitted, giving highly functionalized chains. Polyketides such as orsellinic acid **5** and tetrahydroxynaphthalene **11** undergo no reductive steps during their biosynthesis and are called *non-reduced* polyketides. In other cases, hydroxyl, enoyl, or methylene functionalities are introduced during assembly when the alternative versions of the reductive cycle are followed (Scheme 5.1). The resulting compounds are known as *partially* or *highly* reduced polyketides.

The ability to use different chain starter units (such as acetate, benzoate, cinnamate, and/or amino acids) and alternate extender units by bacterial synthases (malonate, methylmalonate, and ethylmalonate) gives rise to further structural diversity among the polyketides. The assembled polyketide chain can also undergo further modifications such as cyclization, reduction or oxidation, alkylation, and rearrangement after release from the polyketide synthase.

1.3. Types of Polyketide Synthase

By analogy to FAS, the two main categories of PKS are referred to as Type I and Type II (McCarthy and Hardie, 1984). Type I FAS is characteristic of fungi and vertebrates and consists of one large multifunctional enzyme which contains all the necessary enzymatic activities as embedded catalytic domains (Figure 5.2). The Type II FAS, found in bacteria, carries out all the main chemical transformations using discrete mono-functional enzymes. If the same active sites are used more than once (for example, the Type II bacterial FAS and the Type I mammalian FAS) then the synthase is described as *iterative*

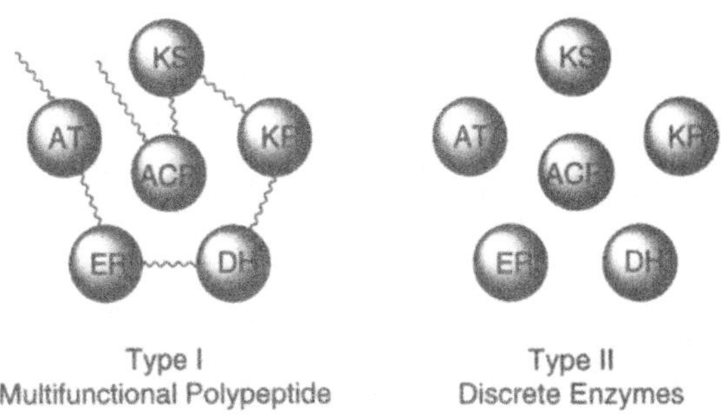

Figure 5.2. Type I and type II PKS architectures.

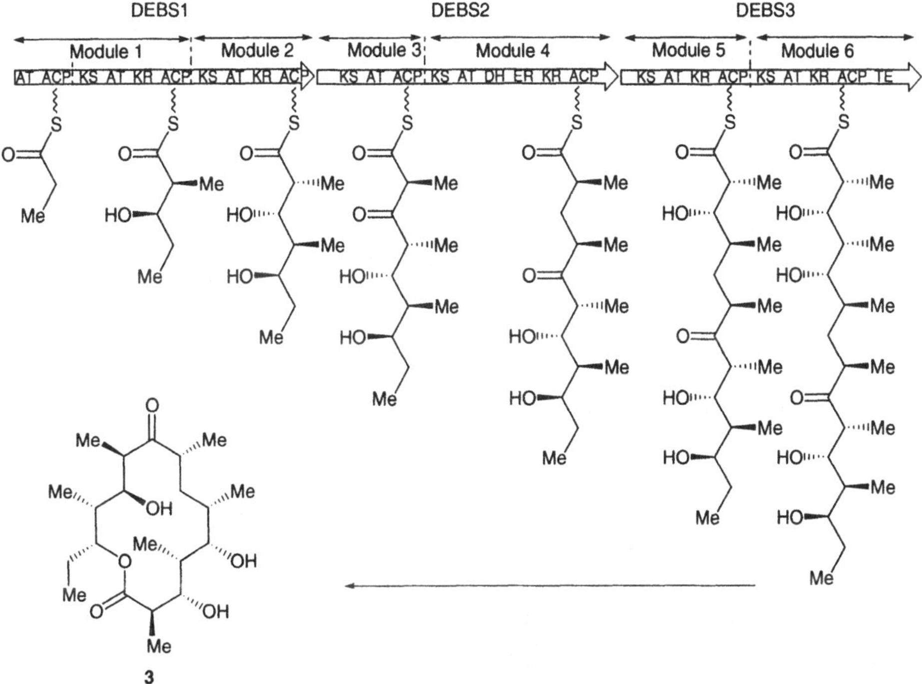

Scheme 5.2. Modular biosynthesis of the erythromycin aglycone 6-dEB **3** by the DEBS proteins. Catalytic domain abbreviations as in Scheme 5.1. TE, thiolesterase.

(Hopwood, 1997). In contrast, systems are known where each set of enzymes acts only once during the biosynthetic cycles (for example, during the biosynthesis of **3**; Scheme 5.2), and these systems are said to be *modular*.

2. Non-Fungal PKS

Most of the knowledge on the functioning of PKS derives from genetic and biosynthetic work in organisms other than fungi. The Type II PKS genes responsible for the synthesis of polycyclic aromatic compounds such as actinorhodin (*act*, **4**) were successfully cloned from bacteria in the late 1980s and early 1990s. Much of the initial genetic knowledge of the bacterial PKS was obtained by Hopwood and coworkers at the John Innes Centre in Norwich while working with *Streptomyces coelicolor* (Hopwood, 1997). Cloning of genomic fragments yielded a large set of genes encoding a Type II PKS involved in *act* **4** biosynthesis.

This work led to an important observation concerning the distribution of the PKS genes on the genome of the organism. It was found that the genes necessary for the biosynthesis of **4** were "clustered" in the same stretch of DNA (Malpartida and Hopwood, 1984; Fernandez-Moreno *et al.*, 1992). This was subsequently confirmed to be the case for all genes of aromatic PKS which meant that entire gene clusters could readily be isolated and sequenced by using fragments of the *act* cluster as gene probes. The sequences of the encoding regions involved in the biosynthesis of numerous aromatic polyketides have since been obtained (Figure 5.3; Hopwood, 1997).

Rather surprisingly, genetic analysis of the synthesis of macrolide PKS, such as those involved in the production of **3** in actinomycetes, did not show an organization similar to the synthases described above (Kao *et al.*, 1995, 1996a, 1996b; Pieper *et al.*, 1996). The early observations made by Peter Leadlay in Cambridge and Leonard Katz at Abbott Laboratories clearly suggested that the biosynthesis of **3** involved very large proteins consisting of repeated multiple sets (modules) of Type I enzyme systems (Scheme 5.2). *Ery* PKS is encoded in three open reading frames (ORFs) producing three large multifunctional proteins (DEBS 1–3), each of which contains two modules. Modular PKS were

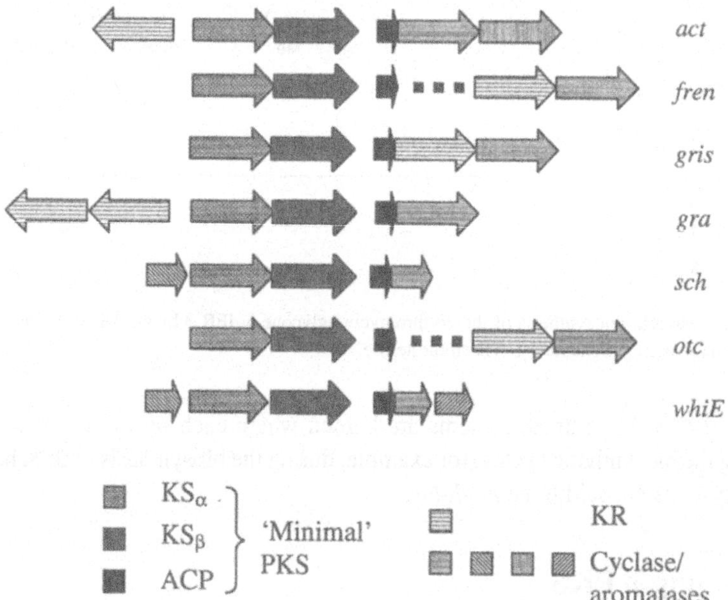

Figure 5.3. Homology of type II PKS genes in actinomycetes: act, actinorhodin; fren, frenolicin; gris, griseusin; gra, granaticin; sch, *Streptomyces halstedii* spore pigment; otc, oxytetracycline; whiE, White E.

subsequently observed to produce macrolide polyketides in other bacteria, where the whole PKS consisted of a number of modules equal to the number of rounds of condensation required to build the polyketide product (Caffrey *et al.*, 1992). By analogy with the gene probing experiments using the *act* probes, *ery* probes were also shown to be effective for cloning other modular PKS genes from other actinomycetes.

The most significant distinction between different types of bacterial PKS, thus, seems to lie in whether the polyketide assembly is controlled by iterative enzyme complexes or by modules where an individual active site acts only once during the biosynthesis. The distinction is relevant to the actual *programming* of PKS. The programming of the macrolide PKS is reflected in the organizational order of the individual modules responsible for the control of a unique cycle in the biosynthesis. Insight into iterative PKS is not possible in such a way since the various active sites are active in all cycles of the biosynthesis (Bohm *et al.*, 1998).

The separation of iterative and modular PKS however, does not seem to be so clear-cut for bacterial taxa other than the actinomycetes. In some synthases, modules within a modular set of PKS genes have been shown to function iteratively (Nowak-Thompson *et al.*, 1997; Gaitatzis *et al.*, 2002). Additionally a single iterative module was shown to be responsible for the dichloro-orsellinic acid fragment of avilamycin **6** in a *Streptomyces* species (Gaisser *et al.*, 1997). The bacterial orsellinic acid synthase (OAS) gene shows considerable amino acid identity (37%) to the fungal *Penicillium patulum* MSAS (*vide infra*). Homologues of OAS were also found in the bacterium *Micromonospora echinospora* (Hosted *et al.*, 2001; Ahlert *et al.*, 2002). An additional iterative Type I PKS responsible for the biosynthesis of the structurally interesting enediyne moiety of the anti-tumor antibiotic C-1027 **7** was found in *M. echinospora* and in *Streptomyces globisporus* (Liu *et al.*, 2002).

The *rppA* gene of *Streptomyces griseus* exemplifies a third type of bacterial PKS, known as Type III. *RppA* encodes a single polypeptide which shows significant homology to the plant chalcone and resveratrol synthases (Ueda *et al.*, 1995; Funa *et al.*, 1999). These synthases operate without the requirement for ACP domains, and since there are no reductive domains, these enzymes consist of simple KS domains alone. Chalcone and resveratrol synthases are highly homologous to each other (greater than 70%), but overall the Type III synthases show very little sequence homology to Type I and Type II PKS and FAS systems (Schroder *et al.*, 1988). Bacteria seem to harbor all three types of PKS and are therefore potentially a useful resource of varied gene probes for isolation of homologues from other genera.

3. Fungal PKS

To date, relatively few fungal PKS genes have been isolated, compared to the large number of bacterial PKS genes. This is due, in part, to the poorer molecular tools available to the fungal researcher. Of late, however, better methods have been developed. To date, a general theme has been observed that fungal PKS are generally iterative Type I enzymes; fungal modular Type I and fungal Type II and III systems have not yet been observed. Unfortunately nomenclature in the area is rather confused, with many fungal PKS genes being named *pks1*, or something similar. To differentiate the various fungal PKS genes we have expanded a system based on the function of the gene. So we refer to *Colletotrichum*

Table 5.1. Fungal PKS genes

Organism	Gene	Synthase component	Metabolite	Type	Accession number
A. parasiticus	*pksL1*	NAS	NA **14**	NR PKS	sw:PKS1_ASPPA
A. nidulans	*pksST*	NAS	NA **14**	NR PKS	sw:STCA_EMENI
A. parasiticus	*pksA*	NAS	NA **14**	PR PKS	em:Z47198
A. nidulans	*wA*	wAS	YWA1 **9**	NR PKS	em:X65866
A. fumigatus	*Alb1*	THNS	Naphthopyrone/THN **11**	NR PKS	em:AF025541
W. dermatitidis	*WdPKS1*	THNS	THN **11**	NR PKS	em:AF130309
C. largenarium	*pks1*	THNS	THN **11**	NR PKS	em:D83643
Nodulisporium sp.	*Pks1*	THNS	THN **11**	NR PKS	em:AF151533
P. patulum	*MSAS*	MSAS	6MSA **1**	PR PKS	sw:MSAS_PENPA
A. terreus	*MSAS*	MSAS	6MSA **1**	PR PKS	em:ATU31329
A. parasiticus	*pksL2*	MSAS	6MSA **1**	PR PKS	em:APU52151
P. griseofulvum	*PKS2*	MSAS	6MSA **1**	PR PKS	em:PGU89769
A. terreus	*lovF*	LDKS	lov diketide **17a**	HR PKS	em:AF141925
A. terreus	*lovB*	LNKS	lov nonaketide **17a**	HR PKS	em:AF151722
C. heterostrophus	*pks1*	TTS	T-toxin **15**	HR PKS	em:CHU68040
G. fujikuroi	*fum5*	FUMS	fumonisin **16**	HR PKS	em:AF155773
P. citrinum	*mlcA*		Compactin diketide **17b**	HR PKS	em:BD013765
P. citrinum	*mlcB*		Compactin nonaketide **17b**	HR PKS	em:BD013765

NR, non-reduced; PR, partially-reduced; HR, highly-reduced.

lagenarium pks1 as THNS (tetrahydroxynaphthalene synthase) and *lovB* from *Aspergillus terreus* as LNKS (lovastatin nonaketide synthase) etc. This nomenclature is shown in full in Table 5.1.

3.1. 6-Methylsalicyclic Acid Synthase

In 1990 the first fungal PKS gene was isolated by Beck and coworkers from the filamentous fungus *Penicillium patulum*. By immunological screening of a genomic expression library and probing a cDNA library, they discovered and characterized the gene associated with 6-methylsalicyclic acid **1** synthase (MSAS, Scheme 5.3) (Beck *et al.*, 1990). A 7.2-kb genomic clone carried a 5,322-bp open reading frame containing a single intron (69-bp) near the 5′ end. The ORF encoded a protein of 1,774 amino acids. Five catalytic domains were found on the Type I PKS and were identified as KS, AT, DH, KR, and ACP by their analogy in amino acid sequence and sequential order to the corresponding sites in rat FAS (Beck *et al.*, 1990; Witkowski *et al.*, 1991). The ER domain was lacking in MSAS as the enzyme does not perform an enoyl reduction. Higher sequence homology was found between MSAS and vertebrate FAS than yeast or *Penicillium* FAS. Additionally significant stretches of homology were detected between MSAS and individual bacterial Type II PKS components.

6-MSA **1** serves as a precursor to the mycotoxin patulin **8** (Forrester and Gaucher, 1972), but no other obvious function has so far been associated with **1** itself. However, genetic homologues of MSAS have since been isolated from numerous fungi including *A. terreus* and *A. parasiticus* (Wang *et al.*, 1991; Walsh, 1993; Fujii *et al.*, 1996; Feng and Leonard, 1998).

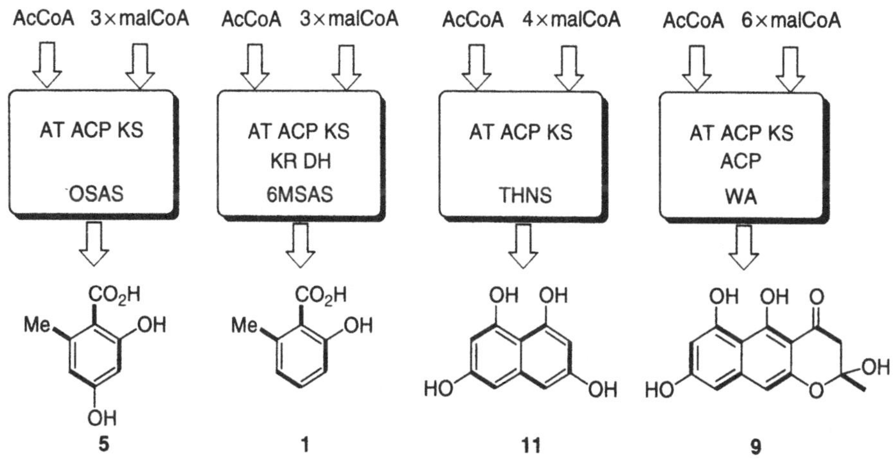

Scheme 5.3. Reactions catalyzed by type I fungal PKS. AcCoA, acetyl CoA; malCoA, malonyl CoA. Other abbreviations as in Scheme 5.1. Bold lines represent intact 2-carbon acetate equivalents as in Scheme 5.1.

MSAS has served as a model for production of fungal polyketides in heterologous hosts such as yeast and *E. coli* (Kealey *et al.*, 1998) and the actinomycete *S. coelicolor* (Bedford *et al.*, 1995).

Interestingly the gene associated with the production of orsellinic acid **5**, a compound very similar structurally to **1** has not yet been cloned in fungi, despite the isolation of its active biosynthetic enzyme (Jordan and Spencer, 1993). Orsellinic acid **5** is the simplest product of an unreduced tetraketide and differs from 6MSA **1** by the lack of reduction at the 4 position (Scheme 5.3). So far, Type I iterative orsellinic acid synthases (OAS) have been found only in bacteria, and it would be interesting to see how the sequence of a fungal OAS would compare to that of the bacterial synthases to assess the suitability of bacterial genes as gene probes in fungi.

3.2. Fungal PKS Involved in Biosynthesis of Conidial Pigment and Melanin

Fungal spore pigmentation contributes both to the survival and pathogenicity of fungal propagules in the environment. The green pigment contained in the asexual spores of the ascomycetous fungus *Aspergillus nidulans* protects them from damage caused by ultraviolet radiation. The *wA* gene product had been shown to be responsible for the production of a yellow intermediate of this pigment (Mayorga and Timberlake, 1990), and in 1992 it was shown that the *wA* gene coded for an iterative Type I PKS. The derived peptide sequence (Scheme 5.3) contained KS, AT, and interestingly, *two* ACP domains. The functional domains show significant sequence similarity with other fungal PKS (*MSAS*), bacterial PKS (*eryA*), and mammalian FAS domains (Mayorga and Timberlake, 1992). Heterologous expression of the *wA* transcript led to the isolation of the novel heptaketide naphthopyrone YWA1 **9** (Schemes 5.3 and 5.4) (Watanabe *et al.*, 1998, 1999).

The Ebizuka group has also isolated *alb1*, a *wA* homologue from the human pathogen *Aspergillus fumigatus* which on heterologous expression in *Aspergillus oryzae*

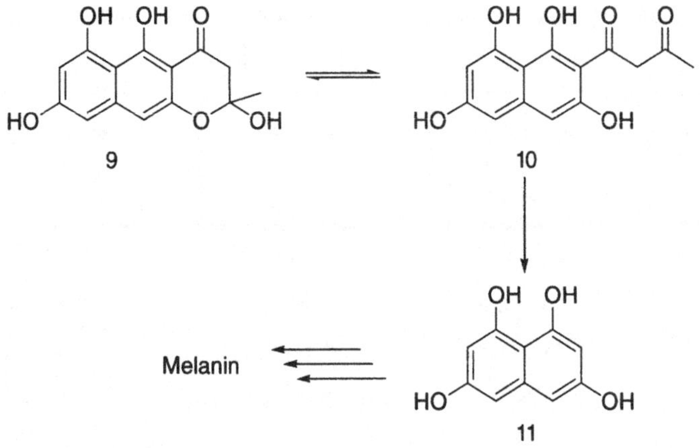

Scheme 5.4. Decarboxylation of Ywa1 **9** to give tetrahydroxynaphthalene **11**.

was shown to produce a yellow pigment naphthopyrone **10** (Watanabe *et al.*, 2000). Interestingly their genetic analysis suggested that the product of *alb1* would catalyze the formation of 1,3,6,8-tetrahydroxynaphtalene **11** (THN) the pentaketide precursor to the brown melanin polymer (Scheme 5.4) (Tsai *et al.*, 1998).

A THNS gene (*pks1*) had previously been cloned from a cosmid library of *Colletotrichum lagenarium* by complementation of a melanin-deficient mutant (Takano *et al.*, 1995). *C. lagenarium*, a plant pathogenic fungus, requires melanin (a polymer of dihydroxynaphthalene) to retain its infectous ability (Vidal-Cros *et al.*, 1994; Takano *et al.* 1995; see Chapter 15). The *pks1* gene contained one open reading frame of 6.7 kb, consisting of 3 exons, separated by two short introns, encoding a multifunctional protein of 2,187 amino acids. THNS shows lower similarity to the *alb1* product (43% identity, 60% similarity) than the latter shows to the *wA* PKS of *A. nidulans* (67% identity, 80% similarity). The gene product of *alb1* has recently been demonstrated to catalyze the chain shortening of the heptaketide naphthopyrone YWA1 **9** to THN **11** in *A. fumigatus* (Scheme 5.4) (Tsai *et al.*, 2001). It therefore appears that two independent routes to THN **11** have evolved in fungi.

Exploiting homology to *C. lagenarium pks1*, other THNS genes showing higher similarity (approx 70%) have since been isolated from a *Nodulisporium* sp. (Fulton *et al.*, 1999) and *Wangiella dermatitidis* (Feng *et al.*, 2001). Watanabe, Ebizuka and coworkers are working on the full characterization of these spore pigment genes and the identification of additional catalytic domains previously undetected in other PKS genes (see Chapter 6).

3.3. Fungal Polyketide Mycotoxins—Norsolorinic Acid Synthase (NAS)

Aflatoxin **12** and sterigmatocystin **13** are potent, polyketide-derived, carcinogenic mycotoxins produced by fungi of the genus *Aspergillus* and constitute a major agricultural problem (Hartley *et al.*, 1963). These biosynthetically related compounds share the same polyketide precursor, norsolorinic acid **14** (NA, Scheme 5.5). The cloning of the norsolorinic synthase (NAS) gene from both the sterigmatocystin pathway of *A. nidulans* (Yu and Leonard, 1995) and the aflatoxin pathway of *A. parasiticus* was achieved in 1995

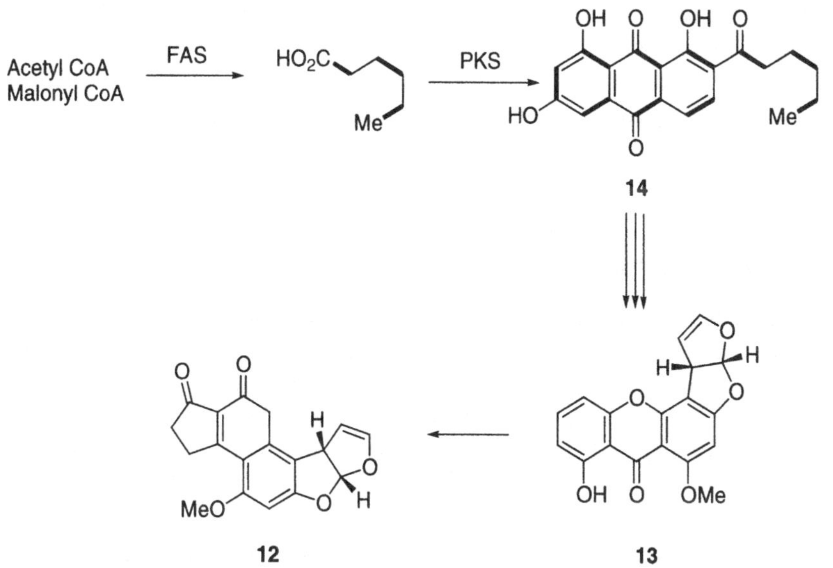

Scheme 5.5. Biosynthetic pathway to the mycotoxins aflatoxin B1 **12** and sterigmatocyctin **13** *via* norsolorinic acid **14**.

(Chang *et al.*, 1995; Feng and Leonard, 1995). In *A. nidulans* a 7.2-kb gene (*pksST*) was isolated by complementation experiments using sterigmatocystin deficient mutants. *PksST* was shown to code for a polypeptide with four catalytic domains (KS, AT, ACP, TE) of significant similarity with the *wA* gene product. Similarly a transcript (*pksL1*) of 6.6 kb was found in *A. parasiticus* encoding a polypeptide which showed sequence similarity with the *pksST* and *wA* gene products. A large number of genes involved in the aflatoxin pathway have been identified and have been shown to be clustered on one chromosome (Minto and Townsend, 1997). This finding is significant as clusters of functionally related genes are generally found in bacteria, but are less common in eukaryotes (Keller and Hohn, 1997).

3.4. Polyketide Synthase in T-Toxin Production

Race T of the fungal pathogen *Cochliobolus heterstrophus* produces T-toxin **15**, a polyketide derived virulence factor responsible for the high virulence of *C. heterstrophus* towards Texas male sterile (T) maize (Yang *et al.*, 1996). Comparison of T-toxin **15** producing race T with non-producing race O of the fungus led to the identification of the *Tox1* locus. This locus was shown to carry a 7.6-kb coding region (*pks1*), containing four introns, which encodes a putative multifunctional PKS of 2,530 amino acid (Yang *et al.*, 1996). The PKS nature of the gene product carrying KS, AT, DH, ER, KR, and an ACP domain was established by gene disruption experiments.

3.5. Polyketide Synthase in Fumonisin Production

Fumonisin **16**, a mycotoxin frequently contaminating maize, is produced by a number of fungi within the *Gibberella fujikuroi* group of crop pests (Proctor *et al.*, 1999). The gene encoding the PKS responsible for the production of the highly reduced polyketide backbone of fumonisin **16** has been isolated from a genomic library. The library was screened with an oligonucleotide probe generated by PCR with degenerate primers on a cDNA template. Genomic clones carried a 7.82-kb ORF (*fum5*) interrupted by five introns, and the translated sequence contained seven domains: KS, AT, DH, *C*MeT (the *C*-methyl transferase was not reported but *vide infra*), ER, KR, and ACP. A BLAST search revealed highest sequence similarity of *fum5* to *C. heterostrophus pks1*.

3.6. Lovastatin Synthases

Lovastatin **17a** is a cholesterol lowering drug naturally produced by *Aspergillus terreus*. This compound consists of two polyketides, a nonaketide and a diketide side-chain. The gene *lovB* encoding LNKS involved in the production of the nonaketide portion of **17a** was identified from a cDNA expression library screened with antiserum raised against a 250-kDa polypetide normally present during **17a** production (Hendrickson *et al.*, 1999). The cDNA clone was then used to isolate a genomic clone from an *A. terreus* genomic library. The 11.6-kb LNKS ORF contains seven short introns and encodes a putative protein of 3,038 amino acids (Scheme 5.6). DNA sequence analysis of the segments surrounding the LNKS led to the identification of a second putative PKS gene, *lovF*. Gene disruption experiments showed *lovF* to be essential for the biosynthesis of the β-methylbutyryl side chain of **17a** and to code for lovastatin diketide synthase (LDKS) (Kennedy *et al.*, 1999). In addition to the catalytic KS, AT, DH, ER, KR, and ACP domains, the two PKS also carried *C*-methyl transferase (*C*MeT) domains expected to be necessary for the addition of the pendant methyl moieties but which had not previously been described in fungal PKS. The LNKS appeared to have an inactive ER domain, and an accessory protein encoded by *lovC* was shown to be required for production of the correct

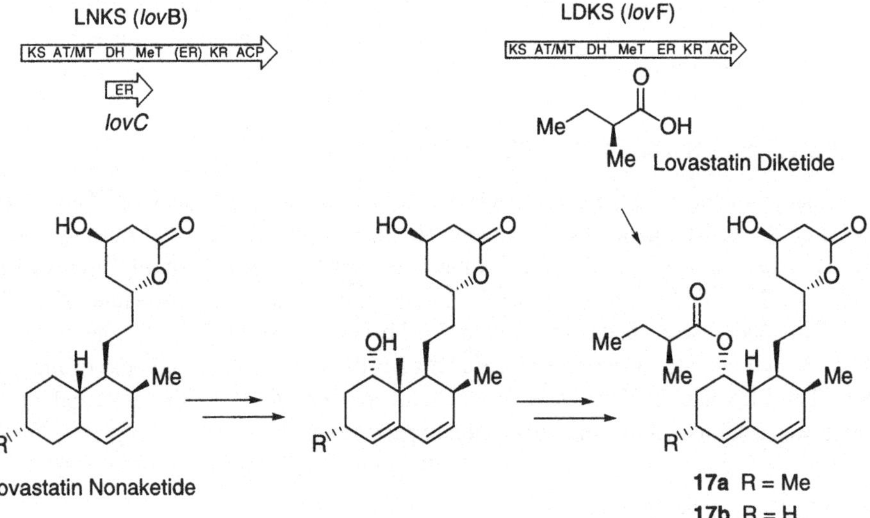

Scheme 5.6. Biosynthesis of lovastatin. LNKS, lovastatin nonaketide synthase; LDKS, lovastatin diketide synthase.

nonaketide precursor to lovastatin (Kennedy *et al.*, 1999). The nature of the large stretch of amino acids at the *C* terminus of LNKS remains unclear, and sequence similarity suggest it may be a peptide synthetase elongation domain.

4. Novel Methods for Accessing PKS Genes

Access to fungal PKS genes described thus far has been achieved by traditional molecular biological techniques targeting those encoding synthases for specific metabolites. However, in most cases the genetic potential of a particular fungus or actinomycete does not seem limited to the production of a unique polyketide. For the remainder of this chapter we will describe recent approaches to accessing all the PKS genes in a microorganism. The methodologies to be reviewed often make use of consensus, or degenerate PCR using primers designed on consensus sequences found in proteins belonging to a specific PKS family (Gould *et al.*, 1989). Two alternative outcomes are desired: either *selective* cloning of a PKS associated with a unique compound or overall assessment of the *total* PKS biosynthetic potential of an organism.

4.1. Problems Associated with Cloning Fungal PKS Genes

Investigations into novel PKS detection methods were sparked by the isolation of irrelevant homologues during the pursuit of a specific PKS gene (Bingle *et al.*, 1999; Santi *et al.*, 2000). For example, in their initial attempts to isolate the gene cluster for the anticancer agent epothilone **18** from *Streptomyces cellulosum*, Santi and coworkers (2000) isolated a set of modular PKS genes associated with another macrolide. Their cloning efforts relied on probing a genomic library with DNA that had been generated by PCR with degenerate primers designed on conserved sequences found in several PKS and FAS. These probes were imperfect as they did not amplify fragments of the PKS of interest. In a second attempt, the degenerate primers were revised so that their sequences were more selective. Thus PCR primers based on KS sequences of soraphen **19** and 6-deoxyerythronolide B **3** synthases were able to generate fragments for several modular PKS including the epothilone PKS. This work prompted a new method to obtain DNA fragments representative of *each and every* modular PKS gene cluster in a bacterial genome. The approach relies on the relatively very large size of modular PKS clusters (40–70 kb) compared to the relatively small size of a bacterial genome (4–10 Mb). When sequencing small fragments (500 bp) of a uniformly-sized random genomic library (1-kb fragments), statistically one fragment of a PKS gene would be found in every 250 clones (approximately 0.4% of a 10-Mb genome). Sequencing of approximately 500 randomly selected clones gave approximately 2.2-fold coverage of the *S. cellulosum* genome, and results suggested that there could be 6–8 PKS gene clusters in this organism (Santi *et al.*, 2000).

It is clear that this partial-sequencing approach is effective only when each gene cluster of interest comprises a significant portion of the genome (0.4–1%). In fungi the iterative Type I PKS genes are only 6–8 kb in size, and the fungal genome is generally at least twice that of an actinomycete, making it unlikely that this library-sequencing method could be efficient. Other problems in fungi center on quite low sequence conservation; for example, the use of PKS probes derived from bacterial sources is not generally successful.

In early studies, our colleagues in Bristol found that fungal PKS genes which are not functional homologues are too highly divergent to make useful probes for cloning novel PKS genes (Bingle *et al.*, 1999).

4.2. Early Efforts to Develop Fungal PKS Probes

To overcome these difficulties, the Bristol group generated homologous probes by PCR using degenerate primers based on known fungal PKS KS domains rather than bacterial PKS or FAS sequences. The four fungal PKS gene sequences available at the time were *wA* from *A. nidulans*, PKS1 encoding THNS from *C. lagenarium*, and the MSAS genes from *P. patulum* and *A. parasiticus* (Walsh, 1993) that is identical to *pksP*1 (Feng and Leonard, 1995). From a comparison of the amino acid sequences of the KS domains, the functional domains of the PKS genes fell into two subclasses designated WA (*wA* and *pks1*) and MSAS (both *MSAS* genes). Significant sequence divergence existed between the condensing domains of the pigment associated PKS genes that lacked reductive domains and the PKS genes responsible for the biosynthesis of the partially reduced 6MSA **1**. This permitted the design of degenerate PCR primers that bind coding sequences conserved within each subclass but not between them. The primer pair LC1 and LC2 was designed for the WA class, and the primer pair LC3 and LC5c for the MSAS class (Table 5.2). PCR experiments using template genomic DNA from a wide range of fungal species showed that these primers could amplify fragments of new PKS genes. Sequence analysis of these fragments indicated that the products fell into two classes closely matching the WA and MSAS types of PKS. Thus, the primer pairs seemed to be selective for different classes of PKS.

Hybridization experiments suggest cross-hybridization occurs only for PCR products generated from the various genomes using the same primer pair. Hybridization was observed to restriction fragments of known MSAS type *pks* exclusively with the LC3/LC5c probes. Data from Southern blot experiments indicated the presence of multiple PKS genes in each fungal genome, and the existence of two subclasses of PKS widely distributed among filamentous fungi.

The PKS genes fall into two distinct clusters in the phylogenetic analysis, where MSAS-type genes cluster with LC3/5c PCR products and where spore pigment, melanin, and norsolorinic acid genes cluster along with the LC1/LC2c products (Figure 5.4). Significantly, the identified PKS enzymes of the WA-type cluster lack KR domains, distinguishing them from the members of the MSAS-type cluster where all the identified gene products possess a KR. Interestingly TTS from *C. heterstrophus*, producing the highly reduced linear polyketide T-toxin **15**, did not cluster significantly with either of the two subgroups.

A similar approach has also been described by Keller *et al.* (1995). They used degenerate PCR to identify enzyme and regulatory genes in the sterigmatocystin **13** pathway of *A. nidulans*. The catalytic domains they targeted were KS and AT. Sequence information on fungal PKS was limited to *P. patulum* MSAS gene and *A. nidulans wA*. They therefore included a series of published bacterial PKS and FAS sequences in their amino acid comparison. Two sets of long (up to 33 bp) and highly degenerate (up to 256-fold) primers were designed on homologous sequences specific for fungal FAS and Type I fungal and modular bacterial PKS (Table 5.2). The sequences did not fall into specific PKS and FAS subgroups for the AT domain. One set of primers was made (Table 5.1) and used on

Table 5.2 5′–3′ Sequences of PCR primers discussed

Primer name	Sequence 5′ to 3′	Domain	Deg.	Dir.	Partner
LC1[a]	GAY CCI MGI TTY TTY AAY ATG	NR PKS KS	32	Fwd	LC2c
LC2c[a]	GT ICC IGT ICC RTG CAT YTC	NR PKS KS	4	Rev	LC1
LC3[a]	GCI GAR CAR ATG GAY CCI CA	PR PKS KS	8	Fwd	LC5c
LC5c[a]	GT IGA IGT IGC RTG IGC YTG	PR PKS KS	4	Rev	LC3
FAS KS Reg.1[b]	GAR GGI GTI GAR ATG GCI TGG ATH ATG GG	FAS KS	24	Fwd	FAS KS Reg. 3
FAS KS Reg.2[b]	GAR WSI TTY ATY ATY ACI ATG WSI GCI TGG G	FAS KS	256	Fwd	FAS KS Reg. 3
FAS KS Reg. 3[b]	CAT ACC IGC NCC IGC NGC ICC YTT IGG RTG TCC	FAS KS	64	Rev	FAS KS Reg.1/Reg.2
PKS KS Reg.1[b]	MGI GAR GCI CAR ATG GAY CCI CAR CAR MG	PKS KS	256	Fwd	PKS KS Reg.2
PKS KS Reg.2[b]	GG RTC NCC IAR YTG IGT ICC IGT ICC RTG IGC	PKS KS	64	Rev	PKS KS Reg.1
PKS AT Reg.1[b]	GGI CAI GGI GYI CAR TGG GYI GGI ATG G	PKS AT	8	Fwd	PKS AT Reg.2
PKS AT Reg.2[b]	GCR ATY TCI CCI ARI SWR TGI CCD ATI ACI GC	PKS AT	96	Rev	PKS AT Reg.1
KS1[c]	GAR GCN GCN GCN YTN GAY CCN CAR CA	PKS KS	4096	Fwd	KS2
KS2[c]	AG NCC NCY NGA RCA NGC NGT RTC NAC	PKS KS	16384	Rev	KS1
FKS1[c]	GCN BHN CAR ATG GAY CCN CA	PKS KS	2058	Fwd	FKS3/FKS4
FKS2[c]	GCN BHN CAR ATG GAY CAR CA	PKS KS	1024	Fwd	FKS3/FKS4
FKS3[c]	GA NGA RCA NGC NGT RTC NAC	PKS KS	1024	Rev	FKS1/FKS2
FKS4[c]	GA NGA RCA NGC NGT RTC RTT	PKS KS	512	Rev	FKS1/FKS2
KS3[d]	TTY GAY GCI GCI TTY TTY AA	HR PKS KS	16	Fwd	KS3
KS4c[d]	RTG RTT IGG CAT IGT IAT ICC	HR PKS KS	4	Rev	KS4c
KR1[d]	YTI ATI ACI GGI GGI YTI GGI	PR PKS KR	4	Fwd	KR2c/3c
KR2c[d]	TA ISW IGC YTG ICC IGG RAA	PR PKS KR	16	Rev	KR1
KR3c[d]	RTT IGC IGC RTA RTT IC	PR PKS KR	8	Rev	KR1
CMeT1[d]	GAR ATI GGI GGI GSI GGI ACI GG	PKS *C*MeT	4	Fwd	CMeT2c/3c
CMeT2c[d]	AT IAR YTT ICC ICC IGG YTT	PKS *C*MeT	8	Rev	CMeT1
CMeT3c[d]	AC CAT YTG ICC ICC IGG YTT	PKS *C*MeT	4	Rev	CMeT1
comKS[e]	GAY ACI GCI TGY AST TC	PKS KS	8	Fwd	AT
comAT[e]	TC ICC IKI RCW GTG ICC	PKS AT	8	Rev	KS
FPKSKSU-2[f]	ATS TCK CCY MRR GAR GC	PKS KS	128	Fwd	FPKSKSD-1
FPKSKSD-1[f]	C HMS RTG RCC CRA YTT KG	PKS KS	512	Rev	FPKSKSU-2

R = A + G; Y = C + T; M = A + T; K = G + T; S = G + C; W = A + T; H = A + C + T; B = G + T + C; D = G + A + T; N = A + C + G + T; I = Inosine.

[a] Bingle *et al.*, 1999.

[b] Keller and Hohn 1997.

[c] Lee *et al.*, 2001.

[d] Nicholson *et al.*, 2001.

[e] Abe *et al.*, 2002.

[f] Sauer *et al.*, 2002.

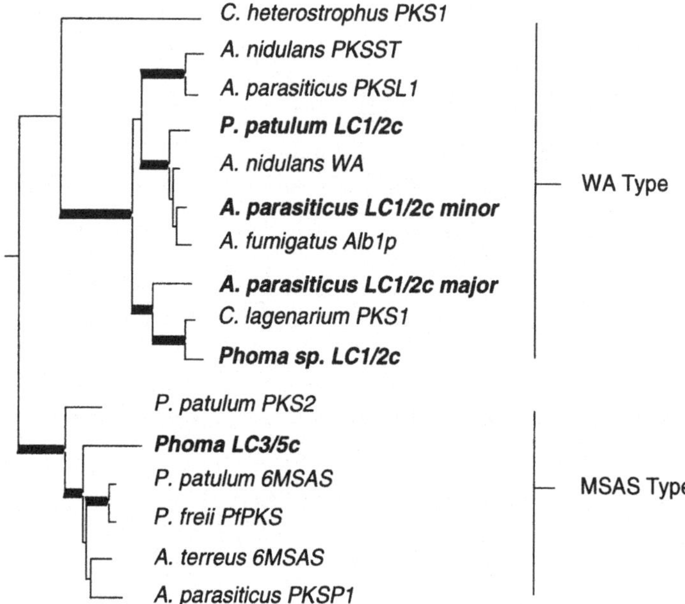

Figure 5.4. Cluster analysis of PCR products (bold) and known PKS genes using the Bingle PCR primers LC1/2c and LC3/5c.

A. nidulans genomic DNA. The resulting PCR amplicons were sequenced and tested for hybridization with cosmids known to carry genes involved in the biosynthesis of sterigamtocystin **13**. The results strongly suggested that fragments of PKS genes associated with sterigamtocystin **13** production had been cloned along with fragments of FAS genes. Given the design of these PKS primers, it is not surprising that they are less selective than the Bingle primers.

Proctor *et al.* (1999) have also used the Keller KS primers in efforts to isolate the PKS associated with fumonisin **16** biosynthesis. PCR experiments using genomic DNA generated seven unrelated PKS KS gene fragments but not the desired fumonisin synthase fragment. On the other hand, PCR with a cDNA template derived from RNA transcripts isolated during **16** biosynthesis in *Gibberella fujikuroi*, generated a single homologous probe which was the desired FUMS component. It appears, therefore, that the Keller primers are suitable for amplification of a great variety, but not the entirety, of putative fungal PKS domain fragments from genomic templates.

It seems that multiple loci within the genome can act as binding sites for the PKS primers designed with the KS domain in mind, and that some regions might be outcompeted during subsequent rounds of amplification. Reduction of the possible competing KS sequencing by generating a favorable cDNA template, during biosynthesis of the compound of interest, does enhance the chances of amplifying a homologous sequence. The subclass-specific primers designed by Bingle, however, could facilitate the cloning of novel PKS genes belonging to the subclasses of non-reduced or partially reduced PKS genes. Competition between KS loci is reduced in this case because the primers, though degenerate, are more specific.

4.3. Assessing Biosynthetic Potential

4.3.1. Prokaryotes

Degenerate primers have provided clear evidence that most fungal strains have the genetic capacity to produce a considerable variety of polyketides. The full potential of a microbe for polyketide production is of interest not only in fundamental research on the biology of the organism and in natural product discovery programs in the pharmaceutical industry, but also to researchers exploring the potential of combinatorial polyketide biosynthesis. Several recent studies in bacteria and fungi have made use of consensus PCR to assess this potential. In 1999 a Finnish group employed the PCR based method to amplify regions of the condensing domain (KS_{α}) of bacterial Type II PKS from *Streptomyces* genomes. Strains producing known aromatic polyketides as well as unidentified strains were used in their study. For most of the strains the single primer pair employed gave a single electrophoretic band, which was sometimes composed of two products. The translated sequences of these products were compared to those of a selection of PKS responsible for the biosynthesis of a variety of polyketide skeletons and differing in the use of starter and extension units. It was concluded that the sequences of putative spore pigment polyketide had high sequence identity to each other, and cluster analysis showed that these separated clearly from PKS antibiotic genes. Amino acid conservation was much lower within the latter group possibly reflecting the structural variety among aromatic antibiotics. Further divisions within this group correlate with the type of starter units employed by the PKS (Metsa-Ketala *et al.*, 1999).

This method, based on the genetic structure of the KS_{α} ORF of Type II PKS, provides a timely means to classify pharmaceutically and chemically interesting *Streptomyces* strains. PCR with degenerate primers has also been used to amplify entire KS_{β} genes from the DNA of uncharacterized organisms that was extracted directly from soil (Seow *et al.*, 1997). KS_{β} forms one of three components (together with KS_{α} and ACP) that encode the so-called minimal PKS or smallest complex of catalytic domains sufficient for polyketide production in bacteria. The approach was feasible only because of conserved organization of the Type II PKS genes coupled with sequence conservation among them. The degenerate primers were based on consensus sequence found in KS_{α} and ACP domains flanking the KS_{β}. To our knowledge no such conservation of genetic organization has been observed in fungal gene clusters to make this powerful technique applicable to components of fungal PKS genes.

4.3.2. Lichens

In a recent review, Miao and coworkers (2001) have described a related genetic approach for revealing the potential for novel polyketide production in lichens. Whereas actinomycete and fungal species are widely employed in pharmaceutical screens for bioactive metabolites, lichens have so far been largely ignored. Lichens, although they have been given systematic Latin names, are in fact symbiotic complexes composed of fungi and either green algae or cyanobacteria or both. Although a large number of biologically active compounds are known to be produced by these symbiotic complexes, they must be isolated from field collected specimens because growth of lichens or the organisms that comprise them is impractical under laboratory conditions (Miao *et al.*, 2001). Recent advances in molecular biology technologies, however, permits access to the cryptic biosynthetic potential of lichens via DNA extracted from lichen thalli (Sinnemann *et al.*, 2000). Heterologous expression of lichen genes in amenable surrogate hosts, with good fermentation characteristics and low profiles of endogenous metabolites, has been achieved (Sinnemann *et al.*, 2000) and is a major step towards making lichen synthetic genes attractive to industry. Earlier Miao and coworkers had tested degenerate KS primers designed by Keller with a limited number of genomic DNA samples from a large collection of lichen DNA. PCR products of various sizes were generated for each strain, with each electrophoretic band often containing multiple KS fragments. Such fragments were amplified from genomes of some strains that were mainly known as producers of shikimate metabolites rather than polyketides, suggesting the ubiquity of PKS genes in lichens. A dendrogram was constructed from the translation of 48 novel KS sequences and known fungal KS domain sequences. Again these highly diverse sequences gave rise to clusters like those described earlier, suggesting correlation between KS domain sequence and metabolite type. To discover new families of compounds the KS tags that do not cluster with the reference domains needed to be explored. Accordingly they have since generated more than 120 new tags from various lichen genomes providing information on the diversity of polyketide pathways in the symbionts and highlighting promising strains likely to produce novel compounds. The authors have employed prioritized KS tags successfully to clone polyketide biosynthetic genes from genomic libraries. Gene clusters containing biosynthetic, tailoring, and regulatory genes were apparent, showing that the genetic architecture found in saprophytic, endophytic, and phytopathogenic fungi applies to lichen fungi as well. Interestingly in studies involving genomic DNA extracted from lichen with cyanobacterial symbionts, KS tags for Type II and modular Type I PKS systems had been generated.

It is generally believed that most of the secondary metabolites in lichens are made by the mycobiont (Culberson and Armaleo, 1992; Hamada *et al.*, 2001). However, approximately 10% of all lichens include cyanobacteria, and research suggests that the prokaryotic partner contributes significantly to the biosynthetic diversity of lichens. Further, a considerable degree of variability exists in the distribution of secondary metabolite genes among specimens of a specific lichen. *Solorian crocoa* collected in five different locations gave differing results in a PCR-based study that employed specific primers for five KS tags that had been obtained from a reference specimen of *S. crocoa* and that were derived from both mycobiont and photobiont. Some PCR reactions did not generate product suggesting that the genes that they were intended to amplify were either absent in a specimen or were markedly divergent from the corresponding gene in the reference specimen. Significant diversity in genotype appears to exist among and within lichen species, and methods are in place to tap into this pool of diverse biosynthetic gene clusters. The PCR screening approach provides an invaluable assessment of the genetic and biosynthetic potential of this largely unexplored group of organisms.

4.3.3. Insect and Nematode Associated Fungi

Polyketide production by insect and nematode associated fungi remains largely undescribed. Gibson and coworkers used a PCR-based approach with degenerate primers to detect and characterize PKS genes in this genetically diverse group of organisms (Lee *et al.*, 2001). For this study the degenerate primers were designed for the KS domain using

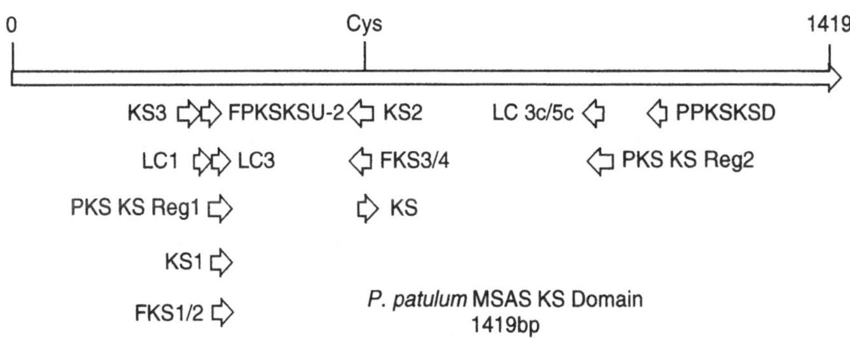

Figure 5.5. Approximate binding positions, and directions, of degenerate KS PCR primers relative to the *P. patulum* MSAS KS domain.

sequence information from known fungal, bacterial, and plant PKS as well as FAS genes. Based on initial results the first primer pair designed (KS1, KS2) was modified to a second set consisting of two forward (FKS1, FKS2) and two reverse primers (FKS3, FKS4, Table 5.2, Figure 5.5). With various primer pairings, PCR amplicons of the expected size (~0.3 kb) were obtained from 92 of 149 fungi examined. At least one PCR product was cloned and sequenced from each of the isolates; PKS sequences were identified in 66 of the 92 PCR products.

Two fungal isolates were analyzed further to determine how many different PKS genes each harbored. More extensive PCR was done with the same primers as well as Bingle's LC primers. For one fungal strain, four individual PKS gene sequences were amplified with the KS/FKS primers, and three additional sequences were found with the LC primers. For the other fungal strain, four KS fragments were detected with the KS/FKS primers, and the LC primers gave one additional sequence. Inferred amino acid sequences for all KS fragments were compared to each other, to 32 KS domain reference sequences from bacteria and fungi, and to a human FAS KS domain. All sequences show high similarity to known PKS within the KS domain region.

A cladistic analysis of the DNA resulted in three equally parsimonious trees showing similar patterns of clades. Two clades were composed of both reference sequences and sequences generated in the studies, whereas the six remaining clades were composed of KS domain sequences from the insect pathogenic fungi. This clustering within clades suggests that this group of fungi may have PKS genes distinctly different from known fungal PKS. Secondary metabolites from the insect and nematode associated fungi have not been well characterized, and it is therefore unclear whether the genetic divergence is reflected in metabolite structure. It is clear, however, that PKS are also widespread in this distinctive group of fungi.

It is noteworthy that the PKS tree generated from these results did not reflect the accepted phylogeny of these organisms, and this raises the possibility that fungal species acquired biosynthetic genes through horizontal transfer. This idea is in accord with the clustering of biosynthetic genes for polyketide and other types of secondary metabolites, an arrangement that provides all the genes for the synthesis of a metabolite in a relatively short stretch of DNA (Walton, 2000).

4.3.4. Endophytic Fungi

Recently a very similar approach was applied to study endophytic fungi (Sauer *et al.*, 2002). A series of degenerate primers based on 12 known fungal PKS genes was designed. The FPKSKSU-2 and FPKSKSD-1 primer pair (Table 5.2, Figure 5.5) was used to screen a number of non-sporulating cranberry endophytes. For 50% of the fungi, the PCR-generated products could be identified as KS fragments on the basis of hybridization with a mixture of 10 probes prepared from known fungal PKS genes. Phylogenetic analysis of the deduced amino acid sequences of the KS domains gave rise to clusters similar to those obtained by Bingle *et al.* (1999). One cluster of sequences was composed of THNS PKS homologues. Within another cluster, two subgroups of aflatoxin and pigment related PKS were found. A third cluster was composed of PKS associated with antibiotic and mycotoxin production, grouping separately from a sub-group of MSAS homologues. There was no suggestion that the phylogenetic relationship of these fungi based on SSU rDNA sequences was mirrored in the sequence relationship of the PKS fragments. This is not surprising because a variety of polyketides are produced by endophytes, and the variety should be reflected in the divergence of their PKS genes. The same primers and approach were recently applied successfully to clone the gene for THNS in *Glarea lozoyensis* (Zhang *et al.*, 2003).

4.4. Biosynthetically Informed Approaches for Accessing Fungal PKS Genes

4.4.1. KS-Specific Primers

Following on our preliminary work (Bingle *et al.*, 1999), we have extended the methods for cloning diverse fungal PKS genes, taking advantage of the growing body of sequence information available for fungal PKS genes to design subclass specific probes (Nicholson *et al.*, 2001). Previous results suggested that subclass discrimination could be based on the extent of reduction in the polyketide produced by a PKS (Bingle *et al.*, 1999). An analysis based on the chemical structure of the polyketide product of a PKS is less subjective than one that utilizes other characteristics such as bioactivity (i.e., antibiotic, anticholesterol, mycotoxin, etc.), species of the producer, or proposed role *in vivo* (i.e., spore pigment, etc.).

Correlating structures with the catalytic domains required for their synthesis, we classified polyketides into three structural types: non-reduced (NR), partially reduced (PR), and highly reduced (HR) (Nicholson *et al.*, 2001). For example, the NR class is represented by 1,3,6,8-tetrahydroxynaphthalene **11**, NA **14**, YWA1 **9**, and orsellinic acid **5** which require no reductive steps during biosynthesis. The PR class is exemplified by 6MSA **1** which shows partial reduction of the polyketide skeleton (i.e., a single round of KR and DH). The HR class contains structurally complex compounds such as T-toxin **15**, lovastatin **17a**, and squalestatin **20** which undergo much higher levels of reductive modification. The overall strategy was to design PCR primers *specific* to these classes of PKS, which could be used to amplify genes from a wide range of fungi. FAS genes were also considered in the primer design to avoid the possibility of recovering endogenous FAS genes, a problem associated with the Keller primers.

The new primers were evaluated in a study limited to a few fungal strains that were known to produce interesting polyketides as well as reference strains *P. patulum*, *A. terreus*, and *C. heterostrophus*. The compactin **17b** producer, *Penicillium citrinum*, and the squalestatin **20** producers, *Phoma* sp. *C2932* and *Phoma MF5453*, were attractive for the following reasons. Compactin **17b** is a structural analogue of lovastatin **17a**, and squalestatin **20** is a potent squalene synthase inhibitor (Blows *et al.*, 1994). Both compounds are composed of a main chain polyketide and an appended side-chain polyketide, which suggests the presence of at least two PKS genes associated with the biosynthesis of each compound. Additionally both polyketides in squalestatin and the side-chain polyketide of compactin possess pendant methyl moieties derived from methionine, implying that their synthases contain CMeT domains.

The LC primer pairs from our earlier work were used for PCR of genomic DNA from the squalestatin producers to provide data for assessing the total number of PKS genes present in the fungi and to test the specificity of the primers with a new target genome. For both strains, at least 12 clones of the PCR products obtained with the NR and PR primer pairs were sequenced. Each primer pair gave only one individual KS fragment

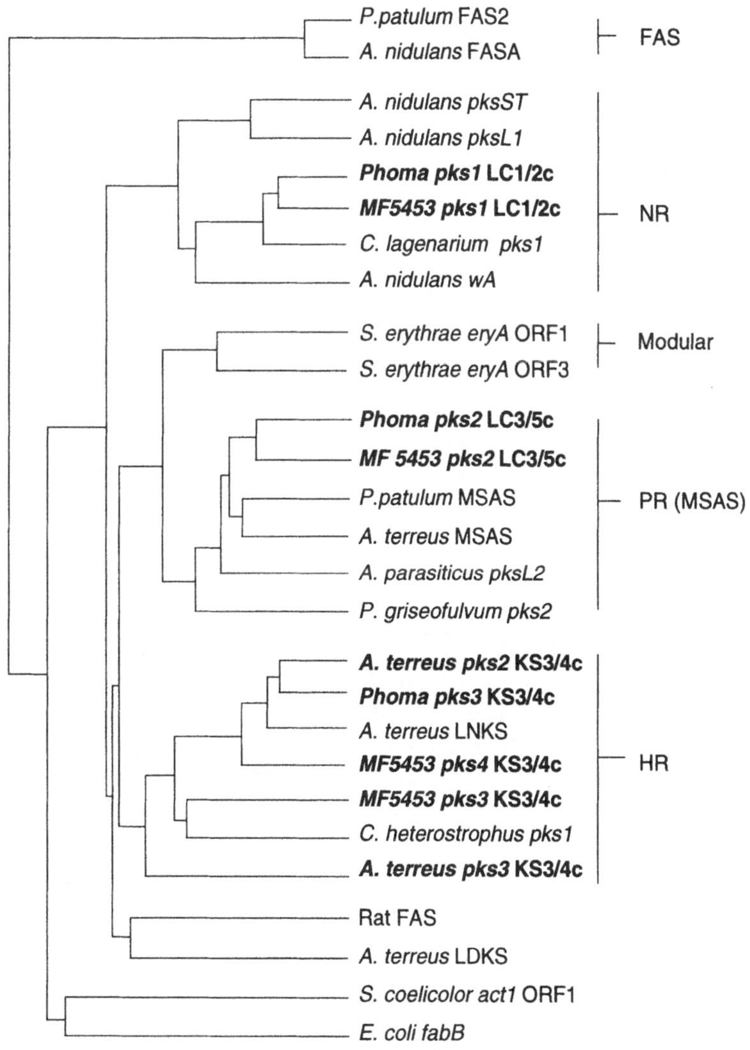

Figure 5.6. Cluster analysis of PCR products (bold) and known PKS genes using the Bristol PCR primers LC1/2c and LC3/5c and KS3/4c.

per strain. Comparison of the deduced polypeptide sequences with reference sequences mirrored the results of Bingle *et al.* (1999) as shown in the dendrogram of the KS domains amplified in this study (Figure 5.4).

The protein sequences of LNKS from *A. terreus* and TTS (T-toxin synthase) from *C. heterostrophus* formed the basis of the design for the third KS primer set. These HR KS primers were designed to bind to the coding sequences of conserved amino acid sequences for the LNKS and TTS which are *not* found in NR and PR PKS or in fungal FAS, and known intron positions were consciously avoided. PCR reactions were predicted to yield products of approximately 750 bp, depending on introns, and were positive with genomic templates from *A. terreus* and *C. heterostrophus* as predicted. The tests were equally successful for *P. citrinum, MF5453* and the *Phoma* sp., where multiple products with at least one of expected size were observed. These products, except for the ones from *P. citrinum*, were cloned and sequenced, and a number of sequences matching HR PKS KS were obtained.

Predicted amino acid sequences were compared and included in the cluster analysis. The result is shown in Figure 5.6. Three different HR PKS genes were cloned for *A. terreus* (including LNKS, but not LDKS), the TTS KS was isolated from *C. heterostrophus*, and one sequence highly homologous to LNKS was isolated from both *Phoma* species. An additional HR KS fragment was observed for *MF5453*. The sequences appear sufficiently divergent from the NR and PR KS tags to group together as a HR KS cluster in the dendrogram (Figure 5.6). The findings were tested by cross-hybridization experiments between members of the subclasses. Paralleling previous findings (Bingle *et al.*, 1999), products from one subclass hybridized well with one another but not to products from the other subclasses. Additionally no cross-hybridization occurred between the KS probes and fungal FAS KS domains.

Southern blotting experiments on genomic digests of *MF5453* and *Phoma* sp. also revealed no overlap between the patterns of restriction fragments detected by the different subclass derived probes. This also holds true for probes belonging to the same subclass where sequence is fairly divergent. Probing genomic digests of *Phoma* sp. with the second HR KS from *MF5453* yielded new hybridization bands, suggesting the absence of a homologue in this species.

4.4.2. KR-Specific Primers

Aiming to distinguish genes of PKS with various reductive capacities, we also considered KR domains. We therefore examined sequence data from known fungal PKS KR domains in conjunction with bacterial modular and Type II PKS KR domains. We also considered fungal, vertebrate, and bacterial FAS KR domains, as well as post-PKS reductive enzymes and reductive enzymes from primary metabolism in order to *select against* these types of enzyme. However, the use of the KR primers was disappointing as only one primer set, selective for the PR MSAS, produced positive results. Another target domain, however, became available with the publication of sequence data for genes associated with the biosynthesis of lovastatin.

4.4.3. *C*meT-Specific Primers

Both LNKS and LDKS possess *C*-methyl transferase (*C*Met) domains that catalyze the introduction methionine-derived methyl branches onto the nascent polyketide chains

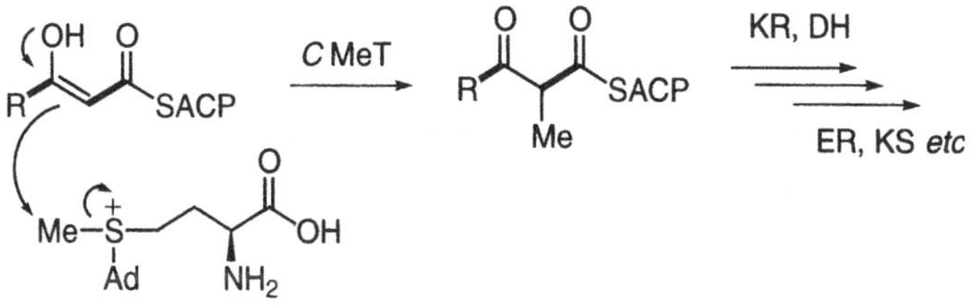

Scheme 5.7. Reaction catalyzed by fungal PKS *C*-methyl transferases (*C*MeT). Ad, Adenosyl; other abbreviations as in Scheme 5.1.

destined to become lovastatin. These domains are thought to add a methyl group (derived from *S*-adenosyl methionine) to β-ketoacyl PKS intermediates (Scheme 5.7). (This strategy for introducing branch points is different from that used by bacterial modular synthases where side-chain groups derive from the incorporation of modified chain extender [malonyl] units.) PCR primers designed to amplify *C*meT domains could provide additional specificity to the cloning approach for methylated polyketides. LNKS and LDKS were the only sequences available for the design of primers; *O*- and *N*-methyl tranferase sequences were also considered to avoid amplification of coding regions from methyl transferases active on intact polyketides or in primary metabolism.

Low overall sequence homology is generally observed for the various *C*meT domains and enzymes; however, Kagan and Clarke (1994) have identified three conserved sequence motifs (Motifs I, II, and III) for non-DNA MeTs. Motif I is involved in *S*-adenosylmethionine (SAM) binding, and Motif III is part of the active site or participates in binding the magnesium cofactor binding. Both are found in all known MeTs, (Kagan and Clarke, 1994; Vidgren *et al.*, 1994), including LNKS and LDKS, the templates for our *C*MeT primer design. A set of primers composed of one forward primer (*C*MeT1) and two reverse primers (*C*MeT2c, *C*MeT3c) were designed where the reverse primers were biased to either LDKS or LNKS, respectively.

As found for the KS reactions, low annealing temperature PCR conditions gave multiple products with both primer combinations and templateDNA from the target and reference (*A. terreus*) genomes. Bands in the predicted size range were cloned and sequenced. PCR with *P. citrinum* genomic DNA was successful with both *C*MeT primer pairs. Amino acid sequence comparison for the other *C*MeT clones revealed two novel fungal PKS *C*MeT domains for both *MF5453* and *Phoma* sp. One of the two domains was amplified by both primer combinations whereas the other could be generated only with *C*MeT1 in conjunction with *C*MeT2c. There was no obvious distinction between these amplicons in terms of their sequence. This was reflected in cross-hybridization experiments where sufficient hybridization occurs with heterologous gene fragments of both types. Heterologous hybridization with *C*MeT probes has also been successful in rapidly isolating entire PKS genes from fungal cDNA libraries (unpublished), and the probes show multiple bands of high intensity in genomic blots. The Bristol group has used this method for rapidly cloning HR PKS from a range of fungal species, and when coupled with temporally specific cDNA libraries, these techniques offer a rapid and very selective way of cloning specific PKS genes.

4.4.4 Lessons and Outlook

The strategy of designing PCR primers based upon the extent to which reduction is required in the formation of the polyketide product and other primers based upon the methylation domains appears to overcome the limitations that were inherent with earlier primers. The methodology appears to be successful in linking protein sequence, or domain fine-structure, in the PKS to the nature of the chemical structure synthesized. Primers designed by this approach appear to be the best means presently available to assess the full complement of PKS in the genome of a fungus. The results for the organisms that produce squalestatin **20** suggest the presence of four to five PKS genes, and in *A. terreus* seven distinct PKS genes were found. Recent RT-PCR experiments using the primers to compare temporal profiles of PKS expression with production of the polyketides themselves suggest that these are underestimates of biosynthetic potential, however. Selective generation of homologous probes for KS domains in conjunction with the use of *C*MeT domain primers have proven invaluable in screening libraries for fungi producing both mycotoxins and pharmaceutically useful polyketides (unpublished).

The cloning of the biosynthetic gene cluster for compactin from *P. citrinum* has recently been accomplished by another laboratory group (Abe *et al.*, 2002). Again, degenerate PCR primers provided gene probes for the isolation of the biosynthetic genes. Three pairs of PCR primers, based on the homology sequence found in the active site region of KS, and AT, successfully amplified four (1- to 2-kb) PKS gene fragments. One primer pair (denoted comKS and comAT in Table 5.2) generated a fragment that, when used in Northern blots, appeared to correlate mRNA expression with the pattern of compactin production. This probe was used successfully to identify a cluster containing LNKS and LDKS homologues as well as genes homologous to other genes in the lovastatin cluster of *A. terreus*. It was interesting that *mlcA*, the LNKS gene homologue, contains a region homologous with the known *C*MeTs that is apparently inactive during the biosynthesis of compactin.

The substantial amount of sequence information generated by the degenerate PCR approaches is undeniable. Sequence tags provide a vital assessment of the polyketide producing potential of both pathogenic and pharmaceutically attractive fungi. The methodologies covered in this chapter provide the tools, in the form of gene probes, for the discovery of pharmaceutical leads from even largely unexplored fungal sources and can help to pinpoint pathogenesis factors, that is, polyketide-derived toxins (T-toxin, fumonisin) and penetration effectors (melanin). The isolation of novel PKS genes and associated clustered ORFs, encoding modifying and tailoring enzymes, increase the pool of PKS components for combinatorial expression in well developed heterologous systems for the production of novel "unnatural" natural products. We envisage the strategies for design and production of probes for the gene of secondary metabolites could be extended to other important classes of secondary metabolites. For non-ribosomal polypeptide synthases (NRPS) many sequences are known, and there is a detailed understanding of the relationship of protein sequence and potential product structure to validate a similar approach (see Chapter 7).

5. The Genomic Era

The advanced genetic tool kit currently available to fungal biologists (Chapter 3) will improve our ability to assess the full range of gene sets devoted to secondary metabolite production in a given organism. Genome sequencing projects and annotated genome databases for yeast and *Streptomyces coelicolor* have opened new avenues to variety of areas, including metabolite production. Certainly analysis of the *S. coelicolor* genome has enabled the identification of polyketide related genes in other species (Hopwood, 1999). Genome-wide analyses for filamentous fungi, however, have started only in recent years. Preliminary sequence assemblies exist for the opportunistic pathogen *Candida albicans* and the mold *Neurospora crassa*. Fungal genomic resources, including public projects for several *Aspergillus* species, *Magnaporthe grisea, Candida glabrata, Cryptococcus neoformans, Pneumocystis carinii*, and other fungal species have been provided in Chapter 2 and elsewhere (Yoder and Turgeon, 2001; Lorenz, 2002). In addition, a comprehensive Fungal Genome Initiative (FGI) proposes to sequence, at a pace of one genome per month, 15 fungi chosen on the basis of their medical, agricultural or purely scientific interest (Birren *et al*., 2002). Aside from other obvious benefits, this project would provide an unprecedented amount of information on the secondary metabolism. From the initial comparisons of fungal genomes, the value of this knowledge is already apparent. The question "how many polyketide metabolites can a fungus produce" remains unanswered, but from the available data one can safely assume it is many! PCR methodologies have sampled different templates including fungal genomic DNA and cDNA generated at different growth stages. The results indicated that a typical filamentous fungus has the genetic potential to produce as many as 10 structurally diverse polyketides.

References

Abe, Y., Suzuki, T., Ono, C., Iwamoto, K., Hosobuchi, M., and Yoshikawa, H. (2002). Molecular cloning and characterization of an ML-236B (compactin) biosynthetic gene cluster in *Penicillium citrinum*. *Mol. Genet. Genomics* **267**, 636–646.

Ahlert, J., Shepard, E., Lomovskaya, N., Zazopoulos, E., Staffa, A., Bachmann, B.O., Huang, K.X., Fonstein, L. *et al*. (2002). The calicheamicin gene cluster and its iterative type I enediyne PKS. *Science* **297**, 1173–1176.

Beck, J., Ripka, S., Siegner, A., Schiltz, E., and Schweizer, E. (1990). The multifunctional 6-methylsalicylic acid synthase gene of *Penicillium patulum*—Its gene structure relative to that of other polyketide synthases. *Eur. J. Biochem.* **192**, 487–498.

Bedford, D.J., Schweizer, E., Hopwood, D.A., and Khosla, C. (1995). Expression of a functional fungal polyketide synthase in the bacterium *Streptomyces coelicolor* A3(2). *J. Bacteriol.* **177**, 4544–4548.

Bennett, J.W. (1995). From molecular genetics and secondary metabolism to molecular metabolites and secondary genetics. *Can. J. Bot.-Revue Canadienne De Botanique* **73**, S917–S924.

Bingle, L.E.H., Simpson, T.J., and Lazarus, C.M. (1999). Ketosynthase domain probes identify two subclasses of fungal polyketide synthase genes. *Fungal Genet. Biol.* **26**, 209–223.

Birren, B., Fink, G., and Lander, E. (2002). Fungal genome initiative: White paper developed by the fungal research community, available online at *http://www-genome.wi.mit.edu/seq/fgi*, 2002.

Blows, W.M., Foster, G., Lane, S.J., Noble, D., Piercey, J.E., Sidebottom, P.J., and Webb, G. (1994). The squalestatins, novel inhibitors of squalene synthase produced by a species of *Phoma*. 5. Minor metabolites. *J. Antibiot.* **47**, 740–754.

Bohm, I., Holzbaur, I.E., Hanefeld, U., Cortes, J., Staunton, J., and Leadlay, P.F. (1998). Engineering of a minimal modular polyketide synthase, and targeted alteration of the stereospecificity of polyketide chain extension. *Chem. Biol. 5*, 407–412.

Caffrey, P., Bevitt, D.J., Staunton, J., and Leadlay, P.F. (1992). Identification of DEBS1, DEBS2 and DEBS3, the multienzyme polypeptides of the erythromycin-producing polyketide synthase from *Saccharopolyspora erythraea*. *Febs Letters,* **304**, 225–228.

Chang, P.K., Cary, J.W., Yu, J.J., Bhatnagar, D., and Cleveland, T.E. (1995). The *Aspergillus parasiticus* polyketide synthase gene *pksA*, a homolog of *Aspergillus nidulans wA*, is required for aflatoxin B-1 biosynthesis. *Mol. Gen. Genet.* **248**, 270–277.

Culberson, C.F. and Armaleo, D. (1992). Induction of a complete secondary product pathway in a cultured lichen fungus. *Exp. Mycol.* **16**, 52–63.

Demain, A.L. (1986). Regulation of secondary metabolism in fungi. *Pure Appl. Chem.* **58**, 219–226.

Demain, A.L. (1999). Pharmaceutically active secondary metabolites of microorganisms. *Appl. Microbiol. Biotechnol.* **52**, 455–463.

Feng, B., Wang, X., Hauser, M., Kaufmann, S., Jentsch, S., Haase, G., Becker, J.M., and Szaniszlo, P.J. (2001). Molecular cloning and characterization of *WdPKS1*, a gene involved in dihydroxynaphthalene melanin biosynthesis and virulence in *Wangiella* (*Exophiala*) *dermatiditis*. *Infect. Immun.* **69**, 1781–1794.

Feng, G.H. and Leonard, T.J. (1995). Characterization of the polyketide synthase gene (*pksL1*) required for aflatoxin biosynthesis in *Aspergillus parasiticus*. *J. Bacteriol.* **177**, 6246–6254.

Feng, G.H. and Leonard, T.J. (1998). Culture conditions control expression of the genes for aflatoxin and sterigmatocystin biosynthesis in *Aspergillus parasiticus* and *A. nidulans*. *Appl. Environ. Microbiol.* **64**, 2275–2277.

Fernandez-Moreno, M.A., Martinez, E., Boto, L., Hopwood, D.A., and Malpartida, F. (1992). Nucleotide sequence and deduced functions of a set of cotranscribed genes of *Streptomycetes coelicolor* A3(2) including the polyketide synthase for the antibiotic actinorhodin. *J. Biol. Chem.* **267**, 19278–19290.

Firn, R.D. and Jones, C.G., (2000). The evolution of secondary metabolism—a unifying model. *Mol. Microbiol.* **37**, 989–994.

Forrester, P.I. and Gaucher, G.M. (1972). m-Hydroxybenzyl alcohol dehydrogenase. *Biochemistry* **11**, 1108–1114.

Fujii, I., Ono, Y., Tada, H., Gomi, K., Ebizuka, Y., and Sankawa, U. (1996). Cloning of the polyketide synthase gene *atX* from *Aspergillus terreus* and its identification as the 6-methylsalicylic acid synthase gene by heterologous expression. *Mol. Gen. Genet.*. **253**, 1–10.

Fulton, T.R., Ibrahim, N., Losada, M.C., Grzegorski, D., and Tkacz, J.S. (1999). A melanin polyketide synthase (PKS) gene from *Nodulisporium* sp that shows homology to the *pks1* gene of *Colletotrichum lagenarium*. *Mol. Gen. Genet.* **262**, 714–720.

Funa, N., Ohnishi, Y., Fujii, I., Shibuya, M., Ebizuka, Y., and Horinouchi, S. (1999). A new pathway for polyketide synthesis in microorganisms. *Nature* **400**, 897–899.

Gaisser, S., Trefzer, A., Stockert, S., Kirschning, A., and Bechthold, A. (1997). Cloning of an avilamycin biosynthetic gene cluster from *Streptomyces viridochromogenes* Tu57. *J. Bacteriol.* **179**, 6271–6278.

Gaitatzis, N., Silakowski, B., Kunze, B., Nordsiek, G., Blocker, H., Hofle, G., and Muller, R. (2002). The biosynthesis of the aromatic myxobacterial electron transport inhibitor stigmatellin is directed by a novel type of modular polyketide synthase. *J. Biol. Chem.* **277**, 13082–13090.

Gould, S.J., Subramani, S., and Scheffler, I.E. (1989). Use of the DNA polymerase chain reaction for homology probing—Isolation of partial cDNA or genomic clones encoding the iron sulfur protein of succinate dehydrogenase from several species. *Proc. Natl. Acad. Sci. U.S.A.* **86**, 1934–1938.

Hamada, N., Tanahashi, T., Miyagawa, H., and Miyawaki, H. (2001). Characteristics of secondary metabolites from isolated lichen mycobionts. *Symbiosis* **31**, 23–33.

Hartley, R.D., Nesbitt, B.F., and O'Kelly, J. (1963). Toxic metabolites of *Aspergillus flavus*. *Nature* **198**, 1056–1058.

Hendrickson, L., Davis, C.R., Roach, C., Nguyen, D.K., Aldrich, T., McAda, P.C., and Reeves, C.D. (1999). Lovastatin biosynthesis in *Aspergillus terreus*: Characterisation of blocked mutants, enzyme activities and a multifunctional polyketide synthase gene. *Chem. Biol.* **6**, 429–439.

Hopwood, D.A. (1997). Genetic contributions to understanding polyketide synthases. *Chem. Rev.* **97**, 2465–2497.

Hopwood, D.A. (1999). Forty years of genetics with *Streptomyces*: From *in vivo* through *in vitro* to *in silico*. *Microbiology-UK* **145**, 2183–2202.

Hosted, T.J., Wang, T.X., Alexander, D.C., and Horan, A.C. (2001). Characterization of the biosynthetic gene cluster for the oligosaccharide antibiotic, evernimicin, in *Micromonospora carbonacea* var. *africana* ATCC39149. *J. Ind. Microbiol. Biotechnol.* **27**, 386–392.

Hunter, I.S. (1992). Function and evolution of secondary metabolites—no easy answers. *Trends Biotechnol.* **10**, 144–146.

Jordan, P.M. and Spencer, J.B. (1993). The biosynthesis of tetraketides—enzymology, mechanism and molecular programming. *Biochem. Soc. Trans.* **21**, 222–228.

Kagan, R.M. and Clarke, S. (1994). Widespread occurrence of three sequence motifs in diverse S- adenosylmethionine-dependent methyltransferases suggests a common structure for these enzymes. *Arch. Biochem. Biophys.* **310**, 417–427.

Kao, C.M., Luo, G.L., Katz, L., Cane, D.E., and Khosla, C. (1995). Manipulation of macrolide ring size by directed mutagenesis of a modular polyketide synthase. *J. Am. Chem. Soc.* **117**, 9105–9106.

Kao, C.M., Luo, G.L., Katz, L., Cane, D.E., and Khosla, C. (1996a). Engineered biosynthesis of structurally diverse tetraketides by a trimodular polyketide synthase. *J. Am. Chem. Soc.* **118**, 9184–9185.

Kao, C.M., Pieper, R., Cane, D.E., and Khosla, C. (1996). Evidence for two catalytically independent clusters of active-sites in a functional modular polyketide synthase. *Biochemistry* **35**, 12363–12368.

Kealey, J.T., Liu, L., Santi, D.V., Betlach, M.C., and Barr, P.J. (1998). Production of a polyketide natural product in nonpolyketide- producing prokaryotic and eukaryotic hosts. *Proc. Natl. Acad. Sci. U.S.A.* **95**, 505–509.

Keller, N.P., Brown, D., Butchko, R.A.E., Fernandes, M., Kelkar, H.S., Nesbitt, T.C., Segner, S., Bhatnagar, D. *et al.* (1995). A onserved polyketide mycotoxin gene cluster in *Aspergillus nidulans*. In M. Eklund, R.L. Richard, and K. Mise (eds) *Molecular Approaches to Food Safety Issues Involving Toxic Microorganisms*. Alaken, Fort Collins, pp. 263–277.

Keller, N.P. and Hohn, T.M. (1997). Metabolic pathway gene clusters in filamentous fungi. *Fungal Genet. Biol.* **21**, 17–29.

Kennedy, J., Auclair, K., Kendrew, S.G., Park, C., Vederas, J.C., and Hutchinson, C.R. (1999). Modulation of polyketide synthase activity by accessory proteins during lovastatin biosynthesis. *Science* **284**, 1368–1372.

Lee, T., Yun, S.H., Hodge, K.T., Humber, R.A., Krasnoff, S.B., Turgeon, G.B., Yoder, O.C., and Gibson, D.M. (2001). Polyketide synthase genes in insect- and nematode-associated fungi. *Appl. Microbiol. Biotechnol.* **56**, 181–187.

Liu, W.S., Christenson, D., Standage, S., and Shen, B. (2002). Biosynthesis of the enediyne antitumor antibiotic C-1027. *Science* **297**, 1170–1173.

Lorenz, M.C. (2002). Genomic appraoches to fungal pathogenicity. *Curr. Opin. Plant Biol.* **5**, 372–378.

Malpartida, F. and Hopwood, D.A. (1984). Molecular-cloning of the whole biosynthetic pathway of a *Streptomyces* antibiotic and its expression in a heterologous host. *Nature* **309**, 462–464.

Mayorga, M.E. and Timberlake, W.E. (1990). Isolation and molecular characterization of the *Aspergillus nidulans wA* Gene. *Genetics* **126**, 73–79.

Mayorga, M.E. and Timberlake, W.E. (1992). The developmentally regulated *Aspergillus nidulans wA* gene encodes a polypeptide homologous to polyketide and fatty acid synthases. *Mol. Gen. Genet.* **235**, 205–212.

McCarthy, A.D. and Hardie, D.G. (1984). Fatty acid synthases—an example of protein evolution by gene fusion. *TIBS* **9**, 60–63.

Metsa-Ketala, M., Salo, V., Halo, L., Hautala, A., Hakala, J., Mantsala, P., and Ylihinko, K. (1999). An efficient approach for screening minimal PKS genes from *Streptomyces*. *FEMS Microbiol. Lett.* **180**, 1–6.

Miao, V., Coeffet-Le Gal, M.F., Brown, D., Sinnemann, S., Donaldson, G., and Davies, J. (2001). Genetic approaches to harvesting lichen products. *Trends Biotechnol.* **19**, 349–355.

Minto, R.E. and Townsend, C.A. (1997). Enzymology and molecular biology of aflatoxin biosynthesis. *Chem. Rev.* **97**, 2537–2555.

Nicholson, T.P., Rudd, B.A.M., Dawson, M., Lazarus, C.M., Simpson, T.J., and Cox, R.J. (2001). Design and utility of oligonucleotide gene probes for fungal polyketide synthases. *Chem. Biol.* **8**, 157–178.

Nowak-Thompson, B., Gould, S.J., and Loper, J.E. (1997). Identification and sequence analysis of the genes encoding a polyketide synthase required for pyoluteorin biosynthesis in *Pseudomonas fluorescens* Pf-5. *Gene* **204**, 17–24.

O'Hagan, D. (1991). *The Polyketide Metabolites*. Ellis Horwood Limited, Chichester.

Pieper, R., Ebert-Khosla, S., Cane, D., and Khosla, C. (1996). Erythromycin biosynthesis: Kinetic studies on a fully active modular polyketide synthase using natural and unnatural substrates. *Biochemistry* **35**, 2054–2060.

Proctor, R.H., Desjardins, A.E., Plattner, R.D., and Hohn, T.M. (1999). A polyketide synthase gene required for biosynthesis of fumonisin mycotoxins in *Gibberella fujikuroi* slating population A. *Fungal Genet. Biol.* **27**, 100–112.

Santi, D.V., Siani, M.A., Julien, B., Kupfer, D. and Roe, B. (2000). An approach for obtaining perfect hybridization probes for unknown polyketide synthase genes: A search for the epothilone gene cluster. *Gene* **247**, 97–102.

Sauer, M., Lu, P., Sangari, R., Kennedy, S., Polishook, J., Bills, G., and An, Z. (2002). Estimating polyketide metabolic potential among non-sporulating fungal endophytes of *Vaccinium macrocarpon*. *Mycol. Res.* **106**, 460–470.

Schroder, G., Brown, J.W.S., and Schroder, J. (1988). Molecular analysis of resveratrol synthase. cDNA, genomic clones and relationship with chalcone synthase. *Eur. J. Biochem.* **172**, 161–169.

Seow, K.T., Meurer, G., Gerlitz, M., Wendt-Pienkowski, E., Hutchinson, C.R., and Davies, J. (1997). A study of iterative type II polyketide synthases, using bacterial genes cloned from soil DNA: A means to access and use genes from uncultured microorganisms. *J. Bacteriol.* **179**, 7360–7368.

Simpson, T.J. (1995). Polyketide biosynthesis. *Chem. Ind. 1995*, 407–411.

Sinnemann, S.J., Andresson, O.S., Brown, D.W., and Miao, V.P.W. (2000). Cloning and heterologous expression of *Solorina crocea pyrG. Curr. Genet.* **37**, 333–338.

Takano, Y., Kubo, Y., Shimizu, K., Mise, K., Okuno, T., and Furusawa, I. (1995). Structural analysis of *pks1*, a polyketide synthase gene involved in melanin biosynthesis in *Colletotrichum lagenarium. Mol. Gen. Genet.* **249**, 162–167.

Tsai, H.F., Chang, Y.C., Washburn, R.G., Wheeler, M.H., and Kwon-Chung, K.J. (1998). The developmentally regulated *alb1* gene of *Aspergillus fumigatus*: Its role in modulation of conidial morphology and virulence. *J. Bacteriol.* **180**, 3031–3038.

Tsai, H.F., Fujii, I., Watanabe, A., Wheeler, A.H., Chang, Y.C., Yasuoka, Y., Ebizuka, Y., and Kwon-Chung, K.J. (2001). Pentaketide melanin biosynthesis in *Aspergillus fumigatus* requires chain-length shortening of a heptaketide precursor. *J. Biol. Chem.* **276**, 29292–29298.

Ueda, K., Kim, K.-M., Beppu, T., and Horinouchi, S. (1995). Overexpression of a gene cluster encoding a chalcone synthase-like protein confers redbrown pigment production in *Streptomyces griseus. J. Antibiot.* **48**, 638–646.

Vidal-Cros, A., Viviani, F., Labesse, G., Bocara, M., and Gaudry, M. (1994). Polyhydroxynapthalene reductase involved in melanin biosynthesis in *Magnaporthe grisea. Eur. J. Biochem.* **219**, 985–992.

Vidgren, J., Svensson, L.A., and Liljas, A. (1994). Crystal structure of catechol O-methyltransferase. *Nature* **368**, 354–358.

Walsh, M., (1993). Cloning fungal polyketide synthase genes, Doctoral Thesis, University of Bristol, UK.

Walton, J.D. (2000). Horizontal gene transfer and the origin of secondary metabolite gene clusters in fungi: An hypothesis. *Fungal Genet. Biol.* **30**, 167–171.

Wang, I.K., Reeves, C., and Gaucher, G.M. (1991). Isolation and sequencing of a genomic DNA clone containing the 3′ terminus of the 6-methylsalicylic acid polyketide synthetase gene of *Penicillium urticae. Can. J. Microbiol.* **37**, 86–95.

Watanabe, A., Fujii, I., Sankawa, U., Mayorga, M.E., Timberlake, W.E., and Ebizuka, Y. (1999). Re-identification of *Aspergillus nidulans wA* gene to code for a polyketide synthase of naphthopyrone. *Tet. Lett.* **40**, 91–94.

Watanabe, A., Fujii, I., Tsai, H.F., Chang, Y.C., Kwon-Chung, K.J., and Ebizuka, Y. (2000). *Aspergillus fumigatus alb1* encodes naphthopyrone synthase when expressed in *Aspergillus oryzae. FEMS Microbiol. Lett.* **192**, 39–44.

Watanabe, A., Ono, Y., Fujii, I., Sankawa, U., Mayorga, M.E., Timberlake, W.E., and Ebizuka, Y. (1998). Product identification of polyketide synthase coded by *Aspergillus nidulans wA* gene. *Tet. Lett.* **39**, 7733–7736.

Witkowski, A., Rangan, V.S., Randhawa, Z.I., Amy, C.M., and Smith, S. (1991). Structural organization of the multifunctional animal fatty acid synthase. *Eur. J. Biochem.* **198**, 571–579.

Yang, G., Rose, M.S., Turgeon, B.G., and Yoder, O.C. (1996). A polyketide synthase is required for fungal virulence and production of the polyketide T-toxin. *Plant Cell* **8**, 2139–2150.

Yoder, O.C. and Turgeon, B.G. (2001). Fungal genomics and pathogenicity. *Curr. Opin. Plant Biol.* **4**, 315–321.

Yu, J.H. and Leonard, T.J. (1995). Sterigmatocystin biosynthesis in *Aspergillus nidulans* requires a novel type-I polyketide synthase. *J. Bacteriol.* **177**, 4792–4800.

Zhang, A., Lu, P., Dahl-Roshak, A.M., Paress, P.S., Kennedy, S., Tkacz, J.S., and An, Z. (2003). Efficient disruption of a polyketide synthase gene (*pks1*) required for melanin synthesis through *Agrobacterium*-mediated transformation of *Glarea lozoyensis. Mol. Genet. Genomics* **268**, 645–655.

6

More Functions for Multifunctional Polyketide Synthases

Isao Fujii, Akira Watanabe, and Yutaka Ebizuka

1. Introduction

More than a hundred years ago, J. N. Collie coined the term "polyketide" to represent natural products derived from simple two-carbon CH_2-CO building blocks (Collie and Myers, 1893). This proposal was later proven experimentally by A. J. Birch who used isotopically labeled acetate in the study of 6-methylsalicylic acid (6MSA) biosynthesis in fungi and showed that it was formed from four acetate units (Birch and Donovan, 1953; Birch *et al.*, 1955). Then, Lynen and his coworkers succeeded in detecting 6-methysalicylic acid synthase (MSAS) activity in a cell-free extract of *Penicillium patulum*, the first demonstration of polyketide synthase function *in vitro* (Dimroth *et al.*, 1971). These chemical and biochemical experiments with fungi established the concepts of "polyketide biosynthesis" and "polyketide synthase" (PKS).

However, clarification of details of polyketide biosynthesis and PKSs was impeded by the difficulty of further biochemical analysis. In 1980s, D. Hopwood and his colleagues developed a molecular genetic approach to polyketide biosynthetic genes in streptomycete bacteria (Malpartida and Hopwood, 1984; Malpartida *et al.*, 1987). Since then, extensive studies on bacterial PKS genes and the clusters in which they occur have been carried out (Hopwood, 1997).

In eukaryotic microorganisms, the MSAS gene from *Penicillium patulum* was the first PKS gene to be cloned (Beck *et al.*, 1990). Since then, a number of fungal PKS genes have been cloned, and analysis of their sequences has revealed a remarkable difference between the bacterial and fungal enzymes. All fungal PKSs presently known are iterative type I PKSs (Fujii *et al.*, 1998). Bacterial PKSs are either modular type I PKSs, producing reduced complex type polyketides such as macrolides, or type II PKSs that make aromatic compounds. Despite these architectural distinctions, the basic reactions involved in these multifunctional enzymes are common in all types of PKSs (Hopwood, 1997). This chapter is intended to

---

Isao Fujii and Yutaka Ebizuka • Graduate School of Pharmaceutical Sciences, The University of Tokyo, 7-3-1 Hongo, Bunkyo-ku, Tokyo, Japan. **Akira Watanabe** • Structural Biology Laboratory, The Salk Institute for Biological Studies, San Diego, CA 92186-5800.

Advances in Fungal Biotechnology for Industry, Agriculture, and Medicine. Edited by Jan S. Tkacz and Lene Lange, Kluwer Academic/Plenum Publishers, 2004.

describe the newly identified functions or additional functions in fungal PKSs other than the basic functions in PKSs. However, it is worthwhile starting with a brief overview on basic PKS functions and the domains that catalyze them.

Polyketide synthesis is initiated when acetate (the starter unit) is first loaded from the 4′-phosphopantetheine arm of coenzyme A (CoA) to an acyl carrier protein (ACP)-bound 4′-phosphopantetheine residue and then is subsequently transferred to a cysteine residue in the active site of a β-ketoacyl synthase (KS) domain. Malonate (the extender unit) is then transferred by an acyltransferase (AT) domain from malonyl-CoA to an ACP-bound 4′-phosphopantetheine residue. Carbon nucleophile formed by decarboxylation of the malonate loaded on ACP then attacks the carbonyl carbon of the starter acyl residue on the KS to form the nascent β-ketoacyl chain on the ACP. These acyltransfers and decarboxylative condensation reactions are the basic reactions in PKSs, and further repetitive condensations lead to the formation of a poly-β-ketoacyl chain.

With a PKS for an aromatic product, the mature β-ketoacyl chain undergoes cyclization (aldol and/or Claisen-type), aromatization, and is then released from the PKS protein. As a typical example, the orsellinic acid synthase (OAS) reactions are shown in Figure 6.1. In the reactions of PKSs leading to reduced complex-type compounds (RD-PKSs), the nascent ACP-anchored β-ketoacyl chain formed in a given condensation round may or may not be subjected to reduction and/or dehydration reactions in that round. Synthesis proceeds in a highly programmed manner (Figure 6.2), with the extent of reduction in each cycle determining the functionality formed: β-keto (no reduction), β-hydroxy (keto reduction), enoyl (keto reduction and dehydration), to alkyl (keto reduction, dehydration, and enoyl reduction). This programmed control of β-keto reduction is the key feature of RD-PKS that differentiates these enzymes from fatty acid synthase (FAS) and that leads to a great structural diversity in polyketide compounds (Hutchinson and Fujii, 1995; Hopwood, 1997).

Most fungal polyketides are derived from tetraketides ($C_2 \times 4$) to decaketides ($C_2 \times 10$). They can be classified into three basic skeletons depending on the cyclization

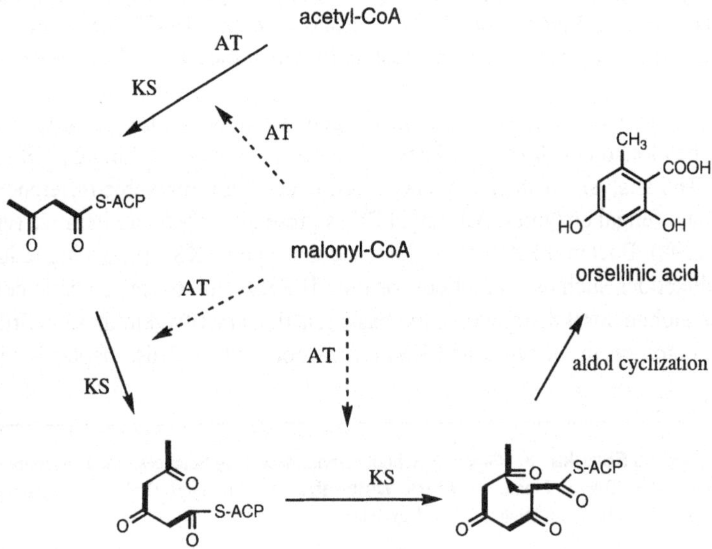

Figure 6.1. Reactions catalyzed by orsellinic acid synthase.

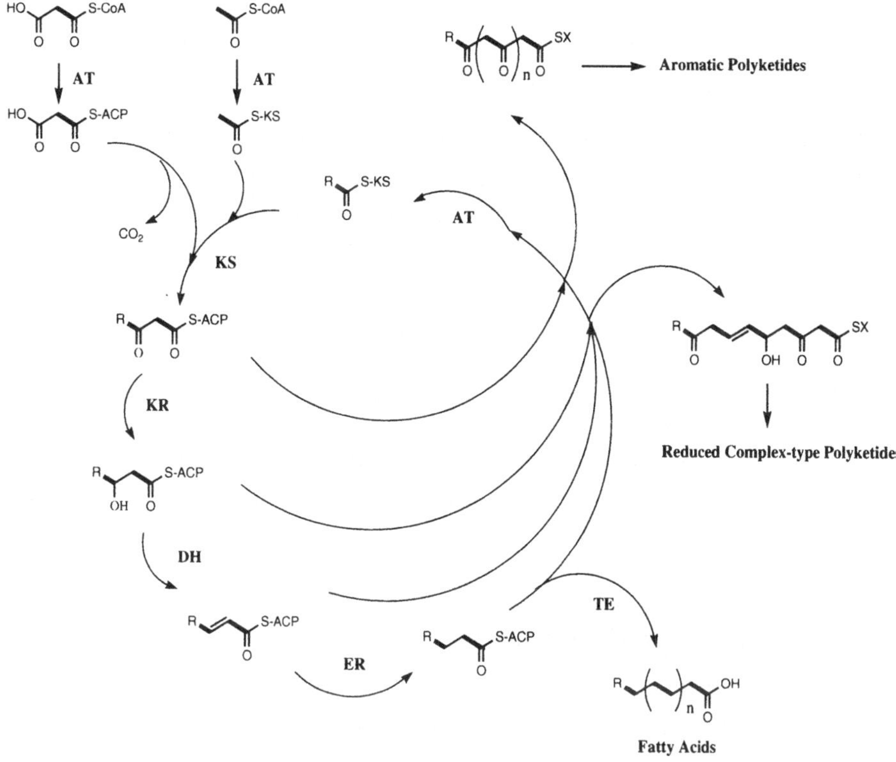

Figure 6.2. The basic pathway of fatty acid and polyketide biosynthesis. KS, β-ketoacyl synthase; KR, β-keto reductase; DH, dehydratase; ER, enoyl reductase; ACP, acyl carrier protein; TE, thioesterase.

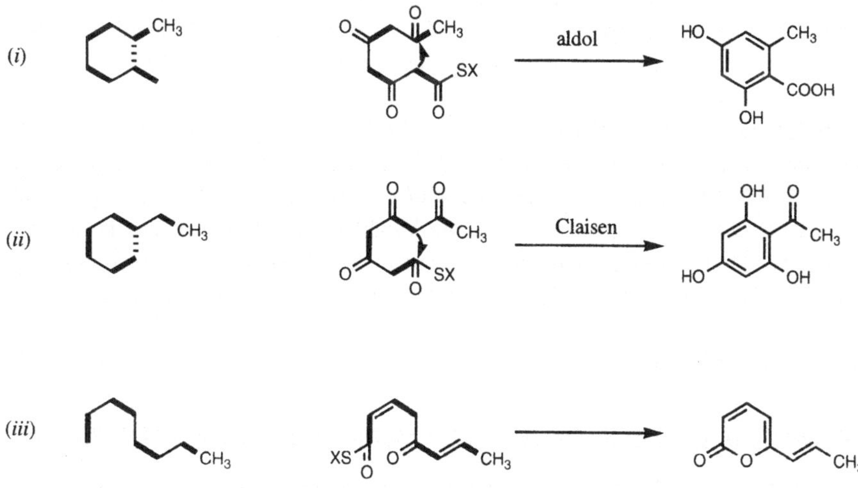

Figure 6.3. Three basic folding patterns in tetraketides. (*i*) Aldol cyclization type. (*ii*) Claisen cyclization type. (*iii*) Linear open chain type.

patterns. These are (i) aldol cyclization type, (ii) Claisen cyclization type, and (iii) linear open-chain type. Typical examples of tetraketides are shown in Figure 6.3. In higher ketides, folding of polyketomethylene chains can vary to form different ring patterns, patterns that nevertheless conform to this classification scheme.

In fungi, extensive modifications following the basic PKS reactions, such as oxidative ring-cleavages, rearrangements, oxidative coupling etc., lead to highly modified carbon skeletons as exemplified by aflatoxin biosynthesis. These secondary modifications are another reason for vast diversity in chemical structures of fungal polyketides.

2. Architecture and Functions of Fungal Polyketide Synthases

All fungal PKSs are large multi-functional polypeptides classified as type I PKSs. In contrast to bacterial modular type I PKSs, fungal PKSs apparently use their functional domains reiteratively (Fujii *et al.*, 1998). Just as we have divided the basic skeleton of the polyketide products into three categories, three patterns of domain organization or architecture are seen in fungal PKSs: (i) MSAS/OAS type, (ii) aromatic multi-ring PKS (AR-PKS), and (iii) reduced complex-type PKS (RD-PKS).

2.1. MSAS/OAS Polyketide Synthases

MSAS/OAS PKSs are the smallest type I PKSs with less than 2,000 amino acid residues (Figure 6.4(*i*)). MSAS genes have been cloned from several fungi and production of 6MSA was confirmed by heterologous expression (Bedford *et al.*, 1995; Fujii *et al.*, 1996). Purification of *P. patulum* MSAS indicated the homotetrameric subunit structure (Beck *et al.*, 1990). Orsellinic acid synthase (OAS) gene has not been cloned from fungi, but *in vitro* enzyme activity of OAS was detected in the cell-free extract of *Penicillium cyclopium* (Woo, 1989), and its polypeptide size was estimated to be ~130 kDa (Jordan and Spencer, 1993). Interestingly, the OAS gene cloned from *Streptomyces viridochromogenes* is an exceptional bacterial PKS, being an iterative type I enzyme (Gaisser *et al.*, 1997). The predicted polypeptide of 1,293 amino acids corresponds well to the size of *P. cyclopium* OAS.

Functionalities necessary for MSAS are β-ketoacyl synthase (KS), acyltransferase (AT), acyl carrier protein (ACP), ketoreductase (KR), and dehydratase (DH), while those for OAS are limited to KS, AT, and ACP. Additional functionalities presumed to be necessary for MSAS/OAS are aldol-type cyclase and/or aromatase for aromatic ring formation and activity to release the product from MSAS/OAS protein. However, these enzyme domains have not been identified nor assigned on the MSAS/OAS polypeptide. Aldol cyclization and aromatization could be a spontaneous chemical reaction, and MSAS/OAS enzyme might just provide a cavity for the stabilization and appropriate folding of tetraketomethylene intermediate.

2.2. Polyketide Synthases for Aromatic Multi-Ring Products (AR-PKSs)

Although a limited number of aromatic multi-ring PKSs (AR-PKS) genes have been cloned, PKSs of this type have additional characteristics in their domain organization when compared with MSAS/OAS (Fujii, 1999). They are polypeptides of 2,100~2,200 amino acids, about 300 amino acid longer than MSASs (Figure 6.4(*ii*)). The KS region is

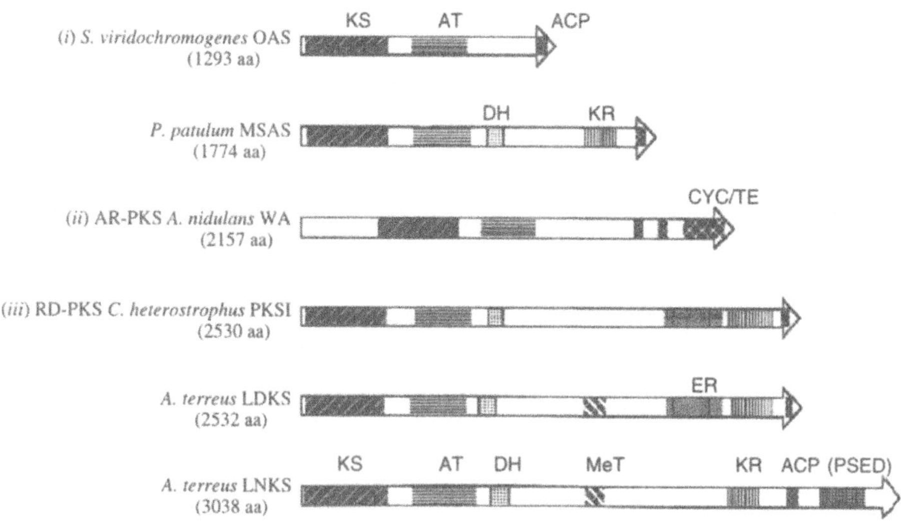

Figure 6.4. Architecture of fungal polyketide synthases. (*i*) MSAS/OAS. *Streptomyces viridochromogenes* OAS coded by *aviM* gene (accession number AF333038) and *Penicillium patulum* MSAS (X55776) are shown. *Aspergillus terreus* ATX (D85860) has similar architecture to that of *P. patulum* MSAS. (*ii*) AR-PKS. As a typical AR-PKS, *Aspergillus nidulans* WA, heptaketide naphthopyrone YWA1 synthase (X65866) is shown. The following PKSs have similar architecture: *Colletotrichum lagenarium* PKS1, 1,3,6,8-tetrahydroxynaphthalene (T4HN) synthase (D83643); *Nodulisporium* sp. PKS1, T4HN synthase (AF151533); *Exophiala dermatitidis* WdPKS1, T4HN synthase (AF130309); *Aspergillus fumigatus* Alb1p, YWA1 synthase (AF025541); *A. nidulans* STCA involved in sterigmatocystin biosynthesis (L39121); *Aspergillus parasiticus* PKSL1 involved in aflatoxin biosynthesis (L42765). (*iii*) RD-PKS. Architecture of *Cochliobolus heterostrophus* PKS1 involved in T-toxin biosynthesis (U68040), *Aspergillus terreus* LDKS, lovastatin diketide synthase (AF141925), and *A. terreus* LNKS lovastatin nonaketide synthase (AF151722) are shown. Following PKSs have similar architecture to that of LDKS: *Gibberella moniliformis* FUM5p involved in fumonisin biosynthesis (AF155773); *Alternaria solani* PKSN, alternapyrone synthase. KS, β-ketoacyl synthase; AT, acyltransferase; ACP, acyl carrier protrein; DH, dehydratase; KR, β-ketoreductase; CYC/TE, Claisen cyclase/thioesterase; ER, enoyl reductase; MeT, methyltransferase; PSED, peptide synthetase elongation domain.

shifted toward the *C*-terminus by ~340 amino acids, and typically, tandem ACPs exist in the *C*-terminus region. Thioesterase-like motifs are found at the *C*-terminus, and in the *Aspergillus nidulans* WA PKS, the *C*-terminus region functions in Claisen-type cyclization (Fujii *et al.*, 2001). Only two of these genes, namely *Colletotrichum lagenarium* T4HN synthase *PKS1* (Fujii *et al.*, 1999) and *A. nidulans wA* (Watanabe *et al.*, 1998; Watanabe *et al.*, 1999), have been expressed in heterologous fungi to study their functions. Details of this analysis are described later in the chapter. Interestingly, their fundamental architectures are the same despite differences in choice of starter units, chain-lengths, folding and cyclization patterns etc., that lead to different products.

2.2.1. Pentaketide 1,3,6,8-Tetrahydroxynaphthalene Synthases

In phytopathogenic fungi, melanization has been recognized as an important virulence factor, and intensive investigations on 1,8-dihydroxynaphthalene (DHN)-melanin biosynthesis have been carried out (Figure 6.5) (Wheeler and Bell, 1988). Most of the PKS

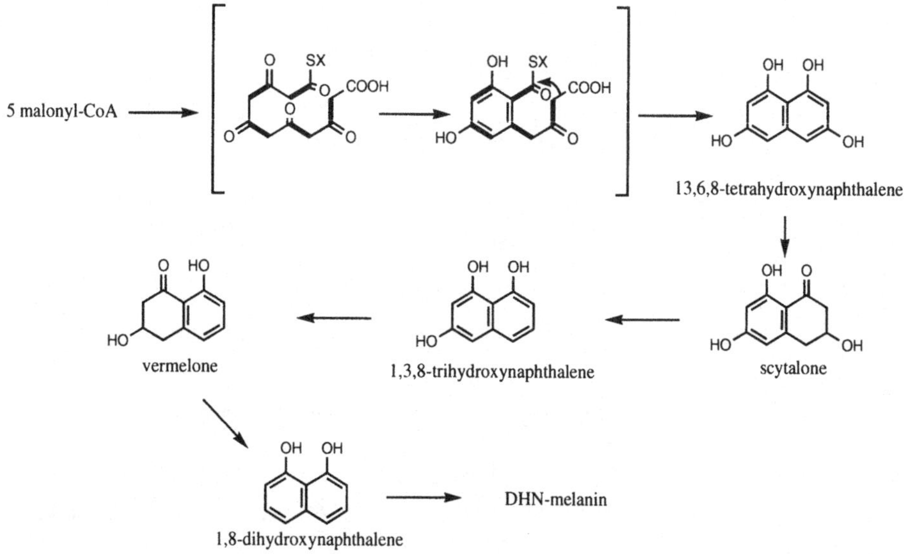

Figure 6.5. 1,3,6,8-Tetrahydroxynaphthalene synthase reaction and DHN-melanin biosynthesis.

genes involved in DHN-melanin biosynthesis were cloned by complementation screening of non-melanizing albino mutants. These are *C. lagenarium PKS1* (Takano *et al.*, 1995), *Nodulisporium* sp. *pks1* (Fulton *et al.*, 1999), *Exophiala dermatitidis WdPKS1* (Ye, Feng *et al.*, 1999), *Glarea lozoyensis pks1* (Zhang *et al.*, 2003), *Xylaria* sp. *pks12* (Punnya *et al.*, 2001), etc. These genes code for polypeptides of ~2,100 amino acid with KS, AT, ACPs, and Claisen-type cyclase (CYC) domains as shown in Figure 6.4(*ii*). The assumed function of a PKS for the formation of 1,3,6,8-tetrahydroxynaphthalene (T4HN) was confirmed for the *C. lagenarium* PKS1 by heterologous expression and identification of the products made in the transformants (Fujii *et al.*, 1999).

2.2.2. Heptaketide Naphthopyrone Synthases

The heptaketide naphthopyrone synthase gene *wA* was originally cloned from *A. nidulans* in the study of fungal differentiation by genetic complementation of the spore pigment-deficient (albino) *wA* mutant (Mayorga and Timberlake, 1990; Mayorga and Timberlake, 1992). Disruption of the *wA* locus confirmed that the gene is required for synthesis of green spore pigment, and the product of WA PKS was assumed to be converted to spore pigment by the *yA*-encoded laccase (Aramayo and Timberlake, 1990). The nucleotide sequence originally reported indicated that the *wA* gene codes for a polypeptide of 1,986 amino acids with a mass of 217 kDa containing KS and AT domains and tandem ACPs (Mayorga and Timberlake, 1992). A revision of the sequence provided for a polypeptide with 2,157 amino acids and additional CYC domain (Watanabe *et al.*, 1999). Its PKS product has been identified as a heptaketide naphthopyrone YWA1 (Figure 6.6) (Watanabe *et al.*, 1999). A similar naphthopyrone synthase gene, *alb1*, was also cloned from *Aspergillus fumigatus* as a gene involved in its spore pigment biosynthesis (Tsai *et al.*, 1998).

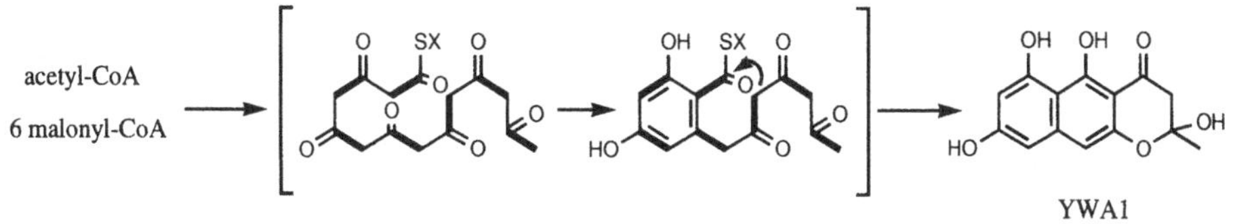

Figure 6.6. WA PKS reaction for heptaketide naphthopyrone YWA1 synthesis.

2.2.3. PKSs Involved in Aflatoxin Biosynthesis

The unusually long and complex biosynthetic pathway of aflatoxins starts from heptaketide 10-deoxynorsolorinic acid (Vederas and Nakashima, 1980). This anthrone compound with a C_6 side-chain has a polyketide origin and was initially considered to be the product of a decaketide synthase (Figure 6.7). However, hexanoate was identified as the starter unit for 10-deoxynorsolorinic acid synthesis by feeding experiments with its *N*-acetylcysteamine thioester (Townsend *et al.*, 1984).

Cloning and sequence analysis of the aflatoxin biosynthetic gene clusters from aflatoxigenic fungi (*Aspergillus flavus, Aspergillus parasiticus*, etc.) and a related cluster for sterigmatocystin from *A. nidulans* identified the presence of PKS genes and the specialized FAS genes responsible for the formation of the hexanoyl precursor which serves as a starter unit of the PKS that makes the 10-deoxynorsolorinic acid carbon skeleton (Yu and Leonard, 1995; Yu *et al.*, 1995; Brown *et al.*, 1996; Mahanti *et al.*, 1996). No direct confirmation of PKS product and/or detection of PKS enzyme activity has yet been carried out, although disruption of PKS genes resulted in mutants unable to produce aflatoxins or any biosynthetic intermediates. Homologous PKS genes, *pksA, pksL1*, and *stcA*, were cloned from *A. parasiticus, A. nidulans*, etc.

2.3. Polyketide Synthases for Reduced Products (RD-PKSs)

Fungal RD-PKSs produce reduced complex-type compounds such as lovastatin, T-toxin, fumonisin, etc. Polyketide chains of RD-PKS products vary in their state of reduction and dehydration as well as in chain-length. The RD-PKSs studied to date are polypeptides of over 2,500 amino acids in length and thus are the largest iterative PKSs known (Figure 6.4(*iii*)). Fungal PKSs of this class have a domain organization quite similar to that of mammalian FASs, possessing DH, ER, and KR domains in addition to KS, AT, and ACP domains. In FASs reactions, β-ketoacyl chain formed by each condensation reaction is converted to a fully saturated acyl chain by successive KR, DH, and ER reactions. On the other hand, RD-PKSs control the extent of reduction at each round of chain extension according to the length of the carbon chain growing on the ACP. A unique functional domain found in some RD-PKSs is the methyltransferase (MeT) responsible for transfer of methyl group from *S*-adenosyl methionine to polyketide chain. MeT domain is not present in FASs and other type of PKSs, but is found in some modular PKSs such as the epothilone PKS from myxobacterium *Sorangium cellulosum* (Julien *et al.*, 2000; Molnár *et al.*, 2000; Tang *et al.*, 2000). Architectures of fungal RD-PKSs are shown in Figure 6.4(*iii*).

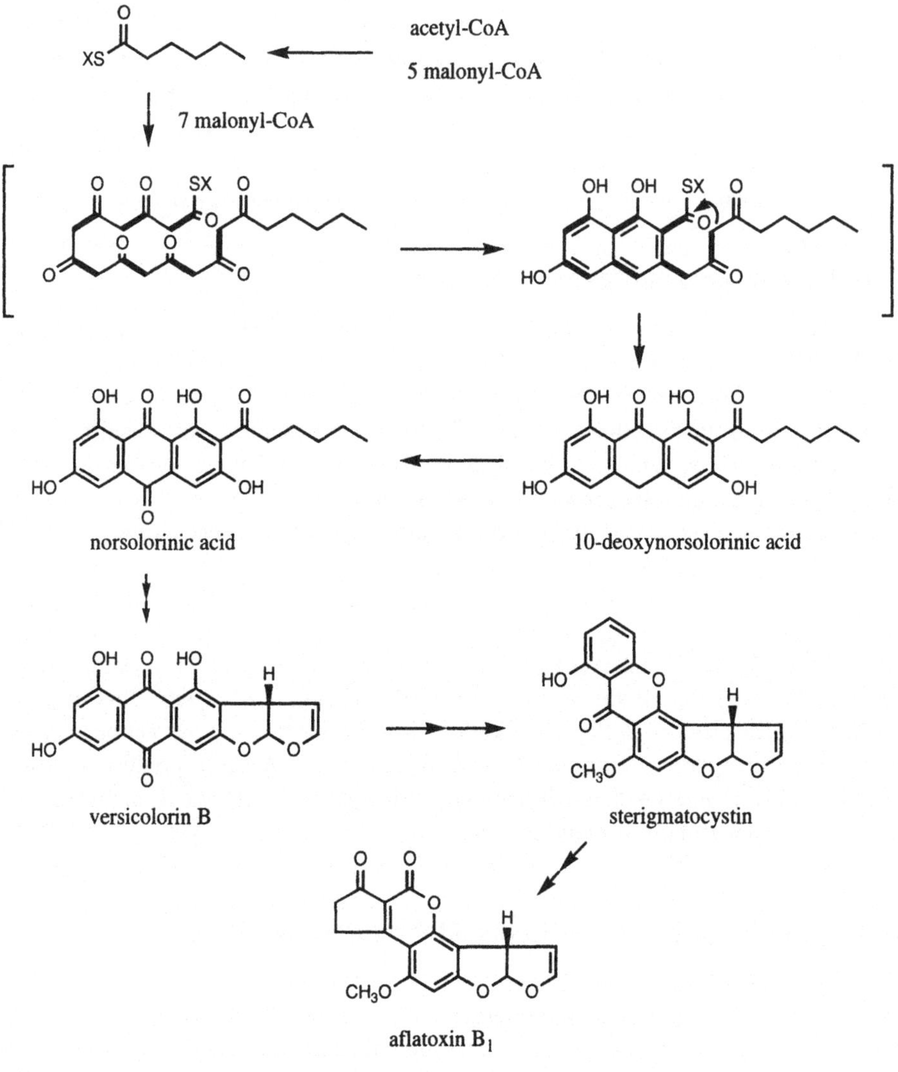

Figure 6.7. Norsolorinic acid synthesis reaction and aflatoxin biosynthesis scheme.

2.3.1. T-toxin PKS

Phytopathogenic fungi, *Cochliobolus heterostrophus* and *Phyllostica maydis* produce T-toxins (Figure 5.1 compound **15**) and related PM-toxins, which are linear long-chain polyketol compounds (Kono and Daly, 1979; Kono *et al.*, 1980; Kono *et al.*, 1983). T-toxin PKS gene was cloned from *C. heterostrophus* by REMI (*r*estriction *e*nzyme *m*ediated *i*ntegration) tagging (Lu *et al.*, 1994). The tagged genomic DNA was recovered from the T-toxin non-producing transformant and its sequence analysis identified the RD-PKS gene *PKS1* that encodes a 2,530 amino-acid polypeptide (Yang *et al.*, 1996). Disruption of the *PKS1* gene abrogated the production of T-toxin and thus confirmed the involvement of the gene in the biosynthesis of T-toxin. However, the direct product of *C. heterostrophus* PKS1 reaction has not been identified.

2.3.2. PKSs Involved in Lovastatin Biosynthesis

Lovastatin and its derivatives are potent (3*S*)-hydroxy-3-methylglutaryl-CoA reductase inhibitors used as cholesterol-lowering drugs (Alberts *et al.*, 1980; Endo and Hasumi, 1993). These compounds are composed of a conjugated decene ring system and 2-methylbutyryl side-chain joined with an ester linkage (Figure 5.1, compounds **17a, 17b**). Studies on the biosynthesis of lovastatin in *Aspergillus terreus* indicate that it is formed by a polyketide pathway (Endo *et al.*, 1985; Moore *et al.*, 1985; Yoshizawa *et al.*, 1994; Wagschal *et al.*, 1996). Decene ring moiety is a nonaketide that undergoes cyclization to a hexahydronaphthalene ring system (Witter and Vederas, 1996) and 2-methylbutyrate is a simple diketide. Identification of a ~250 kDa polypeptide correlated with a production of lovastatin led to the cloning of nonaketide synthase (LNKS) gene *lovB* that encodes a 3,038 amino-acid polypeptide of 269 kDa (Hendrickson *et al.*, 1999). Diketide synthase (LDKS) gene *lovF* was identified in the lovastatin biosynthetic gene cluster adjacent to the *lovB* gene (Kennedy *et al.*, 1999).

The compactin (ML-236B) biosynthetic gene cluster was cloned from *Penicillium citrinum*. Presence of *mlcA* and *mlcB* genes which show high homology with *lovB* and *lovF* genes respectively, were identified in the gene cluster (Abe *et al.*, 2002).

LDKS is considered to catalyze a single condensation followed by ketoreduction, dehydration, enoyl reduction, and *C*-methyltransfer in the synthesis of (2*R*)-2-methylbutyrate. LNKS is responsible for the synthesis of nonaketide chain of decene ring backbone. It consists of fundamental KS, AT, DH, and KR domains and has a MeT domain in the central region and a domain at its *C*-terminus homologous to peptide synthetase elongation domain (PSED) of nonribosomal peptide synthase (von Dören and Kleinkauf, 1997; Zuber and Marahiel, 1997). Its potential ER domain does not contain a full nucleotide binding consensus sequence and has low similarity to the known ERs. The enzyme also does not contain an obvious TE domain.

2.3.3. Fumonisin PKS

Fumonisins are mycotoxins produced by *Fusarium* species that frequently contaminate maize (Nelson *et al.*, 1993). Fumonisins, similar in structure to the sphingolipid intermediate sphinganine, disrupt sphingolipid metabolism via inhibition of the enzyme sphinganine *N*-acyltransferase (Wang *et al.*, 1991). This disruption may account for some of the fumonisin-induced diseases in human and animals (Merrill *et al.*, 1997; Howard *et al.*, 1999). Although there are number of different groups of fumonisins, fumonisin B_1 is the most abundant in nature making up about 70% of the total fumonisin content. Fumonisin B (Figure 5.1, compounds **16**) consists of a linear 20-carbon backbone with an amine group at C-2, methyl group at C-12 and C-16, tricarballylic ester groups at C-14 and C-15, and hydroxyl groups at C-3, C-5, and C-10. Feeding experiments have revealed that C-3 to C-20 of the backbone are derived from acetate (Blackwell *et al.*, 1994), the amine group and C-1 and C-2 are derived from alanine (Branham and Plattner, 1993), the methyl groups are derived from methionine (Plattner and Shackelford, 1992), and the oxygen atoms attached directly to the backbone are derived from molecular oxygen (Caldas *et al.*, 1998).

A polyketide synthase gene, *FUM5*, from *Fusarium verticillioides* was isolated and shown by gene disruption to be required for fumonisin biosynthesis (Proctor and

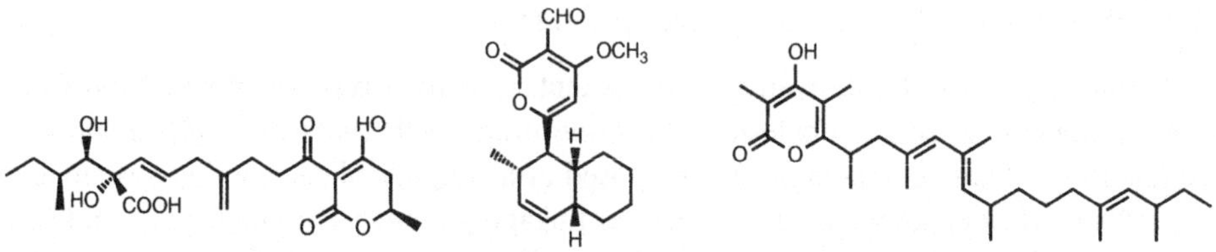

Figure 6.8. Reduced complex-type polyketides from *Alternaria solani.*

Desjaradins, 1999). The predicted PKS has 2,607 amino acids with six basic functional domains characteristic to type I PKS: KS, AT, DH, ER, KR, and ACP. In addition, presence of MeT domain was indicated by comparison with LNKS and LDKS domain organization. There was no amino acid sequence indicative of TE domain similar to those found in other RD-PKSs.

2.3.4. RD-PKS from *Alternaria solani*

Alternaria solani is a causal fungus of early blight disease of potato and tomato. Two kinds of phytotoxins, alternaric acid (Brian *et al.*, 1949; Brian *et al.*, 1951; Brian *et al.*, 1952; Barteles-Keith, 1960) and solanapyrones (Ichihara *et al.*, 1983; Ichihara *et al.*, 1985; Oikawa *et al.*, 1989b), were isolated from the different strains of the fungus (Figure 6.8). Feeding experiments for these reduced complex-type compounds showed the polyketide origin of both compounds (Oikawa *et al.*, 1989a; Ichihara and Oikawa, 1997; Oikawa *et al.*, 1998). Interesting features including C–C bond formation between separate polyketomethylene chains in alternaric acid biosynthesis and involvement of enzymological Diels–Alder reaction in solanapyrone biosynthesis have been shown (Ichihara and Oikawa, 1997).

Cloning of RD-PKS gene was carried out from the solanapyrone producing strain (Fujii *et al.*, unpublished). The cloned PKS gene *PKSN* codes for a 2,551 amino-acid polypeptide with KS, AT, DH, ER, KR, and ACP domains together with a MeT domain. Its domain organization is quite similar to other RD-PKSs, except LNKS that possesses the additional PSED-like domain at its *C*-terminus. Expression of PKSN in *A. oryzae* identified its product to be a novel decaketide pyrone, alternapyrone (Figure 6.8), with multiple *C*-methylations (Fujii *et al.*, unpublished).

3. More Functions for Fungal Polyketide Synthases

In addition to the basic PKS functional domains, KS, AT, KR, DH, ER, and ACP, other capabilities are necessary for the production of polyketide compounds. The key aspects that fungal PKSs should control in their reactions are (a) the selection of a starter unit other than acetate when necessary, (b) the choice of extender units other than malonate when necessary, (c) the extent of reduction, (d) the methylation or alkylation of the growing polyketomethylene chain, (e) the folding conformation of polyketomethylene chain for correct cyclization, (f) the cyclization step (aldol, Claisen, lactonization), and (g) the release of product from the enzyme.

Due to the limited biochemical and mechanistic information on fungal PKSs, these functional aspects have not been fully clarified yet. However, some progress has been attained through heterologous expression, enzymological analysis, and analysis of genetic information, etc. This section of the chapter will summarize these developments.

3.1. Claisen Cyclase Domain in AR-PKSs

In the formation of most aromatic multi-ring polyketides, a Claisen-type cyclization has been assumed to occur as the last stage of the AR-PKS reaction sequence. For example, in 10-deoxynorsolorinic acid synthesis, the third ring is considered to be formed by Claisen-type cyclization as shown in Figures 6.5–6.7.

The Claisen Cyclase domain was identified in WA PKS of *A. nidulans* (Fujii *et al.*, 2001). As mentioned earlier, it was considered that the product of *wA* gene is responsible for production of yellow intermediate which is subsequently converted to the mature green spore pigment by *yA*-encoded laccase (Clutterbuck, 1972; Aramayo and Timberlake, 1990). The structure of the *A. nidulans* green spore pigment and its yellow precursor(s) has not been clarified in spite of some attempts (Brown *et al.*, 1993; Brown and Salvo, 1994). To identify the function of WA PKS, it was expressed in a heterologous fungus that had been successfully used by our group to identify the function of *atX* gene cloned from *Aspergillus terreus* (Fujii *et al.*, 1996). The host was *Aspergillus oryzae* M-2-3, and the expression vector was pTAex3 with a β-amylase promoter (Fujii *et al.*, 1995). The expression plasmid pTA-wA was constructed based on the originally reported wA nucleotide sequence in GenBank. The *A. oryzae* transformant harboring the expression plasmid pTA-wA produced compounds which were not detected in the host *A. oryzae* or even *A. nidulans* itself. The newly produced compounds were identified as citreoisocoumarin and derivatives thereof (Figure 6.9) which are made from heptaketide intermediate with or without reduction of carbonyl in their side-chains (Watanabe *et al.*, 1998).

It was obvious that production of these compounds was directed by the *wA* gene. However, there remained some ambiguity whether the expressed WA PKS functioned correctly because the compounds made were colorless and a yellow spore pigment precursor was expected (Clutterbuck, 1972). Resequencing of the *wA* gene revealed the absence of a single base in the reported nucleotide sequence just upstream of the predicted stop codon of the *wA* ORF. In the corrected WA PKS sequence, the ORF increased from 1,986 amino acids to 2,167 amino acids by inclusion of an additional 181 amino acids at the *C*-terminus. A conserved sequence indicative of a TE domain was found in this *C*-terminal region. When *A. oryzae* was transformed with a new plasmid, pTA-nwA, expressing the full-length WA polypeptide, it produced a yellow compound named YWA1 (Watanabe *et al.*, 1999). The basic carbon skeleton of naphthopyrone YWA1 (Figure 6.10) is that of the *A. parasiticus* spore pigment parasperone A (Brown *et al.*, 1993) and related pigments like fonsecin of

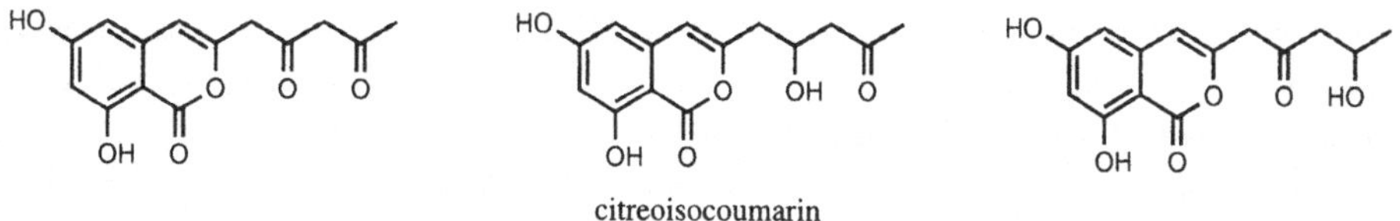

Figure 6.9. Polyketide compounds produced by *Aspergillus oryzae* transformant with pTA-wA.

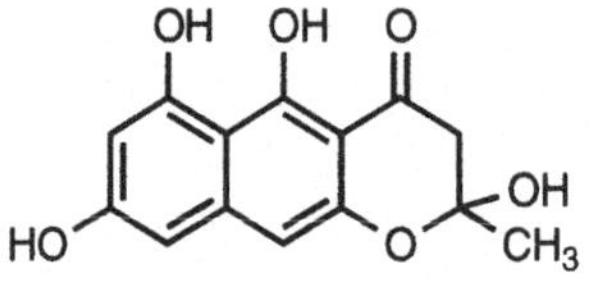

YWA1 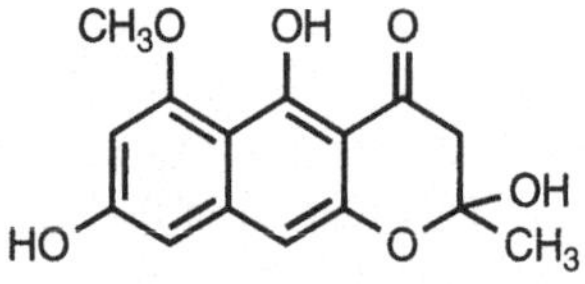 fonsecin (*A. fonsecaeus*)

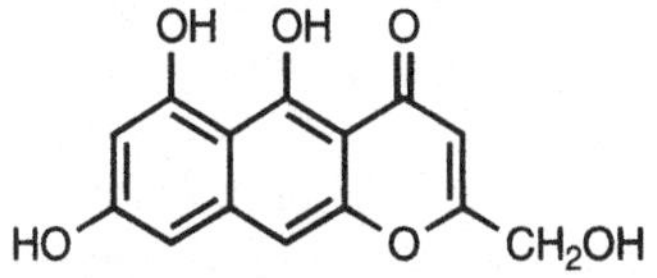

parasperone A (*A. parasiticus*) 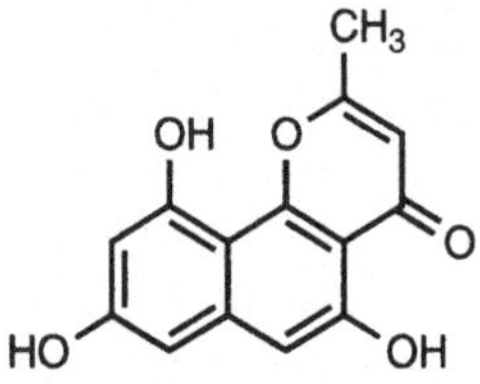 flavasperone (*A. niger*)

Figure 6.10. YWA1 and related fungal naphthopyrones.

Aspergillus fonsecaeus (Galmarini and Stodola, 1965) and rubrofusarin of *Fusarium culmorum* (Stout *et al.*, 1961; Tanaka and Tamura, 1961). This naphthopyrone YWA1 is thus considered to be the true yellow intermediate of green spore pigments of *A. nidulans*.

The result of these two expression experiments with WA PKS indicated that the *C*-terminal region, previously assumed to be a TE domain, has some critical role in cyclization of second aromatic ring of YWA1. To prove this hypothesis, *C*-terminal truncated WA PKS derivatives were expressed in *A. oryzae* (Fujii *et al.*, 2001). Interestingly, a deletion of only 30 amino acids from the *C*-terminus resulted in the production of isocoumarin instead of naphthopyrone. Further truncation up to TE active site motif (-G W S^{1967} A G-) neither abolished the production of heptaketide citreoisocoumarin nor reduced total polyketide production as compared with that of wild type WA.

It is known that mammalian FAS and bacterial modular type I PKSs such as 6-deoxyerythronolide B synthase (DEBS) require TE functionality to release their products (Singh *et al.*, 1984; Aggarwal *et al.*, 1995). In fungal PKSs, the domain that has been termed the TE domain is not always present. MSAS/OASs and RD-PKSs lack this domain. The former enzymes use an aldol cyclization mechanism, and the latter lack cyclization/aromatization steps. To the best of our knowledge, all fungal AR-PKSs with TE domains involve Claisen-type cyclization in their PKS reactions as shown in Figures 6.5–6.7. Thus, the TE domains in these AR-PKSs may not be simple thioesterase acting just to release their products from PKS enzymes but may possibly serve as Claisen Cyclase (CYC) as found in WA PKS (Fujii *et al.*, 2001).

To date, five TEs have been structurally characterized by X-ray crystallography (Lawson *et al.*, 1994; Benning *et al.*, 1998; Bellizzi *et al.*, 2000; Li *et al.*, 2000; Tsai *et al.*, 2001). Of these, the *Vibrio harvey* myristoyl ACP TE (Lawson *et al.*, 1994) and the mammalian palmitoyl protein TE (Bellizzi *et al.*, 2000) contain a classic Ser-His-Asp catalytic triad in their active sites, and a similar catalytic triad would be expected to operate in fungal PKS Claisen-type cyclization. Alignment of *C*-terminus regions of AR-PKSs (Figure 6.11) showed the presence of several highly conserved amino acid residues including Ser^{1967} and His^{2129} (residue numbers in WA PKS). Site-directed mutants of

```
C. lagenarium PKS1        ATKNLWMVPD GSGCATSYTE ISQVSS-NWA VWGLFSPFMK TPEEYKCGVY GMAAKFIEAM   1994
E. dermatitidis WdPKS1    ATXNLFLFPD GSGSATSYVS IPAIDSXNLA VYGLNCPFMK DPTSYTCGIX SVSXLYLEKV   1978
Nodulisporium PKS1        ATRHLFMIPD GSGSATSYTE ISDLGS-DVA VWGMFSPFMR TPEEYKCGVY GMATKFIEEM   1971
A. fumigatus Alb1p        ATKKLFMFPD GSGSASSYAT IPALSP-DVC VYGLNCPYMK TPQNLTCSLD ELTEPYLAEI   1947
A. nidulans WA            ASKTLFLFPD GSGSATSYAT IPGVSP-NVA VYGLNCPYMK APEKLTCSLD SLTTPYLAEI   1952
A. nidulans STCA          AKQILFMLPD GGGSASSYLT IPRLHA-DVA IVGLNCPYAR DPENMNCTHQ SMIQSFCNEI   2013
A. parasiticus PKSL1      ARKTLFMLPD GGGSAFSYAS LPRLKS-DTA VVGLNCPYAR DPENMNCTHG AMIESFCNEI   1922
                          *    *::.** *.*.* **   :. :.  :  .:  *:  .*: :  *  .  .*       :    :

C. lagenarium PKS1        KARQSKGPYS LAGWSAGGVI AYEIVNQLTK AG-------E TVENLIIIDA PCPVTIEPLP   2047
E. dermatitidis WdPKS1    LXRQPNGPYI LXGWSASGVF AYXITXQLXD LQXLHPDKNY TVEKLNLIXS PCPIRLEPLP   2038
Nodulisporium PKS1        KRRQPEGPYA VSGWSAGGVI AYEIVNQLTK AG-------D EVSHLLIIDA PCPITIEPLP   2024
A. fumigatus Alb1p        RRRQPKGPYS FGGWSAGGIC AFDAARQLIL EE------GE EVERLLLLDS PFPIGLEKLP   2001
A. nidulans WA            RRRQPTGPYN LGGWSAGGIC AYDAARKLVL QQ------GE IVETLLLLDT PFPIGLEKLP   2006
A. nidulans STCA          KRRQPEGPYH LGGWSSGGAF AYVTAEALIN AG-------N EVHSLIIIDA PVPQVMEKLP   2066
A. parasiticus PKSL1      RRRQPRGPYH LGGWSSGGAF AYVVAEALVN QG-------E EVHSLIIIDA PIPQAMEQLP   1975
                            **.  ***   .  ***:.*   *:    .   *             *   *  ::  :  * *   :*  **

C. lagenarium PKS1        RSLHAWFASI GLLGEGDDE- --AAKKIPSW LLPHFAASVT ALSNYTAEPI PK------EK   2098
E. dermatitidis WdPKS1    ARLHHFFDEI GLLGTGTG-- -----KTPNW LLPHFEYSIK ALTAYRPELK STRD----FN   2087
Nodulisporium PKS1        AGLHAWFAEI GLLGEGDG-- --EAKKIPSW LLPHFAASVT ALSNYTADPI PK------DK   2074
A. fumigatus Alb1p        PRLYKFFNSI GLFGDGKR-- -----APPDW LLPHFLAFID SLDAYKAVPL PFNDSKWAKK   2054
A. nidulans WA            PRLYSFFNSI GLFGEGKA-- -----APPAW LLPHFLAFID SLDAYKAVPL PFNEQEWKGK   2059
A. nidulans STCA          TSFYEYCNNL GLFSNQPGGT TDGTAQPPPY LIPHFQATVD VMLDYRVAPL KTN------R   2120
A. parasiticus PKSL1      RAFYEHCNSI GLFATQPGAS PDGSTEPPSY LIPHFTAVVD VMLDYKLAPL HAR------R   2029
                            ::     .:  **:.                 *  :  *:***     :     :   *               .

C. lagenarium PKS1        CPNVMAIWCE DGVCHLPTDP RPDPYPTG-- ---HALFLLD NRTDFGPNRW DEYLDVNKFR   2153
E. dermatitidis WdPKS1    APPTLLIWAT DGVCGKPGDP RPPPQADDPK ---SMKWLLE NRTDFGPNGW DKLLGAEVCK   2144
Nodulisporium PKS1        CPKVTTIWCE DGVCKLPTDP RPDPYPTG-- ---HALFLLD NRTDFGPNRW DEYLDIEKMT   2129
A. fumigatus Alb1p        MPKTYLIWAK DGVCGKPGDP RPEPAEDGSE DPREMQWLLN DRTDLGPNKW DTLVGPQNIG   2114
A. nidulans WA            LPKTYLVWAK DGVCPKPGDP WPEPAEDGSK DPREMVWLLS NRTDLGPNGW DTLVGKENIG   2119
A. nidulans STCA          MPKVGIIWAS ETVMDEDNAP ---------- KMKGMHFMVQ KRWDFGPDGW DVVCPGAVFD   2170
A. parasiticus PKSL1      MPKVGIVWAA DTVMDERDAP ---------- KMKGMHFMIQ KRTEFGPDGW DTIMPGASFD   2079
                           *  .   :*.   :  *        *                       ::::.  .*  ::**:  *  *

C. lagenarium PKS1        TRHMPG--NH FSMMHGDYVS QTTLSPYNDD NLTRSF--                          2187
E. dermatitidis WdPKS1    MVTVVG--NH FTMMKPPVAK GVGQYIRESL SMXRA---                          2177
Nodulisporium PKS1        FHHMPG--NH FSMMHGDLAK QLGGFLKEGI HA------                          2159
A. fumigatus Alb1p        GIHVMEDANH FTMTTGQKAK ELSQFMATAM SS------                          2146
A. nidulans WA            GITVIHDANH FTMTKGEKAK ELATFMKNAL GVCERRLV                          2157
A. nidulans STCA          ILRAEG-ANH LR-------- -----------------                            2181
A. parasiticus PKSL1      IVRADG-ANH FTLMQKEHVS IISDLIDRVM A-------                          2109
                                  ** :
```

Figure 6.11. Alignment of *C*-terminus regions of AR-PKSs. Key amino acid residue serine and histidine in Claisen Cyclase domain are shown inverted. Fully or highly conserved amino acids are shown with '*' or ':'/'.' below the alignment.

WA PKS, viz., S1967A and H2129Q, did not produce naphthopyrone at all but produced citreoisocoumarin instead, confirming that both amino acid residues are involved in Claisen cyclization (Fujii *et al.*, 2001).

In the biosynthesis of macrocyclic natural products, such as macrolides, by modular type I PKSs, and macrolactams and macrolactones by non-ribosomal peptide synthetases, a *C*-terminal TE domain is involved in cyclization of a linear precursor. It was shown that *C*-terminal TE domains excised from nonribosomal peptide synthase subunit proteins retain macrocyclization activity (Kohli *et al.*, 2002). Solving the crystal structure of the TE domain of the DEBS PKS provided the first elucidation of a modular PKS domain (Tsai *et al.*, 2001). It belongs to the α/β-hydrolase family with the active site triad comprised of Asp-His-Ser in the middle of the substrate channel. Although the amino acid sequence homology between DEBS TE and WA CYC domains is not very high, a Claisen cyclization mechanism by WA CYC could be assumed based on the structural and mechanistic information of macrocyclization TE domains as shown in Figure 6.12.

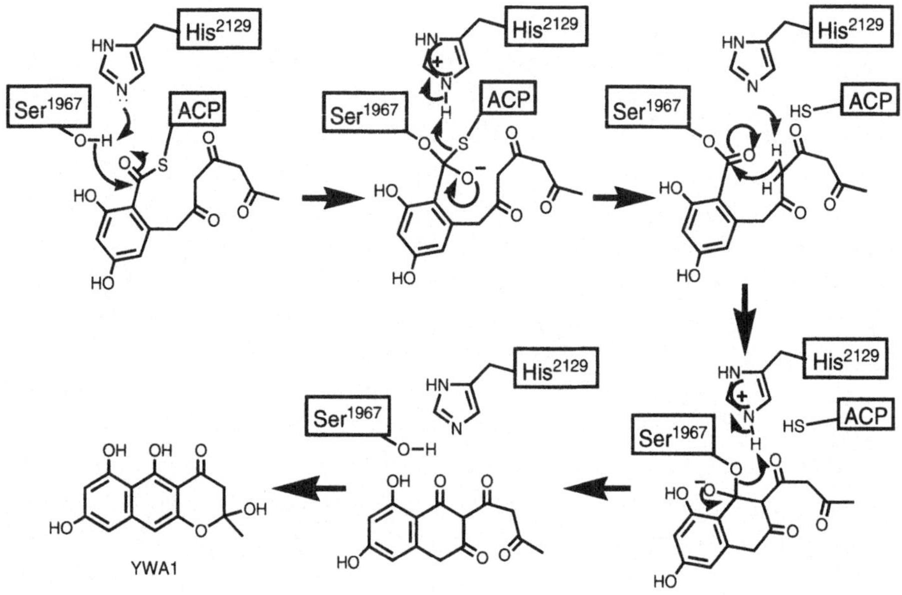

Figure 6.12. Proposed Claisen cyclase mechanism involved in the naphthopyrone second ring formation catalyzed by WA PKS (From Fujii *et al.*, 2001, copyright 2001, with permission from Elsevier Science).

3.2. More Functions for AR-PKSs

3.2.1. Starter Units

Some fungal AR-PKSs utilize starter units other than acetate although their variety is limited compared with bacterial PKSs. *In vitro* studies on starter specificity have been carried out on MSAS and *C. lagenarium* PKS1.

MSAS showed loose starter specificity accepting acetyl-CoA, acetoacetyl-CoA, propionyl-CoA, butyryl-CoA, pentanoyl-CoA, crotony-CoA, hexanoyl-CoA, and heptanoyl-CoA. There is an inverse relationship between the carbon chain-length of the starter unit and the level of incorporation into the 6-alkylsalicylic acids. *In vitro* incorporation of acetoacetyl-CoA into 6MSA was 80% of that of acetyl-CoA, but incorporation of hexanoyl-CoA and heptanoyl-CoA into corresponding 6-alkylsalicylic acids were around 15% and 4%, respectively. On the other hand, MSAS is highly selective for its extender unit, malonyl-CoA. Methylmalonyl-CoA which is often used by bacterial PKSs is not accepted by MSAS (Shoolingin-Jordan and Campuzano, 1999).

In vitro T4HN synthase activity of *C. lagenarium* PKS1 was the first example of AR-PKS detected in a cell-free extract of an *A. oryzae* transformant expressing a heterologous PKS. Fujii *et al.* (2000) tried to identify the starter unit for synthesis of T4HN that apparently lacks a starting unit-derived substituent on its aromatic ring system. However, based on the assumed cyclization mechanism for naphthalene ring formation, malonyl-CoA had been proposed to be a possible starter in the T4HN synthase reaction (Fujii *et al.*, 1999). From the *in vitro* T4HN synthase reaction with ^{14}C-labeled acetyl-CoA and/or ^{14}C-labeled malonyl-CoA as substrates, malonyl-CoA (Figure 6.5) was unambiguously confirmed to serve as the starter as well as an extender unit in the formation of T4HN (Fujii *et al.*, 2000).

The starter unit of PKSs for aflatoxin/sterigmatocystin biosynthesis is considered to be a hexanoyl-CoA (Brobst and Townsend, 1994). Presence of genes for fatty acid synthase α- and β-subunits in the aflatoxin biosynthetic gene clusters strongly supports the idea although no direct *in vitro* evidence has been obtained yet. Alignment of AT domains of AR-PKSs suggests the presence of the conserved amino acid residues specific in aflatoxin PKSs. However, it is not clear enough to differentiate their functions involved in hexanoyl starter loading. Specific protein–protein interaction between FAS and PKS may be necessary in the transfer of the starter hexanoyl unit from the FAS to the PKS (Watanabe *et al.*, 1996) and would explain the starter specificity, but this possibility remains to be explored experimentally.

3.2.2. *N*-Termini

In addition to the conserved KS, AT, ACP, and CYC domains, the *N*-termini of AR-PKSs show relatively high homology to each other (Figure 6.13). This region, upstream of KS domain, is characteristic only to AR-PKSs and is absent in MSAS/OASs and RD-PKSs. If AR-PKSs have homodimeric subunit structures with head-to-tail interaction as shown for mammalian FASs (Smith, 1994), *C*-terminus CYC domains might interact with *N*-termini of the complementary subunit. *N*-Terminus deletion experiments on WA PKS have shown that the deletion of only five amino acids from the *N*-terminus abolished the PKS activity (Fujii *et al.*, 2001), suggesting an important role of *N*-terminus which is yet to be clarified.

3.2.3. Interdomain Regions

AR-PKSs possess interdomain regions connecting AT and ACP domains. In mammalian FAS, the interdomain region between DH and ER is necessary to form a dimeric subunit structure (Chirala *et al.*, 2001). To assess the role of AR-PKS interdomain region, Watanabe (unpublished) constructed the chimeric AR-PKSs from *A. nidulans* WA and *C. lagenarium* PKS1. The SW-2 chimeric PKS consisted of PKS1 *N*-terminus-KS-AT and WA interdomain-ACP-CYC produced a mixture of products (Figure 6.14), tetraketide orsellinic acid, pentaketide α-acetylorsellinic acid, and its isocoumarin derivative, hexaketide isocoumarins, and heptaketide naphthopyrone YWA1. The SW-B chimera consisted of PKS1 *N*-terminus-KS-AT-interdomain and WA ACP-CYC also produced a mixture of products, pentaketide α-acetylorsellinic acid, pentaketide isocoumarin, hexaketide isocoumarins, and hexaketide 2-acetyl-1,3,6,8-tetrahydroxynaphthalene. However, pentaketide T4HN and heptaketide YWA1 were not produced by the SW-B (Watanabe and Ebizuka, 2002). Interestingly, WS chimeras with WA *N*-terminus half and PKS1 *C*-terminus half were apparently not functional. These results indicated that KS-AT of PKS1 could function with ACP-CYC of WA but without strict chain-length control and that WA CYC could cyclize not only heptaketide intermediate but also hexaketide intermediate.

3.2.4. ACP Domains

Typically, AR-PKSs possess tandem duplicated ACP motifs in ACP domains. Similar ACP domain with tandem motifs is also found in *Bacillus subtilis pksX* (Scotti

```
C. lagenarium PKS1        ---MADTMSY LLFGDQSLDT HGFLAEFCRN GNPSILAKTF LEQAGQALRE EIDGLGKLER   57
E. dermatitidis WdPKS1    ------MEEV YVFGDQTADC RAFFTKVFTR KD-NVLLQSF LERAGEAVRF ENQNRS-HPS   52
Nodulisporium PKS1        --MMADQMAF LLFGDQSLDT HGFLADFYRR GNPSVLSKEF LRLTGDALRD EIDRLPREER   58
A. fumigatus Alb1p        ---MEDLHRL YLFGDQTISC DEGLRNLLQA KN-HTIVASF IERCFHALRQ EITRLPPSQR   56
A. nidulans WA            ---MEDPYRV YLFGDQTGDF EVGLRRLLQA KN-HSLLSSF LQRSYHAVRQ EISHLPPSER   56
A. nidulans STCA          MASHAEPTRL FLFGDQTYDF VADLRDLLNI RN-NPILVAF LEQSHHVIRA QMIRELPPKE   59
A. parasiticus PKSL1      ---MAQSRQL FLFGDQTADF VPKLRSLLSV QD-SPILAAF LDQSHYVVRA QMLQSMNTVD   56
                          __________ _:****:_._ ___:__.___ _:___:___* :_____.:*_ :_________

C. lagenarium PKS1        SKLPTFQTLR QLNERYHAQS IKHPGIDSAL LCTTQLAHYI DRTE-KEPQD ACLHDHTFFM  116
E. dermatitidis WdPKS1    KAVPNFSTIQ ELVDRYYRGD AKDAATESAL VCISQFCHFI GAFEERRPSY IQPNSDARLV  112
Nodulisporium PKS1        RKIPIFRTLQ QLNERYHAQQ IKFPGIDSAL LCIAQLAHYI DRSE-KEYED VTDHRNTKLS  117
A. fumigatus Alb1p        TLFPRFTSIA DLLAQHRE-S GTNPALGSAL TCIYQLGCFI DYHG-DRGHP YPSSDDG-LL  113
A. nidulans WA            STFPRFTSIG DLLARHCE-S PGNPAIESVL TCIYQLGCFI NYYG-DLGHT FPSHSQSQLV  114
A. nidulans STCA          HKQARTASLA ELLQKYVD-R KLPSAFQTAL SCVTQIGLFM RQFD-DPRVL YPHANDSYVL  117
A. parasiticus PKSL1      HKLARTADLR QMVQKYVD-G KLTPAFRTAL VCLCQLGCFI REYE-ESGNM YPQPSDSYVL  114
                          ___.____:_ ::__::____ ___..__:.* _*__*:__:: __________ _____.__._

C. lagenarium PKS1        GLCTGLFAAA AIASTPSVST LIPLAVQVVL MAFRTGSHVG SLAERLSPPV GQSEPWTHIL  176
E. dermatitidis WdPKS1    GLCTGLIAAT AVAASDSLTA LIPLAVEAVR IAFRAGAHVG KVAQQTECDS KTQS--WSTI  170
Nodulisporium PKS1        GLCTGLFAAT AIASSYSLST LLPIAIQVVL MAFRTGSHVA SLAERLSPST EKSESWTYVL  177
A. fumigatus Alb1p        GSCTGMLSCT AVSSCKNVGE LLPLAVEIVR LTIHLGLCVM RVREMVDSTE SSSGSWSILV  173
A. nidulans WA            GLCTGLLSCA AVSCASNIGE LLKPAVEVVV VALRLGLCVY RVRKLFGQDQ AAPLSWSALV  174
A. nidulans STCA          GVCTGSLAAA AISCSTSLSE LLPIAVQTVL VAFRLGLWAE KVRDNLEISE TNQTQPWSAV  177
A. parasiticus PKSL1      GFCMGSLAAV AVSCSRSLSE LLPIAVQTVL IAFRLGLCAL EMRDRVDGCS DDRGDPWSTI  174
                          *_*_*_::.. *::.__.:__ *:__*::_*_ ::::_*__._ _:_.______ _________:

C. lagenarium PKS1        PGLKESDAKE ALANFHESNY IS---VASQA YVSAVSSSSL AISGPPATLK ALDDQNVFG-  232
E. dermatitidis WdPKS1    VAADEKSAQE ALDAFHKEXG TS---PINQL WISVSSATSV TISVPPWTKA RLXEESEFFR  227
Nodulisporium PKS1        PAAQEAQAKS ILAEFHESEG IS---LAAQA YVSAVSANNI AISGPPATLQ ALFSKDLFE-  233
A. fumigatus Alb1p        SEINEADATS LIGDFVKKRG IP---PSSQP YISAVGSKGL TISAPPEILD NFIEEGLPKE  230
A. nidulans WA            SGLSESEGTS LIDKFTRRNV IP---PSSRP YISAVCANTL TISGPPVVLN QFLDTFISGK  231
A. nidulans STCA          CHVPPEEVAI AIDRFSHKKV RSPVYRAQRP WITATSAKTT TVSASPDILS QLASQAPFTN  237
A. parasiticus PKSL1      VWG--LDPQQ ARDQIEVFCR TTNVPQTRRP WISCISKNAI TLSGSPSTLR AFCAMP-QMA  231
                          ______.___ ____:_____ _.______:_ :::____.__ ::*_.*____ _:________

C. lagenarium PKS1        -VKSTAIPVY GPYHAAHLHG TADVEKILRL NDPKVGEILA KTKPRSAIMS GTKGIWFAET  291
E. dermatitidis WdPKS1    TQKSAPVSIF APYHASHXHS QSDLDKILRP ---QTKTIFG NTTVRFPVCS SVTGKPFNAE  284
Nodulisporium PKS1        -SRPTAIPVY GPYHAPHLHG LVDIRKMLRL DDEAVTIALA GSKPRSVVMS CVSGEPFRET  292
A. fumigatus Alb1p        YKHFKAPGVS GPYHAPHLYN DREIRNILSF CS---EDVIL RHTPRVPLVS SNTGKLVQVK  287
A. nidulans WA            NKAVMVP-IH GPFHASHLYE KRDVEWILKS CN---VETIR NHKPRIPVLS SNTGELIVVE  287
A. nidulans STCA          SKLWREIPIY VPAHNNHLFS SRDVDDILAT TN---ENPWS TFGAQIPFLS SVTGKLAWVR  294
A. parasiticus PKSL1      QHRTAPIPIC LPAHNGALFT QADITTILDT TP---TTPWE QLPGQIPYIS HVTGNVVQTS  288
                          ______:__* _*____.___ ::__:*____ __________ ____:____* __.*______

C. lagenarium PKS1        DTKSLLQAVT HECLVDVLQF QKGIEGCIET ARDFEGSTCL VVPFGPTHNA ETLCKLIQDR  351
E. dermatitidis WdPKS1    NGFELLQAAL KEIIIDPLRW DKVLKYCAAG ----KASEAK VFAVGPTNLA SSVVSALKAS  340
Nodulisporium PKS1        DTKSLLTAVV HEILNESLIF HKTLSGNANE ARGFKGSRIL VIPYGPTQAA STLANVLRSQ  352
A. fumigatus Alb1p        SMRDLLKVAL EEILLRKICW DKVTESCLSI VQATND-KPW RILPIASNAT QGLVTALQRM  346
A. nidulans WA            NMEGFLKIAL EEILLRQMSW DKVTDSCISI LKSVGDNKPK KLLPISSTAT QSLFNSLKKS  347
A. nidulans STCA          NYRDLLHLAL SQCLIEPIRW DVVEAEVPRL LKDRDGLDTL TIVAFTTVLS KSLSNALVT-  353
A. parasiticus PKSL1      NYRDLIEVAL SETLLEQVRL DLVETGLPRL LQSRQ-VKSV TIVPFLTRMN ETMSNILPD-  346
                          .___::__._ _:_:___:__ .________ __________ .____:____ ._:_._:___
```

Figure 6.13. Alignment of *N*-terminus regions of AR-PKSs. Fully or highly conserved amino acids are shown with '*' or ':'/'.' below the alignment.

et al., 1993) among others. Significance of this feature was probed by site-directed mutagenesis on WA PKS (Fujii *et al.*, 2001). Both S^{1682} and/or S^{1804} in ACP domain were mutated (Figure 6.15). WA mutants S1682C, S1682A, and S1804C functioned as naphthopyrone synthase to produce YWA1, but S1804A produced citreoisocoumarin instead of naphthopyrone probably due to induced change of *C*-terminus (CYC domain) secondary structure. Double mutants S1682C+S1804C and S1682A+S1804A lost PKS activity. These results showed that the presence of at least a single intact Ser residue in ACP domain is good enough for WA PKS to function as naphthopyrone synthase. However, it is unknown whether both or just one of the Ser residues are modified with 4′-phosphopantetheine in wild type WA PKS.

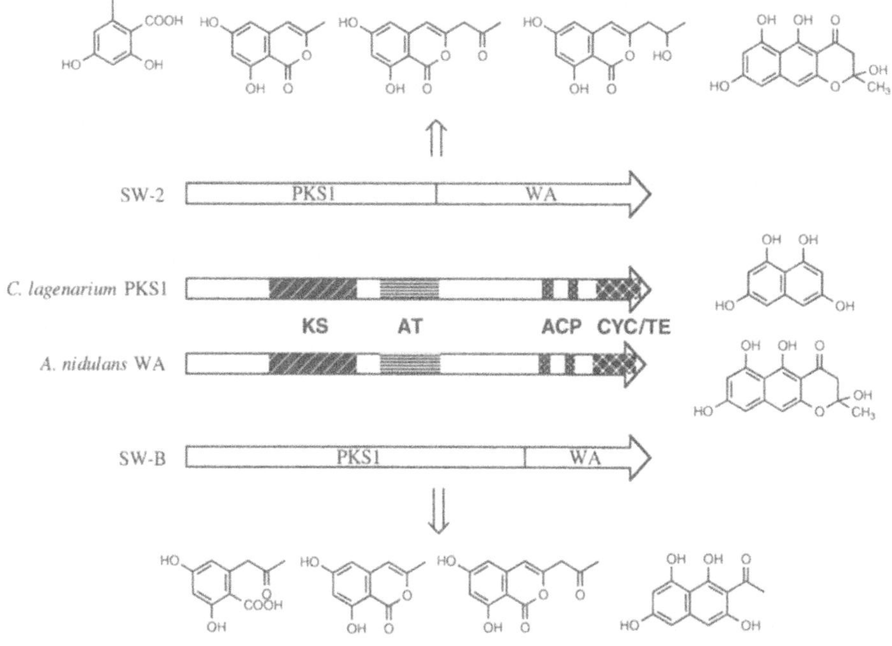

Figure 6.14. PKS1-WA chimeric PKSs and their products.

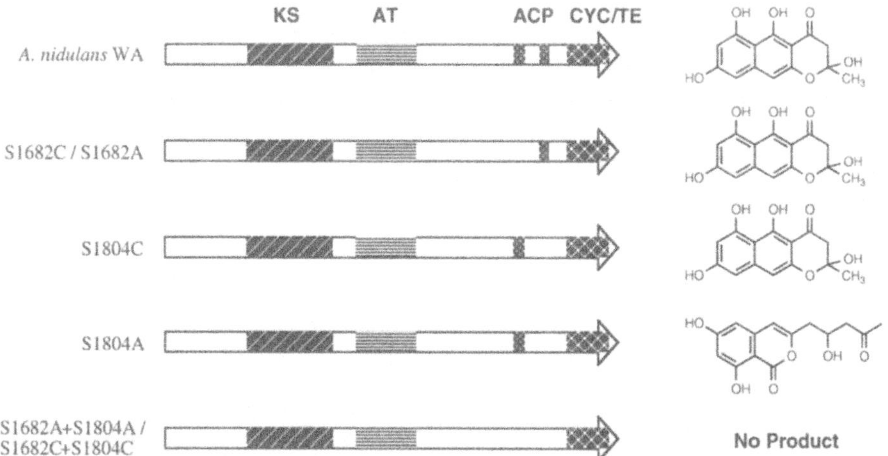

Figure 6.15. WA ACP mutants with site-directed mutagenesis and their products.

3.3. C-Methyltransferase Domains in RD-PKSs

Some fungal reduced complex-type polyketides such as lovastatin and fumonisin have methyl groups attached on the polyketide backbones. The most extreme example is alternapyrone that has eight methyl branches in the molecule (Figure 6.8). Feeding experiments with [^{13}C-methyl]methionine proved that these methyl groups are derived from methionine presumably introduced by *S*-adenosyl-methionine-dependent methyltransferase activity (Fujii, unpublished; Moore *et al.*, 1985; Plattner and Shackelford, 1992). As mentioned earlier, analysis of domain architecture of RD-PKSs showed the presence of methyltransferase (MeT) domains in LNKS, LDKS, FUM5p, and PKSN in the center of multifunctional PKS polypeptides (Figure 6.4(*iii*)). These MeT domains consist of conserved *S*-adenosyl methionine binding motifs, motif I, II, and III as proposed by Kagan and Clarke (1994). It is particularly noteworthy that PKSN MeT controls multi-*C*-methylations with strict regio- and stereospecificities (Fujii *et al.*, unpublished).

MlcA is the counter part of LNKS for compactin (ML-236B) synthesis in *P. citrinum* (Abe *et al.*, 2002). Compactin and lovastatin differ according to the nature of the substitution at C-6: lovastatin has a methyl group at C-6 but this substituent is absent at this position in compactin. MlcA and LNKS show high homology with each other at 59% identity. Alignment of amino acid sequences indicated that MlcA retains conserved MeT motifs but these contain a few substitutions. MlcB for diketide synthesis shows 61% identity with LDKS and also conserves MeT motif I, II, and III. Amino acid substitutions in MlcA could be critical for MeT inactivation. *C. heterostrophus* PKS1 for T-toxin synthesis retains apparent MeT motifs but with several substitutions that also might inactivate its MeT activity. Alignment of MeT domains in RD-PKSs is shown in Figure 6.16. Engineering of conserved MeT motifs in RD-PKSs may become a powerful tool for controlling *C*-methylations of carbon backbones in reduced complex-type polyketides.

3.4. PSED (Peptide Synthetase Elongation Domain)-Like Domains in RD-PKSs

An unusual feature of LNKS and MlcA nonaketide synthases for lovastatin and compactin (ML-236B) is the presence of a domain similar to the condensation domain of nonribosomal peptide synthetase (NRPS) (von Dören and Kleinkauf, 1997; Zuber and

```
                                            Motif I
A. terreus LDKS              SELVRLCC HKNPRARILE IGGGTG---- GCTQLVVDSL GPNPPVGRYDFT  1467
A. terreus LNKS              RELVAQIA HRYQSMDILE IGAGTGG--- ---ATKYVLA TPQLGFNSYTYT  1470
G. moniliformis FUM5p        IDFFATAG HTRPTLRVLE IGAGTGGG-- AQVILEGLTN GKERLFSTYAYT  1496
A. solani PKSN               AEYVGLIA DKQPGLRILE IGAGTGGTTY HVLERLRNPD G-TSKAAQYYFT  1487
C. heterostrophus PKS1       ATYLHLLG HKKPSLRVLT VGPQSGPTSL NLLMLLAELG GGEIPFAVLHHS  1481
                                 .    .      :*  :*  :*                          .:

                                                                       Motif II
A. terreus LDKS              DVSAGFFE AARKRFAGWQ NVMDFRKLDI EDDPEAQG-- FVCGSYDVVLAC  1515
A. terreus LNKS              DISTGFFE QAREQFAPFE DRMVFEPLDI RRSPAEQG-- FEPHAYDLIIAS  1518
G. moniliformis FUM5p        DISAGFFV AAQERFKAYK G-LDFKVLDI TKDPSEQG-- FESGSFDLIIAG  1543
A. solani PKSN               DISPGFLA KAADRFNQDA SIMQFGTLNI ENAPTEQG-- FSPELFDLIVCA  1535
C. heterostrophus PKS1       DAELNIDQ TVRSRFPSWA DSVGFRDVFN ESGASQQNPP IVNETYDIVVAF  1531
                             * . .:   . .:*    .  : *  :      .  *.    :    :*::. 

                                                  Motif III
A. terreus LDKS              QVLHATSN MQRTLTNVRK LLKPGGKLIL VETTRDELDL F-FTFGLLPGWW  1564
A. terreus LNKS              NVLHATPD LEKTMAHARS LLKPGGQMVI LEITHKEHTR LGFIFGLFADWW  1568
G. moniliformis FUM5p        NVIHATPT LNETLANVRK LLAPEGYLFL QELSPKMRMV N-LIMGILPGWW  1592
A. solani PKSN               NVLHATKS IQETLTHCKL LLKPGGRLVL SEVTIKRIFS G-FIMGPLPGWW  1584
C. heterostrophus PKS1       NVLGSSPG FSKTLSAAAP LLNARGKILL VDNSHKSPMA A-LVWGPLPSFL  1580
                             :*: ::   :..*::     ** . * :.:  : : .       :  * :..: 
```

Figure 6.16. Alignment of methyltransferase domains in RD-PKSs. Methyltransferase consensus motifs (Motifs I–III) are indicated by the boxes. Fully or highly conserved amino acids are shown with '*' or ':'/'.' below the alignment.

Marahiel, 1997) in their *C*-terminal regions, downstream of ACP domains. In NRPS such domains catalyze the condensation of two amino acid residues attached as thioesters to adjacent thiolation domains to form a peptide linkage. A histidine residue in the highly conserved motif (HHxxxDG) is considered to serve as the general base promoting nucleophilic attack of the amino group of the attaching amino acid to the carbonyl carbon of the elongating polypeptide (Marahiel *et al.*, 1997). This conserved motif is preserved in both LNKS and MlcA in their *C*-termini which also show overall similarity with the NRPS condensation domain in sequence (Figure 6.17) and in length, ~450 amino acids. Thus, as shown in RD-PKS architecture (Figure 6.4(*iii*)), LNKS (3,038 amino acids) and MlcA (3,032 amino acids) are the largest iterative PKSs presently known.

The PSED domain is also closely related with domains in chloramphenicol acetyltransferase and dihydrolipoyl transacetylase (de Crécy-Lagard and Marliere, 1995). These enzymes with the conserved motif (HxxxDG) catalyze the transfer of an acetyl group onto a hydroxy or thiol moiety. Although the role of LNKS/MlcA *C*-terminal PSED-like domain is still unclear, a similar role in esterification could be assumed. Transesterifcation of diketide acid to the octahydronaphthalene moiety is catalyzed by a separate enzyme LovD/MlcH. Thus, the probable function of the domain could be as a lactonizing thioesterase releasing dihydromonacolin L and acting as an alternative to the TE domain commonly found in type I modular PKS and NRPSs (Figure 6.18). This idea is also supported by the assumed function of RapP which is a specialized multi-domain protein with similarity to PSED involved in the formation of the ester and amide bonds to pipecolic acid in rapamycin biosynthesis (Molnár *et al.*, 1996).

3.5. "Diels–Alderase" in RD-PKSs

The Diels–Alder reaction is one of the pericyclic reactions generating a six-membered ring through the 1,4-addition of the double bond of a dienophile to a conjugated diene (Figure 6.19). A number of potential Diels–Alder adducts are found in polyketides

```
A. terreus LDKS   HHPLSLGQEY SWRIQQGAED PTVFNNTIGM FMKGSIDLKR LYKALRAVLR RHEIFRTGFA  2659
pfam00668         EYPLSPAQER LWFLSKLEGG TSAYNVPFVL RLPGGLDPER LEKALKALVE RHDALRTRFL  63
                   :***  **   * : :        : :*     :   : * :* :* * ***:*::  **:  ** *

A. terreus LDKS   NVDENGMAQL VFGQTKNKVQ TIQVSDRAGA EEG----YRQ LVQTRYNPAA GDTLRLVDFF  2715
pfam00668         R-DEGEPVQV VLEADEPDLK YLEDLSVTAE EEEILEALRQ DLQQPFDLEK GPLFRVALFK  122
                    **     *: *     :  ::  ::       **       **  :*  ::    *   *:  *

A. terreus LDKS   WGQDDHLLVV AYHRLVGDGS TTENIFVEAG QLYDG--TSL SPHVPQFADL AARQRAMLED  2773
pfam00668         LEEDRHRLLF SIHHLIVDGW SWRILLEELA ALYAGLLEGL PPLSPSYKDY AEWQQWYLQS  182
                    :* * *:  : * *: **  :   :  *    ** *    *  *  * : *  *  *:  *:

A. terreus LDKS   GRMEEDLAYW KKMHYRPSSI PVLPLMRPLV GNSSRSDTPN FQHCGPWQQH EAVARLDPMV  2833
pfam00668         DRREKERAYW LEQLSGIEPS LQLPLDRPRP PLQTYDGD-- ---------- RLTFSLSAET  230
                   * *:: ***  :             *** **       :                    *

A. terreus LDKS   AFRIKERSRK HKATPMQFYL AAYQVLLARL TDSTDLTVGL ADTNR--ATV DEMAAMGFFA  2891
pfam00668         TALLRKLAAA YGTTLNDVLL AAWGLVLSRY TGQDDIVVGT PGSGREHPIP DIERMVGWFT  290
                     ::: :    :  *      * **: ::*:*  *   *: **    : *      *    :*:*

A. terreus LDKS   NLLPLRFR  2899
pfam00668         NTLPLRVD  298
                  * ****
```

Figure 6.17. Alignment of *A. terreus* LDKS PSED-like domain with pfam00668 conserved condensation domain. Amino acid residues in active site motif (HHxxxDG) of pfam (protein family database) 00668 are highlighted. Identical or similar amino acids are shown with '*' or ':', respectively, below the alignment.

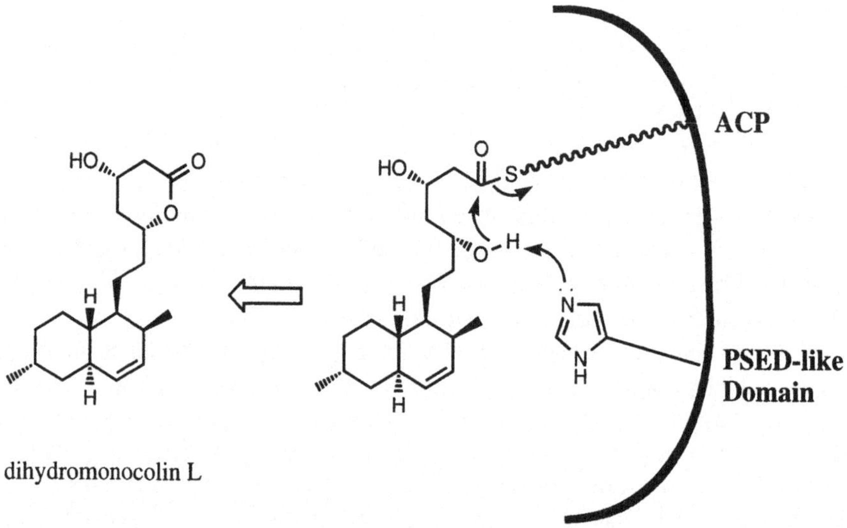

Figure 6.18. Possible function of PSED-like domain in lovastatin biosynthesis.

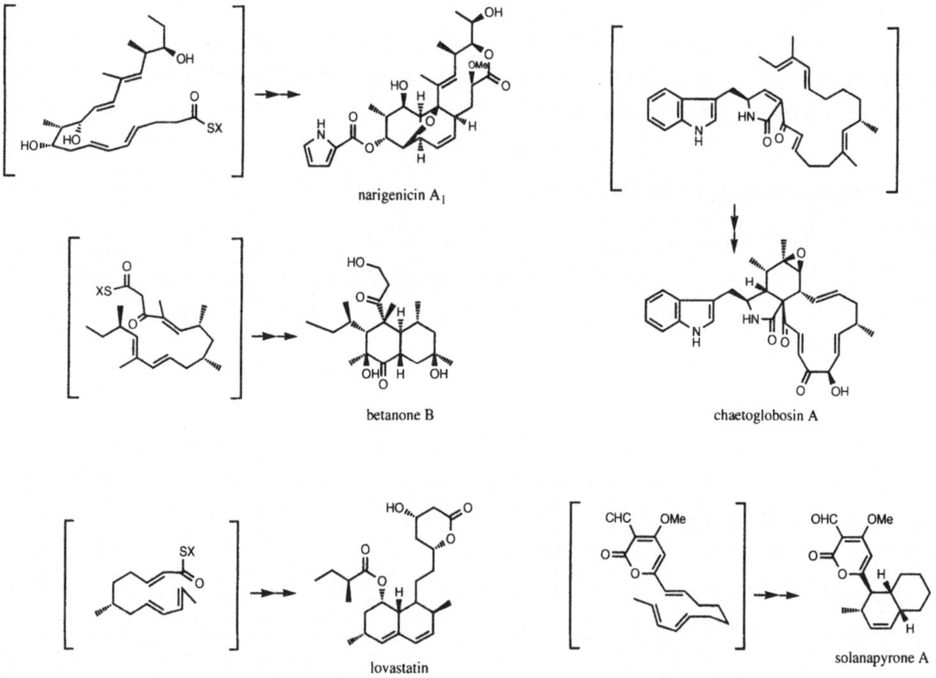

Figure 6.19. Diels–Alder reactions in polyketide biosynthesis.

not only from fungi but also from various microorganisms (Ichihara and Oikawa, 1999). The major class of biological [4 + 2]-adducts in polyketides is a "1,2-dialkyldecalin polyketide." Intramolecular Diels–Alder reaction of the reduced polyketide chain which has a conjugated *E,E*-diene and an *E*-dienophile can explain the stereochemistry found in metabolites of this class.

Lovastatin and related statins are of this class of compounds. Vederas and his coworkers established the origin of carbon, hydrogen, and oxygen atoms by experiments incorporating the stable-isotope labeled precursors (Moore *et al.*, 1985; Yoshizawa *et al.*, 1994). From these results, they proposed that the polyketide synthase may catalyze the [4 + 2]-cycloaddition of hexaketide intermediate. Unfortunately, ^{13}C-labeled hexaketide precursor was not incorporated by the intact fungus (Witter and Vederas, 1996). However, cloning and identification of *lovB* gene coding for a nonaketide synthase LNKS from *A. terreus* enabled them to carry out a more direct approach (Hendrickson *et al.*, 1999). Overexpression of LovB in *A. nidulans* under the control of the *alcA* promoter caused production of two polyketide pyrone compounds (Kennedy *et al.*, 1999). These are hexaketide and heptaketide polyenes with *C*-methylation, respectively, indicating that LNKS malfunctioned in the absence of enoyl reductase activity producing shunt products with shorter carbon chains. The *lovC* gene adjacent to *lovB* encodes a protein of 363 amino acids that has high similarity to the ER domains of PKSs. The *A. terreus* disruptant of *lovC* gene actually produced the same polyene shunt products but not lovastatin or any of the post-PKS intermediates. These results suggested that the LovC protein might be necessary to produce a polyketide of the correct chain-length with correct reduction level and cyclization pattern. This was confirmed by coexpression experiments of LovB and LovC in *A. nidulans*. This double transformant actually produced dihydromonocolin L, an expected cyclized nonaketide product (Figure 6.20). Thus, LovC protein was an accessory enzyme that complexes to LovB protein LNKS and imparts enoyl reductase activity necessary for successful assembly of the normal PKS product, dihydromonocolin L (Kennedy *et al.*, 1999).

Production of dihydromonocolin L by only LovB PKS and enoyl reductase LovC indicated that this PKS system can catalyze the Diels–Alder cyclization as proposed by Vederas

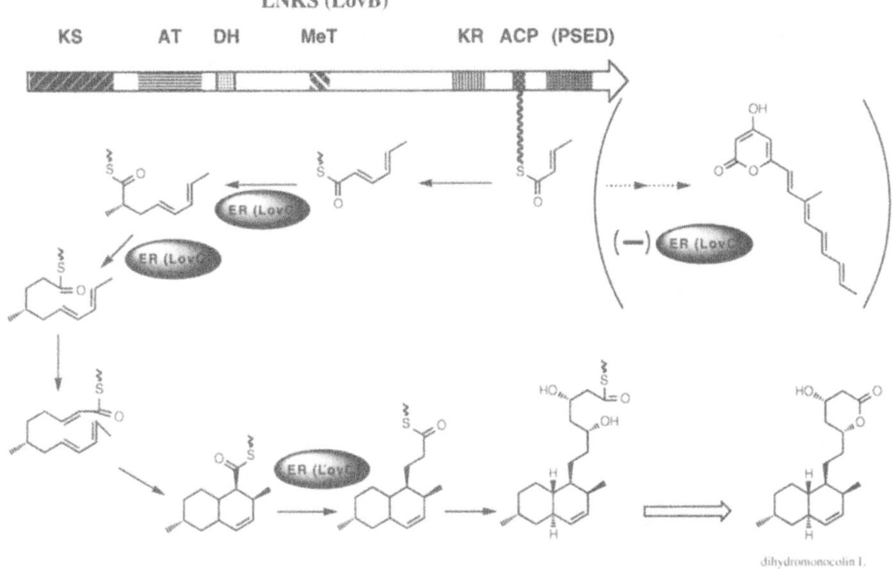

Figure 6.20. LNKS reaction and role of LovC protein in dihydromonocolin L synthesis.

and his coworkers. This idea was finally proven by their effort using the purified LovB LNKS and synthetic substrate analogue, (*E,E,E*)-(*R*)-6-methyldodecatri-2,8,10-enoic acid *N*-acetyl-cysteamine (NAC) thioester. Incubation of this substrate analogue with the purified LNKS caused cyclization to give an *endo* Diels–Alder product with correct stereochemistry (Figure 6.21) that was not detected in non-enzymatic cyclization reaction (Auclair *et al.*, 2000).

Apparently, a key function of LNKS is to bind the substrate in a conformation that resembles the *endo* transition state leading to the product. It is also likely that the protein assists closure through hydrogen bonding of the carbonyl oxygen to make the dienophile more electron-deficient. This "Diels-Alderase" function of LNKS is astonishing as a PKS function. Although Diels–Alder reaction is assumed to occur in biosynthesis of many secondary metabolites, LNKS is the first example of a PKS having Diels–Alderase activity. In solanapyrone biosynthesis (Figure 6.22), it is shown that a separate oxidase catalyzes the formation of solanapyrone by Diels–Alder reaction from prosolanapyrone, that is also a reduced complex-type polyketide (Oikawa *et al.*, 1995).

4. Concluding Remarks

In step with the general progress in fungal genetic analysis, the number of cloned fungal PKS genes is dramatically increasing. As discussed in Chapter 5, sequence information on KS domains can be obtained by PCR using degenerated primers designed based on the highly conserved KS sequences (Nicholson *et al.*, 2001). However, their functions as PKSs, presumed on the basis of homology, has rarely been determined experimentally. Expression of fungal PKSs in heterologous fungal hosts has been successful in identifying their functions, but this approach has been attempted in just a few cases (Fujii *et al.*, 1996; Watanabe *et al.*, 1998; Fujii *et al.*, 1999; Watanabe *et al.*, 1999). Expression in yeast and/or *E. coli* will dramatically help functional analysis of fungal PKSs as has been exemplified in MSAS expression (Kealey *et al.*, 1998).

All fungal PKSs cloned so far can be classified in one of three groups, MSAS/OASs, AR-PKSs, and RD-PKSs depending on their product-types and architecture. Phylogenic analysis of full-length fungal PKSs clearly supports these groupings (Figure 6.23).

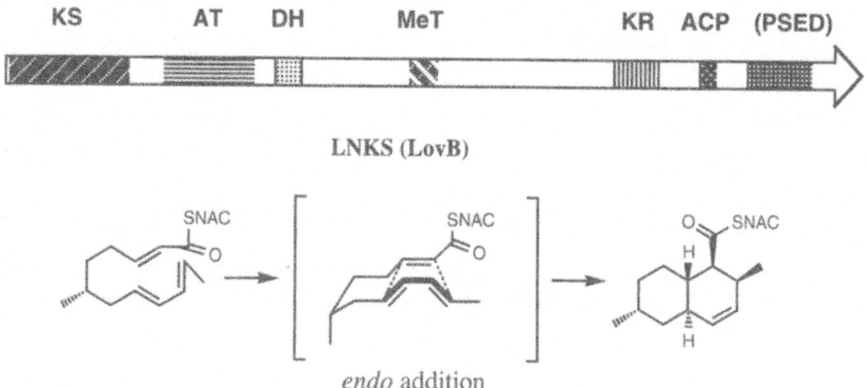

Figure 6.21. LNKS catalized Diels–Alder cyclization with substrate analogue. NAC, *N*-acetyl cysteamine.

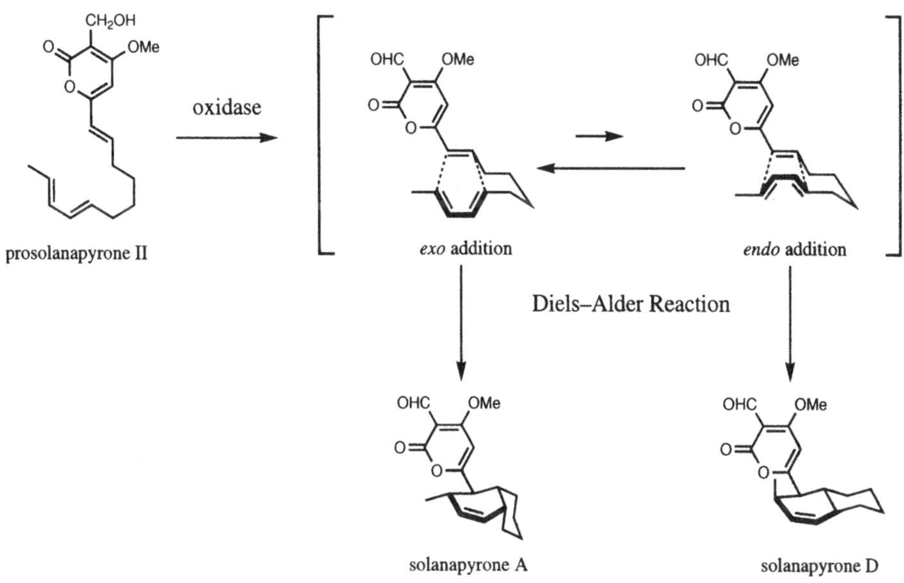

Figure 6.22. Diels–Alder cyclization in solanapyrone biosynthesis.

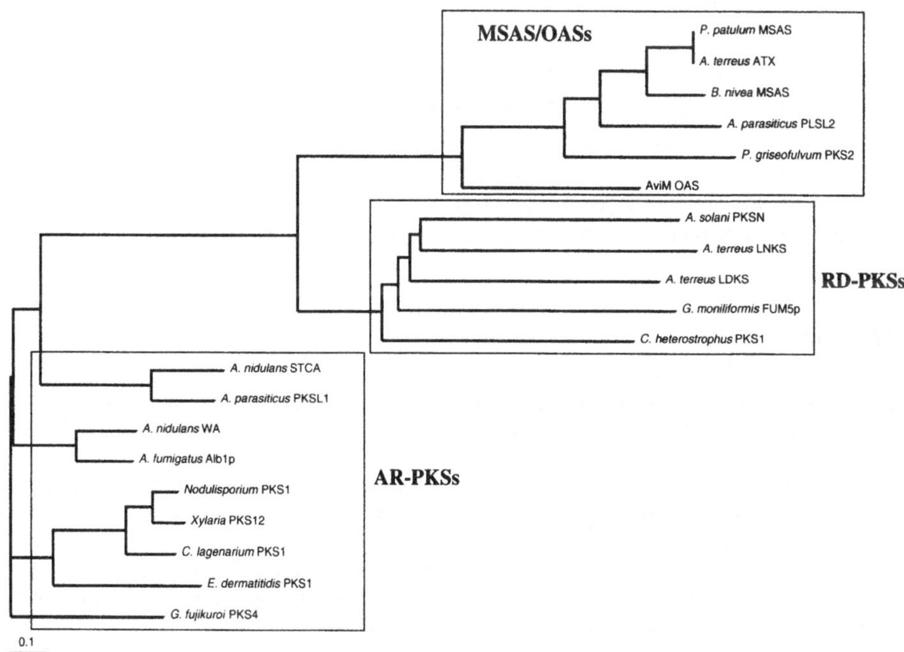

Figure 6.23. Phylogenetic tree of full-length fungal PKSs. In addition to PKSs described in Figure 6.4, *Byssochlamys nivea* MSAS (accession number AF360398), *A. parasiticus* PKSL2 (U52152), *Penicillium griseofulvum* PKS2 (U89769), and *Xylaria* sp. PKS12 (AF395534) are included. The phylogenetic tree was developed with the program CLUSTAL W using the neighbor-joining method and drawn with TREE VIEW. The indicated scale represents 0.1 amino acid substitution per site.

Within AR-PKSs, functionally related PKSs such as pentaketide T4HN synthases, heptaketide naphthopyrone synthases, and 10-deoxynorsolorinic acid synthases form separate phylogenic branches. Interestingly, MSAS/OASs for the simplest single aromatic ring polyketides appear in more descendant position than AR-PKSs.

In addition to basic PKS domains, CYC in AR-PKSs, MeT, and PSED-like domain in RD-PKSs are newly recognized as functional domains although actual function of PSED-like domain is still unidentified. Considering the structural variety in fungal polyketides, more functions and functional domains likely await identification. This information will lead to the design of novel iterative PKSs for production of desired compounds.

Despite this progress, fundamental aspects of fungal PKSs still remain unclarified. Among these are how iterative-type PKSs control the carbon chain-length of their product compounds, that is, the number of condensation cycles, and how PKSs stabilize the unstable active methylene carbonyl intermediates to avoid nonspecific aldol and/or Claisen cyclization. Although answering these fundamental but critical issues seems to be difficult, 3D structural analysis of PKS-intermediate analog complexes might shed some light in the future.

Acknowledgments

We are deeply indebted to our coworkers, H. Tada, Y. Ono, Y. Mori, N. Yoshida, S. Shimomaki, and Professor emeritus U. Sankawa in our laboratory. We are also grateful to our collaborators, Drs M. E. Mayorga, W. E. Timberlake, G. Tsuji, and Professors Y. Kubo and K. Gomi. Our work was supported in part by the Research for the Future Program (JSPS-RFRF96I00302), Japan Society for the Promotion of Science, a Grant-in-Aid for Scientific Research on Priority Area (A) (No. 12045213) from the Ministry of Education, Culture, Sports, Science and Technology, Japan (Y.E.), a Grant-in-Aid for Scientific Research (C) (No. 13836002) from the Ministry of Education, Culture, Sports, Science and Technology, Japan (I.F.), and a JSPS young research fellowship (A.W.).

References

Abe, Y., Suzuki, T., Ono, C., Iwamoto, K., Hosobuchi, M., and Yoshikawa, H. (2002). Molecular cloning and characterization of an ML-236B (compactin) biosynthetic gene cluster in *Penicillium citrinum*. *Mol. Gen. Genet.* **267**, 636–646.

Aggarwal, R., Caffrey, P., Leadlay, P.F., Smith, C.J., and Staunton, J. (1995). The thioesterase of the erythromycin-producing polyketide synthase: Mechanistic studies *in vitro* to investigate its mode of action and substrate specificity. *J. Chem. Soc. Chem. Commun.* 1519–1520.

Alberts, A.W., Chen, J., Curon, G., Hunt, V., Huff, J., Hoffman, C., Rothrock, J., Lopez, M. *et al.* (1980). Mevinolin: A highly potent competitive inhibitor of hydroxymethylglutaryl-coenzyme A reductase and a cholecterol-lowering agent. *Proc. Natl. Acad. Sci. U.S.A.* **77**, 3957–3961.

Aramayo, R. and Timberlake, W.E. (1990). Sequence and molecular structure of the *Aspergillus nidulans yA* (laccase I) gene. *Nucleic Acids Res.* **18**, 3415.

Auclair, K., Sutherland, A., Kennedy, J., Witter, D.J., van den Heever, J.P., Hutchinson, C.R., and Vederas, J.C. (2000). Lovastatin nonaketide synthase catalyzes an intramolecular Diels–Alder reaction of a substrate analogue. *J. Am. Chem. Soc.* **122**, 11519–11520.

Barteles-Keith, J.R. (1960). Alternaric acid. Part III. Structure. *J. Chem. Soc.* 1662–1665.

Beck, J., Ripka, S., Signer, A., Schiltz, E., and Schweizer, E. (1990). The multifunctional 6-methylsalicylic acid synthase gene of *Penicillium patulum*. Its gene structure relative to that of other polyketide synthases. *Eur. J. Biochem.* **192**, 487–498.

Bedford, D.J., Schweizer, E., Hopwood, D.A., and Khosla, C. (1995). Expression of a functional fungal polyketide synthase in the bacterium *Streptomyces coelicolor* A3(2). *J. Bacteriol.* **177**, 4544–4548.

Bellizzi, J.J., Widom, J., Kemp, C., Lu, J.-Y., Das, A.K., Hofman, S.L., and Clardy, J. (2000). The crystal structure of palmitoyl protein thioesterase 1 and the molecular basis of infantile neuronal ceroid lipofuscinosis. *Proc. Natl. Acad. Sci. USA* **97**, 4573–4578.

Benning, M.M., Wesenberg, G., Liu, R., Taylor, K.L., Dunaway-Mariano, D., and Holden, H.M. (1998). The three-dimensional structure of 4-hydroxybenzoyl-CoA thioesterase from *Pseudomonas* sp. strain CBS-3. *J. Biol. Chem.* **273**, 33572–33579.

Birch, A.J. and Donovan, F.W. (1953). Studies in relation to biosynthesis. I. Some possible routes to derivatives of orcinol and phloroglucinol. *Austral. J. Chem.* **6**, 360–368.

Birch, A.J. Massy-Westropp, R.A., and Moye, C.J. (1955). Biosynthesis (VII) 2-hydroxy-6-methyl benzoic acid in *Penicillium griseofulvum. Austral. J. Chem.* **8**, 539–544.

Blackwell, B.A., Miller, J.D., and Savard, M.E. (1994). Production of carbon 14-labeled fumonisin in liquid culture. *J. AOAC Int.* **77**, 506–511.

Branham, B.E. and Plattner, R.D. (1993). Alanine is a precursor in the biosynthesis of fumonisin B_1 by *Fusarium moniliforme. Mycophathologia* **124**, 99–104.

Brian, P.W., Curtis, P.J., Hemming, H.G., Jefferys, E.G., Unwin, C.H., and Wright, J.M. (1951). Alternaic acid—active metabolic product of *Alternaria solani*—production, and antifungal properties. *J. Gen. Microbiol.* **1**, 619–632.

Brian, P.W., Curtis, P.J., Hemming, H.G., Unwin, C.H., and Wright, J.M. (1949). Alternaric acid, a biologically active metabolic product of the fungus *Alternaria solani. Nature* **164**, 534.

Brian, P.W., Elson, G.W., Hemming, H.G., and Wright, J.M. (1952). Phytotoxic properties of alternaric acid in relation to the etiology plant diseases caused by *Alternaria solani. Ann. Appl. Biol.* **39**, 308–321.

Brobst, S. and Townsend, C.A. (1994). The potential role of fatty acid initiation in the biosynthesis of the fungal aromatic polyketide aflatoxin B_1. *Can. J. Chem.* **72**, 200–207.

Brown, D.W., Hauser, F.M., Tommasi, R., Corlett, S., and Salvo, J.J. (1993). Structural elucidation of a putative conidial pigment intermediate in *Aspergillus parasiticus. Tetrahedron Lett.* **34**, 419–422.

Brown, D.W. and Salvo, J.J. (1994). Isolation and characterization of sexual spore pigments from *Aspergillus nidulans. Appl. Environ. Microbiol.* **60**, 979–983.

Brown, D.W., Yu, J.-H., Kelkar, H.S., Fernandes, M., Nesbitt, T.C., Keller, N.P., Adams, T.H., and Leonards, T.J. (1996). Twenty-five coregulated transcripts define a sterigmatocystin gene cluster in *Aspergillus nidulans. Proc. Natl. Acad. Sci. USA* **93**, 1418–1422.

Caldas, E.D., Sadilkova, K., Ward, B.L., Jones, A.D., Winter, C.K., and Gilchrist, D.G. (1998). Biosynthetic studies of fumonisin B_1 and AAL toxins. *J. Agric. Food. Chem.* **46**, 4734–4743.

Chirala, S.S., Jayakumar, A., Gu, Z.-W., and Wakil, S.J. (2001). Human fatty acid synthase: Role of interdomain in the formation of catalytically active synthase dimer. *Proc. Natl. Acad. Sci. USA* **98**, 3104–3108.

Clutterbuck, A.J. (1972). Absence of laccase from yellow-spored mutants of *Aspergillus nidulans. J. Gen. Microbiol.* **70**, 423–435.

Collie, J.N. and Myers, W.S. (1893). The formation of orcinol and other condensation products from dehydracetic acid. *J. Chem. Soc.* 122–126.

de Crécy-Lagard, V. and Marliere, P.S. (1995). Multienzymatic non-ribosomal peptide biosynthesis: Identification of the functional domains catalysing peptide elongation and epimerisation. *C.R. Acad. Sci. II* **318**, 927–936.

Dimroth, P., Walter, H., and Lynen, F. (1971). Biosynthese von 6-methylsalicylsäure. *Eur. J. Biochem.* **13**, 98–110.

Endo, A. and Hasumi, K. (1993). HMG-CoA reductase inhibitors. *Nat. Prod. Rep.* **10**, 541–550.

Endo, A., Negishi, Y., Iwashita, T., Mizukawa, K., and Hirama, M. (1985). Biosynthesis of ML-236B (compactin) and monocolin K. *J. Antibiot.* **38**, 444–448.

Fujii, I. (1999). Polyketide biosynthesis in filamentous fungi. In U. Sankawa (ed.) *Comprehensive Natural Products Chemistry*. Elsevier, Oxford, Vol. 1, pp. 409–441.

Fujii, I., Mori, Y., Watanabe, A., Kubo, Y., Tsuji, G., and Ebizuka, Y. (1999). Heterologous expression and product identification of *Colletotrichum lagenarium* polyketide synthase encoded by the *PKS1* gene involved in melanin biosynthesis. *Biosci. Biotechnol. Biochem.* **63**, 1445–1452.

Fujii, I., Mori, Y., Watanabe, A., Kubo, Y., Tsuji, G., and Ebizuka, Y. (2000). Enzymatic synthesis of 1,3,6,8-tetrahydroxynaphthalene solely from malonyl-coenzyme A by a fungal iterative Type I polyketide synthase PKS1. *Biochemistry* **39**, 8853–8858.

Fujii, I., Ono, Y., Tada, H., Gomi, K., Ebizuka, Y., and Sankawa, U. (1996). Cloning of the polyketide synthase gene *atX* from *Aspergillus terreus* and its identification as the 6-methylsalicylic acid synthase gene by heterologous expression. *Mol. Gen. Genet.* **253**, 1–10.

Fujii, I., Watanabe, A., Mori, Y., and Ebizuka, Y. (1998). Structures and functional analyses of fungal polyketide synthase genes. *Actinomycetology* **12**, 1–14.

Fujii, I., Watanabe, A., Sankawa, U., and Ebizuka, Y. (2001). Identification of Claisen cyclase domain in fungal polyketide synthase WA, a naphthopyrone synthase of *Aspergillus nidulans. Chem. Biol.* **8**, 189–197.

Fujii, T., Yamaoka, H., Gomi, K., Kitamoto, K., and Kuagai, C. (1995). Cloning and nucletide sequence of the ribonuclease T1 Gene (rntA) from *Aspergillus oryzae* and its expression in *Saccharomyces cerevisiae* and *Aspergillus oryzae. Biosci. Biotech. Biochem.* **59**, 1869–1874.

Fulton, T.R., Ibrahim, N., Losada, M.C., Grzegorski, D., and Tkacz, J.S. (1999). A melanin polyketide synthase (PKS) gene from *Nodulisporium* sp. that shows homology to the pks1 gene of *Colletotrichum lagenarium. Mol. Gen. Genet.* **264**, 714–720.

Gaisser, S., Trefzer, A., Stckert, S., Kirshning, A., and Bechthold, A. (1997). Cloning of an avilamycin biosynthetic gene cluster from *Streptomyces viridochromogenes* Tü 57. *J. Bacteriol.* **179**, 6271–6278.

Galmarini, O.L. and Stodola, F.H. (1965). Fonsecin, a pigment from an *Aspergillus fonsecaeus* mutant. *J. Org. Chem.* **30**, 112–115.

Hendrickson, L., Davis, C.R., Roache, C., Nguyen, D.K., Aldrich, T., McAda, P.C., and Reeves, C.D. (1999). Lovastatin biosynthesis in *Aspergillus terreus*: Characterization of blocked mutants, enzyme activities and multifunctional polyketide synthase gene. *Chem. Biol.* **6**, 429–439.

Hopwood, D.A. (1997). Genetic contributions to understanding polyketide synthases. *Chem. Rev.* **97**, 2465–2497.

Howard, P.C., Eppley, R.M., Stack, M.E., Warbritton, A., Voss, K.A., Lorentzen, R.J., Kovach, R., and Bucci, T.J. (1999). Carcinogenicity of fumonisin B_1 in a two-year bioassay with Fischer344 rats and B6C3F1 mice. *Mycotoxins Suppl.* 45–54.

Hutchinson, C.R. and Fujii, I. (1995). Polyketide synthase gene manupilation: A structure–function approach in engineering novel antibiotics. *Annu. Rev. Microbiol.* **49**, 201–238.

Ichihara, A., Miki, M., and Sakamura, S. (1985). Absolute configuration of (-)-solanapyrone A. *Tetrahedron Lett.* **26**, 2453–2454.

Ichihara, A. and Oikawa, H. (1997). Biosynthesis of phytotoxins from *Alternaria solani. Biosci. Biotechnol. Biochem.* **61**, 12–18.

Ichihara, A. and Oikawa, H. (1999). The Diels-Alder reaction in biosynthesis of polyketide phytotoxins. In U. Sankawa (ed.), *Comprehensive Natural Products Chemistry*. Elsevier, Oxford, Vol. 1, pp. 367–408.

Ichihara, A., Tazaki, H., and Sakamura, S. (1983). Solanapyrones A, B and C, phytotoxic metabolites from the fungus *Alternaria solani. Tetrahedron Lett.* **24**, 5373–5376.

Jordan, P.M. and Spencer, J.B. (1993). The biosynthesis of tetraketides: Enzymology, mechanism and molecular programming. *Biochem. Soc. Trans.* **21**, 222–228.

Julien, B., Shah, S., Ziermann, R., Goldman, R., Katz, L., and Khosla, C. (2000). Isolation and characterization of the epothilone biosynthetic gene cluster from *Sorangium cellulosum. Gene* **249**, 153–160.

Kagan, R.M. and Clarke, S. (1994). Widespread occurrence of three sequence motifs in diverse *S*-adenosylmethionine-dependent methyltransferase suggests a common structure for these enzymes. *Arch. Biochem. Biophys.* **310**, 417–427.

Kealey, J.T., Liu, L., Santi, D.V., Betlach, M.C., and Barr, P.J. (1998). Production of a polyketide natural product in nonpolyketide-producing prokaryotic and eukaryotic hosts. *Proc. Natl. Acad. Sci. USA* **95**, 505–509.

Kennedy, J., Auclair, K., Kendrew, S.G., Park, C., Vederas, J.C., and Hutchinson, C.R. (1999). Modulation of polyketide synthase activity by accessory proteins during lovastatin biosynthesis. *Science* **284**, 1368–1372.

Kohli, R.M., Takagi, J., and Walsh, C.T. (2002). The thioesterase domain from a nonribosomal peptide synthetase as a cyclization catalyst for integrin binding peptides. *Proc. Natl. Acad. Sci. USA* **99**, 1247–1252.

Kono, Y. and Daly, J.M. (1979). Characterization of the host-specific pathotoxin produced by *Helminthosporium maydis*, Race T, affecting corn with Texas male sterile cytoplasm. *Bioorg. Chem.* **8**, 391–397.

Kono, Y., Danko, S.J., Suzuki, Y., Takeuchi, S., and Daly, J.M. (1983). Structure of the host-specific pathotoxins produced by *Phyllostica maydis. Tetrahedron Lett.* **24**, 3803–3806.

Kono, Y., Takeuchi, S., Kawarada, A., Daly, J.M., and Knoche, H.W. (1980). Structure of the host-specific pathotoxins produced by *Helminthosporium maydis*, race T. *Tetrahedron Lett.* **21**, 1537–1540.

Lawson, D.M., Derenwenda, U., Serre, L., Ferri, S., Szittner, R., Wei, Y., Meighen, E.A., and Derenwenda, Z.S. (1994). Structure of a myristoyl-ACP-specific thioesterase from *Vibrio harvey. Biochemistry* **33**, 9382–9388.

Li, J., Derenwenda, U., Dauter, Z., Smith, S., and Derenwenda, Z.S. (2000). Crystal structure of the *Escherichia coli* thioesterase II, a homolog of the human Nef binding enzyme. *Nature Struct. Biol.* **7**, 555–559.

Lu, S., Lyngholm, L., Yang, G., Bronson, C., Yoder, O.C., and Turgeon, B.G. (1994). Tagged mutations at the *Tox1* locus of *Cochliobolus heterostrophus* by restriction enzyme-mediated integration. *Proc. Natl. Acad. Sci. U.S.A.* **91**, 12649–12653.

Mahanti, N., Bhatnagar, D., Cary, J.W., Joubran, J., and Linz, J.E. (1996). Structure and function of *fas*-1A, a gene encoding a putative fatty acid synthase directly involved in aflatoxin biosynthesis in *Aspergillus parasiticus. Appl. Environ. Microbiol.* **62**, 191–195.

Malpartida, F., Hallam, S.E., Kieser, H.M., Motamedi, H., Hutchinson, C.R., Butler, M.J., Sugden, D.A., Warren, M. *et al.* (1987). Homology between *Streptomyces* genes coding for synthesis of different polyketides used to clone antibiotic biosynthetic genes. *Nature* **325**, 818–821.

Malpartida, F. and Hopwood, D.A. (1984). Molecular cloning of the whole biosynthetic pathway of a *Streptomyces* antibiotic and its expression in a heterologous host. *Nature* **309**, 462–464.

Marahiel, M.A., Stachelhaus, T., and Mootz, H.D. (1997). Modular peptide synthetases involved in nonribosomal peptide synthesis. *Chem. Rev.* **97**, 2651–2673.

Mayorga, M.E. and Timberlake, W.E. (1990). Isolation and molecular characterization of *the Aspergillus nidulans wA* gene. *Genetics* **126**, 73–79.

Mayorga, M.E. and Timberlake, W.E. (1992). The developmentally regulated *Aspergillus nidulans wA* gene encodes a polypeptide homologous to polyketide and fatty acid synthases. *Mol. Gen. Genet.* **235**, 205–212.

Merrill, A.H., Schmelz, E.-M., Dillehay, D.L., Spiegel, S., Shayman, J.A., Schoroeder, J.J., Riley, R.T., Voss, K.A. *et al.* (1997). Sphingolipids—The enigmatic lipid class: Biochemistry, physiology and pathophysiology. *Toxicol. Appl. Pharmacol.* **142**, 208–225.

Molnár, I., Aparicio. J.F., Haydock, S.F., Khaw, L.E., Schwecke, T., König, A., Staunton, J., and Leadlay, P.F. (1996). Organisation of the biosynthetic gene cluster for rapamycin in *Streptomyces hygroscopicus*: Analysis of genes flanking the polyketide synthase. *Gene* **169**, 1–7.

Molnár, I., Schupp, T., Ono, M., Zirkle, R.E., Milnamov, M., Nowak-Thompson, B., Engel, N., Toupet, C. *et al.* (2000). The biosynthetic gene cluster for the microtubule-stabilizing agents epothilones A and B from *Sorangium cellulosum* So ce90. *Chem. Biol.* **7**, 97–109.

Moore, R.N., Bigman, G., Chan, J.K., Hogg, A.M., Nakashima, T.T., and Vederas, J.C. (1985). Biosynthesis of the hypocholesterolemic agent mevinolin by *Aspergillus terreus*. Determination of the origin of carbon, hydrogen, and oxygen atoms by ^{13}C NMR and mass spectrometry. *J. Am. Chem. Soc.* **107**, 3694–3701.

Nelson, P.E., Desjaradins, A.E., and Plattner, R.D. (1993). Fumonisins, mycotoxins produced by *Fusarium* species: Biology, chemistry, and significance. *Annu. Rev. Phytopathol.* **31**, 233–252.

Nicholson, T.P., Rudd, B.A.M., Dawson, M., Lazarus, C.M., Simpson, T.J. and Cox, R.J. (2001). Design and utility of oligonucleotide gene probes for fungal polyketide synthases. *Chem. Biol.*, **8**, 157–178.

Oikawa, H., Katayama, K., Suzuki, Y., and Ichihara, A. (1995). Enzymatic activity catalyzing *exo*-selective Diels–Alder reaction in solanapyrone biosynthesis. *J. Chem. Soc. Chem. Commun.* 1321–1322.

Oikawa, H., Yokota, T., Abe, T., Ichihara, A., Sakamura, S., Yoshizawa, Y., and Vederas, J.C. (1989a). Biosynthesis of solanapyrone A, a phytotoxin of *Alternaria solani. J. Chem. Soc. Chem. Commun.* 1282–1284.

Oikawa, H., Yokota, T., Ichihara, A., and Sakamura, S. (1989b). Structure and absolute configuration of solanapyrone D: A new clue to the occurrence of biological Diels–Alder reactions. *J. Chem. Soc. Chem. Commun.* 1284–1285.

Oikawa, H., Yokota, T., Sakano, C., Suzuki, Y., Naya, A., and Ichihara, A. (1998). Solanapyrones, phytotoxins produced by *Alternaria solani*: Biosynthesis and isolation of minor components. *Biosci. Biotechnol. Biochem.* **62**, 2016–2022.

Plattner, R.D. and Shackelford, D.D. (1992). Biosynthesis of labeled fumonisins in liquid cultures of *Fusarium moniliforme. Mycophathologia* **117**, 17–22.

Proctor, R.H. and Desjaradins, A.E. (1999). A polyketide synthase gene required for biosynthesis of *Fusarium* mycotoxins in *Gebberella fujikuroi* mating population A. *Fungal Genet. Biol.* **27**, 100–112.

Punnya, J., Cheevadhanak, S., and Saunders, G. (2001). Putative melanin polyketide synthase gene (PKS12) from *Xylaria* sp. BCC 1067. Direct submission to Genbank (accession No. AF395534).

Scotti, C., Piatti, M., Cuzzoni, A., Perani, P., Tognoni, A., Grandi, G., Galizzi, A., and Albertini, A. (1993). A *Bacillus subtilis* large ORF coding for a polypeptide highly similar to polyketide synthases. *Gene* **130**, 65–71.

Shoolingin-Jordan, P.M. and Campuzano, I.D.G. (1999). Biosynthesis of 6-methylsalicylic acid. In U. Sankawa (ed.), *Comprehensive Natural Products Chemistry*. Elsevier, Oxford, Vol. 1, pp. 187–216.

Singh, N., Wakil, S.J., and Stoops, J.K. (1984). On the question of half- or full-site reactivity of animal fatty acid synthetase. *J. Biol. Chem.* **259**, 3605–3611.

Smith, S. (1994). The animal fatty acid synthase: One gene, one polypeptide, seven enzymes. *FASEB J.* **8**, 1248–1259.

Stout, G.H., Dreyer, D.L., and Jensen, L.H. (1961). Structure of rubrofusarin. *Chem. Ind.* 289–290.

Takano, Y., Kubo, Y., Shimizu, K., Mise, K., Okuno, T., and Furusawa, I. (1995). Structual analysis of PKS1, a polyketide synthase gene involved in melanin biosynthesis of *Colletotrichum lagenarium. Mol. Gen. Genet.* **249**, 162–167.

Tanaka, H. and Tamura, T. (1961). The chemical constitution of rubrofusarin. *Tetrahedron Lett.* **4**, 151–155.

Tang, L., Shah, S., Chung, L., Carney, J., Katz, L., Khosla, C., and Julien, B. (2000). Cloning and heterologous expression of the epothilone gene cluster. *Science* **287**, 640–642.

Townsend, C.A., Christensen, S.B., and Trautwein, K. (1984). Hexanoate as a starter unit in polyketide biosynthesis. *J. Am. Chem. Soc.* **106**, 3868–3869.

Tsai, H.-F., Chang, Y.C., Washburn, R.G., Wheeler, M.H., and Kwon-Chung, K.J. (1998). The developmentally regulated *alb1* gene of *Aspergillus fumigatus*: Its role in modulation of conidial morphology and virulence. *J. Bacteriol.* **180**, 3031–3038.

Tsai, S.-C., Miercke, L.J.W., Krucinski, J., Gokhale, R., Chen, J.C.-H., Foster, P.G., Cane, D.E., Khosla, C. *et al.* (2001). Crystal structure of the macrocycle-forming thioesterase domain of the erythromycin polyketide synthase: Versatility from a unique substrate channel. *Proc. Natl. Acad. Sci. USA* **98**, 14808–14813.

Vederas, J.C. and Nakashima, T.T. (1980). Biosynthesis of averufin by *Aspergillus parasiticus*; detection of ^{18}O label by ^{13}C-NMR isotope shifts. *J. Chem. Soc. Chem. Commun.* 183–185.

von Dören, H. and Kleinkauf, H. (1997). Enzymology of peptide synthetases. In W.R. Strohl (ed.), *Biotechnology of Antibiotics*. Mercel Dekker, New York, pp. 217–240.

Wagschal, K., Yoshizawa, Y., Witter, D.J. and Vederas, J.C. (1996). Biosynthesis of ML-236C and the hypocholesterolemic agents compactin by *Penicillium aurantiogreseum* and lovastatin by *Aspergillus terreus*: Determination of the origin of carbon, hydrogen and oxygen atoms by ^{13}C NMR spectrometry and observation of unusual labelling of acetate-derived oxygens by $^{18}O_2$. *J. Chem. Soc. Perkin. Trans.* **1**, 2357–2363.

Wang, E., Norred, W.P., Bacon, C.W., Riley, R.T., and Merrill, A.H. (1991). Inhibition of sphingolipid biosynthesis by fumonisins: Implications for diseases associated with *Fusarium moniliforme. J. Biol. Chem.* **266**, 14486–14490.

Watanabe, A. and Ebizuka, Y. (2002). A novel hexaketide naphthopyrone synthesized by a chimeric polyketide synthase composed of fungal pentaketide and heptaketide synthases. *Tetrahedron Lett.* **43**, 843–846.

Watanabe, A., Fujii, I., Sankawa, U., Mayorga, M.E., Timberlake, W.E., and Ebizuka, Y. (1999). Re-identification of *Aspergillus nidulans wA* gene to code for a polyketide synthase of naphthopyrone. *Tetrahedron Lett.* **40**, 91–94.

Watanabe, A., Ono, Y., Fujii, I., Sankawa, U., Mayorga, M.E., Timberlake, W.E., and Ebizuka, Y. (1998). Product identification of polyketide synthase coded by *Aspergillus nidulans wA* gene. *Tetrahedron Lett.* **39**, 7733–7736.

Watanabe, C.M.H., Wilson, D., Linz, J.E., and Townsend, C.A. (1996). Demonstration of the catalytic roles and evidence for the physical association of type I fatty acid synthases and a polyketide synthase in the biosynthesis of aflatoxin B1. *Chem. Biol.* **3**, 463–469.

Wheeler, M.H. and Bell, A.A. (1988). Melanins and their importance in pathogenic fungi. In M.R. McGinnis (ed.) *Current Topics in Medical Mycology*. Springer-Verlag, New York, Vol. 2.

Witter, D.J. and Vederas, J.C. (1996). Putative Diels–Alder-catalyzed cyclization during the biosynthesis of lovastatin. *J. Org. Chem.* **61**, 2613–2623.

Woo, E.-R., Fujii, I., Ebizuka, Y., Sankawa, U., Kawaguchi, A., Beale, J.M., Shibuya, M., Mocek, U. *et al.* (1989). Nonstereospecific proton removal in the enzymatic formation of orsellinic acid from chiral malonate. *J. Am. Chem. Soc.* **111**, 5498–5500.

Yang, G., Rose, M.S., Turgen, B.G., and Yoder, O.C. (1996). A polyketide synthase is required for fungal virulence and production of the polyketide T-toxin. *Plant Cell* **8**, 2139–2150.

Ye, X., Feng, B., and Szaniszlo, P.J. (1999). A color-selectable and site-specific integrative transformation system for gene expression studies in the dematiaceous fungus *Wangiella* (*Exophiala*) *dermatitidis. Curr. Genet.* **36**, 241–247.

Yoshizawa, Y., Witter, D.J., Liu, Y., and Vederas, J.C. (1994). Revision of the biosynthetic origin of oxygens in mevinolin (Lovastatin), a hypercholesterolemic drug from *Aspergillus terreus* MF4845. *J. Am. Chem. Soc.* **107**, 3694–3701.

Yu, J., Chang, P.-K., Cary, J.W., Wright, M., Bhatnagar, D., Cleveland, T.E., Payne, G.A., and Linz, J.E. (1995). Comparative mapping of aflatoxin pathway gene cluster in *Aspergillus parasiticus* and *Aspergillus flavus*. *Appl. Environ. Microbiol.* **61**, 2365–2371.

Yu, J.-H. and Leonard, T.J. (1995). Sterigmatocystin biosynthesis in *Aspergillus nidulans* requires a novel type I polyketide synthase. *J. Bacteriol.* **177**, 4792–4800.

Zhang, A., Ping, L., Dahl-Roshak, A.M., Paress, P.S., Kennedy, S., Tkacz, J.S., and An, Z. (2003). Efficient disruption of a polyketide synthase gene (*pks*1) required for melanin synthesis through *Agrobacterium*-mediated transformation of *Glarea lozoyensis*. *Mol. Genet. Genomics* **268**, 645–655.

Zuber, P. and Marahiel, M. (1997). Structure, function, and regulation of genes encoding multidomain peptide synthetase. In W.R. Strohl (ed.), *Biotechnology of Antibiotics*. Marcel Dekker, New York, pp. 187–216.

7

Peptide Synthesis without Ribosomes

Jonathan D. Walton, Daniel G. Panaccione, and Heather E. Hallen

1. Introduction

One of the largest and most important groups of microbial secondary metabolites comprises peptides that are synthesized by enzymes instead of ribosomes. Of the thousands of known non-ribosomal peptides (NRPs), many are cyclic, but others are linear or a mixture of cyclic and linear (von Döhren, 1990). A hallmark of NRPs is the presence of unusual amino acids, either proteinogenic ones modified by acylation, epimerization, N-methylation, hydroxylation, etc., or non-proteinogenic amino acids such as ornithine and β-amino acids. NRPs can contain even larger, more complex moieties, such as lysergic acid in the case of ergotamine (Figure 7.1D) or 3-hydroxy-4-methylanthranilic acid in actinomycin (Marahiel *et al.*, 1997). Some NRPs contain ester-linked hydroxy acids in addition to amino acids. Biogenically, there is no fundamental distinction between the ester-containing NRPs, called depsipeptides, and the NRPs that contain just amino acids.

Non-ribosomal peptides are produced by bacteria and fungi. Although there are the usual expected differences in gene organization and expression, bacterial and fungal NRP synthetases and the genes that encode them are very similar in structure and function, as proven by the ability to interchange bacterial and fungal NRP synthetase modules (e.g., Stachelhaus *et al.*, 1995).

Animals and higher plants do contain certain NRPs (e.g., glutathione and various dipeptides), but there is scant evidence that any produce NRPs of the microbial type. There is only one published report of a plant-produced NRP in which the structure was confirmed by modern analytical methods. Jenett-Siems *et al.* (1994) isolated ergobalansine, an ergopeptine similar to ergotamine (Figure 7.1D) but having L-Ala in place of L-Pro, from a South American morning glory, *Ipomoea piurensis*. NRPs have been isolated from marine sponges, but the possibility that the compounds came from associated fungi or bacteria could not be excluded (Clark *et al.*, 1997; Clark *et al.*, 1998). There are no reports of molecular or biochemical characterization of NRP synthetases from any plant or animal.

Jonathan D. Walton and Heather E. Hallen • DOE Plant Research Lab, Michigan State University, E. Lansing MI 48824. **Daniel G. Panaccione** • Division of Plant and Soil Sciences, West Virginia University, 401 Brooks Hall, P.O. Box 6058, Morgantown, WV 26506-6058.

Advances in Fungal Biotechnology for Industry, Agriculture, and Medicine. Edited by Jan S. Tkacz and Lene Lange, Kluwer Academic/Plenum Publishers, 2004.

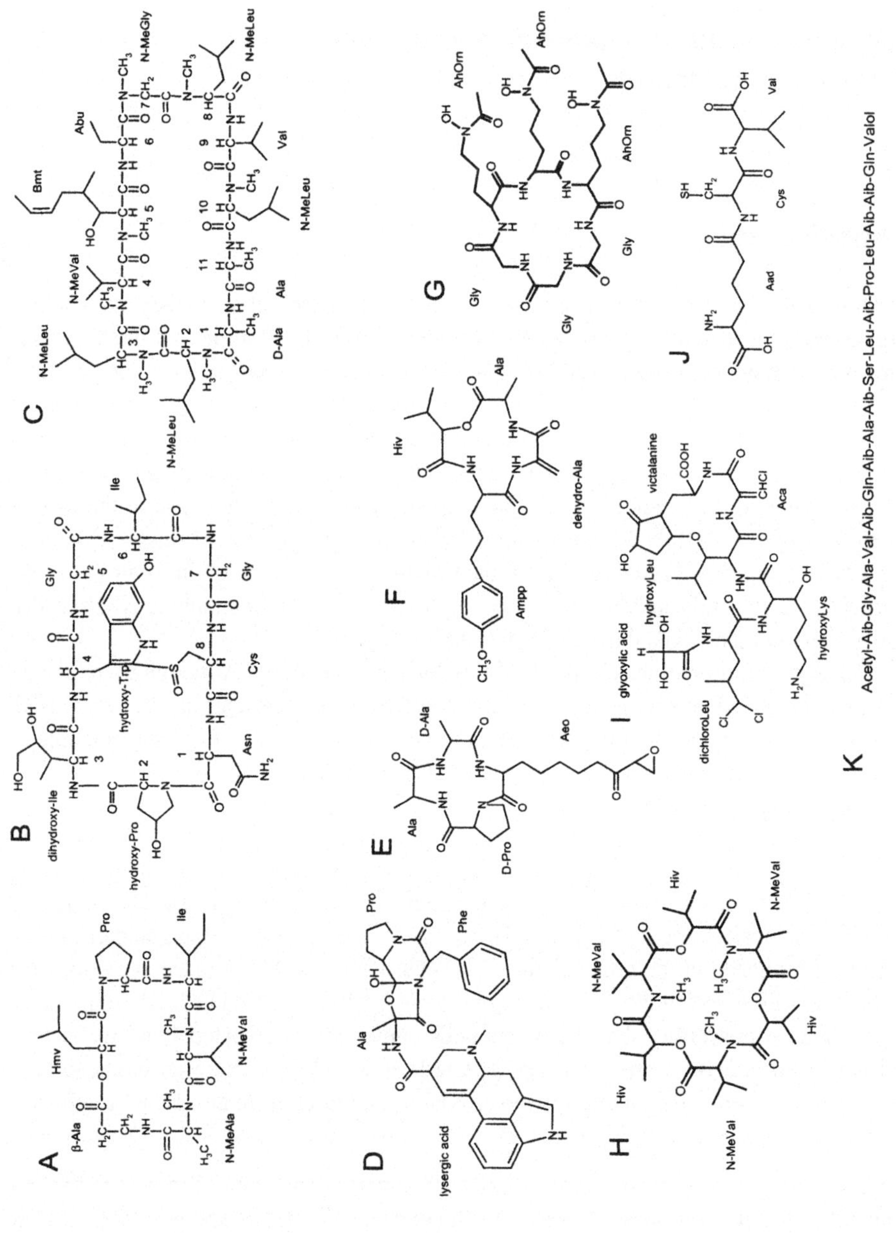
A
β-Ala
Hmv
Pro
Ile
N-MeVal
N-MeAla
B
dihydroxy-Ile
hydroxy-Pro
hydroxy-Trp
Gly
Ile
Gly
Cys
Asn
C
N-MeLeu
N-MeVal
Bmt
Abu
N-MeGly
N-MeLeu
Val
N-MeLeu
Ala
D-Ala
N-MeLeu
D
lysergic acid
Ala
Pro
Phe
E
Ala
D-Ala
Aeo
D-Pro
F
Hiv
Ala
Ampp
dehydro-Ala
G
Gly
AhOrn
AhOrn
AhOrn
Gly
Gly
H
Hiv
N-MeVal
Hiv
N-MeVal
Hiv
N-MeVal
I
glyoxylic acid
hydroxyLeu
victalanine
Aca
dichloroLeu
hydroxyLys
J
Aad
Cys
Val
K
Acetyl-Aib-Gly-Ala-Val-Aib-Gln-Aib-Ala-Aib-Ser-Leu-Aib-Pro-Leu-Aib-Gln-Valol

Some microbial taxa make few if any secondary metabolites, including NRPs, whereas others, such as *Streptomyces*, make a large number and variety. This conclusion is supported by chemical analysis of microbial secondary metabolites and, more recently, by genome analyses, which indicate that microorganisms differ considerably from each other in numbers of predicted genes for secondary metabolites (Kroken *et al.*, 2002). For example, phytotoxic metabolites are clearly an effective virulence strategy for fungal plant pathogens, yet almost all of the more than 20 known host-selective toxins are made by just two genera of fungi (Walton, 1996). Why some microbial isolates, species, or genera produce many more types of secondary metabolites than others is not clear. It could reflect their colonization of an ecological niche in which secondary metabolites are particularly advantageous, for example, exceptionally high competition from other microorganisms or the necessity to interact with insects or plants. It might also reflect the adaptation of certain taxa to the acquisition and genomic integration of genes for novel biochemical pathways.

Most research on NRPs has been rationalized by the tremendous importance of many of them, such as penicillin, cyclosporin, and ergotamine, as pharmaceutical agents. A better understanding of how NRPs are synthesized could lead to improved fermentation yields and the production of novel compounds through rational genetic engineering (Du *et al.*, 2001). Another rationale for studying NRPs is to understand their ecological functions, that is, to determine the selective advantages that they confer on the organisms that produce them. Research in such areas as biocontrol, entomology, and plant pathology have led to the discovery of many NRPs and hence the cognate synthetases and genes.

In this chapter we focus on NRP synthetases from fungi; the gene nomenclatural style adopted by the original authors has been maintained. There is no consistency in the naming of filamentous fungal genes. Researchers studying *Aspergillus nidulans, Neurospora crassa*, yeast, and the numerous plant pathogenic fungi use different conventions, despite attempts at standardization (Yoder *et al.*, 1986). The context should make it clear when wild type versus mutant alleles, or genes versus proteins, are being considered.

2. Overview of Non-Ribosomal Peptide Synthetases

With a few exceptions, notably glutathione, all NRPs are biosynthesized by a well-defined set of enzymes known as non-ribosomal peptide synthetases (NRP synthetases). Several general overviews of NRPs and NRP synthetases have been published recently (Marahiel *et al.*, 1997; von Döhren *et al.*, 1997; Mootz and Marahiel, 1999; Neilan *et al.*,

Figure 7.1. A representative selection of non-ribosomal peptides (NRPs) from fungi. All of them exist as families; only the major or best known member is shown in each case. **A**, destruxin; **B**, amanitin; **C**, cyclosporin; **D**, ergotamine; **E**, HC-toxin; **F**, AM-toxin; **G**, ferrichrome; **H**, enniatin **I**, victorin; **J**, δ-(L-2-aminoadipyl)-L-cysteinyl-D-valine (ACV); **K**, peptaibol. Abbreviations: Aad, L-2-aminoadipic acid; Abu, L-2-aminobutyric acid; Aca, 2-amino-3-chloroacrylic acid; Aeo, L-2-amino-9,10-epoxi-8-oxodecanoic acid; AhOrn, δ-N-acyl-N-hydroxyornithine; Aib, 2-aminoisobutyric acid; Ampp, L-2-amino-5-(p-methyoxyphenyl) pentanoic acid; Bmt, (2S,3R,4R,6E)-2-amino-3-hydroxy-4-methyl-6-octenoic acid, also known as 2-butenyl-4-methyl-L-Thr; Hiv, D-2-hydroxyisovaleric acid; Hmv, D-2-hydroxymethylvaleric acid; Valol, reduced Val. The numbering of cyclosporin reflects the order of modules in cyclosporin synthetase (Figure 7.2); previous numbering schemes started with Bmt.

1999; Dittmann *et al.*, 2001). Although NRP synthetases from fungi share many important characteristics with bacterial NRP synthetases, they also vary from their bacterial counterparts in some key traits. The structures of some fungal NRPs are shown in Figure 7.1.

Non-ribosomal peptide synthetases are large, multifunctional enzymes typically comprised of numerous semiautonomous catalytic domains in a linear series. The domains are arranged in a predictable distance from each other and in a characteristic sequence that reflects the order of their activity in the assembly and tailoring of the peptide or peptide-containing product. Two of the essential functional domains viz., the adenylation (A) domain and thiolation (T) domain, are frequently arranged together as a module of ~650 amino acids. The A domain recognizes the amino acid (or other carboxylic acid) substrate and adenylates it (using ATP as cosubstrate) at the carboxylic acid group. This adenylation step is also catalyzed by many other enzymes constituting a superfamily of adenylate-forming enzymes that includes acyl-CoA synthetases, luciferase, and coumaryl CoA-ligase, among others (Conti *et al.*, 1996). The only essential criterion of a substrate for activation by an NRP synthetase A domain is the presence of a carboxylic acid moiety. Collectively, NRP synthetase A domains can activate molecules as disparate as hydroxy acids, lysergic acid, and 3-hydroxy-4-methylanthranilic acid.

After aminoacyl-adenylate formation (also known as amino acid activation), the amino acid is thioesterified to the thiol group of enzyme-bound 4′-phosphopantetheine (4′-PP) at the thiolation (T) domain (also known as the peptidyl carrier protein, or PCP), which is found immediately downstream from the A domain. The 4′-PP is bound to a universally conserved Ser residue in the T domain, defined by the motif DXFFXXLGG(H/D)S(L/I) (Schlumbohm *et al.*, 1991; Marahiel *et al.*, 1997).

Activation and thioesterification are catalyzed for each constituent of the peptide product at separate modules regularly spaced along the synthetase, resulting in a series of thioesterified amino acids tethered to enzyme-bound 4′-PP at each of the T domains. Condensation (C) domains (Gehring *et al.*, 1998; Stachelhaus *et al.*, 1998), of about 450 amino acids, which are located between the modules, then catalyze peptide bond formation. Condensation begins with the transfer of the amino acid on the first T domain (which will remain the amino-terminal amino acid) to the amino acid bound to the second T domain. The growing chain is then transferred from the second T domain to the amino acid tethered to the third T domain, with the new peptide bond catalyzed by the condensation domain between the second and third modules. The assembly of NRPs resembles polypeptide synthesis via translation in that the chain grows from amino to carboxyl terminus, with the growing chain transferred to the next awaiting amino acid at each step.

The structures of representative fungal NRP synthetases are shown in Figure 7.2. In addition to the A/T domains (which together are also referred to as a "minimal module") and the intervening C domains, some NRP synthetases carry additional domains for specific tailoring functions such as N-methylation and epimerization (see below). Domains catalyzing these modifications act upon the amino acid substrates as they are bound to the enzyme. The N-methylation domain, when present, is typically located between the A and T domain, and the epimerase domain, for those modules that contain one, is found immediately downstream of the T domain (Figure 7.2).

The final step in the biosynthesis of NRPs is the release of the covalently bound nascent peptide from the enzyme or enzyme complex. At this step there are several permutations of the basic modular organization of NRP synthetases, and differences between

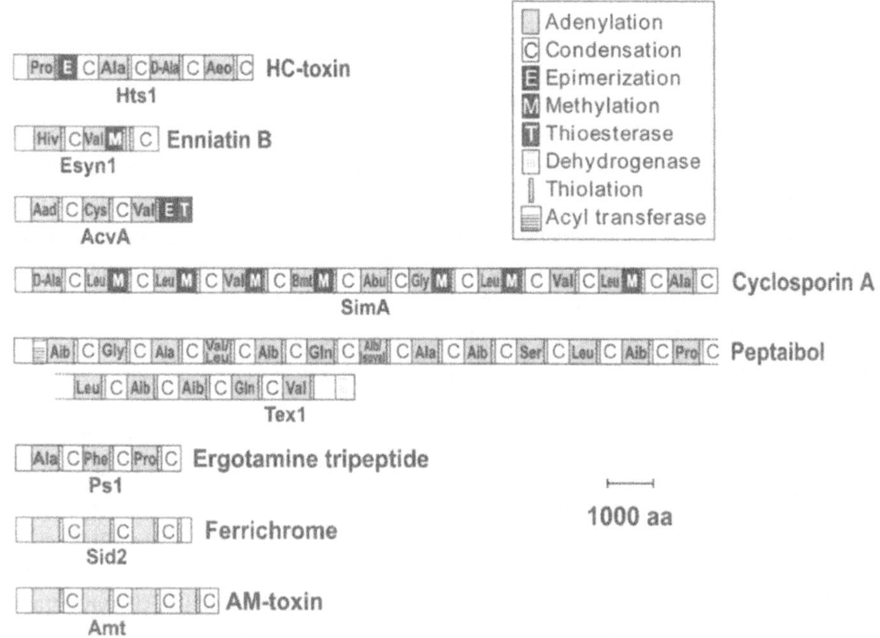

Figure 7.2. Structures of characterized fungal NRP synthetases. The corresponding NRP and the name of the gene product are shown to the right and below, respectively. The wavy line in module 4 of AM-toxin synthetase indicates a gap of ~300 amino acids from the adenylation domain (Johnson *et al.*, 2000).

prokaryotic and eukaryotic enzymes also become apparent. In many bacterial NRP synthetases the final domain is a thioesterase (TE), whose function is to liberate the nascent peptide from its covalent linkage to the enzyme. Keating *et al.* (2001) have proposed that in cases of the release of a carboxylic acid end product (i.e., a linear peptide), the T domain is in an environment in which the thioester bond between the nascent peptide and 4′-PP is susceptible to hydrolysis. In cases where cyclic peptides or cyclic peptides containing one more lactone linkages (depsipeptides) are the end products, the nascent peptide is hypothesized to be bound to the T domain in an environment in which there is limited exposure to water so that the peptide has time for intramolecular condensation, resulting in cyclization during hydrolysis. Fungal NRP synthetases, with a few exceptions, lack recognizable TE domains. (The exceptions are ACV synthetases, and an uncharacterized NRP synthetase in the genome of *Phanerochaete chrysosporium*, see below.) The existence of a TE activity encoded cryptically near the amino terminus of fungal NRP synthetases cannot be ruled out but seems unlikely given the conservation of other functional domains between fungal and bacterial NRP synthetases. In place of a TE domain, some fungal NRP synthetases have an additional condensation domain (Figure 7.2). This final condensation domain is hypothesized to promote peptide bond formation between the amino group at the amino terminus of the nascent peptide, or an internal amino acid, with the carboxyl group attached as a thioester to 4′-PP at the final T domain. This catalysis is hypothetical and has not been demonstrated experimentally.

An alternate mechanism of peptide termination and release has been proposed for enniatin (Glinski *et al.*, 2002) (Figures 7.1H, 7.2). Enniatin has a total of six residues (three carboxylic acids and three amino acids in the order X–Y–X–Y–X–Y), but enniatin synthetase has only two modules, so it must work by an iterative mechanism, analogous to fatty acid synthases and some polyketide synthases. Following the second module of enniatin synthetase, there is a final set of domains that appears to be a fusion of a T domain and a C domain. This fused T–C domain is hypothesized to have a "holding" function for dimer and tetramer intermediates of enniatin during its iterative biosynthesis. By this scheme, the first dimer is transferred to the extra T domain where it waits until the appearance of the second dimer, to which it is then linked by the extra C domain. This step is repeated to form a linear hexadepsipeptide bound to the extra T domain, and, finally, condensation to the final cyclic form is catalyzed by the extra C domain.

Another variation on the basic one module/one amino acid scheme of NRP synthetases occurs in ferrichrome biosynthesis by the product of the *sid2* gene (ferrichrome synthetase) of *Ustilago maydis* (Figures 7.1G, 7.2) (Yuan *et al.*, 2001). Like enniatin, ferrichrome is a cyclic hexadepsipeptide composed of just two components, but, in contrast to enniatin, its pattern is X–X–X–Y–Y–Y. Also in contrast to enniatin synthetase, Sid2 has three modules. The third module is followed by an extra C domain and an extra T domain (Figure 7.2). At this point, there is no good model to explain how a three-module NRP synthetase could synthesize an NRP with the pattern found in ferrichrome. One possibility is that its biosynthesis requires an additional NRP synthetase.

Another variation on termination found in fungal NRP synthetases is that of the peptaibols. These compounds occur as families of differing lengths. For example, a single NRP synthetase (Tex1) in *Trichoderma virens* makes peptaibols of 11, 14, and 18 amino acids (Wiest *et al.*, 2002). The presence of specific shorter peptides may indicate that certain T domains in peptaibol synthetase are not as well sheltered from water as other T domains, permitting premature chain termination (Keating *et al.*, 2001).

3. The "Non-Ribosomal Code" for Fungal NRP Synthetases

The crystalline structure of the Phe-specific adenylation (A) domain of GrsA of *Bacillus brevis* has provided important information on the positions of amino acid residues that form the amino acid substrate-binding pocket of the A domains of NRP synthetase modules (Conti *et al.*, 1997). These residues occur within an approximately 110-amino acid sequence beginning just after conserved motif A3 [AY(V/I)(L/I)FTSGSTGxPKG] and ending just after conserved motif A5 [N(G/A)(Y/W)GP(T/A)E] (motifs as defined in Marahiel *et al.*, 1997). On the basis of amino acids occupying these critical eight or nine positions in modules of known amino acid substrate specificity, Stachelhaus *et al.* (1999) and Challis *et al.* (2000) have proposed "non-ribosomal codes" to predict amino acid substrate specificity of A domains. These independently developed codes differ only by the inclusion of one additional position in the code of Stachelhaus *et al.* (1999). The eight or nine amino acids are considered a signature sequence for the substrate that will be recognized by the A domain. The non-ribosomal codes are based primarily on bacterial NRP synthetase gene sequences, which far outnumber the fungal sequences in public databases and have been validated as effective predictors of the specificity of bacterial NRP synthetase modules (e.g., Shen *et al.*, 2002). A website designed to simplify the task is available at *http://raynam.chm.jhu.edu/~nrps*.

We have investigated the utility of this code for identification of substrates for fungal A domains by querying with sequences of fungal NRP synthetase modules with known specificity. Fungal sequences known to have Ala substrate specificity found their own entries (when available) in BLAST searches of this database but otherwise did not retrieve Ala-activating domains. For example, the Ala-activating modules from ergotamine tripeptide synthetase (Ps1 and LpsA; Figure 7.2) each retrieved the Val-activating module of fengamycin synthetase as their closest match. Both of the Ala-activating modules of cyclosporin synthetase were most similar to Leu-activating modules. The Ala-activating domains of peptaibol synthetase (Tex1) were most similar to Pro-activating domains. Queries with fungal Val-activating domains and Pro-activating domains yielded similar results.

In general, queries of the database with putative fungal amino acid recognition motifs resulted in retrieval of domains that recognize amino acids bearing side chains with similar properties. Identity and similarity scores for domains with the specificity for the same versus similar amino acids were not noticeably different. With the currently available data, it appears that, at least for hydrophobic substrates, the database can provide information about the general chemical nature of the side chain of an amino acid likely to be activated by a fungal module with unknown specificity, but cannot provide an identity for the substrate with a high degree of confidence.

To investigate the possibility that a fungal code may differ in signature amino acid residues at the defined positions or nearby positions, we made multiple alignments of the signature sequences of Stachelhaus *et al.* (1999) for Ala, Val, and Pro, and available fungal A domains known to have specificity for these amino acids. These three amino acids were chosen because they are relatively well represented among the sequenced fungal NRP synthetase genes. We examined multiple alignments for the potential of a slightly modified fungal code using the amino acids at the same or neighboring positions as those derived from the Phe-activating domain of GrsA. For Ala, there is a general similarity in the types of side chains occurring at most of the signature residue positions among the Ala-adenylating domains of lysergyl peptide synthetase (from *Claviceps purpurea* and *Neotyphodium lolii*), cyclosporin synthetase, HC-toxin synthetase, and peptaibol synthetase with the signature residues identified by Stachelhaus *et al.* (1999). However, for other positions, side-chain properties vary widely. The first critical position in the Ala code contains an Asp residue (an acidic residue is highly conserved at this position among all domains that recognize amino acids as opposed to other carboxylic acids), and this residue is believed to interact with the α-amino group of the amino acid. (Thus, the absence of an Asp residue at this position in a domain of unknown specificity indicates that it recognizes a non-amino acid substrate.) For the second, third, fourth, and sixth conserved position, Stachelhaus *et al.* (1999) identified amino acids with hydrophobic residues, and the fungal Ala-activating domains all have residues with hydrophobic side chains at these positions but typically not the specified amino acid. The Gly residue at the fifth position of the code is not well represented in name or property by the amino acids appearing at this position in the fungal Ala-activating domains. Instead, Cys and Tyr residues are more common. Finally, the Val and Ala residues filling the eighth and ninth conserved positions are conserved in property by hydrophobic residues at these positions in the fungal Ala-activating domains. However, similar amino acids occur in all known fungal Val-activating domains and Pro-activating domains as well (with exceptional Cys residues in domains from peptaibol synthetase). Thus, these two positions are not particularly informative in discriminating among the fungal domains, at least those that activate amino acids with non-polar side chains. Analyses of fungal Val-activating domains and Pro-activating domains yielded similar findings. Again, the nature of the amino acid side chain was often conserved at the critical positions but the identity of the amino acid was not.

These analyses focused on three different hydrophobic substrates because the majority of fungal NRP synthetase modules with known substrates activate hydrophobic amino acids. However, it may be more difficult to distinguish domains for hydrophobic amino acids from one another compared to those that adenylate polar amino acids. Challis *et al.* (2000) noted that domains that activate hydrophobic substrates require hydrophobic residues in the recognition sites, but show a lower degree of substrate specificity, and consequently provide a code with lower predictive value. However, domains responsible for the adenylation of substrates with polar side chains were readily identifiable, with highly conserved polar residues present in the signature sequence. For this reason, we also investigated predictive ability of the code for a polar amino acid. Based on several prokaryotic Ser-activating domains (module 1 of Cda1, EntF, and modules 1 and 2 of SyrE), a signature sequence of D(V/L)WH(I/F/L)SL(I/V) was deduced by Challis *et al.* (2000). The His residue at the third signature position and the Ser at the sixth signature position were universally conserved among these prokaryotic sequences. However, the only fungal NRP synthetase module known to activate Ser, module 10 of the Tex1 peptaibol synthetase (Figures 7.1K, 7.2), contains the signature sequence DVGYLAAV (Wiest *et al.*, 2002), which matches the code signature at only 4/8 positions. Moreover, at the two diagnostic positions noted above, Tex1 module 10 has Tyr instead of His at the fourth position and Ala in place of Ser at the sixth position.

To investigate a potential structural basis for amino acid substrate specificity further, we conducted more comprehensive distance-based phylogenetic analyses of 41 fungal NRP synthetase modules. These included modules for Ala, L-2-aminoadipic acid (Aad), Cys, Leu, Val, and Pro that are available from multiple fungi. Additionally, we analyzed multiple sequences for 2-aminoisobutyric acid (Aib) and Glu (Wiest *et al.*, 2002); however, data from these latter two residues are less than ideal because the sequences come from the same organism and gene. It is thus not possible to determine whether the high degree of sequence similarity between Aib-activating modules or Glu-activating modules is due to the specificity requirements of the substrate or to the common evolutionary origin of the modules. The same holds for analyses of the Cys-activating and Aad-activating modules of ACV synthetase.

No clear clustering by substrate emerged in distance-based phylogenetic analyses of 44 fungal NRP synthetase modules conducted with PAUP* (Swofford, 2002), based either on entire A domains (631 aligned positions), the 109-residue region spanning the putative substrate specificity sites, or the nine specificity sites alone (Table 7.1). A tendency to cluster by organism was always observed, although this tendency was diminished in the 109-residue and 9-residue data sets. However, while unconstrained analyses of the 9-residue data set grouped A domains by organism, trees produced by analyses in which Ala, Leu, Pro, Val, or all four were constrained to monophyly did not differ significantly from the unconstrained tree.

In summary, when applied to fungal NRP synthetases, the currently defined codes (Stachelhaus *et al.*, 1999; Challis *et al.*, 2000) can provide a degree of confirmation of suspected substrate specificity but have limited predictive value. Perhaps this will change as more fungal domains with known amino acid substrate specificities are discovered. On the other hand, it is also possible that a firm code will never be realized, due to fungal

Table 7.1. Specificity Residues in Fungal NRP Synthetase Adenylation(A) Domains, as defined by Challis *et al.* (2000) and Stachelhaus, Mootz, and Marahiel (1999)

			Specificity residue position								
Organism	Gene[a]	Substrate[b]	235	236	239	278	299	301	322	330	331
Cochliobolus carbonum	*HTS1*-3	D-Ala	D	L	L	F	G	I	S	V	L
Tolypocladium niveum	*simA*-1	D-Ala	D	L	W	F	Y	I	A	V	V
Claviceps purpurea	*ps1*-1	Ala	D	L	F	F	C	G	G	C	P
Cochliobolus carbonum	*HTS1*-2	Ala	D	A	G	G	C	A	M	V	A
Tolypocladium niveum	*simA*-11	Ala	D	V	F	I	Y	A	A	I	L
Trichoderma virens	*tex1*-3	Ala	D	V	G	F	V	A	G	V	L
Trichoderma virens	*tex1*-8	Ala	D	I	F	V	V	A	G	V	I
Tolypocladium niveum	*simA*-7	Sar	D	I	Q	M	F	V	A	M	Q
Trichoderma virens	*tex1*-5	Aib	D	L	G	W	L	C	G	V	F
Trichoderma virens	*tex1*-1	Aib	D	L	G	Y	L	A	G	V	F
Trichoderma virens	*tex1*-12	Aib	D	L	G	Y	L	A	G	V	F
Trichoderma virens	*tex1*-9	Aib	D	L	G	Y	L	A	G	C	F
Trichoderma virens	*tex1*-15	Aib	D	L	G	F	L	A	G	V	F
Trichoderma virens	*tex1*-16	Aib	D	L	G	F	L	A	G	L	F
Trichoderma virens	*tex1*-7	Aib/isoVal	D	C	G	W	V	V	G	V	V
Trichoderma virens	*tex1*-18	Val	D	A	I	I	I	V	G	V	T
Aspergillus nidulans	*acvA*-3	Val	D	F	E	S	T	A	A	V	Y
Penicillium chrysogenum	*acvA*-3	Val	D	F	E	S	T	A	A	V	Y
Fusarium scirpi	*esyn1*-2	Val	D	G	W	F	I	G	I	I	I
Tolypocladium niveum	*simA*-4	Val	D	A	W	M	F	A	A	I	L
Tolypocladium niveum	*simA*-9	Val	D	A	W	M	F	A	A	V	L
Trichoderma virens	*tex1*-4	Val or Leu	D	M	G	F	L	G	G	V	C
Trichoderma virens	*tex1*-11	Leu	D	F	L	Y	F	G	G	V	V
Trichoderma virens	*tex1*-14	Leu	D	A	A	L	I	G	A	V	F
Tolypocladium niveum	*simA*-2	Leu	D	A	W	L	Y	G	A	V	M
Tolypocladium niveum	*simA*-3	Leu	D	A	W	L	Y	G	A	V	M
Tolypocladium niveum	*simA*-8	Leu	D	A	W	L	Y	G	A	V	M
Tolypocladium niveum	*simA*-10	Leu	D	A	W	L	Y	G	A	V	M
Claviceps purpurea	*ps1*-3	Pro	D	I	T	L	V	A	G	S	L
Cochliobolus carbonum	*HTS1*-1	Pro	D	I	A	V	I	T	V	L	I
Trichoderma virens	*tex1*-13	Pro	D	V	L	F	C	G	L	I	C
Trichoderma virens	*tex1*-6	Gln	D	G	G	M	V	G	G	N	Y
Trichoderma virens	*tex1*-17	Gln	D	G	G	M	V	G	G	N	Y
Aspergillus nidulans	*acvA*-2	Cys	D	H	E	S	D	V	G	I	T
Penicillium chrysogenum	*acvA*-2	Cys	D	H	E	S	D	V	G	I	T
Aspergillus nidulans	*acvA*-1	Aad	E	P	R	N	I	V	E	F	V
Penicillium chrysogenum	*acvA*-1	Aad	E	P	R	N	I	V	E	F	V
Claviceps purpurea	*ps1*-2	Phe	D	L	V	G	M	A	A	S	V
Trichoderma virens	*tex1*-2	Gly	D	I	G	M	V	V	G	V	L
Trichoderma virens	*tex1*-10	Ser	D	V	G	Y	L	A	A	V	Y
Fusarium scirpi	*esyn1*-2	Hiv	G	A	L	H	V	V	G	S	I
Cochliobolus carbonum	*HTS1*-4	Aeo	D	V	L	L	C	T	G	I	M
Tolypocladium niveum	*simA*-5	Bmt	D	A	W	T	Y	G	G	V	I
Tolypocladium niveum	*simA*-6	Abu	D	A	W	F	H	A	V	A	Y

[a]Numbers following gene names refer to the appropriate module of that gene.
[b]For abbreviations of amino acids, see footnote to Figure 7.1.

A domains having an excessively relaxed substrate specificity. All fungal NRPs occur as families, members of which differ in their constituent amino acids, and in every case studied, mutation of the cognate NRP synthetases has indicated that all family members are synthesized by the same and not separate enzymes. It has also been experimentally demonstrated that the spectrum of NRPs made by a particular fungus *in vivo*, or by a particular synthetase *in vitro*, can be easily altered by changing the composition of the available amino acid pool. There are many examples of such "directed biosynthesis" of novel NRPs, including penicillins, cyclosporins, enniatins, ergopeptines, and others (e.g., Mach and Tatum, 1964; Traber *et al.*, 1989; Lawen and Traber, 1993; Baldwin *et al.*, 1994; Riederer *et al.*, 1996; Krause *et al.*, 2001). The ease with which novel NRPs can be made simply by feeding alternate amino acids indicates that NRP synthetase modules have intrinsically low substrate specificity.

4. Parsing Fungal NRP Synthetases

There are two occasions when one would want to predict the amino acid substrate of a particular A domain. The first is when the structures of a peptide and its cognate synthetase gene are known, but the assignment of module to amino acid is not (e.g., AM-toxin and ferrichrome; Figure 7.2). The other is when NRP synthetase genes are available from either PCR cloning or genomic projects, but the NRP product is not known. As discussed in the previous section, the synthetase code can give a rough indication of the nature of the side chain recognized by a particular A domain, but short of the existence of a reliable code for fungal synthetases, the following guidelines are offered for help in defining the function of novel NRP synthetase A domains.

4.1. Guideline 1

The number of modules (A/T domains) in the enzyme determines the number of amino acids and/or hydroxy acids in the NRP product

Exceptions to this rule will occur on two occasions. The first is when the synthetase is composed of two or more subunits, which is common among bacterial but rare among fungal synthetases (the one known fungal example is ergotamine synthetase, which is composed of the one-module lysergic acid synthetase and the three-module ergotamine tripeptide synthetase; Riederer *et al.*, 1996; Tudzynski *et al.*, 1999). Typically, with a two-polypeptide NRP synthetase, the second (or receiving) polypeptide in such a pair should begin with a condensation (C) domain (Marahiel *et al.*, 1997), and therefore when a synthetase begins with a C domain it is likely to be the second (receiving) enzyme in a synthetase composed of multiple polypeptides. Although the published sequence of the lysergyl peptide synthetase gene from *C. purpurea (ps1)* (Tudzynski *et al.*, 1999) does not appear to contain such a C domain, the *N. lolii* homologue of this gene, *lpsA* (Panaccione *et al.*, 2001) does have an additional C domain in the expected position (Damrongkool and Panaccione, unpublished). Although C domains, in general, have conserved motifs, they tend to be more variable than other NRP synthetase domains. Analysis of the suspected area by BLASTP should reveal the presence of a C domain. C domains in NRP synthetases are also identified by the Conserved Domain Database (CDD) function of BLASTP.

The second exception to the one module/one amino acid rule is iterative enzymes. One established case is known in fungi, enniatin synthetase (Figure 7.2). This enzyme has an extra T and an extra C domain at its carboxyl terminus. As mentioned earlier, whether a particular module adenylates (activates) an hydroxy acid or an amino acid appears correlated with a non-acidic residue instead of Asp in the first conserved position of the eight or nine amino acids that form the amino acid substrate-binding pocket in the A domain (Table 7.1). Thus, the absence of this Asp is an indication of a non-amino acid substrate for that A domain.

4.2. Guideline 2

Methylation or epimerase domains embedded within the appropriate module indicate products with N-methylated amino acids or D*-amino acids*

Details of these two "tailoring domains" will be discussed subsequently. The sequences of both types of domains can be extracted from enzymes known to contain them (enniatin synthetase for N-methylation, and HC-toxin synthetase or ACV synthetase for epimerase) and, when present, serve as more reliable indicators that the protein in question is a true NRP synthetase than does the presence of A domains alone, because the latter are found in the entire superfamily of adenylating enzymes. Core domains have been defined for ~350-amino acid epimerase modules by Stachelhaus and Walsh (2000) and for ~420-amino acid N-methylation modules by Haese *et al.* (1993). Motifs (5–15 amino acids) for these two auxiliary domains have been defined by Marahiel *et al.* (1997). The absence of an epimerase domain in a novel NRP synthetase does not indicate the absence of D-amino acids in the NRP product, because in at least two cases (cyclosporin and HC-toxin), D-Ala is made by a separate enzyme instead of the NRP synthetase itself (discussed later). D-amino acids are useful for *in vivo* biosynthetic feeding studies because of their lower tendency to disappear into primary metabolism (Meeley and Walton, 1991). If HC-toxin synthetase (Hts1) is typical, however, an NRP synthetase module that has an attached epimerase domain will activate only the L-amino acid, not the corresponding D isomer (Walton and Holden, 1988).

4.3. Guideline 3

Extra domains can indicate a cyclic, linear, iterative, or N- or C-modified NRP product

A C-domain located on the carboxyl-terminal side of the final minimal module indicates cyclization of the peptide product. "Cyclization" can be of the last amino acid to the first or of the last to an internal amino acid to create a branched product (e.g., the lysergyl peptide lactam product of lysergyl peptide synthetase).

Most, perhaps all, bacterial NRP synthetases contain thioesterase (TE) domains at their C termini, whether the resulting NRP product is linear or cyclic. It would seem to make sense, by analogy to fatty acid synthases and polyketide synthases, that a terminal TE domain would be involved in releasing the nascent peptide from the synthetase, and that this would result in a linear peptide analogous to a linear fatty acid. Continuing with this argument, fungal NRP synthetases do not have TE domains, with the notable exception of ACV synthetases (and possibly the uncharacterized NRP synthetase predicted in the genome of *Phanerochaete chrysosporium*; see below). Except for the peptaibols, whose C termini are modified by reduction, ACV is the only *linear* fungal NRP whose synthetase is known. However, the possible bacterial origin of fungal ACV synthetases confuses the issue, making it unclear if the TE domain of ACV synthetase relates to the linearity of its product (isopenicillin N) or reflects its prokaryotic origin (discussed later).

The presence of an additional T domain (beyond that occurring within the final minimal module) followed by an extra C domain is indicative of an iterative NRP synthetase. This conclusion is based solely on enniatin synthetase (Glinski *et al.*, 2002), as the case of the synthetase (Sid2) for the iterative cyclic peptide ferrichrome in *Ustilago maydis* is still unclear (Figure 7.2).

Tex1 (peptaibol synthetase) contains a reductase domain at its carboxyl terminus and an acyltransferase at its N terminus (Figure 7.2). These, and other domains with similarities to other known enzymes, could help in predicting the NRPs made by novel NRP synthetases. In addition, N-methylated NRPs can be identified by radiolabelling *in vivo* with [^{35}S]-adenosylMet.

The CDD of BLASTP can identify AMP-binding, condensation, thiolation, epimerase, methylation, and thioesterase domains but, in our experience, is not always reliable, especially for auxiliary domains. When analyzing NRP synthetase sequences, any irregularities in module spacing should be reason to do a more intensive search for additional functional domains.

5. Strategies to Identify NRP Synthetases and Genes

The first NRP synthetases were identified by enzyme purification and genetics (Krause *et al.*, 1985; van Liempt *et al.*, 1989; Díez *et al.*, 1990; MacCabe *et al.*, 1990; Smith *et al.*, 1990b; MacCabe *et al.*, 1991; Scott-Craig *et al.*, 1992; Turgay *et al.*, 1992). Other ways in which known or putative NRP synthetases have been identified include chromosome walking from other genes involved in the same pathway (e.g., Tudzynski *et al.*, 1999; Panaccione *et al.*, 2001; Yuan *et al.*, 2001), PCR with primers based on conserved sequences found in NRP synthetases (e.g., Johnson *et al.*, 2000; Wiest *et al.*, 2002), and analysis of fungal genomes (e.g., Kroken *et al.*, 2002; described later).

NRP synthetases can be assayed *in vitro* using ATP/PP$_i$ exchange (Gevers *et al.*, 1969; Lee and Lipmann, 1975), a reaction that measures the initial amino acid activation step (amino acyl-AMP formation). Consequently, the assay does not require complete *in vitro* biosynthesis of the NRP in question or even a completely intact NRP synthetase enzyme, since for amino acid activation, each module functions independently of the others. The substrates are radiolabelled pyrophosphate, unlabelled ATP, and the appropriate substrate amino acid in an unlabelled form. Because formation of the amino acyl-AMP intermediate is reversible, during the course of the reaction ^{32}P is transferred from ^{32}PP$_i$ to ATP. The ATP is selectively removed from the reaction by binding to charcoal, and its radioactivity is quantified.

Other enzymes, notably the aminoacyl-tRNA synthetases, also catalyze amino acid-dependent PP$_i$ exchange and can therefore interfere with identification of NRP synthetases. Since tRNA synthetases recognize only the 20 proteinogenic amino acids as substrates, the use of unusual amino acids such as ornithine or D-amino acids, when possible, will avoid interference. However, for some unusual amino acids the form recognized by the NRP synthetase is not always obvious. D-amino acids, for example, can be synthesized from the L forms either before or after binding to the NRP synthetase (discussed later), and apparently it is not known at what stage hydroxylation occurs for any peptide containing hydroxylated amino acids.

The ATP/PP_i exchange assay has been used to identify and purify many NRP synthetases and thereby identify the encoding genes. In principle, the ATP/PP_i exchange assay could be used to identify the NRP product of novel NRP synthetase genes isolated by PCR or on the basis of genomic sequence data, by heterologous expression of the NRP synthetase. However, there are two problems with this approach. One is the difficulty of expressing NRP synthetases in heterologous systems, and the second is the low substrate specificity of most NRP synthetases, at least *in vitro*. Therefore, assaying an NRP synthetase with different amino acids is not likely to succeed in identifying the true amino acid substrate used *in vivo*.

NRP synthetase genes have been isolated using PCR with primers based on conserved sequences of known NRP synthetases (Turgay and Marahiel, 1994; Nikolskaya *et al.*, 1995; Panaccione, 1996; Stachelhaus *et al.*, 1999; Johnson *et al.*, 2000; Wiest *et al.*, 2002). The major drawback of this approach is the subsequent difficulty in determining the specific function of the recovered NRP synthetase genes. The same difficulty pertains to NRP synthetase genes that can be recognized in the genomic sequences. All of the genomes that have recently been sequenced completely contain predicted NRP synthetase genes of unknown function (described later). The most important method for proof of function is targeted mutation, which is now feasible for most fungi (Johnson *et al.*, 2000; Panaccione *et al.*, 2001; Wiest *et al.*, 2002).

6. Tailoring Enzymes and Auxiliary Domains

Typically, NRPs contain unusual or modified amino acids. D isomers and N-methylated amino acids are common, as are non-proteinogenic amino acids (e.g., ornithine, pipecolic acid), and highly unusual amino acids (e.g., L-2-amino-9,10-epoxi-8-oxodecanoic acid in at least four fungal NRPs, [2S,3R,4R,6E]-2-amino-3-hydroxy-4-methyl-6-octenoic acid in cyclosporin; Figure 7.1). Modifications of linear NRPs include N-acylation and C-terminal reduction. Many bacterial NRPs undergo extensive modifications, including P450-catalyzed oxidative cyclization, β-hydroxylation, heterocyclization, and glycosylation (Walsh *et al.*, 2001). NRP tailoring can be catalyzed by the NRP synthetase itself or by distinct enzymes. Some modifications, such as epimerization to produce D-amino acids, can occur either way.

6.1. N-Methylation

N-Methylation is common in both fungal and bacterial NRPs (Billich and Zocher, 1990; Burmester *et al.*, 1995). In every case known, N-methylation is catalyzed by methylation domains within the synthetase itself. For example, enniatin synthetase (Figure 7.2) contains one methylation domain for modification of Val (Figure 7.1H). Seven of the cyclosporin synthetase modules (Figure 7.2) contain methylation domains, one for methylation of each of the seven N-methylated amino acids found in the product (Figure 7.1C; Weber *et al.*, 1994).

The N-methylation domain of enniatin synthetase is ~430 amino acids in length and contains a motif, VL(E/D)XGXGXG, that is a conserved feature of known S-adenosylMet (SAM)-dependent methyltransferases (Haese *et al.*, 1993; Pieper *et al.*, 1995; Hacker *et al.*, 2000). Binding of D-isovaleric acid to module 1, L-Val to module 2, and SAM to the N-methylation module has been demonstrated experimentally (Burmester *et al.*, 1995; Pieper *et al.*, 1995). Critical amino acids and amino acid motifs in the N-methylation module have been identified by site-directed mutagenesis (Hacker *et al.*, 2000; Velkov and Lawen, 2003).

6.2. Epimerization

Amino acid epimerization is sometimes catalyzed by the NRP synthetase itself and sometimes by a separate enzyme. When the D-amino acid is produced on the synthetase, the enzyme contains an epimerase domain of ~450 amino acids immediately after the thiolation domain of the core module. In such cases, the NRP synthetase recognizes only the L-amino acid and epimerizes it after the thiolation step. The mechanism of epimerization is still not known; pyridoxal phosphate is apparently not involved (Stachelhaus and Walsh, 2000). Early analyses of epimerization domains identified two motifs, HHXXXDXXSW and (S/A)RTXGWFT(T/S) (Scott-Craig *et al.*, 1992; de Crécy-Lagard *et al.*, 1995). The first of these is also found in the condensation domain and therefore is not a reliable indicator, but the second is found only in epimerase domains. Marahiel *et al.* (1997) have named another five epimerase signature motifs. A comparison of 45 fungal and bacterial epimerase domains identified 23 invariant and 30 highly conserved amino acids. Detailed kinetic analyses of site-directed mutants identified five critical residues (Stachelhaus and Walsh, 2000).

Cyclosporin contains one D-amino acid, D-Ala, but cyclosporin synthetase does not contain any epimerase domain. HC-toxin contains two D-amino acids but its synthetase (Hts1) contains only one epimerization domain. The epimerization domain of Hts1 comes right after module 1, which activates L-Pro, so the epimerization domain of Hts1 must epimerize L-Pro (Scott-Craig *et al.*, 1992). It is now known that the D-Ala in cyclosporin and in HC-toxin is synthesized from L-Ala by a separate enzyme, alanine racemase. The D-Ala is then activated by the relevant NRP synthetase and incorporated into the final NRP. The gene for the Ala racemase involved in cyclosporin biosynthesis in *Tolypocladium inflatum* was identified by purification of the enzyme (Hoffmann *et al.*, 1994); it is listed as "sequence 33 from Patent WO9425606" in the GenBank patent database (accession A40406). The Ala racemase gene in *Cochliobolus carbonum*, *TOXG*, was isolated by its physical linkage to other genes involved in HC-toxin biosynthesis. It is part of the *TOX2* gene cluster and is co-regulated by the ToxE transcription factor (described later). The function of *TOXG* was established by gene disruption and complementation of a D-Ala auxotroph of *Escherichia coli* (Cheng and Walton, 2000). The two fungal Ala racemases are almost identical in size and overall have ~43% identity at the amino acid level. They are structurally, and presumably mechanistically, related to other pyridoxal-containing enzymes such as threonine aldolase (e.g., *GLY1* of *Candida albicans*).

TOXG mutants of *C. carbonum* fail to synthesize the D-Ala-containing forms of HC-toxin but can still synthesize a minor form that contains Gly in place of D-Ala (Rasmussen and Scheffer, 1988). *C. carbonum* apparently makes more HC-toxin than it needs for pathogenicity because the *TOXG* mutant is almost as virulent as the wild type. The *TOXG* mutant phenotype can be rescued by supplementing fungal cultures with D-Ala (Cheng and Walton, 2000). The Ala racemase of *T. inflatum* has apparently not been studied genetically.

6.3. Other Tailoring Reactions of Fungal NRP Synthetases

The linear NRPs known as peptaibols are synthesized by several species of *Trichoderma* (*Hypocrea*) and other fungi (discussed later). The peptaibol synthetase gene, *TEX1*, of *Trichoderma virens* has recently been identified (Wiest *et al.*, 2002). The deduced protein product contains 18 modules, consistent with the fact that the major peptaibol produced by *T. virens* contains that number of amino acids, but it also contains a predicted acyl transferase domain at its N terminus and a dehydrogenase domain at its C terminus. Peptaibols are typically acylated at their N termini and reduced at their C termini. Therefore, both the N- and the C-terminal modifications of peptaibols are catalyzed by the synthetase itself. Apparently, the complete biosynthesis of peptaibols requires only a single polypeptide, albeit one of 2.3 MDa.

The biosynthesis of most fungal NRPs requires enzymes and hence genes in addition to the NRP synthetases themselves. These enzymes are involved in the biosynthesis of unusual amino acids or hydroxyacids, transport, and regulation. For example, biosynthesis of δ-N-acyl-N-hydroxyornithine (AhOrn), one of the two components of the siderophore ferrichrome in *U. maydis* (Figure 7.1G), requires ornithine hydroxylase. This enzyme is encoded by *sid1*, which is clustered with the ferrichrome synthetase gene, *sid2* (Mei *et al.*, 1993; Yuan *et al.*, 2001).

Many permutations of NRPs are evident from structural considerations, but little if anything is known about the biosynthetic pathways involved. The host-selective phytotoxins victorin and peritoxin, made by the plant pathogens *Cochliobolus* (*Helminthosporium* or *Bipolaris*) *victoriae* and *Periconia circinata*, respectively, contain chlorinated amino acids (Figure 7.1I; Wolpert *et al.*, 1985; Macko *et al.*, 1992). Amatoxins such as α-amanitin (Figure 7.1B) contain an intramolecular Trp–Cys sulfoxide bridge. It will be interesting to explore the enzymatic mechanisms of these modifications once the relevant NRP synthetases and tailoring genes have been isolated. The biosynthesis and molecular genetics of other fungal NRP constituents are discussed in the individual sections below.

6.4. Pantetheinylation

All NRP synthetases have one molecule of 4′-phosphopantetheine (4′-PP) bound to each module (Schlumbohm *et al.*, 1991; Stein *et al.*, 1994, 1996). Therefore, activity of NRP synthetases depends on successful post-translational addition of 4′-PP from coenzyme A catalyzed by 4′-PP transferase. Several housekeeping enzymes are also pantetheinylated, such as α-aminoadipate semialdehyde dehydrogenase (product of the yeast *lys2* gene) involved in lysine biosynthesis, acyl carrier protein (in bacteria and plants that have the type II fatty acid synthesis pathway), and the fatty acid synthase itself (in fungi and mammals that have the type I pathway). Polyketide synthases, which are involved in the biosynthesis of a large number of bacterial and fungal secondary metabolites (see Chapters 5 and 6), also require post-translational pantetheinylation for their activity. The fatty acid synthase of yeast, and presumably other fungi as well, pantetheinylates itself, but the other known pantetheinylated enzymes do not (Fichtlscherer *et al.*, 2000).

Transferases capable of pantetheinylating NRP synthetases have been identified in several bacteria (Lambalot *et al.*, 1996). For example, *sfp* and *gsp* in *Bacillus subtilis* encode 4′-PP transferases that pantetheinylate the surfactin and gramicidin peptide synthetases, respectively (Lambalot *et al.*, 1996). *sfp* and *gsp* are clustered with their respective NRP synthetase genes (Borchert *et al.*, 1994).

Fungal and bacterial 4′-PP transferases have been identified by complementation of yeast *lys5*, the wild-type product of which pantetheinylates Lys2. *sfp* from *B. subtilis* can complement the *lys5* mutation. A fungal 4′-PP transferase, encoded by *npgA*, was identified by weak sequence similarity to *sfp*, and subsequently shown to encode a functional

4′-PP transferase by complementation of *lys5* (Mootz *et al.*, 2002). *npgA* was originally discovered by its involvement in the biosynthesis of spore pigment in *A. nidulans*. The precursor of this pigment is naphthopyrone which is produced by a polyketide synthase, the *wA* gene product.

The completed genome of *Aspergillus fumigatus* contains a single ortholog of *npgA*, and therefore NpgA is probably responsible for pantetheinylating all enzymes that require it, both in primary and secondary metabolism. *npgA* was shown to be allelic to a previously isolated temperature sensitive mutant of *A. nidulans*, *cfwA2*. Significantly, *npgA* mutants of *A. nidulans* no longer synthesize penicillin, consistent with ACV synthetase being a substrate of NpgA (Keszenman-Pereyra *et al.*, 2003).

Searches of the completed genomes of *N. crassa* and the plant pathogenic ascomycete *Magnaporthe grisea* indicate that, like *A. fumigatus*, both of these fungi have single genes related to *ngpA* (*http://www-genome.wi.mit.edu/annotation/fungi*). The homologues are NCU00581.1 in *N. crassa* and MG03046.1 in *M. grisea*. The *N. crassa* and *M. grisea* proteins are 44% identical, and NpgA is ~25% identical to the other two. Melanin is essential for penetration of host leaves by *M. grisea* (Howard and Valent, 1996), so mutants of its *npgA* ortholog would be nonpathogenic if this fungus truly has only a single 4′-PP transferase gene. They should also be defective in all NRP and polyketide biosynthetic activities. However, insofar as 4′-PP transferases have essential housekeeping functions, the mutant strains might be too unhealthy to be very informative or indeed to survive in nature.

The existence of only a single *npgA*-like gene in filamentous fungi, and the known broad substrate specificity of bacterial 4′-PP transferases, suggests that the same enzyme adds 4′-PP to all enzymes that require it (Lambalot *et al.*, 1996; Mootz *et al.*, 2002; Keszenman-Pereyra *et al.*, 2003). If true, it indicates a significant overlap between primary and secondary metabolism in fungi and bacteria.

One explanation for the difficulty of expressing functional NRP synthetase enzymes and modules in heterologous hosts such as *E. coli* has been failure to obtain efficient pantetheinylation (e.g., Haese *et al.*, 1994). Thus, identification of the cognate 4′-PP transferases might facilitate heterologous expression of fungal NRP synthetases and hence study of their function.

7. Regulation

Regulation of NRP biosynthesis has been studied in only a few fungal systems. Like other secondary metabolites, NRP production can be regulated by environmental signals in complex ways. Penicillin production in *A. nidulans* and *Penicillium chrysogenum* is regulated by pH, carbon source, nitrogen source, and amino acids (Litzka *et al.*, 1999; Martín, 2000). A relaxation of wild-type transcriptional controls was probably important in the anthropogenic evolution of commercial, high-producing strains of *P. chrysogenum* (Litzka *et al.*, 1999). Several DNA-binding proteins that regulate expression of the penicillin biosynthetic genes, including *acv*A encoding ACV synthetase (Figure 7.2), have been characterized, but none of them appear to be specific regulators of the penicillin pathway (Martín, 2000; Schmitt and Kuck, 2000).

Ferrichrome (Figure 7.1G) production by *U. maydis* is suppressed by iron, consistent with the role of siderophores in scavenging this essential nutrient. Under high iron concentrations, repression of *sid1*, encoding L-ornithine-N^5-oxygenase, and *sid2*, encoding ferrichrome synthetase (Figure 7.2), is mediated by *urbs1*. Urbs1 is a zinc finger transcription factor in the GATA family. Transcription of *urbs1* is repressed under low iron conditions, thereby allowing expression of *sid1* and *sid2* and production of siderophores (Mei *et al.*, 1993; Voisard *et al.*, 1993; Yuan *et al.*, 2001).

In nature, production of ergopeptines such as ergotamine (Figure 7.1D) in *Claviceps purpurea* is closely associated with the differentiation of sclerotia (hardened resting structures), indicating developmental regulation (Floss, 1976; Gröger and Floss, 1998). In culture, the production of ergopeptines is repressed by phosphate and induced by tryptophan (Krupinski *et al.*, 1976). Since lysergyl peptide synthetase acts near the end of a long pathway, it may well be that earlier steps in the pathway instead of (or perhaps in addition to) the NRP synthetase are the points of regulation. Based on western blotting, Riederer *et al.* (1996) reported that lysergyl peptide synthetase was expressed constitutively in culture. Tudzynski *et al.* (1999) identified promoter elements indicative of transcriptional regulation in the promoter region of the gene encoding this enzyme.

HC-toxin production by *C. carbonum* is constitutive. Nonetheless, a specific positive-acting transcription factor, ToxE, regulates expression of the HC-toxin biosynthetic genes (Ahn and Walton, 1998). A defined binding site for ToxE (the "tox-box") is present in one or two copies in the promoters of all known HC-toxin biosynthetic genes (Pedley and Walton, 2001). Ahn and Walton (1998) reported that ToxE was not required for expression of Hts1, as assayed enzymatically. However, more recent results indicate that ToxE does regulate *HTS1*, when assayed at the level of *HTS1* mRNA (Pedley and Walton, unpublished). The reason for this discrepancy is still not known, but the latter results are consistent with the presence of two copies of the tox-box in the intergenic region between *TOXA* and *HTS1* (Pedley and Walton, 2001).

The role of *TOXE* appears to be restricted to the regulation of HC-toxin biosynthesis, and *TOXE* is clustered with the other HC-toxin biosynthetic genes. In these two respects, *TOXE* resembles *aflR* that regulates aflatoxin (sterigmatocystin) biosynthesis in *Aspergillus*, and *tri6* that regulates trichothecene biosynthesis in *Fusarium* (Woloshuk *et al.*, 1993; Brown *et al.*, 1996; Hohn *et al.*, 1999). Aflatoxins are polyketides and trichothecenes are sesquiterpenoids.

ToxE has a structure unique among known transcription factors. At its N-terminus it contains a bZIP-like basic DNA binding domain but no leucine zipper. Its C-terminus contains four copies of the ankyrin motif. Although both of these motifs have been found in many other proteins, they have not previously been found in the same protein. A gene with a similar structure, of unknown function, has been reported from the plant pathogenic fungus *Cladosporium fulvum*, and "bANK" has been proposed as the name for this class of proteins (Bussink *et al.*, 2001).

8. Status of Research on Selected Fungal Systems

8.1. AM-Toxin

Like HC-toxin (described later), interest in this compound derives from its involvement in plant disease. Both compounds are "host-selective" toxins, meaning that they are active only against the small subset of plant species or cultivars that also are susceptible to the producing fungus. Structurally, host-selective toxins include polyketides, proteins, cyclic peptides, and isoprenoids (Walton, 1996). Several, such as victorin, HC-toxin, and AM-toxin, are NRPs. Some non-specific phytotoxins are also NRPs, such as tentoxin, a cyclic N-methylated tetrapeptide produced by several species of *Alternaria*. Tentoxin is a specific inhibitor of plastid ATP synthase (Groth, 2002). Tentoxin synthetase has been partially characterized *in vitro*, but the encoding gene has not been reported (Ramm *et al.*, 1994).

The fungus that produces AM-toxin, *Alternaria alternata* apple pathotype (formerly *A. mali*), infects only certain varieties of apple, and AM-toxin is toxic only to those same varieties. Neither the mode of action of AM-toxin nor the basis of its specificity is known in detail (Kohmoto and Otani, 1991).

AM-toxin is a four-membered cyclic depsipeptide (Figure 7.1F). The gene encoding AM-toxin synthetase, called *AMT*, was isolated by PCR with primers based on conserved NRP synthetase motifs. *AMT* is present only in isolates and species of *Alternaria* that produce AM-toxin (Johnson *et al.*, 2000). In this regard it resembles *HTS1*, which is present only in isolates of *C. carbonum* that produce HC-toxin (Panaccione *et al.*, 1992). *AMT* contains an intronless 13.1-kb open reading frame encoding a four-module NRP synthetase. This modular structure is consistent with the structure of AM-toxin. The role of *AMT* in AM-toxin synthesis and pathogenicity was proven by targeted gene disruption. One unusual feature of the gene product of *AMT* is that it is predicted to be missing ~300 amino acids from the adenylation domain of module 4, and therefore it is unlikely to be enzymologically active. A resolution of this paradox might reside in the apparent presence of multiple copies of *AMT* in AM-toxin-producing isolates of *A. alternata*. Perhaps only one copy of *AMT* (fortuitously the disrupted one) is functional, but one of the other, nonfunctional copies was sequenced (Johnson *et al.*, 2000).

A sub-cultured strain of *A. alternata* apple pathotype that spontaneously lost the ability to make AM-toxin had simultaneously lost the entire chromosome (~1.8 Mb) on which *AMT* resides, indicating that this chromosome is dispensable for normal growth (Johnson *et al.*, 2001). Several fungal virulence genes, including those for some other host-selective phytotoxins, are also on dispensable or unstable chromosomes (e.g., Pitkin *et al.*, 2000; Han *et al.*, 2001; Ahn *et al.*, 2002; Hatta *et al.*, 2002).

8.2. Cyclosporin

A question that has long intrigued scientists working on secondary metabolites is to what degree their biosynthesis is compartmentalized. Temporal or developmental compartmentalization is well-known in the case of secondary metabolites such as gramicidin that are made only at defined stages of the producing organism's life cycle. In fact, production at a defined stage in the life cycle was formerly closely tied to the definition of secondary metabolism in general. Physical compartmentalization, in which secondary metabolites are biosynthesized in specialized compartments within the cell, could facilitate biosynthesis by maintaining high local concentrations of intermediates, and could protect the rest of the cell from potentially toxic secondary metabolite intermediates, or both. NRP synthetases themselves constitute a type of metabolic channeling because the intermediate peptides remain attached to the enzyme.

By cell fractionation and immunocytochemistry, it was recently shown that cyclosporin synthetase and alanine racemase (which provides D-Ala to the synthetase) co-localize to the vacuolar membrane (Hoppert *et al.*, 2001). The evidence regarding compartmentalization of the penicillin and cephalosporin enzymes and intermediates has recently been reviewed (van de Kamp *et al.*, 1999). To summarize the conclusion of these authors, there appears to be some degree of compartmentalization of the enzymes of the β-lactam pathways, but there is no strong evidence that the entire pathways are located in particular, specialized compartments.

Cyclosporin contains the non-proteinogenic amino acid 2-butenyl-4-methyl-L-Thr (Bmt), which is methylated by the synthetase itself to produce N-MeBmt. Feeding studies indicate that the C9 backbone of Bmt is synthesized from acetyl-CoA and malonyl-CoA by an as-yet unidentified polyketide synthase (Offenzeller *et al.*, 1996).

The ecological rationale for production of cyclosporin is not known, although it might be related to its weak anti-fungal activity. The perfect (sexual) stage of the producing fungus, *Tolypocladium inflatum* (*T. niveum*), was described as *Cordyceps subsessilis* (Hodge *et al.*, 1996). The fungus is found growing on beetle larvae and has entomopathogenic properties. Cyclosporins have insecticidal properties, so perhaps they facilitate pathogenesis of insect hosts. Cyclosporins are made by several additional fungal species, including *Trichoderma polysporum*, *Beauveria nivea*, and *Cylindrocarpon lucidum*.

8.3. Destruxins

Thirty-five naturally occurring destruxins comprise this family of cyclic hexadepsipeptides containing one hydroxy acid and five amino acids (Figure 7.1A). All destruxins are N-methylated and contain either Pro or pipecolic acid (Pedras *et al.*, 2002). Members of the "E" series are epoxidated. Like some other fungal NRPs, destruxins (and related compounds such as roseotoxin, bursaphelocide, and roseocardin) are made by several, not closely related filamentous fungi, including *Metarhizium anisopliae* (formerly *Oospora destructor*), *Alternaria brassicae*, *Ophiosphaerella herpotricha*, *Nigrosabulum globosum*, and *Trichothecium roseum*. *M. anisopliae* is an insect pathogen; *A. brassicae* and *O. herpotricha* are plant pathogens.

The evidence that destruxins have a role in either insect or plant pathogenesis is suggestive but not conclusive. There is some degree of host-selectivity of destruxins, both against insects (parasitized by *M. anisopliae*) as well as plants (parasitized by *A. brassicae*) (Bains and Tewari, 1987; Buchwaldt and Green, 1992; Shivanna and Sawhney, 1993; Kershaw *et al.*, 1999). Metabolism of destruxin by sequential hydroxylation and glucosylation occurs in *Brassica napus*, the host of *A. brassicae*. The hydroxylated intermediate induces production of antimicrobial phytoalexins in the host plant (Pedras *et al.*, 2001).

An NRP synthetase gene, *pesA* from *M. anisopliae*, has been isolated and sequenced (Bailey *et al.*, 1996; accession number X89442). *pesA* is unlikely to encode destruxin synthetase because it has four modules, whereas the presence of six amino and hydroxy acids in destruxin predicts a synthetase of three or six modules. The *pesA* protein (5,157 amino acids) is nonetheless intriguing in having two predicted epimerase domains, attached to modules 1 and 4. The NRP product of *pesA* is therefore predicted to be a four amino-acid peptide with at least two D-amino acids.

8.4. Enniatins

Several NRPs contain hydroxy acids in addition to amino acids. The model NRP of this type is enniatin, a hexadepsipeptide composed of three units of a dimer of the amino acid N-Me–Val and the hydroxy acid D-hydroxyisovaleric acid (Hiv) (Figure 7.1H).

Members of the enniatin family are made by several fungi, particularly species in the genus *Fusarium*. The related beauvericins, in which Me–Phe replaces Me–Val, are made by fungi in the ascomycetous genera *Beauveria* and *Paecilomyces*, and also by the basidiomycete *Laetiporus* (formerly *Polyporus*) *sulphureus*.

An unidentified non-sporulating fungus ("*mycelia sterilia*") produces a class of N-methylated cyclo-octadepsipeptides, one of which is known as PF1022A. Structurally, PF1022A is related to bassianolide, an insecticidal metabolite from *Beauveria*. The structure of the compound and enzymatic analysis of the cognate NRP synthetase indicates that this compound is made by the sequential condensation of four dimers of a hydroxy acid and an amino acid (Weckwerth *et al.*, 2000).

Many enniatin-producing fungi are pathogens of insects or plants. Enniatins and beauvericins have been detected in wheat and maize infected with *Fusarium* spp., but the toxicological significance of this for mammals, if any, is not known (Logrieco *et al.*, 2002). An engineered enniatin-minus mutant of *F. avenaceum* had reduced ability to colonize potato tuber tissue, indicating a role for enniatins as virulence factors in plant pathogenesis (Herrmann *et al.*, 1996). The role of enniatins in infection of intact, healthy plants has apparently not been evaluated, nor have enniatins been evaluated in diseases caused by other species of *Fusarium*, such as *F. oxysporum* or *F. culmorum*.

Little is known about the biosynthesis of the hydroxy acids found in NRPs. However, a hydroxyisovalerate dehydrogenase that reduces 2-ketoisovalerate (2-KIV) to D-Hiv has been purified from the enniatin producing fungus *Fusarium sambucinum* (Lee et *al.*, 1992).

8.5. Ergopeptines

Ergopeptines are a family of NRPs containing lysergic acid and three amino acids that vary among members of the family. For example, ergotamine, an abundant ergopeptine produced by the ergot fungus *C. purpurea*, contains D-lysergic acid, L-Phe, L-Pro, and L-Ala (Figure. 7.1D). In ergovaline, an ergopeptine produced by several fungal grass endophytes, L-Phe is substituted by L-Val. Numerous other ergopeptines differ by one or two amino acid substitutions. Ergopeptines are the most complex and stable members of the ergot alkaloid family that also contains simple amides of lysergic acid (e.g., ergonovine and ergine) and numerous "clavines," some of which are precursors to lysergic acid (Floss, 1976; Gröger and Floss, 1998). In addition to the ergot fungus, *C. purpurea*, ergopeptines are produced by several other fungi in the Clavicipitaceae (Hypocreales, Ascomycetes), including mutualistic grass endophytes in the genus *Neotyphodium* (Lyons *et al.*, 1986; Christensen *et al.*, 1993; Bush *et al.*, 1997; Schardl and Phillips, 1997).

In the ergot diseases of rye and other grasses, *Claviceps* spp. infect inflorescences and replace the developing seed with a sclerotium, the overwintering structure of the fungus. The sclerotium often contains high concentrations of ergopeptines and other ergot alkaloids. Ergopeptines are antagonists or partial agonists of several important neurotransmitters, including dopamine, 5-hydroxytryptamine, and noradrenaline (Berde and Stürmer, 1978; Groger and Floss, 1998). They have been used or considered as treatments for Parkinson's disease, hypertension, depression, and migraine headaches (e.g., Markstein *et al.*, 1992; Ohno *et al.*, 1994).

Ergotism caused by the consumption of grain contaminated with ergot alkaloid-producing fungi has essentially been eliminated in humans. In the livestock and dairy industries, however, ergopeptines still have a significant impact, as they are present in endophytic fungi frequently colonizing important forage grasses, such as *Neotyphodium coenophialum* in tall fescue and *N. lolii* in perennial ryegrass. These endophytes provide numerous agronomic benefits to their host grasses including increased growth and reproduction and greater tolerance to biotic and abiotic stresses (reviewed in Schardl and Phillips, 1997; Bacon and White, 2000; Malinowski and Belesky, 2000). Roles for the ergopeptines in these beneficial traits and in the toxicoses associated with endophyte infection have been difficult to test because of the complexity of the grass–endophyte symbiosis and because multiple classes of bioactive alkaloids may be produced in different endophyte isolates. The recent production of a genetically defined, ergopeptine-deficient endophyte should facilitate delineation of the contribution of ergopeptines to these various phenotypes (Panaccione *et al.*, 2001).

The NRP synthetase that catalyzes the assembly of ergopeptines is unique among fungal synthetases in that it consists of two separately encoded polypeptides, Lps1 and Lps2 (Riederer *et al.*, 1996). Lps1 catalyzes the adenylation and thiolation of the three amino acids of ergopeptines. Lps2 activates D-lysergic acid. Walzel *et al.* (1997) demonstrated that the synthesis of the lysergyl peptide lactam begins with activation and thiolation of D-lysergic acid by Lps2. D-Lysergic acid is then attached via a peptide bond to thioesterified Ala bound to the first module of Lps1. The growing chain is then passed to the amino acid attached to the second module of Lps1 (L-Phe in the case of ergotamine; L-Val in the case of ergovaline), and then to the amino acid bound to the third module of Lps1 (L-Pro, with one exception). The mechanism of cyclization to release the lysergyl peptide lactam product has been discussed in a previous section.

Genes encoding Lps1 have been isolated from *C. purpurea* (*ps1* described by Tudzynski *et al.*, 1999) and *N. lolii* (*lpsA* described by Panaccione *et al.*, 2001). The Lps1-encoding gene was originally detected as a predicted NRP synthetase gene tightly linked to the gene encoding dimethylallyltryptophan (DMAT) synthase in *C. purpurea* (Tudzynski *et al.*, 1999), and its function has been demonstrated by gene knockout in *Neotyphodium* sp. Lp1 (Panaccione *et al.*, 2001). As expected, the gene sequence predicts a three-module NRP synthetase. An additional condensation domain near the carboxyl terminus (which is not as highly conserved as condensation domains that occur between amino acid-activating modules) may function in formation of the diketopiperazine between the final two amino acids in ergopeptines and in release of the product from the enzyme. The product is then further metabolized to ergopeptines. This unusual cyclization (diketopiperazine formation from the last two amino acids) may occur in place of a complete cyclization (by peptide bond formation between the amino and carboxyl termini) because the N-terminal moiety in ergopeptines (D-lysergic acid) lacks a free amino group. The slow release of product observed in the enzymatic studies of Walzel *et al.* (1997) was consistent with a non-catalytic mechanism. However, the extra carboxyl-terminal condensation domains in the sequences of the *C. purpurea* and *N. lolii* genes suggest enzyme-catalyzed lactam bond formation and peptide release.

At least several genes in addition to those encoding the NRP synthetases are required for the complete synthesis of ergopeptines. These include genes involved in lysergic acid biosynthesis as well as those for the hydroxylating activity required to oxidize the lactam product of the NRP synthetase prior to formation of the final cyclol ring (Quigley and Floss, 1980). Genes encoding DMAT synthase, which catalyzes the prenylation of tryptophan with dimethylallyldiphosphate (the first step in lysergic acid biosynthesis), have been cloned from *Claviceps fusiformis* (Tsai *et al.*, 1995), *C. purpurea* (Tudzynski *et al.*, 1999; Wang, 2001), and various endophytes including *N. lolii* and *N. coenophialum* (Wang, 2001). In *C. purpurea*, the genes encoding DMAT synthase and Lps1 are clustered along with several other predicted genes that appear to form an ergot alkaloid gene cluster (Tudzynski *et al.*, 1999). In *Neotyphodium* spp. there is no evidence yet for a gene cluster, and if there is a cluster, it is arranged differently from that in *C. purpurea* (Schardl and Panaccione, unpublished).

8.6. HC-Toxin

This cyclic tetrapeptide (Figure 7.1E) is an essential virulence determinant for the producing fungus, *C. carbonum*, during pathogenesis of maize (Walton, 1996, 1997). Root growth of maize of genotype *hm1/hm1* is inhibited by HC-toxin at ~1 μM whereas maize of genotype *HM1/HM1* and *HM1/hm1* are affected at ~100 μM. Only HC-toxin-producing (Tox^+) isolates of *C. carbonum* cause disease on maize of genotype *hm1/hm1*. It has been shown biochemically and genetically that *HM1* encodes an enzyme, HC-toxin reductase, that confers insensitivity to HC-toxin (and hence resistance to the fungus) by reducing and thereby detoxifying the 8-carbonyl group in the side chain of L-2-amino-9,10-epoxi-8-oxodecanoic acid (Aeo; Figure 7.1E) (Johal and Briggs, 1992; Meeley *et al.*, 1992). Maize contains a paralog of *HM1*, called *HM2*, that is expressed in, and confers resistance to, adult maize plants (Multani *et al.*, 1998). Other cereals (family Poaceae) have HC-toxin reductase activity and orthologs of *HM1* (Han *et al.*, 1997; Multani *et al.*, 1998).

The central enzyme in HC-toxin biosynthesis is Hts1, an NRP synthetase with four modules. An epimerase domain occurs between the first and second modules (Walton and Holden, 1988; Panaccione *et al.*, 1992; Scott-Craig *et al.*, 1992). Modules 1–4 activate L-Pro, L-Ala, D-Ala, and L-Aeo respectively. This assignment is based on the location within Hts1 of peptide sequences determined from proteolytic fragments that activate either L-Pro, or L-Ala and D-Ala (Walton, 1987; Walton and Holden, 1988), and from the consideration that Hts1 contains only a single epimerase domain, after module 1, that must epimerize Pro (Cheng and Walton, 2000). It is not known if Aeo itself or a precursor of Aeo is the substrate for Hts1.

Several other fungi make compounds related to HC-toxin. *Pochonia* (*Diheterospora*) *chlamydosporia* makes chlamydocin, *Calonectria morgani* (*Cylindrocladium scoparium*) and *Cal. pteridis* (*Cyl. macrosporum*) make Cyl-2, *Petriella guttulata* makes WF-3161, and *Helicoma ambiens* makes trapoxin (Itazaki *et al.*, 1990; Walton, 1990). All of these are cyclic tetrapeptides containing Aeo and an imino acid, either Pro or pipecolic acid. A cyclic tetrapeptide, JM-47, containing the Aeo derivative 2-amino-8-oxo-9-hydroxydecanoic acid (Aoh) was isolated from a marine species of *Fusarium* (Jiang *et al.*, 2002). At least two terrestial *Fusarium* species produce the cyclic tetrapeptide apicidin, which contains 2-amino-8-oxo-decanoic acid in place of Aeo (Darkin-Rattray *et al.*, 1996; Park, *et al.*, 1999). *Cal. pteridis* and *P. chlamydosporia* are pathogens of plants and nematodes respectively, but the involvement of their respective Aeo peptides in virulence has not been investigated.

The biosynthesis of Aeo must also require additional enzymes. Whether Aeo itself, rather than an Aeo precursor, is the substrate for the cognate NRP synthetase is unknown. Although the pathway of Aeo biosynthesis has not been elucidated, some genes with presumed roles in its biosynthesis have been identified in *C. carbonum*. These include *TOXC*, encoding a fatty acid synthase beta subunit, and *TOXF*, encoding a presumed branched-chain amino-acid aminotransferase (Ahn and Walton, 1997; Cheng *et al.*, 1999). Both *TOXC* and *TOXF* are required for, and dedicated to, HC-toxin biosynthesis, and are co-regulated with the other HC-toxin genes. Because they appear to encode biosynthetic enzymes and the other three amino acids of HC-toxin are simple proteinogenic amino acids, *TOXC* and *TOXF* probably have a role in Aeo biosynthesis, ToxC plausibly synthesizing the decanoic acid backbone of Aeo and ToxF subsequently providing the amino group. Aeo biosynthesis is predicted to require additional genes, for example, ones encoding an alpha subunit of fatty acid synthase and one or more enzymes for carbonylation and epoxidation.

The cognate NRP synthetases have not been identified for the other Aeo or Aeo-like-containing peptides, although some NRP synthetase fragments of unknown function have been isolated from some of the fungi (Nikolskaya *et al.* 1995). DNA probes based on *HTS1*, *TOXA*, or *TOXC* did not hybridize at low stringency to DNA in the other fungi that make Aeo-containing peptides (Nikolskaya, *et al.*, 1995). Therefore, despite the structural similarity of the NRPs, the synthetase genes either evolved independently or diverged long ago.

Genetic crosses between HC-toxin-producing (Tox2$^+$) and non-producing (Tox2$^-$) isolates of *C. carbonum* indicated that HC-toxin biosynthesis is under the control of a single Mendelian locus, called *TOX2* (Scheffer *et al.*, 1967). The explanation for this is that the genes for HC-toxin biosynthesis are clustered (Ahn *et al.*, 2002; discussed later). In addition to *HTS1*, *TOXC*, *TOXE*, *TOXF*, and *TOXG*, which have already been discussed, other known genes in the *TOX2* cluster are *TOXA* and *TOXD*. *TOXA* is tightly clustered with *HTS1* and is predicted to encode a membrane transporter of the major facilitator superfamily. Therefore, ToxA might be involved in the export of HC-toxin (Pitkin *et al.*, 1996). Disruption of *TOXD* does not affect HC-toxin biosynthesis (Cheng and Walton, unpublished), but a gene related to *TOXD* has a role in the biosynthesis of the polyketide lovastatin in *Aspergillus terreus* (Kennedy *et al.*, 1999).

All of the Aeo-containing peptides, and also apicidin, are inhibitors of histone deacetylases (HDACs) from a variety of organisms (Kijima *et al.*, 1993; Brosch *et al.*, 1995; Darkin-Rattray *et al.*, 1996). An affinity column based on trapoxin was used for the first isolation of an HDAC (Taunton *et al.*, 1996). HDAC inhibitors such as the Aeo-containing peptides are used in research on HDACs and are being studied as potential anti-cancer drugs (Jung, 2001; Marks *et al.*, 2001; Vigushin and Coombes, 2002). A group of chemically unrelated bicyclic depsipeptides from bacteria (e.g., FR901228 and FR228) are also HDAC inhibitors with potential in cancer treatment (Richon and O'Brien, 2002).

8.7. Penicillin and Cephalosporin

The biosynthesis and regulation of penicillins and related cephalosporins have been reviewed recently (Brakhage, 1998; Demain and Elander, 1999; Litzka *et al.*, 1999; Martín, 2000). Penicillin production is regulated by pH, carbon, and nitrogen. Despite extensive analysis of the regulation of penicillin production, there is no evidence to date for a pathway-specific transcription factor as is found for sterigmatocystin, HC-toxin, and some other fungal secondary metabolites.

In the penicillin and cephalosporin pathways, the initial NRP product, ACV (Figure 7.1J), undergoes extensive modification. Isopenicillin-N synthase (IPNS) catalyzes formation of the β-lactam ring from ACV to form isopenicillin N (IPN). IPN is further modified by transacylases, epimerases, and other enzymes to produce the penicillins and cephalosporins (Brakhage, 1998; Demain and Elander, 1999).

In the case of cephalosporins, the genes are in two clusters in *Acremonium chrysogenum*. The "early cluster" contains the genes for ACV synthetase (*pcb*AB or *acv*A) and IPN synthase (*pcb*C), whereas the "late cluster," which is unlinked to the early cluster, contains the genes for the later stages of cephalosporin biosynthesis, including the expandase-hydroxylase gene (*cef*EF) and deacetylcephalosporin C acetyltransferase (*cef*G) (Gutiérrez *et al.*, 1999).

The mechanism by which isopenicillin N is converted to penicillin N (i.e., epimerization of the alpha carbon of aminoadipic acid) has been recently elucidated in fungi (Ullán, *et al.*, 2002). In contrast to the bacterial pathway, epimerization in *A. chrysogenum* requires two proteins, the products of the *cef*D1 and *cef*D2 genes. These two genes are clustered with *pcb*AB and *pcb*C. The conversion of isopenicillin N to penicillin N occurs by completely different pathways in fungi and bacteria.

8.8. Peptaibols

The peptaibols are a large family of linear NRPs made by many fungi. Peptaibols include alamethicin, antiamoebin, efrapeptin, harzianin, hypelcin, heptaibin, leucinostatin, paracelsin, suzukacillin, trichogin, trichopolin, trichlorzianin, zervamycin, etc. They contain 7–20 amino acids and characteristically have an acylated N-terminus, a reduced C-terminus, and a high content of 2-amino-isobutyric acid (Aib). Unusual amino and imino acids found in peptaibols include isoVal, β-Ala, hydroxyPro, and pipecolic acid. The efrapeptins have a diazobicyclononene derivative (DBN) attached to the reduced C-terminal amino acid (Nagaraj *et al.*, 2001). Peptaibols have been isolated from species in the genera *Trichoderma* (*Hypocrea* or *Gliocladium*), *Paecilomyces, Emericellopsis, Stilbella, Acremonium*, and others.

Some *Trichoderma* species are parasites of other fungi and are therefore used as biocontrol agents to control plant pathogens. Peptaibols act synergistically with extracellular cell wall degrading enzymes to inhibit the growth of pathogenic fungi (Schirmbock *et al.*, 1994). However, the role of peptaibols in mycoparasitism and hence biological control has not yet been rigorously tested.

A significant advance in our understanding of peptaibol biosynthesis came from the recent isolation of a gene encoding a peptaibol synthetase from *T. virens*. The isolated gene, *tex1*, contains a 63-kb open reading frame (Wiest *et al.*, 2002). The predicted translation product has 20,925 amino acids and a molecular weight of 2.3 MDa, making it the largest known polypeptide from any organism. The largest peptaibol made by *T. virens* contains 18 amino acids, and Tex1 contains 18 NRP synthetase modules. Furthermore, at its amino terminus the Tex1 protein contains domains similar to ketoacyl synthases and acyl transferases, which could be responsible for the N-terminal acylation characteristic of peptaibols. Tex1 also contains a domain similar to dehydrogenases (reductases) at its carboxyl terminal, which could plausibly be responsible for reducing the C-terminal carboxyl group to the alcohol. Thus, Tex1 probably catalyzes the complete synthesis of peptaibols in *T. virens* from the constituent amino acids.

Disruption of *TEX1* results in loss of production of the entire family of more than 10 peptaibols made by *T. virens* (Wiest *et al.*, 2002). Synthesis of the chemically unrelated dipeptide, gliotoxin, is not affected. Thus, it appears that all of the peptaibols of *T. virens* are synthesized by Tex1. This is consistent with results obtained with other NRP-producing fungi and bacteria, and contributes to the general conclusion that within an organism all members of a family of NRPs are biosynthesized by a single synthetase, not by multiple synthetases.

An NRP synthetase gene, called *psy1*, had earlier been isolated from *T. virens* and implicated in the biosynthesis of the dipeptide siderophore known as dimerum acid (Wilhite *et al.*, 2001). The sequence of *psy1* is identical to the carboxyl end of *tex1*, and gene knockout studies indicate that there is only a single copy of this gene in *T. virens* (Wilhite, 2001; Wiest *et al.*, 2002). Wiest *et al.* (2002) found no change in dimerum acid production in their *tex1* mutant of *T. virens*. Based on its overall structure, *psy1/tex1* most likely encodes peptaibol and not dimerum acid synthetase. A *psy1* mutant was uncompromised in biocontrol competence against the plant pathogenic fungi *Pythium ultimum* and *Rhizoctonia solani* (Wilhite *et al.*, 2001). This suggests that peptaibols do not play a role in biocontrol of these two plant pathogens by *T. virens*.

9. Evolution of NRPs and NRP Synthetases

9.1. Clustering

The genes for the biosynthesis of a particular secondary metabolite are typically clustered into operons in bacteria. Genes for secondary metabolite pathways are often clustered in fungi as well, but as is typical of eukaryotes, each gene has its own promoter. A dramatic example of a secondary metabolite gene cluster in fungi is that involved in biosynthesis of the polyketide sterigmatocystin (an aflatoxin derivative) in *A. nidulans*, where more than 25 co-regulated genes are found in a 60-kb region (Brown *et al.*, 1996).

There are only a few examples in which fungal NRP synthetase genes are known to be clustered with the other genes involved in biosynthesis of the same NRP. In some cases the genes are known not to be clustered, in other cases the NRP is synthesized entirely by a single gene product, and in other cases the biosynthetic pathway is not yet sufficiently elucidated.

The three genes for the first three enzymes in the pathway of penicillin biosynthesis are clustered both in fungi and in bacteria (Smith *et al.*, 1990b; Gutiérrez *et al.*, 1999). The organization of the three genes is conserved among all fungi and even most bacteria. The sufficiency of these three genes for penicillin biosynthesis has been demonstrated by transfer of the cluster from *P. chrysogenum* to *N. crassa* and to *A. niger* (Smith *et al.*, 1990a). However, the genes for cephalosporin biosynthesis are only partially clustered in fungi.

The known genes for ferrichrome biosynthesis in *U. maydis* are at least partially clustered. *sid1* (encoding L-ornithine N^5-oxygenase) and *sid2* (encoding the ferrichrome NRP synthetase) are clustered and divergently transcribed. Both are regulated by the transcription factor Urbs1. Ferrichrome biosynthesis is predicted to require additional genes. The first genes encountered by sequencing upstream and downstream of *sid1* and *sid2* are housekeeping genes, so any other genes involved in ferrichrome biosynthesis might be physically linked but are not tightly clustered with the known ones (Yuan *et al.*, 2001).

Two genes known to be involved in ergopeptine biosynthesis (encoding DMAT synthase and lysergyl peptide synthetase) are clustered in *C. purpurea* (Tudzynski *et al.*, 1999; Panaccione *et al.*, 2001). Additional open reading frames capable of encoding oxidoreductases, which are activities reasonably associated with biosynthesis of lysergic acid, are also clustered with the ergopeptine biosynthetic genes (Tudzynski *et al.*, 1999).

As mentioned, seven genes involved in HC-toxin biosynthesis have been identified. All of the genes are present in every natural HC-toxin-producing isolate of *C. carbonum* in two or three functional copies, and, with one exception, all copies of all of the genes are physically linked within a ~600-kb region (Ahn and Walton, 1996; Ahn *et al.*, 2002). All of the known HC-toxin genes are absent from natural toxin non-producers. An exception to the clustering of the HC-toxin genes occurs in some isolates of *C. carbonum*, in which one of the two copies of *TOXE* (described earlier) is on a different chromosome from the other HC-toxin genes. The origin of this gene arrangement can be explained by a reciprocal chromosome translocation (Ahn *et al.*, 2002).

9.2. Evolution of Secondary Metabolite Pathways

Clustering of the genes for a particular secondary metabolite is common although not universal in lower eukaryotes. What, if any, adaptive value this might have is unclear (Keller and Hohn, 1997). In contrast to prokaryotic operons, clustering in eukaryotes does not present any obvious advantages in terms of co-regulation. On the contrary, the existence of *trans*-acting pathway-specific transcription factors, as in the case of HC-toxin, should allow adequate co-regulation even for genes that are dispersed throughout the genome. Walton (2000) proposed the "selfish cluster" hypothesis to account for clustering, which states that insofar as traits such as secondary metabolite production are under weak selection pressure, and therefore would tend to mutate to oblivion relatively quickly, they may depend on horizontal gene transfer for their long-term persistence. Clustering would promote survival of the genes and hence the trait during horizontal transfer, whereas it would not, within the limits of our knowledge of intragenomic processes, confer any advantage during normal, vertical descent. The "selfish cluster" model is related to the "selfish operon" theory of Lawrence and Roth (1996), which was proposed for the same reasons to explain the evolution and maintenance of bacterial genes in operons.

Although there is, to date, no strong experimental evidence for the horizontal transfer of genes in fungi, including those for secondary metabolites (Rosewich and Kistler, 2000), there are many intriguing observations for which horizontal transfer is a plausible explanation. One is the case of the Aeo (or Aeo-like)-containing cyclic peptides, which are made by at least five fungi that are not closely related (as already discussed). Possible explanations for this are that the fungi independently evolved the capacity to synthesize Aeo, that Aeo biosynthesis is an ancestral trait subsequently lost from intervening taxa, or that the Aeo biosynthetic genes have been transferred horizontally since taxonomic divergence.

Another intriguing case of possible horizontal gene transfer is that of the enniatins, which are made by several Ascomycetes as well as at least one Basidiomycete. These two major groups of fungi are estimated to have split ~350 million years ago (Berbee and Taylor, 2001). A third possible example of horizontal gene transfer is the bicyclic peptide amatoxins (amanitin) and phallotoxins (phalloidin). Certain species in several dispersed genera within the Agaricales (Basidiomycota), including *Amanita, Galerina, Conocybe*, and *Lepiota*, make amatoxins, or phallotoxins, or both, whereas other species in these genera and in the intervening taxa do not (Hallen *et al.*, 2002). Whether this occurred by common descent or by horizontal transfer cannot be addressed until at least one of the responsible biosynthetic genes has been isolated.

A fourth example of possible horizontal transfer, one that has been much remarked upon, is the case of the β-lactam antibiotics (penicillin and cephalosporin). These compounds are made by both prokaryotes and eukaryotes, and in both groups of organisms the pathways commence with ACV synthetase and IPN synthase. The sequences of the bacterial and fungal ACV synthetases are very similar, and in both bacteria and fungi the two genes are clustered. Furthermore, all fungal ACV synthetases contain a thioesterase domain at their carboxyl termini, a trait typical of bacterial NRP synthetases but not found in other fungal enzymes. (One possible exception is an uncharacterized protein in *Phanerochaete chrysosporium*, discussed later.) There are several possible explanations for the presence of these genes and this pathway in both eukaryotes and prokaryotes, including independent (convergent) evolution, origin in an ancestral lineage before the prokaryote/eukaryote split, or horizontal gene transfer from one to the other after the prokaryote/eukaryote split. The evidence for and against horizontal transfer in the evolution of β-lactam pathways is discussed by Gutiérrez *et al.* (1999).

Although horizontal gene transfer makes an attractive explanation for the similarities between the fungal and bacterial β-lactam biosynthetic genes, other evidence makes the case less clear. A phylogenetic analysis of IPN synthase genes concluded that horizontal gene transfer was not necessary to explain its presence in both eukaryotes and prokaryotes (Smith *et al.*, 1992). Furthermore, a recently discovered aspect of the β-lactam pathway suggests independent evolution of at least some parts of the cephalosporin pathway in prokaryotes and eukaryotes. The mechanism (and hence the enzymes and genes) of epimerization of isopenicillin N to penicillin N, the first committed step in cephalosporin biosynthesis, is completely different in prokaryotes and eukaryotes (Ullán *et al.*, 2002).

In conclusion, there does not seem to be a clearly preferred explanation for the extant distribution and organization of β-lactam biosynthetic genes in diverse organisms. Perhaps the true explanation is a mixture of the various hypotheses, that is, partly horizontal transfer (conceded to have been ~350 million years ago), and partly convergent evolution. A parallel might be drawn to gibberellin biosynthesis. The biosynthetic pathways for this complex group of secondary metabolites/plant hormones in fungi and in plants have certain striking similarities but also a completely different genome organization and, at certain steps, completely different mechanisms (Hedden *et al.*, 2002).

Most fungal NRP synthetase genes lack introns, a fact which has been used as an argument in favor of the prokaryote origin of eukaryotic NRP synthetase genes. However, this argument is vitiated by several additional considerations: (1) many "normal" fungal genes also lack introns, (2) some fungal NRP synthetase genes (e.g., *pesA* from *Metarhizium anosiplae*, and *ps1*/*lpsA* from *C. purpurea*/*N. lolii*) do have introns, and (3) other genes clustered with NRP synthetase genes, such as *TOXA* in *C. carbonum*, contain typical fungal introns (Bailey *et al.*, 1996; Pitkin *et al.*, 1996; Tudzynski *et al.*, 1999; Panaccione *et al.*, 2001). Furthermore, the issue is confounded by our ignorance of the biological significance of introns in the first place, and by the relatively small number of fungal NRP synthetase genes that have been characterized to date.

10. NRP Synthetases in the Genomics Age

The genomic sequences of several fungi are now in the public domain (e.g., *Saccharomyces cerevisiae, Neurospora crassa, Magnaporthe grisea, Cryptococcus neoformans*, and *Phanerochaete chrysosporium*), the genomes of some are in progress (e.g., *Fusarium graminearum*), and others are expected to be sequenced in the next few years (see Chapter 2 and *http://genome.jgi-psf.org/whiterot1/whiterot1.home.html*). The genomes of some fungi (e.g., *U. maydis, Botrytis cinerea*, and *Cochliobolus heterostrophus*) have been sequenced in the private sector and are sometimes available to academic researchers by individual negotiation.

BLAST searching with known NRP synthetase genes as queries indicates that some of these fungi contain novel NRP synthetase genes, ranging from three in *N. crassa* to more than 12 in *M. grisea*. Because many of the contigs in the latter fungus are still <30 kb, however, it is possible that some hits represent large NRP synthetases that are split among two contigs. Some of these genes probably encode members of the adenylating superfamily and not true NRP synthetases. One of the *N. crassa* NRP synthetases (NCU04531.1, on contig 3.231) has an epimerase domain, but none has an N-methylation domain. Two *M. grisea* proteins, MG07858.1 and MG03401.1, contain predicted epimerase domains and one, MG00022.1, has a predicted N-methylation domain. These are therefore probably true NRP synthetases whose NRP products remain to be identified.

The *ACE1* gene of *M. grisea*, mutations in which reduce virulence, is predicted to encode a polyketide synthase fused to a single NRP synthetase module (Böhnert *et al.*, 2002). Such hybrid enzymes have been found in prokaryotes but not previously in fungi (Du *et al.*, 2001). The putative polyketide/NRP product of Ace1 is not known.

The only NRP synthetase from Basidiomycetes that has been characterized to date is ferrichrome synthetase (Sid2) from *U. maydis* (Yuan *et al.*, 2001). The white rot fungus, *P. chrysosporium*, has three predicted proteins with AMP-binding domains. One of these (scaffold 28; 2736 amino acids; *www.jgi.doe.gov/programs/whiterot.htm*), has two adenylation domains associated with thiolation domains as well as single condensation and N-methylation domains. The scaffold 28 protein also contains a carboxy-terminal thioesterase domain, which is unusual for a fungal NRP synthetase. Although there is no reliable gene prediction algorithm, the *P. chrysosporium* NRP synthetase genes apparently contain multiple introns. Based on a TBLASTN search with Hts1 and ACV synthetase, *C. neoformans*, a basidiomycete pathogen of mammals, has at most one NRP synthetase gene.

As more genomes become available and are analyzed for their NRP synthetases genes, it will be interesting to see how the types and numbers of NRP synthetase genes (as well as secondary metabolite genes in general), correlate with ecological niche and taxonomic relatedness (e.g., Kroken *et al.*, 2002). There are indications that diversity in secondary metabolism does not correlate well with taxonomic diversity, suggesting that distinct forces are driving the evolution of secondary metabolite genes (Ward *et al.*, 2002).

Acknowledgments

The authors thank the U.S. Department of Agriculture, the U.S. Department of Energy, and the U.S. National Science Foundation for supporting the research discussed in this review. Figure 7.2 was prepared by Marlene Cameron (Michigan State University).

This chapter is published with the approval of the West Virginia Agriculture and Forestry Experiment Station as Scientific Article #2837.

References

Ahn, J.-H. and Walton, J.D. (1996). Chromosomal organization of *TOX2*, a complex locus required for host-selective toxin biosynthesis in *Cochliobolus carbonum. Plant Cell* **8**, 887–897.

Ahn, J.-H. and Walton, J.D. (1997). A fatty acid synthase gene required for production of the cyclic tetrapeptide HC-toxin, cyclo(D-prolyl-L-alanyl-D-alanyl-L-2-amino-9,10-epoxi-8-oxodecanoyl). *Mol. Plant–Microbe Interactions* **10**, 207–214.

Ahn, J.-H. and Walton, J.D. (1998). Regulation of cyclic peptide biosynthesis and pathogenicity in *Cochliobolus carbonum* by *TOXE*, a gene encoding a novel protein with a bZIP basic DNA binding motif and four ankyrin repeats. *Mol. and Gen. Genet.* **260**, 462–469.

Ahn, J.-H., Cheng, Y.-Q. and Walton, J.D. (2002). An extended physical map of the *TOX2* locus of *Cochliobolus carbonum* required for biosynthesis of HC-toxin. *Fungal Genet. and Bio.* **35**, 31–38.

Bacon, C.W. and White, J.F., Jr. (2000). *Microbial Endophytes*. Marcel Dekker, New York.

Bailey, A.M., Kershaw, M.J., Hunt, B.A., Paterson, I.C., Charnley, A.K., Reynolds, S.E., and Clarkson, J.M. (1996). Cloning and sequence analysis of an intron-containing domain from a peptide synthetase-encoding gene of the entomopathogenic fungus *Metarhizium anisopliae. Gene* **173**, 195–197.

Bains, P.S. and Tewari, J.P. (1987). Purification, chemical characterization and host-specificity of the toxin produced by *Alternaria brassicae. Physiol. Mol. Plant Pathol.* **30**, 259–271.

Baldwin, J.E., Shiau, C.Y., Byford, M.F., and Schofield, C.J. (1994). Substrate specificity of L-δ-(α-aminoadipyl)-L-cysteinyl-D-valine synthetase from *Cephalosporium acremonium*: Demonstration of the structure of several unnatural tripeptide products. *Biochem. J.* **301**, 367–372.

Berbee, M.L. and Taylor, J.W. (2001). Fungal molecular evolution: Gene trees and geologic time. In D.M. McLaughlin, E.G. McLaughlin, and P.A. Lemke (eds.) *The Mycota VII Part B* Springer-Verlag, Berlin, pp. 229–245.

Berde, B. and Stürmer, E. (1978). Introduction to the pharmacology of ergot alkaloids and related compounds as a basis of their therapeutic application. In B. Berde and H.O. Schild (eds.) *Ergot Alkaloids and Related Compounds*, Vol. 49, Springer-Verlag, New York, pp. 1–28.

Billich, A. and Zocher, R. (1990). Formation of N-methylated peptide bonds in peptides and peptidols. In H. Kleinkauf and H. von Döhren (eds.) *Biochemistry of Peptide Antibiotics* W. de Gruyter, Berlin, pp. 57–79.

Böhnert, H.U., Fudal, I., Dioh, W., Tharreau, D., Notteghem, J.L., and Lebrun, M.H. (2002). The hybrid polyketide synthase/nonribosomal peptide synthetase ACE1 controls recognition of avirulent *Magnaporthe grisea* by resistant rice. Abstract, 6th European Conference on Fungal Genetics, Pisa, Italy, pp. 6–9.

Borchert, S., Stachelhaus, T., and Marahiel, M.A. (1994). Induction of surfactin production in *Bacillus subtilis* by *gsp*, a gene located upstream of the gramicidin S operon in *Bacillus brevis. J. Bacteriol.* **176**, 2458–2462.

Brakhage, A.A. (1998). Molecular regulation of beta-lactam biosynthesis in filamentous fungi. *Microbiol. and Mol. Biol. Rev.* **62**, 547–585.

Brosch, G., Ransom, R., Lechner, T., Walton, J.D., and Loidl, P. (1995). Inhibition of maize histone deacetylase by HC-toxin, the host-selective toxin of *Cochliobolus carbonum. Plant Cell* **7**, 1941–1950.

Brown, D.W., Yu, J.H., Kelkar, H.S., Fernandes, M., Nesbitt, T.C., Keller, N.P., Adams, T.H., and Leonard, T.J. (1996). Twenty-five coregulated transcripts define a sterigmatocystin gene cluster in *Aspergillus nidulans. Proc. Natl. Acad. Sci. USA* **93**, 1418–1422.

Buchwaldt, L. and Green, H. (1992). Phytotoxicity of destruxin B and its possible role in the pathogenesis of *Alternaria brassicae. Plant Pathol.* **41**, 55–63.

Burmester, J., Haese, A., and Zocher, R. (1995). Highly conserved N-methyltransferases as an integral part of peptide synthetases. *Biochem. Mol. Bio. Int.* **37**, 201–207.

Bush, L.P., Wilkinson, H.H., and Schardl, C.L. (1997). Bioprotective alkaloids of grass–fungal endophyte symbioses. *Plant Physiol.* **114**, 1–7.

Bussink, H.J., Clark, A., and Oliver, R. (2001). The *Cladosporium fulvum* Bap1 gene: Evidence for a novel class of Yap-related transcription factors with ankyrin repeats in phytopathogenic fungi. *Euro. J. Plant Pathol.* **107**, 655–659.

Challis, G.L., Ravel, J., and Townsend, C.A. (2000). Predictive, structure-based model of amino acid recognition by nonribosomal peptide synthetase adenylation domains. *Chem. Biol.* **7**, 211–224.

Cheng, Y.-Q. and Walton, J.D. (2000). A eukaryotic alanine racemase involved in cyclic peptide biosynthesis. *J. Biol. Chem.* **275**, 4906–5004.

Cheng, Y.-Q., Ahn, J.-H., and Walton, J.D. (1999). A putative branched-chain-amino-acid transaminase gene required for HC-toxin biosynthesis and pathogenicity in *Cochliobolus carbonum. Microbiology* **145**, 3539–3546.

Christensen, M.J., Leuchtmann, A., Rowan, D.D., and Tapper, B.A. (1993). Taxonomy of *Acremonium* endophytes of tall fescue (*Festuca arundinacea*), meadow fescue (*F. pratensis*), and perennial rye-grass (*Lolium perenne*). *Mycol. Res.* **97**, 1083–1092.

Clark, D.P., Carroll, J., Naylor, S., and Crews, P. (1998). An antifungal cyclodepsipeptide, cyclolithistide A, from the sponge *Theonella swinhoei. J. Org. Chem.* **63**, 8757–8764.

Clark, W.D., Corbett, T., Valeriote, F., and Crews, P. (1997). Cyclocinamide A, an unusual cytotoxic halogenated hexapeptide from the marine sponge *Psammocinia. J. Am. Chem. Soc.* **119**, 9285–9286.

Conti, E., Franks, N.P., and Brick, P. (1996). Crystal structure of firefly luciferase throws light on a superfamily of adenylate-forming enzymes. *Structure* **4**, 287–298.

Conti, E., Stachelhaus, T., Marahiel, M.A., and Brick, P. (1997). Structural basis for the activation of phenylalanine in the non-ribosomal biosynthesis of gramicidin S. *EMBO J.* **16**, 4174–4183.

Darkin-Rattray, S.J., Gurnett, A.M., Myers, R.W., Dulski, P.M., Crumley, T.M., Allocco, J.J., Cannova, C., Meinke, P.T. *et al.* (1996). Apicidin: A novel antiprotozoal agent that inhibits parasite histone deacetylase. *Proc. Natl. Acad. Sci. USA* **93**, 13143–13147.

de Crécy-Lagard, V., Marlière, P., and Saurin, W. (1995). Multi-enzymatic non-ribosomal peptide biosynthesis: Identification of the functional domains catalyzing peptide elongation and epimerization. *Life Sci.* **318**, 927–936.

Demain, A.L. and Elander, R.P. (1999). The beta-lactam antibiotics: Past, present, and future. *Antonie van Leeuwenhoek* **75**, 5–19.

Díez, B., Gutiérrez, S., Barredo, J.L., van Solingen, P., van der Voort, L.H., and Martín, J.F. (1990). The cluster of penicillin biosynthetic genes. Identification and characterization of the *pcb*AB gene encoding the α-aminoadipyl-cysteinyl-valine synthetase and linkage to the *pcb*C and *pen*DE genes. *J. Biol. Chem.* **265**, 16358–16365.

Dittmann, E., Neilan, B.A., and Borner, T. (2001). Molecular biology of peptide and polyketide biosynthesis in cyanobacteria. *Appl. Microbiol. Biotechnol.* **57**, 467–473.

Du, L., Sánchez, C., and Shen, B. (2001). Hybrid peptide-polyketide natural products: Biosynthesis and prospects toward engineering novel molecules. *Metab. Eng.* **3**, 78–95.

Fichtlscherer, F., Wellein, C., Mittag, M., and Schweizer, E. (2000). A novel function of yeast fatty acid synthase. Subunit alpha is capable of self-pantetheinylation. *Eur. J. Biochem.* **267**, 2666–2671.

Floss, H.G. (1976). Biosynthesis of ergot alkaloids and related compounds. *Tetrahedron* **32**, 873–912.

Gehring, A.M., Mori, I., Perry, R.D., and Walsh, C.T. (1998). The nonribosomal peptide synthetase HMWP2 forms a thiazoline ring during biogenesis of yersiniabactin, an iron-chelating virulence factor of *Yersinia pestis. Biochemistry* **37**, 11637–11650.

Gevers, W., Kleinkauf, H., and Lipmann, F. (1969). Peptidyl transfers in gramicidin S biosynthesis from enzyme-bound thioester intermediates. *Proc. Natl. Acad. Sci. USA* **65**, 1335–1342.

Glinski, M., Urbanke, C., Hornbogen, T., and Zocher, R. (2002). Enniatin synthetase is a monomer with extended structure: Evidence for an intramolecular reaction mechanism. *Arch. Microbiol.* **178**, 267–273.

Gröger, D. and Floss, H.G. (1998). Biochemistry of ergot alkaloids—achievements and challenges. In G.A. Cordell (ed.) *The Alkaloids*, Vol. 50. Academic Press, New York, pp. 171–218.

Groth, G. (2002). Structure of spinach chloroplast F1-ATPase complexed with the phytopathogenic inhibitor tentoxin. *Proc. Natl. Acad. Sci. USA* **99**, 3464–3468.

Gutiérrez, S., Fierro, F., Casquiero, J., and Martín, J.F. (1999). Gene organization and plasticity of the β-lactam genes in different filamentous fungi. *Antonie van Leeuwenhoek* **75**, 81–94.

Hacker, C., Glinski, M., Hornbogen, T., Doller, A., and Zocher, R. (2000). Mutational analysis of the N-methyltransferase domain of the multifunctional enzyme enniatin synthetase. *J. Biol. Chem.* **275**, 30826–30832.

Haese, A., Schubert, M., Herrmann, M., and Zocher, R. (1993). Molecular characterization of the enniatin synthetase gene encoding a multifunctional enzyme catalysing N-methyldepsipeptide formation in *Fusarium scirpi. Mol. Microbiol.* **7**, 905–914.

Haese, A., Pieper, R., von Ostrowski, T., and Zocher, R. (1994). Bacterial expression of catalytically active fragments of the multifunctional enzyme enniatin synthetase. *J. Mol. Biol.* **243**, 116–122.

Hallen, H.E., Adams, G.C., and Eicker, A. (2002). Amatoxins and phallotoxins in indigenous and introduced South African *Amanita* species. *South African J. Botany* **68**, 322–326.

Han, F., Kleinhofs, A., Kilian, A., and Ullrich, S.E. (1997). Cloning and mapping of a putative barley NADPH-dependent HC-toxin reductase. *Mol. Plant–Microbe Interact* **10**, 234–239.

Han, Y., Liu, X., Benny, U., Kistler, H.C., and van Etten, H.D. (2001). Genes determining pathogenicity to pea are clustered on a supernumerary chromosome in the fungal plant pathogen *Nectria haematococca. Plant J.* **25**, 305–314.

Hatta, R., Ito, K., Hosaki, Y., Tanaka, T., Tanaka, A., Yamamoto, M., Akimitsu, K., and Tsuge, T. (2002). A conditionally dispensable chromosome controls host-specific pathogenicity in the fungal plant pathogen *Alternaria alternata. Genetics* **161**, 59–70.

Hedden, P., Phillips, A.L., Rojas, M.C., Carrera, E., and Tudzynski, B. (2002). Gibberellin biosynthesis in plants and fungi: A case of convergent evolution? *J. Plant Growth Regul.* **20**, 319–331.

Herrmann, M., Zocher, R., and Haese, A. (1996). Enniatin production by *Fusarium* strains and its effect on potato tuber tissue. *Appl. Environ. Microbiol.* **62**, 393–398.

Hodge, K.T., Krasnoff, S.B., and Humber, R.A. (1996). *Tolypocladium inflatam* is the anamorph of *Cordyceps subsessilis. Mycologia* **88**, 715–719.

Hoffmann, K., Schneider-Scherzer, E., Kleinkauf, H., and Zocher, R. (1994). Purification and characterization of eucaryotic alanine racemase acting as key enzyme in cyclosporin biosynthesis. *J. Biol. Chem.* **269**, 12710–12714.

Hohn, T.M., Krishna, R., and Proctor, R.H. (1999). Characterization of a transcriptional activator controlling trichothecene toxin biosynthesis. *Fungal Genet. Biol.* **26**, 224–235.

Hoppert, M., Gentzsch, C., and Schorgendorfer, K. (2001). Structure and localization of cyclosporin synthetase, the key enzyme of cyclosporin biosynthesis in *Tolypocladium inflatum. Arch. Microbiol.* **176**, 285–293.

Howard, R.J. and Valent, B. (1996). Breaking and entering: Host penetration by the fungal rice blast pathogen *Magnaporthe grisea. Annu. Rev. Microbiol.* **50**, 491–512.

Itazaki, H., Nagashima, K., Sugita, K., Yoshida, H., Kawamura, Y., Yasuda, Y., Matsumoto, K., Ishii, K. *et al.* (1990). Isolation and structural elucidation of new cyclotetrapeptides, trapoxins A and B, having detransformation activities as antitumor agents. *J. Antibiot.* **43**, 1524–1532.

Jenett-Siems, K., Kaloga, M., and Eich, E. (1994). Ergobalansine/ergobalansinine, a proline-free peptide-type alkaloid of the fungal genus *Balansia*, is a constituent of *Ipomoea piurensis. J. Nat. Prod.* **57**, 1304–1306.

Jiang, Z., Barret, M.O., Boyd, K.G., Adams, D.R., Boyd, A.S., and Burgess, J.G. (2002). JM47, a cyclic tetrapeptide HC-toxin analogue from a marine *Fusarium. Phytochemistry* **60**, 33–38.

Johal, G. and Briggs, S.P. (1992). Reductase activity encoded by the *HM1* disease resistance gene in maize. *Science* **258**, 985–987.

Johnson, L.J., Johnson, R.D., Akamatsu, H., Salamiah, A., Otani, H., Kohmoto, K., and Kodama, M. (2001). Spontaneous loss of a conditionally dispensable chromosome from the *Alternaria alternata* apple pathotype leads to loss of toxin production and pathogenicity. *Curr. Genetics* **40**, 65–72.

Johnson, R.D., Johnson, L., Itoh, Y., Kodama, M., Otani, H., and Kohmoto, K. (2000). Cloning and characterization of a cyclic peptide synthetase gene from *Alternaria alternata* apple pathotype whose product is involved in AM-toxin synthesis and pathogenicity. *Mol. Plant–Microbe Interact.* **13**, 742–753.

Jung, M. (2001). Inhibitors of histone deacetylase as new anticancer agents. *Curr. Medicinal Chem.* **8**, 1505–1511.

Keating, T.A., Ehmann, D.E., Kohli, R.M., Marshall, C.G., Trauger, J.W., and Walsh, C.T. (2001). Chain termination steps in nonribosomal peptide synthetase assembly lines: Directed acyl-S-enzyme breakdown in antibiotic and siderophore biosynthesis. *Chembiochem.* **2**, 99–107.

Keller, N.P. and Hohn, T.M. (1997). Metabolic pathway gene clusters in filamentous fungi. *Fungal Genet. Biol.* **21**, 17–29.

Kennedy, J., Auclair, K., Kendrew, S.G., Park, C., Vederas, J.C., and Hutchinson, C.R. (1999). Modulation of polyketide synthase activity by accessory proteins during lovastatin biosynthesis. *Science* **284**, 1368–1372.

Kershaw, M.J., Moorhouse, E.R., Bateman, R., Reynolds, S.E., and Charnley, A.K. (1999). The role of destruxins in the pathogenicity of *Metarhizium anisopliae* for three species of insect. *J. Invertebr. Pathol.* **74**, 213–223.

Keszenman-Pereyra, D., Lawrence, S., Twfieg, M., Price, J., and Turner, G. (2003). The *npgA/cfwA* gene encodes a putative 4′ phosphopantetheinyl transferase which is essential for penicillin biosynthesis in *Aspergillus nidulans. Curr. Genet.* **43**, 186–190.

Kijima, M., Yoshida, M., Suita, K., Horinouchi, S., and Beppu, T. (1993). Trapoxin, an antitumor cyclic tetrapeptide, is an irreversible inhibitor of mammalian histone deacetylase. *J. Biol. Chem.* **268**, 22429–22435.

Kohmoto, K. and Otani, H. (1991). Host recognition by toxigenic plant pathogens. *Experientia* **47**, 755–764.

Krause, M., Marahiel, M.A., von Döhren, H., and Kleinkauf, H. (1985). Molecular cloning of an ornithine-activating fragment of the gramicidin S synthetase 2 gene from *Bacillus brevis* and its expression in *Escherichia coli. J. Bacteriol.* **162**, 1120–1125.

Krause, M., Lindemann, A., Glinski, M., Hornbogen, T., Bonse, G., Jeschke, P., Thielking, G., Gau, W. *et al.* (2001). Directed biosynthesis of new enniatins. *J. Antibiot.* **54**, 797–804.

Kroken, S., Glass, N.L., Taylor, J.W., Turgeon, B.G., and Yoder, O. (2002). Evolution of secondary metabolite pathways: Non-ribosomal peptide synthetases and polyketide synthases. Abstract 227 of the Seventh International Mycological Congress, Oslo, Norway, 11–17, August.

Krupinski, V.M., Robbers, J.E., and Floss, H.G. (1976). Physiological study of ergot [*Claviceps* fungi]: Induction of alkaloid synthesis by tryptophan at the enzymatic level. *J. Bacteriol.* **125**, 158–165.

Lambalot, R.H., Gehring, A.M., Flugel, R.S., Zuber, P., LaCelle, M., Marahiel, M.A., Reid, R., Khosla, C. *et al.* (1996). A new enzyme superfamily—the phosphopantetheinyl transferases. *Chem. Biol.* **3**, 923–936.

Lawen, A. and Traber, R. (1993). Substrate specificities of cyclosporin synthetase and peptolide SDZ 214-103 synthetase. Comparison of the substrate specificities of the related multifunctional polypeptides. *J. Biol. Chem.* **268**, 20452–20465.

Lawrence, J.G. and Roth, J.R. (1996). Selfish operons: Horizontal transfer may drive the evolution of gene clusters. *Genetics* **143**, 1843–1860.

Lee, C., Gorisch, H., Kleinkauf, H., and Zocher, R. (1992). A highly specific D-hydroxyisovalerate dehydrogenase from the enniatin producer *Fusarium sambucinum. J. Biol. Chem.* **267**, 11741–11744.

Lee, S.G. and Lipmann, F. (1975). Tyrocidine synthetase system. *Meth. Enzymol.* **43**, 585–602.

Litzka, O., Bergh, K.T., van den Brulle, J., Steidl, S., and Brakhage, A.A. (1999). Transcriptional control of expression of fungal beta-lactam biosynthesis genes. *Antonie van Leeuwenhoek* **75**, 95–105.

Logrieco, A., Rizzo, A., Ferracane, R., and Ritieni, A. (2002). Occurrence of beauvericin and enniatins in wheat affected by *Fusarium avenaceum* head blight. *Appl. Environ. Microbiol.* **68**, 82–85.

Lyons, P.C., Plattner, R.D., and Bacon, C.W. (1986). Occurrence of peptide and clavine ergot alkaloids in tall fescue grass. *Science* **232**, 487–489.

MacCabe, A.P., Riach, M.B.R., Unkles, S.E., and Kinghorn, J.R. (1990). The *Aspergillus nidulans npeA* locus consists of three contiguous genes required for penicillin biosynthesis. *EMBO J.* **9**, 279–287.

MacCabe, A.P., van Liempt, H., Palissa, H., Unkles, S.E., Riach, M.B.R., Pfeifer, E., von Döhren, H., and Kinghorn, J.R. (1991). δ-(L-α-Aminoadipyl)-L-cysteinyl-D-valine synthetase from *Aspergillus nidulans. J. Biol. Chem.* **266**, 12646–12654.

Mach, B. and Tatum, E.L. (1964). Environmental control of amino acid substitutions in the biosynthesis of the antibiotic polypeptide tyrocidine. *Proc. Natl. Acad. Sci. USA* **52**, 876–884.

Macko, V., Stimmel, M.B., Wolpert, T.J., Dunkle, L.D., Acklin, W., Banteli, R., Jaun, B., and Arigoni, D. (1992). Structure of the host-specific toxins produced by the fungal pathogen *Periconia circinata. Proc. Natl. Acad. Sci. USA* **89**, 9574–9578.

Malinowski, D.P. and Belesky, D.P. (2000). Adaptations of endophyte-infected cool-season grasses to environmental stresses: Mechanisms of drought and mineral stress tolerance. *Crop Sci.* **40**, 923–940.

Marahiel, M.A., Stachelhaus, T., and Mootz, H. D. (1997). Modular peptide synthetases involved in nonribosomal peptide synthesis. *Chem. Rev.* **97**, 2651–2673.

Marks, P.A., Richon, V.M., Breslow, R., and Rifkind, R. A. (2001). Histone deacetylase inhibitors as new cancer drugs. *Curr. Opin. Oncol.*, **13**, 477–483.

Markstein, R., Seiler, M.P., Jaton, A., and Briner, U. (1992). Structure activity relationship and therapeutic uses of dopaminergic ergots. *Neurochem. Int.* **20**, S211–S214.

Martín, J.F. (2000). Molecular control of expression of penicillin biosynthesis genes in fungi: Regulatory proteins interact with a bidirectional promoter region. *J. Bacteriol.* **182**, 2355–2362.

Meeley, R.B. and Walton, J.D. (1991). Enzymatic detoxification of HC-toxin, the host-selective cyclic peptide from *Cochliobolus carbonum. Plant Physiol.*, **97**, 1080–1086.

Meeley, R.B., Johal, G., Briggs, S.P., and Walton, J.D. (1992). A biochemical phenotype for a disease resistance gene of maize. *Plant Cell* **4**, 71–77.

Mei, B., Budde, A.D., and Leong, S.A. (1993). *sid1*, a gene initiating siderophore biosynthesis in *Ustilago maydis*: Molecular characterization, regulation by iron, and role in phytopathogenicity. *Proc. Natl. Acad. Sci. USA* **90**, 903–907.

Mootz, H.D., and Marahiel, M.A. (1999). Design and application of multimodular peptide synthetases. *Curr. Opin. Biotechnol.* **10**, 341–348.

Mootz, H.D., Schörgendorfer, K., and Marahiel, M.A. (2002). Functional characterization of 4′-phosphopantetheinyl transferase genes of bacterial and fungal origin by complementation of *Saccharomyces cerevisiae lys5*. *FEMS Microbiol. Lett.*, **213**, 51–57.

Multani, D.S., Meeley, R.B., Paterson, A.H., Gray, J., Briggs, S.P., and Johal, G.S. (1998). Plant-pathogen microevolution: Molecular basis for the origin of a fungal disease in maize. *Proc. Natl. Acad. Sci. USA*. **95**, 1686–1691.

Nagaraj, G., Uma, M.V., Shivayogi, M.S., and Balaram, H. (2001). Antimalarial activities of peptide antibiotics isolated from fungi. *Antimicrobial Agents Chemother.* **45**, 145–149.

Neilan, B.A., Dittmann, E., Rouhiainen, L., Bass, R.A., Schaub, V., Sivonen, K., and Borner, T. (1999). Nonribosomal peptide synthesis and toxigenicity of cyanobacteria. *J. Bacteriol.* **181**, 4089–4097.

Nikolskaya, A.N., Panaccione, D.G., and Walton, J.D. (1995). Identification of peptide synthetase-encoding genes from filamentous fungi producing host-selective phytotoxins or analogs. *Gene* **165**, 207–211.

Offenzeller, M., Santer, G., Totschnig, K., Su, Z., Moser, H., Traber, R., and Schneider-Scherzer, E. (1996). Biosynthesis of the unusual amino acid (4R)-4-[(E)-2-butenyl]-4-methyl-L-threonine of cyclosporin A: Enzymatic analysis of the reaction sequence including identification of the methylation precursor in a polyketide pathway. *Biochemistry* **35**, 8401–8412.

Ohno, S., Koumori, M., Adachi, Y., Mizukoshi, K., Nagasaka, M., and Ichihara, K. (1994). Synthesis and structure-activity relationships of new (5R,8S,10R)-ergoline derivatives with antihypertensive or dopaminergic activity. *Chem. Pharm. Bull.* **42**, 2042–2048.

Panaccione, D. G. (1996). Multiple families of peptide synthetase genes from ergopeptine-producing fungi. *Mycol. Res.* **100**, 429–436.

Panaccione, D.G., Scott-Craig, J.S., Pocard, J.A., and Walton, J.D. (1992). A cyclic peptide synthetase gene required for pathogenicity of the fungus *Cochliobolus carbonum* on maize. *Proc. Natl. Acad. Sci. USA* **89**, 6590–6594.

Panaccione, D.G., Johnson, R.D., Wang, J., Young, C.A., Damrongkool, P., Scott, B., and Schardl, C. L. (2001). Elimination of ergovaline from a grass—*Neotyphodium* endophyte symbiosis by genetic modification of the endophyte. *Proc. Natl. Acad. Sci. USA* **98**, 12820–12825.

Park, J.-S., Lee, K.-R., Kim, J.-C., Lim, S.-H., Seo, J.-A., and Lee, Y.-W. (1999). A hemorrhagic factor (apicidin) produced by toxic *Fusarium* isolates from soybean seeds. *Appl. Environ. Microbiol.* **65**, 126–130.

Pedley, K.F. and Walton, J.D. (2001). Regulation of cyclic peptide biosynthesis in a plant pathogenic fungus by a novel transcription factor. *Proc. Natl. Acad. Sci. USA* **98**, 14174–14179.

Pedras, M.S. C., Zaharia, I.L., Gai, Y., Zhou, Y., and Ward, D.E. (2001). *In planta* sequential hydroxylation and glycosylation of a fungal phytotoxin: Avoiding cell death and overcoming the fungal invader. *Proc. Natl. Acad. Sci. USA* **98**, 747–752.

Pedras, M.S.C., Zaharia, I.L., and Ward, D.E. (2002). The destruxins: Synthesis, biosynthesis, biotransformation, and biological activity. *Phytochemistry* **59**, 579–596.

Pieper, R., Haese, A., Schorder, W., and Zocher, R. (1995). Arrangement of catalytic sites in the multifunctional enzyme enniatin synthetase. *Eur. J. Biochem.* **230**, 119–126.

Pitkin, J.W., Panaccione, D.G., and Walton, J.D. (1996). A putative cyclic peptide efflux pump encoded by the *TOXA* gene of the plant pathogenic fungus *Cochliobolus carbonum*. *Microbiology* **142**, 1557–1565.

Pitkin, J.W., Nikolskaya, A., Ahn, J.-H., and Walton, J.D. (2000). Reduced virulence caused by meiotic instability of the *TOX2* chromosome of the maize pathogen *Cochliobolus carbonum*. *Mol. Plant–Microbe Interact.* **13**, 80–87.

Quigley, F.R. and Floss, H.G. (1980). Mechanism of amino acid α-hydroxylation and formation of the lysergyl moiety in ergotamine biosynthesis. *J. Org. Chem.* **46**, 464–466.

Ramm, K., Ramm, M., Liebermann, B., and Reuter, G. (1994). Studies of the biosynthesis of tentoxin by *Alternaria alternata*. *Microbiology* **140**, 3257–3266.

Rasmussen, J.B. and Scheffer, R.P. (1988). Isolation and biological activities of four selective toxins from *Helminthosporium carbonum*. *Plant Physiol.* **86**, 187–191.

Richon, V.M. and O'Brien, J.P. (2002). Histone deacetylase inhibitors: A new class of potential therapeutic agents for cancer treatment. *Clin. Cancer Res.* **8**, 662–664.

Riederer, B., Han, M., and Keller, U. (1996). D-Lysergyl peptide synthetase from the ergot fungus *Claviceps purpurea*. *J. Biol. Chem.* **271**, 27524–27530.

Rosewich, U.L. and Kistler, H.C. (2000). Role of horizontal gene transfer in the evolution of fungi. *Annu. Rev. Phytopathol.* **38**, 325–363.

Schardl, C.L. and Phillips, T.D. (1997). Protective grass endophytes: Where are they from and where are they going? *Plant Dis.* **81**, 430–437.

Scheffer, R.P., Nelson, R.R., and Ullstrup, A.J. (1967). Inheritance of toxin production and pathogenicity in *Cochliobolus carbonum* and *Cochliobolus victoriae. Phytopathology* **57**, 1288–1289.

Schirmbock, M., Lorito, M., Wang, Y. L., Hayes, C. K., Arisan-Atac, I., Scala, F., Harman, G.E., and Kubicek, C.P. (1994). Parallel formation and synergism of hydrolytic enzymes and peptaibol antibiotics, molecular mechanisms involved in the antagonistic action of *Trichoderma harzianum* against phytopathogenic fungi. *Appl. Environ. Microbiol.* **60**, 4364–4370.

Schlumbohm, W., Stein, T., Ullrich, C., Vater, J., Krause, M., Marahiel, M.A., Kruft, V., and Wittmann-Liebold, B. (1991). An active serine is involved in covalent substrate amino-acid binding at each reaction center of gramicidin S synthetase. *J. Biol. Chem.* **266**, 23135–23141.

Schmitt, E.K. and Kuck, U. (2000). The fungal CPCR1 protein, which binds specifically to beta-lactam biosynthesis genes, is related to human regulatory factor X transcription factors. *J. Biol. Chem.* **275**, 9348–9357.

Scott-Craig, J. S., Panaccione, D.G., Pocard, J.A., and Walton, J.D. (1992). The multifunctional cyclic peptide synthetase catalyzing HC-toxin production in the filamentous fungus *Cochliobolus carbonum* is encoded by a 15.7-kb open reading frame. *J. Biol. Chem.* **67**, 26044–26049.

Shen, B., Du, L., Sanchez, C., Edwards, D.J., Chen, M., and Murrell, J.M. (2002). Cloning and characterization of the bleomycin biosynthetic gene cluster from *Streptomyces verticillus* ATCC15003. *J. Nat. Prod.* **65**, 422–431.

Shivanna, K.R. and Sawhney, V.K. (1993). Pollen selection for *Alternaria* resistance in oilseed brassicas—responses of pollen grains and leaves to a toxin of *A. brassicae. Theor. App. Genet.* **86**, 339–344.

Smith, D.J, Burnham, M.K.R., Edwards, J., Earl, A.J., and Turner, G. (1990a). Cloning and heterologous expression of the penicillin biosynthetic gene cluster from *Penicillium chrysogenum. Biotechnology* **8**, 39–41.

Smith, D.J., Earl, A.J. and Turner, G. (1990b). The multifunctional peptide synthetase performing the first step of penicillin biosynthesis in *Penicillium chrysogenum* is a 421 073 dalton protein similar to *Bacillus brevis* peptide antibiotic synthetases. *EMBO J.* **9**, 2743–2750.

Smith, M.W., Feng, D.-F., and Doolittle, R.F. (1992). Evolution by acquisition: The case for horizontal gene transfers. *Trends Biochem. Sci.* **17**, 489–493.

Stachelhaus, T. and Walsh, C. T. (2000). Mutational analysis of the epimerization domain in the initiation module PheATE of gramicidin S synthetase. *Biochemistry* **39**, 5775–5787.

Stachelhaus, T., Schneider, A., and Marahiel, M. A. (1995). Rational design of peptide antibiotics by targeted replacement of bacterial and fungal domains. *Science* **269**, 69–72.

Stachelhaus, T., Mootz, H.D., Bergendahl, V., and Marahiel, M.A. (1998). Peptide bond formation in nonribosomal peptide biosynthesis catalytic role of the condensation domain. *J. Biol. Chem.* **273**, 22773–22781.

Stachelhaus, T., Mootz, H.D., and Marahiel, M.A. (1999). The specificity-conferring code of adenylation domains in nonribosomal peptide synthetases. *Chem. Biol.* **6**, 493–505.

Stein, T., Vater, J., Kruft, V., Wittmann-Liebold, B., Franke, P., Panico, M., McDowell, R., and Morris, H.R. (1994). Detection of 4′-phosphopantetheine at the thioester binding site for L-valine of gramicidin S synthetase 2. *FEBS Lett.* **340**, 39–44.

Stein, T., Vater, J., Kruft, V., Otto, A., Wittmann-Liebold, B., Franke, P., Panico, M., McDowell, R., *et al.* (1996). The multiple carrier model of nonribosomal peptide biosynthesis at modular multienzymatic templates. *J. Biol. Chem.* **271**, 15428–15435.

Swofford, D. L. (2002). *PAUP*: Phylogenetic Analysis Using Parsimony (and Other Methods)* version 4.0b10. Sinauer Associates, Sunderland, MA.

Taunton, J., Hassig, C.A., and Schreiber, S.L. (1996). A mammalian histone deacetylase related to the yeast transcriptional regulator Rpd3p. *Science* **272**, 408–411.

Traber, R., Hofmann, H., and Kobel, H. (1989). Cyclosporins—new analogues by precursor directed biosynthesis. *J. Antibio.* **42**, 591–597.

Tsai, H.-F., Wang, H., Gebler, J.C., Poulter, C.D., and Schardl, C.L. (1995). The *Claviceps purpurea* gene encoding dimethylallyltryptophan synthase, the committed step for ergot alkaloid biosynthesis. *Biochem. Biophys. Res. Commun.* **216**, 119–125.

Tudzynski, P., Holter, K., Correia, T., Arntz, C., Grammel, N., and Keller, U. (1999). Evidence for an ergot alkaloid gene cluster in *Claviceps purpurea. Mol. Gen. Genet.* **261**, 133–141.

Turgay, K. and Marahiel, M.A. (1994). A general approach for identifying and cloning peptide synthetase genes. *Peptide Res.* **7**, 238–241.

Turgay, K., Krause, M., and Marahiel, M.A. (1992). Four homologous domains in the primary structure of GrsB are related to domains in a superfamily of adenylate-forming enzymes. *Mol. Microbiol.* **6**, 529–546.

Ullán, R.V., Casqueiro, J., Bañuelos, O., Fernández, F.J., Gutiérrez, S., and Martín, J.F. (2002). A novel epimerization system in fungal secondary metabolism involved in the conversion of isopenicillin N into penicillin N in *Acremonium chrysogenum*. *J. Biol. Chem.* **277**, 46216–46225.

van de Kamp, M., Driessen, A. J. M., and Konings, W. N. (1999). Compartmentalization and transport in β-lactam antibiotic biosynthesis by filamentous fungi. *Antonie van Leeuwenhoek* **75**, 41–78.

van Liempt, H., von Döhren, H., and Kleinkauf, H. (1989). δ-(L-α-aminoadipyl)-L-cysteinyl-D-valine synthetase from *Aspergillus nidulans*. *J. Biol. Chem.* **264**, 3680–3684.

Velkov, T. and Lawen, A. (2002). Mapping and molecular modeling of S-adenosyl-L-methionine binding site in N-methyltransferase domains of the multifunctional polypeptide cyclosporin synthetase. *J. Biol. Chem.* **278**, 1137–1148.

Vigushin, D.M. and Coombes, R.C. (2002). Histone deacetylase inhibitors in cancer treatment. *Anticancer Drugs* **13**, 1–13.

Voisard, C., Wang, J., McEvoy, J.L., Xu, P.L., and Leong, S.A. (1993). *URBS1*, a gene regulating siderophore biosynthesis in *Ustilago maydis*, encodes a protein similar to the erythroid transcription factor GATA-1. *Mol. Cell. Biol.* **13**, 7091–7100.

von Döhren, H. (1990). Compilation of peptide structures—a biogenetic approach. In H. Kleinkauf and H. von Döhren (eds.) *Biochemistry of Peptide Antibiotics* W. de Gruyter, Berlin, pp. 411–507

von Döhren, H., Keller, U., Vater, J., and Zocher, R. (1997). Multifunctional peptide synthetases. *Chem. Rev.* **97**, 2675–2705.

Walsh, C. T., Chen, H., Keating, T. A., Hubbard, B. K., Losey, H. C., Luo, L., Marshall, C. G., Miller, D. A. *et al.* (2001). Tailoring enzymes that modify nonribosomal peptides during and after chain elongation on NRPSs assembly lines. *Curr. Opin. Chem. Bio.* **5**, 525–534.

Walton, J. D. (1987). Two enzymes involved in biosynthesis of the host-selective phytotoxin HC-toxin. *Proc. Natl. Acad. Sci. USA* **84**, 8444–8447.

Walton, J. D. (1990). Peptide phytotoxins from plant pathogenic fungi. In H. Kleinkauf and H. von Döhren (eds.) *Biochemistry of Peptide Antibiotics* de Gruyter Berlin, pp. 179–203.

Walton, J.D. (1996). Host-selective toxins: Agents of compatibility. *Plant Cell* **8**, 1723–1733.

Walton, J.D. (2000). Horizontal gene transfer and the origin of secondary metabolite gene clusters in fungi: An hypothesis. *Fungal Genet. Biol.* **30**, 167–171.

Walton, J.D. and Holden, F.R. (1988). Properties of two enzymes involved in the biosynthesis of the fungal pathogenicity factor HC-toxin. *Mol. Plant–Microbe Interact.* **1**, 128–134.

Walton, J.D., Ransom, R., and Pitkin, J.W. (1997). Northern corn leaf spot of maize: Chemistry, enzymology, and molecular genetics of a host-selective phytotoxin. In G. Stacey and N. T. Keen (eds.) *Plant–Microbe Interact.*, Vol. 3. Chapman and Hall, New York, pp. 94–123.

Walzel, B., Riederer, B., and Keller, U. (1997). Mechanism of alkaloid cyclopeptide synthesis in the ergot fungus *Claviceps purpurea*. *Chem. Biol.* **4**, 223–230.

Wang, J. (2001). *dmaW* encoding tryptophan dimethylallyltransferase in ergot alkaloid biosynthesis from Clavicipitaceous fungi. Doctoral dissertation, University of Kentucky, Lexington, KY.

Ward, T.J., Bielawski, J.P, Kistler, H.C., Sullivan, E., and O'Donnell, K. (2002). Ancestral polymorphism and adaptive evolution in the trichothecene mycotoxin gene cluster of phytopathogenic *Fusarium*. *Proc. Natl. Acad. Sci. USA* **99**, 9278–9283.

Weber, G., Schörgendorfer, K., Schneider-Scherzer, E., and Leitner, E. (1994). The peptide synthetase catalyzing cyclosporine production in *Tolypocladium niveum* is encoded by a giant 45.8-kilobase open reading frame. *Curr. Genet.* **26**, 120–125.

Weckwerth, W., Miyamoto, K., Iinuma, K., Krause, M., Glinski, M., Storm, T., Bonse, G., Kleinkauf, H., *et al.* (2000). Biosynthesis of PF1022A and related cyclooctadepsipeptides. *J. Biol. Chem.* **275**, 17909–17915.

Wiest, A., Grzegorski, D., Xu, B.-W., Goulard, C., Rebuffat, S., Ebbole, D. J., Bodo, B., and Kenerley, C. (2002). Identification of peptaibols from *Trichoderma virens* and cloning of a peptaibol synthetase. *J. Biol. Chem.* **277**, 20862–20868.

Wilhite, S.E., Lumsden, R.D., and Straney, D.C. (2001). Peptide synthetase gene in *Trichoderma virens*. *Appl. Environ. Microbiol.* **67**, 5055–5062.

Woloshuk, C.P., Foutz, K.R., Brewer, J.F., Bhatnagar, D., Cleveland, T.E., and Payne, G.A. (1993). Molecular characterization of *aflR*, a regulatory locus for aflatoxin biosynthesis. *Appl. Environ. Microbiol.* **60**, 2408–2414.

Wolpert, T.J., Macko, V., Acklin, W., Jaun, B., Seibl, J., Meili, J., and Arigoni, D. (1985). Structure of victorin C, the major host-selective toxin from *Cochliobolus victoriae*. *Experientia* **41**, 1524–1529.

Yoder, O.C., Valent, B., and Chumley, F. (1986). Genetic nomenclature and practice for plant pathogenic fungi. *Phytopathology* **76**, 383–385.

Yuan, W.M., Gentil, G.D., Budde, A.D., and Leong, S.A. (2001). Characterization of the *Ustilago maydis sid2* gene, encoding a multidomain peptide synthetase in the ferrichrome biosynthetic gene cluster. *J. Bacteriol.* **183**, 4040–4051.

8

Isoprenoids: Gene Clusters and Chemical Puzzles

D. Barry Scott, Geoffrey B. Jameson, and Emily J. Parker

1. Introduction

The isoprenoids are one of the most chemically diverse groups of natural products found in nature, with greater than 23,000 compounds identified (Dewick, 2002). All are derived from linear isoprenoid diphosphates synthesized from isopentenyl diphosphate (IPP) and dimethylallyl diphosphate (DMAPP) by a family of prenyltransferases that catalyze the sequential condensations of IPP with allylic isoprenoid diphosphates (Figure 8.1). The primary isoprenoid substrates can be derived from either the mevalonate pathway or the more recently discovered deoxyxylulose phosphate (mevalonate-independent) pathway (Rohmer *et al.*, 1993), that is found in eubacteria and plants (Dewick, 2002). Most fungi contain both farnesyl diphosphate (FPP) and geranylgeranyl diphosphate (GGPP) synthases, which catalyze the formation of the linear C15 and C20 isoprenoids, FPP and GGPP, respectively. FPP is a key branch point in the isoprene pathway from where additional prenyl transferases catalyze the synthesis of precursors for sterols, ubiquinones, dolichols, sesquiterpenes, and farnesylated proteins. GGPP is the substrate for another group of prenyl transferases involved in the biosynthesis of gibberellins, carotenoids, indole-diterpenes, and geranylgeranylated proteins (Figure 8.1).

While clustering of genes is a common feature of prokaryotes, such organization is less common in eukaryotes. However, gene clusters, are now recognized as a common feature of filamentous fungal genomes, where they are associated with "dispensable" metabolic functions such as the utilization of low molecular weight metabolites (e.g., quinate and proline) or the biosynthesis of low molecular weight secondary metabolites (e.g., isoprenoids) (Keller and Hohn, 1997). Gene clusters, through the novel metabolic functions they encode, can

---

D. Barry Scott • Institute of Molecular BioSciences, College of Sciences, Massey University, Private Bag 11 222, Palmerston North, New Zealand. **Emily J. Parker** • Institute of Fundamental Sciences, College of Sciences, Massey University, Private Bag 11 222, Palmerston North, New Zealand. **Geoffrey B. Jameson** • Institute of Fundamental Sciences, College of Sciences, Massey University, Private Bag 11 222, Palmerston North, New Zealand.

Advances in Fungal Biotechnology for Industry, Agriculture, and Medicine. Edited by Jan S. Tkacz and Lene Lange, Kluwer Academic/Plenum Publishers, 2004.

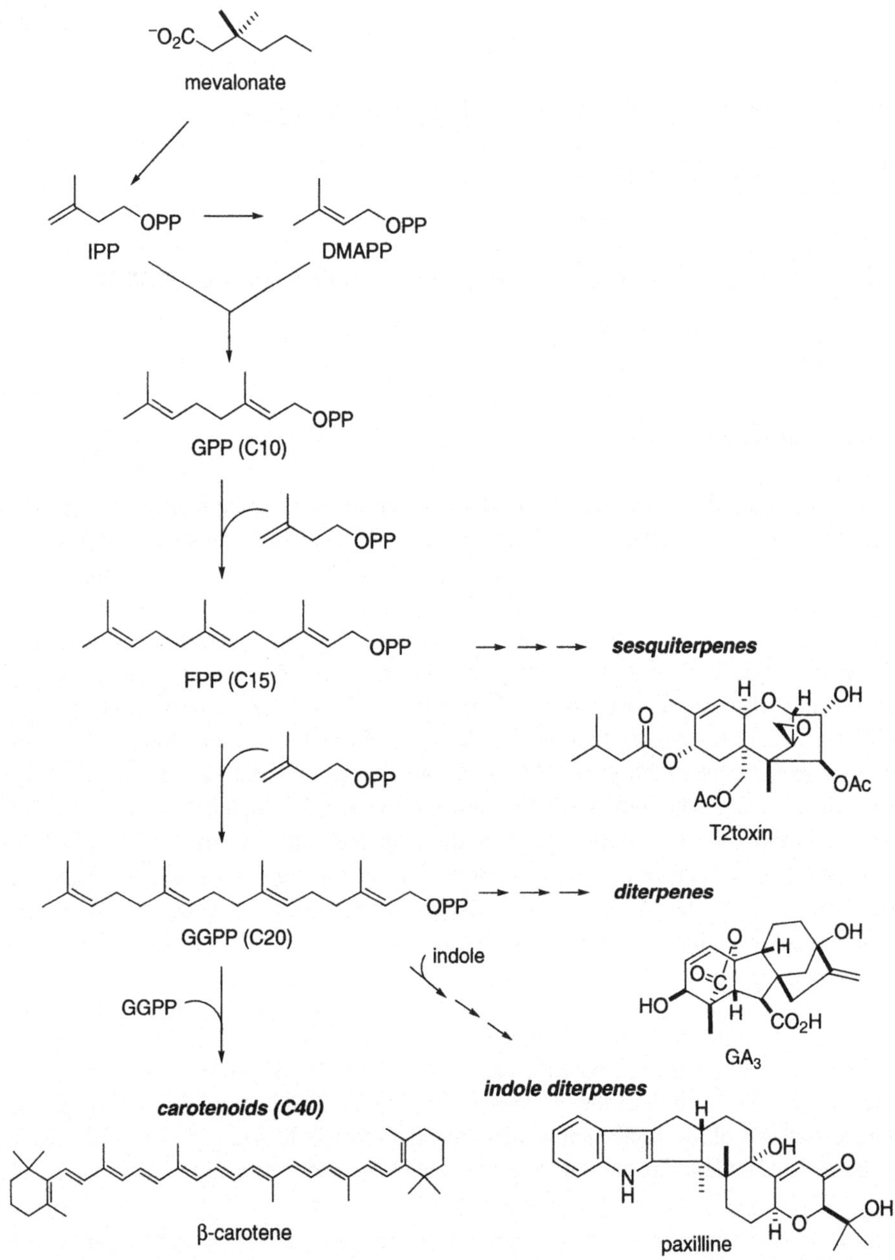

mevalonate

IPP → DMAPP

GPP (C10)

FPP (C15) → → → *sesquiterpenes*

T2toxin

GGPP (C20) → → → *diterpenes*

indole

GA$_3$

GGPP

indole diterpenes

carotenoids (C40)

β-carotene

paxilline

Figure 8.1. Isoprenoid biosynthesis in fungi. Fungal sesquiterpenes (e.g., T2 toxin), diterpenes (e.g., GA3), indole-diterpenes (e.g., paxilline), and carotenoids are derived from mevalonate via the linear isoprenoids farnesyl diphosphate (FPP) or geranylgeranyl diphosphate (GGPP).

be viewed, therefore, as "fitness" attributes for fungal survival in the natural environment. The recent molecular cloning and characterization of gene clusters for the biosynthesis of trichothecenes (Hohn *et al.*, 1993; Brown *et al.*, 2001), gibberellins (Tudzynski and Hölter, 1998), indole-diterpenes (Young *et al.*, 2001), and carotenoids (Verdoes *et al.*, 1999; Velayos *et al.*, 2000; Arrach *et al.*, 2001; Linnemannstöns *et al.*, 2002) has provided new insights into the enzymatic steps that comprise these biosynthetic pathways, how the pathways are regulated, and the molecular basis for the structural diversity.

Much of the isoprenoid structural diversity arises from the chain elongation and cyclization reactions catalyzed by the prenyl transferases (Sacchettini and Poulter, 1997). These enzymes are responsible for both chain elongation and the stereospecific cyclization by electrophilic alkylation. The 3-D structures of fungal trichodiene (Rynkiewicz *et al.*, 2001) and aristolochene (Caruthers *et al.*, 2000) synthases, as well as prenyltransferases of bacterial, plant, and mammalian origin, provide new insights into how this structural diversity arises. Subsequent functionalization of the core cyclic isoprenoids, principally by cytochrome P450 enzymes, adds to the diversity of the isoprenoids that are found in nature.

This chapter will summarize our current knowledge of fungal isoprenoid gene clusters, review proposed biosynthetic schemes for the pathways encoded by cluster genes, and provide some insight into the molecular basis of the structural diversity observed within this class of fungal metabolites. Genes and gene products will be designated according to the convention used for *Aspergillus nidulans* (Bennett and Lasure, 1985). Where discrepancies exist among published structure formulae, we have accepted the Chemical Abstracts depiction as authoritative. Where this solution was not available, we have chosen to show the structure without absolute stereochemistry.

2. Sesquiterpenes

The sesquiterpenes are a diverse group of FPP-derived secondary metabolites found in both plants and fungi. The most well-known fungal sesquiterpenoids are the trichothecenes and aristolochenes.

2.1. Trichothecenes

The trichothecenes form a structurally diverse group of sesquiterpene epoxides produced by a range of fungi including *Fusarium, Myrothecium, Stachybotrys, Cephalosporium, Trichoderma, Trichothecium*, and plants of the genus *Baccharis* (Jarvis *et al.*, 1988; Brown *et al.*, 2001). Their synthesis in fungi appears to be confined to the Hypocreales, suggesting an evolutionary origin within this Order (Berbee, 2001). The trichothecenes are potent inhibitors of eukaryotic protein synthesis and are highly toxic to both plants (phytotoxic) and animals (mycotoxic). The reduced virulence of *Fusarium graminearum* (teleomorph; *Gibberella zeae*) on maize and wheat, associated with mutations affecting trichothecene synthesis, demonstrates that these metabolites are virulence factors (Proctor *et al.*, 1995a; Desjardins *et al.*, 1996; Harris *et al.*, 1999).

2.1.1. Chemical Diversity

The structural diversity of the trichothecenes arises through the pattern of substitution of the trichothecene skeleton and, if present, the macrocyclic ring. *Fusarium sporotrichioides*

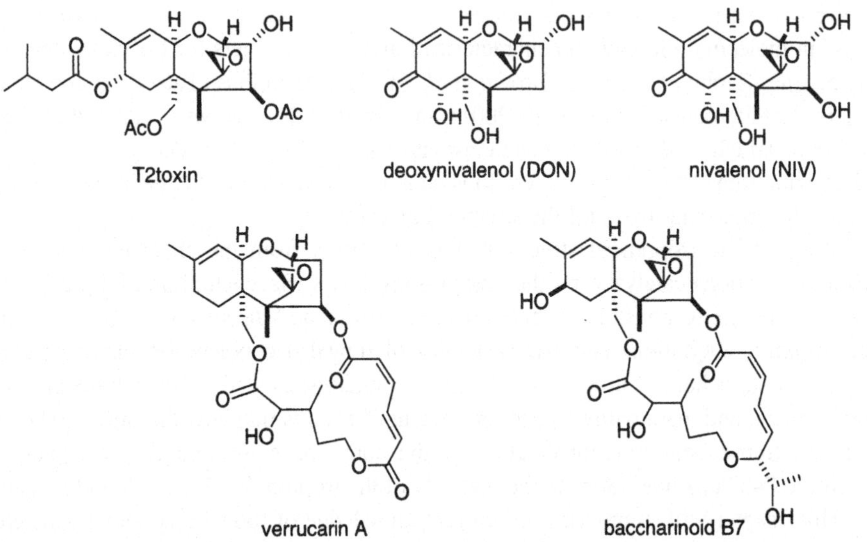

Figure 8.2. Chemical structures of T2 toxin, deoxynivalenol, nivalenol, verrucarin A, and baccarinoid B7.

produces A-type trichothecenes (Figure 8.2), such as T2 toxin or 4,15-diacetoxyscirpenol, which have either a C-8 hydroxyl or an ester function or which lack a C-8 substituent (Brown *et al.*, 2001). *F. graminearum* produces B-type trichothecenes (Figure 8.2), such as deoxynivalenol (DON) and nivalenol (NIV), which have a carbonyl group at C-8. The more structurally complex macrocyclic trichothecenes, which are about 10-fold more toxic than *Fusarium* trichothecenes, characteristically contain a macrolide-ester ring that bridges C-4 and C-15 of the trichothecene skeleton (Figure 8.2).

2.1.2. Gene Clusters

The first trichothecene biosynthetic gene cluster to be identified was that for T2 toxin biosynthesis in *F. sporotrichioides* (Hohn *et al.*, 1993). The cloning of a trichodiene cyclase gene was the breakthrough that led to the subsequent isolation and identification of the set of linked genes required for T2 toxin biosynthesis (Hohn and Beremand, 1989). To date 11 genes have been assigned to the trichothecene gene cluster in *F. sporotrichioides* (Figure 8.3) (Brown *et al.*, 2001). One additional gene required for T2 toxin biosynthesis, *tri101*, resides outside the cluster (Kimura *et al.*, 1998; McCormick *et al.*, 1999). Similar clusters have been identified in *F. graminearum* (Brown *et al.*, 2001), *Myrothecium roridum* (Trapp *et al.*, 1998), and *F. sambucinum* (cited in Tag *et al.*, 2001). The organization of the 11 genes in the *F. graminearum* and *F. sporotrichioides* gene clusters is almost identical (Brown *et al.*, 2001), but the *M. roridum* orthologues of *tri4, tri5*, and *tri6* are organized very differently (Figure 8.3) (Trapp *et al.*, 1998). The proposed function of each of these genes is presented in Table 8.1.

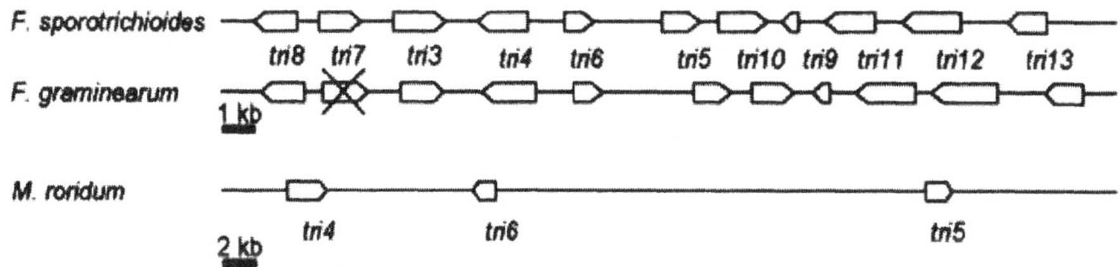

Figure 8.3. Trichothecene biosynthesis gene clusters in *Fusarium sporotrichioides, Fusarium gramineareum,* and *Myrothecium roridum.*

Table 8.1. Proposed Function of Trichothecene Biosynthetic Genes from *Fusarium* Species

Gene	Proposed function	Reference
tri8	Esterase	Brown *et al.*, 2001; McCormick and Alexander, 2002
tri7	Transacetylase	Brown *et al.*, 2001
tri3	Transacetylase	McCormick, Hohn, and Desjardins, 1996
tri4	P450 monooxygenase	Hohn, Desjardins, and McCormick, 1995
tri6	Transcription factor	Proctor, Hohn, McCormick, and Desjardins, 1995
tri5	Trichodiene synthase	Hohn and Beremand, 1989
tri10	Regulation	Tag *et al.*, 2001
tri9	Unknown	Brown *et al.*, 2001
tri11	P450 monooxygenase	Alexander, Hohn, and McCormick, 1998
tri12	Efflux pump	Alexander, McCormick, and Hohn, 1999
tri13	Hydroxylase	Lee, Han, Kim, Yun, and Lee, 2002

2.1.3. Biosynthesis of T2-Toxin

The gene clusters for T2 toxin biosynthesis in *F. sporotrichioides* and for DON formation in *F. graminearum* have provided useful model systems for unravelling the biochemical and genetic basis of trichothecene biosynthesis (Figure 8.4). The biosynthetic sequence of events proceeding from trichodiene to the more complex trichothecenes, such as T2 toxin, has been established by a combination of feeding experiments with mutants blocked at various steps in the pathway and heavy-isotope labeling of pathway precursors.

The first unique step in trichothecene biosynthesis, catalyzed by Tri5, involves stereospecific cyclization of FPP to form bicyclic trichodiene (Figure 8.4). It has been proposed that Tri4, a cytochrome P450 monooxygenase, oxidizes trichodiene to 2-hydroxytrichodiene (Hohn *et al.*, 1995). However, while it is evident that trichodiene is the substrate for Tri4, the product(s) of this enzyme has not yet been experimentally confirmed. Disruption of *tri4* resulted in the loss of both trichothecene and isotrichodiol production and in the accumulation of trichodiene (Hohn *et al.*, 1995). Transformants lacking *tri4* were able to convert isotrichotriol to T2 toxin, suggesting that the enzymes needed beyond this point are present in the mutant, including those for isomerization and cyclization of isotrichotriol to isotrichodermol. Given the multifunctional nature of other cytochrome P450 monooxygenases involved in isoprenoid biosynthesis (Helliwell *et al.*, 2001; Rojas *et al.*, 2001;

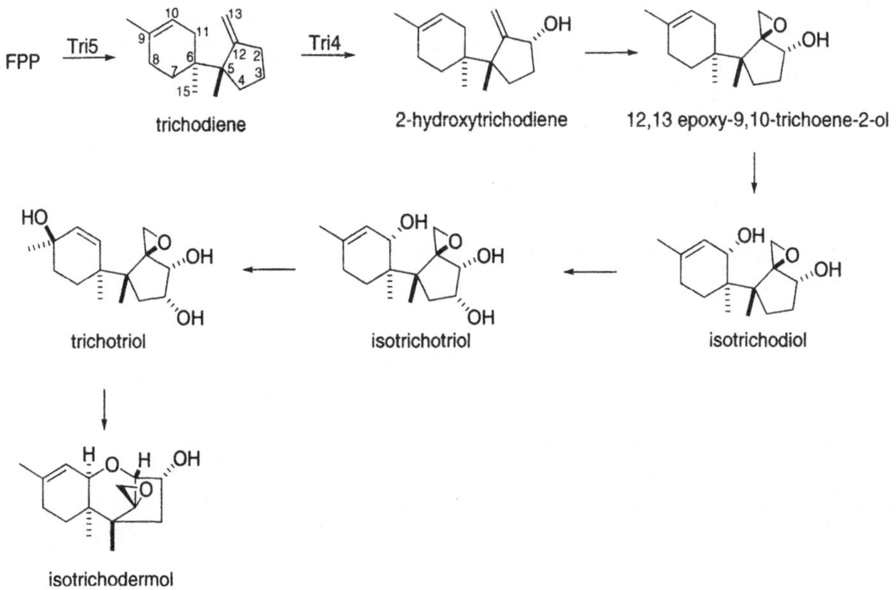

FPP Tri5 trichodiene Tri4 2-hydroxytrichodiene 12,13 epoxy-9,10-trichoene-2-ol

isotrichodiol isotrichotriol trichotriol

isotrichodermol

Figure 8.4(A) Proposed biochemical pathway for trichothecene biosynthesis in *Fusarium* species. This portion of Figure 8.4 shows the pathway to isotrichodermol.

Tudzynski *et al.*, 2002), Tri4 may catalyze all four oxidation steps from trichodiene to isotrichotriol. This hypothesis could be tested by feeding trichodiene either to transformants of a *tri* cluster deletion derivative of *F. sporotrichioides* containing only *tri4*, or to a heterologous host such as *Saccharomyces cerevisiae* expressing *tri4*. Alternatively, other genes, either within or outside the cluster, may be required for these steps.

Following the sequential oxidations on the bicyclic trichodiene, isomerization and cyclization occurs to form the first compound with the trichothecene 12,13-epoxytrichothec-9-ene skeleton, isotrichodermol. The genes that encode the enzymes that catalyze the isomerization and cyclization are as yet unidentified.

A gene located outside the cluster, *tri101*, is required for the next step in the pathway, involving the transacetylation of isotrichodermol to isotrichodermin (Kimura *et al.*, 1998; McCormick *et al.*, 1999). It has been proposed that Tri101 acts as a self-protection or resistance factor during biosynthesis, as a free C-3 hydroxyl group is a key component of *Fusarium* trichothecene phytotoxicity. Following the cloning of *tri101* from *F. graminearum* (Kimura *et al.*, 1998), *in vitro* assays of recombinant enzyme purified from *Escherichia coli* demonstrated that Tri101 catalyzes *O*-acetylation of the trichothecene ring, specifically at the C-3 position, in an acetyl CoA-dependent manner. In addition, Kimura *et al.* (1998) established that *tri101* is separated by at least 35 kb from the *tri* cluster. Subsequently, *tri101* was cloned from *F. sporotrichioides* (McCormick *et al.*, 1999). While expression of *tri101* in yeast conferred resistance to trichothecenes, disruption of *tri101* in *F. sporotrichioides* did not result in a loss of toxin self-protection (McCormick *et al.*, 1999), suggesting that there is at least one other mechanisms for protecting the producer from these toxins, perhaps the efflux pump encoded by *tri12* (Alexander *et al.*, 1999). The *tri101* disruptant accumulates isotrichodermol

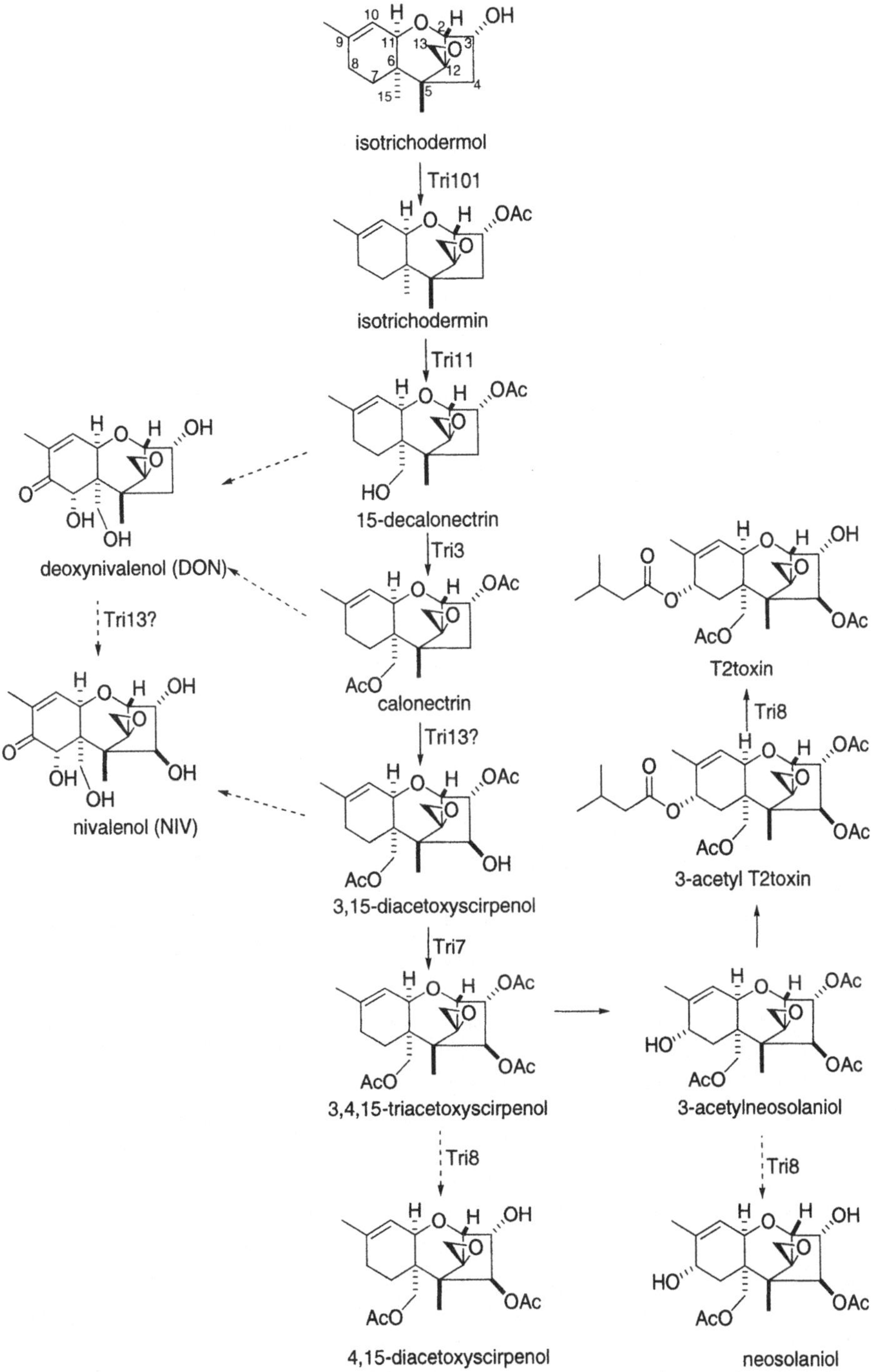

isotrichodermol

Tri101

isotrichodermin

Tri11

15-decalonectrin

Tri3

calonectrin

Tri13?

3,15-diacetoxyscirpenol

Tri7

3,4,15-triacetoxyscirpenol

Tri8

4,15-diacetoxyscirpenol

deoxynivalenol (DON)

Tri13?

nivalenol (NIV)

T2toxin

Tri8

3-acetyl T2toxin

3-acetylneosolaniol

Tri8

neosolaniol

Figure 8.4(B) Proposed biochemical pathway for trichothecene biosynthesis in *Fusarium* species. This portion of Figure 8.4 shows the pathway from isotrichodermol to T2 toxin.

predominantly, along with small amounts of 3,15-didecalonectrin and 3-decalonectrin, trichothecenes that are not observed in culture filtrates of the wild-type strain. The accumulation of isotrichodermol in this mutant provides genetic support for the biochemical evidence that Tri101 is trichothecene 3-*O*-acetyltransferase (Kimura *et al.*, 1998). The formation of 3,15-didecalonectrin and 3-decalonectrin suggests that the P450 monooxygenase and the C-15 acetyltransferase encoded by *tri11* and *tri3*, respectively, are able to accept isotrichodermol and 3,15-didecalonectrin as substrates when isotrichodermol accumulates. Interestingly, chemical feeding studies showed that the *tri101* disruptant was able to convert isotrichodermin and 15-decalonectrin to T2 toxin, but not 3,4,15-triacetoxyscirpenol, suggesting that there may be deacetylation and reacetylation required at position C-3 later in the pathway (McCormick *et al.*, 1999).

The subsequent C-15 hydroxylation of isotrichodermin to form 15-decalonectrin is catalyzed by a second P450 monooxygenase, Tri11 (Figure 8.4) (Alexander *et al.*, 1998). Disruption of the gene encoding this protein resulted in mutants that accumulated isotrichodermin, and whole-cell feeding experiments showed that a *tri11* mutant efficiently converted 15-decalonectrin to T2 toxin. The C-15 hydroxyl group is characteristic of most *Fusarium* trichothecenes and all macrocyclic trichothecenes found to date, suggesting that 15-decalonectrin is the last common intermediate of the various *Fusarium* trichothecene biosynthesis pathways. Synthesis of DON is proposed to branch off the T2 toxin pathway at this intermediate (Brown *et al.*, 2001).

The acetylation of 15-decalonectrin to calonectrin is catalyzed by Tri3 (Figure 8.4) (McCormick *et al.*, 1996). Disruption of the corresponding gene results in the accumulation of deacetylated calonectrins, including 3,15-didecalonectrin and 15-decalonectrin (Figure 8.4). Tri3 shares no sequence similarity to other fungal *O*-acetyltransferases, including Tri101. Moreover, there are no reports to date of a *tri3* homologue in the macrocyclic trichothecene producers such as *Myrothecium* spp.

Recent work on *G. zeae* (anamorph: *F. graminearum*) suggests that *tri13* may encode the enzyme required for the oxidation of calonectrin at C-4, producing 3,15-diacetoxyscirpenol on the pathway to T2 toxin (Figure 8.4). *G. zeae* can be divided into two chemotypes based on the absence or presence of the C-4 hydroxyl group, giving DON and NIV, respectively. Lee *et al.* (2002) have demonstrated that a NIV-producing chemotype can be altered to produce DON by targeted deletion of *tri13*. Likewise, heterologous expression of the *tri13* gene from the NIV-producing chemotype into the DON-producing strain conferred on the latter the ability to synthesize NIV. The expression of a functional *tri13* has been proposed to act as a switch between the two *G. zeae* phenotypes, although the actual substrates and products of Tri13 are yet to be unequivocally identified.

The subsequent acetylation of the C-4 oxygen is performed by Tri7 (Figure 8.4) (Brown *et al.*, 2001). Disruption of *tri7* resulted in the accumulation of HT-2 toxin instead of T2 toxin. The pathway proposed for the synthesis of HT-2 toxin involves C-8 oxidation, C-8 esterification and C-3 deacetylation; steps that are presumably catalyzed by the same enzymes that are used for T2 toxin biosynthesis. The presence of an acetyl group on the C-4 oxygen of T2 toxin and the lack of an acetyl group on the C-4 oxygen of HT-2 toxin suggested that *F. sporotrichioides* Tri7 was responsible for this specific acetylation. This was confirmed by feeding studies, which showed that a C-4 acetyl group was required for the *tri7* deletion strain to convert trichothecenes to T2 toxin (Brown *et al.*, 2001). A key difference between the *F. sporotrichioides* and *F. graminearum* gene clusters is the presence in the latter of a nonfunctional *tri7* pseudogene (Figure 8.3) (Brown *et al.*, 2001). This block is presumably responsible for the formation of DON and NIV as the predominant secondary metabolites in this species, rather than T2 toxin.

The genes encoding enzymes for C-8 oxidation of 3,4,15-triacetoxyscirpenol to 3-acetylneosolaniol, and subsequent esterification with isovalerate, to 3-acetyl T2 toxin, have not yet been identified. Although Tri8 was initially proposed to be involved in one or both of these steps (Brown *et al.*, 2001), McCormack and Alexander (2002) subsequently showed that Tri8 is in fact an esterase, specific for deacetylation of the C-3 position (described later). The isolation of a mutant of *F. sporotrichioides* that is unable to hydroxylate C-8, but is able to form the isovalerate ester at this position, would suggest that two distinct enzymes are required for these reactions. Furthermore, the inability of cosmids that contain trichothecene cluster genes to complement this mutant indicates that the gene required for C-8 oxidation is located outside the core cluster (Hohn *et al.*, 1993; Brown *et al.*, 2001).

The final step in the formation of T2 toxin involves deacetylation of the C-3 acetyl group by Tri8 (McCormick and Alexander, 2002). Disruption of *tri8* in *F. sporotrichioides* gave rise to mutants that accumulated 3-acetyl T2 toxin, 3-acetyl neosolaniol, and 3,4,15-triacetoxyscirpenol rather than T2 toxin, neosolaniol, and 4,15-diacetoxyscirpenol, strongly suggesting that *tri8* encodes an esterase specific for the C-3 position. *F. graminearum*, which has a non-functional *tri7* (acetyltransferase), accumulated 3,15-acetyl-DON, 7,8-dihydroxycalonectrin, and calonectrin instead of 15-acetyl-DON. Interestingly, these same three metabolites accumulate in wild-type cultures of *F. culmorum*, raising the possibility that this species may naturally have non-functional *tri7* and *tri8* genes. When seven possible substrates for Tri8, all with C-3 acetyl groups, were fed to cells of *S. cerevisiae* transformed with both *tri8* and *tri12* (efflux pump), all were deacetylated at position C-3. Five were rapidly and efficiently converted to the corresponding 3-hydroxyl derivative, but two were relatively poor substrates. Similar results were obtained when the same substrates were fed to cell-free extracts of both *F. sporotrichioides* and *F. graminearum*. Taken together, these data provide strong support for Tri8 acting as an esterase for the removal of the C-3 acetyl group from trichothecenes. Since C-3 hydroxyl trichothecenes are more phytotoxic than the corresponding acetylated intermediates, Tri8 could be considered to be a toxicity factor (McCormick and Alexander, 2002).

Biosynthesis of DON and NIV by *F. graminearum, G. zeae*, and *F. culmorum* apparently requires additional enzyme-catalyzed transformations not observed in T2 toxin biosynthesis. The formation of DON has been proposed to branch from T2 biosynthesis at 15-decalonectrin, but the accumulation of 15-acetyl-DON precursors by *F. graminearum* following disruption of Tri8 suggests that branching may actually occur at calonectrin. Biosynthesis of NIV is proposed to branch from the T2 toxin pathway after C-4 hydroxylation of calonectrin to produce 3,15-diacetoxyscirpenol. As this compound is acetylated at C-15 and NIV is not, it is clear an enzyme with C-15 deacetylase activity is required to produce NIV. The fact that NIV or DON chemotypes are defined by the presence or absence of a C-4 hydroxylase, suggests that C-15 deacetylation occurs in organisms that produce either compound.

The two other features that distinguish these mycotoxins from T2 toxin are the presence of a C-7 hydroxyl and a C-8 carbonyl group. The accumulation of 7,8-dihydroxycalonectrin in *tri8* disruptants suggests the C-8 oxidation is a two-step process via the 8-hydroxyl species. It is possible that an as yet unidentified gene encoding the C-8 oxidation has a role in both T2 toxin biosynthesis and DON and NIV biosynthesis. The results with the *tri8* disruptants detailed above suggest Tri8 is responsible for C-3 deacetylation to produce both DON and NIV.

The biosynthetic pathway to macrocyclic trichothecenes is closely related to the other trichothecene pathway, despite the organization and orientation of the *tri* cluster genes differing markedly from the *tri* cluster of *F. sporotrichioides* (Figure 8.3). To date, *tri5* (encoding a trichodiene synthase) and *tri4* (encoding a P450 monooxygenase for the first oxidation) have been identified in *M. roridum* (Trapp *et al.*, 1998). The deduced sequences of *M. roridum* Tri5 and Tri4 display overall identities of 55% and 63% with their respective *F. sporotrichioides* counterparts. Complementation of a *F. sporotrichioides tri4* mutant with the orthologue from *M. roridum* demonstrated that these genes are functionally similar, but there are some subtle differences in the regulation of the pathway, in as much as the chemical phenotype of the heterologous transformants was different from that of transformants containing the homologous wild-type gene (Trapp *et al.*, 1998).

2.1.4. Regulation

The four remaining genes within the *tri* cluster encode a Cys_2His_2 zinc finger transcription factor, *tri6* (Proctor *et al.*, 1995b), an efflux pump, *tri12* (Alexander *et al.*, 1999), a novel regulatory gene, *tri10* (Tag *et al.*, 2001), and a gene, *tri9* of unknown function that encodes a 43-amino acid polypeptide (Brown *et al.*, 2001).

Disruption of *tri6* resulted in a mutant that was unable to produce trichothecenes but that did accumulate low levels of the trichothecene precursor, trichodiene (Proctor *et al.*, 1995b). This mutant was unable to convert six trichothecene intermediates to T2 toxin, and its transcription of both *tri4* and *tri5* were down regulated. Fusion of *tri6* to the DNA-binding region of *GAL4* in *S. cerevisiae* activated expression of a *GAL1–lacZ* fusion, confirming that Tri6 is a transcriptional activator. A consensus binding sequence (TNAGGCCT) for Tri6 has been identified in the upstream region of all the *tri* biosynthetic genes, with the exception of *tri10* (Hohn *et al.*, 1999; Brown *et al.*, 2001). This motif was shown to be specific for Tri6 binding *in vitro* (Hohn *et al.*, 1999). These results indicate that Tri6 is a pathway-specific regulator of trichothecene biosynthetic genes in *F. sporotrichioides* and acts at the level of transcription. An orthologue of the *F. sporotrichioides tri6* has also been found in *M. roridum* (Trapp *et al.*, 1998).

The gene product of *tri12* appears to function as a trichothecene efflux pump (Alexander *et al.*, 1999). Disruption of Tri12 resulted in reduced growth of *F. sporotrichioides* on complex media and reduced levels of trichothecenes. Furthermore, yeast transformed with both *tri3* and *tri12* converted 15-decalonectrin, a toxic intermediate, to calonectrin very efficiently compared with yeast transformed with *tri3* alone. These results are consistent with an efflux role for Tri12 in providing self-protection for *F. sporotrichioides* against trichothecenes.

Disruption of *tri10* in *F. sporotrichioides* abolishes T2 toxin synthesis and dramatically decreases transcript levels of *tri4, tri5, tri6, tri101*, and a putative FPP synthase, suggesting it has a regulatory role in trichothecene biosynthesis (Tag *et al.*, 2001). Although orthologues are found in *F. graminearum* and *F. sambucinum, tri10* has no similarity to any other gene. Tag *et al.* (2001) propose that Tri10 acts upstream of the cluster-specific transcription factor Tri6 and is required for full expression of trichothecene genes, both within and outside the cluster, as well as genes for primary metabolism, whose products form the precursors that feed the trichothecene pathway. The demonstration that Tri10 regulates FPP synthase is the first evidence of a regulatory connection between genes required for the production of precursors for both primary and secondary metabolism and the cluster genes encoding the enzymes of the secondary pathway itself. As more fungal genomes are characterized and microarray experiments become feasible, it will be possible to explore fully the regulatory connection between expression of secondary metabolite genes and the genes required for the synthesis of the necessary precursors in primary metabolism.

2.2. Aristolochenes

Aristolochene is an eremophilane-type sesquiterpenoid (Figure 8.5) that is produced by fungi, insects, and plants (Proctor and Hohn, 1993). The −/− enantiomer has been identified in plants and insects, whereas the +/− enantiomer has been identified in extracts from the fungi *Aspergillus terreus* and *Penicillium roqueforti* (Proctor and Hohn, 1993), and another diastereoisomer, 5-*epi*-aristolochene, has been identified in tobacco (*Nicotiana tabacum*) (Ralston, 2001). Aristolochene is also a likely precursor for a range of sesquiterpenoid toxins produced by filamentous fungi (Hohn *et al.*, 1991), including PR-toxin (Figure 8.5) produced *by P. roqueforti*. PR-toxin has been implicated in incidents of mycotoxicoses resulting from the consumption of contaminated grain, but unlike the trichothecenes little is known about the mode of action or the functional groups required for toxicity.

As some strains of *P. roqueforti* produce large amounts of PR-toxin, this organism provides an excellent experimental system for studying aristolochene-derived biosynthesis. Purification and characterization of the sesquiterpene cyclase, aristolochene synthase, made it possible to clone the corresponding gene, *ari1* (Proctor and Hohn, 1993). Expression of *ari1* in *E. coli* as a Protein A/Ari1 fusion confirmed that the enzyme is a sesquiterpene cyclase. As with the trichothecene biosynthesis genes, *ari1* is upregulated in stationary phase cultures of *P. roqueforti*. Southern analysis confirmed that *ari1* homologues are present in *Penicillium camembertii* and *Penicillium chrysogenum*, but no hybridization was observed with other *Penicillium* species or more distantly related fungi that produce aristolochene-like sesquiterpenoids (Hohn *et al.*, 1991). More recently, Ari1 was purified from *A. terreus*, and cDNA and genomic clones were isolated by PCR (Cane and Kang, 2000). The

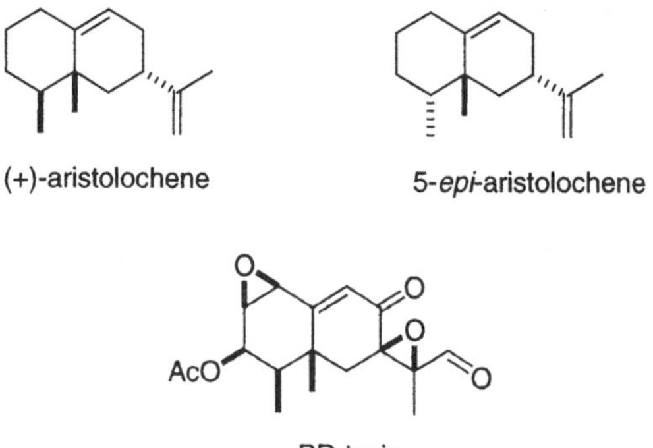

Figure 8.5. Structures of (+)-aristolochene, 5-*epi*-aristolochene, and PR-toxin.

A. terreus and *P. roqueforti* genes are highly homologous. To date there is no evidence that either of them forms part of a gene cluster. However, as *A. terreus* is one of 15 fungal genomes soon to be sequenced (Birren *et al.*, 2002), the definitive answer to this question should be forthcoming.

3. Diterpenes

3.1. Gibberellins

Gibberellins are a large family of tetracyclic diterpenoid carboxylic acids, some of which function as plant hormones. They were first identified as secondary metabolites of the rice pathogen, *Gibberella fujikuroi*, but other fungi, including *Sphaceloma manihoticola* and *Phaeosphaeria* sp., synthesize these metabolites (Tudzynski, 1999). *G. fujikuroi* is used for commercial production of these metabolites and has been the organism of choice for dissecting the biosynthetic pathway. The diterpenoid nature of gibberellins was established by incorporation of radioactivity from [2-^{14}C] mevalonic acid into GA_3 (Figure 8.6), the end product of the fungal pathway.

3.1.1. Chemical Diversity

There are over 120 gibberellins identified from green plants, fungi, and bacteria (Dewick, 2002). While GA_3 is the major gibberellin produced by *G. fujikuroi* (Tudzynski, 1999), it is a minor component in most plant species (Dewick, 2002). These compounds all share the same tetracyclic skeleton but vary in their functionalization. Indeed the range of functionalization patterns observed for known gibberellins means that their biosynthesis is perhaps best referred to as a metabolic grid rather than a pathway. Closely related to

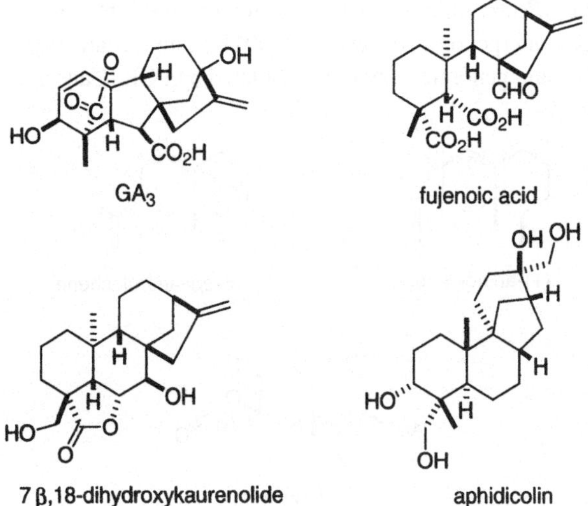

Figure 8.6. Structures of GA_3, 7β, 18-dihydroxykaurnolide, fujenoic acid, and aphidicolin.

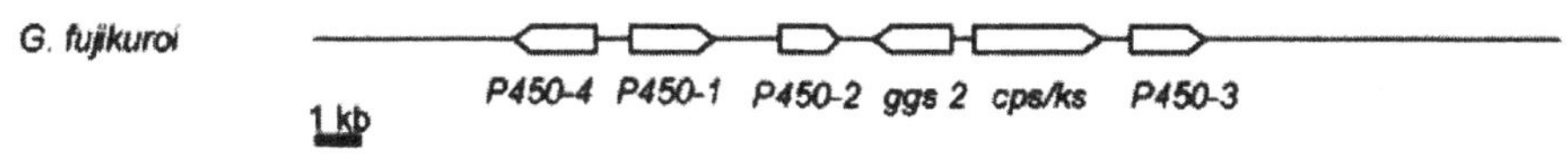

Figure 8.7. The *Gibberella fujikuroi* gibberellin gene cluster.

the gibberellins are the kaurenolides and the fujenoic acids, based on six-membered rather than five-membered rings (Figure 8.6).

3.1.2. Gene Cluster

In *G. fujikuroi* the gibberellin biosynthetic genes are organized in a cluster of at least six genes (Figure 8.7), including a GGPP synthase (*ggs-2*), a bi-functional terpene cyclase (*cps/ks*), and four cytochrome P450 monooxygenases (P450-1, P450-2, P450-3, and P450-4) (Tudzynski and Hölter, 1998; Tudzynski *et al.*, 2001). These genes were identified by a combination of differential cDNA screening and chromosome walking (Tudzynski and Hölter, 1998; Tudzynski *et al.*, 2001). Two key strategies have been employed by Tudzynski and colleagues to determine the function of the genes in the gibberellin cluster. The first involved identifying the intermediates that accumulate in gene-specific disruptants. For the second, the gene cluster was deleted and the intermediates formed when specific pathway genes were reintroduced, were identified (Linnemannstöns *et al.*, 1999; Rojas *et al.*, 2001; Tudzynski *et al.*, 2002). The chemical feeding has proved to be particularly useful as an *in vivo* assay for the reaction catalyzed by each enzyme.

3.1.3. Biosynthesis of GA_3

The proposed first committed step in the pathway to GA_3 (Figure 8.8) is catalyzed by GGPP synthase. Interestingly, there are two copies of this gene in *G. fujikuroi: ggs-1*, possibly required for primary metabolism (Mende *et al.*, 1997), and *ggs-2*, specific for gibberellin biosynthesis (Tudzynski and Hölter, 1998). After GGPP is synthesized, the key cyclization event in gibberellin biosynthesis occurs, setting in place the tetracyclic skeleton with the appropriate stereochemistry. This is a two-step process whereby GGPP is first converted to bicyclic −/− copalyl diphosphate [−/− CDP] in a type-B cyclization (initiated by double bond protonation), followed by a type-A cyclization (initiated by diphosphate loss) and skeletal rearrangement to give tetracyclic *ent*-kaurene (Figure 8.8). This process is catalyzed by a bifunctional copalyl diphosphate synthase/kaurene synthase (Cps/Ks) in both *G. fujikuroi* and a *Phaeosphaeria* sp. (Kawaide *et al.*, 1997, 2000). In plants *ent*-kaurene is also synthesized via CDP but by two independent enzymes, Cps and Ks. The two cyclization reactions catalyzed by Cps/Ks lead to the generation of five new stereocentres in *ent*-kaurene. Control of stereochemistry in both cyclization steps is essential to gibberellin biosynthesis. Another bifunctional enzyme from the fungus *Phoma betae* is involved in the initial steps leading to aphidicolin, an antiviral tetracyclic diterpene (Figure 8.6) (Oikawa *et al.*, 2001). Aphidicolan-16β-ol synthase has remarkable sequence similarity to Cps/Ks and catalyzes a two-step cyclization via a diastereoisomer of −/−CDP, +/−*syn*-CDP. This suggests that only minor sequence changes in the enzyme

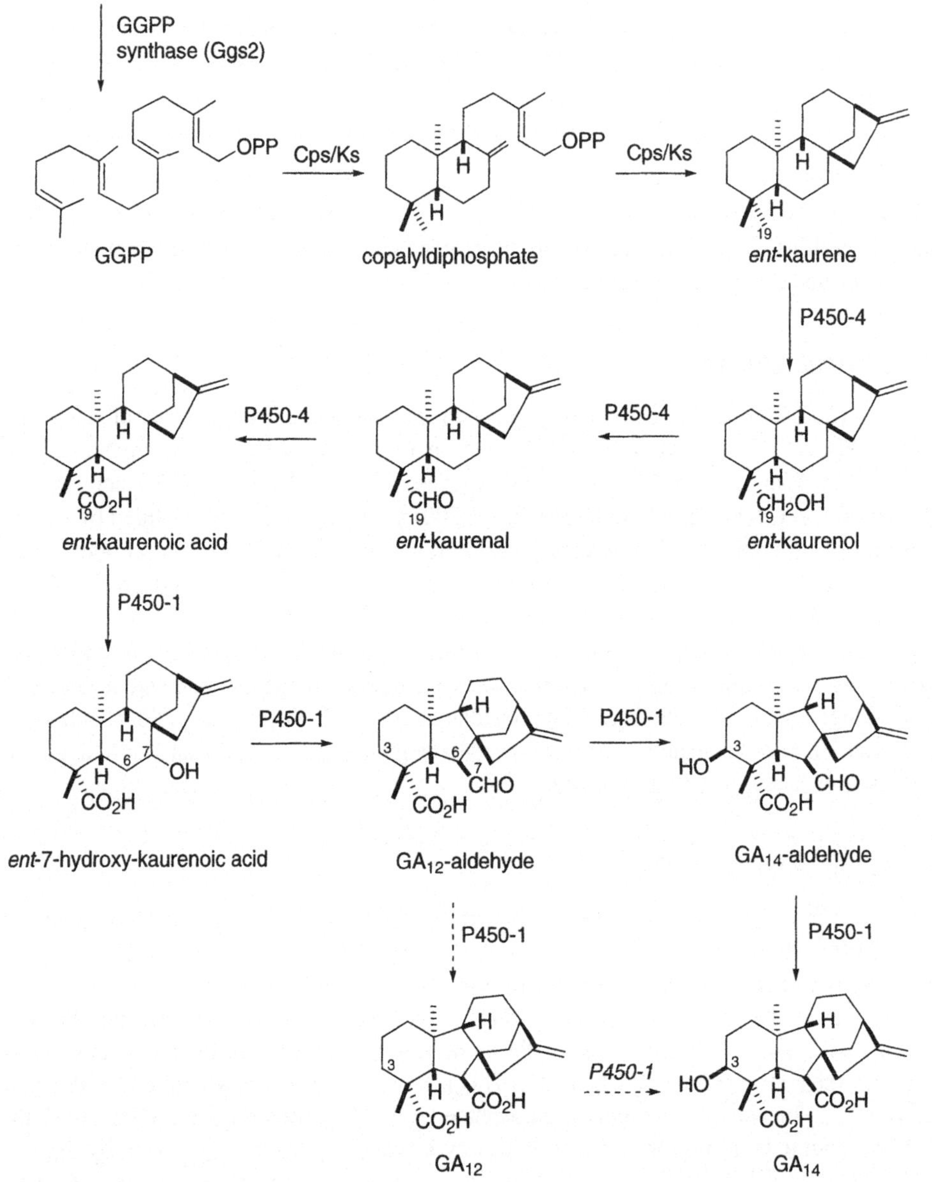

GGPP synthase (Ggs2)

GGPP → Cps/Ks → copalyldiphosphate → Cps/Ks → *ent*-kaurene

P450-4

ent-kaurenoic acid ← P450-4 ← *ent*-kaurenal ← P450-4 ← *ent*-kaurenol

P450-1

ent-7-hydroxy-kaurenoic acid → P450-1 → GA_{12}-aldehyde → P450-1 → GA_{14}-aldehyde

P450-1

GA_{12} → *P450-1* → GA_{14}

Figure 8.8(A) Biosynthetic pathway for GA_3 synthesis in *Gibberella fujikuroi*. This part of Figure 8.8 shows the pathway to GA_{12} and GA_{14}.

are sufficient to alter the diterpene skeleton. The bifunctional abietadiene synthase from grand fir (*Abies grandis*), catalyzes diterpene cyclization via another stereoisomer of −/− CDP (Peters and Croteau, 2002).

Formation of *ent*-kaurene in *G. fujikuroi* is followed by sequential oxidation at C-19 to produce *ent*-kaurenoic acid. Using the approaches described above, *P450-4* was shown to encode a multi-functional *ent*-kaurene oxidase that performs all three oxidation steps

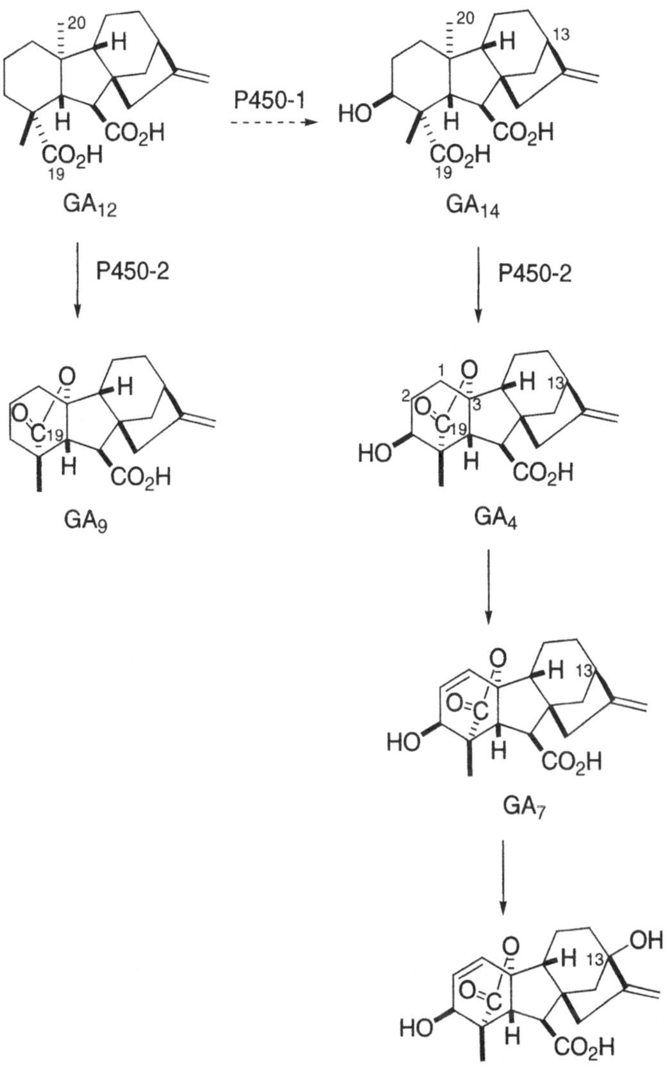

Figure 8.8(B) Biosynthetic pathway for GA$_3$ synthesis in *Gibberella fujikuroi*. This part of Figure 8.8 shows the pathway from GA$_{12}$ and GA$_{14}$ to GA$_3$.

from *ent*-kaurene to *ent*-kaurenoic acid, via *ent*-kaurenol and *ent*-kaurenal (Tudzynski *et al.*, 2001) (Figure 8.8). The conversion of 4-[^{14}C]kaurene to *ent*-kaurenoic acid by a *P450-4* transformant in a cluster-deletion background supports this conclusion. Disruption of *P450-4* results in the accumulation of *ent*-kaurene as the only identifiable intermediate, suggesting that this is the substrate for P450-4. Plants also contain a multifunctional *ent*-kaurene oxidase capable of catalyzing all three oxidation steps, but the fungal and plant enzymes belong to different P450 families (Swain *et al.*, 1995; Helliwell *et al.*, 1999). When a *P450-4* disruptant was fed ^{14}C-labelled *ent*-kaurene, *ent*-kaurenol, *ent*-kaurenal, or *ent*-kaurenoic acid, only the latter was converted to GA_4. Surprisingly, ^{14}C-labelled GA_{12}-aldehyde and GA_{14} were metabolized only as far as GA_4 in this mutant, suggesting that the enzyme, or enzymes, required for the conversion of GA_4 through GA_7 to GA_3 are not expressed. By contrast another mutant with a point mutation in *P450-4* is able to convert GA_4 through to GA_3. Transcript analysis showed that disruption of *P450-4* in either mutant background negatively influenced the expression of *cps/ks*, suggesting that *P450-4* is under positive feedback control by a product of the pathway. The later pathway enzyme or enzymes may be under similar control.

In *G. fujikuroi* the next four steps in the pathway from *ent*-kaurenoic acid to GA_{14} are conducted by the multifunctional P450-1 enzyme (Rojas *et al.*, 2001) (Figure 8.8). Targeted disruption of this gene resulted in a mutant that accumulated *ent*-kaurenoic acid suggesting that this is the substrate of P450-1. However, incubation of this mutant with *ent*-[^{14}C]kaurenoic acid or [^{14}C]GA_{12}-aldehyde showed that, in contrast to wild-type, neither compound could be further metabolized, suggesting that these compounds were also substrates and intermediates for P450-1. This was confirmed by feeding a number of pathway intermediates to a cluster-deletion strain, containing a functional *P450-1*: when *ent*-[^{14}C]kaurenoic acid, *ent*-7α-hydroxy[^{14}C]kaurenoic acid, [^{14}C]GA_{12}-aldehyde, or [^{14}C]GA_{12} were fed, the main product that accumulated was [^{14}C]GA_{14}, demonstrating *in vivo* that P450-1 can catalyze all four oxidation steps in this pathway. In addition P450-1 may be involved in two branch pathways to the fujenoic acids and the C-18 oxidized kaurenolides (Figure 8.6). These results demonstrate that P450-1 is capable not only of sequential oxidations at a single carbon atom, C-7, but also of oxidations at carbon atoms C-6, C-3, and possibly C-18. Additionally, the sequence of some of these oxidations may be interchanged. While the main pathway from GA_{12}-aldehyde is C-3 oxidation to produce GA_{14}-aldehyde followed by further oxidation at C-7 to give GA_{14} (all catalyzed by P450-1), it appears that C-7 oxidation can precede C-3 oxidation giving GA_{12}. Similar multifunctional P450 *ent*-kaurenoic acid oxidases from *Arabidopsis thaliana* and barley have recently been identified and characterized (Helliwell *et al.*, 2001). Interestingly, *Phaeosphaeria* sp. produces significant amounts of GA_1 through a pathway that is similar to that in higher plants; namely, 3β-hydroxylation of GA_9 and GA_{20} occurs to form GA_4 and GA_1, respectively (Kawaide *et al.*, 1997).

In *G. fujikuroi* C-20 GA_{14} and GA_{12} are then converted to the C-19 lactones, GA_4 and GA_9 respectively, by oxidative removal of C-20 in a step mediated by the enzyme encoded by *P450-2* (Figure 8.8) (Tudzynski *et al.*, 2002). Disruption of *P450-2* gave rise to mutant strains that accumulated GA_{14}. This result indicated that P450-2 is required for oxidation of GA_{14} at C-20. This was confirmed by feeding proposed substrates GA_{14} and GA_{12} to a cluster-deletion strain containing an ectopic copy of *P450-2*. Incubation with [^{14}C]GA_{14} and [^{14}C]GA_{12} resulted in efficient conversion of these substrates to [^{14}C]-GA_4 and [^{14}C]-GA_9, respectively. To investigate whether the C-20 oxidation involved the participation of alcohol and aldehyde intermediates, the C-20 alcohol and aldehyde of both GA_{12} and 3β-hydroxylated GA_{14} were incubated with this strain. Only partial conversion of the alcohol into the acid was observed. These results suggest that, unlike the situation found in plants, the C-20 oxidation catalyzed by P450-2 does not involve the formation of free alcohols or aldehydes and that the intermediates remain enzyme-bound. The intermediates for the *ent*-kaurene oxidase (P450-4) and GA_{14} synthase (P450-1) also fail to accumulate.

In plants the series of progressive oxidations required to remove the C-20 of GA_{20} to form the C_{19}-GAs are catalyzed by a multifunctional 2-oxoglutarate-dependent dioxygenase, rather than by a P450 as in *G. fujikuroi* (Lange *et al.*, 1994; Phillips *et al.*, 1995). By contrast, no dioxygenase genes are found in the *G. fujikuroi* gene cluster.

The final steps in GA_3 biosynthesis in *G. fujikuroi* are C-1/2 desaturation followed by C-13 hydroxylation. While the enzyme(s) responsible for these steps have not yet been identified, it is tempting to speculate that *P450-3* may encode an enzyme responsible for both these oxidation reactions. Experiments similar to those described above may establish a role of this enzyme in the pathway.

Are additional genes required for GA biosynthesis? No additional genes that would appear to have a role in gibberellin biosynthesis have been identified near those already described. Adjacent to *P450-4* is an unidentified ORF, *orf3*, a gene similar to a major facilitator transporter, *smt*, and a dehydrogenase, *alc-dh* (Voss *et al.*, 2001). Deletion of *smt* had no effect on the yield of GA.

3.1.4. Regulation

Unlike the aflatoxin and trichothecene gene clusters (Woloshuk *et al.*, 1994; Proctor *et al.*, 1995b; Yu *et al.*, 1996), no pathway-specific transcription factor has been identified for the GA gene cluster. However, all genes appear to be under the control of the global nitrogen regulator, AreA (Tudzynski *et al.*, 1999). Classic motifs recognized by AreA are found in the upstream region of all gibberellin biosynthetic genes, and deletion of the *G. fujikuroi areA* represses expression of this pathway. Gibberellin biosynthesis in *G. fujikuroi* is specifically reduced by high concentrations of ammonium or glutamine.

In plants, GA 20-oxidase is an important regulatory enzyme for GA biosynthesis, its expression being negatively feedback-regulated by biologically active GAs. However, there is no evidence for a similar mechanism operating in *G. fujikuroi* (Tudzynski *et al.*, 2002).

3.2. Indole-Diterpenes

Indole-diterpenes are a large, structurally diverse group of fungal secondary metabolites, many of which are potent tremorgenic mammalian mycotoxins (Steyn and Vleggaar, 1985). Metabolites of this class have been reported in ascomycetous fungi of the subphyla Plectomycetes (*Aspergillus* and *Penicillium*) and Pyrenomycetes (*Claviceps* and *Epichloë*). Indole-diterpenes are commonly synthesized by fungi that form endophytic associations with grasses (Scott, 2001) and woody plants (Ondeyka *et al.*, 1997; Bills *et al.*, 1992). In species of the graminicolous Clavicipitaceae (sub-family Clavicipitideae) they are associated with various animal disorders such as 'ryegrass staggers' syndrome (Fletcher and Harvey, 1981). The recent report of indole-diterpenes in *Chaunopycnis pustulata* and *C. alba* (Bills *et al.*, 2002) extends their distribution to insect or fungal parasitic lineages of the Clavicipitaceae (subfamily Cordycipitoideae). The ecological affinities of these fungi for plants, insects, or other fungi may provide a clue to the benefits of this biosynthetic ability. Synthesis of indole-diterpenes by the fungus may confer an ecological advantage for both parasite and host, and be a selective pressure for selection and maintenance of gene clusters for the synthesis of these metabolites (Schardl, 1996).

The biological effects most commonly associated with indole-diterpenes are tremorgenicity in mammals and toxicity for insects. The biochemical mechanisms for these toxicities appear to involve multiple effects on neurotransmitter release in the central and peripheral nervous systems. Mechanisms identified to date include modulation of mammalian γ-aminobutyric acid (GABA)-gated ion channels (Yao *et al.*, 1989), inhibition of mammalian calcium-activated K^+ (maxi-K) channels (Knaus *et al.*, 1994), and inhibition of glutamate-gated chloride channels of insects (Smith *et al.*, 2000b).

3.2.1. Chemical Diversity

The indole-diterpenes have been somewhat arbitrarily classified into six structural groups, namely the penitrems, janthitrems, lolitrems, aflatrem, paxilline, and the paspaline/paspalinine/paspalitrems (Figure 8.9) (Steyn and Vleggaar, 1985). To these groups could be added the more recently discovered terpendoles (Gatenby *et al.*, 1999; Huang *et al.*, 1995; Tomoda *et al.*, 1995), shearinines (Belofsky *et al.*, 1995), sulpinines (Laakso *et al.*, 1992), and nodulisporic acid (Ondeyka *et al.*, 1997) (Figure 8.9). The most recently discovered fungal indole-diterpenes, the thiersinines, possess a unique spirocyclic structure (Li *et al.*, 2002). All these compounds have a cyclic diterpene skeleton derived from four isoprene units and an indole moiety derived from tryptophan or a tryptophan precursor, but very little is known about the pathways for their biosynthesis. Biosynthetic schemes have been proposed on the basis of chemical identification of likely intermediates, from the organism of interest and related filamentous fungi (Gatenby *et al.*, 1999; Mantle and Weedon, 1994; Munday-Finch *et al.*, 1996), but until recently none of the proposed steps had been validated by biochemical or genetic studies.

3.2.2. Gene Cluster

The recent cloning of a cluster of genes from *Penicillium paxilli* that are necessary for the biosynthesis of paxilline (Figure 8.10), a potent mammalian tremorgen (Cole *et al.*, 1974), has provided for the first time an insight into the enzymes required for indole-diterpene biosynthesis (Young *et al.*, 2001). Although the precise boundaries of the cluster are still to be defined, at least 6 genes required for paxilline biosynthesis have been identified on the basis of gene inactivation studies. The first committed step in the pathway is catalyzed by a geranylgeranyl diphosphate synthase, encoded by *paxG* (Figure 8.11). *P. paxilli*, like *G. fujikuroi*, contains two GGPP synthase enzymes, one for primary metabolism, *ggs1*, and one for secondary metabolism, *paxG*. The recent discovery of two GGPP synthase genes in the perennial ryegrass endophyte, *Neotyphodium lolii* (Young and Scott, unpublished results), which synthesizes the indole-diterpene lolitrem B, would suggest that a "signature" for identifying fungi that have the capability to produce diterpenes is the presence of two copies of the GGPP synthase gene. Deletion of *paxG* in *P. paxilli* resulted in a paxilline-negative phenotype, consistent with this gene being necessary for paxilline biosynthesis (Young *et al.*, 2001). The complete absence of paxilline and other indole-diterpenes in this deletion, and other deletion derivatives isolated in this species, would suggest that there is subcellular compartmentalization of the two GGPP synthases, as the GGPP product of Ggs1 is unable to complement deletion derivatives of *paxG* (Young *et al.*, 2001).

The identification of a GGPP synthase in this cluster suggests that IPP and DMAPP, rather than GGPP (Munday-Finch *et al.*, 1996), are the carbon precursors from primary metabolism that support indole-diterpene biosynthesis. The high rates of incorporation of radiolabeled anthranilic acid, compared to tryptophan, into nodulisporic acid by a

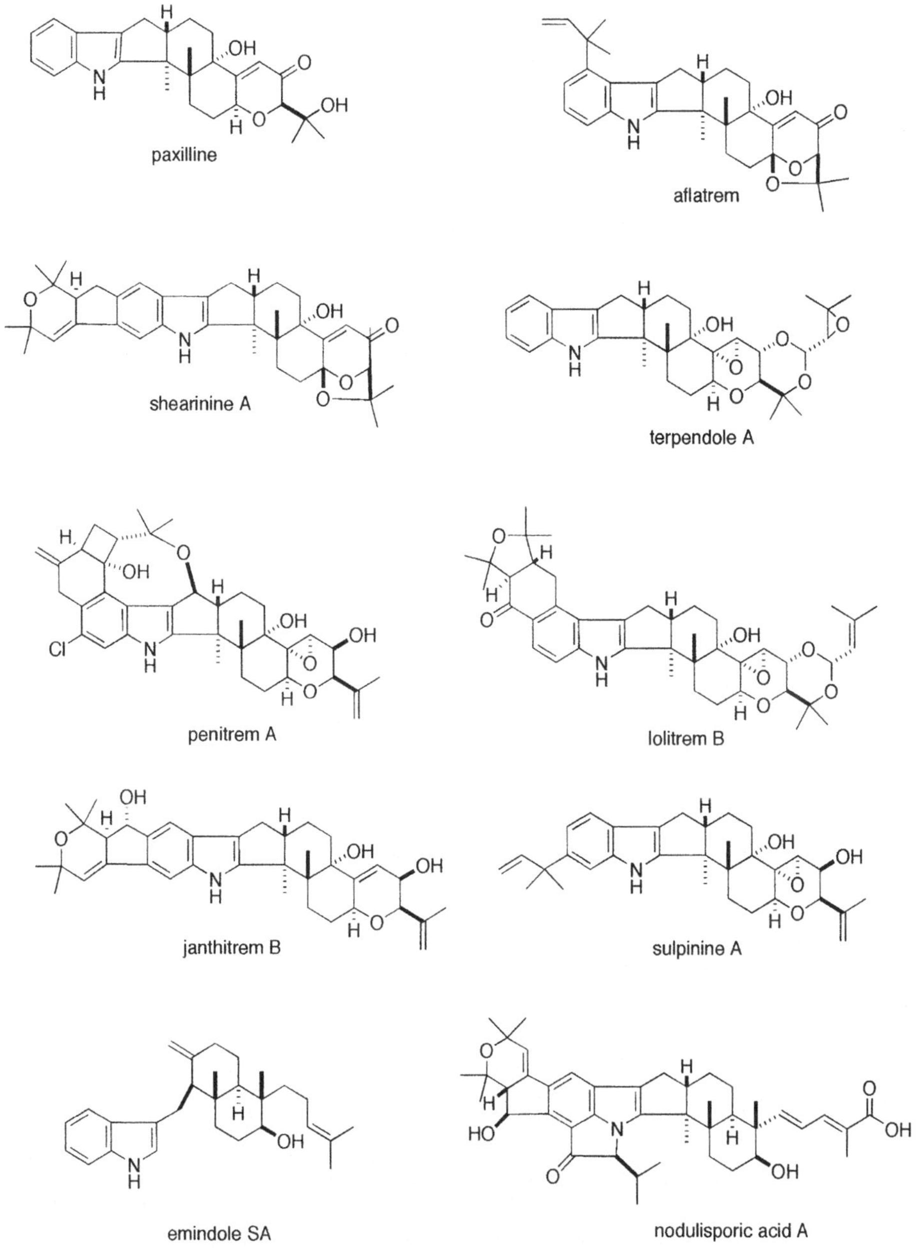

paxilline
aflatrem
shearinine A
terpendole A
penitrem A
lolitrem B
janthitrem B
sulpinine A
emindole SA
nodulisporic acid A

Figure 8.9. Structural diversity of the indole-diterpenes.

P. paxilli
1 kb
paxG
paxM
paxC
paxP
paxQ

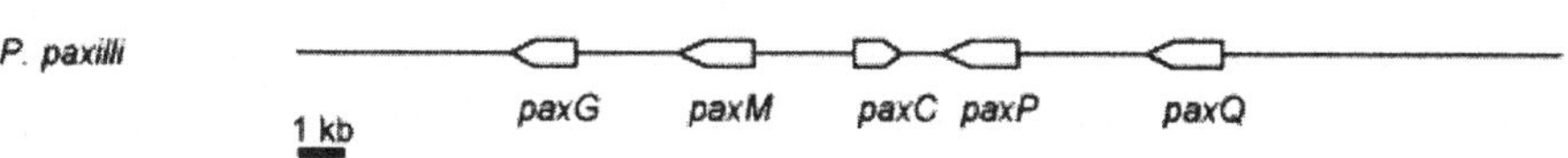

Figure 8.10. Paxilline biosynthesis gene cluster in *Penicillium paxilli*.

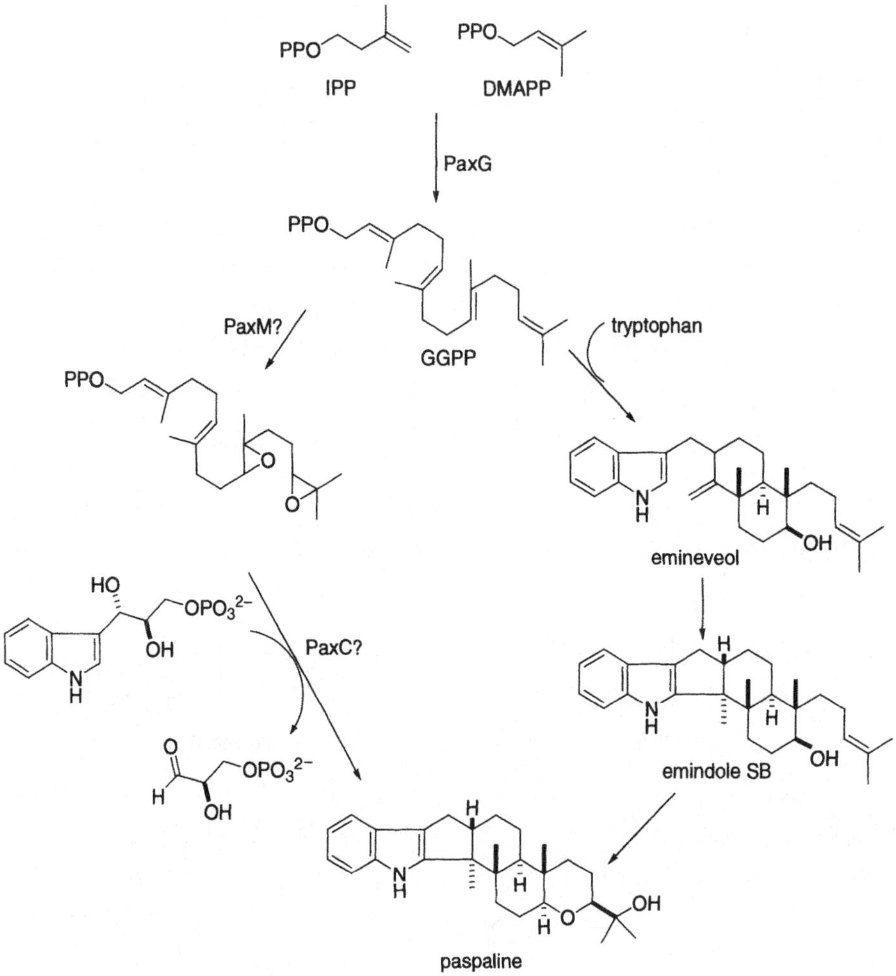

Figure 8.11(A) Formation of an indole-diterpene from GGPP and an indole source. This part of Figure 8.11 shows schemes proposed by ourselves and Munday-Finch, Wilkins, and Miles (1996) for the route to paspaline, with indole 3 glycerol phosphate as the proposed indole source in the former and tryptophan in the latter.

Nodulisporium sp. (Byrne *et al.*, 2002) would suggest that indole-3-glycerol phosphate is the primary source of the indole group for this class of compounds (Figure 8.11). Candidate genes for the early steps in indole-diterpene biosynthesis are *paxM*, a FAD-dependent monooxygenase, and *paxC*, a prenyl transferase (cyclase). Deletion of either *paxM* or *paxC* (Scott *et al.*, unpublished) results in mutants that lack the ability to synthesize any identifiable indole-diterpene. On the basis of these results, we propose that the first stable indole-diterpene product formed is paspaline and that production requires the action of PaxG, M, and C. Figure 8.11 outlines a proposed chemical scheme for the incorporation of indole-3-glycerol phosphate into the C-20 terpenoid and cyclization (including skeletal rearrangement) of the product to form a stable indole-diterpene. Formation of the first stable cyclic intermediate is clearly under precise enzymatic control giving only the paspaline PaxP paspaline B PC-M6 13-desoxypaxilline PaxQ PaxQ paxilline β-paxitriol

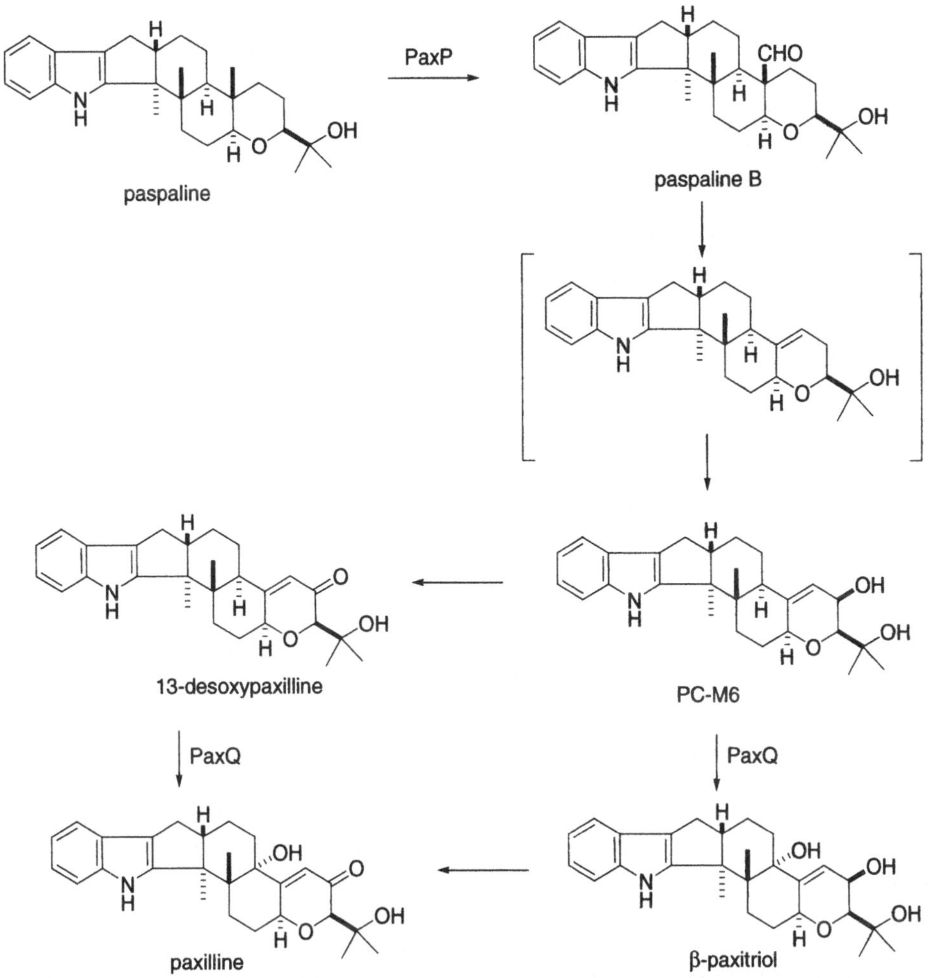

Figure 8.11(B) Formation of an indole-diterpene. Proposed pathway for the formation of paxilline from paspaline.

desired ring structure of the paxilline-related indole-diterpenes. A number of bicyclic and tricyclic isomers of paspaline have been isolated from *Emericella* species, indicating alternative cyclization pathways of indole and GGPP adducts are possible without skeletal rearrangement. A low-yielding biomimetic chemical synthesis of emindole SA has been achieved by Smith and coworkers (Rainier and Smith, 2000), emphasising the power of enzyme catalysis in the control of the correct backbone folding in the formation of paxilline-related indole-diterpene compounds. Considerable effort has gone into the total chemical synthesis of several indole-diterpenes (Smith and Mewshaw, 1985; Smith *et al.*, 2000a). Products of partial cyclization have been also isolated from *Emericella* species, leading to the inclusion of emeniveol and emindole SB in the biosynthetic pathway proposed by Munday-Finch *et al.* (1996). To date there is no evidence from gene cluster analysis that these compounds are intermediates in the biosynthesis of paxilline by *P. paxilli*.

Emindole SB, however, may yet prove to be an intermediate in the biosynthesis of nodulisporic acid A. Apart from the recently discovered thiersinines, all other known indole-diterpenes have the tetracyclic skeleton of paspaline.

At least three other genes are required for the biosynthesis of paxilline (Figure 8.11). Deletion of *paxP* or *paxQ*, both coding for cytochrome P450 monooxygenase enzymes, gave rise to mutants that accumulate paspaline and 13-desoxypaxilline, respectively (McMillan *et al.*, 2003). These compounds are proposed intermediates in a metabolic grid for the biosynthesis of paxilline and other indole-diterpenes (Munday-Finch *et al.*, 1996). The isolation of mutants that accumulate these compounds confirms that *paxP* and *paxQ* are essential for paxilline biosynthesis and that paspaline and 13-desoxypaxilline are the most likely substrates for the corresponding enzymes (Figure 8.11). However, identification of the products of the reactions catalyzed by these enzymes will require either *in vivo* chemical feeding experiments or *in vitro* assays using microsomal fractions or recombinant enzyme over-expressed in a suitable host.

Given that a single cytochrome P450 enzyme is required to oxidatively demethylate the C-14 position of lanosterol in ergosterol biosynthesis (Trzaskos *et al.*, 1986; Fischer *et al.*, 1989), PaxP alone may catalyze the conversion of paspaline to PC-M6, via paspaline B. This hypothesis is supported by the demonstration that P450 enzymes required for gibberellin biosynthesis, in *Gibberella fujikuroi*, are capable of multiple catalytic steps (Rojas *et al.*, 2001; Tudzynski *et al.*, 2002). Thus, fewer genes/enzymes may be required for paxilline biosynthesis than originally proposed (Young *et al.*, 2001). The metabolic grid as proposed suggests that the penultimate substrate for the formation of paxilline could be either 13-desoxypaxilline or β-paxitriol (Munday-Finch *et al.*, 1996). If this is correct, then PC-M6 may also be a substrate for PaxQ, as the conversion from PC-M6 to β-paxitriol also involves a C-13 hydroxylation. Similarly, the conversion of β-paxitriol to paxilline and PC-M6 to 13-desoxypaxilline both involve oxygenation at position C-10, suggesting that a single dehydrogenase enzyme may perform both reactions. A further prediction from this scheme is that double mutants of *paxQ* and this dehydrogenase will accumulate PC-M6.

The cloning of these genes now opens the way to the cloning of other fungal indole-diterpene clusters. Because GGPP synthase is a relatively conserved gene, it will be possible to design degenerate primers to conserved regions of *P. paxilli paxG* and other fungal GGPP synthases and amplify *paxG* orthologues from other fungi. Given that the genes for the biosynthesis of most fungal secondary metabolites are organized in clusters (Keller and Hohn, 1997), genes for other indole-diterpene pathways will be located by chromosome walking from *paxG* orthologues.

Production of the more complex indole-diterpenes requires further functionalization of the paxilline skeleton. Interestingly, this appears to frequently involve the addition of further isoprene units.

4. Tetraterpenes

4.1. Carotenoids

Carotenoids are a structurally diverse group of tetraterpenoid pigments synthesized by photosynthetic and non-photosynthetic bacteria, fungi, and higher plants (Goodwin, 1980).

The majority of the >600 different carotenoids known are C-40 tetraterpenes. They play a crucial role in photooxidative protection and photosynthesis, and although they are not synthesized by animals, play a key role in mammalian nutrition, vision, and cellular differentiation. The numerous benefits of carotenoids for human health and nutrition identify this group of compounds as important industrial products. The most valuable carotenoids commercially are β-carotene and astaxanthin. Both can be synthesized chemically but are also produced by fungal fermentation and in algal cultures (Nonomura, 1989). One of the most promising natural sources of astaxanthin is from *Xanthophyllomyces dendrorhous* (Verdoes *et al.*, 1999).

4.1.1. Chemical Diversity

Many fungi synthesize carotenoid pigments (Goodwin, 1980). The most abundant fungal carotenoid is β-carotene, which accumulates as the end product in carotenogenic species of the Zygomycotina, Ascomycotina, and Basidiomycotina. β-carotene and neurosporaxanthin are the main carotenoids that accumulate in the ascomycetes, *G. fujikuroi* and *Neurospora crassa*. Astaxanthin is the main carotenoid in the basidiomycete yeast, *Xanthophyllomyces dendrorhous*. The Mucorales fungi, *Blakeslea trispora, Mucor circinelloides*, and *Phycomyces blakesleeanus* all accumulate β-carotene. Typical fungal carotenoids have carbonyl groups, may be mono- or bi-cyclic, and possess between 11 and 13 conjugated double bonds.

4.1.2. Biosynthetic Pathway

Even though the end products of carotenoid biosynthesis are very diverse, a common pathway leads to the synthesis of the C-40 cyclic-β-carotene in many prokaryotic and eukaryotic organisms (Figure 8.12) (Sandmann, 2001). The first committed step, catalyzed by a phytoene synthase (CrtB/Psy), involves the head-to-head condensation of two molecules of GGPP to form prephytoene diphosphate, a cyclopropylcarbinyl intermediate. Prephytoene diphosphate is then converted to phytoene by a 1′–1 rearrangement. In plants 15-*cis* phytoene is formed by elimination of the diphosphate and stereospecific proton abstraction. In carotenogenic bacteria and fungi, 15-*cis* phytoene is converted to lycopene in four desaturation steps that are accomplished by a single enzyme, a phytoene desaturase (CrtI) (Sandmann, 2001). As no specific isomerase for the conversion of 15-*cis* phytoene to all-*trans* phytoene has been identified, this reaction is assumed to occur non-enzymatically, possibly by photoisomerization. Alternatively, isomerization may occur on route from phytoene to lycopene. In cyanobacteria and higher plants these steps are catalyzed by three enzymes, phytoene desaturase (CrtP/Pds), ζ-carotene desaturase (CrtQ/Zds), and a recently discovered *cis/trans* isomerase (CrtH/CrtISO) (Masamoto *et al.*, 2001; DellaPenna and Pogson, 2002; Park *et al.*, 2002). Lycopene is the main substrate for cyclization. Two main types of ionone rings are formed: the β-ring and the ε-ring. β-carotenes contain a β-ring on both ends of the molecule whereas α-carotenes have a β-ring at one terminus and an ε-ring on the other. The mechanism of cyclization, substantially supported by D_2O-labelling experiments, involves proton attack at C-2 (and C-2′) of lycopene and addition of the C-1 carbocation to the C5–C6 double bond. The resulting carbocation is quenched by loss of a proton either from the C-1 or C-4 to yield a β- or ε-ring, respectively. β-arotene

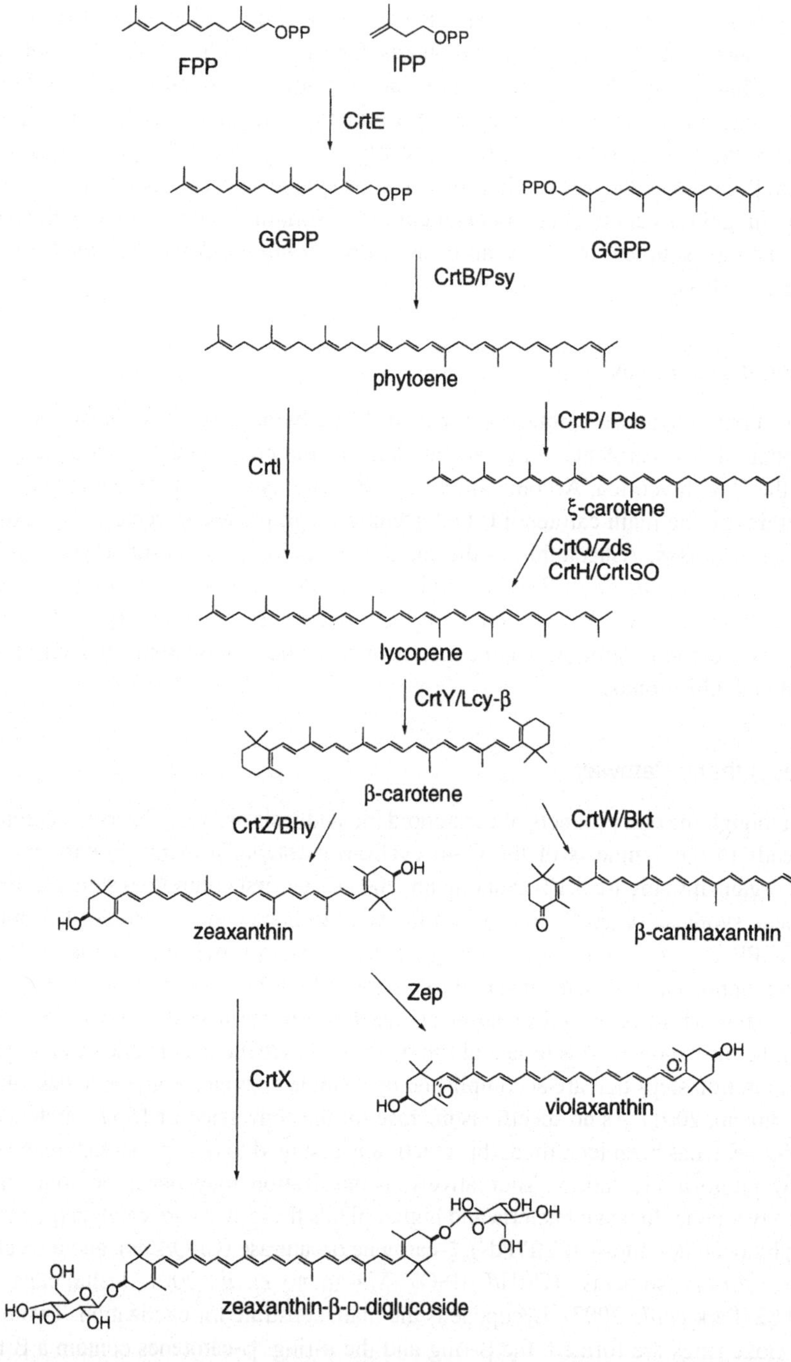

FPP IPP

CrtE

GGPP GGPP

CrtB/Psy

phytoene

CrtI

CrtP/ Pds

ξ-carotene

CrtQ/Zds
CrtH/CrtISO

lycopene

CrtY/Lcy-β

β-carotene

CrtZ/Bhy CrtW/Bkt

zeaxanthin β-canthaxanthin

Zep

CrtX

violaxanthin

zeaxanthin-β-D-diglucoside

Figure 8.12. Biosynthesis of β-carotene and its oxidation products. The enzymes catalyzing the individual reactions have been assigned according to the corresponding gene: CrtE, GGPP synthase; CrtB/Psy, bacterial/eukaryotic phytoene synthase; CrtI, bacterial phytoene desaturase; CrtP/Pds, cyanobacterial/plant phytoene desaturase; CrtQ/Zds, cyanobacterial/plant ζ-carotene desaturase; CrtH/CrtISO cyanobacterial/plant carotene *cis-trans* isomerase; CrtY/Lcp-β bacterial/plant lycopene cyclase; CrtZ/Bhy, bacterial/plant β-carotene hydroxylase; CrtW/Bkt bacterial/plant β-carotene ketolase; Zep, zeaxanthin epoxidase; CrtY, bacterial β-D-diglucoside synthase. Phytoene and ζ-carotene are drawn in the *all*-trans configuration even though *cis*-phytoene may be the first product formed. Adapted from (Sandmann, 2001).

is subsequently converted to various oxidized products (xanthophylls) such as β-canthaxanthin (CrtW/Bkt), violaxanthin (Zep), zeaxanthin (CrtZ/Bhy), and zeaxanthin-β-D-diglucoside (CrtX) (Figure 8.12).

Genes for the synthesis of carotenoids were first cloned from the soil bacteria *Erwinia uredovora* (Misawa *et al*., 1990) and *Erwinia herbicola* (Armstrong *et al*., 1990) and the photosynthetic bacterium *Rhodobacter capsulatus* (Armstrong *et al*., 1989). In these species, as in most carotenogenic eubacteria, the genes are organized in gene clusters (Misawa *et al*., 1990). The *E. uredovora* gene cluster is comprised of six genes, *crtBEIYZX* (Figure 8.12). *E. coli* cells transformed with the entire *crt* cluster from *E. uredovora* are able to accumulate the yellow xanthophylls, zeaxanthin, and zeaxanthin-β-D-diglucoside (Misawa *et al*., 1990; Sandmann *et al*., 1993). The development of this *in vivo* carotenoid biosynthesis assay, using various combinations of these genes expressed in *E. coli*, has been a powerful tool for analyzing the function of carotenoid biosynthesis genes from many organisms, including the fungi (Sandmann *et al*., 1993).

The model ascomycetous fungus, *N. crassa*, has been used extensively to study many biological problems including carotenoid biosynthesis. This organism has a branched carotenoid pathway leading to the dicyclic β-carotene via lycopene and the monocyclic carotenoid, torulene, via 3,4-didehyrolycopene (Figure 8.13). The synthesis of the acyclic carotenoids requires *al-3*, a GGPP synthase (Nelson *et al*., 1989), *al-2*, a phytoene synthase (Schmidhauser *et al*., 1994) and *al-1*, a phytoene desaturase (Schmidhauser *et al*., 1990). *al-1* and *al-2* are linked on chromosome I, but separated by other genes, whereas *al-3* is located on chromosome V (Perkins *et al*., 2001). The completed genome sequence of *N. crassa* reveals the presence of a single GGPP synthase, in contrast to the two copies found in fungi that synthesize gibberellins and indole-diterpenes (*Neurospora* sequencing project; Whitehead Institute/MIT Center for Genome Research; *www-genome.wi.mit.edu*). Carotenogenesis in *N. crassa* is under both light and developmental regulation, at the level of transcription and post-transcription (Ballario and Macino, 1997). Functional analysis of the cDNA for *al-3* in *E. coli* using both *in vitro* and *in vivo* assays demonstrated that Al-3 is a GGPP synthase (Sandmann *et al*., 1993). Substrate specificity assays indicate that DMAPP rather than FPP is the preferred allylic substrate for Al-3, in contrast to the GGPP synthase (CrtE) identified in *E. uredovora* which utilizes FPP as its allylic substrate. In a heterologous complementation assay, co-expression of *N. crassa al-3* with a plasmid, pACCAR25Δ*crtE*, containing all of the *E. uredovora crt* genes except *crtE*, resulted in the synthesis in *E. coli* of yellow carotenoid pigments. This result confirmed that *al-3* is a functional GGPP synthase that is able to substitute for *crtE* in the biosynthesis of carotenoids in *E. coli* (Figure 8.12). This conclusion was further supported by the synthesis of 15-*cis*-phytoene and its all-*trans* isomer in *E. coli* containing *al-3* and *crtB* (Sandmann *et al*., 1993).

The next steps in the carotenoid biosynthesis pathway (Figure 8.13) require the action of Al-2 (Schmidhauser *et al*., 1994) and Al-1 (Schmidhauser *et al*., 1990). *al-2* encodes a phytoene synthase that condenses two molecules of GGPP to form phytoene (Schmidhauser *et al*., 1994) whereas *al-1* encodes a phytoene desaturase that performs the four-step desaturation of phytoene to lycopene (Figure 8.13). Mutations in *al-2* and *al-1* prevent the synthesis of phytoene and lycopene respectively (Hausmann and Sandmann, 2000). In contrast to the structurally related bacterial enzymes, which require FAD for desaturation, Al-1 uses NAD as the cofactor. *In vitro* experiments also showed that Al-1 can accept as substrates many of the intermediates that form in the conversion of phytoene to lycopene (Hausmann and Sandmann, 2000). Interallelic complementation experiments

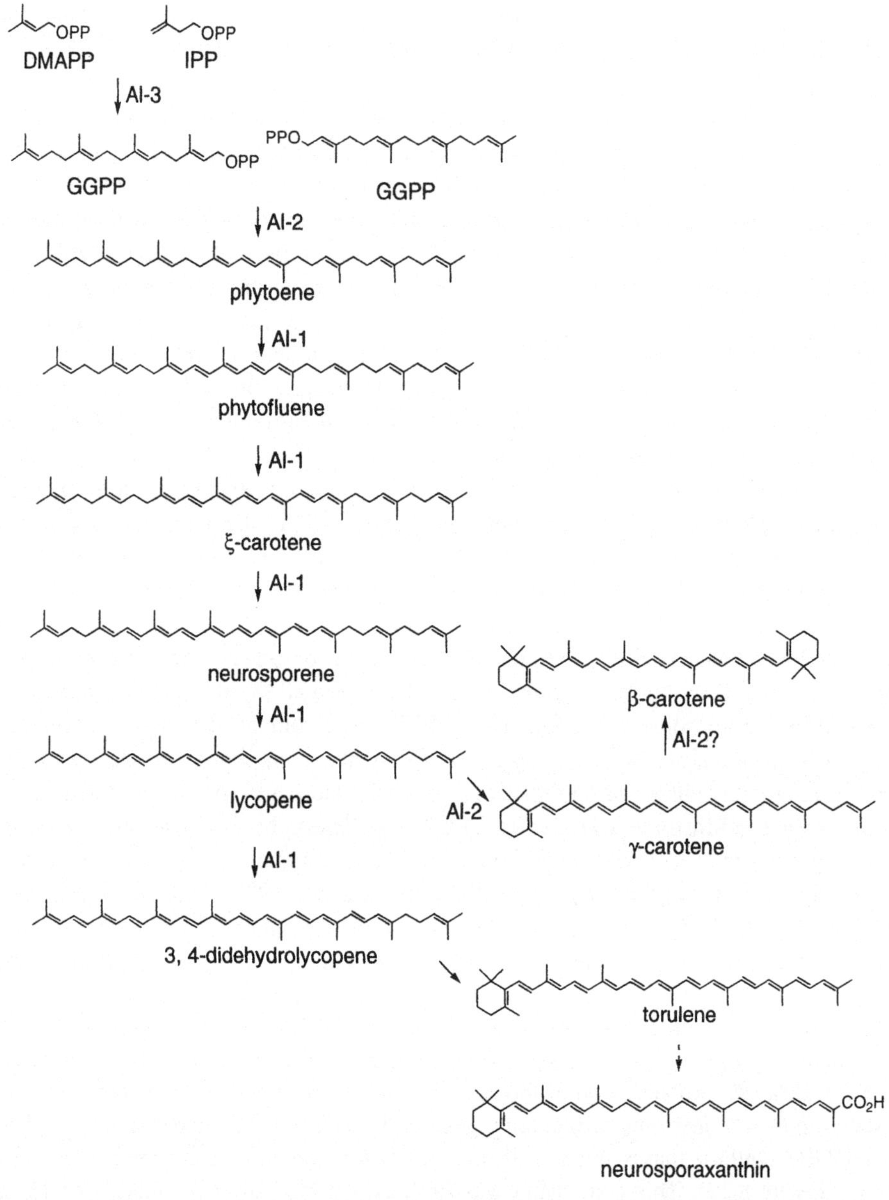

DMAPP IPP

Al-3

GGPP GGPP

Al-2

phytoene

Al-1

phytofluene

Al-1

ξ-carotene

Al-1

neurosporene

β-carotene

Al-2?

Al-1

lycopene

Al-2

γ-carotene

Al-1

3, 4-didehydrolycopene

torulene

neurosporaxanthin

Figure 8.13. Carotenoid biosynthesis pathway in *Neurospora crassa*. The figure is adapted from that of Arrach *et al.* (2002). The enzymes catalyzing the individual reactions have been assigned according to the corresponding gene: Al-3, GGPP synthase; Al-2, bifunctional phytoene synthase, and lycopene cyclase; Al-1 phytoene desaturase. The question mark indicates a possible reaction catalyzed by Al-2. Phytoene and ζ-carotene are drawn in the *all*-trans configuration even though *cis*-phytoene may be the first product formed.

in *Phycomyces blakesleeanus* have recently provided genetic evidence for the multimeric organization of the phytoene desaturase (Sanz *et al.*, 2002).

Until recently the enzyme cyclizing lycopene and 3,4-didehydrolycopene to β-carotene and torulene had not been identified in *N. crassa*. However, recent work in the basidiomycete, *Xanthophyllomyces dendrorhous* (*Phaffia rhodozyma*) demonstrated that *crtYB*, encodes a bifunctional enzyme that has both phytoene synthase (*crtB*) and lycopene cyclase (*crtY*) functions (Verdoes *et al.*, 1999). The dual function associated with *crtYB* has also been found for the homologous genes, *carRA* and *carRP*, from the Zygomycetes, *P. blakesleeanus* (Arrach *et al.*, 2001) and *Mucor circinelloides* (Velayos *et al.*, 2000), and *carRA* in the Ascomycete, *G. fujikuroi* (Linnemannstöns *et al.*, 2002). However, the CarRA protein from *P. blakesleeanus* may be post translationally cleaved into two independent proteins (Arrach *et al.*, 2001). The *crtYB, carRA*, and *carRP* genes in these four fungi are all physically linked with their corresponding phytoene dehydrogenase (*carB*) to form small carotenoid gene clusters; both genes being coordinately regulated by blue light. Alignment of the Al-2 polypeptide sequence from *N. crassa* with the other fungal sequences demonstrated that this fungus also appears to contain a bifunctional enzyme (Arrach *et al.*, 2002). Taking advantage of the fact that *N. crassa* accumulates predominantly neurosporaxanthin when illuminated at low temperatures, Arrach *et al.* (2002) were able to isolate UV-induced mutants that accumulated a pale red polar carotenoid pigment, probably corresponding to the oxidized product of 3,4-didehydrolycopene (Figure 8.13). Transformation of the wild-type *al-2* gene into the mutants restored the wild-type pigmentation but in just a small proportion of the transformants (Arrach *et al.*, 2002). This low frequency of restoration was probably due to quelling, a vegetative gene-silencing phenomenon associated with the presence of duplicate copies of the gene (Romano and Macino, 1992). The mutations responsible for the red pigmentation are located in the N-terminal domain of Al-2, the domain attributed with cyclase activity in other fungi (Verdoes *et al.*, 1999; Velayos *et al.*, 2000; Arrach *et al.*, 2001). The cyclase activity in the fungal enzyme is associated with an N-terminal extension that is absent in the phytoene synthases from plants and bacteria. These organisms have a separate gene for the cyclase, which shows some similarity with the N-terminal domain of the fungal cyclase. The presence of a bifunctional phytoene synthase/lycopene cyclase in representative species of the Basidiomycetes, Ascomycetes, and Zygomycetes would suggest that this gene must have been acquired very early in the evolution of the fungi, possibly from a bacterial ancestor of the extant Actinomycetales (Krubasik and Sandmann, 2000). By contrast the plant β- and ε-cyclases and the β-cyclase from cyanobacterium appear to have evolved from the classical monomeric β-cyclase gene, *crtY* (Krubasik and Sandmann, 2000).

5. Proteins of Isoprenoid Biosynthetic Pathways

The chemical diversity of the terpene family originates ultimately in the regio- and stereospecific-cyclization of linear terpenes. Terpene synthesis and cyclization are catalyzed by enzymes from three distinct superfamilies of prenyl transferases (Murzin *et al.*, 1995). Although the three-dimensional structures are known for two fungal terpene cyclases (Caruthers *et al.*, 2000; Rynkiewicz *et al.*, 2001), much about the mechanism of action of these enzymes, and the oxidoreductases that functionalize and modify the core structure, remains to be determined.

5.1. Initiation of Prenyl Transfer

The initial step common to all terpene chain elongations and cyclizations characterized to date is the formation of an electrophilic center (carbocation or epoxide). The final step is the quenching of the carbocation and release from the enzyme of a diphosphate group. The initial carbocation can be formed in two distinct ways: cleavage of the diphosphate group, designated class A, or protonation of a C=C double bond or an epoxide, designated class B (Sacchettini and Poulter, 1997). Class A reactions require a divalent metal ion for activity. Following carbocation formation, there are three distinct reaction pathways: head-to-tail addition of IPP to make linear terpene diphosphates, intramolecular cyclization, and head-to-head joining of FPP and GGPP moieties to make C30 and C40 terpenoids.

5.2. Prenyl Transferase Structure and Classification

Many prenyl transferases share the "prenyl fold" of the *terpenoid synthases* superfamily, a multiple-helix bundle, first identified in avian FPP synthase (Tarshis *et al.*, 1994). The relative orientations of the six inner helices are strongly conserved and form a substantially hydrophobic cleft for the substrate(s) (Figure 8.14). All sesquiterpene cyclases structurally characterized to date belong to this family. Other structurally characterized members of this superfamily include human squalene synthase, a C-30 synthase, and FPP synthase. In all structurally characterized members of this superfamily, the initial carbocation is formed by cleavage of the diphosphate. A diphosphate-Mg^{2+} binding site located between a pair of helices of the substrate-binding pocket with sequence motifs (N/D)Dxx(D/E) on one helix followed about 130 residues later by (N/D)Dxx(S/T)xxxE on a second helix (Figure 8.14). In FPP synthase, which binds two diphosphate moieties, the

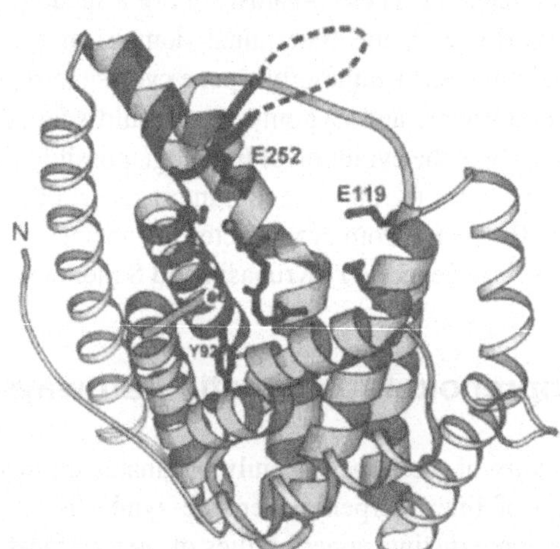

Figure 8.14. Cartoon representation of the structure of aristolochene synthase from *P. roqueforti*. The core prenyl fold is shown with wide helices; peripheral structure is shown with narrow helices. The key side chains of the conserved DDxxD motif (here ^{115}DxxxE), and the (N,D)Dxx(S,T)xxxE (here ^{244}NxxxSxxxE) are shown in ball-and-stick representation. The side chain of Tyr 92, which protonates the C6=C7 double bond in the second cyclization step, is similarly displayed. The bound farnesol is shown in the substrate-binding pocket, but not necessarily in the conformation adopted by farnesyl diphosphate. The helices and loop that undergo major conformational change on binding of diphosphate to trichodiene synthase are depicted with reverse shading. The 8-residue part of this loop that is disordered in the structure of aristolochene synthase is shown as a dashed line.

first four residues of the latter motif are replaced by DDxxD. For squalene synthase (Pandit *et al.*, 2000), the motifs appear as DxxxD. A second distinguishing characteristic of these prenyl transferases is the presence of polarizable tryptophan and asparagine side chains in the substrate binding cleft, presumably to direct proper folding and stabilize the carbocations. While these motifs have been observed in all structurally characterized examples of this superfamily, copalyl diphosphate synthases of both plant and fungi lack these motifs, characteristic of Type A cyclization. Instead they feature a DxDD motif.

Although their involvement in fungal secondary metabolism has not been observed to date, there are two other superfamilies of prenyl transferases, exemplified by squalene–hopene cyclase from the bacterium *Alicyclobacillus acidocaldarius* (Wendt *et al.*, 1997) and by undecaprenyl synthase from the bacterium *Micrococcus luteus* (Fujihashi *et al.*, 2001). The $(\alpha\alpha)_6$ toroidal fold of squalene cyclase is shared by protein–prenyl transferases, such as rat Rab geranylgeranyl transferase (Zhang *et al.*, 2000).

There is some confusion in the literature on prenyl transferase categorization. Most recently, they were classified into three classes (I, II, III) on the basis of substrate (Liang *et al.*, 2002), but head-to-head condensation was not considered. Previously, prenyl transferases were classified on the basis of carbocation origin (Wendt and Schulz, 1998), on the assumption (now probably incorrect) that all Class I (Type A, described earlier) shared the prenyl fold (Figure 8.14).

With structural, functional, and sequence information now at hand, we propose to categorize prenyl transferases firstly according to superfamily membership (Types I, II, III, ...) and secondly according to reaction type (A, B, mixed, ...). Type I prenyl transferases have the α-helical terpene fold (Figure 8.14) and are exemplified by fungal enzymes trichodiene synthase (Rynkiewicz *et al.*, 2001) and aristolochene synthase (Caruthers *et al.*, 2000), plant 5-*epi*-aristolochene synthase (Starks *et al.*, 1997), bacterial enzyme pentalenene synthase (Lesburg *et al.*, 1997), and the vertebrate FPP synthase (Tarshis *et al.*, 1994), and squalene synthase (Pandit *et al.*, 2000). Type II prenyl transferases have the αβ fold found in *cis*-undecaprenyl synthase (Fujihashi *et al.*, 2001). Type III prenyl transferases have an all α-helical toroidal fold, as found in squalene cyclase and several protein prenyl transferases (Wendt *et al.*, 1997; Zhang *et al.*, 2000).

5.3. Trichodiene Synthase

The first step in the trichothecene biosynthetic pathway is trichodiene synthase. The structure of the trichodiene synthase (Tri5) from *F. sporotrichioides* has been determined in apo and diphosphate-Mg^{2+} bound forms for wild-type (Rynkiewicz *et al.*, 2001) and for the weakly active D100E mutant (Rynkiewicz *et al.*, 2002). Binding of one diphosphate and three Mg^{2+} ions to wild-type enzyme are sufficient by themselves to trigger substantial conformational changes in the protein. Whether this conformational change is merely to sequester substrate more effectively from the solvent or whether the conformational change promotes rupture of the C–O (phosphate) bond and separation of the diphosphate group from the cation awaits structures with bound inhibitor or pseudo-substrate. Evidence of diphosphate cleavage as the initial step in the cyclization cascade is found in the formation of the intermediate species (3*R*)-nerolidyl diphosphate, resulting from attack of the diphosphate on the nascent cationic C3 position (Cane and Ha, 1988). In addition to binding to three magnesium ions, the diphosphate moiety binds also to the side chains of tyrosine, lysine, and arginine. Moreover, on binding the diphosphate-Mg^{2+} moiety, ^{101}D of the ^{100}DDSKD moiety changes its hydrogen-bonding partner to close the mouth of the active-site pocket. The D100E mutant in its apo form is negligibly different in structure from the wild-type. However, on diphosphate-Mg^{2+} binding, only minor localized conformational changes occur. Just two magnesium ions coordinate, and these no longer interact with the Arg and Lys side chains. This absence of structural change and weakened electropositive environment for the diphosphate group are consistent with the observed reduced activity and loss of product specificity. The consensus (L/V)(V/L/A)(N/D)Dxx(S/T)xxxE motif, which forms part of an Mg^{2+} binding site is observed here as ^{223}WV**NDLMSFYKE.** All fungal trichiodiene synthases identified to date have very highly conserved sequences.

5.4. Aristolochene Synthase

The structure of aristolochene synthase from *P. roqueforti* conforms to the canonical terpene synthase structure (Caruthers *et al.*, 2000). The fungal structure studied was a Sm^{3+} derivative; the cation binds to the DDxxD motif (here ^{115}DDVLE). The other conserved motif appears here as ^{242}VV**ND**IY**S**YD**KE**. The initial cyclization forms a 10-membered monocyclic species, germacrene A. Formation of the bicyclic +/− aristolochene is initiated by protonation at C-6 of the C6=C7 bond by a nearby tyrosine, Tyr 92 (Calvert *et al.*, 2002a). Mutation of this tyrosine to valine led to formation of β-(*E*)-farnesene, indicating that not only did Tyr function as the acid, but it also directed folding of FPP into the conformation needed for germacrene A formation (Calvert *et al.*, 2002b). Extensive molecular modelling has given a structural basis to the regiospecificity of this enzyme by comparison with 5-*epi*-aristolochene synthase from *N. tabacum*. In the presence of inhibitors farnesyl hydroxyphosphonate or trifluoro-FPP, three magnesium ions are bound to 5-*epi*-aristolochene synthase (Starks *et al.*, 1997), in a manner similar to those in the diphosphate-Mg^{2+} derivative of trichodiene synthase, although Lys and Arg side chains are provided from different secondary structure elements.

6. Final Remarks

The discovery of fungal gene clusters as a co-regulated set of dispensable genes for the production of secondary metabolites has coincided with a rapidly increasing interest in biosynthetic pathways of terpenoid species. To date gene clusters have been discovered for the biosynthesis of polycyclic sesquiterpenes (trichothecenes), diterpenes (gibberellin), indole-diterpenes (paxilline), and tetraterpenes (carotenoids). Evidence to date would suggest that, for the biosynthesis of gibberellins, paxilline, and some of the carotenoids, all the enzymes for the pathway are part of a gene cluster. For the biosynthesis of the trichothecenes, the gene for at least one enzymatic step lies outside the cluster. However, this gene is coordinately regulated by a pathway-specific transcription factor.

The key enzyme encoded by an isoprenoid cluster gene is a prenyl transferase, in this case a terpene cyclase, which directs in a regio- and stereospecific manner a complex sequence of carbon–carbon bond scission and formation, selecting one product from the many thousands of possible sesqui- or diterpene products that could be produced upon cyclization. In some clusters, an additional and essential second prenyl transferase, for example, GGPP synthase, produces the universal precursor. As the products may be self-toxic, additional enzymes, which add protecting groups and proteins to mediate export of the product from the cell, may also be found. A second source of chemical diversity is generated by additional cluster enzymes that mediate functionalization of the cyclic hydrocarbon skeleton. These cytochrome P450 monooxygenases, FAD-dependent oxidoreductases, ester synthases, and esterases may accept substrate variants (although the target atom or group usually, but not always, remains constant), leading to different groups of products. In addition, cytochrome P450 enzymes may exhibit multiple functionality, oxidizing saturated carbon atoms and mediating ring contractions and demethylations. Finally, inactivation of one enzyme, either in natural or engineered fungal strains may lead to a new set of products.

There is still much that is not known about regulation of gene expression, metabolic advantage (if any) of clustering genes, cellular localization of proteins encoded by cluster genes, product channelling of intermediates from one enzyme to another, export machinery, and detailed enzymology, including 3-D structure analyses, of many of the enzymes involved. Thus, there is risk in making generalizations about isoprenoid gene clusters when so few have been characterized. Some of these gaps will be addressed in the next few years when more than a dozen additional fungal genomes become sequenced.

Remarkably few enzymes are required in nature to synthesize compounds, which if synthesized by classical organic methods may require in excess of 50 separate synthetic steps. Harnessing these enzymes will be a major focus of biotechnology in the future.

Acknowledgments

The authors would like to thank Rohan Lowe for preparation of three of the figures, and Susan McCormick and Bettina Tudzynski for comments on the trichothecene and gibberellin sections of the manuscript. This research was supported by research grants MAU-010 and MAUX-0127 from the Royal Society of New Zealand and the Foundation for Research, Science and Technology, respectively.

References

Alexander, N.J., Hohn, T.M., and McCormick, S.P. (1998). The *TRI11* gene of *Fusarium sporotrichioides* encodes a cytochrome P-450 monooxygenase required for C-15 hydroxylation in trichothecene biosynthesis. *Appl. Environ. Microbiol.* **64**, 221–225.

Alexander, N.J., McCormick, S.P., and Hohn, T.M. (1999). TRI12, a trichothecene efflux pump from *Fusarium sporotrichioides*: Gene isolation and expression in yeast. *Mol. Gen. Genet.* **261**, 977–984.

Armstrong, G.A., Alberti, M., and Hearst, J.E. (1990). Conserved enzymes mediate the early reactions of carotenoid biosynthesis in nonphotosynthetic and photosynthetic prokaryotes. *Proc. Natl. Acad. Sci. USA* **87**, 9975–9979.

Armstrong, G.A., Alberti, M., Leach, F., and Hearst, J.E. (1989). Nucleotide sequence, organization, and nature of the protein products of the carotenoid biosynthesis gene cluster of *Rhodobacter capsulatus. Mol. Gen. Genet.* **216**, 254–268.

Arrach, N., Fernandez-Martin, R., Cerda-Olmedo, E., and Avalos, J. (2001). A single gene for lycopene cyclase, phytoene synthase, and regulation of carotene biosynthesis in *Phycomyces. Proc. Natl. Acad. Sci. USA* **98**, 1687–1692.

Arrach, N., Schmidhauser, T.J., and Avalos, J. (2002). Mutants of the carotene cyclase domain of *al-2* from *Neurospora crassa. Mol. Genet. Genomics* **266**, 914–921.

Ballario, P. and Macino, G. (1997). White collar proteins: PASsing the light signal in *Neurospora crassa. Trends Microbiol.* **5**, 458–462.

Belofsky, G.N., Gloer, J.B., Wicklow, D.T., and Dowd, P.F. (1995). Antiinsectan alkaloids: Shearinines A-C and a new paxilline derivative from the ascostromata of *Eupenicillium shearii. Tetrahedron* **51**, 3959–3968.

Bennett, J.W. and Lasure, L.L. (1985). Conventions for gene symbols. In J.W. Bennett and L.L. Lasure (eds.) *Gene Manipulations in Fungi* Academic Press, London, pp. 537–544.

Berbee, M.L. (2001). The phylogeny of plant and animal pathogens in the Ascomycota. *Physiol. Mol. Plant Pathol.* **59**, 165–187.

Bills, G.F., Giacobbe, R.A., Lee, S.H., Paláez, F., and Tkacz, J.S. (1992). Clavine tremorgenic mycotoxins, paspalitrem A and C, from a tropical *Phomopsis. Mycol. Res.* **96**, 977–983.

Bills, G.F., Polishook, J.D., Goetz, M.A., Sullivan, R.F., and White, J.J.F. (2002). *Chaunopycnis pustulata* sp. nov., a new clavicipitalean anamorph producing metabolites that modulate potassium ion channels. *Mycol. Prog.* **1**, 3–17.

Birren, B., Fink, G., and Lander, E. (2002). *Fungal genome initiative: white paper developed by the fungal research community (http://www-genome.wi.mit.edu/).* Centre for Genome Research, Cambridge, MA, USA.

Brown, D.W., McCormick, S.P., Alexander, N.J., Proctor, R.H., and Desjardins, A.E. (2001). A genetic and biochemical approach to study trichothecene diversity in *Fusarium sporotrichioides* and *Fusarium graminearum. Fungal Genet. Biol.* **32**, 121–133.

Byrne, K.M., Smith, S.K., and Ondeyka, J.G. (2002). Biosynthesis of nodulisporic acid A: Precursor studies. *J. Am. Chem. Soc.* **124**, 7055–7060.

Calvert, M.J., Ashton, P.R., and Allemann, R.K. (2002a). Germacrene A is a product of the aristolochene synthase-mediated conversion of farnesylpyrophosphate to aristolochene. *J. Am. Chem. Soc.* **124**, 11636–11641.

Calvert, M.J., Taylor, S.E., and Allemann, R.K. (2002b). Tyrosine 92 of aristolochene synthase directs cyclisation of farnesyl pyrophosphate. *J. Chem. Soc. Chem. Commun.* 2384–2385.

Cane, D.E. and Ha, H.-J. (1988). Trichodiene biosynthesis and the role of nerolidyl pyrophosphate in the enzymic cyclization of farnesyl pyrophosphate. *J. Am. Chem. Soc.* **110**, 6865–6870.

Cane, D.E. and Kang, I. (2000). Aristolochene synthase: Purification, molecular cloning, high-level expression in *Escherichia coli*, and characterisation of the *Aspergillus terreus* cyclase. *Arch. Biochem. Biophys.* **376**, 354–364.

Caruthers, J.M., Kang, I., Rynkiewicz, M.J., Cane, D.E., and Christianson, D.W. (2000). Crystal structure determination of aristolochene synthase from the blue cheese mold, *Penicillium roqueforti. J. Biol. Chem.* **275**, 25533–25539.

Cole, R.J., Kirksey, J.W., and Wells, J.M. (1974). A new tremorgenic metabolite from *Penicillium paxilli. Can. J. Microbiol.* **20**, 1159–1162.

Desjardins, A.E., Proctor, R.H., Bai, G.-H., McCormick, S.P., Shaner, G., Buechley, G., and Hohn, T.M. (1996). Reduced virulence of trichothecene-nonproducing mutants of *Gibberella zeae* in wheat field tests. *Mol. Plant Microbe Interact.* **9**, 775–781.

Dewick, P.M. (2002). The biosynthesis of C5–C25 terpenoid compounds. *Nat. Prod. Rep.* **19**, 181–222.

Fischer, R.T., Stam, S.H., Johnson, P.R., Ko, S.S., Magolda, R.L., Gaylor, J.L., and Trzaskos, J.M. (1989). Mechanistic studies of lanosterol 14 alpha-methyl demethylase: Substrate requirements for the component reactions catalysed by a single P-450 isozyme. *J. Lipid Res.* **30**, 1621–1632.

Fletcher, L.R. and Harvey, I.C. (1981). An association of a *Lolium* endophyte with ryegrass staggers. *N Z Vet. J.* **29**, 185–186.

Fujihashi, M., Zhang, Y.W., Higuchi, Y., Li, X.Y., Koyama, T., and Miki, K. (2001). Crystal structure of *cis*-prenyl chain elongating enzyme, undecaprenyl diphosphate synthase. *Proc. Natl. Acad. Sci. USA* **98**, 4937–4942.

Gatenby, W.A., Munday-Finch, S.C., Wilkins, A.L., and Miles, C.O. (1999). Terpendole M, a novel indolediterpenoid isolated from *Lolium perenne* infected with the endophytic fungus *Neotyphodium lolii. J. Agric. Food Chem.* **47**, 1092–1097.

Goodwin, T.W. (1980). *The Biochemistry of the Carotenoids* (2nd edn. Vol. 1). Chapman & Hall, London.

Harris, L.J., Desjardins, A.E., Plattner, R.D., Nicholson, P., Butler, G., Young, J.C., Weston, G., Proctor, R.H. *et al.* (1999). Possible role of trichothecene mycotoxins in virulence of *Fusarium graminearum* on maize. *Plant Dis.* **83**, 954–960.

Hausmann, A. and Sandmann, G. (2000). A single five-step desaturase is involved in the carotenoid biosynthesis pathway to beta-carotene and torulene in *Neurospora crassa. Fungal Genet. Biol.* **30**, 147–153.

Helliwell, C.A., Chandler, P.M., Poole, A., Dennis, E.S., and Peacock, W.J. (2001). The CYP88A cytochrome P450, *ent*-kaurenoic acid oxidase, catalyzes three steps of the gibberellin biosynthesis pathway. *Proc. Natl. Acad. Sci. U.S.A.* **98**, 2065–2070.

Helliwell, C.A., Poole, A., Peacock, W.J., and Dennis, E.S. (1999). Arabidopsis *ent*-kaurene oxidase catalyzes three steps of gibberellin biosynthesis. *Plant Physiol.* **119**, 507–510.

Hohn, T.M. and Beremand, P.D. (1989). Isolation and nucleotide sequence of a sesquiterpene cyclase gene from the trichothecene-producing fungus *Fusarium sporotrichiodes. Gene* **79**, 131–138.

Hohn, T.M., Desjardins, A.E., and McCormick, S.P. (1995). The *Tri4* gene of *Fusarium sporotrichioides* encodes a cytochrome P450 monooxygenase involved in trichothecene biosynthesis. *Mol. Gen. Genet.* **248**, 95–102.

Hohn, T.M., Krishna, R., and Proctor, R.H. (1999). Characterization of a transcriptional activator controlling trichothecene toxin biosynthesis. *Fungal Genet. Biol.* **26**, 224–235.

Hohn, T.M., McCormick, S.P., and Desjardins, A.E. (1993). Evidence for a gene cluster involving trichothecene-pathway biosynthetic genes in *Fusarium sporotrichioides. Curr. Genet.* **24**, 291–295.

Hohn, T.M., Proctor, R.H., and Desjardins, A.E. (1991). *Biosynthesis of sesquiterpenoid toxins by fungal pathogens*. Paper presented at the EMBO workshop on the molecular biology of filamentous fungi, Berlin.

Huang, X.-H., Tomoda, H., Nishida, H., Masuma, R., and Omura, S. (1995). Terpendoles, novel ACAT inhibitors produced by *Albophoma yamanashiensis*. I. Production, isolation and biological properties. *J. Antibio.* **48**, 1–4.

Jarvis, B.B., Midiwo, J.O., Bean, G.A., Aboul-Nasr, M.B., and Barros, C.S. (1988). The mystery of trichothecene antibiotics in *Baccharis* species. *J. Nat. Prod.* **51**, 736–744.

Kawaide, H., Imai, R., Sassa, T., and Kamiya, Y. (1997). *Ent*-kaurene synthase from the fungus *Phaeosphaeria* sp. L487. cDNA isolation, characterization, and bacterial expression of a bifunctional diterpene cyclase in fungal gibberellin biosynthesis. *J. Biol. Chem.* **272**, 21706–21712.

Kawaide, H., Sassa, T., and Kamiya, Y. (2000). Functional analysis of the two interacting cyclase domains in *ent*-kaurene synthase from the fungus *Phaeosphaeria* sp. L487 and a comparison with cyclases from higher plants. *J. Biol. Chem.* **275**, 2276–2280.

Keller, N.P. and Hohn, T.M. (1997). Metabolic pathway gene clusters in filamentous fungi. *Fungal Genet. Biol.* **21**, 17–29.

Kimura, M., Kaneko, I., Komiyama, M., Takatsuki, A., Koshino, H., Yoneyama, K., and Yamaguchi, I. (1998). Trichothecene 3-*O*-acetyltransferase protects both the producing organism and transformed yeast from related mycotoxins. Cloning and characterization of *Tri101*. *J. Biol. Chem.* **273**, 1654–1661.

Knaus, H.-G., McManus, O.B., Lee, S.H., Schmalhofer, W.A., Garcia-Calvo, M., Helms, L.M.H., Sanchez, M., Giangiacomo, K. *et al.* (1994). Tremorgenic indole alkaloids potently inhibit smooth muscle high-conductance calcium-activated channels. *Biochemistry* **33**, 5819–5828.

Krubasik, P. and Sandmann, G. (2000). Molecular evolution of lycopene cyclases involved in the formation of carotenoids with ionone end groups. *Biochem. Soc. Trans.* **28**, 806–810.

Laakso, J.A., Gloer, J.B., Wicklow, D.T., and Dowd, P.F. (1992). Sulpinines A-C and secopenitrem B: New anti-insectan metabolites from the sclerotia of *Aspergillus sulphureus. J. Org. Chem.* **57**, 2066–2071.

Lange, T., Hedden, P., and Graebe, J.E. (1994). Expression cloning of a gibberellin 20-oxidase, a multifunctional enzyme involved in gibberellin biosynthesis. *Proc. Natl. Acad. Sci. USA* **91**, 8552–8556.

Lee, T., Han, Y.-K., Kim, K.-H., Yun, S.-H., and Lee, Y.-W. (2002). *Tri13* and *Tri7* determine deoxynivalenol- and nivalenol-producing chemotypes of *Gibberella zeae. Appl. Environ. Microbiol.* **68**, 2148–2154.

Lesburg, C.A., Zhai, G., Cane, D.E., and Christianson, D.W. (1997). Crystal structure of pentalenene synthase: Mechanistic insights on terpenoid cyclization reactions in biology. *Science* **277**, 1820–1824.

Li, C., Gloer, J.B., Wicklow, D.T., and Dowd, P.F. (2002). Thiersinines A and B: Novel antiinsectan indole diterpenoids from a new fungicolous *Penicillium* species (NRRL 28147). *Org. Lett.* **4**, 3095–3098.

Liang, P.-H., Ko, T.-P., and Wang, A.H.J. (2002). Structure, mechanism and function of prenyltransferases. *Eur. J. Biochem.* **269**, 3339–3354.

Linnemannstöns, P., Prado, M.M., Fernández-Martín, R., Tudzynski, B., and Avalos, J. (2002). A carotenoid biosynthesis gene cluster in *Fusarium fujikuroi*: The genes *carB* and *carRA*. *Mol. Genet. and Genomics.* **267**, 593–602.

Linnemannstöns, P., Voss, T., Hedden, P., Gaskin, P., and Tudzynski, B. (1999). Deletions in the gibberellin biosynthesis gene cluster of *Gibberella fujikuroi* by restriction enzyme-mediated integration and conventional transformation-mediated mutagenesis. *Appl. Environ. Microbiol.* **65**, 2558–2564.

Mantle, P.G. and Weedon, C.M. (1994). Biosynthesis and transformation of tremorgenic indole-diterpenoids by *Penicillium paxilli* and *Acremonium lolii. Phytochemistry* **36**, 1209–1217.

Masamoto, K., Wada, H., Kaneko, T., and Takaichi, S. (2001). Identification of a gene required for *cis*-to-*trans* carotene isomerization in carotenogenesis of the cyanobacterium *Synechocystis* sp. PCC 6803. *Plant Cell Physiol.* **42**, 1398–1402.

McCormick, S.P. and Alexander, N.J. (2002). *Fusarium Tri8* encodes a trichothecene C-3 esterase. *Appl. Environ. Microbiol.* **68**, 2959–2964.

McCormick, S.P., Alexander, N.J., Trapp, S.E., and Hohn, T.M. (1999). Disruption of *TRI101*, the gene encoding trichothecene 3-*O*-acetyltransferase from *Fusarium sporotrichioides*. *Appl. Environ. Microbiol.* **65**, 5252–5256.

McCormick, S.P., Hohn, T.M., and Desjardins, A.E. (1996). Isolation and characterization of *Tri3*, a gene encoding 15-*O*-acetyltransferase from *Fusarium sporotrichioides*. *Appl. Environ. Microbiol.* **62**, 353–359.

McMillan, L.K., Carr, R.C., Young, C.A., Astin, J.W., Lowe, R.G.T., Parker, E.J., Jameson, G.B. Finch, S.C., Miles, C.O., McManus, O.B., Schmalhofer, W.A., Garcia, M.L., Kaczorowski, G.J., Goetz, M., Tkacz, J.S., and Scott, B. (2003). Molecular analysis of two cytochrome P450 monooxygenase genes required for paxilline biosynthesis in Penicillium paxilli and effects of paxilline intermediates on mammalian maxi-K ion channels. *Mol. Gen. Genom.* **270**, 9–23.

Mende, K., Homann, V., and Tudzynski, B. (1997). The geranylgeranyl diphosphate synthase gene of *Gibberella fujikuroi*: Isolation and expression. *Mol. Gen. Genet.* **255**, 96–105.

Misawa, N., Nakagawa, M., Kobayashi, K., Yamano, S., Izawa, Y., Nakamura, K., and Harashima, K. (1990). Elucidation of the *Erwinia uredovora* carotenoid biosynthetic pathway by functional analysis of gene products expressed in *Escherichia coli*. *J. Bacteriol.* **172**, 6704–1612.

Munday-Finch, S.C., Wilkins, A.L., and Miles, C.O. (1996). Isolation of paspaline B, an indole-diterpenoid from *Penicillium paxilli*. *Phytochemisty* **41**, 327–332.

Murzin, A.G., Brenner, S.E., Hubbard, T., and Chothia, C. (1995). SCOP: A structural classification of proteins database for the investigation of sequences and structures. *J. Mol. Biol.* **247**, 536–540.

Nelson, M.A., Morelli, G., Carattoli, A., Romano, N., and Macino, G. (1989). Molecular cloning of a *Neurospora crassa* carotenoid biosynthetic gene (*albino-3*) regulated by blue light and the products of the *white collar* genes. *Mol. Cell. Biol.* **9**, 1271–1276.

Nonomura, A.M. (1989). Industrial biosynthesis of carotenoids. In N.I. Krinski, M.M. Mathews-Roth, and R.F. Taylor (eds.) *Carotenoid chemistry and biology*. Plenum Press, New York, pp. 365–375.

Oikawa, H., Toyomasu, T., Toshima, H., Ohashi, S., Kawaide, H., Kamiya, Y., Ohtsuka, M., Shinoda, S. *et al.* (2001). Cloning and functional expression of cDNA encoding aphidicolan-16 beta-ol synthase: A key enzyme responsible for formation of an unusual diterpene skeleton in biosynthesis of aphidicolin. *J. Am. Chem. Soc.* **123**, 5154–5155.

Ondeyka, J.G., Helms, G.L., Hensens, O.D., Goetz, M.A., Zink, D.L., Tsipouras, A., Shoop, W.L., Slayton, L. *et al.* (1997). Nodulisporic acid A, a novel and potent insecticide from a *Nodulisporum* sp. isolation, structure determination and chemical transformations. *J. Am. Chem. Soc.* **119**, 8809–8816.

Pandit, J., Danley, D.E., Schulte, G.K., Mazzalupo, S., Pauly, T.A., Hayward, C.M., Hamanaka, E.S., Thompson, J.F. *et al.* (2000). Crystal structure of human squalene synthase. A key enzyme in cholesterol biosynthesis. *J. Biol. Chem.* **275**, 30610–30617.

Park, H., Kreunen, S.S., Cuttriss, A.J., DellaPenna, D., and Pogson, B.J. (2002). Identification of the carotenoid isomerase provides insight into carotenoid biosynthesis, prolamellar body formation, and photomorphogenesis. *Plant Cell* **14**, 321–332.

Perkins, D.D., Radford, A., and Sachs, M.S. (2001). *The Neurospora compendium: Chromosomal loci*. Academic Press, San Diego.

Peters, R.J. and Croteau, R.B. (2002). Abietadiene synthase catalysis: Conserved residues involved in protonation-initiated cyclization of geranylgeranyl diphosphate to (+)-copalyl diphosphate. *Biochemistry* **41**, 1836–1842.

Phillips, A.L., Ward, D.A., Uknes, S., Appleford, N.E.J., Lange, T., Huttly, A.K., Gaskin, P., Graebe, J.E. *et al.* (1995). Isolation and expression of three gibberellin 20-oxidase cDNA clones from *Arabidopsis*. *Plant Physiol.* **108**, 1049–1057.

Proctor, R.H. and Hohn, T.M. (1993). Aristolochene synthase. Isolation, characterization, and bacterial expression of a sesquiterpenoid biosynthetic gene (*Ari1*) from *Penicillium roqueforti*. *J. Biol. Chem.* **268**, 4543–4548.

Proctor, R.H., Hohn, T.M., and McCormick, S.P. (1995a). Reduced virulence of *Gibberella zeae* caused by disruption of a trichothecene toxin biosynthetic gene. *Mol. Plant–Microbe Interact.* **8**, 593–601.

Proctor, R.H., Hohn, T.M., McCormick, S.P., and Desjardins, A.E. (1995b). *Tri6* encodes an unusual zinc finger protein involved in regulation of trichothecene biosynthesis in *Fusarium sporotrichioides*. *Appl. Environ. Microbiol.* **61**, 1923–1930.

Ralston, L.F. (2001). *Cloning, heterologous expression, and functional characterization of 5-epi-aristolochene-1,3-dihydroxylase and a related gene from tobacco*. University of Kentucky, Lexington.

Rainier, J.D. and Smith, A.B., III, (2000). Polyene cyclizations to indole diterpenes. The first synthesis of (+)-emindole SA using a biomimetic approach. *Tetrahedron Lett.* **41**, 9419–9423.

Rohmer, M., Knani, M., Simonin, P., Sutter, B., and Sahm, H. (1993). Isoprenoid biosynthesis in bacteria: A novel pathway for the early steps leading to isopentenyl diphosphate. *Biochem. J.* **295**, 517–524.

Rojas, M.C., Hedden, P., Gaskin, P., and Tudzynski, B. (2001). The *P450–1* gene of *Gibberella fujikuroi* encodes a multifunctional enzyme in gibberellin biosynthesis. *Proc. Natl. Acad. Sci. USA* **98**, 5838–5843.

Romano, N. and Macino, G. (1992). Quelling: Transient inactivation of gene expression in *Neurospora crassa* by transformation with homologous sequences. *Mol. Microbiol.* **6**, 3343–3353.

Rynkiewicz, M.J., Cane, D.E., and Christianson, D.W. (2001). Structure of trichodiene synthase from *Fusarium sporotrichioides* provides mechanistic inferences on the terpene cyclization cascade. *Proc. Natl. Acad. Sci. USA* **98**, 13543–13548.

Rynkiewicz, M.J., Cane, D.E., and Christianson, D.W. (2002). X-ray crystal structures of D100E trichodiene synthase and its pyrophosphate complex reveal the basis for terpene product diversity. *Biochemistry* **41**, 1732–1741.

Sacchettini, J.C. and Poulter, C.D. (1997). Creating isoprenoid diversity. *Science* **277**, 1788–1789.

Sandmann, G. (2001). Carotenoid biosynthesis and biotechnological application. *Arch. Biochem. Biophys.* **385**, 4–12.

Sandmann, G., Misawa, N., Wiedemann, M., Vittorioso, P., Carattoli, A., Morelli, G., and Macino, G. (1993). Functional identification of *al-3* from *Neurospora crassa* as the gene for geranylgeranyl pyrophosphate synthase by complementation with *crt* genes, in vitro characterization of the gene product and mutant analysis. *J. Photochem. Photobiol.* **18**, 245–251.

Sanz, C., Alvarez, M.I., Orejas, M., Velayos, A., Eslava, A.P., and Benito, E.P. (2002). Interallelic complementation provides genetic evidence for the multimeric organization of the *Phycomyces blakesleeanus* phytoene dehydrogenase. *Eur. J. Biochem.* **269**, 902–908.

Schmidhauser, T.J., Lauter, F.R., Russo, V.E., and Yanofsky, C. (1990). Cloning, sequence, and photoregulation of *al-1*, a carotenoid biosynthetic gene of *Neurospora crassa. Mol. Cell. Biol.* **10**, 5064–5070.

Schmidhauser, T.J., Lauter, F.R., Schumacher, M., Zhou, W., Russo, V.E., and Yanofsky, C. (1994). Characterization of *al-2*, the phytoene synthase gene of *Neurospora crassa*. Cloning, sequence analysis, and photoregulation. *J. Biol. Chem.* **269**, 12060–12066.

Schardl, C.L. (1996). *Epichloë* species: Fungal symbionts of grasses. *Annu. Rev. Phytopathol.* **34**, 109–130.

Scott, B. (2001). *Epichloë* endophytes: Fungal symbionts of grasses. *Curr. Opin. Microbiol.* **4**, 393–398.

Smith, A.B., III, Kanoh, N., Ishiyama, H., and Hartz, R.A. (2000a). Total synthesis of (−)-penitrem D. *J. Am. Chem. Soc.* **122**, 11254–11255.

Smith, A.B., III, and Mewshaw, R. (1985). Total synthesis of (−)-paspaline. *J. Am. Chem. Soc.* **107**, 1769–1771.

Smith, M.M., Warren, V.A., Thomas, B.S., Brochu, R.M., Ertel, E.A., Rohrer, S., Schaeffer, J., Schmatz, D. *et al.* (2000b). Nodulisporic acid opens insect gluatamate-gated chloride channels: Identification of a new high-affinity modulator. *Biochemistry* **39**, 5543–5554.

Starks, C.M., Back, K., Chappell, J., and Noel, J.P. (1997). Structural basis for cyclic terpene biosynthesis by tobacco 5-*epi*-aristolochene synthase. *Science* **277**, 1815–1820.

Steyn, P.S. and Vleggaar, R. (1985). Tremorgenic mycotoxins. *Fortschr. Chem. Org. Naturst.* **48**, 1–80.

Swain, S.M., Ross, J.J., Reid, J.B., and Kamiya, Y. (1995). Gibberellins and pea seed development. Expression of the *lhi, ls* and *le* mutations. *Planta* **195**, 426–433.

Tag, A.G., Garifullina, G.F., Peplow, A.W., Ake, C., Jr., Phillips, T.D., Hohn, T.M., and Beremand, M.N. (2001). A novel regulatory gene, *Tri10*, controls trichothecene toxin production and gene expression. *Appl. Environ. Microbiol.* **67**, 5294–5302.

Tarshis, L.C., Yan, M., Poulter, C.D., and Sacchettini, J.C. (1994). Crystal structure of recombinant farnesyl diphosphate synthase at 2.6-Å resolution. *Biochemistry* **33**, 10871–10877.

Tomoda, H., Tabata, N., Yang, D., Takayanagi, H., and Omura, S. (1995). Terpendoles, novel ACAT inhibitors produced by *Albophoma yamanashiensis*. III. Production, isolation and structure elucidation of new components. *J. Antibio.* **48**, 793–804.

Trapp, S.C., Hohn, T.M., McCormick, S., and Jarvis, B.B. (1998). Characterization of the gene cluster for biosynthesis of macrocyclic trichothecenes in *Myrothecium roridum. Mol. Gen. Genet.* **257**, 421–432.

Trzaskos, J., Kawata, S., and Gaylor, J.L. (1986). Microsomal enzymes of cholesterol biosynthesis. Purification of lanosterol 14 alpha-methyl demethylase cytochrome P-450 from hepatic microsomes. *J. Biol. Chem.* **261**, 14651–14657.

Tudzynski, B. (1999). Biosynthesis of gibberellins in *Gibberella fujikuroi*: Biomolecular aspects. *Appl. Microbiol. Biotechnol.* **52**, 298–310.

Tudzynski, B., Hedden, P., Carrera, E., and Gaskin, P. (2001). The P450-4 gene of *Gibberella fujikuroi* encodes ent-kaurene oxidase in the gibberellin biosynthesis pathway. *Appl. Environ. Microbiol.* **67**, 3514–3522.

Tudzynski, B. and Hölter, K. (1998). Gibberellin biosynthetic pathway in *Gibberella fujikuroi*: Evidence for a gene cluster. *Fungal Genet. Biol.* **25**, 157–170.

Tudzynski, B., Honman, B., Feng, B., and Marzluf, G.A. (1999). Isolation, characterisation and disruption of the areA nitrogen regulatory gene of *Gibberella fujikuroi. Mol. Gen. Genet.* **261**, 106–114.

Tudzynski, B., Rojas, M.C., Gaskin, P., and Hedden, P. (2002). The gibberellin 20-oxidase of *Gibberella fujikuroi* is a multifunctional monooxygenase. *J. Biol. Chem.* **277**, 21246–21253.

Velayos, A., Eslava, A.P., and Iturriaga, E.A. (2000). A bifunctional enzyme with lycopene cyclase and phytoene synthase activities is encoded by the *carRP* gene of *Mucor circinelloides. Eur. J. Biochem.* **267**, 5509–5519.

Verdoes, J.C., Krubasik, K.P., Sandmann, G., and van Ooyen, A.J. (1999). Isolation and functional characterisation of a novel type of carotenoid biosynthetic gene from *Xanthophyllomyces dendrorhous. Mol. Gen. Genet.* **262**, 453–461.

Voss, T., Schulte, J., and Tudzynski, B. (2001). A new MFS-transporter gene next to the gibberellin biosynthesis gene cluster of *Gibberella fujikuroi* is not involved in gibberellin secretion. *Curr. Genet.* **39**, 377–383.

Wendt, K.U., Poralla, K., and Schulz, G.E. (1997). Structure and function of a squalene cyclase. *Science* **277**, 1811–1815.

Wendt, K.U. and Schulz, G.E. (1998). Isoprenoid biosynthesis: Manifold chemistry catalyzed by similar enzymes. *Structure* **6**, 127–133.

Woloshuk, C.P., Fount, K.R., Brewer, J.F., Bhatnagar, D., Cleveland, T.E., and Payne, G.A. (1994). Molecular characterisation of *aflR*, a regulatory locus for aflatoxin biosynthesis. *Appl. Environ. Microbiol.* **60**, 2408–2414.

Yao, Y., Peter, A.B., Baur, R., and Sigel, E. (1989). The tremorgen aflatrem is a positive allosteric modulator of the aminobutyric acid receptor channel in *Xenopus* oocytes. *Mol. Pharmacol.* **35**, 319–323.

Young, C.A., McMillan, L., Telfer, E., and Scott, B. (2001). Molecular cloning and genetic analysis of an indole-diterpene gene cluster from *Penicillium paxilli. Mol. Microbiol.* **39**, 1–13.

Yu, J.-H., Butchko, R.A.E., Fernandes, M., Keller, N.P., Leonard, T.J., and Adams, T.H. (1996). Conservation of structure and function of the aflatoxin regulatory gene *aflR* from *Aspergillus nidulans* and *A. flavus. Curr. Genet.* **29**, 549–555.

Zhang, H., Seabra, M.C., and Deisenhofer, J. (2000). Crystal structure of Rab geranylgeranyltransferase at 2.0 Å resolution. *Structure Fold. Des.* **8**, 241–251.

III

Enzymes and Green Chemistry

9

Heterologous Expression and Protein Secretion in Filamentous Fungi

Wendy Thompson Yoder and Jan Lehmbeck

1. Introduction

The first reports of transformation of filamentous fungi began appearing in the scientific literature in the 1960s and 1970s (e.g., Shamoian *et al.*, 1961; Shockley and Tatum, 1962; Sen *et al.*, 1969; Mishra and Tatum, 1973). However, no recombinant DNA methods were available at that time to confirm the integration of donor DNA, and it was only in 1979 that Case, Schweizer, Kushner, and Giles reported the transformation of *Neurospora crassa* with a plasmid vector. As a result of these and subsequent breakthroughs (including the expression of heterologous genes), a multi-million dollar per year industry has emerged, whose core technology is based on the expression and secretion of heterologous proteins by filamentous fungi. The industry supplies a diverse array of products with applications in a wide range of market sectors including detergents, food and beverages, textiles, starch, animal feed, personal care, pulp and paper, leather, fats and oils, and biocatalysis. In many cases these products are being used in manufacturing and industrial process development in lieu of harsh and toxic chemicals (Nedwin, 1997). In 2001, the world market for industrial enzymes was ~1.6 billion US dollars (Novozymes, 2002), with products derived from filamentous fungi accounting for approximately 50% of this market.

Protein secretion is a very complex process, and our understanding of the underlying molecular mechanism in fungi is still at a relatively early developmental stage. A number of detailed reviews of heterologous protein production and secretion have been published, most recently by members of the TNO group (Conesa *et al.*, 2001a; Punt *et al.*, 2002). Other useful reviews include those of Radzio and Kuck (1997), Archer and Peberdy (1997), Jarai (1997), and Kruszewska (1999). In addition, while Pentilla focused on *Trichoderma* in a 1998 review covering heterologous protein production, Maras *et al.* (1999) reviewed the use of filamentous fungi specifically as production organisms for glycoproteins of biomedical interest.

Rather than reiterate what these authors have recently discussed in detail, this chapter will summarize, in tabular form, the current status of heterologous expression in filamentous

Wendy T. Yoder • Novozymes Biotech Inc., 1445 Drew Avenue, Davis, CA 95616. **Jan Lehmbeck** • Novozymes A/S, Krogshoejvej 36, DK-2880, Bagsvaerd, Denmark.

Advances in Fungal Biotechnology for Industry, Agriculture, and Medicine. Edited by Jan S. Tkacz and Lene Lange, Kluwer Academic/Plenum Publishers, 2004.

fungi, and highlight selected examples. A section describing the development of a new fungal host (*Fusarium venenatum*) will follow, and the final section will discuss the future directions that this burgeoning field is likely to take in response to the world population's needs, to our customer's demands, and to our scientific curiosity in understanding the molecular basis of secretion. In this regard, some of the emergent technologies which may be expected to advance our understanding of heterologous gene expression and protein secretion will be considered.

2. The Past Decade

The broad range of different classes of enzymes that have been expressed in filamentous fungi, as well as the variety of sources from which the heterologous genes originated are illustrated in Tables 9.1 and 9.2. Table 9.1 exemplifies heterologous fungal gene expression, while Table 9.2 summarizes the non-fungal proteins that have been expressed and secreted, with varying degrees of success, in filamentous fungal hosts. Both Tables note the promoters used to drive expression of the heterologous genes, and references describing the various strategies devised to increase expression levels have also been included.

When presented in this unidimensional format, the list reveals little about the effort involved in achieving the cited goals. For example, in Table 9.2 the commercially viable yields of chymosin in *Aspergillus awamori* (1.3 g/L) were achieved only after incremental yield improvements were made over a period of several years (Dunn-Coleman *et al.*, 1991). A multiplicity of techniques were employed, including targeted deletion of the host *pepA* gene (encoding the major extracellular protease), construction of glucoamylase–chymosin gene fusions, insertion of an additional N-glycosylation site into the protein, classical mutagenesis with nitrosoguanidine, isolation of 2-deoxyglucose resistant mutants, and the utilization of robotic screening to aid identification of improved strains. In contrast, the higher yields of some of the heterologous fungal enzymes shown in Table 9.1 resulted simply (and in many cases by virtue of serendipity) from the generation of highly expressing primary transformants, which warranted little or no yield improvement.

Clearly the recent use of filamentous fungi for large-scale production of proteins has been very successful. However, while a significant number of heterologous fungal proteins have been efficiently expressed, the attainment of high yields of proteins, whose corresponding genes originate from outside the fungal kingdom, seems in general to be more problematic (see Table 9.2). Transcriptional efficiency does not appear to be a major contributing factor to this situation (Gouka *et al.*, 1996b). Rather, low heterologous protein levels seem to be primarily affected by translational inefficiency, degradation by host proteases, delays within the secretion pathway, re-direction to the vacuole for degradation, or a combination of these factors. The negative influence of host protease(s) on the yields of secreted proteins has been remedied in many cases either by selection of mutant strains that secrete very low levels of proteases (Mattern *et al.*, 1992; van den Hombergh *et al.*, 1995), or by disruption of the gene(s) encoding the protease(s) (van den Hombergh *et al.*, 1997; Lehmbeck, 1998; Moralejo *et al.*, 2002). Another approach has been to prevent expression of host proteases in fermentations by addition of amino acid-rich supplements, ensuring a high ammonia concentration in the medium, or by changing the pH of the medium (Mackenzie *et al.*, 1994; Bartking *et al.*, 1996). Hintz *et al.* (1995) used *alcA*

Table 9.1. Heterologous Production of Fungal Proteins in Filamentous Fungi

Production host	Protein[a]	Promoter[b]	Yield[c]	Reference
Acremonium chrysogenum	Alkaline protease		1.5 g/L	Morita *et al.*, 1994
	(*Fusarium* sp.)		4.0 g/L[e]	
Aspergillus awamori	α-Amylase (*Aspergillus oryzae*)		10.3 U/g	Ruttkowski *et al.*, 1989
	Aspartic protease (*Mucor miehei*)	*glaA*	3.0 g/L[d]	Ward and Kodama, 1991
	Cutinase (*Fusarium solani f.* sp. *pisi*)	*exlA*	64 mg/L	van Gemeren *et al.*, 1996
	Glucoamylase (*Aspergillus niger*)		4.6 g/L	Finkelstein *et al.*, 1989
	Glucoamylase (*Humicola grisea*)		0.66 g/L[e]	Berka *et al.*, 1988
Aspergillus nidulans	α-Amylase (*A. oryzae*)		2070 U/ml	Lachmund *et al.*, 1993
	α-L-Arabino-furanosidase (*A. niger*)		2.64 U/ml	Filipphi *et al.*, 1993
	Aspartic protease (*Mucor miehei*)		2 mg/ml	Gray *et al.*, 1986
	Glucoamylase	*gpdA*	0.2 mg/g	Punt *et al.*, 1990
	(*A. niger*)	*alcA*	1.2 g/L[e]	Devchand *et al.*, 1989
	Glucose oxidase		5.4 U/mg	Whittington *et al.*, 1990
	(*A. niger*)		2.7 U/mg	Witteveen *et al.*, 1993
	Pectin lyase A, B, & C		NR	Kusters-van Someren, 1991
	Polygalacturonase I		510 U/ml	Bussink *et al.*, 1992
	Polygalacturonase C (*A. niger*)		33 U/ml	
Aspergillus niger	Chloroperoxidase (*Caldariomyces fumago*)	*glaA*	500 mg/L	Conesa *et al.*, 2001b
	Glucoamylase (*A. niger*)	*amy*	NR	Nielsen *et al.*, 2002
	Laccase (*Pycnoporus cinnabarinus*)	*gpdA*	1 mg/L	Record *et al.*, 2002
	Lignin peroxidase (*Phanerochaete chrysosporium*)	*glaA*	NA	Conesa *et al.*, 2000
	Lipase (*Humicola lanuginosa*)	*amy*	NR	Boel and Huge-Jensen, 1989
	Maganese peroxidase (*P. chrysosporium*)	*glaA*	400 mg/L[e]	Conesa *et al.*, 2002b
	Polygalacturonase II (*Aspergillus tubingensis*)		NR	Bussink *et al.*, 1991
	Xylanase (*Aspergillus* awamori)		140 kU/ml	Hessing *et al.*, 1994
Aspergillus oryzae	Acid Phosphatase (*F. venenatum*)	*amy*	68 U/L	Yaver *et al.*, 2001
	α-1,3-glucanase (*Trichoderma harzianum*)	*amy*	NR	Fuglsang *et al.*, 1999
	α-1,3-glucanase (*Penicillium purpurogenum*)	*amy na2/tpi*	NR	Berka *et al.*, 1998a
	Aspartic protease	*amy*	3.3 g/L[e]	Christensen *et al.*, 1988;
	(*Mucor miehei*)	*glaA*	0.4 g/L[e] NR	Boel *et al.*, 1987.

Table 9.1. (*Continued*)

Production host	Protein[a]	Promoter[b]	Yield[c]	Reference
Aspergillus oryzae	β-1,4-galactanase (*Aspergillus aculeatus*)	*amy*	NR	Christgau *et al.*, 1995
	Carbohydrate oxidase (*Microdochium nivale*)	*amy*		Xu *et al.*, 2001b
	Cellobiose dehydrogenase (*Humicola insolens*)	*amy*	NR	Xu *et al.*, 2001a
	Galactose oxidase (*Dactylium dendroides*)	*amy*	0.1 g/L	Xu *et al.*, 2000
	Glucoamylase (*Aspergillus shirousami*)		4970 U/g	Shibuya *et al.*, 1990
	Glucoamylase (*Thielavia terrestris*)	*amy*	++	Rey *et al.*, 2002
	Haloperoxidase (*Curvularia verruculosa*)	*amy* *na2/tpiI*	8,390 mOD_{600} per min per ml	Fuglsang *et al.*, 1999
	Laccase (Lcc1) (*Coprimus cinereus*)	*amy*	135 mg/L	Yaver *et al.*, 1999
	Laccase (*Myceliophthora thermophila*)	*amy*	19 mg/L	Berka *et al.*, 1997
	Laccase (*Scytalidium thermophilum*)	*amy*	1.8 g/L	Berka *et al.*, 1998b
	Laccases (Lcc1,2 & 4) (*Rhizoctonia solani*)	*amy*	NR	Wahleithner *et al.*, 1996
	Laccase (Lcc1) (*Trametes villosa*)	*amy*	NR	Yaver *et al.*, 1996
	Lipase (*Absidia corymbifera*)	*amy* *na2/tpi*	NR	Berka *et al.*, 1998c
	Lipase (*Absidia sporophora-variabilis*)	*amy* *na2/tpi*	NR	Berka *et al.*, 1998c
	Lipase (*Acremonium berkeleyanum*)	*na2/tpi*	NR	Tsatsumi *et al.*, 2002
	Lipase (*Fusarium culmorum*)	*na2/tpi*	NR	Tsatsumi *et al.*, 2002
	Lipase (*Fusairum solani*)	*na2/tpi*	NR	Tsatsumi *et al.*, 2002
	Lipase (*Fusarium sulphureum*)	*na2/tpi*	130 LU/ml	Tsatsumi *et al.*, 2002
	Lipase (*H. lanuginosa*)	*amy*	NR	Boel and Huge-Jensen, 1989
	Maganese Peroxidase (*P. chrysosporium*)	*amy*	5 mg/L	Stewart *et al.*, 1996
	Pectin Methyl Esterase (*Aspergillus aculeatus*)	*amy*	NR	Christgau *et al.*, 1996
	Peroxidase (*C. cinereus*)	*amy*	1000 PODU/ml	Andersen *et al.*, 1992
	Peroxidase (*Geotrichum candidum*)	*amy*	NR	Sugano *et al.*, 2000
	Phospholipase (*Fusarium solani*)	*na2/tpi*	NR	Tsatsumi *et al.*, 2002
	Phospholipase B (*A. oryzae*)	*na2/tpi*	NR	Harris and Brown, 2000
	Rhamnoglacturonan Acetylesterase (*A. aculeatus*)	*amy*	NR	Kauppinen *et al.*, 1995
	Triglyceride Lipase (*Mucor miehei*)	*amy*	2 mg/L	Boel and Huge-Jensen, 1989
Chrysosporium lucknowense	Hydrolytic enzymes (*Trichoderma*)	*cbh1 (source unknown)*	NR	van Zeijl *et al.*, 2001
Cladosporium fulvum	Tomatinase (*Septoria lycopersici*)	*gpd (A. nidulans)*	++	Melton *et al.*, 1998
Cochliobolus heterostrophus	Cutinase (*Nectria haematococca*)		18.6 U/μg	Oeser and Yoder, 1994

Fusarium venenatum	Acid Phosphatase (*F. venenatum*)	*Trypsin (F. oxysporum)*	97 U/L	Yaver *et al.*, 2001
	Carbohydrate Oxidase (*Microdochium nivale*)	*Trypsin (F. oxysporum)*	NR	Xu *et al.*, 2001b
	Cellobiose Dehydrogenase (*Humicola insolens*)	*Trypsin (F. oxysporum)*	NR	Xu *et al.*, 2001a
	Endo-glucanase (*Humicola insolens*)	*Trypsin (F. oxysporum)*	NR	Royer *et al.*, 1995
	Galactose Oxidase (*F. venenatum*)	*Trypsin (F. oxysporum)*	NR	Golightly *et al.*, 2000
	Glucoamylase (*A. niger*)	*Trypsin (F. oxysporum)*	46 mg/g biomass	Weibe *et al.*, 1999
	Glucose Oxidase (*Cladosporium oxysporum*)	*Trypsin (F. oxysporum)*	NR	Cherry *et al.*, 1999
	Lipase (*Thermomyces lanuginosus*)	*Trypsin (F. oxysporum)*	NR	Royer *et al.*, 1995
	Lipase (*Thermomyces lanuginosus*)	*Daria glaA Quinn*	1967 LU/ml 2731 LU/ml 1990 LU/ml	Berka *et al.*, 2002
	Lipase I (*F. venenatum*)	*Trypsin (F. oxysporum)*	NR	Rey and Golightly, 2002
	Phytase (*T. lanuginosus*)	*Trypsin (F. oxysporum)*	NR	Berka *et al.*, 1998d
	Serine Carboxy-peptidase (*A. oyzae*)	*Trypsin (F. oxysporum)*	NR	Blinkovsky *et al.*, 1999
	Tripeptide amino-peptidase (*A. oryzae*)	*Trypsin (F. oxysporum)*	NR	Rey and Golightly, 1999
	Tryptic protease (*F. oxysporum*)	*Trypsin (F. oxysporum)*	NR	Royer *et al.*, 1995
Mucor circinelloides	Aspartic protease (*Mucor miehei*)		1.2 mg/L	Dickinson *et al.*, 1987
Trichoderma reesei	Acid phosphatase (*A. niger*)	*cbh1*	0.5 g/L	Saloheimo *et al.*, 1993
	Endo-chitinase (*Trichoderma harzianum*)	*cbh1*	150 mg/L	Margolles-Clark *et al.*, 1996
	Glucoamylase		NP	Joutsjoki and Torkkeli, 1992
	(*Hormoconis resinae*)	*cbh1*	0.7 g/L	Joutsjoki *et al.*, 1993
	Invertase (*A. niger*)		0.47 U/mg	Bergès *et al.*, 1993
	Lignin oxidase (*Phlebia radiata*)	*cbh1*	20 mg/L[e]	Saloheimo *et al.*, 1991
	Lignin peroxidase (*P. radiata*)	*cbh1*	NP	Saloheimo *et al.*, 1989
	Phytase (A. *niger*)	*cbh1*	2 g/L	Saloheimo *et al.*, 1993
Trichoderma viride	α-Amylase		4 mg/L	Cheng *et al.*, 1990
	(*A. oryzae*)	*cbh1*	1 g/L	Cheng and Udaka, 1991

[a]The source of the gene is indicated in parentheses.

[b]Unless otherwise noted, the gene's native promoter was used. The heterologous promoters originate from the host organism, unless indicated by superscript symbols. The promoters are *alcA*, alcohol dehydrogenase I; *alp*, alkaline protease; *amy*, alpha-amylase; *aphA*, repressible acid phosphatase; *cbhI*, cellobiohydrolase I; *cesB*, B2 wide-spectrum esterase from *Acremonium chrysogenum*; *exlA*, 1,4-ß-endoxylanase; *glaA*, glucoamylase; *gdh1*, glyceraldehydes 3 phosphate dehydrogenase; *gpdA*, glyceraldehydes-3-phosphate dehydrogenase; *na2*, *A. niger* neutral amylase 2; *tpi*, *A. nidulans tpi* leader sequence; *xyl*, xylanase.

[c]The amount of protein produced by the best transformants is indicated. NA, no activity, but a band on a SDS-page gel; ND, not detectable; NP, no protein product; NR, the amount of protein produced is not reported.

[d]High levels obtained after mutagenesis.

[e]Submerged fermentations.

++ Activity detected: see reference for details.

Table 9.2. Heterologous Production of Non-Fungal Proteins in Filamentous Fungal Hosts

Production host	Protein[a]	Promoter[b]	Yield[c]	Reference
Acremonium chrysogenum	Lysozyme (Human)	*alp (Fusarium* sp.)	40 mg/L	Morita *et al.*, 1995
Aspergillus awamori	α-galactosidase (*Cyamopsis tetragonoloba*) (Plant)	*exlA*	12 mg/L[i]	Gouka *et al.*, 1996a, 1997
	Chymosin (Bovine)	*glaA (A. niger)*	1.3 g/L	Dunn-Coleman *et al.*, 1991
	Interleukin-6 (Human)	*exlA*	10 mg/L[g]	Gouka *et al.*, 1996b
	Lactoferrin (Human)	*glaA*	250 mg/L 2 g/L[e]	Ward *et al.*, 1995
	Prochymosin (Bovine)	*glaA*	140 mg/L[g]	Ward *et al.*, 1990
	Prochymosin (Bovine)	*amy*	50 mg/L[g]	Korman *et al.*, 1990
	scFv-LYS	*xyl*	109 mg/L[g]	Sotiriadis *et al.*, 2001
	Thaumatin (*Thaumatococcus daniellii*) (Plant)	*gdhA*	11 mg/L[g,h,i,k]	Moralejo *et al.*, 1999
		cesB, gdhA	14 mg/L[g,h,i]	
		cesB, gdhA	100 mg/L[g,h,i]	
Aspergillus nidulans	Interferon alpha2	*alcA*	1 mg/L	Gwynne *et al.*, 1989
	Interferon alpha2	*aphA (A. niger)*	0.2 mg/L	MacRae *et al.*, 1993
	Interleukin-6 (Human)	*glaA (A. niger)*	<1 μg/L	Carrez *et al.*, 1990
			5 mg/L[g]	Contreras *et al.*, 1991
	Lactoferrin (Human)	*alcA*	5 mg/L	Ward *et al.*, 1992
Aspergillus niger	Chymosin (Bovine)	*glaA (A. oryzae & TAKA signal)*	100 mg/L	Christensen, 1995
	Enterokinase (Bovine)	*glaA*	5 mg/L[g,h]	Svetina *et al.*, 2000
	Insulin (Human)	*glaA*	776 mU/L	Mestric *et al.*, 1996
	Interleukin-6 (Human)	*gpdA (A. nidulans)*	ND, 40 mg/L[g,h]	Broekhuijsen *et al.*, 1993
	Lymphotoxin (Human)	*glaA*	50 pg/L[g,h]	Krasevec *et al.*, 2000
	Lysozyme (Equine)	*glaA*	150 mg/L[g,i]	Spencer *et al.*, 1999
	Lysozyme (Hen)	*glaA*	40 mg/L[g]	Spencer *et al.*, 1999
	Lysozyme (Hen)	*glaA*	50 mg/L 1 g/L[g]	Jeenes *et al.*, 1993
	Lysozyme (Human)	*glaA*	140 mg/L[g]	Spencer *et al.*, 1999
	Pancreatic prophospholipase A2 (Porcine)	*glaA*	ND 10 mg/L[g]	Roberts *et al.*, 1992
	Tisue plasminogen activator (Mammalian)	*glaA*	10 mg/L[g,h]	Wiebe *et al.*, 2001
Aspergillus oryzae	Chymosin (Bovine)	*glaA*	0.16 mg/L	Tsuchiya *et al.*, 1993
	Der f1 (*Dermatophagoides farinae* allergen)	*glaA*	8 mg/L	Shoji *et al.*, 1999

	Lactoferrin (Human)	*gla (A. awamori)*	2 g/L	Ward *et al.*, 1995
	Polygalacturonase PG1 (*Cucumis melo*)			Hadfield *et al.*, 1998
	Prochymosin (Bovine)	*glaA* 150 mg/L[g,j]	70 μg/L	Tsuchiya *et al.*, 1994
	Prochymosin (Bovine)	*glaA (A. niger) + amy*	100 mg/L	Christensen, 1995
Neurospora crassa	Chymosin (Bovine)	*Beta-tubulin, grg-1*	0.5 mg/L 1 mg/L	Nakano *et al.*, 1993
	Kappa light chain &	*gla1*	mg/L range	Buczynski *et al.*, 1995
	Gamma heavy chain (Human)	*gla1*	mg/L range	
Tolypocladium geodes	Lysozyme (*Human*)	*Tr1*[d]	2 mg/L[f] 150 mg/L[g]	Baron *et al.*, 1992
Trichoderma reesei	Chymosin (Bovine)	*cbh1*	100 mg/L[g]	Penttila, 1998
	Fab antibody fragments	*cbh1*	1 mg/L	Nyyssönen *et al.*, 1993
			150 mg/L[f,g]	Nyyssönen and Keränen, 1995
	Interleukin-6 (Mammalian)	*cbh1*	5 mg/L	Demolder *et al.*, 1994
	Single chain antibodies (Murine)	*gpd*	1 mg/L	Nyyssonen, 1993

[a]The source of the gene is indicated in parentheses.

[b]Unless otherwise noted, the gene's native promoter was used. The heterologous promoters originate from the host organism, unless indicated by superscript symbols. The promoters are *alcA*, alcohol dehydrogenase I; *alp*, alkaline protease; *amy*, alpha-amylase; *aphA*, repressible acid phosphatase; *cbhI*, cellobiohydrolase I; *cesB*, B2 wide-spectrum esterase from *Acremonium chrysogenum*; *exlA*, 1,4-β-endoxylanase; *glaA*, glucoamylase; *gdh1*, glyceraldehydes 3 phosphate dehydrogenase; *gpdA*, glyceraldehydes-3-phosphate dehydrogenase; *grg-1*, glucose repressible promoter from *N. crassa*; *xyl*, xylanase.

[c]The amount of protein produced by the best transformants is indicated. NA, no activity, but a band on a SDS-page gel; ND, not detectable; NP, no protein product; NR, the amount of protein produced is not reported.

[d]*Trichoderma reesei* promoter.

[e]High levels obtained after mutagenesis.

[f]Submerged fermentations.

[g]Expression by gene-fusion strategy.

[h]Protease deficient strain.

[i]Synthetic gene with optimized fungal codon usage.

[j]Double transformant generated using two different promoters.

[k]Synthetic gene with optimized yeast codon usage.

++ Activity detected: See reference for details.

promoter variants which were insensitive to glucose repression to ensure earlier expression and secretion of heterologous product (although in their case host proteases still posed a problem late in the fermentations).

One somewhat successful approach for improving the expression of heterologous proteins has been to make translational fusions (reviewed by Gouka *et al.*, 1997). This involves fusion of the heterologous protein to a highly secreted homologous protein, the rationale being that the homologous protein facilitates passage through the secretion pathway, although the molecular basis of this process is still poorly understood. In other examples the signal sequence or the signal and pro-sequences have been exchanged with the corresponding sequences from highly secreted proteins. This has been effective in some cases, but the general conclusion from such studies is that a number of different combinations of signal/pro-sequences must be tested in order to elicit increased production yields (Conesa *et al.*, 2001a, and references therein).

Over the last 5 years attempts to modify the levels of chaperones such as BiP, PDI (protein disulfide isomerase), and calnexin, which promote folding and maturation, have been tested. Recent results obtained from fungal strains in which the levels of each of these chaperones were altered have implicated a role for these proteins, but the results are again far from conclusive. Increased levels of secreted protein have been obtained by overexpression of calnexin or PDI in *Aspergillus niger* strains producing either *Phanerochaete chrysosporium* manganese peroxidase (Conesa, 2002a) or *Thaumatococcus danielli* thaumatin (Moralejo *et al.*, 2002), respectively. In the latter case the best thaumatin-producing strains were associated with intermediate levels of PDI, while strains having high levels of PDI produced lower levels of thaumatin. In another case overexpression of BiP increased intracellular levels, but not extracellular levels, of two fusion proteins (Punt *et al.*, 1998). Implementation of the various improvement strategies mentioned has increased the number of proteins that can be produced in significant amounts, but many proteins remain whose production at commercially-relevant levels is problematic. The future approaches that will probably be applied to this problem will be discussed in the final section of this chapter.

3. Development of a New Fungal Expression Host: *Fusarium venenatum* Nirenberg

Novozymes A/S, a major supplier of industrial enzymes and microorganisms for industrial use, employs *Aspergillus oryzae, A. niger* and *Bacillus* species as the workhorse organisms for enzyme production. In 1992 the company initiated the process of identifying an alternative fungal host because certain enzymes were not well expressed in *A. oryzae* or *A. niger*, and it was considered likely that expression levels might be higher in different host backgrounds (with a different spectrum of host proteases, different pH optima, etc.).

3.1. Selection Criteria

The search for a new fungal host began with over one hundred candidates including mesophilic and thermophilic ascomycetous saprotrophs and basidiomycetous saprotrophs. These candidates were screened for their (1) ease of transformation, (2) low secreted protease levels, (3) low total spectrum of secreted proteins, (4) capacity for high level heterologous expression, (5) fermentation morphology, and (6) current or potential "GRAS" (generally recognized as safe) or equivalent status. At the conclusion of the evaluations two candidates clearly stood out from the rest. One was *Fusarium venenatum*, which enjoyed the added benefit of having been approved in England (17 years previously) as a food component safe for human consumption (marketed as "Quorn™" mycoprotein) (Trinci, 1992). The Quorn™ fungus had originally been identified as *Fusarium graminearum*, but convincing molecular, phylogenetic, morphological, and mycotoxin evidence has since shown that the A3/5 Quorn™ strain should be re-classified as *F. venenatum*. This work was published by two independent research groups (O'Donnell *et al.*, 1998; Yoder and Christianson, 1998), 3 years after Nirenberg (1995) had published strong cultural and morphological evidence for splitting the *Fusarium sambucinum sensu lato* group into three separate species; *F. sambucinum sensu stricto, F. torulosum*, and *F. venenatum*.

3.2. Heterologous Expression

The first heterologous product expressed in *F. venenatum* was a tryptic protease from *Fusarium oxysporum* (Rypniewski *et al.*, 1993). Expression of this gene in *Aspergillus* had been problematic, the data indicating that a proteolytic activity required for efficient processing of the trypsin was lacking in *A. oryzae* (Royer *et al.*, 1995). In *F. venenatum*, however, the tryptic protease was efficiently expressed, processed, and secreted by the host strain (Royer *et al.*, 1995), suggesting that *F. venenatum* possesses the maturing activity necessary for the correct processing of this protein. Using the promoter and terminator sequences of the *F. oxysporum* trypsin gene, other heterologous genes from less closely related fungi were also expressed successfully in the *Fusarium* host. These included a lipase from *Thermomyces lanuginosus* and an endoglucanase from *Humicola insolens* (Royer *et al.*, 1995).

3.3. Improved Morphological Mutants

Initially in continuous flow cultures used for the production of Quorn™ myoprotein, the A3/5 strain would be supplanted, after ~600 hr, by different classes of recessive, colonial, highly branched mutants, following shear-induced separation of hyphal fragments (Wiebe *et al.*, 1991). These mutants displayed radial colonial growth rates (K_r; $\mu m\ h^{-1}$) significantly lower than that of the A3/5 parent strain (Wiebe *et al.*, 1991, 1992, 1996). Since the morphologies of these mutants were of interest from a secretion perspective, a number of them were screened and their heterologous expression levels compared to those of A3/5. The *T. lanuginosus* lipase gene was used as one of two reporter genes, and the *A. nidulans* acetamidase (*amdS*) gene, that confers the ability to grow on acetamide as sole nitrogen source (Kelly and Hynes, 1985), was employed as the selectable marker. On the basis of these comparisons, a morphological mutant designated CC1-3 (Wiebe *et al.*, 1991) had the most favorable characteristics.

3.4. Selectable Markers

Two selectable marker genes were tested for their potential to support the generation of transformants yielding high levels of heterologous proteins. Comparisons were made between the heterologous protein yields of transformants generated in MLY3 (a spontaneous mutant of the CC1-3 mutant) using either the *Streptomyces hygroscopicus bar* gene or the *amdS* gene as selectable markers. The *bar* gene encodes phosphinothricin acetyltransferase that confers resistance to the herbicide bialaphos (whose toxicity is caused by the glutamic acid analog phosphinothricin) (Thompson *et al.*, 1987). From these comparisons the *bar* gene emerged as the superior marker. On further testing the combination of the MLY3 host and the *bar* gene consistently supported high level expression, by primary transformants, of a range of heterologous enzymes (listed in Table 9.1).

3.5. Targeted Gene Deletions

Prior to its approval in the UK as a food for human consumption, the A3/5 strain underwent extensive toxicity testing (Trinci, 1992). To eliminate any potential production of diacetoxyscirpenol (DAS), a trichothecene made by certain *F. venenatum* strains, the *tri5* (= *tox5*) gene encoding the trichodiene synthase (the first enzyme in the trichothecene biosynthetic pathway) was deleted in the *Fusarium* MLY3 strain (Royer *et al.*, 1999). The resulting strain, designated LyMC1A, was subsequently shown to be incapable of producing DAS or modified trichothecenes, even under inducing conditions (Miller and MacKenzie, 2000).

Although the LyMC1A strain did not produce detectable levels of enniatin B (another potentially toxic secondary metabolite produced by some *Fusarium* species) (Miller and MacKenzie, 2000), the *dps1* (depsipeptide synthase) gene, a putative *Fusarium torulosum* enniatin synthase (*esyn1*) homolog, was identified as a second candidate for deletion and was subsequently removed (Yoder, Rey, Yaver, and Berka, unpublished). Yields of heterologous proteins were not negatively impacted in the *tri5*-deleted MLY-3 strain or the *tri5*- and *dps1*-deleted MLY-3 strain.

3.6. GRAS Status for the First Heterologous Enzyme Produced in *F. venenatum*

A *T. lanuginosus* endo-1,4-beta-xylanase was the first *F. venenatum*-derived heterologous protein submitted to the FDA as a candidate for GRAS status. This enzyme is used as a processing aid in baked goods, serving to degrade the insoluble fraction of arabinoxylans in flour and thereby improving handling and stability of the dough. The strain expressing this enzyme underwent rigorous safety testing, including oral toxicity testing in rats, Ames mutagenicity tests, cytogenicity tests, and plant pathogenicity tests, prior to submission of the application. The FDA granted GRAS status in February 2001 (FDA, 2001; Notice Number GRN 000054).

3.7. The First Commercial Recombinant *F. venenatum* Product

The first microbially produced alternative to animal trypsin, with applications in cell culture and bio-production, is being manufactured in the current *F. venenatum* host (a relative of the xylanase host strain) for distribution by Invitrogen ("Gibco Cell Culture Division," 2002; Invitrogen, 2002).

3.8. *Fusarium venenatum* Genomics

A detailed discussion of *F. venenatum* genomics is beyond the scope of this chapter. However, the reader is directed to a recent overview (Berka, 2003), which provides one of the first exposés of the genetic complement of *F. venenatum*, based on analysis of its karyotype, expressed sequence tags (ESTs), genomic DNA sequences, and cDNA in microarray format.

4. "To Infinity and Beyond"

Some of the difficulties inherent in achieving high level expression of non-fungal genes in filamentous fungal hosts will undoubtedly be overcome as we continue to increase our understanding of the detailed functioning of the secretion pathway in these organisms. Processing and analyzing the data from the filamentous species currently being sequenced (*Ashbya gossypii, Aspergillus flavus, A. fumigatus, A. nidulans, A. oryzae, F. graminearum, Magnaporthe grisea, Nectria haematococca, N. crassa*, and *P. chrysosporium*) and the new developments in molecular biology, collectively known as the "-omics" techniques (i.e., genomics, proteomics, transcriptomics, metabolomics, and phenomics) will be key tools in the elucidation of the problematic steps involved in high-level, inter-kingdom protein production and secretion. For example, Phenotype Microarray (PM™) technology (Biolog, 2002) allows easy and efficient testing of thousands of cellular traits simultaneously and has been used in studies of gene discovery and gene function assignment in *M. grisea* and *M. graminicola* (Hamer *et al.*, 2001; Paradigm Genetics, 2002). The continued refinement of PM™ for use with filamentous fungi would provide a tremendous boost to the expression and secretion fields, bridging the gap between, and facilitating interpretation of, data derived from nucleic acid microarrays and proteomics.

Genomics approaches allow insight into entire pathways (e.g., Chu *et al.*, 1998; De Gregorio *et al.*, 2001), and they provide possibilities for elucidating the function of unknown genes (e.g., Eisen *et al.*, 1998). Collection and detailed analysis of data from experiments where, for example, a strain which secretes high yields of a heterologous protein is compared with a strain that is apparently incapable of secreting high levels of a heterologous protein, should allow identification of those genes which play an important role in the secretion of heterologous proteins. But, even with this new knowledge of the molecular basis of the functioning of the secretion pathway, it is likely to be many years before such information can be implemented and used in the generation of new and optimized production strains. This is due to the complexity of the processes and the rather slow, laborious way in which most production-relevant filamentous fungal strains are manipulated genetically today.

The much anticipated advances that will ultimately facilitate efficient, routine, large scale, and safe production of therapeutic proteins in filamentous fungi are currently hindered by the inherent differences in glycosylation pathways in filamentous fungi and mammals. The glycosylation patterns of therapeutic proteins are known to modulate a number of critical characteristics such as activity, clearance rate from the circulation, and immunogenicity (Varki, 1993). Maras *et al.* (1999) reviewed the current state of our understanding of protein glycosylation in filamentous fungi and summarized the attempts that have been made to modify the process so that more mammalian-like glycosylation patterns will result. A number of strategies for engineering such modifications have been proposed. One involves manipulating the fungal host by disrupting host genes and/or by the introduction of mammalian glycosyl-transferases. The introduction of the mammalian cDNA encoding N-acetyl-glucosaminyltransferase I (GlcNAc T1) into *Trichoderma reesei*, for example, resulted in the predicted formation of GlcNAcMan5GlcNAc2, as shown by NMR (Maras *et al.*, 1997a). Another approach involves deglycosylation of the protein *in vitro* followed by *in vitro* re-glycosylation to form the desired substitution pattern (e.g., Maras *et al.*, 1997b). This and other examples demonstrate that such strategies are feasible. The development of recombinant protein drugs that require effective glycosylation has essentially forced the issue of understanding and controlling glycosylation. For example, because of inadequate glycosylation, Amgen was forced to discard up to 80% of their recombinant erythropoietin until a method was developed for the addition of two sugar residues to create the longer-acting version of the drug, known as Aranesp (Lesney, 2002). "Glycochip" arrays have recently been developed (Boguslavsky, 2002; Glycominds, 2002; Perkel, 2002), allowing the simultaneous analysis of glycan–protein interactions *en masse*. There is no doubt that complementation of gene-only approaches with "glycomics" era technology will be essential if the complex metabolic, structural, and regulatory aspects of glycans are to be elucidated (Lesney, 2002).

In the forthcoming decade, the application of the "omics" platforms will undoubtedly result in significant advances in our understanding of the many processes and molecular mechanisms that enable high (i.e., commercially viable) levels of heterologous proteins to be produced and secreted. New options and opportunities will arise for the control and regulation of certain genes and pathways involved directly or indirectly with heterologous expression and secretion. In addition to the post-translational glycosylation processes mentioned previously, such advances should include an improved understanding of (1) the molecular basis of cellular/nuclear competence for transformation, (2) optimized vector design (e.g., inclusion of matrix attachment regions, upstream activating sequences and other functional elements, transcriptional hot spot sequences, distal regulatory sequences, etc.), (3) optimized host development (increased secretion capacity, improved fermentation morphology, etc.), (4) fungal promoter functioning (e.g., the identity and mode of action of additional proteins involved in transcriptional regulation, effects of overproduction of positively acting transcription factors), and (5) modifications of enzymes involved in protein folding and formation of disulfide bridges. We expect that fermentation optimization in addition to pathway engineering based on the recognition and understanding of the metabolic burdens imposed on host systems by heterologous expression will also play crucial and complementary roles in achieving these goals.

5. Conclusions

The production of heterologous proteins in filamentous fungi has in numerous cases been very successful, although expression of many proteins remains problematic. Much of the recent success represents the culmination of a number of strategies that have been developed and applied over the last decade. Nevertheless it is clear that these strategies alone are inadequate to address all the problems which hinder further progress in this field. As we enter a new technological era, we anticipate that the necessary major breakthroughs will ultimately be made, although some of these (e.g., "glycomics") will necessitate the development of highly sophisticated techniques and instrumentation. Future advances will offer new possibilities enabling development of large scale, efficient, commercially viable, and safe processes for the manufacture of industrial, agricultural, neutraceutical, and pharmaceutical products.

References

Andersen, H.D., Jensen, E.B., and Welinder, K.G. (1992). A process for producing heme proteins. *European Patent Application EP 0505311-A2.*

Archer, D.B. and Peberdy, J.F. (1997). The molecular biology of secreted enzyme production by fungi. *Crit. Rev. Biotechnol.* **17**, 273–306.

Baron, M., Tiraby, G., Calmels, T., Parriche, M., and Durand, H. (1992). Efficient secretion of human lysozyme fused to the Sh *ble* phleomycin resistance protein by the fungus *Tolypocladium geodes. J. Biotechnol.* **24**, 253–266.

Bartking, S., van den Hombergh, J-P., Olsen, O., von Wetstein, D., and Visser, J. (1996). Expression of an *Erwinia* pectate lyase in three species of *Aspergillus. Curr. Genet.* **29**, 474–481.

Bergès, T., Barreau, C., Peberdy, J.F., and Boddy, L.M. (1993). Cloning of an *Aspergillus niger* invertase gene by expression in *Trichoderma reesei. Curr. Genet.* **24**, 53–59.

Berka, R.M., Boominathan, K.C., and Sandal, T. (1998c). Nucleic acids encoding polypeptides having lipase activity. *US Patent 5,821,102.*

Berka, R.M., Christgau, S., Halkier, T., Shuster, J.R., and Fuglsang, C.C. (1998a). *Penicillium purpurogenum* mutanases and nucleic acids encoding same. *US Patent 5,853,702.*

Berka, R.M., Nelson, B.A., Zaretsky, E.J., Yoder, W.T., and Rey, M.W. (2003). Genomics of *Fusarium venenatum*: An alternative fungal host to make enzymes. In D.K. Arora and G.G. Khachatourians (eds.) *Applied Mycology and Biotechnology, Volume 4: Fungal Genomics*. Elsevier Science, Amsterdam. (In Press).

Berka, R.M., Rey, M.W., Brown, K.M., and Brown, S.H. (2002). Promoters for expressing genes in a fungal cell. *US Patent 6,361,973 B1.*

Berka, R.M., Rey, M.W., Brown, K.M., Byun, T., and Klotz, A.V. (1998d). Molecular characterization and expression of a phytase gene from the thermophilic fungus *Thermomyces lanuginosus. Appl. Environ. Microbiol.* **64**, 4423–4427.

Berka, R.M., Schneider, P., Golightly, E.J., Brown, S.H., Madden, M., Brown, K.M., Halkier, T., Mondorf, K., and Xu, F. (1997). Characterization of the gene encoding an extracellular laccase of *Myceliophthora thermophila* and analysis of the recombinant enzyme expressed in *Aspergillus oryzae. Appl. Environ. Microbiol.* **63**, 3151–3157.

Berka, R.M., Thompson, S.A., and Xu, F. (1998b). Purified *Scytalidium* laccases and nucleic acids encoding same. *US Patent 5,843,745.*

Berka, R., Ward, M., Thompson, C., Rey, M., Fong, K., Wilson, L., Lamsa, M., and Gray, G. (1988). Foreign protein secretion in *Aspergillus*; Current status. In K. Brew, F. Ahmad, H. Bialy, S. Black, R.E. Fenna, D. Puett, W.A. Scott, J. van Brunt, R.W. Voellmy, W.J. Whelan, and J.F. Woessner (eds.) *Advances in gene technology: Protein engineering and production. Proceedings of the 1988 Miami Biotechnology Winter Symposium.* IRL, Oxford, pp. 50–51.

Biolog (2002). Hayward, California (September, 2002); *http://www.biolog.com.*

Blinkovsky, A.M., Byun, T., Brown, K.M., and Golightly, E.J. (1999). Purification, characterization and heterologous expression in *Fusarium venenatum* of a novel serine carboxypeptidase from *Aspergillus oryzae. Appl. Environ. Microbiol.* **65**, 3298–3303.

Boel, E., Christensen, T., and Wöldike, H.F. (1987). Process for the production of protein products in *Aspergillus oryzae* and a promoter for the use in *Aspergillus. European Patent Application EP 0238023.*

Boel, E. and Huge-Jensen, I.B. (1989). Recombinant *Humicola* lipase and process for the production of recombinant *Humicola* lipase. *European Patent Application EP 0305216.*

Boguslavsky, J. (2002). Novel microarrays push ingenuity. *Genomics & Proteomics* **2002**, 33–36.

Broekhuijsen, M.P., Mattern, I.E., Contreras, R., Kinghorn, J.R., and van den Hondel, C.A.M.J.J. (1993). Secretion of heterologous proteins by *Aspergillus-niger*-production of active human interleukin-6 in a protease-deficient mutant by kex2-like processing of a glucoamylase- IL-6 fusion protein. *J. Biotechnol.* **31**, 135–145.

Buczynski, S., Schneck, D., Vann, D., Kato, E., and Dorsey, W. (1995). *Eighteenth Fungal Genetics Conference*, Pacific Grove, CA, USA.

Bussink, H.J., Buxton, F.P., Fraaye, B.A., de Graaff, L.H., and Visser, J. (1992). The polygalacturonases of *Aspergillus niger* are encoded by a family of diverged genes. *Eur. J. Biochem.* **208**, 83–90.

Bussink, H.J., Buxton, F.P., and Visser, J. (1991). Expression and sequence comparison of the *Aspergillus niger* and *Aspergillus tubingensis* genes encoding polygalacturonase II. *Curr. Genet.* **19**, 467–474.

Carrez, D., Janssens, W., Degrave, P., van den Hondel, C.A., Kinghorn, J.R., Fiers, W., and Contreras, R. (1990). Heterologous gene expression by filamentous fungi: Secretion of human interleukin-6 by *Aspergillus nidulans*. *Gene* **94**, 147–154.

Case, M.E., Schweizer, M., Kushner, S.R., and Giles, N.H. (1979). Efficient transformation of *Neurospora crassa* utilizing hybrid plasmid DNA. *Proc. Natl. Acad. Sci. USA* **76**, 5259–5263.

Cheng, C., Tsukagoshi, N., and Udaka, S. (1990). Transformation of *Trichoderma viride* using the *Neurospora crassa pyr4* gene and its use in the expression of a Taka-amylase A gene from *Aspergillus oryzae*. *Curr. Genet.* **18**, 453–456.

Cheng, C. and Udaka, S. (1991). Efficient production of Taka-amylase A by *Trichoderma* viride. *Agric. Biol. Chem.* **55**, 1817–1822.

Cherry, J.R., Berka, R.M., and Halkier, T. (1999). Recombinant expression of a glucose oxidase from a *Cladosporium* strain. *US Patent 5,879,921*.

Christensen, T. (1995). A process for producing chymosin. *European Patent Specification 0575462*.

Christensen, T., Woeldike, H., Boel, E., Mortensen, S.B., Hjortshoej, K., and Hansen, M.T. (1988). High level expression of recombinant genes in *Aspergillus oryzae*. *Bio/Technology* **6**, 1419–1422.

Christgau, S., Kofod, L.V., Halkier, T., Andersen, L.N., Hockauf, M., Dörreich, K., Dalbøge, H., and Kauppinen, S. (1996). Pectin methyl esterase from *Aspergillus aculeatus*: Expression cloning in yeast and characterization of the recombinant enzyme. *Biochemical J.* **319**, 705–712.

Christgau, S., Sandal, T., Kofod, L.V., and Dalbøge, H. (1995). Expression cloning, purification and characterization of a beta-1,4-galactanase from *Aspergillus aculeatus*. *Curr. Genet.* **27**, 135–141.

Chu, S., Derisi, J., Eisen, M., Mulholland, J., Botstein, D., Brown, P.O., and Herskowitz, I. (1998). The transcriptional program of sporulation in budding yeast. *Science* **282**, 699–705.

Conesa, A., Jeenes, D., Archer, D.B., van den Hondel, C.A.M.J.J., and Punt, P.J. (2002a). Calnexin overexpression increases manganese peroxidase production in *Aspergillus niger*. *Appl. Environ. Microbiol.* **68**, 846–851.

Conesa, A., Punt, P.J., and van den Hondel, C.A.M.J.J. (2002b). Fungal peroxidases: Molecular aspects and applications. *J. Biotechnol.* **93**, 143–158.

Conesa, A., Punt, P.J., van Luijk, N., and van den Hondel, C.A.M.J.J. (2001a). The secretion pathway in filamentous fungi: A biotechnological view. *Fungal Genet. Biol.* **33**, 155–171.

Conesa, A., van den Hondel, C.A.M.J.J., and Punt, P.J. (2000). Studies on the production of fungal peroxidases in *Aspergillus niger*. *Appl. Environ. Microbiol.* **66**, 3016–3023.

Conesa, A., van de Velde, F., van Rantwijk, F., Sheldon, R.A., van den Hondel, C.A.M.J.J., and Punt, P.J. (2001b). Expression of the *Caldariomyces fumago* chloroperoxidase in *Aspergillus niger* and characterization of the recombinant enzyme. *J. Biol. Chem.* **276**, 17635–17640.

Contreras, R., Carrez, D., Kinghorn, J.R., van den Hondel, C.A.M.J.J., and Fiers, W. (1991). Efficient KEX2-like processing of a glucoamylase-interleukin-6 fusion protein by *Aspergillus nidulans* and secretion of mature interleukin-6. *Bio/Technology* **9**, 378–381.

De Gregorio, E., Spellman, P.T., Rubin, G.M., and Lemaitre, B. (2001). Genome-wide analysis of the *Drosophila* immune response by using oligonucleotide microarrays. *Proc. Natl. Acad. Sci. U.S.A.* **98**, 12590–12595.

Demolder, J., Saelens, X., Penttila, M., Fiers, W., and Contreras, R. (1994). KEX-2-like processing of glucoamylase-interleukin 6 and cellobiohydralase-interleukin 6 fusion proteins in *Trichoderma reesei*. *Second European Conference on Fungal Genetics*, Lunteren, The Netherlands. Abstract B38.

Devchand, M., Williams, S.A., Johnstone, J.A., Gwynne, D.I., Buxton, F.P., and Davies, R.W. (1989). Expression systems for *Aspergillus nidulans*. In C.L. Hershberger, S.W. Queener, G. Hegeman (eds.) *Genetics and molecular biology of industrial microorganisms*. American Society of Microbiology, Washington DC, pp. 301–303.

Dickinson, L., Harboe, M., van Heeswijck, R., Stroman, P., and Jepsen, L.P. (1987). Expression of active *Mucor miehei* aspartic protease in *Mucor circinelloides*. *Carlsberg Res. Commun.* **52**, 243–252.

Dunn-Coleman, N.S., Bloebaum, P., Barka, M., Bodie, E., Robinson, N., Armstrong, G., Ward, M., Przetak, M., Carter, G., Lamsa, M., and Hiensohn, H. (1991). Commercial levels of chymosin production by *Aspergillus*. *Mol. Gen. Genet.* **230**, 288–294.

Eisen, M.B., Spellman, P.T., Brown, P.O., and Botstein, D. (1998). Cluster analysis and display of genome-wide expression patterns. *Proc. Natl. Acad. Sci. U.S.A.* **95**, 14863–14868.

FDA. (2001). Washington (September, 2002); *http://vm.cfsan.fda.gov/~rdb/opa-g054.html*.

Finkelstein, D.B., Rambosek, J., Crawford, M.S., Soliday, C.L., McAda, P.C., and Leach, J. (1989). Protein secretion in *Aspergillus niger*. In C.L. Hershberger, S.W. Queener, G. Hegeman (eds.) *Genetics and molecular biology of industrial microorganisms*. American Society of Microbiology, Washington DC, pp. 295–300.

Filipphi, M.J.A., Visser, J., van der Veen, P., and De Graaff, L.H. (1993). Cloning of the *Aspergillus niger* gene encoding alpha-L-arabinofuranosidase A. *Appl. Microbiol. Biotechnol.* **39**, 335–340.

Fuglsang, C.C., Oxenboll, K., Halkier, T., Berka, R.M., and Cherry, J.R. (1999). Haloperoxidases from *Curvularia verruculosa* and nucleic acids encoding same. *US Patent 5,965,418.*

"Gibco cell culture division of Invitrogen reports that Invitrogen has signed a worldwide exclusive agreement with Novozymes." (2002, November 7). Metropolitan Daily Business Report, San Diego.

Glycominds Ltd., Lod, Israel (October 2002); *http://www.glycominds.com.*

Golightly, E.J., Berka, R.M., and Rey, M.W. (2000). Polypeptides having galactose oxidase activity and nucleic acids encoding same. *US Patent 6,090,604.*

Gouka, R.J., Hessing, J.G., Punt, P.J., Stam, H., Musters, W., and Van den Hondel, C.A.M.J.J. (1996a). An expression system based on the promoter region of the *Aspergillus awamori* 1,4-beta-endoxylanase A gene. *Appl. Microbiol. Biotechnol.* **46**, 28–35.

Gouka, R.J., Punt, P.J., Hessing, G.M., and van den Hondel, C.A.M.J.J. (1996b). Analysis of heterologous protein production in defined recombinant *Aspergillus awamori* strains. *Appl. Environ. Microbiol.* **62**, 1951–1957.

Gouka, R.J., Punt, P.J., and van den Hondel, C.A.M.J.J. (1997). Glucoamylase gene fusions alleviate limitations for protein production in *Aspergillus awamori* at the transcriptional and (post)translational levels. *Appl. Environ. Microbiol.* **63**, 488–497.

Gray, G.L., Hayenga, K., Cullen, D., Wilson, L.J., and Norton, S. (1986). Primary structure of *Mucor miehei* aspartyl protease: Evidence for a zymogen intermediate. *Gene* **48**, 41–53.

Gwynne, D.L., Buxton, F.P., Williams, S.A., Garven, S., and Davies, R.W. (1989). Genetically engineered secretion of active human interferon and a bacterial endoglucanase from *Aspergillus nidulans. Bio/Technology* **5**, 713–719.

Hadfield, K.A., Rose, J.K.C., Yaver, D.S., Berka, R.M., and Bennett, A.B. (1998). Polygalacturonase gene expression in ripe melon fruit supports a role for polygalacturonase in ripening-associated pectin disassembly. *Plant Physiol.* **117**, 363–373.

Hamer, L., Adachi, K., Montenegro-Chamorro, M.V., Tanzer, M.M., Sanjoy K. Mahanty, S.K., Lo, C., Tarpey, R.W., Skalchunes, A.R. *et al.* (2001). Gene discovery and gene function assignment in filamentous fungi. *Proc. Natl. Acad. Sci. USA* **98**, 5110–5115.

Harris, P. and Brown, K.M. (2000). Polypeptides having phospholipase B activity and nucleic acids encoding same. *US Patent 6,146,869.*

Hessing, H.J.G.M., van Rotterdam, C., Verbakel, J.A.M., Roza, M., Maat, J., Gorcom, R.F.M., and van den Hondel, C.A.M.J.J. (1994). Isolation and characterization of a 1,4-beta-endoxylanase gene of *A. awamori. Curr. Genet.* **26**, 228–232.

Hintz, W.E., Kalsner, I., Plawinski, E., Guo, Z., and Lagosky, P.A. (1995). Improved gene expression in *Aspergillus nidulans. Can. J. Bot.* **73**, S876–S884.

Invitrogen (2002). "Invitrogen Signs Exclusive Worldwide Distribution Agreement with Novozymes for Bioengineered Trypsin Alternative." *http://www.invitrogen.com.* Press Release, November 6th.

Jarai, G. (1997). Heterologous gene expression in filamentous fungi. In T. Anke (ed.) *Fungal Biotechnology*, Chapman & Hall, Weinheim, pp. 251–261.

Jeenes, D.J., Marczinke, B., Mackenzie, D.A., and Archer, D.B. (1993). A truncated glucoamylase gene fusion for heterologous protein secretion from *Aspergillus niger. FEMS Microbiol. Lett.* **107**, 267–271.

Joutsjoki, V.V., Kuittien, M., Torkkeli, T.K., and Suominen, P.L. (1993). Secretion of the *Hormoconis resinae* glucoamylase enzyme from *Trichoderma reesei* directed by the natural and the *cbh1* gene secretion signal. *FEMS Microbiol. Lett.* **112**, 281–286.

Joutsjoki, V.V. and Torkkeli, T.K. (1992). Glucoamylase P gene of *Hormoconis resinae*: Molecular cloning, sequencing and introduction into *Trichoderma reesei. FEMS Microbiol. Lett.* **78**, 237–243.

Kauppinen, S., Christgau, S., Kofod, L.V., Halkier, T., Dörreich, K., and Dalbøge, H. (1995). Molecular cloning and characterization of a rhamnogalacturonan acetylesterase from *Aspergillus aculeatus.* Synergism between rhamnogalacturonan degrading enzymes. *J. Biol. Chem.* **270**, 27172–27178.

Kelly, J.M. and Hynes, M. (1985). Transformation of *Aspergillus niger* by the *amdS* gene from *Aspergillus nidulans. EMBO J.* **4**, 475–479.

Korman, D.R., Bayliss, F.T., Barnett, C.C., Carmona, C.L., Kodama, K.H., Royer, T.J., Thompson, S.A., Ward, M. *et al.* (1990). Cloning, characterization, and expression of two alpha-amylase genes from *Aspergillus niger* var. *awamori. Curr. Genet.* **17**, 203–212.

Krasevec, N., van de Hondel, C.A.M.J.J., and Komel, R. (2000). Expression of human lymphotoxin alpha in *Aspergillus niger. Pflugers Archiv.* **440**, R83.

Kruszewska, J.S. (1999). Heterologous expression of genes in filamentous fungi. *Acta Biochim. Pol.* **46**, 181–195.

Kusters-van Someren, M. (1991). Characterization of an *Aspergillus niger* pectin lyase gene family. Doctoral dissertation. Agricultural University Wageningen, The Netherlands.

Lachmund, A., Urmann, U., Minol, K., Wirsel, S., and Ruttkowski, E. (1993). Regulation of alpha-amylase formation in *Aspergillus oryzae* and *Aspergillus nidulans* transformants. *Curr. Microbiol.* **26**, 47–51.

Lehmbeck, J. (1998). New modified host cells—are modified to express reduced levels of metallo-protease and alkaline protease, used to increase production of heterologous protein products. *World Patent Appl. WO 9812300.*

Lesney, M.S. (2002). Next stop glycomics. *Modern Drug Discovery*, October 2002, 35–39.

Mackenzie, D.A., Gendron, L.C.G., Jeenes, D.J., and Archer, D.B. (1994). Physiological optimization of secreted protein-production in *Aspergillus niger. Enzyme Microb. Technol.* **16**, 276–280.

MacRae, W.D., Buxton, F.P., Gwynne, D.I., and Davies, R.W. (1993). Heterologous protein secretion directed by a repressible acid phosphatase system. *Gene* **132**, 193–198.

Maras, M., Debruyn, A., Schraml, J., Herdewijn, P., Claeyssens, M., Fiers, W., and Contreras, R. (1997a). Structural characterization of n-linked oligosaccharides from cellobiohydrolase I secreted by the filamentous fungus *Trichoderma reesei* rutc-30. *Eur. J. Biochem.* **245**, 617–625.

Maras, M., Saelens, X., Laroy, W., Piens, K., Claeyssens, M., Fiers, W., and Contreras, R. (1997b). *In-vitro* conversion of the carbohydrate moiety of fungal glycoproteins to mammalian-type oligosaccharides—evidence for n-acetylglucosaminyltransferase I-accepting glycans from *Trichoderma reesei. Eur. J. Biochem.* **249**, 701–707.

Maras, M., van Die, I., Contreras, R., and van den Hondel, C.A.M.J.J. (1999). Filamentous fungi as production organisms for glycoproteins of bio-medical interest. *Glycoconjugate J.* **16**, 99–107.

Margolles-Clark, E., Hayes, C.K., Harman, G.E., and Penttila, M. (1996). Improved production of *Trichoderma harzianum* endochitinase by expression in *Trichoderma reesii. Appl. Environ. Microbiol.* **62**, 2145–2151.

Mattern, I.E., van Noort, J.M., Berg, P., Archer, D.B., Roberts, I.N., and van den Hondel, C.A.M.J.J. (1992). Isolation and characterization of mutants of *Aspergillus niger* deficient in extracellular proteases. *Mol. Gen. Genet.* **234**, 332–336.

Melton, R.E., Flegg, L.M., Brown, J.K.M., Oliver, R.P., Daniels, M.J., and Osbourn, A.E. (1998). Heterologous expression of *Septoria lycopersici* tomatinase in *Cladosporium fulvum*: Effects on compatible and incompatible interactions with tomato seedlings. *Mol. Plant Microbe Interact.* **11**, 228–236.

Mestric, S., Punt, P.J., Valinger, R., and van den Hondel, C.A.M.J.J. (1996). Expression of human insulin gene in *Aspergillus niger. Fungal Genet. Newsl.* **43B**, 25.

Miller, J.D. and MacKenzie, S. (2000). Secondary metabolites of *Fusarium venenatum* strains with deletions in the *tri5* gene encoding trichodiene synthetase. *Mycologia* **92**, 764–771.

Mishra, N.C. and Tatum, E.L. (1973). Non-Mendelian inheritance of DNA induces inositol independence in *Neurospora crassa. Proc. Natl. Acad. Sci. USA* **70**, 3873–3879.

Moralejo, F.J., Cardoza, R.C., Gutierrez, S., Lombraña, M., Fierro, F., and Martín, J.F. (2002). Silencing of the aspergillopepsin B (*pepB*) gene of *Aspergillus awamori* by antisense RNA expression or protease removal by gene disruption results in a large increase in Thaumatin production. *Appl. Environ. Microbiol.* **68**, 3550–3559.

Moralejo, F.J., Cardoza, R.E., Gutierrez, S., and Martín, J.F. (1999). Thaumatin production in *Aspergillus awamori* by use of expression cassettes with strong fungal promoters and high gene dosage. *Appl. Environ. Microbiol.* **65**, 1168–1174.

Morita, S., Kuriyama, M., Nakatsu, M., and Kitano, K. (1994). High level expression of *Fusarium* alkaline protease gene in *Acremonium chrysogenum. Biosci. Biotechnol. Biochem.* **58**, 627–630.

Morita, S., Kuriyama, M., Nakatsu, M., Suzuki, M., and Kitano, K. (1995). Secretion of active human lysozyme by *Acremonium chrysogenum* using a *Fusarium* alkaline protease promoter system. *J. Biotechnol.* **42**, 1–8.

Nakano, E.T., Fox, R.D., Clements, D.E., Woo, K., Stuart, W.D., and Ivy, J.M. (1993). Expression vectors for *Neurospora crassa* and expression of a bovine preprochymosin cDNA. *Fungal Genet. Newsl.* **40**, 54–56.

Nedwin, G.E. (1997). Green chemistry: Using enzymes as benign substitutes for synthetic chemicals and harsh conditions in industrial processes. In G.S. Sayler (ed.) *Biotechnology in the Sustainable Environment,* Plenum Press, New York, pp. 13–32.

Nielsen, B.R., Lehmbeck, J., and Frandsen, T.B. (2002). Cloning, heterologous expression, and enzymatic characterization of a thermo-stable glucoamylase from *Talaromyces emersonii. Protein Expression and Purification* **26**(1), 1–8.

Nirenberg, H.I. (1995). Morphological differentiation of *Fusarium sambucinum* Fuckel *sensu stricto, F. torulosum* (Berk. & Curt.)Nirenberg comb. nov. and *F. venenatum* Nirenberg sp. nov. *Mycopathologia* **129**, 131–141.

Novozymes (2002). Annual Report (2002). Financial Highlights, p. 6.

Nyyssönen, E. (1993). Monoclonal antibodies: Production and use in studies of the rate-limiting steps in heterologous protein production by the filamentous fungus *Trichoderma reesei*. Doctoral dissertation. University of Helsinki, p. 92.

Nyyssönen, E. and Keränen, S. (1995). Multiple roles of the cellulase CBHI in enchancing production of fusion antibodies by the filamentous *Trichoderma reesei. Curr. Genet.* **28**, 71–79.

Nyyssönen, E., and Penttilä, M., Harkki, A., Saloheimo, A., Knowles, J.K.C., and Keränen, S. (1993). Efficient production of antibody fragments by the filamentous fungus *Trichoderma reesei. Bio/Technology* **11**, 591–595.

O'Donnell, K. Cigelnik, E., and Casper, H.H. (1998). Molecular phylogenetic and mycotoxin data support re-identification of the Quorn™ mycoprotein fungus as *Fusarium venenatum. Fungal Genet. Biol.* **23**, 57–67.

Oeser, B. and Yoder, O.C. (1994). Pathogenesis by *Cochliobolus heterostrophus* transformants expressing a cutinase-encoding gene from *Nectria haematococca. Mol. Plant Microbe Interact.* **7**, 282–288.

Paradigm Genetics (2002). North Carolina (September, 2002); *http://www.paradigmgenetics.com.*

Pentilla, M. (1998). Heterologous protein production in *Trichoderma*. In G.E. Harman, and C.P. Kubicek (eds.) *Trichoderma and Gliocladium Volume 2; Enzymes, biological control and commercial applications*. Taylor & Francis Ltd, London, pp. 365–382.

Perkel, J.M. (2002). Glycobiology goes to the ball. *The Scientist* **16**, 32.

Punt, P.J., van Biezen, N., Conesa, A., Albers, A., Mangnus, J., and van den Hondel, C.A.M.J.J. (2002). Filamentous fungi as cell factories for heterologous protein production. *Trends Biotechnol.* **20**, 200–206.

Punt, P.J., van Gemeren, I.A., Drint-Kuijvenhoven, J., Hessing, J.G.M., van Muijlwijk-Harteveld, G.M., Beijersbergen, A., and Verrips, C.T. (1998). Analysis of the role of the gene *bipA*, encoding the major endoplasmic reticulum chaperone protein in the secretion of homologous and heterologous proteins in black Aspergilli. *Appl. Microbiol. Biotechnol.* **50**, 447–454.

Punt, P.J., Zegers, N.D., Busscher, M., Pouwels, P.H., and van den Hondel, C.A.M.J.J. (1990). Intracellular and extracellular production of proteins in *Aspergillus* under control of expression signals of the highly expressed *Aspergillus nidulans gpdA* gene. *J. Biotechnol.* **17**, 19–34.

Radzio, R. and Kuck, U. (1997). Synthesis of biotechnologically relevant heterologous proteins in filamentous fungi. *Process Biochem.* **32**, 529–539.

Record, E., Punt, P.J., Chamkha, M., Labat, M., van den Hondel, C.A.M.J.J., and Asther, M. (2002). Expression of the *Pycnoporus cinnabarinus* laccase gene in *Aspergillus niger* and characterization of the recombinant enzyme. *Eur. J. Biochem.* **269**, 602–609.

Rey, M.W. and Golightly, E.J. (1999). Nucleic acids encoding polypeptides having tripeptide aminopeptidase activity. *US Patent 5,989,889.*

Rey, M.W. and Golightly, E.J. (2002). Polypeptides having lipase activity and nucleic acids encoding same. *US Patent 6,432,898.*

Roberts, I.N., Jeenes, D.J., MacKenzie, D.A., Wilkinon, A.P., Sumner, I.G., and Archer, D.B. (1992). Heterologous gene expression in *Aspergillus niger*: A glucoamylase-porcine pancreatic prophospholipase A2 fusion protein is secreted and processed to yield mature enzyme. *Gene* **122**, 155–161.

Royer, J.C., Christainson, L.M., Yoder, W.T., Gambetta, G.A., Klotz, A.V., Brody, H., and Otani, S. (1999). Deletion of the trichodiene synthase gene of *Fusarium venenatum*: Two systems for repeated gene deletions. *Fungal Genet. Biol.* **28**, 68–78.

Royer, J.C., Moyer, D.M., Reiwitch, S.G., Madden, M.S., Jensen, E.B., Brown, S.H., Yonker, C.C., Johnstone, J.A. *et al.* (1995). *Fusarium* graminearum A3/5 as a novel host for heterologous protein production. *Bio/Technology* **13**, 1479–1483.

Ruttkowski, E., Khanh, N.Q., Wirsel, S., Wildhardt, G., and Gottschalk, M. (1989). Expression of the alpha-amylase gene from *Aspergillus oryzae* in *Aspergillus awamori. DECHEMA Biotechnology Conferences 4.* VCH Verlagsgesellschaft Weinhei, Germany, pp. 325–328.

Rypniewski, W.R., Hastrup, S., Betzel, C.H., Dauter, M., Dauter, Z., Papendorf, G., Branner, S., and Wilson, K.S. (1993). The sequence and X-ray structure of the trypsin gene from *Fusarium oxysporum. Prot. Eng.* **6**, 341–348.

Saloheimo, M., Barajas, V., Niku-Paavola, M.L., and Knowles, J.K. (1989). A lignin peroxidase-encoding cDNA from the white-rot fungus *Phlebia radiata*: Characterization and expression in *Trichoderma reesei. Gene* **85**, 343–351.

Saloheimo, M., Meittinen-Oinonen, A., Torkkeli, T., Nevalainen, H., and Suominen, P. (1993). Enzyme production in *T. reesei* using the *cbh1* promoter. In P. Suominen and T. Reinikainen (eds.) *Proceedings of the 2nd Tricel Symposium, Majvic, Finland. Foundation for Biotechnical and Industrial Fermentation Research, Helsinki*, **8**, 229–237.

Saloheimo, M., Niku-Paavola, M.L., and Knowles, J.K. (1991). Isolation and structural analysis of the laccase gene from the lignin-degrading fungus *Phlebia radiata*. *J. Gen. Microbiol.* **137**, 1537–1544.

Sen, K., Nandi, P., and Mishra, A.K. (1969). Transformation of nutritionally deficient mutants of *Aspergillus niger*. *J. Gen. Microbiol.* **55**, 195–200.

Shamoian, C.A., Canzanelli, A., and Melrose, J. (1961). Back mutation of a *Neurospora crassa* mutant by a nucleic acid complex from wild strain. *Biochim. Biophys. Acta* **47**, 208–211.

Shibuya, I., Gomi, K., Iimura, Y., Takahashi, K., Tamura, G., and Hara, S. (1990). Molecular cloning of the glucoamylase gene of *Aspergillus shirousami* and its expression in *Aspergillus oryzae*. *Agric. Biol. Chem.* **54**, 1905–1914.

Shoji, H., Horiuchi, H., and Takagi, M. (1999). Production of recombinant Der f1 (a major mite allergen) by *Aspergillus oryzae*. *Biosci. Biotechnol. Biochem.* **63**, 703–709.

Shockley, T.E. and Tatum, E.L. (1962). A search for genetic transformation in *Neurospora crassa*. *Biochim. Biophys. Acta* **61**, 567–572.

Sotiriadis, A., Keshavarz, T., and Keshavarz-Moore, E. (2001). Factors affecting the production of a single-chain antibody fragment by *Aspergillus awamori* in a stirred tank reactor. *Biotechnol. Prog.* **17**, 618–623.

Spencer, A., Morozov-Roche, L.A., Noppe, W., MacKenzie, D.A., Jeenes, D.J., Joniau, M., Dobson, C.M., and Archer, D.B. (1999). Expression, purification, and characterization of the recombinant calcium-binding equine lysozyme secreted by the filamentous fungus *Aspergillus niger*: Comparisons with the production of hen and human lysozymes. *Protein Expr. Purif.* **16**, 171–180.

Stewart, P., Whitwam, R.E., Kersten, P.J., Cullen, D., and Tien, M. (1996). Efficient expression of a *Phanerochaete chrysosporium* manganese peroxidase gene in *Aspergillus oryzae*. *Appl. Environ. Microbiol.* **62**, 860–864.

Sugano, Y., Nakano, R., Sasaki, K., and Shoda, M. (2000). Efficient heterologous expression in *Aspergillus oryzae* of a unique dye-decolorizing peroxidase, DyP, of *Geotrichum candidum*. *Appl. Environ. Microbiol.* **66**, 1754–1758.

Svetina, M., Krasevec, N., Gaberc-Porekar, V., and Komel, R. (2000). Expression of catalytic subunit of bovine enterokinase in the filamentous fungus *Aspergillus niger*. *J. Biotechnol.* **76**, 245–251.

Thompson, C.J., Movva, N.R., Tizard, R, Crameri, R., Davies, J.E., Lauwereys, M., and Botterman, J. (1987). Characterization of the herbicide-resistance gene from *Streptomyces hygroscopicus*. *EMBO J.* **6**, 2519–2523.

Trinci, A.P.J. (1992). Presidential Address 1991. Myco-protein: A twenty year, overnight success story. *Mycol. Res.* **96**, 1–13.

Tsatsumi, N., Sasaki, Y., Rey, M.W., Zeretsky, E.J., Spendler, T., and Vind, J. (2002). Lipolytic enzyme. *World Patent Application WO 02/00852*.

Tsuchiya, K., Gomi, K., Kitamoto, K., Kumagai, C., and Tamura, G. (1993). Secretion of calf chymosin from the filamentous fungus *Aspergillus oryzae*. *Appl. Microbiol. Biotechnol.* **40**, 327–332.

Tsuchiya, K., Nagashima, T., Yamamoto, Y., Gomi, K., Kitamoto, K., Kumagai, C., and Tamura, G. (1994). High level secretion of calf chymosin using a glucoamylase–prochymosin fusion gene in *Aspergillus oryzae*. *Biosci. Biotechnol. Biochem.* **58**, 895–899.

van den Hombergh, J.P.T.W., Gelpke, M.D.S., van de Vondervoot, P.J., Buxton, F.P., and Visser, J. (1997). Disruption of three acid proteases in *Aspergillus niger*: Effects on protease spectrum, intracellular proteolysis, and degradation of target proteins. *Eur. J. Biochem.* **247**, 605–613.

van den Hombergh, J.P.T.W., van de Vondervoot, P.J., van der Heijden, N.C.B.A., and Visser, J. (1995). New protease mutants in *Aspergillus niger* result in strongly reduced *in vitro* degradation of target proteins: Genetical and biochemical characterization of seven complementation groups. *Curr. Genet.* **28**, 299–308.

van Gemeren, I.A., Beijersbergen, A., Musters, W., Gouka, R.J., and van den Hondel, C.A.M.J.J. (1996). The effect of pre- and pro-sequences and multicopy integration on heterologous expression of the *Fusarium solani* var *pisi* cutinase gene in *Aspergillus awamori*. *Appl. Microbiol. Biotechnol.* **45**, 755–763.

van Zeijl, C., Bartels, J., Punt, P., Emelfarb, M., Burlingame, R., Olson, P.T., Pynnonen, C.M., Sinitsyn, A., and van den Hondel, C.A.M.J.J. (2001). *Chrysosporium lucknowense*, a new fungal host for protein production. *Twenty-first Fungal Genetics Conference*, Asilomar, CA.

Varki, A. (1993). Biological roles of oligosaccharides—all of the theories are correct. *Glycobiology* **3**, 97–130.

Wahleithner, J., Xu, F., Brown, S.H., Brown, K.M., Golightly, E.J., Kauppinen, S., Pedersen, A., and Schneider, P. (1996). The identification and characterization of four lacccases from the plant pathogenic fungus *Rhizoctonia solani*. *Curr. Genet.* **29**, 395–403.

Ward, M. and Kodama, K.H. (1991). Introduction to fungal proteinases and expression in fungal systems. In B.M. Dunn (ed.) *Advances in Experimental Medicine and Biology*, 306 Plenum, New York, pp. 149–160.

Ward, P.P., May G.S., Headon, D.R., and Conneely, O.M. (1992). An inducible expression system for the production of human lactoferrin in *Aspergillus nidulans. Gene* **122**, 219–223.

Ward, P.P., Piddington, C.S., Cunningham, G.A., Zhou, X., Wyatt, R.D., and Conneely, O.M. (1995). A system for production of commercial quantities of human lactoferrin: A broad spectrum natural antibiotic. *Bio/Technology* **13**, 498–503.

Ward, M., Wilson, L.J., Kodama, K.H., Rey, M.W., and Berka, R.M. (1990). Improved production of chymosin in *Aspergillus* by expression as a glucoamylase–chymosin fusion. *Bio/Technology* **8**, 435–440.

Weibe, M.G., Blakebrough, M.L., Craig, S.H., Robson, G.D., and Trinci, A.P.J. (1996). How do highly branched (colonial) mutants of *Fusarium graminearum* A3/5 arise during Quorn^{TM} myco-protein fermentations? *Microbiology* **142**, 525–532.

Weibe, M.G., Robson, G.D., Trinci, A.P.J., and Oliver, S.G. (1992). Characterization of morphological mutants generated spontaneously in glucose-limited, continuous flow cultures of *Fusarium graminearum* A3/5. *Mycol. Res.* **96**, 555–562.

Wiebe, M.G., Karandikar, A., Robson, G.D., Trinci, A.P., Candia, J.L., Trappe, S., Wallis, G., Rinas, U. *et al.* (2001). Production of tissue plasminogen activator (t-PA) in *Aspergillus niger. Biotech. Bioeng.* **76**, 164–174.

Wiebe, M.G., Robson, G.D., Shuster, J.R., and Trinci, A.P.J. (1999). pH regulation of recombinant glucoamylases production in *Fusarium venenatum* JeRS325, a transformant with a *Fusarium oxysporum* alkaline (trypsin-like) protease promoter. *Biotechnol. Bioeng.* **64**, 368–372.

Wiebe, M.G., Trinci, A.P.J., Cunliffe, B., Robson, G.D., and Oliver, S.G. (1991). Appearance of morphological (colonial) mutants in glucose-limited, continuous flow cultures of *Fusarium graminearum. Mycol. Res.* **95**, 1284–1288.

Whittington, H., Kerry-Williams, S., Bidgood, K., Dodsworth, N., Peberdy, J., Dobson, M., Hinchliffe, E., and Balance, D.J. (1990). Expression of the *Aspergillus niger* glucose oxidase gene in *A. niger, A. nidulans* and *Saccharomyces cerevisiae. Curr. Genet.* **18**, 531–536.

Witteveen, F.B., van de Vondervoort, P.J., van den Broeck, H.C., van Engelenburg, A.C., de Graaff, L.H., Hillebrand, M.H., Schaap, P.J., and Visser, J. (1993). Induction of glucose oxidase, catalase, and lactonase in *Aspergillus niger. Curr. Genet.* **24**, 408–416.

Xu, F., Golightly, E.J., Duke, K.R., Lassen, S.F., Knusen, B., Christensen, S., Brown, K.M., Brown, S.H., *et al.* (2001a). *Humicola insolens* cellobiose dehydrogenase: Cloning, redox chemistry and logic gate-like dual functionality. *Enz. Microbiol. Technol.* **28**, 744–753.

Xu, F., Golightly, E.J., Schneider, P., Berka, R.M., Brown, K.M., Johnstone, J.A., Baker, D.H., Fuglsang, C.C. *et al.* (2000). Expression and characterization of a recombinant *Fusarium* sp. galactose oxidase. *Appl. Biochem. Biotechnol.* **88**, 23–32.

Xu, F., Golightly, E.J., Schneider, P., Fuglsang, C.C., Christensen, S., Duke, K.R., Graeve-Kampfenkel, K.H., Dybdal, L. *et al.* (2001b). A novel carbohydrate: Acceptor oxidoreductase from *Microdochium nivale. Eur. J. Biochem.* **268**, 1136–1142.

Yaver, D.S., Berka, R.M., and Rey, M.W. (2001). Polypeptides having acid phosphatase activity and nucleic acids encoding same. *US Patent* 6,309,869.

Yaver, D.S., Overjero, M.D., Xu, F., Nelson, B.A., Brown, K.M., Halkier, T., Bernauer, S., Brown, S.H. *et al.* (1999). Molecular characterization of laccase genes from the basidiomycete *Coprinus cinereus* and heterologous expression of the laccase lcc1. *Appl. Environ. Microbiol.* **65**, 4943–4948.

Yaver, D.S., Xu, F., Golightly, E.J., Brown, K.M., Brown, S.H., Rey, M.W., Schneider, P., Halkier, T. *et al.* (1996). Purification, characterization, molecular cloning, and expression of two laccase genes from the white rot basidiomycete *Trametes villosa. Appl. Environ. Microbiol.* **62**, 834–841.

Yoder, W.T. and Christianson, L.M. (1998). Species-specific primers resolve members of *Fusarium* section *Fusarium*: Taxonomic status of the edible Quorn^{TM} fungus re-evaluated. *Fungal Genet. Biol.* **23**, 68–80.

10

Artificial Evolution of Fungal Proteins

Jesper Vind

1. Introduction

Numerous techniques have been developed over the last two decades to change characteristics of proteins, and with these techniques it has become feasible to customize proteins for specific commercial and environmental applications. For instance, a fungal lipase from *Thermomyces lanuginosus* has been improved to increase its performance in detergent solutions (Borch *et al.*, 1998). The same lipase has been changed into a phospholipase that can be applied in bread making to generate emulsifiers from the lipids present in the dough, eliminating the need to add emulsifiers (Bojsen *et al.*, 1998). This chapter will provide a general overview of the advances within artificial molecular evolution. It will focus on the *in vitro* and *in vivo* procedures that have been applied to fungal proteins, some techniques having been developed specifically for filamentous fungi. Finally, thoughts on the future direction of artificial evolution of fungal proteins will be offered.

2. Artificial Evolution in General

2.1. Idea Generation

The first step in protein engineering is to establish the question to be answered. Often there are biochemical data (e.g., thermostability, substrate specificity, pH profile) giving rise to a hypothesis on which questions can be based. These data may be complemented by an X-ray structure of the protein of interest. Such a structure can be used in various *in silico* experiments involving docking of the substrate (reviewed by Halperin *et al.*, 2002), electrostatic calculations (reviewed by Kumar and Nussinov, 2002), or molecular dynamics (reviewed by Bizzarri and Cannistraro, 2002; Daggett, 2002). If the X-ray structure is not available, it might be possible to build a model based on a known structure of a homologous protein (reviewed by Sanchez and Sali, 1997). Alignment of homologous genes can often give hints for mutations that might be interesting to generate. Based on the biochemical and *in silico* data, it is often possible to generate many ideas for new mutations to test the hypothesis or improve the characteristics of a protein.

Jesper Vind • Novozymes, NOVO Alle 1, 1U.1.20, 2880 Bagsvaerd Denmark.

Advances in Fungal Biotechnology for Industry, Agriculture, and Medicine. Edited by Jan S. Tkacz and Lene Lange, Kluwer Academic/Plenum Publishers, 2004.

2.2. *In Vitro* Generation of Gene Variants

One way to generate protein variants is to modify *in vitro* the gene encoding the protein and then introduce the DNA into a host cell for expression. Since the late 1970s it has been feasible to perform DNA oligonucleotide-directed mutagenesis (Hutchison *et al.*, 1978), and for the last decade a number of commercial kits for making site-directed variants have been available (reviewed by Ling and Robinson, 1997). As these tools became available, they were applied to the analysis of protein structure and function (Dalbadie-McFarland *et al.*, 1982).

In the 1980s it was difficult, as it is today, to predict the outcome when mutations were introduced. Thus, Site-Saturation mutagenesis was established to explore the consequences of replacing a particular residue in the protein sequence with each of the other natural amino acids (Schultz and Richards, 1986). To accomplish this, DNA oligonucleotides were synthesized with the codon of interest in the form NNS where N = A, T, C, or G and S = G or C. This did not solve the difficulty in prediction, but allowed the possibility of arriving at an answer faster. Since then the approach has become slightly more sophisticated; oligonucleotide synthesis can be biased toward specific amino acid codons by manipulation of the percentages of A, T, G, or C in the reaction mixtures for each position within a given codon (Jensen *et al.*, 1998). Various techniques were soon developed to introduce multiple mutations in a gene simultaneously (Rasmussen and Nickoloff, 1992; Rouwendal *et al.*, 1993; Nickoloff *et al.*, 1996). An alternative to using DNA oligonucleotides was to introduce mutations in the gene of interest randomly with chemicals, for example, hydroxylamine (Busby *et al.*, 1982; Myers *et al.*, 1985). Simpler methods to generate random point mutations became available with the introduction of PCR (Leung *et al.*, 1989), which made it feasible easily to produce numerous random mutations concurrently. With a good screen, it was possible to identify a transformant expressing a protein variant with improved characteristics, which might then be subjected to iterative mutagenesis cycles for further improvement. The technique was refined and made more efficient by mixing improved protein variants from the first round of mutagenesis, rather than continuing with just a single variant, so that the chances for improvement would not be limited by the initial choice (Fuchs *et al.*, 1993). This can be done either by shuffling *in vitro*, where the DNA encoding improved variants is mixed, cut with a restriction enzyme, and reassembled by ligation or PCR (Stemmer, 1994; Vind, 1997; Dupret *et al.*, 1998), or by *in vivo* shuffling, where intact genes are subjected to enzymes involved in DNA recombination (reviewed by Wang, 2000; Kurtzman *et al.*, 2001). Further refinements have made possible the shuffling of genes with little or no homology, although these methodologies still have some drawbacks, such as giving rise to only a single crossover event (Sieber *et al.*, 2001). New *in vitro* and *in vivo* shuffling formats take advantage of the DNA repair system. Heteroduplexes of homologous genes are created by annealing opposite DNA strands, and these are then treated *in vivo* or *in vitro* with DNA repair enzymes (Volkov *et al.*, 1999; Bernstein *et al.*, 2000; Vind, 2000).

3. *In Vitro* Mutagenesis and Expression of Fungal Proteins

3.1. Characterization of Protein Variants Expressed in Yeast

From the early 1980s yeast was used as the host for expressing and studying protein variants, such as a cytochrome *c* or functional oxidation-resistant mutants of human

α_1-antitrypsin (Rosenberg *et al.*, 1984; Pielak *et al.*, 1985). Later Naumovski and Friedberg (1986) used random mutagenesis, along with site-directed protein engineering to identify important domains in the *RAD3*-encoded protein from *Saccharomyces cerevisiae*. Because of the high frequency of homologous recombination in this organism, protein variants could be generated simply by transforming them with DNA oligonucleotides containing the mutation of interest (Moerschell *et al.*, 1988). Subsequently a number of techniques have been developed to generate and screen libraries in yeast, for example, the two-hybrid protein system. With this system it is possible to screen for protein–protein interaction *in vivo* (reviewed by Legrain and Selig, 2000). Recently, methods have been developed that provide for the display of proteins on the yeast surface and for screening on the basis of affinity to a substrate or an interacting protein (Boder and Wittrup, 2000). The advantage of this approach compared with protein display on *Escherichia coli*-specific phage is that it allows evaluation of post-translational modifications, including disulfide isomerization and glycosylation (Rodi *et al.*, 2002). Besides improving the characteristics of a protein, artificial molecular evolution can also be used to improve the expression level of a protein in yeast (van der Linden *et al.*, 2000).

Specific examples of artificial evolution will illustrate two strategies by which different fungal enzymes are thermostabilized. The first strategy is to create a consensus protein (Lehmann, Pasamontes, Lassen, and Wyss, 2000b) and is based on the premise that during evolution, there has been a selection for specific amino acid residues in each position of a given protein. By alignment of homologous genes from different species, it is possible to identify the most commonly found amino acid residue at each position, an amino acid that might represent the optimal residue for that position within the protein family. Building a consensus sequence is illustrated in Figure 10.1. The strategy has been applied to phytase. Thirteen homologous wild-type phytases from filamentous fungi were used to deduce a consensus sequence (Lehmann, Kostrewa, Wyss, Brugger, D'Arcy, Pasamontes, and van Loon, 2000a). A synthetic consensus gene was made based on the codon usage of highly expressed yeast genes, and the expressed consensus phytase was more stable by 15–26°C than any of the 13 "parental" phytases as analyzed by differential scanning calorimetry. With further alignments and refinements, the stability was actually increased further by 10°C.

```
aay39897    ATFPLNATLYADFSHDSNLVSIFWALGLYNGTAPLSQTSVESVSQTDGYAAAWTVPFAAR

aar46234    ATFPLNSTLYADFSHDNGIISILFALGLYNGTKPLSTTTVENITQTDGFSSAWTVPFASR

aay39905    ATFPLNATMYVDFSHDNSMVSIFFALGLYNGTEPLSRTSVESAKELDGYSASWVVPFGAR

aay69554    ATFPLDRKLYADFSHDNSMISIFFAMGLYNGTQPLSMDSVESIQEMDGYAASWTVPFGAR

aaw84355    ATFPLNATLYADFSHDNTMTSIFAALGLYNGTAKLSTTEIKSIEETDGYSAAWTVPFGGR

aay69556    RTFPLGRPLYADFSHDNDMMGVLGALGAYDGVPPLDKTARRDPEELGGYAASWAVPFAAR

             ****    * *****         * * * *  *            *     * *** *

consensus:  ATFPLNATLYADFSHDNSMISIFFALGLYNGTAPLSTTSVESIEETDGYSASWTVPFAAR
```

Figure 10.1. Sequence alignment. Part of a sequence alignment of six hypothetical fungal phytase enzymes, which can be used to generate a consensus sequence. Identities are marked with *.

Phytase is added to pig feed to catalyze the release of phosphate that is naturally bound to phytic acid in the feed. Its use eliminates the need to supplement the feed with inorganic phosphate, which if not absorbed by the animal, enriches the manure and can accelerate eutrophication in the environment. In the process of making feed pellets, the raw mixture reaches 60–90°C; the engineered thermostable phytase survives where the wild-type enzymes would be inactivated (Lehmann *et al.*, 2000a).

The second approach is illustrated by the thermostabilization of the *Aspergillus niger* glucoamylase (Frandsen *et al.*, 1998). Highly mobile regions of the glucoamylase were identified through molecular dynamic simulation performed on the X-ray structure (color Figure 10.2). These regions might represent initiation sites for unfolding, and consequently, they were chosen for random mutagenesis. The substitutions were directed to residues that could influence the dynamic behavior by increasing the contacts with nearby residues or by simply occupying space needed for motion by the mobile peptide. DNA oligonucleotides encoding desired changes were mixed with DNAse I-degraded *A. niger* glucoamylase PCR fragments, and open reading frames (ORFs) were assembled by PCR. The assembled fragments were cloned by *in vivo* recombination in yeast, and the library screened for stability. Through a number of iterative cycles, a variant was identified with a functional half-life at 70°C of 115 min, as compared with the 7-min half-life of the wild-type enzyme in the same conditions.

Glucoamylase catalyzes the release of D-glucose from the nonreducing ends of starch or related oligo- and polysaccharides and is used to produce high-fructose corn syrup. Thermostable glucoamylase was developed to allow the corn syrup process to run at an elevated temperature (Frandsen *et al.*, 1998).

3.2. Characterization of Protein Variants Expressed in Filamentous Fungi

Before suitable systems were available for the transformation of filamentous fungi, most filamentous fungal proteins were expressed in yeast or *E. coli* (Sierks *et al.*, 1989; Grunert *et al.*, 1991). However these hosts generate difficulties associated with glycosylation: in yeast the foreign secreted proteins tend to be hyperglycosylated whereas in *E. coli* the proteins lack glycosylation (Romanos *et al.*, 1991). Conducting experiments with incorrectly glycosylated or non-glycosylated proteins can lead to misinterpretation of the results, because glycosylation can affect the activity of enzymes, either directly or indirectly, by altering protein folding during secretion or by changing the stability to pH, protease, or temperature (Steed *et al.*, 1998).

The development of transformation systems for filamentous fungi provided the means to express manipulated fungal genes in fungal hosts (Boel *et al.*, 1987; Christensen *et al.*, 1988; Clausen *et al.*, 1990). Besides removing the glycosylation issue, filamentous fungal hosts, as compared to yeast, were advantageous with respect to potential levels of expression, reaching as high as 30 g/L (Punt *et al.*, 2002). Adequate expression aids purification of enzyme in quantities that are sufficient for a full battery of characterization experiments.

*Pichia pastor*is is a yeast host that is an alternative to *S. cerevisiae*. *Pichia* gives high levels of expression (1–2 g/L) and generally does not hyperglycosylate secreted protein

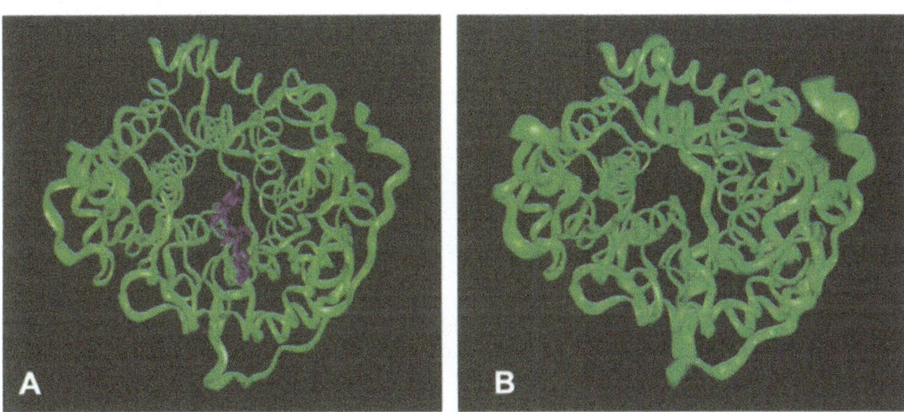

Figure 10.2. Ribbon presentations of structural mobile areas in the glucoamylase structure from *Aspergillus niger*. Molecular dynamics simulation of a modeled structure of the glucoamylase from *A. niger* (modeled from the *Aspergillus awamori* 3D structure 3GLY) at two temperatures; 300K (A) and 400K (B). The ribbon displays the root mean square (RMS) values for each residue's carbon alpha (CA) atom. The increase in temperature results for certain residues in a higher mobility of the atoms as illustrated by the thickness of the ribbons.

(reviewed by Cereghino and Cregg, 2000). Since it is not well suited for making libraries, they can be made and screened in *S. cerevisiae* and the best candidates can be expressed in *P. pastoris* on a large scale (Morawski *et al.*, 2001).

3.3. Library Generation in Filamentous Fungi

Most of the library screening work on fungal proteins has been done in yeast as the host (mainly *S. cerevisiae*) rather than in filamentous fungi. The disadvantage posed by hyperglycosylation is outweighed by several favorable features of the organisms: it grows rapidly, can be manipulated easily, is amenable to library construction, is transformed by episomal plasmids that are easily rescued, and has expressed a number of eukaryotic proteins.

Nevertheless, for the development of commercially viable proteins, it is preferable to screen the host that will be used in industrial process. However with filamentous fungi, there are generally no useful, naturally occurring, nuclear, self-replicating plasmids. When DNA is introduced into a fungus, it must integrate into a chromosome to be stably inherited, and with traditional transformation methods this occurs at random sites in the genome and at various copy numbers. Therefore expression levels vary greatly among the transformants, and this situation is unsuitable for library screening. Further, the transformation frequency is low (10–100 transformants/μg DNA). One exception is the *AMA1* sequence of *Aspergillus nidulans* (Gems *et al.*, 1991). This sequence was recognized in a library of other *A. nidulans* genomic fragments by the fact that the transformants containing the sequence grew as smaller and sectored colonies, a phenotype suggesting that the transforming DNA was maintained in a replicative but unstable form rather than being incorporated into the genome. With the *AMA1* sequence, it is possible to construct plasmids that replicate independently in the nuclei of certain filamentous fungi. Transformation rates with such plasmids are increased 250-fold, but the plasmids tend to be unstable, being easily lost during growth.

An *AMA1*-based plasmid with improved stability has been developed, and it has become feasible to create libraries with it for use in filamentous fungi (Vind, 1999). Although the *pyrG*-encoding orotidine-5′-phosphate (OMP) decarboxylase has been used for dominant selection, incorporation of a mutant *pyrG* that either is poorly expressed or encodes an unstable form of the enzyme affords an *AMA1* plasmid with unusually stringent selection pressure that is maintained at increased copy number. Libraries made in a plasmid of this type can be screened for products with both improved characteristics and the ability to express well in an industrially relevant host. When independent *Aspergillus* clones transformed with such an *AMA1–pyrG* plasmid containing a reporter gene (*T. lanuginosus* lipase) are grown in microtiter dishes, they gave nearly the same lipase expression level with a relative standard deviation of 20%. This low standard deviation makes the system suitable for screening of libraries in filamentous fungi. A small library was made in which the propeptide of the *T. lanuginosus* lipase gene was mutated and the transformants were screened for activity in the presences of detergent. A number of positive transformants were identified, and as shown in Table 10.1, one among them exhibited lipase activity in detergent, which was 15-fold higher than that of the wild-type lipase (Vind, 1999).

Table 10.1. Screening of a Library of *Thermomyces lanuginosus* Lipase Variants Expressed in *Aspergillus*[a]

	PNP-butyrate activity (%)	PNP-palmitate activity in detergent (%)
Wild type	102 ± 16	1.6 ± 80
Clone A	106 ± 11	49 ± 12
Clone B	80 ± 1	29 ± 5
Clone C	117 ± 7	28 ± 1

[a]Media in which different clones from the library had grown were assayed for relative *p*-nitrophenyl-butyrate activity and *p*-nitrophenyl-palmitate activity measured in presence of detergent. The clones show increased activity in presence of detergent (data from Patent WO200024883).

4. *In Vivo* Mutagenesis in Fungi

4.1. *In Vivo* Shuffling in Yeast

For a number of years, yeast has been used for *in vivo* shuffling with positive results (Pompon and Nicolas, 1989; Okkels, 1995). One example covering a number of other techniques is the directed evolution of a *Coprinus cinereus* peroxidase leading to improvements in thermostability and resistance to oxidation (Cherry *et al.*, 1999). Based on the X-ray structure (Petersen *et al.*, 1994), a number of specific residues were chosen for mutagenesis. Methionines and tyrosines, potential oxidation sites, were mutated, as was a glutamic acid residue (E239) that was involved in stability. In the latter case, the Site-Saturation approach was used (Figure 10.3). At the same time error-prone PCR was performed on the entire peroxidase gene. Successive rounds of mutagenesis led finally to 10 candidates for *in vivo* shuffling in yeast. PCR fragments from the 10 variants were introduced into yeast along with a linearized vector. From this *in vivo* shuffling, two improved variants were selected, and these were shuffled further in yeast along with 28 improved variants identified in another error-prone PCR library. One of the peroxidase variants had increased its thermostability nearly 200-fold in the screening assay and its stability in hydrogen peroxide approximately 100-fold (Table 10.2).

4.2. *In Vivo* Shuffling in *Neurospora*

A system based on the hot spot recombination site *cog*^{L} and the *his3* locus (for selection) is being developed for the generation of libraries by *in vivo* shuffling in *Neurospora* (Rasmussen *et al.*, 2002). The *his3* gene is near the *cog*^{L} locus, and recombinations that initiate at *cog*^{L} and correct mutations in the *his3* gene might also include events within the intervening DNA region. The influence of intervening DNA has been assessed by placing the gene for immunoglobulin kappa chain between *cog*^{L} and alleles of *his3*. The *his3* gene encodes a protein with three enzymatic functions required for histidine synthesis. Some mutations in this gene affect only one of the functions, and genetic complementation is possible between functionally distinct alleles. The K26 and K480 alleles do not

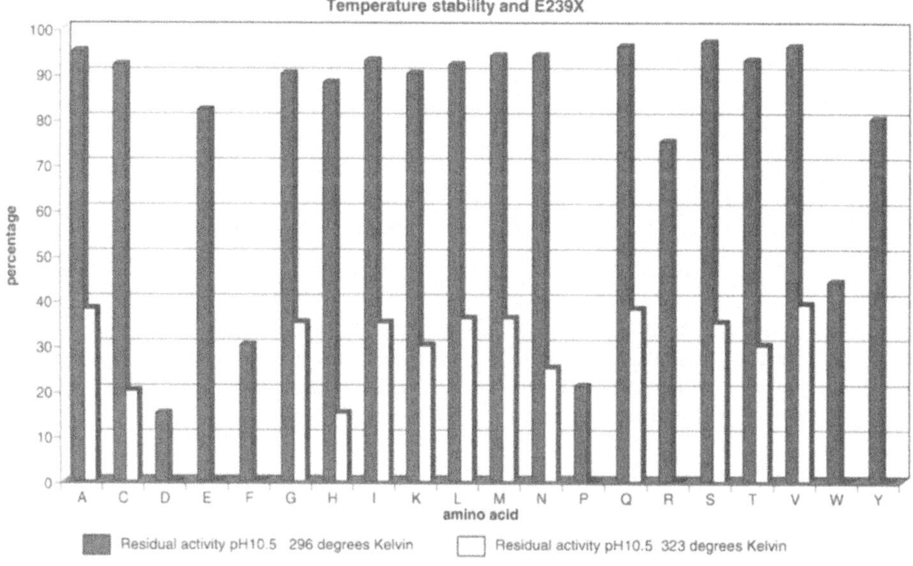

Figure 10.3. Site saturation at position E239 and measurement of residual activity at pH 7.0 after 20-min incubation at pH 10.5 and either 296 K or 323 K (data extracted from Cherry *et al.*, 1999).

Table 10.2. Peroxidase Variants[a]

Mutations	Thermostability (%)	H_2O_2 stability	Specific activity at pH 7	Source
CiP wt	0.5	1.00	1.00	
M166F	0	2.41	0.94	SD
M242I	3	1.21	0.92	SD
Y272F	9	1.28	1.05	SD
E239G	73	1.13	0.96	SS
V53A, V319A	1	2.93	1.13	R
E239G, M242I, Y272F, 149S, V53A, T121A, M166F	87	100	1.15	SH

[a]Improved *Coprinus cinereus* peroxidase stability by a combination of site-directed mutagenesis (SD), site saturation (SS), error-prone random PCR (R), and *in vivo* shuffling (SH). Data is expressed relative to wild type (data extracted from Cherry *et al.*, 1999).

complement each other but both complement the K458 allele, that is, heterokaryons of K26/K458 or of K480/K458 behave as histidine prototrophs. Introduction of a K26–IgκB–cog^L DNA plasmid construct into a K458 strain yielded a His$^+$ heterokaryon, from which K26 monokaryons can be recovered. In these the K458 allele has been replaced by the incoming construct (recombination between the K26 and K480 alleles was not hampered by the presence of the 1.75-kb IgκB sequence). The same procedure was

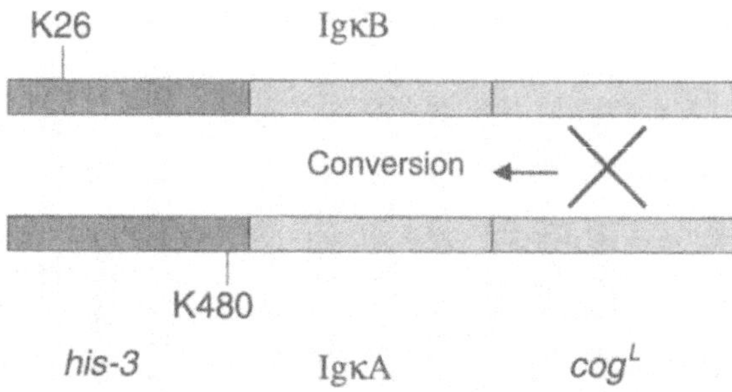

Figure 10.4. *In vivo* shuffling in *Neurospora*. Mating of the two *Neurospora* strains containing the constructs K26–IgκB–cog^L and K480–IgκA–cog^L, respectively, leads to the initiation of recombination at the hot spot recombination site cog^L, which is followed by a conversion passing through the Igκ locus and the *his*-3 locus, leading to the shuffling of the loci and the generation of a wild-type *his-3* gene.

carried out with a K480–IgκA–cog^L construct and a K458 strain of the opposite mating type. Crossing the two auxotrophic monokaryotic strains yielded approximately 300 His^+ recombinants per 10^5 viable spores (Figure 10.4). A control cross, without the Igκ sequence, gave 100 His^+ recombinants per 10^5 viable spores. This indicates that an exogenous DNA sequence does not interfere with the progression of the recombination events initiated at the cog^L locus and encompassing the *his3* gene. Any exogenous sequence placed between the markers should be subject to shuffling.

4.3. *In Vivo* Mutagenesis with the RIP System

A *Neurospora*-based *in vivo* mutagenesis method takes advantage of the repeat-induced point (RIP) mutation process (Barbato *et al.*, 1996; Cambareri and Kato, 2000). Duplicated DNA is subject to G:C to A:T transition mutations during the premeiotic dikaryon phase in the life cycle of this fungus, which can lead to silencing in the case of *Neurospora* genes but which can be used to generate functional diversity in heterologous genes. Only duplicated sequences are subject to this process. During the dikaryon phase, intrachromosomal recombination can take place, leading to a deletion of one of the duplicated DNA sequences and to a cessation of the RIP process. Recombination, itself, also gives rise to diversity, if the duplicated DNA sequences are not entirely homologous.

In vivo RIP mutagenesis have been performed on an albino gene (*al-3*) in *Neurospora* (Barbato *et al.*, 1996). Among 32,000 ascospores from 3 different crosses of *Neurospora*, 70 leaky *al-3* mutants (0.4%) were found in a phenotypic screen for individuals with impaired carotinoid production. Consistent with the leaky phenotype, the number of mutations recovered in the 1400-bp gene in each individual ranged from two to six. Higher rates of RIP mutation would have generated inactive rather than partially active products.

4.4. *In Vivo* Mutagenesis with the Mismatch Repair System

As mentioned earlier, the high rate of homologous recombination in yeast makes it feasible to generate genetic variation by transforming the organisms directly with DNA oligonucleotides. The limitation of this method is that there is no inherent difference to distinguish or select recombinants in the population of unmodified cells. To remedy this, the method can be enhanced by mixing the oligonucleotide with a second oligonucleotide capable of correcting a lethal defect present in the recipient's *cycl* gene (Yamamoto *et al.*, 1992). Recipients rendered viable by the incorporation of this second oligonucleotide are likely to have also taken up and possibly incorporated the oligonucleotide that introduces the mutation into the gene of interest.

Recently, a different method named *delitto perfetto* has been developed (Storici *et al.*, 2001). In this approach, a DNA fragment (CORE) containing counter-selectable genes (such as *URA3* and *kan*) is installed within a chromosomal gene of interest (in this case *trp5*). When the yeast strain is presented with one or two DNA oligonucleotides, ranging in size from 50 to 90 bp, which are targeted to the chromosomal locus flanking the CORE, recombination leads to eviction of the markers and the generation of transformants selectable on the basis of resistance to 5-fluoroorotic acid. The transformants are also sensitive to G418 (i.e., *kan*$^-$). In the published report, 19 of 20 clones examined had the expected change and lacked additional mutations (Storici *et al.*, 2001). This gene replacement could not take place in a *rad52*$^-$ strain, indicating the importance of the Rad52p protein in oligonucleotide recombination.

Another type of oligonucleotide-based mutagenesis system that has been developed for *in vivo* nucleotide exchange in yeast involves chimeric RNA/DNA oligonucleotides folded into double-hairpin conformations (Rice *et al.*, 2001). *In vitro* incubation of a cell-free extract of wild-type yeast with the plasmid pK^{s}m4021 (containing an inactivating mutation in its kanamycin gene) and chimeric oligonucleotides directed to the mutation results in repair of the defective gene. Since the plasmid also carried a functional ampicillin resistance gene, the repair was seen by comparing transformation rates to kanamycin resistance and to ampicillin resistance in *E. coli* (1 kan^r in 4 $\times$ 10^5 amp^r). Transformation of yeast with the same RNA/DNA chimera could repair a similar plasmid-borne defect *in vivo* and produce G418-resistant clones. The conversion was verified by sequencing. A 25-fold increase in conversion was seen when using extract from a *rad52* deleted strain *in vitro* or transforming a *rad52* deleted yeast strain harboring the plasmid (Rice *et al.*, 2001).

DNA oligonucleotide transformation is feasible in filamentous fungi such as *Neurospora* (Calissano and Macino, 1995) and *Aspergillus*. At the Novozymes laboratories, a strain of *Aspergillus oryzae* was identified with an insertion in its *pyrG* gene that disrupts the ORF (AGG GG<u>G</u>C CTT, insertion underlined). When sense or antisense DNA oligonucleotides of 38 bp were introduced into the strain, *pyrG*$^+$ transformants could be selected on minimal medium in which the *pyrG* reading frame had been restored by exact integration of the DNA oligonucleotides into the chromosome. This was confirmed by sequencing, which showed that the *pyrG* insertion was no longer present and that a silent mutation marking the DNA oligonucleotide had been introduced (AG<u>**A**</u> GGC CTT, silent mutation underlined). The drawback with all of the DNA oligonucleotide transformation systems is that they all require a selection screen, which is feasible to create but laborious to perform.

The mismatch repair system used for integrating DNA oligonucleotides may also be employed to generate random mutations *in vivo*. The *E. coli* strain XL1-red having mutations in *mutS*, *mutD*, and *mutT* has been developed for this purpose (Greener and Callahan, 1994; Bornscheuer *et al.*, 1999). However, with this method the mismatch repair system is permanently inactive, and this may lead to additional undesirable mutations. This problem has been circumvented by having the gene encoding a dominant MutD5 protein on a plasmid with a temperature sensitive origin of replication (Selifonova *et al.*, 2001). Upon introduction into the bacterium, the plasmid gives rise to mutations due to the dominant MutD5 mutation, but when the temperature is changed, the plasmid is lost.

A comparable system has been developed for filamentous fungi based upon an *msh2* homolog cloned from *A. oryzae* (Borchert *et al.*, 1999). In humans, some colorectal cancer is caused by dominant mutations in the gene encoding the DNA repair protein Msh2 (Alas *et al.*, 1998). Variants of the human Msh2 protein (G674A, K675A, and S676A) bind very poorly to DNA containing mismatches and exhibit weak ATPase activity compared with wild-type protein (Whitehouse *et al.*, 1996). It is thought that these variants behave as dominant negative mutations because the gene product they produce, though defective, retains the ability to complex with the Msh6 protein, thereby creating an inactive MSH complex unable to bind heteroduplex DNA. Alleles of the *A. oryzae msh2* gene were made (G674A, K675A, and S676A) and cloned down stream of the TAKA-amylase-promoter in an *AMA1* replicative plasmid containing the *pyrG* gene (Vind, 2001). The resulting three plasmids, an identical plasmid containing the wild-type *msh2* gene, and a control plasmid lacking an *msh2* gene were introduced into a *pyrG*$^-$ *A. oryzae* strain. Spore suspensions were made from the *Aspergillus* transformants that grew on minimal medium containing 2% maltose to induce expression of the Msh2 variants. The spores were plated either on minimal medium (where all the transformants were capable of growing) or on medium containing 5% chlorate (to select transformants that acquired mutations in nitrate utilization; chlorate is toxic if metabolized by the *nia*-gene products). A high frequency of chlorate-resistant cells would reflect a high *in vivo* mutation rate. The results of such an experiment are summarized in Table 10.3. As expected, an increase in the level of the wild-type MSH2 protein makes no difference in the mutation frequency *in vivo* as indicated by an insignificant difference between the control (pENI1902) and wild-type MSH2 (pENI2039). However, expression of the dominant mutant *msh2* alleles lead to increases in the frequency of mutations to chlorate resistance ranging from 100-fold (K649A) to 500-fold (G648A or S650A). Thus, providing a dominant inactive variant of a protein involved in mismatch repair is a desirable way to impede or inactivate the

Table 10.3. Msh2 Dominant Mutation. The Chlorate Resistance Frequency of the *Aspergillus oryzae* Transformants Carrying the Indicated Plasmids[a]

Plasmid transformed	Frequency of chlorate resistance
pENI1902 (control W/O MSH2)	5×10^{-6}
pENI2039 (wild-type MSH2)	8×10^{-6}
pENI2040 (G648A MSH2)	3×10^{-3}
pENI2041 (K649A MSH2)	5×10^{-4}
pENI2042 (S650A MSH2)	2×10^{-3}

[a]Expression of dominant variants of MSH2 leads to increased mutation rate seen as increase in chlorate resistances.

DNA repair system in this fungus. The benefits of this mutagenesis technique are that it is fast and easy, it does not require any modification of the host genome, it is independent of genomic copy number, and it can be controlled with a regulatable promoter or by removing the selective pressure needed for maintenance of the plasmid bearing the dominant MSH2 allele.

5. Future in Artificial Evolution of Fungal Proteins

A number of methods have been already developed to perform qualified protein engineering and to generate the diversity when creating libraries. The choice of method depends on the available biochemical data, the knowledge of the protein structure, and the availability of a suitable screen that correlates with desired characteristics. Improving proteins by artificial evolution in the future will be facilitated by better prediction tools for use with knowledge as it becomes available (from genomics, for instance), by better screens to identify the desired variants within the libraries, and by better ways of creating the desired diversity in a relevant host.

More accurate predictions based on knowledge (X-ray structure, *in silico* simulations, biochemical data, genomics, etc.) lead to more focused libraries or improved selection for site-directed variants. A more effective screen giving data that correlates with the characteristics that are to be improved is always a desired goal, but its development is often a difficult task. You get what you screen for, but you also get surprises. Prediction and screening are major issues, though not specific to fungal proteins.

More efficient tools to generate higher quality diversity are needed. The general methods have to be robust and allow tight control of bias, for example, codon bias. With regard to shuffling, a format for shuffling genes of low homology and for getting multiple crossover events would be a very useful tool in artificial evolution.

When it comes to fungal proteins, there are several issues to be solved. There is a need for systems to generate large libraries in filamentous fungi. The transformation frequency of the fungi is often low, and the expression levels are often low and uneven among independent transformants. A different way to generate large fungal libraries is to perform *in vivo* mutagenesis using error-prone polymerases, possibly combined with a mismatch repair deficient host (Borchert and Ehrlich, 1996).

A practical but nevertheless critical point is how to control and confine the growth of filamentous fungi. Fungal transformants often vary greatly in morphology and colony size, and these variations must be taken into account in the context of a plate screen. Approaches used for *E. coli* and yeast are often difficult to apply to or simply inappropriate for filamentous fungi. A way to confine growth is to propagate the fungal transformants in microtiter plates, but to screen a million clones is very time consuming and expensive. Encapsulating the fungi in alginate beads, for example, can circumvent this problem (Beck *et al.*, 2000). Alternative ways of compartmentalizing individual cells and their progeny are being developed and might be useful for filamentous fungi.

Another issue with relation to diversity generation is the post-translational modification of the protein. As discussed in Chapter 9, attempts are being made to modify the fungal glycosylation pattern so that it will resemble more closely the human glycosylation pattern. Thus, in the future, it might be feasible to produce proteins glycosylated in a human-like fashion for pharmaceutical use in filamentous fungi, which could lower production costs significantly (Maras *et al.*, 1999). One desire is for a human-type high mannose glycan. By expressing a potent ER-retained 1,2-α-mannosidase in *P. pastoris*, it was possible to secrete the *Trypanosome cruzi trans*-sialidase glycoprotein with an altered, human-type like N-glycan pattern, that is, mainly $Man_5GlcNAc_2$ rather than the naturally occurring $Man_8GlcNAc_2$ (Callewaert *et al.*, 2001).

Beyond glycosylation, there are other post-translational modifications of the protein to take into consideration such as, acylation, phosphorylation, and the introduction of non-natural or D-amino acids. These modifications will also lead to changes in the protein characteristics, and are likely to become targets for artificial evolution of fungal proteins in the future.

References

Alas, M., Bruin, R., Eyck, L.T., Los, G., and Howell, S. (1998). Prediction-based threading of the hMSH2 DNA mismatch repair protein. *FASEB J.* **12**, 653–663.

Barbato, C., Calissano, M., Pickford, A., Romano, N., Sandmann, G., and Macino, G. (1996). Mild rip—an alternative method for *in-vivo* mutagenesis of the albino-3 gene in *Neurospora crassa. Mol. Gen. Genet.* **252**, 353–361.

Bernstein, J., McCarthy, J.K., and Moore, J.C. (2000). Method for generating recombinant polynucleotides. *World Patent WO200224953-A1.*

Beck, T.C., Ernst, S., Frisner, H., Hansen, P.K., Husum, T.L., Joergensen, B.R., Kongsbak, L., Lamsa, M. *et al.* (2000). Microtiter plate (MTP) based high throughput screening (HTS) assays. *World Patent WO200132844.*

Bizzarri, A.R. and Cannistraro, S. (2002). Molecular dynamics of water at the protein–solvent interface. *J. Phys. Chem. B* **106**, 6617–6633.

Boder, E.T. and Wittrup, K.D. (2000). Yeast surface display for directed evolution of protein expression, affinity, and stability. *Methods Enzymol.* **328**, 430–444.

Boel, E., Christensen, T., Woeldike, H.F., and Huge-Jensen, I.B. (1987). Process for the production of protein products in *Aspergillus oryzae* and a promoter for use in *Aspergillus. European Patent EP238023-B1, US Patent 5536661.*

Bojsen, K., Svendsen A., Fuglsang K.C., Shamkant A.P., Borch K., Vind J., Petri A., Glad S.S. *et al.* (1998). Lipolytic enzyme variants. *World Patent WO200032758-A1.*

Borch, K., Vind, J., Svendsen, A., Petersen, D.A., Patkar, S.A., and Bojsen, K. (1998). Lipase variant. *World Patent WO9942566-A1.*

Borchert, T.V., Christiansen, L., and Vind, J. (1999). Fungal cells with inactivated DNA mismatch repair system. *World Patent WO200050567.*

Borchert, T.V. and Ehrlich, S.D. (1996). A method for *in vivo* production of a mutant library in cells. *World Patent WO9725410.*

Bornscheuer, U.T., Altenbuchner, J., and Meyer, H.H. (1999). Directed evolution of an esterase: Screening of enzyme libraries based on pH-indicators and a growth assay. *Bioorg. Med. Chem.* **7**, 2169–2173.

Busby, S., Irani, M., and Crombrugghe, B. (1982). Isolation of mutant promoters in the *Escherichia coli* galactose operon using local mutagenesis on cloned DNA fragments. *J. Mol. Biol.* **154**, 197–209.

Calissano, M. and Macino, G. (1995). *In vivo* site-directed mutagenesis of *Neurospora crassa* beta-tubulin gene by spheroplasts transformation with oligonucleotides. *Fung. Genet. Newslett.* **43**, 15–16.

Cambareri, E.B. and Kato, E.E. (2000). Methods for diversification of single genes *in vivo. Patent WO200170946-A2.*

Callewaert, N., Laroy, W., Cadirgi, H., Geysens, S., Saelens, X., Min Jou, W., and Contreras, R. (2001). Use of HDEL-tagged *Trichoderma reesei* mannosyl oligosaccharide 1,2-α-D-mannosidase for N-glycan engineering in *Pichia pastoris. FEBS Lett.* **503**, 173–178.

Cereghino, J.L. and Cregg, J.M. (2000). Heterologous protein expression in the methylotrophic yeast *Pichia pastoris. FEMS Microbiol. Rev.* **24**, 45–66.

Cherry, J.R., Lamsa, M.H., Schneider, P., Vind, J., Svendsen, A., Jones, A., and Pedersen, A.H. (1999). Directed evolution of a fungal peroxidase. *Nature Biotechnol.* **17**, 379–384.

Christensen, T., Woeldike, H., Boel, E., Mortensen, S.B., Hjortshoej, K., Thim, L., and Hansen, M.T. (1988). High level expression of recombinant genes in *Aspergillus oryzae. Biotechnology* **6**, 1419–1422.

Clausen, I.G., Gormsen, E., Patkar, S.A., Svendsen, A., Okkels, J.S., and Thellersen, M. (1990). Lipase variants. *World Patent WO9205249-A*.

Daggett, V. (2002). Molecular dynamics simulations of the protein unfolding/folding reaction. *Accounts Chem. Res.* **35**, 422–429.

Dalbadie-McFarland, G., Cohen, L.W., Riggs. A.D., Morin, C., Itakura, K., and Richards, J.H. (1982). Oligonucleotide directed mutagenesis as a general and powerful method for studies of protein function. *Proc. Natl. Acad. Sci. USA* **79**, 6409–6464.

Dupret, D., Lefevre, F., and Masson, J.M. (1998). Method for obtaining *in vitro* recombined polynucleotide sequences, sequence banks and resulting sequences. *World Patent WO200009679*.

Frandsen, T.P., Hendriksen, H.V., Nielsen, B.R., Pedersen, H., Svendsen, A., and Vind, J. (1998). Glucoamylase variants. *World Patent WO200004136-A1*.

Fuchs, M., Henco, K., Lindemann, B., and Schwienhorst, A. (1993). Process for the evolutive design and synthesis of functional polymers based on designer elements and codes. *World Patent WO9517413-A1*.

Gems, D., Johnstone, I.L., and Clutterbuck, J. (1991). An autonomously replicating plasmid transforms *Aspergillus nidulans* at a high frequency. *Gene* **98**, 61–68.

Greener, A. and Callahan, M. (1994). XL1-red: A highly efficient random mutagenesis strain. *Strategies* **7**, 32–34.

Grunert, H.P., Zouni, A., Beineke, M., Quaas, R., Georgalis, Y., Saenger, W., and Hahn, U. (1991). Studies on RNase T1 mutants affecting enzyme catalysis. *Eur. J. Biochem.* **197**, 203–207.

Halperin, I., Ma, B.Y., Wolfson, H., and Nussinov, R. (2002). Principles of docking: An overview of search algorithms and a guide to scoring functions. *Protein Struct. Funct. Genet.* **47**, 409–443.

Hutchison, C.A., Philips, S., Edgell, M.H., Gillam, S., Jahnke, P., and Smith, M. (1978). Mutagenesis at a specific position in a DNA sequence. *J. Mol. Biol.* **253**, 6551–6560.

Jensen, L.J., Andersen, K.V., Svendsen, A., and Kretzschmar, T. (1998). Scoring functions for computational algorithms applicable to the design of spiked oligonucleotides. *Nucl. Acids Res.* **26**, 697–702.

Kumar, S. and Nussinov, R. (2002). Close-range electrostatic interactions in proteins. *Chembiochem* **3**, 604–617.

Kurtzman, A.L., Govindarajan, S., Vahle, K., Jones, J.T., Heinrichs, V., and Patten, P.A. (2001). Advances in directed protein evolution by recursive genetic recombination: Applications to therapeutic proteins. *Curr. Opin. Biotechnol.* **12**, 361–370.

Legrain, P. and Selig, L. (2000). Genome-wide protein interaction maps using two-hybrid systems. *FEBS Lett.* **480**, 32–36.

Lehmann, M., Kostrewa, D., Wyss, M., Brugger, R., D'Arcy, A., Pasamontes, L., and van Loon, A.P.G.M. (2000a). From DNA sequence to improved functionality: Using protein sequence comparisons to rapidly design a thermostable consensus phytase. *Protein Eng.* **13**, 49–57.

Lehmann, M., Pasamontes, L., Lassen, S.F., and Wyss, M. (2000b). The consensus concept for thermostability engineering of proteins. *Biochim. Biophys. Acta—Protein Struct. Mol. Enzymol.* **1543**, 408–415.

Leung, D.W., Chen, E., and Goeddel, D.V. (1989). A method for random mutagenesis of a defined DNA segment using a modified polymerase chain reaction. *Technique* **1**, 11–15.

Ling, M.M. and Robinson, B.H. (1997). Approaches to DNA mutagenesis: An Overview. *Anal. Biochem.* **254**, 157–178.

Maras, M., van Die, I., Contreras, R., and van den Hondel, C.A.M.J.J. (1999). Filamentous fungi as production organisms for glycoproteins of bio-medical interest. *Glycoconj. J.* **16**, 99–107.

Moerschell, R.P., Tsunasawa, S., and Sherman, F. (1988). Transformation of yeast with synthetic oligonucleotides. *Proc. Natl. Acad. Sci. USA* **85**, 524–528.

Morawski, B., Quan, S., and Arnold, F.H. (2001). Functional expression and stabilization of horseradish peroxidase by directed evolution in *Saccharomyces cerevisiae. Biotechnol. Bioeng.* **76**, 99–107.

Myers, R.M., Lerman, L.S., and Maniatis, T. (1985). A general method for saturation mutagenesis of cloned DNA fragments. *Science* **229**, 242–247.

Naumovski, L. and Friedberg, E.C. (1986). Analysis of the essential and excision repair functions of the *rad-3* gene of *Saccharomyces cerevisiae* by mutagenesis. *Mol. Cell. Biol.* **6**, 1218–1227.

Nickoloff, J.A., Deng, W.P., Miller, E.M., and Ray, F.A. (1996). Site-directed mutagenesis of double-stranded plasmids, domain substitution, and marker rescue by co-mutagenesis of restriction enzyme sites. *Methods Mol. Biol.*, **58**, 455–468.

Okkels, J.S. (1995). Method for preparing polypeptide variants. *World Patent WO9707205-A1.*

Pielak, G.J., Mauk, A.G., and Smith, M. (1985). Site-directed mutagenesis of cytochrome C shows that an invariant phe is not essential for function. *Nature* **313**, 152–154.

Petersen, J.F.W., Kadziola, A., and Larsen, S. (1994). Three-dimensional structure of a recombinant peroxidase from *Coprinus cinereus* at 2.6 Å resolution. *FEBS Lett.* **339**, 291–296.

Pompon, D. and Nicolas, A. (1989). Protein engineering by complementary DNA recombination in yeasts shuffling of mammalian cytochrome P-450 functions. *Gene* **83**, 15–24.

Punt, P.J., van Biezen, N., Conesa, A., Albers, A., Mangnus, J., and van den Hondel, C. (2002). Filamentous fungi as cell factories for heterologous protein production. *Trends Biotechnol.* **20**, 200–206.

Rasmussen, J.P., Bowring, F.J., Yeadon, P.J., and Catcheside, D.E.A. (2002). Targeting vectors for gene diversification by meiotic recombination in *Neurospora crassa. Plasmid* **47**, 18–25.

Rasmussen, F.A. and Nickoloff, J.A. (1992). Site-specific mutagenesis of almost any plasmid using a PCR-based version of unique site elimination. *Biotechniques* **13**, 342–348.

Rice, M.C., Bruner, M., Czymmek, K., and Kmiec, E.B. (2001). *In vitro* and *in vivo* nucleotide exchange directed by chimeric RNA/DNA oligonucleotides in *Saccharomyces cerevisiae. Mol. Microbiol.* **40**, 857–868.

Rodi, D.J., Makowski, L., and Kay, B.K. (2002). One from column A and two from column B: The benefits of phage display in molecular-recognition studies. *Curr. Opin. Chem. Biol.*, **6**, 92–96.

Romanos, M.A., Makoff, A.J., Fairweather, N.F., Beesley, K.M., Slater, D.E., Rayment, F.B., Payne, M.M., and Clare, J.J. (1991). Expression of tetanus toxin fragment C in yeast: Gene synthesis is required to eliminate fortuitous polyadenylation sites in AT-rich DNA. *Nucl. Acids Res.* **19**, 1461–1467.

Rosenberg, S., Barr, P.J., Najarian, R.C., and Hallewell, R.A. (1984). Synthesis in yeast of a functional oxidation-resistant mutant of human α_1-antitrypsin. *Nature* **312**, 77–80.

Rouwendal, G.J., Wolbert, E.J., Zwiers, L.H., and Springer, J. (1993). Simultaneous mutagenesis of multiple sites: Application of the ligase chain reaction using PCR products instead of oligonucleotides. *Biotechniques* **15**, 68–76.

Sanchez, R. and Sali, A. (1997). Advances in comparative protein-structure modelling. *Curr. Opin. Struct. Biol.* **7**, 206–214.

Schultz, S.C. and Richards, J.H. (1986). Site-saturation studies of the beta lactamase production and characterization of mutant beta lactamase with all possible amino-acid substitutions at residue 71. *Proc. Natl. Acad. Sci. USA* **83**, 1588–1592.

Selifonova, O., Valle, F., and Schellenberger, V. (2001). Rapid evolution of novel traits in microorganisms. *Appl. Env. Microbiol.* **67**, 3645–3649.

Sieber, V., Martinez, C.A., and Arnold, F.H. (2001). Libraries of hybrid proteins from distantly related genes. *Nature Biotechnol.* **29**, 456–460.

Sierks, M.R., Ford, C., Reilly, P.J., and Svensson, B. (1989). Site-directed mutagenesis at the active site Trp120 of *Aspergillus awamori* glucoamylase. *Protein Eng.* **2**, 621–625.

Steed, P.M., Lasala, D., Liebman, J., Wigg, A., Clark, K., and Knap, A.K. (1998). Characterization of recombinant human cathepsin-b expressed at high-levels in baculovirus. *Protein Sci.* **7**, 2033–2037.

Stemmer, W.P.C. (1994). Rapid evolution of a protein *in vitro* by DNA shuffling. *Nature* **370**, 389–391.

Storici, F., Lewis, L.K., and Resnick, M.A. (2001). *In vivo* site-directed mutagenesis using oligonucleotides. *Nature Biotechnol.* **19**, 773–776.

van der Linden, R.H., de Geus, B., Frenken, G.J., Peters, H., and Verrips, C.T. (2000). Improved production and function of llama heavy chain antibody fragments by molecular evolution. *J. Biotechnol.* **80**, 261–270.

Vind, J. (1997). An *in vitro* method for construction of a DNA library. *World Patent WO9841653-A1.*

Vind, J. (1999). Constructing and screening a DNA library of interest in filamentous fungal cells. *World Patent WO200024883-A1.*

Vind, J. (2000). Method for producing a polynucleotide library. *World Patent WO 2002046396.*

Vind, J. (2001). *In vivo* mutation and recombination in filamentous fungi. *World patent WO200259331.*

Volkov, A.A., Shao, Z., and Arnold, F.H. (1999). Recombination and chimeragenesis by *in vitro* heteroduplex formation and *in vivo* repair. *Nucl. Acids Res.* **27**, E18.

Wang, P.L. (2000). Creating hybrid genes by homologous recombination. *Dis. Markers* **16**, 3–13.

Whitehouse, A., Parmar, R., Deeble, J., Taylor, G.R., Phillips, S.E., Meredith, D.M., and Markham, A.F. (1996). Mutational analysis of the nucleotide binding domain of the mismatch repair enzyme hMSH-2. *Biochim. Biophys. Res. Commun.* **229**, 147–153.

Yamamoto, T., Moerschell, R.P., Wakem, L.P., Ferguson, D., and Sherman, F. (1992). Parameters affecting the frequencies of transformation and co-transformation with synthetic oligonucleotides in yeast. *Yeast* **8**, 935–948.

11

Biocatalysis and Biotransformation

Frieder Schauer and Rainer Borriss

1. Preface

Enzymatic reactions and biotransformations catalyzed by fungal enzymes and used in industry, agriculture, food technology, and medicine have increased in importance tremendously in recent years. Many efforts have been made to detect new sources of enzymes and to adapt these biological catalysts, for example, by methods of gene technology and protein engineering, to new applications for human or commercial benefit. This will be the subject of approximately half of this chapter. On the other hand, enzymatic activities of fungi can damage various products of human endeavor. Fungi may use these products for their growth and development, degrading, destroying, or inactivating many substances and products in the process. This phenomenon is the cause of serious losses such as the destruction of food supplies and plant stocks, the ruination of wood structures, the damage of leather or textile goods, or the inactivation of food preservatives, biocides, and fungicides. These fungal activities and the enzymes involved will be covered in the second part of this chapter.

In addition, several examples will be given illustrating how enzymatic activities of fungi can be used for degradation of environmental pollutants, for bioremediation processes, or the synthesis of chemical compounds in an environmentally friendly way. Lastly, some trends and future developments in the use of fungal enzymes will be summarized.

2. Fungal Enzymes and Biotransformations—An Introduction

Enzymes are remarkably effective catalysts, responsible for the thousands of coordinated chemical reactions involved in the biological processes of living systems. An outstanding feature of enzymes in comparison with chemical catalysts is substrate and reaction specificity that promotes only one reaction with the respective substrate and ensures the synthesis of a specific biomolecular product without the concomitant production of by-products.

Frieder Schauer • Institute of Microbiology, E.-M.-Arndt-University Greifswald, F.-L.-Jahn-Strasse 15, D-17487 Greifswald, Germany. **Rainer Borriss** • Institute of Biology, Humboldt-University Berlin, Chausseestrasse 117, D-10055 Berlin, Germany.

Advances in Fungal Biotechnology for Industry, Agriculture, and Medicine. Edited by Jan S. Tkacz and Lene Lange, Kluwer Academic/Plenum Publishers, 2004.

An advantage of using enzymes in technology lies in the multitude of biochemical reactions available, each offering a biotechnical route with the benefits of high selectivity, mild reaction conditions, and minor environmental and toxicological impact. Since ancient times, enzymes have played a central role in many manufacturing processes, such as the production of wine, cheese, and bread and the modification of starch, and so on. Today, enzymes are commonly used in a wide variety of industrial applications, and the demand for more stable, highly active, and specific enzymes is growing rapidly. By 1983, there were approximately 30 different classes of enzymes in commercial use, of which approximately half were of fungal origin (Bennet, 1998). According to Demain (2000), industrial enzymes have already reached a market size of US$ 1.6 billion.

Enzymes are divided into six major classes according to the types of reactions they catalyze. Only a few of the more than 2,500 enzymes presently known are produced commercially (Borriss, 1987). As can be seen in Table 11.1, most enzymes with importance in technology are hydrolases, accounting for more than 80% of all the commercial enzymes produced. Today, as the world enzyme market grows, an increasing number of representatives from other enzyme classes, especially oxidoreductases, isomerases, transferases, and lyases, are being added to the market.

Biotransformations describe special biochemical or metabolic reactions that are not connected with a complete degradation or mineralization of a substance and therefore result in an accumulation of metabolic products. In a broader sense, biotransformations encompass all catabolic and anabolic reactions that result in an accumulation of intermediates or dead-end products. However, the observed enrichment of substances does not lead to a commercial exploitation in each case.

A *bioconversion* represents the changing of organic molecules by microorganisms or other living cells or by their enzymes with the aim to produce defined chemical substances. Advantages in comparison with chemical syntheses are the following:

- high specificity of the reaction (with respect to different parent substances)
- high regio- and stereoselectivity of the reaction (with respect to a given molecule)
- high enantioselectivity of the reaction
- minimal generation of by-products (in case of enzymes)
- catalysis under mild conditions (neutrality, room temperature, normal pressure, and unaltered atmosphere)
- limited emission of dangerous environmental pollutants (chlorine chemistry and heavy metal catalysts)
- multistep reactions possible nearly without loss of intermediate substances (in case of whole cells)
- tailoring of protein catalysts to a given problem by the methods of gene technology and protein engineering.

Increasingly animal and plant enzymes are being supplanted by microbial enzymes, so that today most commercial enzymes are of microbial origin, produced by bacteria and fungi (yeasts and filamentous fungi). Fungal enzymes compete favorably in this respect with bacterial enzymes; despite the fast growth and the high diversity of metabolic routes of bacteria, fungal enzymes have gained a reliable position in biotechnology and industry. Fungal amylases and endo-β-glucanases, for example, can substitute partially or completely for the enzymes present in malted barley and wheat in the beer, distillery, baking, and textile

Table 11.1. Representative Fungal Enzymes with Biotechnological Importance from the Six Major Classes According to Compilation of the IUPAC Commission on Enzyme Nomenclature

Enzyme	Organism and cellular state[a]	Application	Reference
1. *Oxidoreductases*[b]			
1.1.3.4: Glucose oxidase	*Aspergillus niger*, IC	Diagnostics, baking	Frost and Moss, 1987
	Penicillium notatum, IC		Huber, 1994
1.10.3.2: Phenoloxidases, laccases	*Pycnoporus cinnabarinus*, EC	Lignin degradation	Eggert *et al.*, 1998; Berka *et al.*, 1997
	Myceliophthora thermophila, EC		
	Trametes versicolor, EC	Organic synthesis	
1.11.1.7: Peroxidase			
Lignin peroxidase	LiP, *Phanerochaete chrysosporium*, EC	Lignin degradation	Conesa *et al.*, 2002
Manganese peroxidase	MnP, *P. chrysosporium*, EC	Lignin degradation	Conesa *et al.* 2002
			Reddy and D'Souza, 1994
2. *Transferases*[c]			
2.4: Glycosyltransferases	Bacteria		
3. *Hydrolases*[d]			
3.1.1.1: Lipases	*Candida rugosa*, EC	Organic synthesis	Frost and Moss, 1987; Jaeger and Reetz, 1998
	Candida antarctica, EC	Organic synthesis	
	Thermomyces lanuginosus	Detergent	
	Rhizomucor miehei, EC	Food processing	
3.1.1.3: Carboxylester hydrolase	*Orpinomyces* sp., IC	Xylan degradation	Bornscheuer, 2002
	Aspergillus awamori, IC	Wheat bran	
3.1.3.8: Phytase	PhyA, *Agrocybe pediades*	Feed additive	Lassen *et al.*, 2001
	PhyA, *Ceriporia* sp.		
	PhyA, *Trametes pubescens*		
	PhyA, *Peniophora lycii*		Shimizu, 1993; Mitchell *et al.*, 1997; Berka *et al.*, 1998
	PhyA, *A. niger*, EC		
	PhyB, *A. niger*, EC		
	PhyA, *Aspergillus terreus*, EC		
	PhyA, *M. thermophila*, EC		
	PhyA, *Aspergillus oryzae*, EC		
	PhyA, *T. lanuginosus*, EC		
3.2.1.1: α-Amylase	*Aspergillus oryzae*, EC	Starch hydrolysis, baking	Huber, 1994

(*Continued*)

Table 11.1. *(Continued)*

Enzyme	Organism and cellular state[a]	Application	Reference
3.2.1.2: β-Amylase	*Rhizopus oryzae*, EC	Starch hydrolysis	
3.2.1.3: Glucoamylase	*A. awamori*, EC *A. niger*, EC *Endomycopsis bispora*, EC *Endomycopsis fibuliger*, EC	Manufacture of glucose and fructose syrup Manufacture of glucose and fructose syrup	Huber, 1994
3.2.1.4: 1,4-β-Glucan endohydrolase	EGI, *Trichoderma reesei*, EC EG1, *Trichoderma viride* *Humicola insolens* *Penicillium funiculosum*	Food, fruit juices, textile, and laundry biotechnology	Bennet, 1998
3.2.1.6: 1,3(4)-β-Glucan endohydrolase	*R. oryzae* *Phaffia rhodozyma* *Cochliobolus carbonum*	Brewing	Planas, 2000
3.2.1.8: Xylanase	*Aspergillus kawachii* *A. awamori*, EC *Fusarium oxysporum*, EC *T. lanuginosus*, EC *Trichoderma harzianum*, EC *T. reesei*, EC	Fermentation of shochu Chlorine-free bleaching; improvement in the texture, quality, and shelf life of bakery products; animal feed: additive for poultry diet	Ito *et al.*, 1992 Poutanen, 1997
3.2.1.11: Dextranase	*P. funicolosum*, IC		Jaeger and Reetz, 1998
3.2.1.14: Exochitinase	*Aspergillus fumigatus*	Cell wall digestion	Xia *et al.*, 2001
3.2.1.21: β-Glucosidase	*T. reesei*, EC, IC	Cellulose degradation	Jaeger and Reetz, 1998
3.2.1.22: α-Galactosidase	*A. niger*, IC	Sugar refining	Jaeger and Reetz, 1998
3.2.1.23 β-Galactosidase, lactase	*Kluyveromyces fragilis*, IC	Dairy	Lehmann *et al.*, 2000
3.2.1.26: β-Fructosidase, invertase	Yeast, IC *A. niger* *A. oryzae*	Confectionery	Bennet, 1998
3.2.1.39: 1,3-β-Glucan endohydrolase	*T. reesei* *T. harzianum*	Brewing, biocontrol of plant pathogen, animal feed	Planas, 2000
3.2.1.73: 1,3-1,4-β-Glucan endohydrolase	*Orpinomyces, Talaromyces, Cochliobolus*	Brewery, animal feed: additive for poultry and piglet diet	

3.2.1.91: Cellobiohydrolase	CBHI, *T. reesei*, EC CBHII, *T. reesei*, EC	Textile and laundry Biotechnology, bio-stoning of denim fabrics, bio-polishing of cotton, washing powders	Bhat, 2000
3.4.21: Serine protease	*A. niger* *A. oryzae* *Engydontium album*	Leather industry, manufacture of soy sauce Proteinase K	Huber, 1994 Kulkarni *et al.*, 1999
3.4.22: Thioproteases	*A. oryzae*, EC	Baking	
3.4.23: Aspartic protease, rennet, chymosin	*R. miehei*, EC *Mucor pusillus*, EC *Endothia parasitica*, EC *Rhizopus niveus*	Manufacture of cheese	
3.4.24: Metalloproteinases	*A. fumigatus* *Aspergillus flavus*	Baking	
Pectinase complex containing 3.1.1.11: pectinmethylesterase, 3.2.1.15: polymethylgalacturonases (PMG) with endo-PMG and exo-PMG activities and polygalacturonases (PG) with endo-PG and exo-PG activities	*A. niger, A. oryzae, T. reesei, H. insolens*	Improving in pressing and extraction of juice from fruits and oil from olives, clarification of fruit juices, production of high-viscosity fruit purées, tomato ketchup	Grassin and Fauquembergue, 1996 Heldt-Hansen, 1997
4. *Lyases*[e]			
Pectinase complex containing 4.2.2.2: polymethylgalacturonate lyase and polygalacturonate lyase	*A. niger, T. reesei*, EC	Fruit and vegetable juice	Grassin and Fauquembergue, 1996
5. *Isomerases*[f]			
5.3.1.5: Xylose (glucose) isomerase	Occurs mainly in bacteria		
6. *Ligases*[g]			

[a]IC, cell bound; EC, extracellular.
[b]Enzymes of this group catalyze oxidation–reduction reactions involving oxygenation or overall removal or addition of hydrogen atom equivalents.
[c]These enzymes mediate the transfer of a group, such as aldehydic or ketonic, acyl, sugar, phosphoryl, methyl, or a sulphur-containing one, from one molecule to another.
[d]The range of functional groups hydrolyzed by such enzymes is very broad. It includes esters, anhydrides, peptides, and others. C–O, C–N, and C–C bonds may be cleaved as well as some others.
[e]The types of reactions catalyzed are additions to, or formation of, double bonds such as C=C, C=O, and C=N.
[f]A variety of isomerizations, including racemization, can be affected.
[g]These are often called synthetases, and catalyze the formation of C–O, C–S, and N–C bonds with accompanying adenosine triphosphate (ATP) or other nucleoside triphosphate cleavage.

industries; plant and animal proteases are being displaced by *Aspergillus* protease for meat tenderization and for chill-proofing beer, in detergent preparations, and for softening leather. Application of recombinant DNA technology has drastically strengthened this trend (see Chapter 9). One impressive example is the substitution of recombinant rennet expressed in fungal or yeast cells for the calf chymosin used in cheese ripening. Over the past decade, the industrial use of enzymes is gathering increasing attention in solving environmental problems (Ogawa and Shimizu, 1999). This is illustrated by the recent applications of *Aspergillus* phytase in reducing the phosphorous pollution resulting from animal farming and of *Trichoderma* xylanase in kraft pulp bleaching in paper industry to reduce harmful chlorine pollution.

Today fungi, along with extremophilic and mesophilic archaea and bacteria, are preeminent sources of biocatalysts. The main eukaryotic production organisms for biocatalysts are the molds *Aspergillus niger* (including strains known under the names *A. aculeatus, A. awamori, A. ficuum, A. foetidus, A. japonicus, A. phoenicis, A. saitoi*, and *A. usami*), *Aspergillus oryzae* (including strains known under the names *A. sojae* and *A. effusus*), *Trichoderma reesei, Trichoderma viride*, and the yeasts *Saccharomyces cerevisiae* and *Kluyveromyces fragilis*. These fungi are generally recognized by the Food and Drug Administration (FDA) as safe (GRAS). A list of fungi traditionally used in food or in food processing also includes *Mucor javanicus, Rhizopus arrhizus, Rhizopus oligosporus, Rhizopus oryzae*, and *Kluyveromyces lactis*. Different safety aspects of the production organisms employed in enzyme technology have been reviewed (Bennet, 1998).

A vast body of experience in fermentation technology with filamentous fungi and yeast has been accumulated in past decades. The main advantage provided by fungi, especially representatives of *Aspergillus* spp. and *Trichoderma* spp., is their capacity to secrete enormous amounts of extracellular proteins into the culture medium. Mutant strains of *T. reesei* produce up to 40 g cellulases/L; 50% of the cellulase was accounted for by cellobiohydrolase (Penttila, 1998). Recombinant *Aspergillus* phytase has been expressed in methylotrophic yeast, *Pichia angusta* (syn. *Hansenula polymorpha)*, with a yield up to 13.5 g/L (Mayer *et al*., 1999). Moreover, cultivation of fungi in simple culture medium with high yield is often possible. Compared with bacterial cultures, fungal cultures are easily processed to provide microbe-free enzyme preparations. Culture media rich in secreted enzyme can be readily separated from the mycelium after fermentation. Crude concentrates of such filtrates may display remarkable stability. *A. niger* amyloglucosidase, for example, stored for 1 year at 0–4°C retained 100% of its activity (Frost and Moss, 1987).

3. Glycosyl Hydrolases

Glycosyl hydrolases (EC 3.2.1.N–3.2.3.N) are a widespread group of hydrolases cleaving the glycosidic bond between two or more saccharides or between a saccharide and a non-carbohydrate moiety. Among them are enzymes with great importance in biotechnology, such as the α- and β-glucan hydrolyzing enzymes (degrading starch and cellulose) and various hemicellulases.

The glycosidic bond, particularly that between two glucose residues, is the most stable linkage within naturally occurring biopolymers, the half-life for spontaneous hydrolysis of cellulose and starch being in the range of 5 million years (Wolfenden *et al*., 1998).

Enzymes responsible for the hydrolysis of these materials therefore face a challenging task, yet they accomplish it with rate constants of up to 1,000 s^{-1}, earning them a reputation as some of the most proficient catalysts.

In general, glycosyl hydrolases cleave glycosidic bond between sugar residues through general acid catalysis, requiring a proton donor and a nucleophile. There are two possible stereochemical outcomes of hydrolysis: inversion or retention of anomeric configuration (Figure 11.1). In either case, protonation of the glycosidic oxygen is the first step. In the inverting mechanism this is followed by the attack of the anomeric carbon molecule by an activated water molecule, leading to the inversion of the absolute configuration. Alternatively, a second nucleophilic attack by a water molecule results in overall retention of the anomeric configuration (retaining mechanism). Typically, active site aspartate or glutamate residues participate in catalysis; sometimes tyrosine appears to be involved in transition-state stabilization. The distance between the two active carboxyl groups varies depending on the stereochemical course of the reaction. A distance of about 5.5 Å is found in retaining glycosidases, whereas a larger value around 10 Å is found in inverting xylanases. It is still an open question whether the nucleophile forms a covalent bond with the anomeric carbon or provides its negative charge simply to stabilize the positive charge of the reaction intermediate (Sinnott, 1990). To accelerate the hydrolysis of glycosidic bonds by factors approaching 1017-fold, glycosylases have evolved finely tuned active sites optimally configured for transition-state stabilization.

Glycosyl hydrolases have been classified on the basis of amino acid sequence similarities. Some of them can be grouped further according to their active-site folding. At present 89 families, tentatively grouped in clans A–M, are distinguished (*http://afmb.cnrs-mrs.fr/CAZY/GH_intro.html*). According to their structure, at least three different types of active-site topologies exist in these enzymes. Pockets are optimal for monosaccharidases

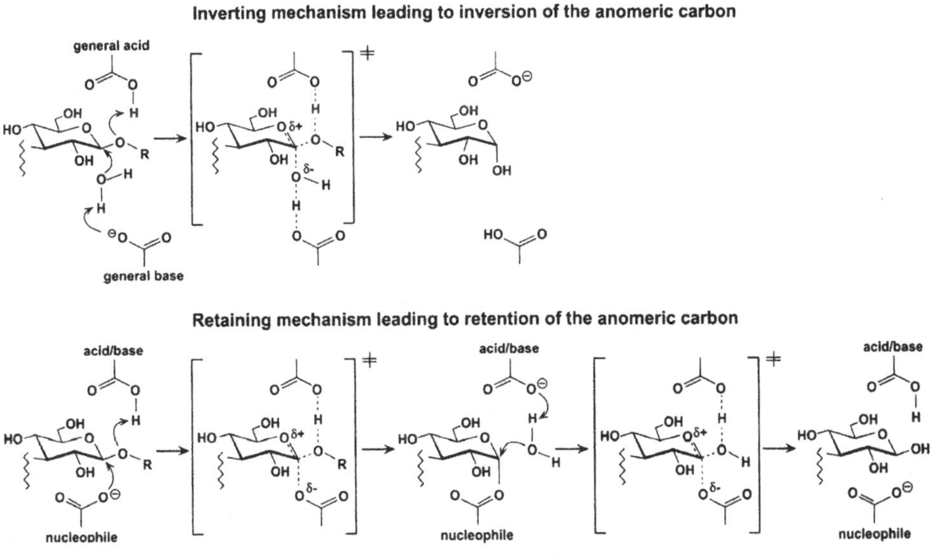

Figure 11.1. Inverting and retaining mechanisms of hydrolysis of the β-1,4 glycosidic bond by glycosyl hydrolases.

and enzymes adapted to branched substrates with large numbers of chain ends available for cleavage (Davies and Henrissat, 1995). A tunnel was observed in cellobiohydrolases (Rouvinen *et al.*, 1990). This topology requires the polysaccharide chain to be threaded through, which is optimal for fibrous substrate such as native cellulose. This appears to be a special case of the third topology, the cleft or groove. Clefts allow the binding of several consecutive sugar units in linear or branched polymeric substrates, and they are commonly observed in endo-acting polysaccharidases (Gruber *et al.*, 1998).

Binding of the polymeric substrate at the active site of polysaccharide hydrolases occurs at multiple subsites, permitting fine tuning of specific protein–substrate interactions. Such a subsite system is encountered not only in glycosyl hydrolases but also in proteases and nucleases. Actually, substrate reactivity is influenced by polymer units that are bound at a distance from the point of catalysis. Characterization of a substrate binding region (subsite mapping) includes experimental determination of the number of subsites flanking the glycosidic bond to be hydrolyzed (Davies *et al.*, 2002). In the following sections, some examples of hydrolases degrading different kinds of macromolecular carbohydrates and possessing industrial importance are presented.

3.1. Starch Hydrolysis: Amylases and Glucoamylases

Commercial amylases used in baking industry are fungal α-amylases, specifically hydrolyzing internal α-1,4 glycosidic linkages of amylose and amylopectin to release soluble dextrins with a degree of polymerization of 2–12. α-Amylase from *A. oryzae* was the first microbial enzyme to be manufactured for sale (Takamine, 1894). In 1958 the first commercial method for glucose production by an enzyme from *R. oryzae* was developed. Other good producers of glucoamylase (GA), such as strains of *A. niger*, have subsequently been discovered. Today, the enzyme from *A. niger* is preferred for industrial use (Harada, 1984).

GAs (1,4-α-D-glucan glucohydrolase, EC 3.2.1.3) catalyze the hydrolysis of α-1,4 and α-1,6 glucosidic linkages to release β-D-glucose from the nonreducing ends of starch. Fungal GA is widely used in the manufacture of glucose and fructose syrups, permitting a glucose yield of up to 96%. The activity toward the α-1,6 linkage is only 0.2% of that for the α-1,4 linkage, and this adversely affects the yield in industrial saccharification. This property together with a need for other developments such as enhanced thermostability motivated both fundamental and applied protein engineering of GAs.

The GAs constituting glycosyl hydrolase family 15 catalyze the hydrolysis of glucosidic linkages with inversion of the anomeric configuration. Seven subsites participating in substrate recognition were identified kinetically. The general acid catalyst and proton donor Glu179 and the catalytic base Glu400 in GA of *A. niger* were identified by mutational analysis. These residues are 9.2 Å from each other, which is consistent with an inverting mechanism of hydrolysis.

The catalytic domain in the GA of *A. awamori* contains 13 α-helixes, 12 of which form an $(\alpha/\alpha)_6$ barrel. In this folding, six outer and six inner α-helixes surround the funnel-shaped active site (Sauer *et al.*, 2000). In addition to a catalytic domain, the GA of *A. awamori*, but not the GA of *Saccharomycopsis fibuligera*, has a C-terminal starch binding domain (SBD). This substrate binding domain consists of eight β-strands organized in two β-sheets, forming a twisted β-barrel structure. The SBD is required for degradation of raw starch by GA and is attached to the catalytic domain by a Ser/Thr-rich linker region.

Industrial saccharification is currently performed at 60°C. Development of a more thermostable GA, capable of performing industrial saccharification at elevated temperature, would thus be of significant importance to the starch processing industry. Attempts to increase thermostability of GAs of *A. niger* and *A. awamori* by protein engineering include replacement of glycine in α-helixes, elimination of fragile Asp–X bonds, and substitution of asparagine in Asn–Gly sequences. The most successful strategy proved to be the introduction of additional disulfide bonds into the molecule, which increased the T_m value of GA up to 4°C (Sauer *et al.*, 2000; see also Chapter 10).

3.2. Cellulose and Cellulases

Cellulose, is a crystalline polymer consisting of linear chains of 1,4-β-linked glucose, in which each residue is rotated with respect to its neighboring residues 180° about the main axis. Thus, the basic recurring unit is cellobiose. Cellulose chains associate to form insoluble fibrils in which the chains are held together by hydrogen bonds. Bundles of fibrils aggregate to form the inert insoluble fibers characteristic of the primary and secondary cell walls of higher plants. Cellulose is the most abundant biopolymer on earth: about 7×10^{11} tons exist. Moreover, it is a renewable resource, about 4×10^{10} tons being synthesized annually as a result of photosynthesis (Coughlan, 1985). In natural environments, cellulose is almost always associated with hemicellulose and lignin. Despite the relative simplicity of its chemical structure, the biodegradation of cellulose with its variable degree of crystallinity represents a major biochemical challenge with practical implications.

Cellulose research started more than 50 years ago. Early in World War II, the U.S. Army was concerned with the rate that uniforms and cellulosic material decayed in the tropics. Indeed, *T. viride* QM6a, from which the most powerful cellulolytic mutants have since been derived, was isolated from a catridge belt rotting in the jungles of New Guinea. Extensive basic and applied research during the 1970s and 1980s demonstrated that enzyme-mediated bioconversion of lignocellulose to soluble sugars was rather difficult and uneconomical (Coughlan, 1985). Nevertheless, continued research on cellulases demonstrated their biotechnical potential in various industries (Table 11.1). Together with hemicellulases and pectinases, cellulases have steadily grown in industrial importance (Rabinovich *et al.*, 2002). Today these enzymes, mostly from *Trichoderma* and *Aspergillus*, account for approximately 20% of the world enzyme market (Bhat, 2000).

Complete and efficient degradation of crystalline cellulose to glucose requires the synergistic action of three cellulolytic enzymes: endoglucanase (EC 3.2.1.4), cellobiohydrolase (EC 3.2.1.91), and β-glucosidase (EC 3.2.1.21). In contrast with bacterial cellulases that often associate into multienzymatic complexes (cellulosomes), fungal cellulose-degrading enzymes function as individual enzymes. Each consists of a catalytic domain, responsible for the hydrolysis reaction, and a cellulose binding domain (CBD), mediating binding of the enzymes to the substrate. The two domains are joined by a linker peptide that must be sufficiently long and flexible to allow proper orientation and efficient operation of both domains.

Endoglucanases act randomly on internal bonds, significantly decreasing the chain length and the viscosity of the cellulose polymer and are thought to be more active on amorphous cellulose. Exocellobioydrolase cleaves cellobiosyl units from the newly exposed nonreducing ends of the cellulose chains, attacking preferentially crystalline cellulose in a processive manner. β-Glucosidase cleaves short cello-oligosaccharides to glucose. The ascomycete *T. reesei* secretes several cellulolytic enzymes comprising a multigene family of enzymes from these three classes. The main enzymes CBHI, CBHII, and EG I have been cloned and characterized; however, by themselves these enzymatic activities are not sufficiently high to make bioconversion of lignocellulose to soluble sugars economically feasible in the near future.

3.2.1. Cellulases in Textile and Laundry Biotechnology

Nevertheless, cellulases have found successful industrial application. Even though cellulases were introduced in the textile and laundry industries only a decade ago, they have now become the third largest group of enzymes used in these industries. Cellulases are able to modify cellulosic fibres in a controlled and desired manner; bio-polishing and bio-stoning constitute the best known illustrations. Cellulases are also increasingly used in household washing powders because they enhance detergent performance, allow the removal of small, fuzzy fibrils from fabric surfaces, and improve the appearance and color brightness (Bhat, 2000).

3.3. Hydrolysis of Hemicellulose: Xylanases and Mixed-Linked β-Glucanases

Current developments in the degradation of very complex polysaccharides in the cell walls of unlignified, green plants are more promising than the efforts to degrade woody materials. The term *hemicellulose* is generally applied to the fraction extracted or isolated from plant materials with dilute alkali. Hemicellulose includes xylan, mannan, galactan, β-1,3-1,4 mixed linked glucan, and arabinan as the main heteropolymers. In plants, hemicelluloses are situated between lignin and the underlying cellulose fibers. An extensive overview of possible applications of fungal cellulases, hemicellulases, and pectinases in the food and feed industries, as well as in the brewery, wine, textile, and pulp and paper industries was published recently (Bhat, 2000; see also Table 11.1).

3.3.1. Xylanases

Xylan is a complex polysaccharide comprised of a backbone of xylopyranose residues linked by β-1,4 glycosidic bonds. Xylan is the most common hemicellulosic polysaccharide in cell walls of land plants, representing as much as 30–35% of the total dry weight (Beg *et al.*, 2001). Most xylans occur as heteropolysaccharides, containing substituent groups such as acetyl, arabinosyl, or glucuronysyl residues on the backbone and on side chains.

Xylanolytic enzyme systems consist of an arsenal of polymer degrading enzymes: β-1,4-endoxylanase, β-xylosidase, α-L-arabinofuranosidase, α-glucuronidase, acetyl xylan esterase, and phenolic acid (ferulic and *p*-coumaric acid) esterases. All these enzymes act cooperatively to convert xylan into its constituent sugars. The ability to produce such multicomponent xylanolytic systems is widespread among fungi (Sunna and Antranikian, 1997).

Endo-β-1,4-xylanases (EC 3.2.1.8) belong to family 10 (former family F) and family 11 (former family G) of glycosyl hydrolases, as evidenced by sequence alignments and hydrophobic cluster analysis. The greater catalytic versatility of family 10 xylanases can be ascribed to differences in their tertiary structures. Three-dimensional (3-D) structures are available for several members of family 11 fungal xylanases: *Trichoderma harzianum,*

T. reesei (xylanases I and II), and *Thermomyces lanuginosus*, previously called *Humicola lanuginosa* (Gruber *et al.*, 1998). The structure of all family 11 xylanases appears smaller and more well packed compared with their family 10 counterparts. Their structures are dominated by two twisted β-sheets. Sheet A forms the outer surface of the enzyme and consists of five antiparallel strands. Its hydrophilic, solvent-accessible surface contains a large number of serine and threonine residues. Sheet B consists of nine, mostly antiparallel strands. One face forms the active site of the enzyme, while the other is packed against sheet A to form the hydrophobic core of the protein. The overall shape of the molecule resembles a right hand with the two β-sheets and an α-helix forming the fingers and the palm and two loop regions forming the thumb and its web connection with the palm. This picture was first introduced in the description of the structure of xylanase II from *T. reesi* (Törrönen *et al.*, 1994). The similarity of this structure with that of bacterial 1,3-1,4 β-glucanases is striking (Keitel *et al.*, 1993). The catalytic groups are present in a cleft that accommodates a chain of 5–7 xylopyranosyl residues. Glu86 and Glu178 are thought to be involved in catalysis by the xylanase of *T. lanuginosus* (Gruber *et al.*, 1998). Short distances between active site carboxylates, typical of retaining hydrolases, are found in the xylanases of *T. reesei* (6.7 Å) and *T. harzianum* (6.4 Å).

The family 10 xylanases usually have both catalytic and substrate binding domains. The tertiary fold of catalytic domain is a typical 8-fold α/β barrel $(\alpha/\beta)_8$ imparting a bowl shape to the molecule. The substrate binds to a shallow groove at the bottom of the bowl. The relative shallowness of the substrate binding region together with possibly higher flexibility of the larger enzymes, may account for the lower substrate specificity of family 10 endo-β-1,4-xylanases (Biely *et al.*, 1997).

3.3.2. Application: Delignification of Kraft Pulps by *Trichoderma* Xylanases

Xylanases have found commercial application in the pulp and paper, food, and animal feed industries. In the food industry, xylanases are used to accelerate the baking of cookies, cakes, crackers, and other foods, by breaking down polysaccharides in the dough. Xylanase from *Aspergillus kawachi* is utilized in the fermentation of shochu, a traditional Japanese spirit (Ito *et al.*, 1992).

Currently, the most effective application of xylanase is in the pre-bleaching kraft pulp that minimizes the chlorine needed in subsequent processing steps. Limited hydrolysis of hemicellulose in pulps by fungal xylanases increases the extractability of lignin from the kraft pulps and reduces the amount of chlorine required for bleaching. It is assumed that the xylanase hydrolyzes the reprecipitated xylan, thereby allowing enhanced leaching of entrapped lignin from the fiber wall and making the pulp more susceptible to the chemical bleaching. The xylanase from *T. reesei* acts uniformly on all accessible surfaces of kraft pulp and is effective for bio-bleaching (Bhat, 2000). Presently research is focused on discovery of enzymes that are more robust with respect to pH stability and temperature kinetics (Kulkarni *et al.*, 1999).

3.3.3. Mixed Linked β-Glucan Hydrolyzing Enzymes

The mixed linked 1,3-1,4-glucans form the major part of cell walls of cereals like oat, barley, rye, sorghum, rice, and wheat. Structurally, they are linear glucans of D-glucosyl residues linked through β-1,3 and β-1,4 glycosidic linkages. Due to their high viscosity, β-glucans can impede brewing processes, reducing the yield of extract and lowering the rates of wort separation or beer filtration. Residual β-glucans in the finished beer can form hazes and gelatinous precipitates (Borriss, 1994a). Enzymatic depolymerization of this polysaccharide is an early event in seed germination, catalyzed by endogenous glycosidases with three different specificities: 1,4-β-glucan-4-glucanohydrolase (EC 3.2.1.4), 1,3-β-glucan-3-glucanohydrolase (EC 3.2.1.39) and 1,3-1,4-β-glucan-4-glucanohydrolase (EC 3.2.1.73). Despite the action of this spectrum of endogenous enzymes, a considerable proportion of the high molecular weight β-glucan remains in the germinated barley seeds (malt) used in brewing process. In particular, if low quality malt is processed or if the malt is partly or completely replaced by ungerminated cereal adjuncts, addition of exogenous microbial enzymes becomes necessary (Borriss, 1994b). 1,3-1,4-β-Glucanases are also important aids in the animal feed industry, especially for broiler chickens and piglets; the addition of β-glucanases improves digestibility of barley diets and reduces sanitary problems (malodorous droppings). By far the most efficient enzyme for hydrolyzing 1,3-1,4-β-glucan is licheninase (EC 3.2.1.73), which acts specifically on 1,4-β-linkages adjacent to 1,3-β-linkage yielding 3-O-β-D-cellobiosyl-D-glucose and 3-O-β-D-cellotriosyl-D-glucose as final products. Representatives of this enzyme class belong to glycosyl hydrolase family 16 and are found mainly in bacteria. Their 3-D structures show all-antiparallel β-sheet, jelly role architecture. The core of the protein is formed by two β-sheets that consist of seven antiparallel strands creating a substrate binding cleft crossing one side of the protein. This concave cleft is lined mainly with aromatic residues along its walls and with acidic residues at its bottom. The convex side of the molecule, remote from the active site, binds a calcium ion, which stabilizes the protein's structure (Keitel *et al.*, 1993).

Recently, enzymes homologous to bacterial licheninases have been detected in fungi, and this enzymatic activity also occurs in plants. The plant counterparts do not share any similarity with the microbial enzymes at the level of primary sequence or tertiary folding. In fact they belong to glycosyl hydrolase family 17 and serve as impressive examples of convergent evolution. The first fungal gene encoding a lichenase with high homology to the bacterial genes was cloned from the anaerobic fungus *Orpinomyces*. Since the gene does not contain introns and its flanking regions are AT-rich, it is likely to have had a bacterial origin. Other fungal lichenases belonging to glycosyl hydrolase family 16 were found in the plant pathogen, *Cochliobolus carbonum*, and in *Talaromyces emersonii* (Planas, 2000).

Despite the recent detection of fungal lichenases, most of the fungal enzymes used in brewing processes are 1,4(3)- or 1,4-β-glucan endohydrolases (EC 3.2.1.4 and EC 3.2.1.6) prepared from *Trichoderma, Aspergillus*, or *Penicillium* species and exhibiting some activity toward mixed linked β-glucans. The β-glucanase from *Trichoderma* appears to be especially suitable for the production of high quality beer from poor quality barley (Saxena *et al.*, 2001).

Enzyme preparations containing 1,3-β-glucan hydrolyzing activity derived from *T. harzianum* are useful for making wine from grapes infected with *Botrytis cinerea*. This fungus generally attacks nearly ripe grapes and produces a soluble, high molecular weight 1,3-β-glucan with short β-(1,6)-linked side chains that makes filtration of the wine difficult (Galante *et al.*, 1998).

3.3.4. Application of Cellulases and Hemicellulases in Animal Feed Biotechnology

The animal feed industry is an important sector of agro-biotech with an annual production of more than 600 million tons of feed, worth more than US$ 50 billion. Of the total feed produced, the major share is consumed by poultry, pigs, and ruminants (up to 90%), while pet foods and fish farming account for approximately 10%. Mixed linked 1,3-1,4-β-glucanases and xylanases have been successfully used in diets to hydrolyze non-starchy polysaccharides such as barley β-glucan and arabinoxylans, which are only poorly degraded by the digestive tracts of young monogastric animals, namely, chickens and piglets (Bhat, 2000).

3.4. Cell Wall Lytic Enzymes

3.4.1 Macerating Enzymes in Fruit and Vegetable Processing

Degradation of plant cell wall polysaccharides is of major importance in the food and feed, beverage, textile, and paper industries and is needed in other industrial processes. Members of the genus *Aspergillus* and *Penicillium* are often used for the production of polysaccharide-degrading enzymes (Mukherje and Majumdar, 1971; Acunaarguelles *et al.*, 1995; Cary *et al.*, 1995; Ryazanova *et al.*, 1996; Barensi *et al.*, 2001; de Vries and Visser, 2001; Molina *et al.*, 2001). *A. aculeatus* produces an enzyme complex able to degrade the rhamnogalacturonan backbone of the pectic substances that are embedding the cellulosic fibers of plant material and linked with protein. The enzyme complex, consisting of 10–15 distinct enzymes, is useful in a large number of separation and liquefaction processes for vegetable raw materials, including the recovery of protein and starch and the liquefaction of apples (Adler-Nissen, 1987).

The production of fruit and vegetable juices requires methods for extraction, clarification, and stabilization. Currently, a combination of pectinases (pectin lyase, pectin methylesterase, endo- and exopolygalacturonases, pectinacetylesterase, rhamnogalacturonase, and endo- and exoarabinases), cellulases (endoglucanases, exoglucanases, and cellobiases), and hemicellulases (endo- and exoxylanases, galactanases, xyloglucanases, and mannanases), which are collectively called macerating enzymes, are used to extract and clarify fruit and vegetable juices (Bhat, 2000). Culture filtrates of the food-grade microorganisms, *A. niger* and *Trichoderma* spp., are the sources of these complex enzyme mixtures.

In recent years, olive oil has attracted attention on the world market because of its purported health benefits. Use of a commercial *Aspergillus* enzyme preparation, Cytolase O, consisting mainly of macerating enzymes has proven advantageous for efficient extraction of olive oil under cold processing conditions that preserve antioxidants and vitamin E (Galante *et al.*, 1998).

4. Phosphorous Mobilization: Phytases

Phytic acid (myo-inositol hexakisphosphate) is the primary phosphorus stock in cereal grains, legumes, and oilseeds accounting for 60–80% of the organically bound phosphorus (Brinch-Pedersen *et al.*, 2002). It is generated in plants by a stepwise phosphorylation of myo-inositol-3-phosphate, a product of conversion of glucose-6-phosphate by D-3-myo-inositol-3-phosphate synthase (Loewus and Murthy, 2000). Almost all phytic acid occurs as phytate, a mixed salt (usually with K^+, Ca^{2+}, Mg^{2+}, or Zn^{2+}) that is deposited as globoid crystals together with protein in membrane-bound vesicles (Barrientos *et al.*, 1994).

In general, phosphatases are not able to hydrolyze phytate. However, a special group of phosphomonoesterases, phytases (myo-inositol hexakisphosphate[myo-inositol P6] phosphohydrolase), catalyze the sequential hydrolysis of myo-inositol P6 to a series of lower phosphoric esters of myo-inositol or, in some cases, to free myo-inositol. The International Union of Biochemistry (1979) recognizes two phytases: (1) 3-phytase, EC 3.1.3.8, that hydrolyzes the ester bond at the 3 position of myo-inositol P6 producing myo-inositol 1,2,4,5,6-pentakisphosphate and orthophosphate (Figure 11.2), and (2) 6-phytase, EC 3.1.3.26, that hydrolyzes the 6-position of myo-inositol P6 releasing D-myo-inositol 1,2,3,4,5-pentakisphosphate and orthophosphate (Wodzinski and Ullah, 1996). Remaining ester bonds in the substrate are subsequently hydrolyzed at different rates yielding different meso-inositol phosphate esters (IP3 to IP1) as the final products of enzymatic action. Based on sequence homology, phytases can be divided into three groups that evolved independently: (1) histidine acid phosphatases, mainly isolated from fungi, (2) *Bacillus* phytases with a six-bladed propeller architecture reminiscent of *Bacillus* alkaline phosphatase (Idriss *et al.*, 2002), and (3) plant purple acid phosphatases.

Phytic acid and phytases are of major economic and ecological importance. The yearly global production of phytic acid is estimated to be more than 51 million metric tons, almost 65% of the elemental phosphorus sold worldwide for use in mineral fertilizers (Brinch-Pedersen *et al.*, 2002). The annual world market for phytase as a feed additive was estimated to be US$500 million in 1999 and is expected to increase steadily (Abelson, 1999). The enormous industrial interest in phytases is based mainly on the fact that its

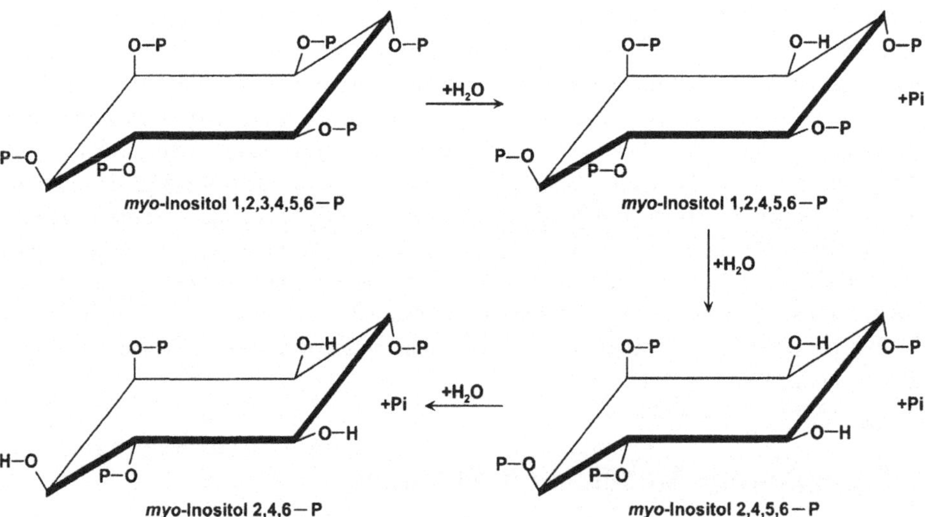

Figure 11.2. Degradation of phytate by 3-phytases starts with dephosphorylation of the phosphate ester bond at C3 and leads to either myo-inositol 2,4,6-P (myo-inositol triphosphate) or myo-inositol-2-P (myo-inositol monophosphate).

activity is very low or entirely absent in the digestive tracts of monogastric animals, for example, poultry and pigs. This means that inorganic phosphorous has to be added to the feed to meet the phosphorous requirements of these animals. However, supplementation imposes ecological problems (eutrophication). Accordingly the substitution of phytase for inorganic phosphate is desirable, because the enzyme releases the phosphate that is already present in the feed as phytic acid. Since the manure accumulated worldwide has increased as a by-product of intensified livestock production, phytic acid that is not metabolized by monogastric animals poses a disposal problem. Environmental pollution with high-phosphate manure has also become an issue in various locations around the world, and as a consequence, animal feed distributors in Europe have begun to formulate feed products supplemented with phytase to improve feedlot productivity and decrease phosphate waste (Berka *et al.*, 1998). Phytic acid can also act as an antinutrient in animal feed by chelating minerals, and addition of phytase increases the feed value by diminishing this factor.

Fungal phytases belong to the family of histidine acid phosphatases, all utilizing a phosphohistidine intermediate in their phosphoryl transfer reaction. *Aspergillus* phytase preferentially hydrolyzes phytic acid at position D-3, generating D-myo-inositol-(1,2,4,5,6)-pentakisphosphate first and subsequently D-myo-inositol-(1,2,5,6)-tetrakisphosphate as the major tetrakisphosphate product (Irvin and Cosgrove, 1972). Like other acid phosphatases, fungal phytases are characterized by RHGXRXP active site motifs that are believed to be the phosphate receptor regions. One exception is the acid histidine phosphatase from *Peniophora lycii* that is a 6-phytase. Simultaneous application of phytases from *A. niger* and *P. lycii* lead to synergistic effects resulting in increased hydrolysis of phytate (Lassen *et al.*, 2000).

X-ray analysis of the phytase from *A. niger* var. *ficuum* provided a model for substrate binding and attack, which features formation of hydrogen bonds to the C3-phosphate group of the substrate and attack by His59 (Kostrewa *et al.*, 1997). Binding of the negatively charged phytic acid is facilitated by a cluster of basic amino acid residues located close to the active site region. The crystal structure of *A. niger* 3-phytase reveals the presence of a large α/β-domain, showing similarity with, but with a smaller α-domain than, the high molecular weight acid phosphatase of rats.

Highest phytase productivity is found in aspergilli, especially within the *A. niger* group. *A. niger* strain NRRL 3135 is able to produce two different phytate hydrolyzing enzymes: PhyA phytase (EC 3.2.1.8), unique in possessing two pH optima at 5.5 and at 2.5, and PhyB (EC 3.2.1.2), a nonspecific acid phosphomonoesterase (Dvorakova, 1998). Genes encoding PhyA homologous were isolated from *Aspergillus terreus* and *Myceliophthora thermophila* and were claimed to form a phytase subgroup distinct from other acid phosphatases (Mitchell *et al.*, 1997; Table 11.1). Most of the known phytases have previously been isolated from *Aspergillus* species and other ascomycetes, but phytase genes have been cloned very recently from fungi belonging to four orders of basidiomycetes, suggesting that PhyA phytases are widely distributed in the fungal kingdom (Lassen *et al.*, 2001). Interestingly these phytases, though homologous to *Aspergillus* PhyA phytase, act as 6-phytases. Yeasts, such as *Saccharomyces cerevisiae, Candida tropicalis, Candida krusei, Debaryomyces occidentalis* (formerly *Schwanniomyces castellii*), and *Kluyveromyces fragilis* are known to produce phytase (Quan *et al.*, 2001). The current commercial feed supplement (Phytase Novo) is a recombinant *A. niger* (previously *A. ficuum*) phytase produced in *A. niger* or *A. oryzae*.

4.1. Engineering of Improved Functionality in *Aspergillus* Phytase

Presently phytase is one of the most important industrial enzymes, and a number of its properties (thermostability, catalytic activity, and pH optimum) have been altered to improve performance in the field. Protein engineering to increase thermostability is often a primary goal in enzyme improvement efforts. Despite extensive knowledge of the general mechanisms governing protein stability, stabilization of a given protein by rational design is not yet easily realized. An impressive example of successful enzyme engineering is the optimization of fungal phytase achieved by the members of Hoffmann-La Roche Ltd. Laboratory in Basel, who applied the consensus approach (Lehmann *et al.*, 2000a) that has been detailed in Chapter 10 (Figure 10.1). As with other thermophilic enzymes, higher phytase thermostability resulted from greater rigidity of the enzyme, but this simultaneously had a negative affect on catalytic activity at ambient temperatures. Another disadvantage of the consensus phytase was a loss of the optimum at pH 2.5, which is beneficial for enzyme activity in the digestive tracts of animals. To overcome these drawbacks while retaining enhanced stability, the Roche researchers replaced all divergent active site residues in the consensus phytase with the respective residues from the highly active *A. niger* enzyme. Although this exchange of the active site was associated with a decrease in intrinsic thermostability relative to consensus "phytase-1," the new consensus "phytase-7" was still over 7°C more thermostable than any of the "parental" phytases. However, more importantly, the second optimum at pH 2.5 was restored, and this was accompanied by a moderate improvement in the specific activity (63.7 U/mg as compared with 44.1 U/mg in consensus "phytase-1" and 102.5 U/mg in *A. niger* PhyA) (Lehmann *et al.*, 2000b). This experience suggests that the route to stabilize a thermolabile enzyme established by the Roche Laboratory, namely combining a consensus strategy with retention of active site characteristics, may be generally applicable.

Another engineering strategy was developed by the same laboratory group to develop a more active version of *Aspergillus fumigatus* phytase. This phytase displays a series of favorable characteristics including a broad pH optimum, a broad substrate specificity, and the capacity to refold properly after heat denaturation. Its low specific activity (26.5 U/mg) precluded its commercialization. By comparing its sequence with that of the most active fungal phytase known to date from *A. terreus* (196.5 U/mg), attention was focused on the substrate binding residue Gln27 as a possible cause of low activity in the *A. fumigatus* phytase. Replacement of this residue with Leu, which is present at the equivalent position in *A. terreus* phytase, resulted in a dramatic increase in catalytic activity to 92.1 U/mg, without affecting the favorable properties of the *A. fumigatus* enzyme (Tomschy *et al.*, 2000).

In developing a fungal phytase maximally adapted to its application in animal feeding, the Roche Laboratory group was also mindful of the digestive tract pH in chickens and piglets. Inspection of the amino acid sequence alignment of phytases together with the known 3-D structure of *A. niger* phytase led to the identification of residues possibly responsible for the activity at pH 2.5, which is a feature of the PhyA and PhyB phytases of *A. niger* and the *P. lycii* 6-phytase. To engineer a protein with a suitable pH profile, the consensus phytase and the *A. fumigatus* phytase were used as starting points. Appropriate substitutions in both enzymes gave rise to enzymes with optima at pH values between 2.8 and 3.4. In addition, a single K68A substitution decreased the pH optima in both enzymes by 0.5–1.0 pH units, with either no change or a slight decrease in specific activity (Tomschy *et al.*, 2002).

5. Lipases (Triacylglycerol Hydrolases, EC 3.1.1.3)

Esterases (EC 3.1.1.N) represent a diverse group of hydrolases catalyzing the cleavage and formation of ester bonds. Esterases show high regio- and stereospecificity that makes them attractive biocatalysts in fine-chemical synthesis of optically pure compounds. Two major classes of hydrolases are of utmost importance: lipases (EC 3.1.1.1, triacylglycerol hydrolases) and "true" esterases (EC 3.1.1.3, carboxyl esterhydrolases). The 3-D structures of both enzymes show the characteristic α/β-hydrolase fold, a distinctive order of α-helices and β-sheets. The catalytic triad is usually composed of Ser–Asp–His, the mechanism for ester hydrolysis or formation being essentially the same for lipases and esterases. The substrate is first bound to the active Ser and the resulting enzyme–substrate intermediate is stabilized by the catalytic His and Asp residues. Then the alcohol releases, and an acyl–enzyme complex is formed. Attack of a nucleophile leads again to an intermediate that is resolved to the product and free enzyme (Bornscheuer, 2002).

Esterases can be employed where regioselectivity is of interest. The most prominent example of a commercial esterase application is the use of carboxyl esterases from *Pycnoporus cinnabarinus* in the release of ferulic acid from pectin or xylan. In xylans, ferulic acid is attached to arabinose residues that are bound to the xylan backbone; in pectins, it can be linked to galactosyl or arabinosyl side chains. The ferulic acid obtained in this process has commercial value as a precursor of vanillin, a major flavor compound (Lesage-Meessen *et al.*, 1996; see also Chapter 13).

As mentioned, lipases catalyze both the hydrolysis and the synthesis of esters formed from glycerol and long-chain fatty acids. After an extended screening of lipase-producing microorganisms, lipases from *Fusarium oxysporum* and *Ovadendron sulphureo-ochraceum* gave the best yields and selectivity in the resolution of racemic ibuprofen and 1-phenylethanol (Cardenas *et al.*, 2001a, b). Apart from numerous applications such as transesterification and ester synthesis in fine-chemical production, and use in laundry detergents, these enzymes also play a wide role in the food and dairy industries.

Lipases have traditionally been obtained from animal pancreas and were originally used as a digestive aid intended for human consumption. Initial interest in microbial lipases was a result of a shortage of pancreas tissue and the difficulties in collecting available material (Frost and Moss, 1987). In as much as the primary effect of adding lipases to fat containing foods is the production of a rancid flavor, limitation of endogenous lipase activity is often important.

The reasons for the enormous biotechnological potential of microbial lipases include their stablity in organic solvents (conditions which favor formation rather than hydrolysis of ester bonds), their activity in the absence of cofactors, their broad substrate specificity, and their high enantioselectivity. A number of fungal lipases have been produced commercially (Jaeger and Reetz, 1998; Jaeger and Eggert, 2002).

Lipolytic reactions occur at the lipid–water interface. Lipases often exhibit a phenomenon known as "interfacial activation," that is, the enhancement of lipase activity towards an insoluble substrate in the form of an emulsion (Cajal *et al.*, 2000). Lipases are considered carboxylesterases capable of acting on emulsified substrates. Although distinguishing them from "true" esterases in this way may seem arbitrary, there is a structural justification. Crystallographic analysis of lipases has shown that access to the active site is blocked by a surface loop, called the lid. Upon binding at an interface, this lid moves away, turning the "closed" enzyme into an "open" form with the active site now accessible to the solvent. At the same time, a large hydrophobic surface is exposed that facilitates binding of the lipase at the interface. These structurally related events are the basis of interfacial activation. However, some lipases, for example, lipase from *Candida antarctica*, do not show interfacial activation, suggesting that this elegant mechanism does not exist in all members of this enzyme family. Simply defined, a lipase is a carboxylesterase that catalyzes the hydrolysis of long chain acylglycerols (Jaeger and Reetz, 1998).

Some enzymes of fungal lipid metabolism are used to alter plant lipids (Budziszewski *et al.*, 1996). Lipases from thermophilic fungi are of increasing interest. Genes encoding lipase from thermophilic fungi have been cloned and overexpressed in heterologous fungi (Maheshwari *et al.*, 2000). Furthermore, fungal phospholipases play a role in virulence and fungal pathogenesis, and they constitute possible therapeutic and diagnostic targets (Ghannoum, 2000).

In commercial terms the most important application of hydrolytic lipases is to fortify laundry and dishwasher detergents. Each year approximately 1,000 tons of lipase are added to 13 billion tons of detergent products, and lipase sales in 1995 were estimated to be US$30 million. Novo Nordisk introduced Lipolase, the first commercial lipase, in 1994; the enzyme originated from the fungus *T. lanuginosus* and was expressed in *A. oryzae*. How the characteristics of this enzyme were improved is described in Chapter 10.

6. Proteases

Proteases are degradative enzymes that catalyze the total hydrolysis of proteins by cleaving their peptide bonds. They represent one of the three largest groups of industrial enzymes and account for about 60% of the total worldwide sale of enzymes. Fungal proteases are mainly used in food processing and, more recently, in the leather industry for dehairing, bating, and tanning hides, replacing toxic chemicals, for example, sodium sulfide. Exo- and endopeptidases are distinguished on the basis of their mode of action. Exopeptidases (EC 3.4.1) act only near the ends of polypeptide chains, and depending on their site of action, are classified as amino- or as carboxypeptidases. Endopeptidases (EC 3.4.21–3.4.34) are characterized by their preferential action on the peptide bonds in the inner region of the polypeptide chain and are divided into four subgroups based on catalytic mechanism: serine proteases, aspartic proteases, cysteine proteases, and metalloprotease. Kulkarni *et al.* (1999) have published an extensive review of the many biochemical and biotechnological aspects of microbial proteases.

Fungi elaborate a wide variety of proteases. For example, *A. oryzae* produces acid, neutral, and alkaline proteases. *A. oryzae* protease preparations are used in bread making due to their lower temperature stability (allowing inactivation) and also their amylase content (Saxena *et al.*, 2001).

Traditionally, "chymosin," an acid protease (EC 3.4.23.4) obtained from the fourth stomach of unweaned calves, has been used to precipitate milk casein in the cheese ripening process. The enzyme cleaves the specific peptide bond Phe105–Met106 to yield para-κ-casein and macropeptides. During cheese making it is important to avoid proteases that hydrolyze milk casein to small bitter-tasting peptides. Fungal acidic proteases, with high milk clotting activity and the ablity to substitute for the expensive animal chymosin, were detected and commercially exploited from *Mucor pusillus* (Noury-Lab, Mucorpepsin, EC 3.4.23.23), *Rhizomucor miehei* (Rennilase, Mucorpepsin, EC 3.4.23.73), and *Cryphonectria parasitica* (Sure curd, Suparen, Endothiapepsin, EC 3.4.23.22). Nevertheless the production of recombinant calf chymosin itself in *K. lactis* represents the most industrially important approach for the preparation of cheese ripening enzyme to date. This yeast is part of the natural flora of kefir. Calf chymosin is also expressed in *T. reesei* (Harkki *et al.*, 1989) or *Aspergillus* cells (Ward *et al.*, 1990). The secreted prochymosin is processed to chymosin under the controlled acidification of the fermentation medium that also serves to precipitate the producing yeast cells. Maxiren (Gist Brocades) was the first microbial chymosin preparation successfully used on an industrial scale in the cheese ripening process (Behnke and Täufel, 1994).

7. Degradation of Lignocellulose: Ligninolytic Enzymes

The cell walls of woody tissues in higher plants have a structure that is physically robust and chemically stable. This material, termed lignocellulose, contains the three main components, lignin, hemicellulose, and cellulose, which are associated closely and sometimes covalently to form a variable and very complex material. Because of the importance of lignocellulose in the paper production, the possibility of its biodegradation has been intensively researched for decades.

In natural terrestrial environments, the decay of dead wood is largely dependent on enzymes from aerobic white-rot fungi (basidiomycetes) and soft-rot fungi (often ascomycetes). In addition to the specific hydrolases discussed in previous sections of this chapter (e.g., cellulases and hemicellulases), redoxenzymes, including the copper-containing phenoloxidase laccase, lignin peroxidase (LiP), manganese peroxidase (MnP), hydrogen peroxide (H_2O_2)-producing oxidases, and cellobiose dehydrogenase, are believed to take part in wood degradation (Nerud and Misurcova, 1996; Henriksson *et al.*, 2000). These enzymes are distributed in ligninolytic fungi in a variable pattern (Table 11.2).

Table 11.2. Distribution of Ligninolytic Enzymes in Selected Fungi

Species	Laccase	Lignin peroxidase	Manganese peroxidase	Reference
Pycnoporus cinnabarinus	+	–	–	Eggert *et al.*, 1996
Pleurotus pulmonaris	+	–	–	Zilly *et al.*, 2002
Phlebia ochraceofulva	+	+	–	Vares *et al.*, 1993
Dichomitus squalens	+	–	+	Perie & Gold, 1991
Panus tigrinus	+	–	+	Golovleva *et al.*, 1993
Agaricus bisporus	+	–	+	Bonnen *et al.*, 1994
Lentinula edodes	+	–	+	Buswell *et al.*, 1995
Pleurotus ostreatus	+	–	+	Ha *et al.*, 2001
Trametes versicolor	+	+	+	Vyas *et al.*, 1994b
Phlebia radiata	+	+	+	Vares *et al.*, 1995
Bjerkandera adusta	+	+	+	Nakamura *et al.*, 1999
Phanerochaete chrysoporium	–	+	+	Datta *et al.*, 1991
Phanerochaete sordida	–	–	+	Ruttimann-Johnson *et al.*, 1994

Despite attempts to apply these enzymes for lignocellulose degradation in woody materials, an economically feasible process is not yet available (Sun and Cheng, 2002). Indeed it is difficult to see the benefit of hydrolyzing cellulose to glucose, considering the immense agricultural surplus of the competing substrate, starch. Nevertheless, ligninolytic enzymes are used for many purposes in biotechnology and industry.

7.1. Lignin Peroxidase and Manganese Peroxidase

LiP and MnP, together with laccase and some other enzymes, belong to the ligninolytic enzyme system of fungi. These peroxidases are heme-containing glycoproteins that require H_2O_2 as the oxidant (Conesa *et al.*, 2002).

The lignin-degrading capability of the white-rot basidiomycete *Phanerochaete chrysosporium* is due mainly to the LiP activity that it produces as a family of extracellular proteins. As many as 15 LiP isozymes, ranging in *M*(*r*) values from 38,000 to 43,000, are produced depending on the strain and culture conditions. Although the major LiP isozymes are encoded by separate genes, they are similar in architecture. The mature form of each protein is 343–344 amino acids long, containing 1 putative N-glycosylation site and a number of putative O-glycosylation sites, and is produced from a precursor that included a 27–28 amino acid leader peptide having a Lys–Arg cleavage site (Reddy and D'Souza, 1994). Additional properties of LiP have been summarized by Piontek (2002).

MnPs are a second family of extracellular heme proteins produced by *P. chrysosporium* that are likewise considered important in lignin degradation (Wariishi *et al.*, 1989). MnP oxidizes Mn (II) to Mn (III), which in turn converts nearby phenolic rings to phenoxy radicals, leading to the decomposition of these phenolic compounds (Glenn *et al.*, 1986; Mester and Tien, 2000; Hakala *et al.*, 2002). A broad spectrum of aromatic substances are partially mineralized by the MnP system of white-rot fungi. The oxidation of hydrophobic substrates such as polycyclic aromatic hydrocarbons is connected with the release of highly polar products and carbon dioxide (CO_2; Scheibner *et al.*, 1997). Therefore, the mineralization of compounds by MnP, occuring outside the fungal cell and leading to CO_2 production, has been considered "enzymatic combustion" (Kirk and Farrell, 1987; Hofrichter *et al.*, 1998).

MnP has been used to degrade various environmental pollutants, such as 2,4,6-trinitrotoluene (van Aken *et al.*, 1999; van Aken and Agathos, 2002), polyvinyl alcohol (Huang *et al.*, 2002), polycyclic aromatic hydrocarbons (Bezalel *et al.*, 1996; Bogan and Lamar, 1996; Sack *et al.*, 1997b; Günther *et al.*, 1998), and organochlorine pesticides (Tekere *et al.*, 2002). It has also been applied in the bleaching of wood (Wong and Mansfield, 1999) and the decolorization of industrial dyes (Conneely *et al.*, 2002; Mielgo *et al.*, 2002).

7.2. Laccase

Laccase (EC 1.10.3.2) represents a family of copper-containing oxidases. Having versatility and broad substrate specificity, laccases could become one of the most important biocatalysts in fungal biotechnology, and consequently their properties and biotechnological applications will be described here in some detail.

7.2.1. Distribution

Laccase, first detected in the Japanese lac tree *Toxicodendron verniciflua*, has been found in certain other plants, in more than 15 genera of insects, and in a variety of fungi, including yeasts (e.g., *Cryptococcus*), molds (e.g., *Penicillium*), mushrooms (e.g., *Agaricus*), and white-rot fungi (e.g., *Pleurotus*). In bacteria its occurrence seems to be restricted to a few species from the genera *Azospirillum* (Faure *et al.*, 1995, 1996; Alexandre and Bally, 1999), *Sinorhizobium* (Castro-Sowinski *et al.*, 2002), *Marinomonas* (Solano *et al.*, 1997, 2000; Sanchez-Amat *et al.*, 2001; Lucas-Elio *et al.*, 2002), *Streptomyces* (Endo *et al.*, 2002), and *Bacillus* (Martins *et al.*, 2002).

7.2.2. Biological Function of Laccase

The biological roles of laccase are diverse. It is wide-spread among ligninolytic basidiomycetes (*Phanerochaete, Pleurotus, Trametes, Pycnoporus, Nematoloma, Sporotrichum, Stropharia, Clitocybula*, etc.). As a result of laccase oxidation, radicals (cationic) can be generated in lignin, resulting in subsequent aliphatic or aromatic C–C bond cleavage and lignin depolymerization (Ander and Eriksson, 1976; Elisashvili, 1993; Hatakka, 1994; Bourbonnais *et al.*, 1995; Eggert *et al.*, 1997; Leonowicz *et al.*, 1999, 2001; ten Have and Teunissen, 2001). Laccase also seems to be involved in depolymerization of coal, humus, and black-colored solutions of humic acid (Cohen and Gabriele, 1982; Fakoussa and Hofrichter, 1999) as well as in local depolymerization of plant cell wall material in mycorrhizal symbiosis (Burke and Cairney, 2002).

On the other hand, monoaromatic, phenolic, and methoxyphenolic compounds such as catechol, methoxyhydroquinone, vanillic acid, syringic acid, protocatechuic acid, sinapinic acid, and ferulic acid are also subject to laccase oxidation (Fahreus and Ljundgreen, 1961; Yaropolov *et al.*, 1994; D'Annibale *et al.*, 1996; Smirnov *et al.*, 2001). The one-electron oxidation of phenolic substrates is accompanied by reduction of molecular oxygen to water by transfer of four electrons. The oxidation of a reducing substrate by laccase typically involves the loss of a single electron and the generation of a free radical (Xu, 1999) that can undergo further laccase-catalyzed oxidation and nonenzymatic reactions (quinone formation from phenol, hydration, disproportion, or polymerization). As a result laccase may participate in polymerization processes such as lignification of plant cell walls (O'Malley *et al.*, 1993; Richardson *et al.*, 2000), regeneration of tobacco protoplasts (de Marco and RoubelakisAngelakis, 1997), insect sclerotization (Anderson, 1985), and pigment formation in fungal and bacterial spores (Tsai *et al.*, 1999; Hullo *et al.*, 2001; Martins *et al.*, 2002). Additionally, laccase has an integral role in melanin synthesis in bacteria (Castro-Sowinski *et al.*, 2002; Lucas-Elio *et al.*, 2002), filamentous fungi (Tanaka *et al.*, 1992), and yeasts of the genus *Cryptococcus* (Williamson *et al.*, 1998; Ikeda *et al.*, 2002). In the opportunistic pathogen *Cryptococcus neoformans*, laccase has been implicated as a virulence factor because of its participation in melanin formation (Perfect *et al.*, 1998; McFadden and Casadevall, 2001; Petter *et al.*, 2001), its ability to oxidize brain catecholamines (Zhu *et al.*, 2001), and its role in protection against alveolar macrophage mediated antifungal activity (Liu *et al.*, 1999). Laccase from the phytopathogenic fungus *Botrytis cinerea* detoxifies antimicrobial substances produced by the plant cell (Slomczynski *et al.*, 1995), but is also able to change the phytoalexin and profungicide, resveratrol, to a more toxic phytoalexin in the grapevine (Schouten *et al.*, 2002). In *Escherichia coli*, a laccase-like protein seems to be an integral part of copper resistance and the protection of enzymes against copper-induced damage (Grass and Rensing, 2001; Roberts *et al.*, 2002).

7.2.3. Isoenzymes

Most fungal laccases are extracellular proteins, but intracellular laccases have been described in plants, insects, and some species of fungi (Thomas *et al.*, 1989; Rigling and van Alfen, 1993; Williamson, 1994; Gavnholt *et al.*, 2002). Laccases often occur as groups of isoenzymes. Whereas *P. cinnabarinus* seems to express only one laccase gene (Eggert *et al.*, 1996), two genes encoding laccase isoenzymes are present in fungi such as *Aspergillus nidulans* (Scherer and Fischer, 2001), *Agaricus bisporus* (Ohga *et al.*, 1999), *B. cinerea* (Schouten *et al.*, 2002), *Dichomitus squalens* (Perie *et al.*, 1998), *Lentinula edodes* (Zhao and Kwan, 1999), *Trametes hispida* (Rodriguez *et al.*, 1999), and *Trametes versicolor* (Paice *et al.*, 1996; Cassland and Jonsson, 1999). Three isoenzymes were detected in *Pleurotus ostreatus* (Giardina *et al.*, 1999), *Ganoderma lucidum* (Ko *et al.*, 2001), and *Trametes* sp. (Mansur *et al.*, 1997). *Rhizoctonia solani* and *Pleurotus sajor-caju* form four isoenzymes (Wahleithner *et al.*, 1996; Soden and Dobson, 2001), and *Trametes villosa* and *Trametes sanguinea* have five genes (Yaver and Golightly, 1996; Hoshida *et al.*, 2001). *Flavodon flavus* expresses 10 isoenzymes (Raghukumar *et al.*, 1999), but even this number is small relative to plants such as ryegrass (*Lolium perenne*) that contain 25 different laccase genes (Gavnholt *et al.*, 2002). Some isoenzymes are constitutively expressed (Slomczynski *et al.*, 1995; Eggert *et al.*, 1996; Farnet *et al.*, 1999; Vasconcelos *et al.*, 2000; Soden and Dobson, 2001; Levin and Forchiassin, 2001), whereas others are inducible under various conditions.

7.2.4. Characterization and Some Biochemical Properties

Laccases from more than 30 species of fungi have been purified and characterized, and the primary structures of many have been determined by protein or DNA sequencing (Ong *et al.*, 1997; Temp *et al.*, 1999b; Xaver *et al.*, 1999; Dedeyan *et al.*, 2000; Jung *et al.*, 2002; Kwon and Anderson, 2002). Crystallization was achieved for laccases of *T. sanguinea* (Nishizawa *et al.*, 1995), *T. versicolor* (Antorini *et al.*, 2002; Bertrand *et al.*, 2002a, 2002b), *Coprinus cinereus* (Ducros *et al.*, 1997), *P. cinnabarinus* (Antorini *et al.*, 2002), and *Melanocarpus albomyces* (Hakulinen *et al.*, 2002).

Some fungal laccases may have quaternary structure as judged from their behavior during electrophoresis under nondenaturating conditions; in SDS–PAGE most laccases exhibit a molecular mass of 50–90 kDa. From 7% to 50% of their molecular weight may be attributed to covalently bound carbohydrate. Glycosylation seems to be important for folding of the nascent peptide, providing thermal or proteolytic stability to the mature protein, and possibly playing a role in copper retention. The absorption spectrum of purified laccase shows a maximum at 605 nm, typical of blue-copper oxidases. In the holoenzyme form, most laccases have four copper atoms per monomer, three of them constituting a trinuclear copper cluster center (Cu II-type center). Some laccases, however, contain other metal atoms, for example, one copper, one manganese, two zinc, as shown for *Phellinus ribis* (Min *et al.*, 2001). If blue laccase from *Phlebia* species react with low-molecular-weight lignin decomposition products, it may appear to be a "yellow" laccase (Leontievsky *et al.*, 1997, 1999).

7.2.5. Regulation of Laccase Production

The synthesis of laccase is regulated by many factors. Some are produced at the end of the growth phase or under acidic culture conditions (Kim *et al.*, 2001), and others are associated with the vegetative growth of the fungus (Das *et al.*, 2001). The yield of inducible laccases may be boosted by aromatic inducers such as 2,5-xylidine (Rogalski *et al.*, 1991), *p*-anisidine (Shuttleworth *et al.*, 1986), syringaldazine, *o*-toluidine (Skorobogatko *et al.*, 1996), aniline, nonylphenol (Mougin *et al.*, 2002), ferulic acid (Leonowicz *et al.*, 1972), veratrylalcohol (Niku-Paavola *et al.*, 1990; Dekker *et al.*, 2001), coniferyl alcohol (Farnet *et al.*, 1999), 1-hydroxybenzotriazole (Collins and Dobson, 1997), or anthraquinone. Even various agrochemicals, industrial compounds, or environmental pollutants and their oxidation products can enhance laccase production (Jonas *et al.*, 1998; Hundt *et al.*, 1999; Mougin *et al.*, 2002). Distinct laccase isoenzymes are induced by different aromatic compounds or induction conditions (Farnet *et al.*, 1999; Soden and Dobson, 2001; Temp *et al.*, 1999b). In *Pleurotus ostreatus*, an extracellular subtilisin-like protease seems to play a role in the regulation of laccase activity by degrading and/or activating different isoenzymes (Palmieri *et al.*, 2001).

Unlike the production of cellulases and xylanases, laccase synthesis is usually not subject to glucose repression (Dekker *et al.*, 2001). Media containing 4% glycerol are also suitable for laccase production by *Trametes multicolor* (Hess *et al.*, 2002). Production is stimulated by high concentrations of oxygen, ammonium ions as nitrogen source (Collins and Dobson, 1997), traces of ethanol (Dekker *et al.*, 2001), and appropriate incubation temperature (Lang *et al.*, 2000; Zhang and Williamson, 2001). Formation of extracellular laccase can be stimulated considerably by the addition of copper (Cu II, 0.5–5 mM) or cadmium (1–5 mM), whereas Ag, Hg, Pb, Zn ions and H_2O_2 decreased the measured activity in the culture medium (Collins and Dobson, 1997; Farnet *et al.*, 1999; Baldrian and Gabriel, 2002; Galhaup *et al.*, 2002; Hess *et al.*, 2002). For some fungi, higher copper concentrations (>1 mM) can inhibit growth and decrease the accompanying manganese peroxidase production (Lenin *et al.*, 2002).

7.2.6. Laccase Mediator Systems

The concerted action of fungal laccases with oxidizable low-molecular-weight compounds (called mediators) can extend laccase action to refractory substrates (Majcherczyk *et al.*, 1999; Johannes and Majcherczyk, 2000). Additionally, a mediator can diminish the rate of phenol polymerization so that oxidative phenol degradation is favored (Duran and Esposito, 2000). Use of laccase mediator systems leads to more effective degradation of polycyclic aromatic hydrocarbons and other organic compounds. These systems have also received widespread attention for their potential in bleaching kraft pulp, an application that can significantly extend the potential of laccase biotechnology. Typical mediators are 2,2′-azino-bis-(3-ethyl-benzothiazoline-6-sulphonic acid) (ABTS) (Majcherczyk *et al.*, 1999), 1-hydroxybenzotriazole (Bohmer *et al.*, 1998; Srebotnik and Hammel, 2000), and violuric acid (Li *et al.*, 1999), though several other compounds may also be used (Fabbrini *et al.*, 2002).

A new function for laccase was found by Schlosser and Hofer (2002), who described an oxidation of Mn^{2+} to Mn^{3+} by purified laccase in the presence of oxalate or malonate as mediator. The resulting Mn^{3+} causes an abiotic decomposition of organic molecules, the formation of superoxide radicals, and the subsequent reduction of superoxide to H_2O_2. This may contribute to an increased substrate oxidation rate and represents a novel type of cooperation between laccase and the H_2O_2-utilizing ligninolytic peroxidases.

7.2.7. Delignification of Lignocellulosics by Laccase

The pulp and paper industry is facing increasing pressure to replace the conventional, chlorine-based, pulp-bleaching technique with an environmentally benign process. Enzymatic bleaching methods have drawn much attention in this respect, and laccase has been the focus for the bio-bleaching of kraft pulp because it can be produced in large amounts at reasonable cost and uses oxygen as electron acceptor (Paice *et al.*, 1995; Sealey *et al.*, 1999; Wong and Mansfield, 1999; Balakshin *et al.*, 2001). As a "bio-pulping" agent, laccase could also be applied to wood chips before pulping to partially degrade the lignin and loosen the lignocellulose structure, so that pulping can be performed more efficiently (Patel *et al.*, 1994; Xu, 1999).

7.2.8. Purification of Colored Waste Waters

Laccase is suitable for the degradation of many phenolic or aromatic compounds. Phenolic effluents occur as waste streams in the pulp and paper, coal conversion, dying, textile, and olive oil industries. The coloration and, in some cases, toxicity of these effluents pose serious environmental issues, because colored lignin compounds and their oxidation products are extremely resistant to microbial attack. Conventional water purification methods with activated sludge are ineffective in removing the color. Therefore, alternative low-cost biotreatment processes are being considered, most of which are based on laccase and other ligninolytic enzymes (Font *et al.*, 1997; D'Annibale *et al.*, 1999; Garg and Modi, 1999; Aggelis *et al.*, 2002; Tsioulpas *et al.*, 2002).

7.2.9. Textile Dye Decolorization

Waste water from the textile industry can contain a variety of polluting substances including dyes. Microorganisms and their enzymes are used for degradation of these dyestuffs (Banat *et al.*, 1996; Fu and Viraraghavan, 2001; McMullan *et al.*, 2001; Robinson *et al.*, 2001). Several dyes can be oxidized by fungal laccases or laccase mediator systems (Chivukula and Renganathan, 1995; Abadulla *et al.*, 2000; Soares *et al.*, 2002). Immobilized laccase showing a higher thermal stability could be repeatedly used for decolorization of industrial dyes (Reyes *et al.*, 1999).

7.2.10. Transformation and Inactivation of Toxic Environmental Pollutants

Laccase and other ligninolytic enzymes can be used to detoxify or remove various aromatic pollutants and xenobiotics found in industrial waste (Karam and Nicell, 1997;

Kahraman and Yesilada, 2001), contaminated soil (Pointing, 2001; Tanaka *et al.*, 2001), or water (Duran and Esposito, 2000). The action of laccase results in either direct degradation and oxidation or in polymerization and immobilization of pollutants and toxic chemicals. Examples include degradation of polycyclic aromatic hydrocarbons (Collins *et al.*, 1996; Johannes *et al.*, 1998; Johannes and Majcherczyk, 2000), 2,4-dichlorophenol (Ahn *et al.*, 2000), 2,4,6-trichlorophenol (Leontievsky *et al.*, 2001), pentachlorophenol (Ruttimann-Johnson and Lamar, 1996), 2,4,6-trinitrotoluene (Thiele *et al.*, 2002), phosphorothiolates (Amitai *et al.*, 1998), and dibenzothiophene (Bressler *et al.*, 2000), and various other pollutants. Further examples of biotransformation processes and dechlorination reactions catalyzed by laccase and other fungal enzymes are given later in this chapter.

7.2.11. Beverage and Food Treatment

It is assumed that phenolic saccharides and other phenolic compounds are responsible for browning and haze formation during production and storage of fruit juice or wine. Laccase is to be employed in the processing of juices to make clear and stable products (Cliffe *et al.*, 1994; Piacquadio *et al.*, 1997; Gokmen *et al.*, 1998), and the enzyme has been subjected to extensive toxicological evaluations that have documented its safety (Brinch and Pedersen, 2002). Laccase seems to be useful also for cross-linking ferulic acid during enzymatic gelation of sugar beet pectin in food products (Norsker *et al.*, 2000).

7.2.12. Laccase-Based Biosensors

The development of laccase-based electrochemical biosensors is underway. These biosensors will enable the amperometric determination of phenolic compounds in waste streams, for example, paper mill effluents (Freire *et al.*, 2002a, 2002b), or the quantitative estimation of a wide variety of aromatic substances (catechols, aniline, etc.) and reducing compounds (Ghindilis *et al.*, 1992; Zouari *et al.*, 1994; Yaropolov *et al.*, 1995).

Oxygen determinations can also be made with laccase biosensors (Gardiol *et al.*, 1996). "Wired" laccase cathodes allow the electroreduction of oxygen to water (Yaropolov *et al.*, 1994; Barton *et al.*, 2001) and may be used in bio-fuel cells (Atanassov, 2002). With phenol-derivatized graphite electrodes, it was shown that laccase can interact with phenolic groups immobilized on a solid surface (Scholz *et al.*, 2000).

7.2.13. Synthesis of New Chemicals by Laccase

Laccase is used for bioconversion of aromatic compounds and several other chemicals to oxidized products (Yaropolov *et al.*, 1994). Methyl aromatic compounds are converted to the corresponding aromatic aldehydes (Fritz-Langhals and Kunath, 1998). Although laccase often mediates effective degradation of various chemical substances, it also has an impressive ability for coupling or oligomerizing molecules. This activity can be harnessed for the synthesis of new compounds, to derivatize bioactive compounds, and to couple substances. The formation of radicals catalyzed by laccase can result in a reaction of the parent molecules themselves (self-coupling) or of the parent molecule with a second type of molecule (cross-coupling). Compounds with aromatic structures are very suitable for transformations and bioconversions by laccase.

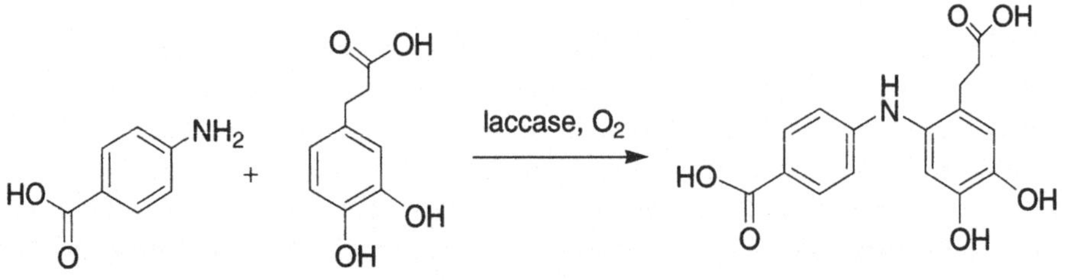

laccase, O_2

4-amino-benzoic acid + 3-(3,4-dihydroxyphenyl) propionic acid → 4-[2-(2-carboxyethyl)-4,5-dihydroxy-phenylamino] benzoic acid

Figure 11.3. Heteromolecular coupling of the bioactive compound 3-(3,4-dihydroxyphenyl) propionic acid with 4-aminobenzoic acid catalyzed by laccase of *Pycnoporus cinnabarinus* (according to Mikolasch *et al.*,

Laccase have been used, therefore, in the polymerization of bisphenol A (Uchida *et al.*, 2001), 1-naphthol (Aktas *et al.*, 2001), and vinyl monomers (Tsujimoto *et al.*, 2001), in the synthesis of phenolic polymers (Uyama, 2001), poly (1,4-oxyphenylene) (Kobayashi and Uyama, 1998), and lignin graft copolymers (Mai *et al.*, 2001a), or in the copolymerization of phenolics with acrylates (Mai *et al.*, 2001b). In the course of such coupling reactions decarboxylation and demethylation of aromatic and phenolic substrates can occur releasing carboxyl groups, methanol, or CO_2 (Dec *et al.*, 2001).

Other homomolecular coupling reactions and biotransformations were used for the synthesis of new imidazole compounds (Schäfer *et al.*, 2001). Of special interest are heteromolecular coupling reactions that can be used as a gentle and specific method for modifying or derivatizing bioactive compounds or for chemical syntheses (Figure 11.3) (Bhalerao *et al.*, 1994; Lindequist and Schauer, 2002; Mikolasch *et al.*, 2002).

7.2.14. Desulfurization and Solubilization of Coal

The combustion of fossil fuels such as coal, petroleum, and natural gas produces harmful emissions, some of which have their origins in sulfur-containing compounds. Conventional physical and chemical fuel desulfurization methods are relatively inefficient and expensive, requiring extreme reaction conditions (temperature and pressure) and corrosion-resistant equipment. An enzymatic desulfurization method would be a welcomed alternative, and earlier results on the application of laccase hold promise (Dordick *et al.*, 1991; Ichinose *et al.*, 1999; Xu, 1999). Additionally, ligninolytic enzymes can be used to solubilize coal and decolorize coal-derived humic acids (Ralph *et al.*, 1996; Hofrichter *et al.*, 1997; Willmann and Fakoussa, 1997; Fakoussa and Frost, 1999; Temp *et al.*, 1999a; Holker *et al.*, 2002).

8. Utilization of Aromatic and Aliphatic Compounds and Hydrocarbons

The capability of fungi to grow on monoaromatic compounds is well known (Gibson, 1968; Halsall *et al.*, 1969); certain species are able to grow on phenol (Zimmermann, 1958; Harris and Ricketts, 1962; Neujahr and Varga, 1970; Hofmann and Schauer, 1988; Krivobok *et al.*, 1994), catechol or orcinol (Gaal and Neujahr, 1979;

Kocwahaluch, 1995; Harwood and Parales, 1996), cresol or toluene (Kennes and Lema, 1994; Prenafeta-Boldu *et al.*, 2001), benzoic or hydroxybenzoic acid (Hofrichter and Scheibner, 1993; Middelhoven, 1993), protocatechuic, vanillic, ferulic, or syringic acid (Rahouti *et al.*, 1999), or aniline (Emtiazi *et al.*, 2001).

Additionally, there is a significant body of data available on the oxidation of polycyclic aromatic hydrocarbons by fungi (summarized by Bumpus, 1989; Cerniglia *et al.*, 1992; Field *et al.*, 1992; Cerniglia, 1993; Sack *et al.*, 1997a; Wolter *et al.*, 1997; Braun-Lullemann *et al.*, 1999; Gramss *et al.*, 1999; Juhasz and Naidu, 2000; Ravelet *et al.*, 2000). These oxidations are mediated either by cytochrome P-450 proteins (Maspahy *et al.*, 1999) or ligninolytic enzymes, such as LiP (Haemmerli *et al.*, 1986; Hammel *et al.*, 1986; Vazquezduhalt *et al.*, 1994; Bogan *et al.*, 1996), MnP (Sack *et al.*, 1997b; Hofrichter *et al.*, 1998), and laccase (Collins *et al.*, 1996; Bohmer *et al.*, 1998; Pickard *et al.*, 1999; Johannes and Majcherczyk, 2000). However, other eukaryotic peroxidases like soybean peroxidase can also catalyze the oxidation of polycyclic aromatic hydrocarbons (Kraus *et al.*, 1999).

Other types of hydrocarbons, namely the aliphatic n-alkanes, are oxidized very quickly, and many fungi find these chemically inert substances suitable as carbon sources for growth (Komagata *et al.*, 1964; Klug and Markovetz, 1967; Hoffmann and Rehm, 1976; Atlas, 1984; April *et al.*, 2000). The key mediators of n-alkane and fatty acid ω-oxidation are alkane cytochrome P-450 monooxygenases (Lebeault *et al.*, 1971), and most fungi capable of using aliphatic hydrocarbons produce multiple forms of microsomal cytochrome P-450. Corresponding *P-450* genes or cDNAs have been isolated from *Candida maltosa* (Schunck *et al.*, 1989; Takagi *et al.*, 1989; Ohkuma *et al.*, 1995), *C. tropicalis* (Sanglard and Loper, 1989; Seghezzi *et al.*, 1992), and *Candida apicola* (Lottermoser *et al.*, 1996). A single purified P-450, 52A3, is able to catalyze (in presence of the corresponding reductase) a cascade of sequential mono- and diterminal monooxygenation reactions from n-alkanes via n-alkanols to α,ω-dioic acids with high regioselectivity (Scheller *et al.*, 1998). The involvement of fungal cytochrome P450 enzymes in many complex bioconversion processes and their use as catalysts for biohydroxylation has been reviewed by several authors (Sheldon, 1994; Lewis, 1996; van den Brink *et al.*, 1998; Lehman and Stewart, 2001).

Growing fungi on water-immiscible hydrocarbons in two-phase, non-glycolytic fermentations can lead to the formation of metabolites that do not appear during growth on conventional substrates. These products include fatty acids (Iida *et al.*, 1980; Bühler and Schindler, 1984; Ratledge, 1988), dicarboxylic acids (Krauel and Weide, 1978; Kaneyuki *et al.*, 1980; Picataggio *et al.*, 1992; Hara *et al.*, 2001), other organic acids such as α-ketoglutarate, citrate, isocitrate, fumarate, malate, anglyceric acid, as well as polyols, ergosterols, and several vitamins (summarized by Fukui and Tanaka, 1981; Sahasrabudhe and Sankpal, 2001). In addition, different exolipids, glycolipids, and biosurfactants may be produced (Inoue and Itoh, 1982; Cirigliano and Carman, 1985; Hommel *et al.*, 1987; Radwan and Sorkhoh, 1993; Davila *et al.*, 1994; Kim *et al.*, 1999).

9. Inactivation of Fungal Biocontrol Agents

9.1. Creosote

The impregnation of wood products with creosote, a complex mixture of polycyclic aromatic hydrocarbons, is a time-honored and effective means of protecting the wood from decay (Creffield *et al.*, 2000). However, inadequate impregnation cannot halt the growth of wood-decomposing fungi, particularly *Lentinus lepideus* (Collett, 1992). Conversely, wood or soil contaminated with creosote can be detoxified by different methods of bioremediation (Pollard *et al.*, 1994). As discussed in the previous section, ligninolytic enzymes of many fungi are able to degrade polycyclic aromatic hydrocarbons, and these include those in creosote-treated or creosote-contaminated materials (Duncan and Deverall, 1964; Cragg and Eaton, 1997; Blakely *et al.*, 2002). Organisms such as *P. chrysosporium, Phanerochaete sordida*, or *P. ostreatus* are very effective for decontaminating creosote-polluted soils (Davis *et al.*, 1993; Bogan and Lamar, 1995; Eggen, 1999; Eggen and Sveum, 1999). Additionally, fungi from the genera *Bjerkandera, Heterobasidion, Schizophyllum, Trametes, Lentinus, Ceratocystis, Stereum, Paecilomyces*, and *Trichoderma* grow with creosote (Schmidt *et al.*, 1991).

9.2. Pentachlorophenol

The antifungal substance and biocide, pentachlorophenol, which has been the most prevalent wood preservative used worldwide for many years, can be dehalogenated and degraded by fungi from the genera *Phanerochaete* (Aiken and Logan, 1996; McAllister *et al.*, 1996; Laugero *et al.*, 1997), *Trametes* (Lamar *et al.*, 1993; Tuomela *et al.*, 1999), *Lentinula* (Okeke *et al.*, 1996), *Armillaria* (Chiu *et al.*, 1998), *Gloephyllum* (Fahr *et al.*, 1999), *Ganoderma, Inonotus* (Logan *et al.*, 1994), *Rhizopus* (Cortes *et al.*, 2002), *Cerrena*, and *Abortiporus* (Cho *et al.*, 2001), as well as a number of other micromycetes and deuteromycetes, especially strains from the genera *Calcarisporium* and *Oidiodendron* (Seigle-Murandi *et al.*, 1995). A biotransformation of pentachlorophenol to pentachloroanisole was found in *Trichoderma* sp. (Rigot and Matsumura, 2002). These transformations are brought about by fungal peroxidases and phenoloxidases (Seigle-Murandi *et al.*, 1993; Ricotta *et al.*, 1996). Laccase can oxidize pentachlorophenol to chloranils and benzoquinones (Ricotta *et al.*, 1996), and also mediate the polymerization of pentachlorophenol (Ullah *et al.*, 2000). Depletion of free pentachlorophenol can be greatly intensified by the addition of 2,5-dimethylaniline (Cho *et al.*, 2001), which contributes to a formation of heteropolymers. Several bioreactors have been developed to degrade pentachlorophenol through the action of fungi and their enzymes (Alleman *et al.*, 1995; Pallerla and Chambers, 1998).

9.3. Inorganic Wood Preservatives

Although brown-rot and white-rot fungi are sensitive to the copper salts used to impregnate wood, many such as *Fibroporia vaillantii, Gloephyllum trabeum, Schizophyllum commune*, and *T. versicolor* can excrete oxalic acid which detoxifies copper (II) sulfate by transforming it into insoluble copper oxalate (Humar *et al.*, 2002). On the other hand, oxalic acid production can convert bound inorganic wood preservatives to water-soluble compounds (leaching), as is the case for chromium and arsenic preservatives (Stephan *et al.*, 1996). Oxalic acid leaching seems to be a suitable process for recycling chromated copper arsenate (CCA)-treated wood (Kartal and Clausen, 2001). The copper-tolerant, wood degrading fungus, *Wolfiporia cocos* (syn. *Poria cocos*), is capable of accumulating copper from copper-based wood preservatives in its mycelium (de Groot and Woodward, 1998, 1999). Organoiodine wood preservatives can be degraded by fungi of the genera *Tyromyces, Serpula*, and *Trametes* (Lee *et al.*, 1992).

9.4. Disinfectants and Deodorants

The disinfectant and antimicrobial substance, cresol, can be degraded by bacteria and fungi. Fungi from the genera *Aspergillus* (Jones *et al.*, 1993) and *Penicillium* (Hofrichter and Scheibner, 1993), as well as fungi such as *Scedosporium apiospermum* and *P. chrysosporium* (Claussen and Schmidt, 1998) can utilize *p*-cresol for growth.

Natural antimicrobial compounds such as essential oils from *Lavandula, Rosmarinus, Artemisia*, or other plants have antifungal properties and are used in cosmetic products and creams as preservatives (Muyima *et al.*, 2002). Nevertheless, these oils can be degraded by several fungi (Nawas and Alkofahi, 1994).

Chlorinated synthetics such as triclosan (2,4,4′-trichloro-2′-hydroxydiphenyl ether) have been used as antimicrobial compounds in deodorants, soaps, and dentifrices for many years (Bhargava and Leonard, 1996). Due to its widespread use and resistance to bacterial and fungal degradation, triclosan can be detected in environmental samples (Okumura and Nishikawa, 1996). The fungi *T. versicolor* and *P. cinnabarinus* transform this substance into two more hydrophilic products that have been identified as carbohydrate conjugates, as well as into a methylated triclosan (Hundt *et al.*, 2000). The conjugates containing glucose or xylose moieties have distinctly lower microbicidal activity than triclosan, and the glycosylation process, probably catalyzed by uracil diphosphate (UDP)-glycosyltransferases, may be considered a detoxification mechanism.

9.5. Fungicides in Agriculture and Medicine

The development of fungicide resistance is an increasing problem in the control of pathogenic fungi. Many fungi have become resistant to fungicides such as guazatine, benomyl, dicarboximide, or others (Pommer and Lorentz, 1982; Wild, 1983; Cui *et al.*, 2002). Benomyl and carbendazim can be degraded by fungi via catabolic enzymes (Ali and Wainwright, 1994; Silva *et al.*, 1999). A single hydroxylation step can result in the partial inactivation of antifungal substances, as shown for itraconazole (Mikami *et al.*, 1994) and rustmicin (Shafiee *et al.*, 2001). Consequently, hydroxylating enzymes from fungi such as the alkane-hydroxylating P-450 protein, CaALK8, from *Candida albicans* may be involved in resistance to fluconazole and some other drugs (Panwar *et al.*, 2001).

Resistance mechanisms that do not involve modification or destruction of the fungicide are discussed elsewhere (Dekker and Georgopoulos, 1982; Dekker, 1995; Georgopoulos, 1995; Steffens *et al.*, 1996; del Sorbo *et al.*, 2000; Morschhäuser, 2002).

9.6. Food Preservatives

Prepared foods are often fortified with sorbic and benzoic acid to ensure safety and extend shelf-life, but certain fungi are capable of degrading these preservative agents (Brul and Coote, 1999). Mollapour and Piper (2001) have characterized a gene from the food spoilage yeast *Zygosaccharomyces bailii* that mediates the catabolism of sorbate and benzoate. Although the gene product seems to be essential for benzoate utilization, it has little homology to fungal benzoate-4-hydroxylases that are microsomal cytochrome P-450 proteins.

10. Biotransformation of Biphenyls by Fungi

10.1. Biphenyl

Biphenyls are aromatic compounds now ubiquitously distributed in the environment due in part to their formation during incomplete combustion of mineral oils and coal. They also have commercial use as fungistatic agents in the packing of citrus fruits and in foods, as components of heat transfer fluids, as dye stuff carriers (textile or copying paper), and as reagents for chemical syntheses. Additionally, biphenyls and their derivatives are natural constituents of plant cells, helping to protect against phytopathogenic fungi (Mori *et al.*, 1997; Cortez *et al.*, 1998). Biphenyl phytoalexins have been found that are produced by plants in response to fungal infection (Kokubun and Harbone, 1995; Kokubun *et al.*, 1995a; Borejsza-Wysocki *et al.*, 1999). Halogenated derivatives of biphenyl are especially toxic. Exposure of humans to biphenyl and highly halogenated biphenyls has been associated with eye and skin disorders and with toxic effects on the liver, kidneys, and nervous system, which are suspected to lead to immune dysfunction, behavioral changes, and cancer. The substances may also have teratogenic effects.

Most research on the microbial inactivation and degradation of biphenyls as poorly soluble environmental pollutants has been performed with aerobic and anaerobic bacteria. Nevertheless, certain fungi also seem to possess remarkable capabilities to oxidize and biotransform these important pollutants. Although fungal transformation of biphenyl has been studied mainly to determine how this antifungal substance may be inactivated, this research serves also as a model for mammalian biphenyl metabolism (Smith and Rosazza, 1974). Additionally, biphenyl and several halogenated derivatives have attracted attention in screens that employ recombinant *S. cerevisiae* cells expressing human or animal estrogen receptors to find hormonally active xenobiotics or estrogenic substances (Collins *et al.*, 1997; Jin *et al.*, 1997; Petit *et al.*, 1997; Schultz *et al.*, 1998). They are also known as inducers of microsomal glutathione S-transferase in fungi and mammals (Datta *et al.*, 1994).

Although fungi constitute the bulk of microbial biomass in soil, reports on fungal metabolism of biphenyl are limited to a few yeasts and filamentous fungi. These fungi seem to possess the capability to monohydroxylate the inert biphenyl molecule to form 2-hydroxybiphenyl and/or 4-hydroxybiphenyl. This is true for filamentous fungi from the genera *Absidia* (Dodge *et al.*, 1979; Schwartz *et al.*, 1980), *Aspergillus* (Smith *et al.*, 1980; Golbeck *et al.*, 1983; Cox and Golbeck, 1985; Mobley *et al.*, 1993), *Cunninghamella* (Smith and Rosazza, 1974; Dodge *et al.*, 1979; Smith *et al.*, 1980), *Fusarium* (Gesell, 2001), *Gliocladium, Thamnostylum* (originally *Helicostylum*) (Smith *et al.*, 1980), *Paecilomyces* (Gesell *et al.*, 2001), and *Pycnoporus* (Jonas, 1997), as well as for yeasts from the genera *Candida* (Wiseman *et al.*, 1975; Cerniglia and Crow, 1981; Romero *et al.*, 2001), *Debaryomyces* (Cerniglia and Crow, 1981; Lange *et al.*, 1998), *Saccharomyces* (Kärenlampi and Hynninen, 1981; Layton *et al.*, 2002), *Yarrowia, Cryptococcus* (Romero *et al.*, 2001), and *Trichosporon* (Sietmann *et al.*, 2000, 2001, 2002). The metabolite 3-hydroxybiphenyl is accumulated only in minor amounts by selected fungi (Dodge *et al.*, 1979; Cerniglia and Crow, 1981; Sietmann *et al.*, 2000, 2002; Gesell *et al.*, 2001). Monohydroxylated biphenyls can be further oxidized to dihydroxylated derivatives in a second oxidation step that cannot be performed by all fungi that are able to monohydroxylate the substrate. Whereas one or two dihydroxylated products were detectable in most cases (Smith *et al.*, 1980; Cerniglia and

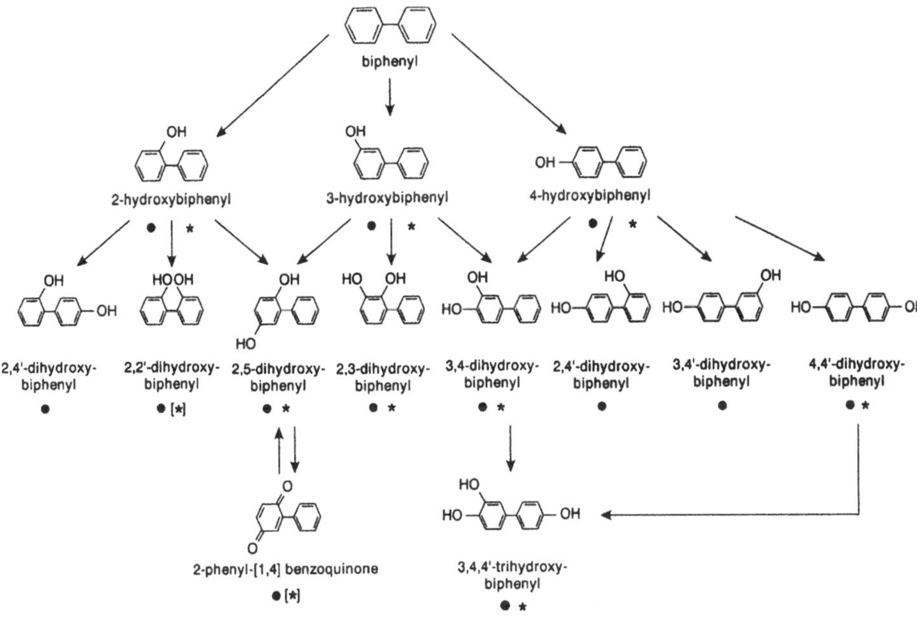

Figure 11.4. Primary oxidation products of biphenyl in the fungi *Trichosporon mucoides* (•) and *Paecilomyces lilacinus* (*) (according to Sietmann *et al.*, 2000; Gesell, 2001; Gesell *et al.*, 2001).

Crow, 1981; Lange *et al.*, 1998), six and nine di- and trihydroxylated intermediates could be identified by high performance liquid chromatography (HPLC), gas chromatography–mass spectrometry (GC–MS), and nuclear magnetic resonance (NMR)-techniques in biotransformations with *Paecilomyces lilacinus* (Gesell *et al.*, 2001) and *Trichosporon mucoides* (Sietmann *et al.*, 2000), respectively (Figure 11.4). The hydroxylation reactions depend upon molecular oxygen, as shown by ^{18}O incorporation (Sietmann *et al.*, 2000), and in most cases appear to be catalyzed by fungal P-450 proteins because cytochrome P-450 inhibitors (metyrapone or 1-aminobenzotriazole) prevent the formation of most hydroxylated intermediates. Only the oxidation of 2- and 4-hydroxybiphenyl to 2,2′- and 4,4′-dihydroxybiphenyl, respectively, and the further transformation of 3,4-dihydroxybiphenyl could not be inhibited by 1-aminobenzotriazole in *T. mucoides* (Sietmann *et al.*, 2000).

Hydroxylated biphenyls are more water soluble than their parental compounds, and also possess increased cellular toxicity (Lange *et al.*, 1998; Gesell *et al.*, 2001) and estrogenic activity (Layton *et al.*, 2002). Some fungi are able to detoxify these hydroxylated compounds either by (1) conjugate formation (sulfate conjugates in *Aspergillus* strains [Golbeck *et al.*, 1983]; glucuronide conjugates in *Cunninghamella* strains [Dodge *et al.*, 1979; Cerniglia *et al.*, 1982]), (2) oligomerization through the action of ligninolytic enzymes (Jonas *et al.*, 2000), or (3) enzymatic splitting of the ring system in di- or trihydroxylated biphenyls (Mobley *et al.*, 1993; Lange *et al.*, 1998; Gesell *et al.*, 2001; Sietmann *et al.*, 2001). In *P. lilacinus* (Gesell *et al.*, 2001) and *T. mucoides* (Sietmann *et al.*, 2001) 6–8 different ring cleavage products could be identified (Figure 11.5). Nevertheless, no growth of these soil fungi on biphenyls or their oxidation products occurs, and consequently, some of these dead-end products accumulate in significant amounts. For biotechnological

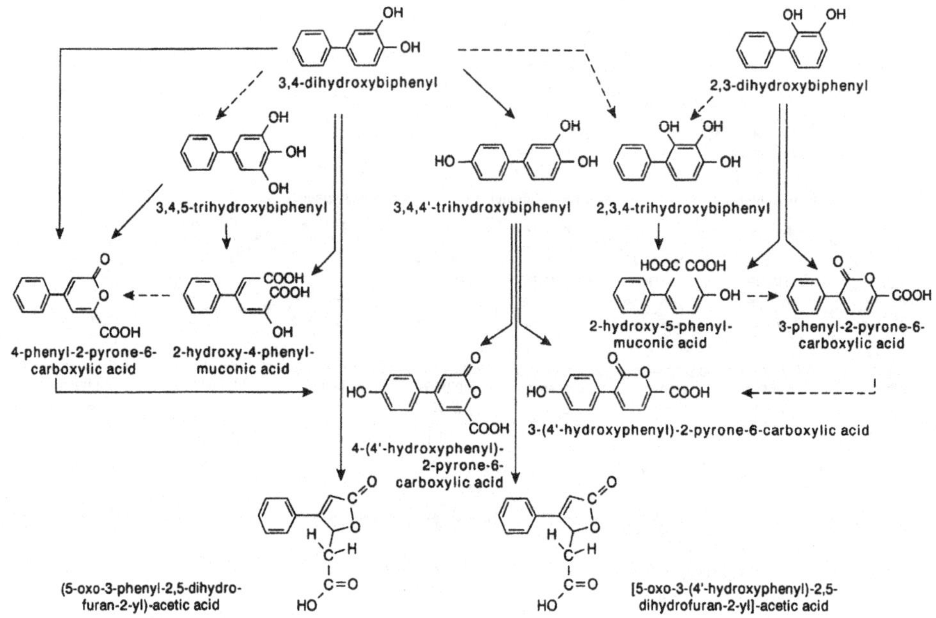

Figure 11.5. Transformation of di- and trihydroxylated biphenyls to ring cleavage products by the yeast *Trichosporon mucoides* (according to Sietmann *et al.*, 2001; Sietmann, 2002). → product formation proved; -→ product formation assumed.

syntheses of special biphenyl derivatives, fungi of the genera *Absidia* (Schwartz *et al.*, 1980), *Aspergillus* (Mobley *et al.*, 1993), *Geotrichum*, and *Saccharomyces* (Fujisawa *et al.*, 1998) have been used.

10.2. Polychlorinated Biphenyls

Polychlorinated biphenyls (PCBs) were widely used in various industrial applications in the past and persist in the environment due to their hydrophobicity and high chemical stability. Some of these chlorinated biphenyl derivatives can be transformed by fungi to form mono- and dihalogenated derivatives analogous to those produced from biphenyl. This was demonstrated for 4-chlorobiphenyl and 4,4′-dichlorobiphenyl with *Rhizopus japonicus* (Wallnöfer *et al.*, 1973), for 4-chlorobiphenyl with *P. lilacinus* (Gesell *et al.*, 2001), for 2-,3-,4-chlorobiphenyl and 2,2′-dichlorobiphenyl with *T. mucoides* (Sietmann, 2002), for a mixture of chlorinated biphenyls with *A. niger* (Dmochewitz and Ballschmiter, 1988), and for 4-fluorobiphenyl with the mycorrhizal fungus *Tylospora fibrilosa* (Green *et al.*, 1999). In *T. mucoides* dihydroxylated derivatives of monochlorinated biphenyls are also formed (Sietmann, 2002). Safe *et al.* (1976) showed that tyrosinase in the mushroom *Agaricus arvensis* is involved in the hydroxylation of 4′-chloro-4-hydroxybiphenyl. However, the product 4′-chloro-2,4-dihydroxybiphenyl could also be found in *Trichosporon* (Sietmann, 2002) and in several mammals (Safe *et al.*, 1975a, 1975b). Monochlorinated ring cleavage products were first characterized in *Paecilomyces* (Gesell *et al.*, 2001) and *Trichosporon* strains (Sietmann, 2002). During this transformation, both the unchlorinated and the chlorinated ring system can be cleaved if the chlorine atom is located at the 2 position, but no oxidation products were found in *T. mucoides* with 4,4′-, 3,4-, 2,4′-, or 2,3-dichlorobiphenyl or with 2,4,4′-trichloro- and 2,2′,3,5′-tetrachlorobiphenyl (Sietmann, 2002). Thus, as also shown for aerobic bacteria, an increasing number of chlorine atoms in the molecule diminishes its susceptibility to biotransformation by fungi (Thomas *et al.*, 1992; Zeddel *et al.*, 1993; Kubatova *et al.*, 2001).

For the white-rot fungus *P. chrysosporium*, a degradation of PCBs was reported for 4,4′-dichlorobiphenyl (Dietrich *et al.*, 1995), 2,2′,4,4′-tetrachlorobiphenyl (Thomas *et al.*, 1992), 3,3′4,4′-tetrachlorobiphenyl (Vyas *et al.*, 1994a), and several mixtures of PCBs ranging from four to seven chlorine atoms per biphenyl molecule (Eaton, 1985; Bumpus and Aust, 1987; Viney and Bewly, 1990; Yadav *et al.*, 1995; Krcmar and Ulrich, 1998; Fernandez-Sanchez *et al.*, 2001). However, in the absence of chemical identification of the metabolites formed, other explanations for the observed results cannot be excluded. In the few studies where metabolites were analyzed, chlorobenzoate, a degradation intermediate typical for bacteria, was sometimes found, for example, in cultures of *P. chrysosporium* and in *A. niger* (Dmochewitz and Ballschmiter, 1988; Dietrich *et al.*, 1995). In other cases the degradation rates showed no significant differences in the decrease of di-, tri-, tetra-, penta-, hexa-, hepta- or octachlorinated congeners (Krcmar and Ulrich, 1998; Krcmar *et al.*, 1999).

Zeddel *et al.* (1993, 1994) demonstrated that *P. ostreatus* or *T. versicolor* are very suitable for the degradation of tri- or tetrachlorinated PCBs; after 5 weeks of incubation with these organisms in a wood chip/soil system, 95% of PCBs were no longer detectable. However, 2,2′,4,4′,5,5′-hexachlorobiphenyl was resistant to degradation, and *P. chrysosporium* was able to degrade only mono- and dichlorinated biphenyls under these conditions. Also in comparison with *Bjerkandera adusta, P. ostreatus*, and *T. versicolor, P. chrysosporium* gave the lowest mineralization and biodegradation rates with di- to hexachlorobiphenyls (Beaudette *et al.*, 1998). In contaminated soil, *P. ostreatus* showed degradation rates with a mixture of chlorinated PCBs that were better than those seen with *P. chrysosporium* or *T. versicolor* (Kubatova *et al.*, 2001). The release of CO_2 from 2,4′,5-trichlorobiphenyl by *T. versicolor* could be stimulated by the presence of the detergent Triton X-100 (Beaudette *et al.*, 2000).

11. Oxidation of Dibenzofurans and Dibenzodioxins

Dibenzofuran is a constituent of coal tar and also occurs in Late Permian sediments (Sephton *et al.*, 1999). Derivatives of dibenzofuran have been found in nature as biologically active substances of streptomycetes (Iwatsuki *et al.*, 1993; Kato *et al.*, 1993), in fungal pigments (Read and Vining, 1959; Gripenberg, 1960; Turner and Aldridge, 1983; Gill and Steglich, 1987), and in lichens (Wachtmeister, 1956; Turner and Aldridge, 1983; Stenroos, 1989; Miyagawa *et al.*, 1993; Kon *et al.*, 1997a, 1997b). The lichen metabolite, usnic acid, is known as an antibiotic and a selective herbicide (Cocchietto *et al.*, 2002). Finally, plants make dibenzofuran phytoalexins in response to fungal attack as part of their defense against fungal invasion (Kokubun *et al.*, 1995b, c, d; Hrazdina *et al.*, 1997; Borejsza-Wysocki *et al.*, 1999).

Particular attention has been given to dibenzofurans in the form of toxic highly chlorinated derivatives, which have been identified as trace contaminants in commercial preparations of PCBs, pentachlorophenol, and hexachlorobenzene and as combustion products of several chlorinated chemical substances (Onodera and Saitoh, 1997). Although polychlorinated dibenzofurans (PCDFs) have not been commercially produced and efforts have been made to reduce chlorinated dibenzofuran and dioxin emissions from the combustion of dust or bleaching of pulp in paper mills, these compounds remain widespread environmental contaminants. In light of their strong toxicity several attempts were made over the past two decades to biodegrade these substances, and much information has accumulated on the activity of bacteria (Armengaud and Timmis, 1997; Wittich, 1998a; Nojiri *et al.*, 2001; Stope *et al.*, 2002). Fungal oxidation and biotransformation of dibenzofuran was first shown by Cerniglia *et al.* (1979) for *Cunninghamella elegans*, which hydroxylates it to 2- and 3-hydroxydibenzofuran as well as 2,3-dihydroxydibenzofuran. It was later demonstrated that filamentous fungi from several genera, including *Circinella, Mortierella, Mucor, Acremonium, Chrysosporium, Fusarium, Paecilomyces, Papulaspora, Penicillium*, and *Trichoderma*, are able to hydroxylate dibenzofuran, forming 1-, 2-, 3-, or 4-hydroxybenzofuran and some other more hydrophilic products (Hammer and Schauer, 1997). Further oxidation of monohydroxylated dibenzofuran via 2,3-dihydroxydibenzofuran leading to a ring cleavage wherein one aromatic ring system is opened to a hydroxymuconic acid structure was shown for the yeast *T. mucoides* (Hammer *et al.*, 1998). Alternatively, the filamentous fungus *Penicillium canescens*, unable to form dihydroxylated dibenzofurans, transforms monohydroxylated derivatives (2-, 3-, or 4-hydroxydibenzofuran) to their corresponding glucoside conjugates (Hammer *et al.*, 2001). Thus the filamentous fungus tends to biotransform the xenobiotic as would a mammalian cell; the monohydroxylated derivatives (phase I reaction) are transformed to the glucosyl conjugates, probably by a highly active UDP-glucosyltransferase (phase II reaction). This path overcomes the toxic effects of hydroxylated derivatives and increases the water solubility of the xenobiotic. In white-rot fungi, ligninolytic enzymes provide a third possibility: hydroxylated dibenzofurans are detoxified by laccase-mediated oligomerization (Jonas *et al.*, 1998). During this process different radicals can be formed (Figure 11.6), causing dimeric and trimeric products to form (Figure 11.7).

There are few reports of dioxin degradation by fungi. Dibenzodioxins share with dibenzofurans the properties of water insolubility, persistence in the environment, and bioaccumulation. Like dibenzofurans they are formed as by-products during synthesis of chlorinated chemicals (chlorophenols, chlorobenzens, and herbicides) or as unwanted and toxic products during combustion of PCBs, polyvinylchloride, or pesticide-treated wastes (Hutzinger and Blumich, 1985; Arthur and Frea, 1989). Only a few bacteria have been found that can grow on unchlorinated dibenzodioxin as the sole carbon source (Fortnagel *et al.*, 1989; Wittich *et al.*, 1992; Ishiguro *et al.*, 2000). The highly halogenated derivatives seem to be completely resistant to degradation by most aerobic bacteria (Wittich, 1998b), although a slow dehalogenation can be achieved by some anaerobic bacteria (Adriaens and Grbic-Galić, 1994; Beurskens *et al.*, 1995; Ballerstedt *et al.*, 1997; Bunge *et al.*, 2001). Therefore the search for other types of microorganisms able to transform toxic halogenated dioxins, such as 2,3,7,8-tetrachlorodibenzo-*p*-dioxin, to products of lower toxicity seems to be a potentially rewarding task.

Some fungi are able to hydroxylate dibenzo-*p*-dioxin. *Trichosporon domesticum* produced one hydroxylation product (Henning, 1993) and *T. mucoides* made three (Sietmann, 2002). However, no ring cleavage products have hitherto been detected with yeasts or

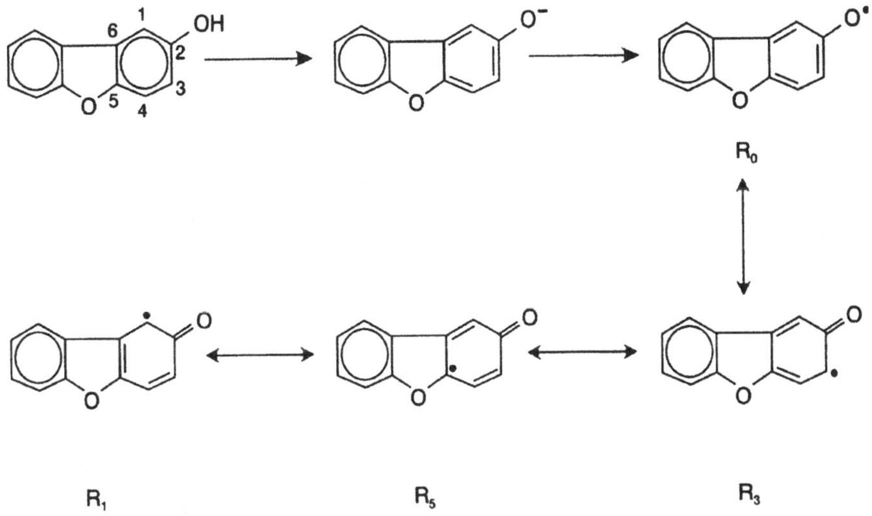

Figure 11.6. Proposed reaction mechanism for the formation of radicals of 2-hydroxydibenzofuran by laccase; R_0, R_1, R_3, and R_5: radicals formed at the free oxygen and carbon 1, 3, and 5, respectively (according to Jonas *et al.*, 1998).

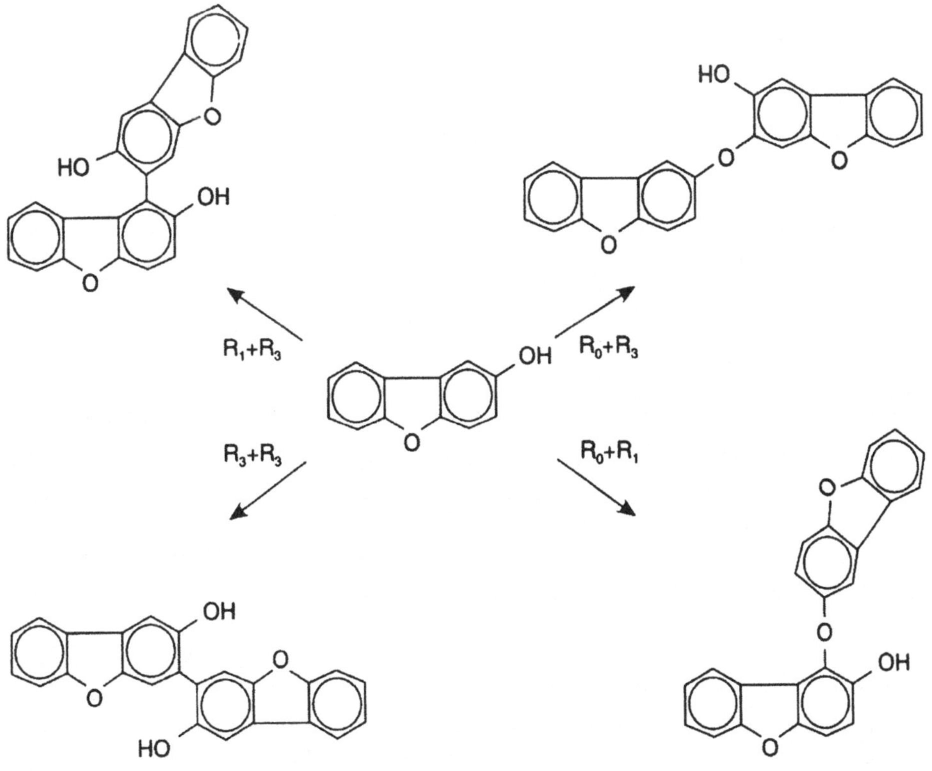

Figure 11.7. Possible structure of dimers formed during oligomerization of 2-hydroxydibenzofuran by a laccase from *Pycnoporus cinnabarinus* (based on Jonas *et al.*, 1998).

molds. Hammel *et al.* (1986) have shown that the ligninase of *P. chrysosporium* in the presence of H_2O_2 produces a dibenzo-*p*-dioxin cation radical, as measured by ultraviolet (UV)-absorption changes and by direct electron spin resonance (ESR) spectrometry. The expected initial product, dibenzo-*p*-dioxin-2,3-dione, however, could not be detected by GC/MS.

Rosenbrock *et al.* (1997) investigated the mineralization rate of [U-^{14}C]dibenzo-*p*-dioxin in soils contaminated with polychlorinated dibenzo-*p*-dioxins and dibenzofurans (PCDD/F). In soils deficient in organic matter, the mineralization rate could be increased from about 20% in 70 days (endogenous dibenzo-*p*-dioxin mineralization) to rates of 28–36% by amending the soil with compost/humus to enhance its organic content or to 50% by inoculating it with ligninolytic white-rot fungi of the genera *Phanerochaete, Pleurotus,* or *Dichomitus*. According to Tanaka *et al.* (1992) *P. sordida* seems able to degrade a mixture of tetra- to octachlorodibenzo-*p*-dioxins and dibenzofurans (chlorination in 2-,3-,7-, and 8-positions) to 4,5-dichlorocatechol and tetrachlorocatechol, respectively (Takada *et al.*, 1996). Unfortunately, these results could not be confirmed as yet by other workers. Production of chlorinated dibenzo-*p*-dioxins from chlorinated monoaromatic precursors by dimerization can be achieved with microorganisms, especially white-rot fungi, or with peroxidases of various origins (Svenson *et al.*, 1989; Öberg *et al.*, 1990; Wagner *et al.*, 1990; Jonas *et al.*, 1998).

12. Biotransformation of Diphenyl Ethers and Phenoxy Herbicides

Diphenyl ethers are constituents of humic substances and of coal, and as methoxylated forms, they serve as lignin precursors. Some fungi produce diphenyl ether derivatives as secondary metabolites (Takahashi *et al.*, 1992, 1993; Adeboya *et al.*, 1996; Hargreaves *et al.*, 2002). Additionally, diphenyl ethers are produced industrially for various applications. Diphenyl ethers and especially their halogenated derivatives have become increasingly significant environmental pollutants that are highly resistant to microbial attack. This is particularly troublesome because chlorinated diphenyl ethers serve as the basic structure for herbicides such as nitrofen (2,4-dichlorophenyl-4′-nitrophenylether) and bifenox [5-(2,4-dichlorophenoxy)-2-nitrobenzoic acid], and are formed as by-products during the synthesis of chlorinated phenols and phenoxy acids (Ahlborg and Thunberg, 1980). Brominated diphenyl ethers are used as flame retardants in electronic equipment. Because of their high persistence in the environment, increasing concentrations are being detected in sea sediments, sewage sludge, and animal tissue (Pijnenburg *et al.*, 1995).

A variety of biotransformation reactions have been characterized for fungi, and are summarized in Table 11.3. After hydroxylation, several ring cleavage products may be formed, such as 2-hydroxy-4-phenoxymuconic acid, 2-hydroxy-4-(4-hydroxyphenoxy)muconic acid, and 6-carboxy-4-phenoxy-2-pyrone (Figure 11.8). The fungus *T. versicolor* is able to cleave the unhalogenated ring of 4-bromo- and 4-chlorodiphenyl ether. In addition to hydroxylated and ring cleavage products, 4-bromo- and 4-chlorophenol were detected as monoaromatic intermediates. Furthermore, in 2-bromo- and 2-chlorodiphenyl ether as well as in 2,4-dichlorodiphenyl ether, the unchlorinated ring system was cleaved, but in 2,4′-dichlorodiphenyl ether and in 2′,3,4-trichlorodiphenyl ether cleavage of the chlorinated

Table 11.3. Range of Biotransformation Reactions Provided by Fungi[a]

Diphenyl ether substrate	Transformation	Fungus	Reference
Diphenyl ether	4-Hydroxylation	*Cunninghamella echinulata*	Seigle-Murandi *et al.*, 1991
	Hydroxylation and ring cleavage	*Trichosporon domesticum*	Schauer *et al.*, 1995
		Trametes versicolor	Hundt *et al.*, 1999
		Paecilomyces lilacinus	Gesell, 2001
		Trichosporon (13 species)	Sietmann *et al.*, 2002
		Cryptococcus curvatus	Sietmann *et al.*, 2002
2-Hydroxydiphenyl ether	Dimerization	*Pycnoporus cinnabarinus*	Jonas *et al.*, 2000
Triclosan[b]	Xylosylation	*T. versicolor*	Hundt *et al.*, 2000
	Glucosylation	*T. versicolor*	Hundt *et al.*, 2000
		P. cinnabarinus	Hundt *et al.*, 2000
	Methylation	*P. cinnabarinus*	Hundt *et al.*, 2000
Chloronitrofen[c]	Hydroxylation	*T. versicolor*	Hiratsuka *et al.*, 2001
	Nitroreduction	*T. versicolor*	Hiratsuka *et al.*, 2001
	Oxidative dehalogenation	*T. versicolor*	Hiratsuka *et al.*, 2001
	Reductive dehalogenation	*T. versicolor*	Hiratsuka *et al.*, 2001
2',3,4-Trichlorodiphenyl ether	Hydroxylation and ring cleavage	*T. versicolor*	Hundt, 2001

[a]Examples for diphenyl ethers and some of their derivatives are shown.
[b]Triclosan: 2,4,4'-trichloro-2'-hydroxydiphenyl ether.
[c]Chloronitrofen: 2,4,6-trichloro-4'-nitrodiphenyl ether.

ring could be observed. 2-Chlorodiphenyl ether was also possibly susceptible to cleavage of its chlorinated ring.

4,4′-Dibromo- and 4,4′dichlorodiphenyl ether were only hydroxylated (without ring cleavage), and for 2,2′,4,4′-tetrabromo- and 3,3′,4,4′-tetrachlorodiphenyl ether no transformation products could be observed at all (Hundt, 2001). All the biotransformation reactions described for diphenyl ether seem to be realized by *T. versicolor* without the participation of its ligninolytic enzyme system, because (1) the hydroxylation reactions could be inhibited by P-450-inhibitors, (2) they are independent of the rate of ligninolytic enzyme expression, and (3) yeasts of the genus *Trichosporon*, which do not have ligninolytic enzymes, perform the same reactions.

The chlorinated diphenyl ether nitrofen can be transformed by *T. versicolor* via hydroxylation, nitroreduction, or oxidative and reductive dechlorination (Figure 11.9).

Whereas the monoaromatic phenoxymuconic acid and pyrone structures formed during biotransformation of diphenyl ethers seem not to support the growth of the fungi investigated for diphenyl ether biotransformation, a few other fungi can grow on some phenoxy acid derivatives (Lottmann *et al.*, 1999). Of particular interest is the ability of soil fungi to degrade phenoxy acid herbicides such as 2,4-dichlorophenoxy acetic acid (2,4-D), 2,4,5-trichlorophenoxy acetic acid (2,4,5-T), and 4-chloro-2-methylphenoxy acetic acid (MCPA) and their esters or amines. These herbicides can cause serious health effects in animals or humans (Inomata *et al.*, 1991; Smith and Christophers, 1992; Venkov *et al.*, 2000). Effects on fungi which show only a certain degree of bioaccumulation of the herbicides (Benoit *et al.*, 1998) are small

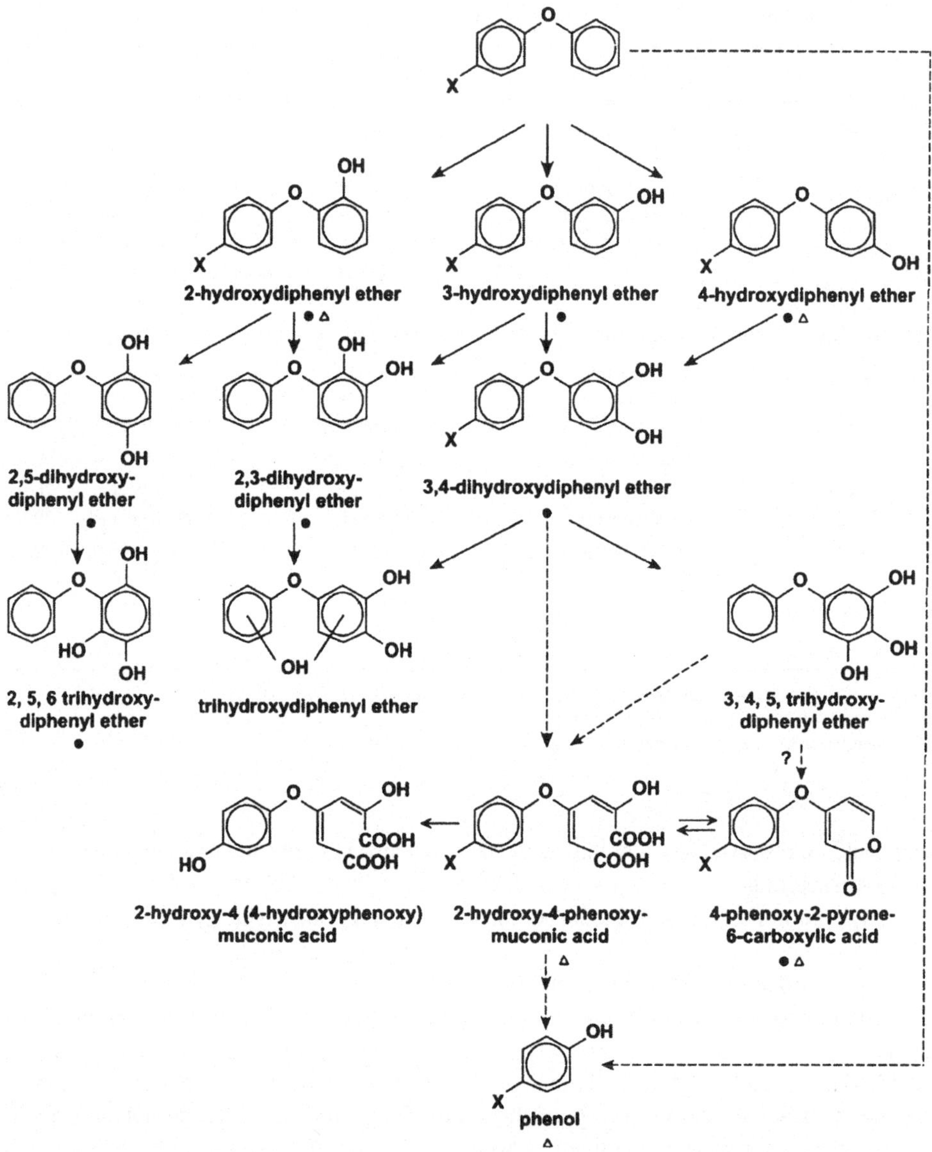

Figure 11.8. Transformation of diphenyl ether by fungi. •, products accumulated by *Trichosporon domesticum* SBUG 752 (Henning, 1993; Schauer *et al.*, 1995); Δ, products accumulated by *Trametes versicolor* (Hundt *et al.*, 1999); *, products accumulated by *Paecilomyces lilacinus* SBUG-M 1093 (Gesell, 2001). X = H, Cl, or Br for *T. versicolor*; X = H for *T. domesticum* and *P. lilacinus* (other substrates not tested). --▸ product formation assumed; → product formation proved.

(Sherry, 1994; Rothemund *et al.*, 1996): only the impairment of spore germination was found (Michailides and Spotts, 1991). *A. niger* seems unable to cleave the ether linkage of 2,4-D to produce the corresponding phenols (Shailubhai *et al.*, 1983) or to dehalogenate chlorinated derivatives of phenoxyacetic acid (Sahasrabudhe *et al.*, 1987). The ligninolytic fungus

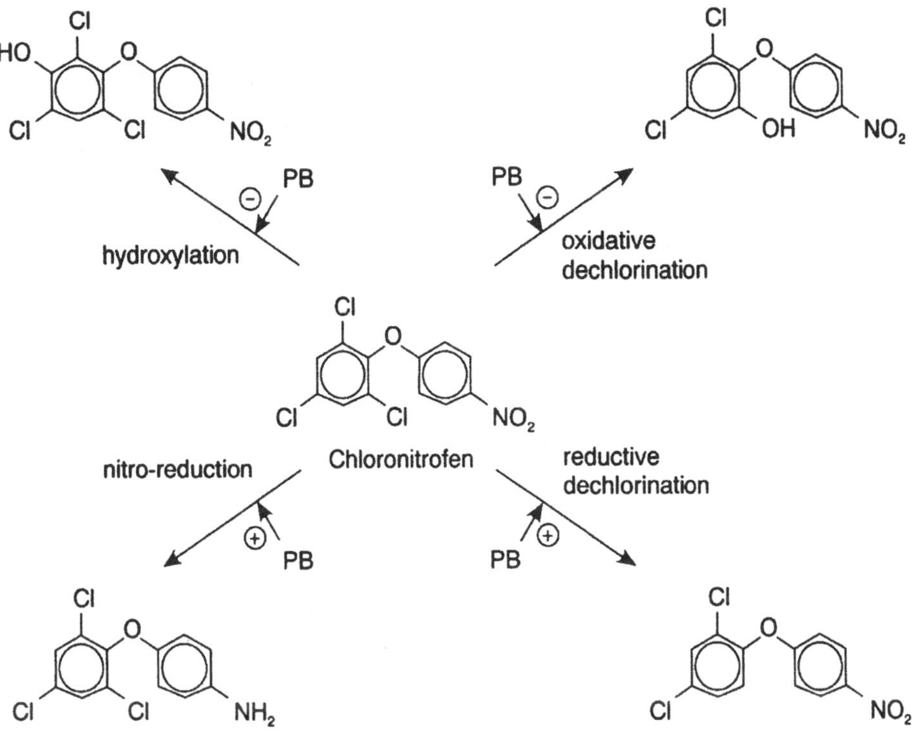

Figure 11.9. Transformation reactions of the herbicide chloronitrofen by *Trametes versicolor* (according to Hiratsuka *et al.*, 2001). (−) Inhibition or (+) stimulation of the reactions by the cytochrome P-450 inhibitor piperonyl butoxide (PB).

D. squalens, like *P. chrysosporium*, initially cleaves the side chain by a mechanism not involving the lignin degradation system; the resulting chlorophenol intermediate can be xylosylated or oxidatively dechlorinated to 2-chloro-*p*-benzoquinone that is susceptible to further dechlorination and ring opening (Reddy *et al.*, 1997).

The addition of wood chips inoculated with *P. chrysosporium* or *Trappea darkeri* can increase the rate of degradation of 2,4-D in forest soils (Entry *et al.*, 1996) and in a solid substrate fermentation system. *P. chrysosporium* grown on straw degrades 75% of MCPA within 20 days (Castillo *et al.*, 2001). Degradation of 2,4-D was observed in soil with several mycorrhizal fungi (Donnelly and Fletcher, 1994; Donnelly *et al.*, 1993), and its removal from an aqueous cultivation medium was shown with *C. elegans, Cunninghamella echinulata, R. solani*, and *Verticillium leccani* (Vroumsia *et al.*, 1999). In comparison with bacteria that degrade 2,4-D, fungi seem to be more active only in dry soil (Han and New, 1994).

13. Dehalogenation of Aromatic Xenobiotics

No doubt exists about the importance of fungi for dechlorination of chlorinated aromatic compounds. Fungi are known to dechlorinate several monoaromatic compounds such as 3- and 4-chlorophenol or their metabolites (*C. maltosa*; Polnisch *et al.*, 1992),

2,4-dichlorophenol (*P. chrysosporium*; Valli and Gold, 1991), 2,4,6-trichlorophenol (*P. chrysosporium* [Reddy *et al.*, 1998]; *T. versicolor* and *Panus tigrinus* [Leontievsky *et al.*, 2000, 2001]), 2,4,5-trichlorophenoxyacetic acid (*D. squalens*; Reddy *et al.*, 1997), or pentachlorophenol (*P. chrysosporium* [Lamar *et al.*, 1990; Milewski *et al.*, 1988]; *L. edodes* [Okeke *et al.*, 1997]; *Cerrena unicolor* and others [Cho *et al.*, 2001]). In comparison with detailed studies on the dechlorination of monoaromatics, descriptions of dehalogenation reactions of biarylic compounds are rare. Although the loss of chlorinated organic molecules during degradation of halogenated biphenyls (PCBs) by white-rot fungi was observed, clearly defined dechlorinated metabolites could only be identified if the substrates were hydroxylated chlorobiphenyls. For example, 3-chloro-2-hydroxybiphenyl was transformed by the yeast *T. mucoides* to 3-chloro-2,4′-dihydroxybiphenyl and the unchlorinated 2,5-dihydroxybiphenyl (Sietmann, 2002), and 5-chloro-2-hydroxybiphenyl was transformed by the laccase of *T. versicolor* to an unchlorinated dimerization product (Schultz *et al.*, 2001). Different mechanisms have been discussed for these dehalogenations, including (1) the cycloisomerization of chlorinated ring cleavage products (Polnisch *et al.*, 1992), (2) the involvement of LiP (Hammel and Tardone, 1988; Reddy *et al.*, 1998), MnP (Reddy *et al.*, 1998; Leontievsky *et al.*, 2000), or laccase (Leontievsky *et al.*, 2000, 2001; Schultz *et al.*, 2001) or (3) radical formation by cellobiose dehydrogenase (Cameron and Aust, 1999). For glutathione-dependent dehalogenation of dichloroacetic acid to glyoxylic acid the participation of a zeta-class glutathione transferase (present in plants, fungi and animals) was assumed (Dixon *et al.*, 2000). Also, a theta-class glutathione S-transferase (occurring in various organisms) has a dehalogenase activity toward several halogenated compounds, such as dichloromethane (Landi, 2000).

The involvement of peroxidase and laccase in the mechanism of dehalogenation is illustrated in several studies. The action of LiP or MnP dechlorinates 2,4,6-trichlorophenol via 2,6-dichloro-1,4-benzoquinone, 2,6-dichloro-1,4-dihydroxybenzene, and 2-chloro-1,4-dihydroxybenzene to 1,4-hydroquinone or to 5-chloro-1,2,4-trihydroxybenzene and 1,2,4-trihydroxybenzene (Reddy *et al.*, 1998; Figure 11.10). Soluble and immobilized laccase degrades 2,4,6-trichlorophenol to 2,6-dichloro-1,4-hydroquinone and 3,5-dichlorocatechol (Leontievsky *et al.*, 2001). An alternative dechlorination mechanism for laccase was described by Schultz *et al.* (2001), who found the loss of chlorine during the dimerization of 3-chloro-4-hydroxybiphenyl to yield 3,3′-di(4-hydroxybiphenyl) (Figure 11.11).

14. Trends and Future Developments

We now turn our attention to some recent and future developments for improving fungal enzymes and adapting them to function optimally under the conditions where they would be commercially useful. These recent trends stem from further exploitation of biodiversity in the fungal world including basidiomycetes, progress in fungal expression and cloning systems, application of directed evolution systems, and rational protein design to tailor enzymes for specific industrial applications. Some of these aspects have been extensively discussed in the Chapters 9 and 10 and will only be briefly mentioned here. Considering the recent trends overall, it seems that the environmental benefits of fungal enzyme technology will drive future developments. Examples of such goals already cited in this chapter are the avoidance or reduction environmental pollution by the application

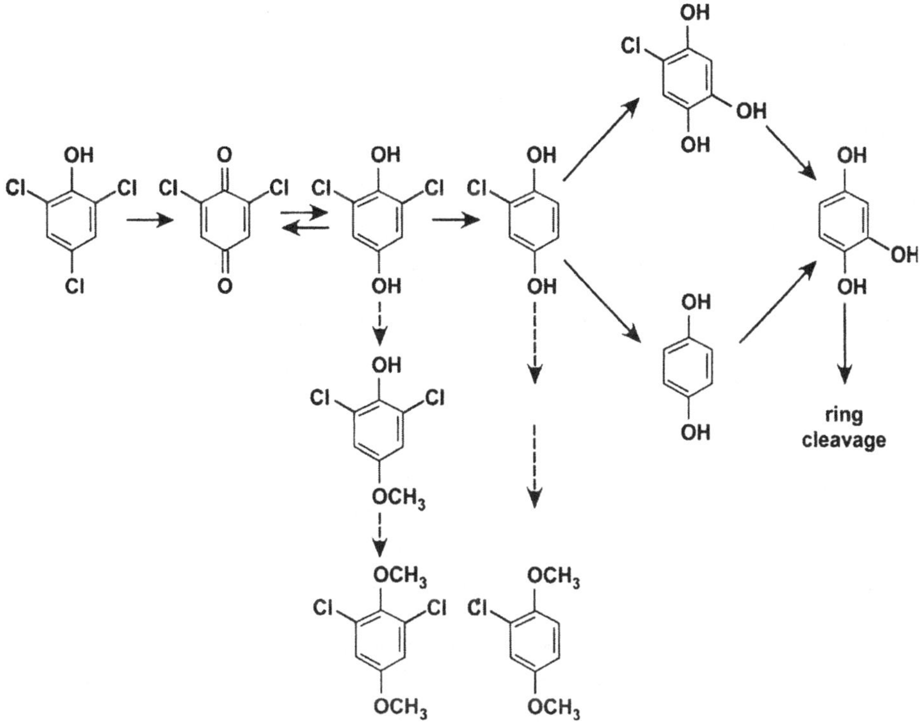

Figure 11.10. Dechlorination of 2,4,6-trichlorophenol by ligninolytic enzymes from *Phanerochaete chrysosporium* (according to Reddy *et al.*, 1998).

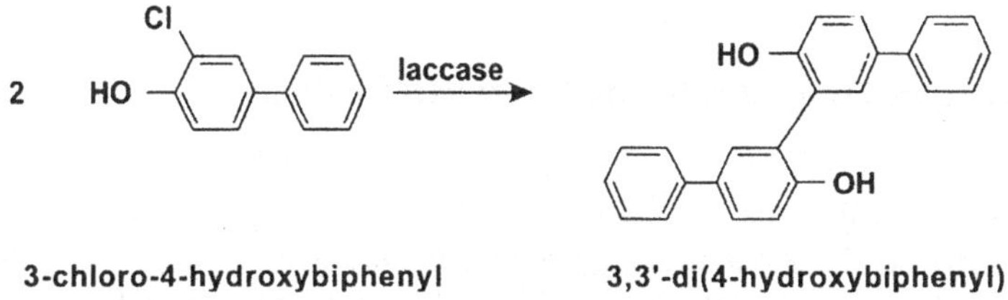

3-chloro-4-hydroxybiphenyl

3,3'-di(4-hydroxybiphenyl)

Figure 11.11. Dechlorination of 3-chloro-4-hydroxybiphenyl by laccase (according to Schultz *et al.*, 2001).

of improved variants of fungal phytases and xylanases and the use of *Trichoderma* strains as fungal antagonists for the biocontrol of plant pathogens.

14.1. Novel Fungal Enzymes: Screening, Development, and Specific Features

One of the most efficient and successful means of finding enzymes with novel activities or new representatives with known activities but perhaps with improved characteristics is to

Table 11.4. Novel Enzymes Found through Screening of Fungal Isolates

Product	Enzyme	Origin	Reference
Ethyl *R*-4-chloro-3-hydroxybutanoate	Aldehyde reductase	*Sporobolomyces salmonicolor*	Ogawa and Shimizu, 1999
	Aldehyde reductase	*Candida magnoliae*	Ogawa and Shimizu, 1999
D-Pantoyl lactone	Carbonyl reductase	*Candida parapsilosis*	Ogawa and Shimizu, 1999
D-Pantoic acid	Lactonase	*Fusarium oxysporum*	Ogawa and Shimizu, 1999
Adenosylmethionine	Adenosylmethionine synthetase	*Saccharomyces sake*	Ogawa and Shimizu, 1999
Phosphate mobilization	Thermostable phytase	*Thermomyces lanuginosus*	Berka *et al.*, 1998
Lignin degradation	Thermostable laccase	*T. lanuginosus*	Berka *et al.*, 1997

screen large numbers of microorganisms, because this group of organisms is biologically diverse and metabolically versatile. Today initial screening can be based either upon whole cell processes, the enzyme activity itself or, in some cases, upon DNA homology. New enzymes can be useful catalysts for the synthesis of many biologically and chemically interesting compounds (Table 11.4).

Several properties of an enzyme may need to be altered to make its application in a given industrial setting economically favorable. The substrates in many industrial processes are synthetic compounds, and enzymes able to perform the desired transformations with them may still be unknown. Therefore, screening for novel enzymes that are capable of catalyzing novel reactions is, and will continue to be, an integral part of enzyme technology. The classical screening philosophy based on microbial diversity is still the most sound approach for finding novel enzymes, and their discovery will provide insight for designing new enzymatic processes (Ogawa and Shimizu, 1999).

14.2. Screening of Fungi Producing Improved Phytases

Since the application of fungal phytase in the formulation of animal feeds, the protein has become one of the most important commercial enzymes. Despite the growing use of *Aspergillus* phytase, there is a pressing need for second-generation phytases displaying improvements such as enhanced thermostability and catalytic activity and being amenable to production at commercially viable levels. Thermal stability is relevant in animal feed because the enzyme is normally incorporated into the grain before the mixture is heated to temperatures of 85–90°C for processing into pellets. Following this step the commercial phytase must be able to function effectively at 37°C in the animal's digestive tract. A screening program examining thermophilic fungi that included gene cloning and expression was successfully applied for this goal. The strategy led to the cloning of PhyA homologs encoding thermostable extracellular phytases from thermophilic fungi such as *A. fumigatus* (Pasamontes *et al.*, 1997b), *Talaromyces thermophilus*, and *T. lanuginosus* (Pasamontes *et al.*, 1997a). The *Thermomyces* phytase expressed in *Fusarium venenatum* displays a high catalytic activity at 37°C, releasing 110 μmol of inorganic phosphate per min per mg protein, and a thermostability superior to that of any known fungal phytase (Berka *et al.*, 1998).

14.3. Diversity of Microbial Enzymes Catalyzing Stereoselective Reactions

Biological reduction of carbonyl compounds presents a good example of the diversity of microbial activity. Prochiral compounds are transformed by stereoselective reactions into optically pure alcohols. Ketopantoyl lactone, for example, can be converted to pantoyl lactone by several yeasts, those in the genera *Candida, Hansenula*, and *Sporobolomyces* producing predominantly ethyl-D-pantothenate and those in the genera *Saccharomyces, Pichia*, and *Rhodotorula* yielding the L enantiomer.

The reduction of 4-chloroacetoacetate ethyl ester to 4-chloro-3-hydroxybutanate ethyl ester is also usually stereospecific. *Candida magnoliae* has high levels of an enzyme that generates the *S* enantiomer, whereas the enzyme in *Sporobolomyces salmonicolor* produces predominantly the *R* enantiomer (Ogawa and Shimizu, 1999).

14.4. Lactonase in D-Pantothenic Acid Production

A novel lactone hydrolyzing enzyme, lactonase, that can stereospecifically hydrolyze pantoyl lactone was isolated from the filamentous fungus *F. oxysporum*. This enzyme will be useful for the optical resolution necessary to obtain D-pantoyl lactone, an intermediate in the commercial production of D-pantothenic acid. As the lactonase reaction involves intermolecular ester-bond hydrolysis to D-pantoic acid and L-pantolactone, the pantoyl-lactone substrate need not be modified for the resolution, making the process much simpler than a chemical resolution.

A simple and convenient assay system is generally considered to be essential in the screening of a large number of microorganisms. For example, when screening for a novel lactonase, particular attention was paid to the specific characteristics of the product, pantoic acid; the accumulation of pantoic acid lowers the pH of the reaction mixture, and a drop in pH could be an indicator of lactonase activity. In fact, it was possible to isolate a superior strain (*F. oxysporum* AKU3702) through an assay based simply upon pH (Ogawa and Shimizu, 1999).

14.5. Aldehyde Reductase in the Production of Chiral Alcohols

Optically active *R*- and *S*-4-chloro-3-hydroxybutanate ethyl esters (*R*- and *S*-CHBE) are promising chiral building blocks for the preparation of a variety of optically active compounds such as L-carnitine (*R*-form), 3-hydroxy-3-methylglutaryl (HMG)-CoA reductase inhibitors, and 1,4-dihydropyridine β-blockers (*S*-form). *S. salmonicolor* produces *R*-CHBE with molar conversion rates through asymmetric reduction of the prochiral carbonyl compound, 4-chloroaceto-acetate ethyl ester. The aldehyde reductase involved resembles a member of the mammalian aldo-keto-reductase superfamily. Similarly, the homologous enzyme from *C. magnoliae* is a useful catalyst for the production of *S*-CHBE (Ogawa and Shimizu, 1999).

14.6. Laccase-Catalyzed Heteromolecular Coupling of Molecules

Heteromolecular coupling reactions (cross-coupling) mediated by laccase can be used to synthesize various chemical substances, such as substituted benzothiadizine-8-ones

(Bhalerao *et al.*, 1994), or to derivatize hydrocaffeic acid by N-coupling of aromatic and aliphatic amines (Figure 11.3) (Mikolasch *et al.*, 2002). Similar coupling reactions may be used for rapid derivatization of various bioactive compounds and drugs under mild conditions (Lindequist and Schauer, 2002). The synthesis of new antibiotics by laccase has been achieved by Schauer *et al.* (2001). In suitable bioreactors, heteromolecular laccase-catalyzed coupling processes can be optimized (Pilz *et al.*, 2003). Entrapment of laccase in reversed micelles allows oxidative and biotransformation reactions to be conducted in various organic solvents (Michizoe *et al.*, 2001).

14.7. Heterologous Expression of Fungal Ligninolytic Enzymes

Most wood-rotting fungi grow very slowly in liquid cultures, but this limitation for the production of ligninolytic enzymes can be circumvented by expressing the genes encoding these enzymes in microorganisms adapted for rapid growth in production fermentors. *P. cinnabarinus* laccase gene (lac1) was overexpressed in *Pichia pastoris* (Otterbein *et al.*, 2000) and in *A. niger*, a host often used to produce large quantities of industrial enzymes (Record *et al.*, 2002). Laccase genes from the filamentous fungi *T. versicolor* and *T. sanguinea* were expressed in the yeasts *S. cerevisiae* (Cassland and Jonsson, 1999; Hoshida *et al.*, 2001) and *P. pastoris* (Jonsson *et al.*, 1997; Brown *et al.*, 2002; O'Callaghan *et al.*, 2002). The production of recombinant laccase in *P. pastoris* was improved by optimization of cultivation condition and with methanol as the carbon source (Hong *et al.*, 2002). Additionally, transgenic plants can be used for laccase production (Hood, 2002). Efforts on heterologous expression of lignin-degrading enzymes have been summarized by Cullen (1997).

14.8. Impact of DNA Recombinant Techniques

At present both homologous and heterologous enzymes are produced commercially in fungi. However, the number of heterologous enzymes approved for industrial applications is expected to increase significantly in the near term. One reason for this is that filamentous fungi, which have been used for more than a half century in the production of industrial enzymes, secrete several of their own enzymes simultaneously, and the traditional products have provided the desired activity together with a number of others (although as we have seen with macerating enzymes, a medley of related activities can be useful in certain applications). The second reason for a rise in the production of heterologous enzymes has to do with the difficulty of achieving economically viable production of homologous proteins by traditional strain improvement strategies.

Acordingly more and more emphasis is being placed on the development of recombinant DNA technologies for production organisms. Efficient gene expression systems are needed to achieve this goal. Further progress can be expected by the introduction of new host systems with improved capacities for enzyme secretion. To date, the *cbhl* promoter directing expression of cellobiohydrolase 1 in *T. reesei* seems to be one of the strongest eukaryotic promoters; constructs with this promoter have been used to produce several heterologous enzymes in milligram and gram quantities (Penttila, 1998). Examples are the lignin peroxidase and laccase from *Phlebia radiata*, GA from *Hormoconis resinae*, the phytase and acid phosphatase from *A. niger*, and the endochitinase from *T. harzianum* (Bhat, 2000).

Table 11.5. Recombinant Enzymes, Derived from a Fungal Host, Approved for Use in Food (after Archer, 2000)

Enzyme	Host	Donor
Catalase	*Aspergillus niger*	*Aspergillus* sp.
Cellulase	*Aspergillus oryzae*	*Humicola* sp.
Cellulase	*Trichoderma reesei (longibrachiatum)*	*Trichoderma* sp.
Chymosin	*A. niger* var. awamori	Calf
α-Galactosidase	*A. oryzae*	*Aspergillus* sp.
β-Glucanase	*T. reesei (longibrachiatum)*	*Trichoderma* sp.
Glucose oxidase	*A. niger*	*Aspergillus* sp.
Lipase	*A. oryzae*	*Candida* sp.
		Rhizomucor sp.
		Thermomyces sp.
Phytase	*A. niger A. oryzae*	*Aspergillus niger* var. ficuum
Protease	*A. oryzae*	*Rhizomucor* sp.
Xylanase	*A. niger* (and var. awamori)	*Aspergillus* sp.
	A. oryzae	*Aspergillus* sp.
		Thermomyces sp.
	T. reesei (longibrachiatum)	*Trichoderma* sp.

The further enhancement of fungal genetic transformation systems will facilitate the manipulation of production strains for efficient expression of homologous and heterologous proteins. The successful use of *Agrobacterium tumefaciens* Ti plasmid for introducing DNA into *A. awamori* and for its integration at a predetermined locus in the genome without concomitant introduction of non-fungal DNA represents a significant advancement for fungal expression systems (Gouka *et al.*, 1999; see also Chapter 4). A selection of recombinant enzymes, derived from fungal hosts and approved for use in food, is given in Table 11.5.

14.9. Expression of *Aspergillus* Phytase in Transgenic Plants

Moving beyond the addition of microbial phytase to feed, alternative strategies have been devised to improve phosphate bioavailability in animal feed and reduce the environmental burden posed by the need to fortify feed with phosphate (Brinch-Pedersen *et al.*, 2002). In the long term, transgenic approaches seem to have favorable prospects. Recently for example, pigs have been engineered to produce phytase in their salivary glands (Golovan *et al.*, 2001). Several attempts have also been made to design transgenic plants with increased phytase production in their seeds. Use of fodder plants expressing recombinant fungal phytase could largely eliminate the present problems linked with the use of feed containing most of its phosphate immobilized in phytic acid.

In a seminal study, tobacco was transformed with a DNA construct comprised of the *phyA* gene from *A. niger* controlled by the potent 35S promoter. When the transgenic seeds were tested under simulated digestive conditions, they released Pi from standard feed more efficiently than an equivalent amount of phytase added exogenously (Pen *et al.*, 1993). Subsequently PhyA was shown to be functionally expressed in soybean, oilseed rape, rice, and wheat (for review see Brinch-Pedersen *et al.*, 2002). Detailed feeding studies have been performed with transgenic soybean and canola (*Brassica napus*) expressing the *A. niger* phytase. By all parameters measured, there were significant improvements for chickens and pigs fed the transgenic diet compared to their counterparts raised on the conventional low phosphate diet; in addition, no adverse effects due to the recombinant enzyme were observed (Zhang *et al.*, 2000a, b).

In the future, attention will be focused on targeting phytase expression in specific compartments of transgenic plant cells. For example, bioengineering plants to secrete heterologous phytase from their roots might be an important strategy for mobilizing phosphate reserves in soil. *Arabidopsis* lines bioengineered to secrete an *A. niger* phytase had a 20-fold increase in root phytase activity and grew better than the wild type on medium with phytic acid as the sole phosphate source (Richardson *et al.*, 2001).

14.10. Gene Libraries

Recently, the discovery of new industrial enzymes has focused on novel microbial sources isolated from extreme environment (extremophiles). In many cases the genes encoding these interesting enzymes can be cloned without prior isolation of the donor microbe, directly from DNA isolated from environmental samples. However, heterologous expression of the novel genes often results in disappointing yields, delaying efforts to commercialize the product.

Screening gene libraries of thermophilic fungi has provided a successful alternative to avoid this disadvantage and obtain biocatalysts with high industrial potential (Berka *et al.*, 1997). One efficient approach to shorten the time needed to clone the desired gene is an expression cloning procedure, which combines the ability of *S. cerevisiae* to express heterologous genes with the utilization of sensitive and reliable enzyme assays. A library of cDNA prepared from the fungus of interest is constructed in a shuttle vector in *E. coli*. Plasmid DNA isolated from subpools of the library is introduced into yeast cells, and transformants expressing the desired gene product are detected on agar plates with the respective assay. High expression of the isolated fungal gene can be subsequently accomplished in *A. oryzae* (Dalboge, 1997).

14.11. Biomolecular Engineering

Rapid progress in genetic engineering has provided the opportunity to design, modify, and engineer enzymes and other natural biomolecules. For industrial application, the enzymes should (1) display high specificity in catalyzing the desired reaction, (2) produce high yield, and (3) be stable in the nonbiological environment of a bioprocess that might require organic solvents, elevated temperatures, and extremes of pH. Site-directed mutagenesis was the first method used to design enzymes on a rationale basis according to available structure/function information. Six mutations increasing enzyme thermostability without changing enzyme kinetic properties were introduced in *A. awamori* GA. The replacement of a very thermolabile extended loop region on the catalytic domain surface gave rise to a free energy of thermoinactivation increase of about 4 kJ/mol (Liu *et al.*, 2000).

Most lipases used in cleaning applications are activated at the oil–water interface by a conformational change in which a portion known as the lid is shifted to expose the hydrophobic binding pocket of the enzyme. Two point mutations in this lid region alter activity of the *T. lanuginosus* lipase (Lipolase). Other mutants with improved laundry performance were also obtained (Ryu and Nam, 2000).

The methodological arsenal of *in vitro* techniques enabling the generation of novel molecules with tailor-made properties has been extended tremendously in recent years. The emerging field of biomolecular engineering will make use of (1) bioinformatics, including functional genomics and proteomics, (2) protein chemistry and protein engineering, and (3) recombinant techniques including random mutation, DNA shuffling, and phage display technology (Ryu and Nam, 2000). Although most of the novel tools have been applied to improve industrially important enzymes of bacterial origin, there are also some examples for their successful application in fungal enzymes, as in case of *Aspergillus* GA and phytase and *Trichoderma* xylanase.

14.12. Concept of Directed Evolution

Molecular diversity is typically created by random mutagenesis of a target gene and/or recombination of a family of related genes. Directed evolution is a technology to generate desired enzyme variants by using a simple iterative Darwinian optimization algorithm (Petrounia and Arnold, 2000). Prerequisites for directed evolution are the availability of the gene(s) encoding the enzyme of interest, and a suitable expression and selection system. Usually random mutagenesis of the gene encoding the protein is performed by error-prone PCR, leading to a huge library of mutants. Alternatively and often more efficiently, methods based on random recombination of DNA fragments such as gene shuffling are used to combine mutations (Arnold and Volkov, 1999). A highly advanced technology introduced by Stemmer (1994) is DNA shuffling that combines repetitive cycles of mutation and recombination of DNA sequences followed by screening for improved protein variants. A variety of enzymatic properties, such as dependence of cofactors (Joo *et al.*, 1999), protein solubility and expression (Waldo *et al.*, 1999), enantioselectivity (May *et al.*, 2000), stability in unusual environments (Miyazaki *et al.*, 2000), and allosteric interactions (Legendre *et al.*, 1999) were successfully altered through various *in vitro* strategies. The evolution of novel activities has also been achieved (Altamirano *et al.*, 2000). Approaches for directed evolution rapidly access a much larger proportion of the sequence space than the more conventional mutagenesis techniques and are therefore more powerful for the systematic engineering of catalytic properties. However, their success depends heavily on the choice of appropriate environment and on efficient screening assays that are never available for every enzyme or every catalytic property. For that reason it is not surprising that most efforts to alter enzymatic properties have been concentrated on bacterial enzymes, and significant examples for improving fungal enzymes by directed evolution are rare. Further progress in developing these techniques might change this situation soon.

References

Abadulla, E., Tzanov, T., Costa, S., Robra, K.-H., Cavaco-Paulo, A., and Gübitz, G.M. (2000). Decolorization and detoxification of textile dyes with a laccase from *Trametes hirsuta. Appl. Environ. Microbiol.* **66**, 3357–3362.

Abelson, P.H. (1999). A potential phosphate crisis. *Science* **283**, 2015.

Acunaarguelles, M.E., Gutierrezrojas, M., Viniegragonzales, G., and Favelatorres, E. (1995). Production and properties of 3 pectinolytic activities produced by *Aspergillus niger* in submerged and solid-state fermentation. *Appl. Microbiol. Biotechnol.* **43**, 808–814.

Adler-Nissen, J. (1987). Newer uses of microbial enzymes in food processing. *Trends Biotechnol.* **5**, 170–174.

Adeboya, M.O., Edwards, R.L., Lassoe, T., Maitland, D.J., Shields, L., and Whalley, A.J.S. (1996). Metabolites of the higher fungi. 29. Maldoxin, maldoxone, dihydromaldoxin, isodihydromaldoxin and dechlorodihydromaldoxin. A spirocyclohexadienone, a depsidone and three diphenyl ethers: Keys in the depsidone biosynthetic pathway from a member of the fungus *Xylaria. J. Chem. Soc. Perk. Trans. I 1996*, 1419–1425.

Adriaens, P. and Grbic-Galić, D. (1994). Reductive dechlorination of PCDD/F by anaerobic cultures and sediments. *Chemosphere* **29**, 2253–2259.

Aggelis, G., Ehaliotis, C., Nerud, F., Stoychev, I., Lyberatos, G., and Zervakis, G.I. (2002). Evaluation of white-rot fungi for detoxification and decolorization of effluents from the green olive debittering process. *Appl. Microbiol. Biotechnol.* **59**, 353–360.

Ahlborg, U.G. and Thunberg, T.M. (1980). Chlorinated phenols: Occurrence, toxicity, metabolism and environmental impact. *CRC Crit. Rev. Toxicol.* **7**, 1–36.

Ahn, M.Y., Dec, J., Kim, J.E., and Bollag, J.M. (2000). Use of free and immobilized laccase for the decontamination of soil polluted with 2,4-dichlorophenol. *Abstr. Pap. Am. Chem. Soc.* **220**, 308-ENVR Part 1.

Aiken, B.S. and Logan, B.E. (1996). Degradation of pentachlorophenol by the white rot fungus *Phanerochaete chrysosporium* grown in ammonium lignosulfonate media. *Biodegradation* **7**, 175–182.

Aktas, N., Cicek, H., Unal, A.T., Kibarer, G., Kolankaya, N., and Tanyolac, A. (2001). Reaction kinetics for laccase-catalyzed polymerization of 1-naphthol. *Biores. Technol.* **80**, 29–36.

Alexandre, G. and Bally, R. (1999). Emergence of laccase-positive variant of *Azospirillum lipoferum* occurs via a two-step phenotypic switching process. *FEMS Microbiol. Lett.* **174**, 371–378.

Ali, T.A. and Wainwright, A.M. (1994). Growth of *Phanerochaete chrysosporium* in soil and its ability to degrade the fungicide benomyl. *Biores. Technol.* **49**, 197–201.

Alleman, B.C., Logan, B.E., and Gilbertson, R.L. (1995). Degradation of pentachlorophenol by fixed films of white-rot fungi in rotating tube bioreactors. *Water Res.* **29**, 61–67.

Altamirano, M.M., Blackburn, J.M., Aguayo, C., and Fersht, A.R. (2000). Directed evolution of new catalytic activity using α/β-barrel scaffold. *Nature* **403**, 617–622.

Amitai, G., Adani, R., Sod-Moriah, G., Rabinovitz, I., Vincze, A., Leader, H., Chefetz, B., Leibovitz-Persky *et al.* (1998). Oxidative biodegradation of phosphorothiolates by fungal laccase. *FEBS Lett.* **438**, 195–200.

Ander, P. and Eriksson, K.-E.L. (1976). The importance of phenol oxidase activity in lignin degradation by the white-rot fungus *Sporotrichum pulverulentum. Arch. Microbiol.* **109**, 1–8.

Anderson, S.O. (1985). Sclerotization and tanning in cutide. In G.A. Kerkut and L.I. Gillert (eds.) *Comparative insect physiology, biochemistry, and pharmacology* (Vol. 3). Pergamin Press, Oxford, pp. 59–64.

Antorini, M., Herpoel-Gimpert, I., Choinowski, T., Sigoillot, J.C., Aster, M., Winterhalter, K., and Piontek, K. (2002). Purification, crystallisation and X-ray diffraction study of fully functional laccases from two ligninolytic fungi. *Biochim. Biophys. Acta* **1594**, 109–114.

April, T.M., Foght, J.M., and Currah, R.S. (2000). Hydrocarbon-degrading filamentous fungi isolated from flare pit soils in northern and western Canada. *Can. J. Microbiol.* **46**, 38–49.

Archer, D.B. (2000). Filamentous fungi as microbial cell factories for food use. *Curr. Opin. Biotechnol.* **11**, 478–483.

Armengaud, J. and Timmis, K.N. (1997). Biodegradation of dibenzofuran-*p*-dioxin and dibenzofuran by bacteria. *J. Microbiol.* **35**, 241–252.

Arnold, F.H. and Volkov, A.A. (1999). Directed evolution of biocatalysts. *Curr. Opin. Chem. Biol.* **3**, 54–59.

Arthur, M.F. and Frea, J.I. (1989). 2,3,7,8-Tetrachlorodibenzo-*p*-dioxin: Aspects of its important properties and its potential biodegradation in soils. *J. Environ. Qual.* **18**, 1–11.

Atlas, R.M. (1984). *Petroleum microbiology*. Macmillan Publ. Co., New York.

Atanassov, P. (2002). Laccase-catalyzed direct electron transfer: Application in bio-fuel cell cathode. *Abstr. Pap. Am. Chem. Soc.* **223**, 378-COLL Part 1.

Balakshin, M., Capanema, E., Chen, C.L., Gratzl, J., Kirkman, A., and Gracz, H. (2001). Biobleaching of pulp with dioxygen in the laccase-mediator system—reaction mechanisms for degradation of residual lignin. *J. Mol. Catal. B—Enzym.* **13**, 1–16.

Baldrian, P. and Gabriel, J. (2002). Copper and cadmium increase laccase activity in *Pleurotus ostreatus. FEMS Microbiol. Lett.* **206**, 69–74.

Ballerstedt, H., Kraus, A., and Lechner, U. (1997). Reductive dechlorination of 1,2,3,4-tetrachlorodibenzo-*p*-dioxin and its products by anaerobic mixed cultures from Saale river sediment. *Environ. Sci. Technol.* **31**, 1749–1753.

Banat, I.M., Nigam, P., Singh, D., and Marchant, R. (1996). Microbial decolorization of textile-dye-containing effluents: A review. *Biores. Technol.* **58**, 217–227.

Barensi, R.I., Chellegatti, M.A.D.C., Fonseca, M.J.V., and Said, S. (2001). Partial purification and characterization of exopolygalacturonase II and III of *Penicillium frequentans. Braz. J. Microbiol.* **31**, 327–330.

Barrientos, L., Scott, J.J., and Murthy, P.P.N. (1994). Specificity of hydrolysis of phytic acid by alkaline phytase from lily pollen. *Plant Physiol.* **106**, 1489–1495.

Barton, S.C., Kim, H.H., Binyamin, G., Zhang, Y.C., and Heller, A. (2001). Electroreduction of O_2 to water on the "wired" laccase cathode. *J. Phys. Chem. B* **105**, 11917–11921.

Beaudette, L.A., Davies, S., Fedorak, P.M., Ward, O.P., and Pickard, M.A. (1998). Comparison of gas chromatography and mineralization experiments for measuring loss of selected polychlorinated biphenyl congeners in cultures of white-rot fungi. *Appl. Environ. Microbiol.* **64**, 2020–2025.

Beaudette, L.A., Ward, O.P., Pickard, M.A., and Fedorak, P.M. (2000). Low surfactant concentration increases fungal mineralization of a polychlorinated biphenyl congener but has no effect on overall metabolism. *Lett. Appl. Microbiol.* **30**, 155–160.

Beg, Q.K., Kapoor, M., Mahajan, L., and Hoondal, G.S. (2001). Microbial xylanases and their industrial application: A review. *Appl. Microbiol. Biotechnol.* **56**, 326–338.

Behnke, U. and Täufel, A. (1994). Peptidases. In H. Ruttloff (ed.) *Industrial enzymes (German).* Behr's Verlag, Hamburg, pp. 779–831.

Bennet, J.W. (1998). Mycotechnology: The role of fungi in biotechnology. *J. Biotechnol.* **66**, 101–107.

Benoit, P., Barriuso, E., and Calvet, R. (1998). Biosorption characterization of herbicides, 2,4-D and atrazine, and two chlorophenols on fungal mycelium. *Chemosphere* **37**, 1271–1282.

Berka, R.M., Rey, M.W., Brown, K.M., Byun, T., and Klotz, A.V. (1998). Molecular characterization and expression of a phytase gene from the thermophilic fungus *Thermomyces lanuginosus. Appl. Environ. Microbiol.* **64**, 4423–4427.

Berka, R.M., Schneider, P., Golightly, E.J., Brown, S.H., Madden, M., Brown, K.M., Halkier, T., Mondorf, K. *et al.* (1997). Characterization of the gene encoding an extracellular laccase of *Myceliophthora thermophila* and analysis of the recombinant enzyme produced in *Aspergillus oryzae. Appl. Environ. Microbiol.* **63**, 3151–3157.

Bertrand, T., Jolivalt, C., Briozzo, P., Caminade, E., Joly, N., Madzak, C., and Mougin, C. (2002a). Crystal structure of a four-copper laccase complexed with an arylamine: Insights into substrate recognition and correlation with kinetics. *Biochemistry* **41**, 7325–7333.

Bertrand, T., Jolivalt, C., Caminade, E., Joly, N., Mougin, C., and Briozzo, P. (2002b). Purification and preliminary crystallographic study of *Trametes versicolor* laccase in its native form. *Acta Crystallogr. D—Biol. Crystallogr.* **58**, 319–321.

Beurskens, J.E.M., Toussaint, M., de Wolf, J., van der Steen, J.M.D., Lot, P.C., Commandeur, L.C.M., and Parson, J.R. (1995). Dehalogenation of chlorinated dioxins by an anaerobic microbial consortium from sediment. *Environ. Toxicol. Chem.* **14**, 939–943.

Bezalel, L., Hadar, Y., and Cernigila, C.E. (1996). Mineralization of polycyclic aromatic hydrocarbons by the white rot fungus *Pleurotus ostreatus. Appl. Environ. Microbiol.* **62**, 292–295.

Bhalerao, U.T., Muralikrishna, C., and Rani, B.R. (1994). Laccase enzyme-catalyzed efficient synthesis of 3-substituted-1,2,4-triazolo(4,3-b)(4,1,2)benzothiadiazine-8-ones. *Tetrahedron* **50**, 4019–4024.

Bhargava, H.N. and Leonard, P.A. (1996). Triclosan: Applications and safety. *Am. J. Infect. Control* **24**, 209–218.

Bhat, M.K. (2000). Cellulases and related enzymes in biotechnology. *Biotechnol. Adv.* **18**, 355–383.

Biely, P., Vrsanska, M., Tenkanen, M., and Kluepfel, D. (1997). Endo-β-1,4-xylanase families: Differences in catalytic properties. *J. Biotechnol.* **57**, 151–166.

Blakely, J.K., Neher, D.A., and Spongberg, A.L. (2002). Soil invertebrate and microbial communities, and decomposition as indicators of polycyclic aromatic hydrocarbon contamination. *Appl. Soil Ecol.* **21**, 71–88.

Bogan, B.W. and Lamar, R.T. (1995). One-electron oxidation in the degradation of creosote polycyclic aromatic-hydrocarbons by *Phanerochaete chrysosporium. Appl. Environ. Microbiol.* **61**, 2631–2635.

Bogan, B.W. and Lamar, R.T. (1996). Polycyclic aromatic hydrocarbon-degrading capabilities of *Phanerochaete laevis* HHB-1625 and its extracellular ligninolytic enzymes. *Appl. Environ. Microbiol.* **62**, 1597–1603.

Bogan, B.W., Schoenike, B., Lamar, R.T., and Cullen, D. (1996). Expression of lip genes during growth in soil and oxidation of anthracene by *Phanerochaete chrysosporium. Appl. Environ. Microbiol.* **62**, 3697–3703.

Bohmer, S., Messner, K., and Srebotnik, E. (1998). Oxidation of phenanthrene by a fungal laccase in the presence of 1-hydroxybenzotriazole and unsaturated lipids. *Biochem. Biophys. Res. Comm.* **244**, 233–238.

Bonnen, A.M., Anton, L.H., and Orth, A.B. (1994). Lignin-degrading enzymes of the commercial button mushroom *Agaricus bisporus. Appl. Environ. Microbiol.* **60**, 960–965.

Borejsza-Wysocki, W., Lester, C., Attygalle, A.B., and Hrazdina, G. (1999). Elicited cell suspension cultures of apple (Malus × domestica) cv. Liberty produce biphenyl phytoalexins. *Phytochemistry* **50**, 231–235.

Bornscheuer, U.T. (2002). Microbial carboxyl esterases: Classification, properties and application in biocatalysis. *FEMS Microbiol. Rev.* **26**, 73–81.

Borriss, R. (1987). Biology of enzymes. In H.J. Rehm and G. Reed (eds.) *Biotechnology*. VCH Verlagsgesellsch, Weinheim, pp. 35–62.

Borriss, R. (1994a). β-glucan hydrolyzing enzymes. In H. Ruttloff (ed.) *Industrial enzymes (German)*. Behr's Verlag, Hamburg, pp. 728–757.

Borriss, R. (1994b). Structure and function of the genes encoding for bacterial endo 1,3-1,4-β-glucanases. *Curr. Top. Mol. Genet. (Life Sci. Adv.)*, 163–188.

Bourbonnais, R., Paice, M.G., Reid, I.D., Lanthier, P., and Yaguchi, M. (1995). Lignin oxidation by laccase isoenzymes from *Trametes versicolor* and role of the mediator 2,2′-azinobis (3-ethylbenzthiazoline-6-sulfonate) in kraft lignin depolymerization. *Appl. Environ. Microbiol.* **61**, 1876–1880.

Braun-Lullemann, A., Hüttermann, A., and Majcherczyk, A. (1999). Screening of ectomycorrhizal fungi for degradation of polycyclic aromatic hydrocarbons. *Appl. Microbiol. Biotechnol.* **53**, 127–132.

Bressler, D.C., Fedorak, P.M., and Pickard, M.A. (2000). Oxidation of carbazole, N-ethylcarbazole, fluorene, and dibenzothiophene by the laccase of *Coriolopsis gallica. Biotechnol. Lett.* **22**, 1119–1125.

Brinch, D.S. and Pedersen, P.B. (2002). Toxicological studies on *Polyporus pinsitus* laccase expressed by *Aspergillus oryzae* intended for use in food. *Food Addit. Contam.* **19**, 323–334.

Brinch-Pedersen, H., Dahl-Sorensen, L., and Holm, P.B. (2002). Engineering crop plants: Getting a handle on phosphate. *Trends Plant Sci.* **7**, 118–125.

Brown, M.A., Zhao, Z.W., and Mauk, A.G. (2002). Expression and characterization of a recombinant multicopper oxidase: Laccase IV from *Trametes versicolor. Inorg. Chim. Acta* **331**, 232–238.

Brul, S. and Coote, P. (1999). Preservative agents in foods—mode of action and microbial resistance mechanisms. *Int. J. Food Microbiol.* **50**, 1–17.

Budziszewski, G.J., Croft, K.P.C., and Hildebrand, D.F. (1996). Uses of biotechnology in modifying plant lipids. *Lipids* **31**, 557–569.

Bühler, M. and Schindler, J. (1984). Aliphatic hydrocarbons. In K. Kieslich (ed.) *Biotransformations (Biotechnology Vol. 6*, Series ed. Rehm, H.-J. and Reed, G.). Verlag Chemie, Weinheim, pp. 329–385.

Bumpus, J.A. (1989). Biodegradation of polycyclic aromatic hydrocarbons by *Phanerochaete chrysosporium. Appl. Envrion. Microbiol.* **55**, 154–158.

Bumpus, J.A. and Aust, S.D. (1987). Biodegradation of chlorinated organic compounds by *Phanerochaete chrysosporium* a wood rotting fungus. *ACS Symp. Ser.* **338**, 340–349.

Bunge, M., Ballerstedt, H., and Lechner, U. (2001). Regiospecific dechlorination of spiked tetra- and trichlordibenzo-*p*-dioxins by anaerobic bacteria from PCDD/F-contaminated Spittelwasser sediments. *Chemosphere* **43**, 675–681.

Burke, R.M. and Cairney, J.W.G. (2002). Laccases and other polyphenol oxidases in ecto- and ericoid mycorrhizal fungi. *Mycorrhiza* **12**, 105–116.

Buswell, J.A., Cai, Y.J., and Chang, S.T. (1995). Effect of nutrient nitrogen and manganese on manganese peroxidase and laccase production by *Lentinula* (*Lentinus*) *edodes. FEMS Microbiol. Lett.* **128**, 81–87.

Cajal, Y., Svendsen, A., de Bolos, J., Patkar, S.A., and Alsina, M.A. (2000). Effect of the lipid interface on the catalytic activity and spectroscopic properties of a fungal lipase. *Biochimie* **82**, 1053–1061.

Cameron, M.D. and Aust, S.D. (1999). Degradation of chemicals by reactive radicals produced by cellobiose dehydrogenase from *Phanerochaete chrysosporium. Arch. Biochem. Biophys.* **367**, 115–121.

Cardenas, F., Alvarez, E., de Castro-Alvarez, M.S., Sanchez-Monteri, J.M., Valmaseda, M., Elson, S.E., and Sinisterra, J.V. (2001a). Screening and catalytic activity in organic synthesis of novel fungal and yeast lipases. *J. Mol. Catal. B—Enzym.* **14**, 111–123.

Cardenas, F. de Castro, M.S., Sanchez-Montero, J.M., Sinisterra, J.V., Valmaseda, M., Elson, S.W., and Alvarez, E. (2001b). Novel microbial lipases: Catalytic activity in reactions in organic media. *Enzyme Microb. Technol.* **28**, 145–154.

Cary, J.W., Brown, R., Cleveland, T.E., Whitehead, M., and Dean, R.A. (1995). Cloning and characterization of a novel polygalacturonase-encoding gene from *Aspergillus parasiticus. Gene* **153**, 129–133.

Cassland, P. and Jonsson, L.J. (1999). Characterization of a gene encoding *Trametes versicolor* laccase A and improved heterologous expression in *Saccharomyces cerevisiae* by decreased cultivation temperature. *Appl. Microbiol. Biotechnol.* **52**, 393–400.

Castillo, M.D., Andersson, A., Ander, P., Stenstrom, J., and Torstensson, L. (2001). Establishment of the white rot fungus *Phanerochaete chrysosporium* on unsterile straw in solid substrate fermentation systems intended for degradation of pesticides. *World J. Microbiol. Biotechnol.* **17**, 627–633.

Castro-Sowinski, S., Martinez-Drets, G., and Okon, Y. (2002). Laccase activity in melanin-producing strains of *Sinorhizobium meliloti. FEMS Microbiol. Lett.* **209**, 119–125.

Cerniglia, C.E. (1993). Biodegradation of polycyclic aromatic hydrocarbons. *Curr. Opin. Biotechnol.* **4**, 331–338.

Cerniglia, C.E. and Crow, S.A. (1981). Metabolism of aromatic hydrocarbons by yeasts. *Arch. Microbiol.* **129**, 9–13.

Cerniglia, C.E., Freeman, J.P., and Mitchum, R.K. (1982). Glucuronide and sulfate conjugation in the fungal metabolism of aromatic hydrocarbons. *Appl. Environ. Microbiol.* **43**, 1070–1075.

Cerniglia, C.E., Morgan, J.C., and Gibson, D.T. (1979). Bacterial and fungal oxidation of dibenzofuran. *Biochem. J.* **180**, 175–185.

Cerniglia, C.E., Sutherland, J.B., and Crow, S.A. (1992). Fungal metabolism of aromatic hydrocarbons. In G. Winkelmann (ed.) *Microbial degradation of natural products.* VCH Verlagsgesellschaft, Weinheim, pp. 193–217.

Chiu, S.W., Ching, M.L., Fong, K.L., and Moore, D. (1998). Spent oyster mushroom substrate performs better than many mushroom mycelia in removing the biocide pentachlorophenol. *Mycol. Res.* **102**, 1553–1562.

Chivukula, M. and Renganathan, V. (1995). Phenolic azo dye oxidation by laccase from *Pyricularia oryzae. Appl. Environ. Microbiol.* **61**, 4374–4377.

Cho, N.S., Nam, J.H., Park, J.M., Koo, C.D., Lee, S.S., Pashenova, N., Ohga, S., and Leonowicz, A. (2001). Transformation of chlorophenols by white-rot fungi and their laccase. *Holzforschung* **55**, 579–584.

Cirigliano, M.C. and Carman, G.M. (1985). Purification and characterization of liposan, a bioemulsifier from *Candida lipolytica. Appl. Environ. Microbiol.* **50**, 846–850.

Claussen, M. and Schmidt, S. (1998). Biodegradation of phenol and *p*-cresol by the hyphomycete *Scedosporium apiospermum. Res. Microbiol.* **149**, 399–406.

Cliffe, S., Fawer, M.S., Maier, G., Takata, K., and Ritter, G. (1994). Enzyme assays for the phenolic content of natural juices. *J. Agric. Food Chem.* **42**, 1824–1828.

Cocchietto, M., Skert, N., Nimis, P.L., and Sava, G. (2002). A review on usnic acid, an interesting natural compound. *Naturwissenschaften* **89**, 137–146.

Cohen, M.S. and Gabriele, P.D. (1982). Degradation of coal by the fungi *Polyporus versicolor* and *Poria monticola. Appl. Environ. Microbiol.* **44**, 23–27.

Collett, O. (1992). Aromatic-compounds as growth substrates for isolates of the brown-rot fungus *Lentinus lepideus* (Fr ex Fr.). Fr. *Mater. Organismen* **27**, 67–77.

Collins, B.M., McLachlan, J.A., and Arnold, S.F. (1997). The estrogenic and antiestrogenic activities of phytochemicals with human estrogen receptor expressed in yeast. *Steroids* **62**, 365–372.

Collins, P.J. and Dobson, A.D.W. (1997). Regulation of laccase gene transcription in *Trametes versicolor. Appl. Environ. Microbiol.* **63**, 3444–3450.

Collins, P.J., Kotterman, M.J.J., Field, J.A., and Dobson, A.D.W. (1996). Oxidation of anthracene and benzo[a]pyrene by laccase from *Trametes versicolor. Appl. Environ. Microbiol.* **62**, 4563–4567.

Conesa, A., Punt, P.J., and van den Hondel, C.A.M.J.J. (2002). Fungal peroxidases: Molecular aspects and applications. *J. Biotechnol.* **93**, 143–158.

Conneely, A., Smyth, W.F., and McMullan, G. (2002). Study of the white-rot fungal degradation of selected pthalocyanine dyes by capillary electrophoresis and liquid chromatography. *Anal. Chim. Acta* **451**, 259–270.

Cortes, D., Barrios-Gonzales, J., and Tomasini, A. (2002). Pentachlorophenol tolerance and removal by *Rhizopus nigricans* in solid-state culture. *Process Biochem.* **37**, 881–884.

Cortez, D.A.G., Young, M.C.M., Marston, A., Wolfender, J.L., and Hostettmann, K. (1998). Xanthones, triterpenes and a biphenyl from *Kielmeyera coriacea. Phytochemistry* **47**, 1367–1374.

Coughlan, M.P. (1985). The properties of fungal and bacterial cellulases with comment on their production and application, Biotechnol. Genet. Eng. 3. In G.E. Russell (ed.) *Biotechnology and genetic engineering* Intercept, Newcastle Upon Tyne, pp. 39–109.

Cox, J.C. and Golbeck, J.H. (1985). Hydroxylation of biphenyl by *Aspergillus parasiticus*: Approaches to yield improvement in fermentor cultures. *Biotechnol. Bioeng.* **27**, 1395–1402.

Cragg, S.M. and Eaton, R.A. (1997). Evaluation of creosote fortified with synthetic pyrethroids as wood preservatives for use in the sea. II. Effects on wood-degrading micro-organisms and fouling invertebrates. *Mater. Organismen* **31**, 197–216.

Creffield, J.W., Greaves, H., Chew, N., and Nguyen, N.K. (2000). A field trial of pigment-emulsion creosote: 11 year data. *Forest Prod. J.* **50**, 77–82.

Cui, W., Beever, R.E., Parkes, S.L., Weeds, P.L., and Templeton, M.D. (2002). An osmosensing histidine kinase mediators dicarboximide fungicide resistance in *Botryotinia fuckeliana* (*Botrytis cinerea*). *Fungal Genet. Biol.* **36**, 187–198.

Cullen, D. (1997). Recent advances on the molecular genetics of ligninolytic fungi. *J. Biotechnol.* **53**, 273–289.

Dalboge, H. (1997). Expression cloning of fungal enzyme genes; a novel approach for efficient isolation of enzyme genes of industrial interest. *FEMS Microbiol. Rev.* **21**, 29–42.

D'Annibale, A., Celetti, D., Felici, M., DiMattia, E., and Sermanni, G.G. (1996). Substrate specificity of laccase from *Lentinus edodes*. *Acta Biotechnol.* **16**, 257–270.

D'Annibale, A., Stazi, S.R., Vinciguerra, V., DiMattia, E., and Sermanni, G.G. (1999). Characterization of immobilized laccase from *Lentinula edodes* and its use in olive-mill wastewater treatment. *Process Biochem.* **34**, 697–706.

Das, N., Chakraborty, T.K., and Mukherjee, M. (2001). Purification and characterization of a growth-regulating laccase from *Pleurotus florida*. *J. Basic Microbiol.* **41**, 261–267.

Datta, A., Bettermann, A., and Kirk, T.K. (1991). Identification of a specific manganese peroxidase among ligninolytic enzymes secreted by *Phanerochaete chrysosporium* during wood decay. *Appl. Environ. Microbiol.* **57**, 1453–1460.

Datta, J., Dutta, T.K., and Samanta, T.B. (1994). Microsomal glutathione-S-transferase (GST) isoenzymes in *Aspergillus ochraceus* TS—induction by 3-methylcholanthrene. *Biochem. Biophys. Res. Commun.* **203**, 1508–1514.

Davila, A.M., Marchal., R., and Vandecasteele, J.P. (1994). Sophorose lipid production from lipidic precursors—predictive evaluation of industrial substrates. *J. Ind. Microbiol.* **13**, 249–257.

Davies, G. and Henrissat, B. (1995). Structures and mechanisms of glycosyl hydrolases. *Structure* **3**, 853–859.

Davies, G.J., Wilson, K.S., and Henrissat, B. (2002). Nomenclature for sugar binding subsites in glycosyl hydrolases. *Biochem. J.* **321**, 557–559.

Davis, M.W., Glaser, J.A., Evans, J.W., and Lamar, R.T. (1993). Field-evaluation of the lignin-degrading fungus *Phanerochaete sordida* to treat creosote-contaminated soil. *Environ. Sci. Technol.* **27**, 2572–2576.

Dec, J., Haider, K., and Bollag, J.M. (2001). Decarboxylation and demethoxylation of naturally occuring phenols during coupling reactions and polymerization. *Soil Sci.* **166**, 660–671.

Dedeyan, B., Klonowska, A., Tagger, S., Tron, T., Iacazio, G., Gil, G., and Le Petit, J. (2000). Biochemical and molecular characterization of a laccase from *Marasmius quercophilus*. *Appl. Environ. Microbiol.* **66**, 925–929.

de Groot, R.C. and Woodward, B. (1998). *Wolfiporia cocos*—a potential agent for composting or bioprocessing Douglas-fir wood treated with copper-based preservatives. *Mater. Organismen* **32**, 195–215.

de Groot, R.C. and Woodward, B. (1999). Using copper-tolerant fungi to biodegrade wood treated with copper-based preservatives. *Int. Biodeter. Biodegr.* **44**, 17–27.

Dekker, J. (1995). Development of resistance to modern fungicide and strategies for its avoidance. In H. Lyr (ed.) *Modern selective fungicides* (2nd edn.). Fischer, Jena, pp. 23–38.

Dekker, J. and Georgopoulos, S.G. (eds.) (1982). *Fungicide resistance in crop protection*. Pudoc, Wageningen.

Dekker, R.F.H., Vasconcelos, A.F.D., Barbosa, A.M., Giese, E.C., and Paccola-Meirelles, L. (2001). A new role for veratryl alcohol: Regulation of synthesis of lignocellulose-degrading enzymes in the ligninolytic ascomycetous fungus, *Botryosphaeria* sp.; influence of carbon source. *Biotechnol. Lett.* **23**, 1987–1993.

del Sorbo, G., Schoonbeek, H., and de Waard, M.A. (2000). Fungal transporters involved in efflux of natural toxic compounds and fungicides. *Fungal Genet. Biol.* **30**, 1–15.

de Marco, A. and RoubelakisAngelakis, K.A. (1997). Laccase activity could contribute to cell-wall reconstitution in regeneration protoplasts. *Phytochemistry* **46**, 421–425.

Demain, A.L. (2000). Microbial biotechnology. *Trends Biotechnol.* **18**, 26–31.

de Vries, R.P. and Visser, J. (2001). *Aspergillus* enzymes involved in degradation of plant cell wall polysaccharides. *Microbiol. Mol. Biol. Rev.* **65**, 497–522.

Dietrich, D., Hickey, W.J., and Lamar, R. (1995). Degradation of 4,4′-dichlorobiphenyl, 3,3′,4,4′-tetrachlorobiphenyl, and 2,2′,4,4′,5,5′-hexachlorobiphenyl by the white rot fungus *Phanerochaete chrysosporium*. *Appl. Environ. Microbiol.* **61**, 3904–3909.

Dixon, D.P., Cole, D.J., and Edwards, R. (2000). Characterisation of a zeta class gluthatione transferase from *Arabidopsis thaliana* with a putative role in tyrosine catabolism. *Arch. Biochem. Biophys.* **384**, 407–412.

Dmochewitz, S. and Ballschmiter, K. (1988). Microbial transformation of technical mixtures of polychlorinated biphenyls (PCB) by the fungus *Aspergillus niger. Chemosphere* **17**, 111–121.

Dodge, R.H., Cerniglia, C.E., and Gibson, D.T. (1979). Fungal metabolism of biphenyl. *Biochem. J.* **178**, 223–230.

Donnelly, P.K. and Fletcher, J.S. (1994). Potential use of mycorrhizal fungi as bioremediation agents. *ACS Symp. Ser.* **563**, 93–99.

Donnelly, P.K., Entry, J.A., and Crawford, D.L. (1993). Degradation of atrazine and 2,4-dichlorophenoxyacetic acid by mycorrhizal fungi at 3 nitrogen concentrations *in-vitro. Appl. Environ. Microbiol.* **59**, 2642–2647.

Dordick, J.S., Ryu, K., and Mc Eldoon, J.P. (1991). Enzymatic cytalysis on coal-related compounds in organic media—kinetics and potential commercial applications. *Res. Conserv. Recycl.* **5**, 195–209.

Ducros, V., Davies, G.J., Lawson, D.M., Wilson, K.S., Brwon, S.H., Ostergaard, P., Pedersen, A.H., Schneider, P. *et al.* (1997). Crystallization and preliminary X-ray analysis of the laccase from *Coprinus cinereus. Acta Crystallogr. D—Biol. Crystallogr.* **53**, 605–607.

Duncan, C.G. and Deverall, F.J. (1964). Degradation of wood preservatives by fungi. *Appl. Microbiol.* **12**, 57–68.

Duran, N. and Esposito, E. (2000). Potential applications of oxidative enzymes and phenoloxidase-like compounds in wastewater and soil treatment: A review. *Appl. Catal. B—Environ.* **28**, 83–99.

Dvorakova, J. (1998). Phytase: Source, preparation and exploitation. *Folia Microbiol.* **43**, 323–338.

Eaton, D.C. (1985). Mineralization of polychlorinated biphenyls by *Phanerochaete chrysosporium*, a ligninolytic fungus. *Enzyme Microb. Technol.* **7**, 194–196.

Eggen, T. (1999). Application of fungal substrate from commercial mushroom production—*Pleurotus ostreatus*—for bioremediation of creosote contamined soil. *Int. Biodeter. Biodegr.* **44**, 117–126.

Eggen, T. and Sveum, P. (1999). Decontamination of aged creosote polluted soil: The influence of temperature, white rot fungus *Pleurotus ostreatus*, and pretreatment. *Int. Biodeter. Biodegr.* **43**, 125–133.

Eggert, C., Lafayette, P.R., Temp, U., Eriksson, K.-E.L., and Dean, J.F.D. (1998). Molecular analysis of a laccase gene from the white rot fungus *Pycnoporus cinnabarinus. Appl. Environ. Microbiol.* **64**, 1766–1772.

Eggert, C., Temp, U., and Eriksson, K.-E.L. (1996). The ligninolytic system of the white rot fungus *Pycnoporus cinnabarinus*: Purification and characterization of the laccase. *Appl. Environ. Microbiol.* **62**, 1151–1158.

Eggert, C., Temp, U., and Eriksson, K.-E.L. (1997). Laccase is essential for lignin degradation by the white-rot fungus *Pycnoporus cinnabarinus. FEBS Lett.* **407**, 89–92.

Elisashvili, V.I. (1993). Physiological regulation of ligninolytic activity in higher basidium fungi. *Microbiology* **62**, 480–487.

Emtiazi, G., Satarii, M., and Mazaherion, F. (2001). The utilization of aniline, chlorinated aniline, and aniline blue as the only source of nitrogen by fungi in water. *Water Res.* **35**, 1219–1224.

Endo, K., Hosono, K., Beppu, T., and Ueda, K. (2002). A novel extracytoplasmic phenol oxidase of *Streptomyces*: Its possible involvement in the onset of morphogenesis. *Microbiology* **148**, 1767–1776.

Entry, J.A., Donnelly, P.K., and Emmingham, W.H. (1996). Mineralization of atrazine and 2,4-D in soils inoculated with *Phanerochaete chrysosporium* and *Trappea darkeri. Appl. Soil Ecol.* **3**, 85–90.

Fabbrini, M., Galli, C., and Gentili, P. (2002). Comparing the catalytic efficiency of some mediators of laccase. *J. Mol. Catal. B—Enzym.* **16**, 231–240.

Fahr, K., Wetzstein, H.G., Grey, R., and Schlosser, D. (1999). Degradation of 2,4-dichlorophenol and pentachlorophenol by two brown rot fungi. *FEMS Microbiol. Lett.* **175**, 127–132.

Fahreus, G. and Ljundgreen, H. (1961). Substrate specificity of purified fungal laccase. *Biochim. Biophys. Acta* **46**, 22–32.

Fakoussa, R.M. and Frost, P.J. (1999). *In vivo*-decolorization of coal-derived humic acids by laccase-excreting fungus *Trametes versicolor. Appl. Microbiol. Biotechnol.* **52**, 60–65.

Fakoussa, R.M. and Hofrichter, M. (1999). Biotechnology and microbiology of coal degradation. *Appl. Microbiol. Biotechnol.* **52**, 25–40.

Farnet, A.M., Tagger, S., and Le Petit, J. (1999). Effects of copper and aromatic inducers on the laccases of the white-rot fungus *Marasmius quercophilus. CR Acad. Sci. III–VIE* **322**, 499–503.

Faure, D., Bouillant, M.L., and Bally, R. (1995). Comparative-study of substrates and inhibitors of *Azospirillum lipoferum* and *Pyricularia-oryzae* laccases. *Appl. Environ. Microbiol.* **61**, 1144–1146.

Faure, D., Bouillant, M.L., Jacoud, C., and Bally, R. (1996). Phenolic derivatives related to lignin metabolism as substrates for *Azospirillum* laccase activity. *Phytochemistry* **42**, 357–359.

Fernandez-Sanchez, J.M., Rodriguez-Vazquez, R., Ruiz-Aguilar, G., and Alvarez, P.J.J. (2001). PCB biodegradation in aged contaminated soil: Interactions between exogenous *Phanerochaete chrysosporium* and indigenous microorganisms. *J. Environ. Sci. Health* **36**, 1145–1162.

Field, J.A., de Jong, E., Costa, G.F., and de Bont, J.A.M. (1992). Biodegradation of polycyclic aromatic hydrocarbons by new isolates of white rot fungi. *Appl. Environ. Microbiol.* **58**, 2219–2226.

Font, X., Caminal, G., Gabarrell, X., Lafuente, J., and Vicent, M.T. (1997). One-line enzyme activity determination using the stopped-flow technique: Application to laccase activity in pulp mill waste-water treatment. *Appl. Microbiol. Biotechnol.* **48**, 168–173.

Fortnagel, P., Harms, H., Wittich, R.-M., Krohn, S., Meyer, H., and Francke, W. (1989). Cleavage of dibenzofuran and dibenzo-*p*-dioxin ring systems by a *Pseudomonas* bacterium. *Naturwissenschaften* **76**, 222–223.

Freire, R.S., Duran, N., and Kubota, L.T. (2002a). Development of a laccase-based flow injection electrochemical biosensor for the determination of phenolic compounds and its application for monitoring remediation of Kraft E1 paper mill effluent. *Anal. Chim. Acta* **463**, 229–238.

Freire, R.S., Duran, N., Wang, J., and Kubota, L.T. (2002b). Laccase-based screen printed electrode for amperometric detection of phenolic compounds. *Anal. Lett.* **35**, 29–38.

Fritz-Langhals, E. and Kunath, B. (1998). Synthesis of aromatic aldehydes by laccase-mediator assisted oxidation. *Tetrahedron Lett.* **39**, 5955–5956.

Frost, G.M. and Moss, D.A. (1987). Production of enzymes by fermentation. In J.F. Kennedy (ed.) *Enzyme technology*. VCH, Weinheim, pp. 65–211.

Fu, Y.Z. and Viraraghavan, T. (2001). Fungal decolorization of dye wastewaters: A review. *Biores. Technol.* **79**, 252–262.

Fujisawa, T., Onogawa, Y., Sato, A., Mitsuya, T., and Shimuzu, M. (1998). Asymmetric reductions of (trifluoroacetyl)biphenyl derivatives with bakers' yeast and with *Geotrichum candium* acetone powder. *Tetrahedron* **54**, 4267–4276.

Fukui, S. and Tanaka, A. (1981). Production of useful compounds from alkane media in Japan. In A. Fiechter (ed.) *Products from alkanes, cellulose and other feedstocks*. Akademie-Verlag, Berlin, pp. 1–35.

Gaal, A. and Neujahr, H.Y. (1979). Metabolism of phenol and resorcinol in *Trichosporon cutaneum*. *J. Bacteriol.* **137**, 13–21.

Galante, Y.M., de Conti, A., and Montevedi, R. (1998). Application of *Trichoderma* enzymes in food and feed industries. In G.E. Hamnn and C. Kubicek (eds.) *Trichoderma and Gliocladium—enzymes, biological control and commercial applications*. Taylor & Francis, London, pp. 327–342.

Galhaup, C., Goller, S., Peterbauer, C.K., Strauss, J., and Haltrich, D. (2002). Characterization of the major laccase isoenzyme from *Trametes pubescens* and regulation of its synthesis by metal ions. *Microbiology* **148**, 2159–2169.

Gardiol, A.E., Hernandez, R.J., Reinhammer, B., and Harte, B.R. (1996). Development of a gas-phase oxygen biosensor using a blue copper-containing oxidase. *Enzyme Microb. Technol.* **18**, 347–352.

Garg, S.K. and Modi, D.R. (1999). Decolorization of pulp-paper mill effluents by white-rot fungi. *Crit. Rev. Biotechnol.* **19**, 85–112.

Gavnholt, B., Larsen, K., and Rasmussen, S.K. (2002). Isolation and characterisation of laccase cDNAs from meristematic and stem tissues of ryegrass (*Lolium perenne*). *Plant Sci.* **162**, 873–885.

Georgopoulos, S.G. (1995). The genetics of fungicide resistance. In H. Lyr (ed.) *Modern selective fungicides* (2nd edn). Fischer, Jena, pp. 39–52.

Gesell, M. (2001). Biotransformation von Biarylverbindungen durch Pilze der Gattungen *Paecilomyces* und *Fusarium* unter besonderer Berücksichtigung des Stammes *Paecilomyces lilacinus*. Doctoral dissertation, University of Greifswald.

Gesell, M., Hammer, E., Specht, M., Francke, W., and Schauer, F. (2001). Biotransformation of biphenyl by *Paecilomyces lilacinus* and characterization of ring cleavage products. *Appl. Environ. Microbiol.* **67**, 1551–1557.

Ghannoum, M.A. (2000). Potential role of phospholipases in virulence and fungal pathogenesis. *Clin. Microbiol. Rev.* **13**, 122–135.

Ghindilis, A.L., Gavrilova, V.P., and Yaropolov, A.I. (1992). Laccase-based biosensor for determination of polyphenols—determination of catechols in tea. *Biosens. Bioelectron.* **7**, 127–131.

Giardina, P., Palmieri, G., Scaloni, A., Fontanella, B., Faraco, V., Cennamo, G., and Sannia, G. (1999). Protein and gene structure of a blue laccase from *Pleurotus ostreatus*. *Biochem. J.* **341**, 655–663.

Gibson, D.T. (1968). Microbial degradation of aromatic compounds. *Science* **161**, 1093–1097.

Gill, M. and Steglich, W. (1987). *Progress in the chemistry of organic natural products 51*. Springer-Verlag, Wien.

Glenn, J.K., Akileswaran, L., and Gold, M.H. (1986). Mn (II) oxidation in the principal function of the extracellular Mn-peroxidase from *Phanerochaete chrysopsorium*. *Arch. Biochem. Biophys.* **251**, 688–696.

Gokmen, V., Borneman, Z., and Nijhuis, H.H. (1998). Improved ultrafiltration for color reduction and stabilization of apple juice. *J. Food Sci.* **63**, 504–507.

Golbeck, J.H., Albaugh, S.A., and Radmer, R. (1983). Metabolism of biphenyl by *Aspergillus toxicarius*: Induction of hydroxylating activity and accumulation of water-soluble conjugates. *J. Bacteriol.* **156**, 49–57.

Golovan, S.P., Meidinger, R.G., Ajakaiye, A., Cottrill, M., Wiederkehr, M.Z., Barney, D.J., Plante, C., Pollard, J.W. *et al.* (2001). Pigs expressing salivary phytase produce low-phosphorous manure. *Nat. Biotechnol.* **19**, 741–745.

Golovleva, L.A., Leontievsky, A.A., Maltseva, O.V., and Myasoedova, N.M. (1993). Ligninolytic enzymes of the fungus *Panus tigrinus* 8/18—biosynthesis, purification and properties. *J. Biotechnol.* **30**, 71–77.

Gouka, R.J., Gerk, C., Hooykaas, P.J.J., Bundock, P., Musters, W., Verrips, C.T., and de Groot, M.J.A. (1999). Transformation of *Aspergillus awamori* by *Agrobacterium tumefaciens*-mediated homologous recombination. *Nat. Biotechnol.* **6**, 598–601.

Gramss, G., Kirsche, B., Voigt, K.D., Günther, T., and Fritsche, W. (1999). Conversion rates of five polycyclic aromatic hydrocarbons in liquid cultures of fifty-eight fungi and the concomitant production of oxidative enzymes. *Mycol. Res.* **103**, 1009–1018.

Grass, G. and Rensing, C. (2001). CueO is a multi-copper oxidase that confers copper tolerance in *Escherichia coli*. *Biochem. Biophys. Res. Commun.* **286**, 902–908.

Grassin, C. and Fauquembergue, P. (1996). Fruit juices. In S. West (ed.) *Industrial enzymology*. Macmillan Press, London, pp. 226–240.

Green, N.A., Meharg, A.A., Till, C., Troke, J., and Nicholson, J.K. (1999). Degradation of 4-fluorobiphenyl by mycorrhizal fungi as determined by F-19 nuclear magnetic resonance spectroscopy and ^{14}C radiolabelling analysis. *Appl. Environ. Microbiol.* **65**, 4021–4027.

Gripenberg, J. (1960). Fungus pigments 12. The structure and synthesis of telephoric acid. *Tetrahedron* **10**, 135–143.

Gruber, K., Klintschar, G., Hayn, M., Schlacher, A., Steiner, W., and Kratky, C. (1998). Thermophilic xylanase fom *Thermomyces lanuginosus*: High resolution X-ray structure and modeling studies. *Biochemistry* **37**, 13475–13485.

Günther, T., Sack, U., Hofrichter, M., and Latz, M. (1998). Oxidation of PAH and PAH-derivatives by fungal and plant oxidoreductases. *J. Basic Microbiol.* **38**, 113–122.

Ha, H.C., Honda, Y., Watanabe, T., and Kuwahara, M. (2001). Production of manganese peroxidase by pellet culture of the lignin-degrading basidiomycete, *Pleurotus ostreatus*. *Appl. Environ. Microbiol.* **55**, 704–711.

Haemmerli, S.D., Leisola, M.S.A., Sanglard, D., and Fiechter, A. (1986). Oxidation of benzo[a]pyrene by extracellular ligninase of *Phanerochaete chrysosporium*. *J. Biol. Chem.* **261**, 6900–6903.

Hakala, T., Lundell, T., Hofrichter, M., and Maijala, P. (2002). Manganese peroxidase—the key enzyme in lignin biodegradation and biopulping by white-rot fungi? *Abstr. Pap. Am. Chem. Soc.* **223**, 028-CELL Part1.

Hakulinen, N., Kiiskinen, L.L., Kruus, K., Saloheimo, M., Paananen, A., Koivula, A., and Rouvinen, J. (2002). Crystal structure of a laccase from *Melanocarpus albomyces* with an intact trinuclear copper site. *Nat. Struct. Biol.* **9**, 601–605.

Halsall, B.E., Darrah, J.A., and Cain, R.B. (1969). Regulation of enzymes of aromatic-ring fission in fungi—organisms using both catechol and protocatechuate pathways. *Biochem. J.* **114**, P75.

Hammel, K.E. and Tardone, P.J. (1988). The oxidative 4-dechlorination of polychlorinated phenols is catalyzed by extracellular fungal lignin peroxidases. *Biochemistry* **27**, 6563–6568.

Hammel, K.E., Kalyanaraman, B., and Kirk, T.K. (1986). Oxidation of polycyclic aromatic hydrocarbons and dibenzo-*p*-dioxins by *Phanerochaete chrysosporium* ligninase. *J. Biol. Chem.* **262**, 16948–16952.

Hammer, E., Krowas, D., Schäfer, A., Specht, M., Francke, W., and Schauer, F. (1998). Isolation and characterization of a dibenzofuran-degrading yeast: Identification of oxidation and ring cleavage products. *Appl. Environ. Microbiol.* **64**, 2215–2219.

Hammer, E. and Schauer, F. (1997). Fungal hydroxylation of dibenzofuran. *Mycol. Res.* **101**, 433–436.

Hammer, E., Schoefer, L., Schäfer, A., Hundt, K., and Schauer, F. (2001). Formation of glucoside conjugates during biotransformation of dibenzofuran by *Penicillium canescens* SBUG-M 1139. *Appl. Microbiol. Biotechnol.* **57**, 390–394.

Han, S. and New, P.B. (1994). Effect of water availability on degradation of 2,4-dichlorophenoxyacetic acid (2,4-D) by soil-microorganisms. *Soil Biol. Biochem.* **26**, 1689–1697.

Hara, A., Ueda, M., Matsui, T., Arie, M., Saeki, H., Matsuda, H., Furuhashi, K., Kanai, T., *et al.* (2001). Repression of fatty-acyl-CoA oxidase-encoding gene expression is not necessarily a determination of high-level production of dicarboxylic acids in industrial dicarboxylic-acid-producing *Candida tropicalis. Appl. Microbiol. Biotechnol.* **56**, 478–485.

Harada, T. (1984). Isoamylase and its industrial significance in the production of sugars from starch, Biotechnol. Genet. Eng. 1. In G.E. Russel (ed.) *Biotechnology and genetic engineering reviews* Intercept, Newcastle Upon Tyne, pp. 39–63.

Hargreaves, J., Park, J.O., Ghisalberti, E.L., Sivasithamparam, K., Skelton, B.W., and White, A.H. (2002). New chlorinated diphenyl ethers from *Aspergillus* species. *J. Nat. Prod.* **65**, 7–10.

Harkki, A., Uusitalo, J., Bailey, M., Penttila, M., and Knowles, J.K.C. (1989). A novel fungal expression system: Secretion of active calf chymosin from the filamentous fungus *Trichoderma reesei. Bio-Technology* **7**, 596–603.

Harris, G. and Ricketts, R.W. (1962). Metabolism of phenolic compounds by yeast. *Nature* **195**, 473–474.

Harwood, C.S. and Parales, R.E. (1996). The beta-ketoadipate pathway and the biology of self-identity. *Annu. Rev. Microbiol.* **50**, 553–590.

Hatakka, A. (1994). Lignin-modifying enzymes from selected white-rot fungi—production and role in lignin degradation. *FEMS Microbiol. Rev.* **13**, 125–135.

Heldt-Hansen, H.P. (1997). Development of enzymes for food application. In K. Poutanen (ed.) *Biotechnology in the food chain. New tools and applications for future foods* Technical Research Centre of Finland, Espoo, Symposium 1998, Helsinki, pp. 45–55.

Henning, K. (1993). Oxidation of diphenyl ether by the yeast *Trichosporon beigelii* (German). Doctoral dissertation, University of Greifswald.

Henriksson, G., Johansson, G., and Petterson, G. (2000). A critical review of cellobiose dehydrogenases. *J. Biotechnol.* **78**, 93–113.

Hess, J., Leitner, C., Galhaup, C., Kulbe, K.D., Hinterstoisser, B., Steinwender, M., and Haltrich, D. (2002). Enhanced formation of extracellular laccase activity by the white-rot fungus *Trametes multicolor. Appl. Biochem. Biotechnol.* **98**, 229–241.

Hiratsuka, N., Wariishi, H., and Tanaka, H. (2001). Degradation of diphenyl ether herbicides by the lignin-degrading basidiomycete *Coriolus versicolor. Appl. Microbiol. Biotechnol.* **57**, 563–571.

Hoffmann, B. and Rehm, H.J. (1976). Degradation of long chain n-alkanes by *Mucorales. Eur. J. Appl. Micobiol.* **3**, 19–30.

Hofmann, K.H. and Schauer, F. (1988). Utilization of phenol by hydrocarbon assimilating yeasts. *Antonie van Leeuwenhoek* **54**, 179–188.

Hofrichter, M., Bublitz, F., and Fritsche, W. (1997). Fungal attack on coal II. Solubilization of low-rank coal by filamentous fungi. *Fuel Proc. Technol.* **52**, 55–64.

Hofrichter, M. and Scheibner, K. (1993). Utilization of aromatic compounds by the *Penicillium* strain BI-7/2. *J. Basic Microbiol.* **33**, 227–232.

Hofrichter, M., Scheibner, K., Schneegass, I., and Fritsche, W. (1998). Enzymatic combustion of aromatic and aliphatic compounds by manganese peroxidase from *Nematoloma frowardii. Appl. Environ. Microbiol.* **64**, 399–404.

Holker, U., Schmiers, H., Grosse, S., Winkelhofer, M., Polsakiewicz, M., Ludwig, S., Dohse, J., and Hofer, M. (2002). Solubilization of low-rank coal by *Trichoderma atroviride*: Evidence for the involvement of hydrolytic and oxidative enzymes by using ^{14}C-labelled lignite. *J. Ind. Microbiol. Biotechnol.* **28**, 207–212.

Hommel, R., Stüwer, O., Stuber, W., Haferburg, D., and Kleber, H.-P. (1987). Production of water soluble surface-active exolipids by *Torulopsis apicola. Appl. Microbiol. Biotechnol.* **26**, 199–205.

Hong, F., Meinander, N.Q., and Jonsson, L.J. (2002). Fermentation strategies for improved heterologous expression of laccase in *Pichia pastoris. Biotechnol. Bioeng.* **79**, 438–449.

Hood, E.E. (2002). From green plants to industrial enzymes. *Enzyme Microb. Technol.* **30**, 279–283.

Hoshida, H., Nakao, M., Kanazawa, H., Kubo, K., Hakukawa, K., Morimasa, K., Akada, R., and Nishizawa, Y. (2001). Isolation of five laccase gene sequences from the white-rot fungus *Trametes sanguinea* by PCR, and cloning, characterization and expression of the laccase cDNA in yeasts. *J. Biosci. Bioeng.* **92**, 372–380.

Hrazdina, G., Borejsza-Wysocki, W., and Lester, C. (1997). Phytoalexin production in an apple cultivar resistant to *Venturia inaequalis. Phytopathology* **87**, 868–876.

Huang, M.H., Shih, Y.P., and Liu, S.M. (2002). Biodegradation of polyvinyl alcohol by *Phanerochaete chrysosporium* after pretreatment with Fenton's reagent. *J. Environ. Sci. Health A—Toxic/Hazardous Substances Environ. Eng.* **37**, 29–41.

Huber, J. (1994). Production of microbial enzyme preparations: Biology and biochemistry. In H. Ruttloff (ed.) *Industrial enzymes (German)*. Behr's Verlag, Hamburg, pp. 193–244.

Hullo, M.F., Moszer, I., Danchin, A., and Martin-Verstraete, I. (2001). CotA of *Bacillus subtilis* is a copper-dependent laccase. *J. Bacteriol.* **183**, 5426–5430.

Humar, M., Petric, M., Pohleven, F., Sentjurc, M., and Kalan, P. (2002). Changes in EPR spectra of wood impregnated with copper-based preservatives during exposure to several wood-rotting fungi. *Holzforschung* **56**, 229–238.

Hundt, K.F. (2001). Biotransformation von halogenierten Diphenylethern durch Pilze unter besonderer Berücksichtigung von *Trametes versicolor*. Doctoral dissertation, University of Greifswald.

Hundt, K., Jonas, U., Hammer, E., and Schauer, F. (1999). Transformation of diphenyl ethers by *Trametes versicolor* and characterization of ring cleavage products. *Biodegradation* **10**, 279–286.

Hundt, K., Martin, D., Hammer, E., Jonas, U., Kindermann, M.K., and Schauer, F. (2000). Transformation of Triclosan by *Trametes versicolor* and *Pycnoporus cinnabarinus*. *Appl. Environ. Microbiol.* **66**, 4157–4160.

Hutzinger, O. and Blumich, M.J. (1985). Sources and fate of PCDDs and PCDFs: An overview. *Chemosphere* **14**, 581–600.

Ichinose, H., Wariishi, H., and Tanaka, H. (1999). Bioconversion of recalcitrant 4-methyldibenzothiophene to water-extractable products using lignin-degrading basidiomycete *Coriolus versicolor. Biotechnol. Progr.* **15**, 706–714.

Idriss, E.E., Makarewicz, O., Farouk, A., Rosner, K., Greiner, R., Bochow, H., Richter, T., and Borriss, R. (2002). Extracellular phytase activity of *Bacillus amyloliquefaciens* FZB45 contributes to its plant growth promoting effect. *Microbiology* **148**, 2097–2109.

Iida, M., Kobayashi, H., and Iizuka, H. (1980). Cellular fatty acids derived from normal alkanes by *Candida rugosa*. *Z. Allg. Mikrobiol.* **20**, 449–457.

Ikeda, R., Sugita, T., Jacobson, E.S., and Shinoda, T. (2002). Laccase and melanization in clinically important *Cryptococcus* species other than *Cryptococcus neoformans*. *J. Clin. Microbiol.* **40**, 1214–1218.

Inomata, N., Yosgida, H., Aoki, A., Tsunoda, M., and Yamamoto, M. (1991). Effects of MCPA and other phenoxyacid compounds on hepatic xenobiotic metabolism in rats. *Tohoku J. Exp. Med.* **165**, 171–182.

Inoue, S. and Itoh, S. (1982). Sophorolipids from *Torulopsis bombicola* as microbial surfactants in alkane fermentations. *Biotechnol. Lett.* **4**, 3–8.

Irvin, G.C.J. and Cosgrove, D.J.J. (1972). Inositol phosphate phosphatases of microbiological origin: The inositol pentaphosphate products of *Aspergillus ficuum* phytases. *J. Bacteriol.* **112**, 434–438.

Ishiguro, T., Ohtake, Y., Nakayama, S., Inamori, Y., Amagai, T., Soma, M., and Matsusita, H. (2000). Biodegradation of dibenzofuran and dioxins by *Pseudomonas aeruginosa* and *Xanthomonas maltophilia*. *Environ. Technol.* **21**, 1309–1316.

Ito, K., Ikemasu, T., and Ishikawa, T. (1992). Cloning and sequencing of the *xynA*-gene encoding xylanase A of *Aspergillus kawachii*. *Biosci. Biotechnol. Biochem.* **56**, 906–912.

Iwatsuki, M., Niki, E., and Kato, S. (1993). Antioxidant activities of natural and synthetic carbazoles. *Biofactors* **4**, 123–128.

Jaeger, K.E. and Eggert, T. (2002). Lipases for biotechnology. *Curr. Opin. Biotechnol.* **13**, 390–397.

Jaeger, K.-E. and Reetz, M. (1998). Microbial lipases form versatile tools for biotechnology. *Trends Biotechnol.* **16**, 396–403.

Jin, L.Z., Tran, D.Q., Ide, C.F., McLachlan, J.A., and Arnold, S.F. (1997). Several synthetic chemicals inhibit progesterone receptor-mediated transactivation in yeast. *Biochem. Biophys. Res. Commun.* **233**, 139–146.

Johannes, C. and Majcherczyk, A. (2000). Natural mediators in the oxidation of polycyclic aromatic hydrocarbons by laccase mediator systems. *Appl. Environ. Microbiol.* **66**, 524–528.

Johannes, C., Majcherczyk, A., and Hüttermann, A. (1998). Oxidation of acenaphthylene by laccase of *Trametes versicolor* in a laccase-mediator system. *J. Biotechnol.* **61**, 151–156.

Jonas, U. (1997). Biotransformation von Biarylverbindungen durch Weißfäulepilze unter besonderer Berücksichtigung des ligninolytischen Enzymsystems von *Pycnoporus cinnabarinus* und *Trametes versicolor*. Doctoral dissertation, University of Greifswald.

Jonas, U., Hammer, E., Haupt, E.T.K., and Schauer, F. (2000). Characterisation of coupling products formed by biotansformation of biphenyl and diphenyl ether by the white rot fungus *Pycnoporus cinnabarinus*. *Arch. Microbiol.* **174**, 393–398.

Jonas, U., Hammer, E., Schauer, F., and Bollag, J.-M. (1998). Transformation of 2-hydroxydibenzofuran by laccases of white rot fungi *Trametes versicolor* and *Pycnoporus cinnabarinus* and characterization of oligomerization products. *Biodegradation* **8**, 321–328.

Jones, K.H., Trudgill, P.W., and Hopper, D.J. (1993). Metabolism of *p*-cresol by the fungus *Aspergillus fumigatus*. *Appl. Environ. Microbiol.* **59**, 1125–1130.

Jonsson, L.J., Saloheimo, M., and Penttila, M. (1997). Laccase from the white-rot fungus *Trametes versicolor*: cDNA cloning of *lcc1* and expression in *Pichia pastoris*. *Curr. Gen.* **32**, 425–430.

Joo, H., Lin, Z., and Arnold, F.H. (1999). Laboratory evolution of peroxide-mediated cytochrome P450 hydroxlation. *Nature* **399**, 670–673.

Juhasz, A.L. and Naidu, R. (2000). Bioremediation of high molecular weight polycyclic aromatic hydrocarbons: A review of the microbial degradation of benzo[a]pyrene. *Int. Biodeter. Biodegr.* **45**, 57–88.

Jung, H.C., Xu, F., and Li, K.C. (2002). Purification and characterization of laccase from wood-degrading fungus *Trichophyton rubrum* LKY-7. *Enzyme Microb. Technol.* **30**, 161–168.

Kaneyuki, H., Deno, H., Hiratsuka, J., Matsuyoshi, T., and Furukawa, T. (1980). Production of sebacic acid from n-decane by mutants derived from *Torulopsis candida*. *J. Ferment. Technol.* **58**, 405–410.

Kahraman, S. and Yesilada, O. (2001). Industrial and agricultural wastes as substrates for laccase production by white-rot fungi. *Folia Microbiol.* **46**, 133–136.

Karam, J. and Nicell, J.A. (1997). Potential applications of enzymes in waste treatment. *J. Chem. Technol. Biotechnol.* **69**, 141–153.

Kärenlampi, S.O. and Hynninen, P.H. (1981). Formation of benzoic acid from biphenyl in the yeast *Saccharomyces cerevisiae*. *Chemosphere* **10**, 391–396.

Kartal, S.N. and Clausen, C.A. (2001). Leachability and decay resistance of particleboard made from acid extracted and bioremediated CCA-treated wood. *Int. Biodeter. Biodegr.* **47**, 183–191.

Kato, S., Kawasaki, T., Urata, T., and Mochizuku, J. (1993). *In vitro* and *ex vivo* free radical scavenging activities of carazostatin, carbazomycin B and their derivatives. *J. Antibiot.* (*Tokyo*) **46**, 1859–1865.

Keitel, T., Simon, O., Borriss, R., and Heinemann, U. (1993). Molecular and active-site structure of a *Bacillus* 1,3–1,4-ß-glucanase. *Proc. Natl. Acad. Sci. U.S.A.* **90**, 5287–5291.

Kennes, C. and Lema, J.M. (1994). Simultaneous biodegradation of *p*-cresol and phenol by the basidiomycete *Phanerochaete chrysosporium*. *J. Ind. Microbiol.* **13**, 311–314.

Kim, H.S., Yoon, B.D., Choung, D.H., Oh, H.M., Katsuragi, T., and Tani, Y. (1999). Characterization of a biosurfactant, mannosylerythritol lipid produced from *Candida* sp SY16. *Appl. Microbiol. Biotechnol.* **52**, 713–721.

Kim, S., Leem, Y., Kim, K., and Choi, H.T. (2001). Cloning of an acidic laccase gene (*clac2*) from *Coprinus congregatus* and its expression by external pH. *FEMS Microbiol. Lett.* **195**, 151–156.

Kirk, T.K. and Farrell, R.L. (1987). Enzymatic combustion—the microbial-degradation of lignin. *Annu. Rev. Microbiol.* **41**, 465–505.

Klug, M.J. and Markovetz, A.J. (1967). Degradation of hydrocarbons by members of the genus *Candida*. *Appl. Microbiol.* **15**, 690–693.

Ko, E.M., Leem, Y.E., and Choi, H.T. (2001). Purification and characterization of laccase from the white-rot basidiomycete *Ganoderma lucidum*. *Appl. Microbiol. Biotechnol.* **57**, 98–102.

Kobayashi, S. and Uyama, H. (1998). Enzymatic polymerization for synthesis of polyester and polyaromatics. *Enzymes Polym. Synth. ACS Symp. Ser.* **684**, 58–73.

Kocwahaluch, R. (1995). Easy and inexpensive diffusion tests for detecting the assimilation of aromatic-compounds by yeast-like fungi. 1. Assimilation of dihydroxyphenols. *Chemosphere* **30**, 209–213.

Kokubun, T. and Harborne, J.B. (1995). Phytoalexin induction in the sapwood of plants of the Maloideae (Rosaceae)—biphenyls or dibenzofurans. *Phytochemistry* **40**, 1649–1654.

Kokubun, T., Harborne, J.B., Eagles, J., and Waterman, P.G. (1995a). Antifungal biphenyl compounds are the phytoalexins of the sapwood of *Sorbus aucuparia*. *Phytochemistry* **40**, 57–59.

Kokubun, T., Harborne, J.B., Eagles, J., and Waterman, P.G. (1995b). Dibenzofuran phytoalexins from the sapwood of *Cotoneaster acutifolius* and 5 related species. *Phytochemistry* **38**, 57–60.

Kokubun, T., Harborne, J.B., Eagles, J., and Waterman, P.G. (1995c). Dibenzofuran phytoalexins from the sapwood tissue of *Photinia, Puracantha* and *Crataegus* species. *Phytochemistry* **39**, 1033–1037.

Kokubun, T., Harborne, J.B:, Eagles, J., and Waterman, P.G. (1995d). 4-dibenzofuran phytoalexins from the sapwood of *Mespilus germanica*. *Phytochemistry* **39**, 1039–1042.

Komagata, K., Nakase, T., and Katsuya, N. (1964). Assimilation of hydrocarbons by yeast. I. Preliminary screening. *J. Gen. Appl. Microbiol.* **10**, 313–321.

Kon, Y., Iwashina, T., Kashiwadini, H., and Wardlaw, J.H. (1997a). A new benzofuran, isostreptsilic acid, produced by cultured mycobiont of the *Usnea orientalis*. *J. Jpn. Botany* **72**, 67–71.

Kon, Y., Kashiwadani, H., Wardlaw, J.H., and Elix, J.A. (1997b). Effects of culture conditions on dibenzofuran production by cultured mycobionts of lichens. *Symbiosis* **23**, 97–106.

Kostrewa, D., Gruninger-Leitch, F., D'Arcy, A., Broger, C., Mitchell, D.B., and van Loon, A.P.G.M. (1997). Crystal structure of phytase from *Aspergillus ficuum* at 2.5 Å; resolution. *Nat. Struct. Biol.* **4**, 185–190.

Krauel, H. and Weide, H. (1978). Dicarbonsäurebildung durch *Candida guilliermondii*, Stamm H17, beim Abbau von n-Tridecan in Batch-Kultur. *Z. Allg. Mikrobiol.* **18**, 47–54.

Kraus, J.J., Munir, I.Z., McEldoon, J.P., Clark, D.S., and Dordick, J.S. (1999). Oxidation of polycyclic aromatic hydrocarbons catalyzed by soybean peroxidase. *Appl. Biochem. Biotechnol.* **80**, 221–230.

Krcmar, P. and Ulrich, R. (1998). Degradation of polychlorinated biphenyl mixtures by the lignin-degrading fungus *Phanerochaete chrysosporium. Folia Microbiol.* **43**, 79–84.

Krcmar, P., Kubatova, A., Votruba, J., Erbanova, P., Novotny, C., and Sasek, V. (1999). Degradation of polychlorinated biphenyls by extracellular enzymes of *Phanerochaete chrysosporium* produced in a perforated plate bioreactor. *World J. Microbiol. Biotechnol.* **15**, 269–276.

Krivobok, S., Benoit-Guyod, J.L., Seigle-Murandi, F., Steiman, R., and Thiault, G.A. (1994). Diversity in phenol-metabolizing capability of 809 strains of micromycetes. *Microbiologica* **17**, 51–60.

Kubatova, A., Erbanova, P., Eichlerova, I., Homolka, L., Nerud, F., and Sasek, V. (2001). PCB congener selective biodegradation by the white rot fungus *Pleurotus ostreatus* in contaminated soil. *Chemosphere* **43**, 207–215.

Kulkarni, N., Shendye, A., and Rao, M. (1999). Molecular and biotechnological aspects of xylanases. *FEMS Microbiol. Rev.* **23**, 411–456.

Kwon, S.I. and Anderson, A.J. (2002). Genes for multicopper proteins and laccase activity: Common features in plant-associated *Fusarium* isolates. *Can. J. Botany* **80**, 563–570.

Lamar, R.T., Evans, J.W., and Glaser, J.A. (1993). Solid-phase treatment of a pentachlorophenol-contaminated soil using lignin-degrading fungi. *Environ. Sci. Technol.* **27**, 2566–2571.

Lamar, R.T., Glase, J.A., and Kirk, T.K. (1990). Fate of pentachlorophenol (PCP) in sterile soils inoculated with the white-rot basidiomycete *Phanerochaete chrysosporium*: Mineralization, voletilization and depletion of PCP. *Soil Biol. Biochem.* **22**, 433–440.

Landi, S. (2000). Mammalian class theta GST and differential susceptibility to carcinogens: A review. *Mut. Res.—Rev. Mut. Res.* **463**, 247–283.

Lang, E., Gonser, A., and Zadrazil, F. (2000). Influence of incubation temperature on activity of ligninolytic enzymes in sterile soil by *Pleurotus* sp and *Dichomitus squalens. J. Basic Microbiol.* **40**, 33–39.

Lange, J., Hammer, E., Specht, M., Francke, W., and Schauer, F. (1998). Biotransformation of biphenyl by the ascomycetous yeast *Debaryomyces vanrijiae. Appl. Microbiol. Biotechnol.* **50**, 364–368.

Lassen, S.F., Breinholti, J., Fuglsang, C.C., Ohmann, A., and Stergaard, P.R. (2000). *Peniophora* phytase. *World Patent WO9828408*.

Lassen, S.F., Breinholti, J., Ostergaard, P.R., Brugger, R., Bischoff, A., Wyss, M., and Fuglsang, C.C. (2001). Expression, gene cloning and characterization of five novel phytases from four basidiomycete fungi: *Peniophora lycii, Agrocybe pediades, Ceriporia* sp., and *Trametes pubescens. Appl. Environ. Microbiol.* **67**, 4701–4707.

Laugero, C., Mougin, C., Sigoillot, J.C., Moukha, S., and Asther, M. (1997). Comparison of static and agitated immobilized cultures of *Phanerochaete chrysosporium* for the degradation of pentachlorophenol and its metabolite pentachloroanisole. *Can. J. Microbiol.* **43**, 378–383.

Layton, A.C., Sanseverino, J., Gregory, B.W., Easter, J.P., Sayler, G.S., and Schultz, T.W. (2002). *In vitro* estrogen receptor binding of PCBs: Measured activity and detection of hydroxylated metabolites in a recombinant yeast assay. *Toxicol. Appl. Pharmacol.* **180**, 157–163.

Lebeault, J.-M., Lode, E.T., and Coon, M. (1971). Fatty acid and hydrocarbon hydroxylation in yeast. Role of cytochrome P-450 in *Candida tropicalis. Biochem. Biophys. Res. Commun.* **42**, 413–419.

Lee, D.H., Takahashi, M., and Tsunoda, K. (1992). Fungal detoxification of organoiodine wood preservatives. 2. Fungal metabolism in the decomposition of the chemicals. *Holzforschung* **46**, 467–469.

Legendre, D., Soumillon, P., and Fastrez, J. (1999). Engineering a regulatable enzyme for homogenous immunoassays. *Nat. Biotechnol.* **17**, 67–72.

Lehman, L.R. and Stewart, J.D. (2001). Filamentous fungi: Potentially useful catalysts for the biohydroxylations of non-activated carbon centers. *Curr. Org. Chem.* **5**, 439–470.

Lehmann, M., Kostrewa, D, Wyss, M., Brugger, R., D'Arcy, A., Pasamontes, L., and van Loon, A.P.G.M. (2000a). From DNA sequence to improved functionality: Using protein squence comparisons to rapidly design a thermostable consensus phytase. *Protein Eng.* **13**, 49–57.

Lehmann, M., Lopez-Ulibarri, R., Loch, C., Viarouge, C., Wyss, M., and van Loon, A.P.G.M. (2000b). Exchanging the active site between phytases for altering the functional properties of the enzyme. *Protein Sci.* **9**, 1866–1872.

Lenin, L., Forchiassin, F., and Ramos, A.M. (2002). Copper induction of lignin-modifying enzymes in the white-rot fungus *Trametes trogii. Mycologia* **94**, 377–383.

Leonowicz, A., Cho, N.S., Luterek, J., Wilkolazka, A., Wojtas-Wasilewska, M., Matuszewska, A., Hofrichter, M., Wesenberg, D. *et al.* (2001). Fungal laccase: Properties and activity on lignin. *J. Basic Microbiol.* **41**, 185–227.

Leonowicz, A., Matuszewska, A., Luterek, J., Ziegenhagen, D., Wojtas-Wasilewska, M., Cho, N.S., Hofrichter, M., and Rogalski, J. (1999). Biodegradation of lignin by white-rot fungi. *Fungal Genet. Biol.* **27**, 175–185.

Leonowicz, A., Trojanow, J., and Nowak, G. (1972). Ferulic acid as inductor of messenger-RNA synthesis related to laccase formation in wood rotting fungus *Pleurotus ostreatus. Microbios* **6**, 23–31.

Leontievsky, A., Myasoedova, N., Pozdnyakova, N., and Golovleva, L. (1997). "Yellow" laccase of *Panus trigrinus* oxidizes non-phenolic substrates without electron-transfer mediators. *FEBS Lett.* **413**, 446–448.

Leontievsky, A.A., Myasoedova, N.M., Baskunov, B.P., Evans, C.S., and Golovleva, L.A. (2000). Transformation of 2,4,6-trichlorophenol by the white rot fungi *Panus tigrinus* and *Coriolus versicolor. Biodegradation* **11**, 331–340.

Leontievsky, A.A., Myasoedova, N.M., Baskunov, B.P., Golovleva, L.A., Bucke, C., and Evans, C.S. (2001). Transformation of 2,4,6-trichlorphenol by free and immobilized fungal laccase. *Appl. Microbiol. Biotechnol.* **57**, 85–91.

Leontievsky, A.A., Myasoedova, N.M., Baskunov, B.P., Pozdnyakova, N.N., Vares, T., Kalkkinen, N., Hatakka, A.I., and Golovleva, L.A. (1999). Reactions of blue and yellow fungal laccases with lignin model compounds. *Biochemistry* **64**, 1150–1156.

Lesage-Meessen, G., Delattre, M., Haon, M., Thibault, J.F., Ceccaldi, B.C., Brunerie, P., and Asther, M. (1996). A two-step conversion process for vanillin production from ferulic acid combining *Aspergillus niger* and *Pycnoporus cinnabarinus. J. Biotechnol.* **50**, 107–113.

Levin, L. and Forchiassin, F. (2001). Ligninolytic enzymes of the white rot basidiomycete *Trametes trogii. Acta Biotechnol.* **21**, 179–186.

Lewis, D.F.V. (1996). *Cytochromes P450. Structure, Function and Mechanism.* Taylor & Francis, London.

Li, K., Xu, F., and Eriksson, K.-E.L. (1999). Comparison of fungal laccases and redox mediators in oxidation of a nonphenolic lignin model compound. *Appl. Environ. Microbiol.* **65**, 2654–2660.

Lindequist, U. and Schauer, F. (2002). Bioactive natural compounds—new possibilities for their derivatization. *Screening* **3**, 48–49.

Liu, H.L., Doleyres, Y., Coutinho, P.M., Ford, C., and Reilly, P.J. (2000). Replacement and deletion mutations in the catalytic domain and belt region of *Aspergillus awamori* glucoamylase to enhance thermostability. *Protein Eng.* **13**, 655–659.

Liu, L., Tewari, R.P., and Williamson, P.R. (1999). Laccase protects *Cryptococcus neoformans* from antifungal activity of alveolar macrophages. *Infect. Immun.* **67**, 6034–6039.

Loewus, F.A. and Murthy, P.P.N. (2000). Myo-inositol metabolism in plants. *Plant Sci.* **150**, 1–19.

Logan, B.E., Alleman, B.C., Amy, G.L., and Gilbertson, R.L. (1994). Adsorption and removal of pentachlorophenol by white-rot fungi in batch cultures. *Water Res.* **28**, 1533–1538.

Lottermoser, K., Schunck, W.-H., and Asperger, O. (1996). Cytochromes P450 of the sophorose lipid-producing yeast *Candida apicola*: Heterogeneity and polymerase chain reaction-mediated cloning of two genes. *Yeast* **12**, 565–575.

Lottmann, J., Hammer, E., and Schauer, F. (1999). Methyl ketone formation during degradation of phenoxybutyric acid by *Penicillium canescens* SBUG-M 1139. *Arch. Microbiol.* **172**, 417–420.

Lucas-Elio, P., Solano, F., and Sanchez-Amat, A. (2002). Regulation of polyphenol oxidase activities and melanin synthesis in *Marinomonas mediterranea*: Identification of *ppoS*, a gene encoding a sensor histidine kinase. *Microbiology* **148**, 2457–2466.

Maheshwari, R., Bharadwaj, G., and Bhat, M.K. (2000). Thermophilic fungi: Their physiology and enzymes. *Microbiol. Mol. Biol. Rev.* **64**, 461–476.

Mai, C., Schormann, W., and Hüttermann, A. (2001a). The effect of ions on the enzymatic induced synthesis of lignin graft copolymers. *Enzyme Microb. Technol.* **28**, 460–466.

Mai, C., Schormann, W., and Hüttermann, A. (2001b). Chemo-enzymatically induced copolymerization of phenolics with acrylate compounds. *Appl. Microbiol. Biotechnol.* **55**, 177–186.

Majcherczyk, A., Johannes, C., and Hüttermann, A. (1999). Oxidation of aromatic alcohols by laccase from *Trametes versicolor* mediated by 2,2'-azino-bis-(3-ethyl-benzothiazoline-6-sulphonic acid) cation radical and dication. *Appl. Microbiol. Biotechnol.* **51**, 267–276.

Mansur, M., Suarez, T., Fernandez-Larrea, J.B., Brizuela, M.A., and Gonzalez, A.E. (1997). Identification of a laccase gene family in the new lignin-degrading basidiomycete CECT 20197. *Appl. Environ. Microbiol.* **63**, 2637–2646.

Martins, L.O., Soares, C.M., Pereira, M.M., Teixeira, M., Costa, T., Jones, G.H., and Henriques, A.O. (2002). Molecular and biochemical characterization of a highly stable bacterial laccase that occurs as a structural component of the *Bacillus subtilis* endospore coat. *J. Biol. Chem.* **277**, 18849–18859.

Maspahy, S., Lamb, D.C., and Kelly, S.L. (1999). Purification and characterization of a benzo[a]pyrene hydroxylase from *Pleurotus pulmonarius. Biochem. Biophys. Res. Commun.* **266**, 326–329.

May, O., Nguyen, P.T., and Arnold, F.H. (2000). Inverting enantioselectivity and increasing total activity of a key enzyme in a multi-enzyme synthesis creates a viable process for production of L-methionine. *Nat. Biotechnol.* **18**, 317–320.

Mayer, A.F., Hellmuth, K., Schlieker, H., Lopez-Ulibarri, R., Oertel, S., Dahlems, U., Strasser, A.W.M., and van Loon, A.P.G.M. (1999). An expression system matures: A highly efficient and cost-effective process for phytase production by recombinant strains of *Hansenula polymorpha. Biotechnol. Bioeng.* **63**, 373–381.

McAllister, K.A., Lee, H., and Trevors, J.T. (1996). Microbial degradation of pentachlorophenol. *Biodegradation* **7**, 1–40.

McFadden, D.C. and Casadevall, A. (2001). Capsule and melanin synthesis in *Cryptococcus neoformans. Med. Mycol.* **39** (Suppl. 1), 10–30.

McMullan, G., Meehan, C., Conneely, A., Kirby, N., Robinson, T., Nigam, P., Banat, I.M., Marchant, R. *et al.* (2001). Microbial decolourisation and degradation of textile dyes. *Appl. Microbiol. Biotechnol.* **56**, 81–87.

Mester, T. and Tien, M. (2000). Oxidation mechanism of ligninolytic enzymes involved in the degradation of environmental pollutants. *Int. Biodeter. Biodegr.* **46**, 51–59 (Sp. Issue).

Michailides, T.J. and Spotts, R.A. (1991). Effects of certain herbicides on the fate of sporangiospores of *Mucor piriformis* and conidia of *Botrytis cinerea* and *Penicillium expansum. Pesticide Sci.* **33**, 11–22.

Michizoe, J., Goto, M., and Furusaki, S. (2001). Catalytic activity of laccase hosted in reversed micelles. *J. Biosci. Bioeng.* **92**, 67–71.

Middelhoven, W.J. (1993). Catabolism of benzene compounds by ascomycetous and basidiomycetous yeasts and yeast-like fungi—a literature-review and an experimental approach. *Antonie van Leeuwenhoek* **63**, 125–144.

Mielgo, I., Moreira, M.T., Feijoo, G., and Lema, J.M. (2002). Biodegradation of a polymeric dye in a pulsed bed reactor by immobilised *Phanerochaete chrysosporium. Water Res.* **36**, 1896–1901.

Mikami, Y., Sakamoto, T., Yazawa, K., Gonoi, T., Ueno, Y., and Hasegawa, S. (1994). Comparison of *in vitro* antifungal activity of itraconazole and hydroxy-itraconazole by colorimetric MTT assay. *Mycoses* **37**, 27–33.

Mikolasch, A., Hammer, E., Jonas, U., Popowski, K., Stielow, A., and Schauer, F. (2002). Synthesis of 3-(3,4-dihydroxy-phenyl)-propionic acid derivatives by N-coupling of amines using laccase. *Tetrahedron* **58**, 7598–7593.

Milewski, G.J., Bumpus, J.A., Jurek, M.A., and Aust, S.D. (1988). Biodegradation of pentachlorophenol by the white rot fungus *Phanerochaete chrysosporium. Appl. Environ. Microbiol.* **54**, 2885–2888.

Min, K.L., Kim, Y.H., Kim, Y.W., Jung, H.S., and Hah, Y.C. (2001). Characterization of a novel laccase produced by the wood-rotting fungus *Phellinus ribis. Arch. Biochem. Biophys.* **392**, 279–286.

Mitchell, D.B., Vogel, K., Weimann, J., Pasamontes, L., and van Loon, A.P.G.M. (1997). The phytase subfamily of histidine acid phosphatases; isolation of genes for two novel phytases from the fungi *Aspergillus terreus* and *Myceliophthora thermophila. Microbiology* **143**, 247–252.

Miyagawa, H., Hamada, N., Sato, M., and Ueno, T. (1993). Hypostrepsilic acid, a new dibenzofuran from the cultured lichen mycobiont of *Evernia esorodia. Phytochemistry* **34**, 589–591.

Miyazaki, K., Wintrode, P.L., Grayling, R.A., Rubingh, D.N., and Arnold, F.H. (2000). Directed evolution study of temperature adaptation in a psychrophilic enzyme. *J. Mol. Biol.* **297**, 1015–1026.

Mobley, D.P., Finkbeiner, H.L., Lockwood, S.H., and Spivack, J. (1993). Synthesis of 3-aryl muconolactones using biphenyl metabolism in *Aspergillus. Tetrahedron* **49**, 3273–3280.

Molina, S.M.G., Pelissari, F.A., and Vitorello, C.B.M. (2001). Screening and genetic improvement of pectinolytic fungi for degumming of textile fibers. *Braz. J. Microbiol.* **32**, 320–326.

Mollapour, M. and Piper, P.W. (2001). The *ZbYME2* gene from the food spoilage yeast *Zygosaccharomyces bailii* confers not only YME2 functions in *Saccharomyces cerevisiae*, but also the capacity for catabolism of sorbate and benzoate, two major weak organic acid preservatives. *Mol. Microbiol.* **42**, 919–930.

Mori, M., Aoyama, M., and Doi, S. (1997). Antifungal constituents in the bark of *Magnolia obovata* Thunb. *Holz Roh. Werkst.* **55**, 275–278.

Morschhäuser, J. (2002). The genetic basis of fluconazole resistance development in *Candida albicans. Biochim. Biophys. Acta* **18**, 240–248.

Mougin, C., Kollmann, A., and Jolivalt, C. (2002). Enhanced production of laccase in the fungus *Trametes versicolor* by the addition of xenobiotics. *Biotechnol. Lett.* **24**, 139–142.

Mukherje, S.K. and Majumdar, S.K. (1971). Fermentative production of pectinases by fungi—screening of organisms and production of enzymes by *Aspergillus niger. J. Ferment. Technol.* **49**, 759–761.

Muyima, N.Y.O., Zulu, G., Bhengu, T., and Popplewell, D. (2002). The potential application of some novel essential oils as natural cosmetic preservatives in an aqueous cream formulation. *Flavour Frag. J.* **17**, 258–266.

Nakamura, Y., Sungusia, M.G., Sawada, T., and Kuwahara, M. (1999). Lignin-degrading enzyme production by *Bjerkandera adusta* immobilized on polyurethane foam. *J. Biosci. Bioeng.* **88**, 41–47.

Nawas, T. and Alkofahi, A. (1994). Microbial contamination and preservative efficacy of topical creams. *J. Clin. Pharm. Ther.* **19**, 41–46.

Nerud, F. and Misurcova, Z. (1996). Distribution of lignonolytic enzymes in selected white-rot fungi. *Folia Microbiol.* **41**, 264–266.

Neujahr, H.Y. and Varga, J.M. (1970). Degradation of phenols by intact cells and cell-free preparations of *Trichosporon cutaneum. Eur. J. Biochem.* **13**, 37–44.

Niku-Paavola, M.L., Karhunen, E., Kantelinen, A., Viikari, L., Lundell, T., and Hatakka, A. (1990). The effect of culture conditions on the production of lignin modifying enzymes by the white-rot fungus *Phlebia radiata. J. Biotechnol.* **13**, 211–222.

Nishizawa, Y., Nakabayashi, K., and Shinagawa, E. (1995). Purification and characterization of laccase from white-rot fungus *Trametes sanguinea* M85-2. *J. Ferment. Bioeng.* **80**, 91–93.

Nojiri, H., Habe, H., and Omari, T. (2001). Bacterial degradation of aromatic compounds via angular dioxygenation. *J. Gen. Appl. Microbiol* **47**, 279–305.

Norsker, M., Jensen, M., and Adler-Nissen, J. (2000). Enzymatic gelation of sugar beet pectin in food products. *Food Hydrocolloids* **14**, 237–243.

Öberg, L.G., Glas, B., Swanson, S.E., Rappe, C.P., and Paul, K.G. (1990). Peroxidase-catalyzed oxidation of chlorophenols to polychlorinated dibenzo-*p*-dioxins and dibenzofurans. *Arch. Environ. Contam. Toxicol.* **19**, 930–938.

O'Callaghan, J., O'Brien, M.M., McClean, K., and Dobson, A.D.W. (2002). Optimisation of the expression of a *Trametes versicolor* laccase gene in *Pichia pastoris. J. Ind. Microbiol. Biotechnol.* **29**, 55–59.

Ogawa, J. and Shimizu, S. (1999). Microbial enzymes: New industrial applications from traditional screening methods. *Trends Biotechnol.* **17**, 13–20.

Ohga, S., Smith, M., Thurston, C.F., and Wood, D.A. (1999). Transcriptional regulation of laccase and cellulase genes in the mycelium of *Agaricus bisporus* during fruit body development on a solid substrate. *Mycol. Res.* **103**, 1557–1560.

Ohkuma, M., Muraoka, S., Tanimoto, T., Fujii, M., Ohta, A., and Takagi, M. (1995). *Cyp 52* (cytochrome-P450 alk) multigene family in *Candida maltosa*. Identification and characterization of 8 members. *DNA Cell Biol.* **14**, 163–173.

Okeke, B.C., Paterson, A., Smith, J.E., and Watson-Craik, I.A. (1997). Comparative biotransformation of pentachlorophenol in soils by solid substrate cultures of *Lentinula edodes. Appl. Microbiol. Biotechnol.* **48**, 563–569.

Okeke, B.C., Smith, J.E., Paterson, A., and Watson-Craik, I.A. (1996). Influence of environmental parameters on pentachlorophenol biotransformation in soil by *Lentinula edodes* and *Phanerochaete chrysosporium. Appl. Microbiol. Biotechnol.* **45**, 263–266.

Okumura, T. and Nishikawa, Y. (1996). Gas chromatography–mass spectrometry determination of triclosans in water, sediment and fish samples via methylation with diazomethane. *Anal. Chim. Acta* **325**, 175–184.

O'Malley, D.M., Whetten, R., Bao, W.L., Chen, C.L., and Sederoff, R.R. (1993). The role of laccase in lignification. *Plant J.* **4**, 751–757.

Ong, E., Pollock, W.B.R., and Smith, M. (1997). Cloning and sequence analysis of two laccase complementary DNAs from the ligninolytic basidiomycete *Trametes versicolor. Gene* **196**, 113–119.

Onodera, S. and Saitoh, K. (1997). Formation of chlorodibenzofurans upon thermo-chemical reactions of diphenyl ether herbicide (CNP). *Jpn. J. Tox. Environ. Health* **43**, 293–299.

Otterbein, L., Record, E., Longhi, S., Asther, M., and Moukha, S. (2000). Molecular cloning of the cDNA encoding laccase from *Pycnoporus cinnabarinus* I-937 and expression in *Pichia pastoris*. *Eur. J. Biochem.* **267**, 1619–1625.

Paice, M.G., Archibald, F.S., Bourbonnais, R., Jurasek, L., Reid, I.D., Charles, T., and Dumonceaux, T. (1996). Enzymology of kraft pulp bleaching by *Trametes versicolor. ACS Symp. Ser.* **655**, 151–164.

Paice, M.G., Bourbonnais, R., Reid, I.D., Archibald, F.S., and Jurasek, L. (1995). Oxidative bleaching enzymes—a review. *J. Pulp Paper Sci.* **21**, J280–J284.

Pallerla, S. and Chambers, R.P. (1998). Reactor development for biodegradation of pentachlorophenol. *Catal. Today* **40**, 103–111.

Palmieri, G., Bianco, C., Cennamo, G., Giardina, P., Marino, G., Monti, M., and Sannia, G. (2001). Purification, characterization, and functional role of a novel extracellular protease from *Pleurotus ostreatus. Appl. Environ. Microbiol.* **67**, 2754–2759.

Panwar, S.L., Krishnamurthy, S., Gupta, V., Alarco, A.M., Raymond, M., Sanglard, D., and Prasad, R. (2001). CaALK8, an alkane assimilating cytochrome P450, confers multidrug resistance when expressed in a hypersensitive strain of *Candida albicans. Yeast* **18**, 1117–1129.

Pasamontes, L., Haiker, T., Wyss, M., Henriquez-Huecas, M., Mitchell, D.B., and van Loon, A.P.G.M (1997a). Cloning of phytases from *Emericella nidulans* and the thermophilic fungus *Talaromyces thermophilus. Biochim. Biophys. Acta* **1353**, 217–223.

Pasamontes, L., Haiker, T., Wyss, M., Tessier, M., and van Loon, A.P.G.M (1997b). Gene cloning, purification, and characterization of a heat stable phytase from the fungus *Aspergillus fumigatus. Appl. Environ. Microbiol.* **63**, 1696–1700.

Patel, R.N., Thakker, G.D., and Rao, K.R. (1994). Potential use of a white-rot fungus *Antrodiella* sp. RK1 for biopulping. *J. Biotechnol.* **36**, 19–23.

Pen, J., Verwoerd, T.C., Vanparidon, P.A., Beudeker, R.F., van den Elzen, P.J.M., Geerse, U., van der Klis, J.D., Versteegh, H.A.J. *et al.* (1993). Phytase-containing transgenic seeds as novel feed additive for improved phosphorous utilization. *Biotechnology* **11**, 811–814.

Penttila, M. (1998). Heterologous protein production in *Trichoderma*. In G.E. Harmann and C. Kubicek (eds.) *Trichoderma and Gliocladium—enzymes, biological control and commercial applications*. Taylor & Francis, London, pp. 365–382.

Perfect, J.R., Wong, B., Chang, Y.C., Kwon-Chung, K.J., and Williamson, P.R. (1998). *Cryptococcus neoformans*: Virulence and host defences. *Med. Mycol.* **36** (Suppl. 1), 79–86.

Perie, F.H. and Gold, M.H. (1991). Manganese regulation of manganese peroxidase expression and lignin degradation by the white rot fungus *Dichomitus squalens. Appl. Environ. Microbiol.* **57**, 2240–2245.

Perie, F.H., Reddy, G.V.V., Blackburn, N.J., and Gold, M.H. (1998). Purification and characterization of laccases from the white-rot basidiomycete *Dichomitus squalens. Arch. Biochem. Biophys.* **353**, 349–355.

Petit, F., Le Goff, P., Cravedi, J.P., Valotaire, Y., and Pakdel, F. (1997). Two complementary bioassays for screening the estrogenic potency of xenobiotics: Recombinant yeast for trout estrogen receptor and trout hepatocyte cultures. *J. Molec. Endocrinol.* **19**, 321–335.

Petrounia, I.P. and Arnold, F.H. (2000). Designed evolution of enzymatic properties. *Curr. Opin. Biotechnol.* **11**, 325–330.

Petter, R., Kang, B.S., Boekhout, T., Davis, B.J., and Kwon-Chung, K.J. (2001). A survey of heterobasidiomycetous yeasts for the presence of the genes homologous to virulence factors of *Filobasidiella neoformans*, CNLAC1 and CAP59. *Microbiology* **147**, 2029–2036.

Piacquadio, P., De Stefano, G., Sammartino, M., and Sciancalepore, V. (1997). Phenols removal from apple juice by laccase immobilized on Cu^{2+}-chelate regenerable carrier. *Biotechnol. Techn.* **11**, 515–517.

Picataggio, S., Rohrer, T., Deanda, K., Lanning, D., Reynolds, R., Mielenz, J., and Eirich, L.D. (1992). Metabolic engineering of *Candida tropicalis* for the production of long-chain dicarboxylic acids. *Bio-Technology* **10**, 894–898.

Pickard, M.A., Roman, R., Tinoco, R., and Vazquez-Duhalt, R. (1999). Polycyclic aromatic hydrocarbon metabolism by white rot fungi and oxidation by *Coriolopsis gallica* UAMH 8260 laccase. *Appl. Environ. Microbiol.* **65**, 3805–3809.

Pijnenburg, A.M.C.M., Everts, J.W., de Boer, J., and Boon, J.P. (1995). Polybrominated biphenyl and diphenyl ether flame retardants: Analysis, toxicity and environmental occurrence. *Rev. Environ. Contam. Toxicol.* **141**, 1–26.

Pilz, R., Hammer, E., Schauer, F., and Kragl, U. (2003). Laccase-catalyzed synthesis of coupling products of phenolic substrates in different reactors. *Appl. Microbiol. Biotechnol.* **60**, 708–712.

Piontek, K. (2002). New insights into lignin peroxidase. *Ind. J. Chem. A—Inorg. Bio-Inorg. Phys. Theoret. Analyt. Chem.* **41**, 46–53.

Planas, A. (2000). Bacterial 1,3-1,4-ß-glucanases: Structure, function and protein engineering. *Biochim. Biophys. Acta* **1543**, 361–382.

Pointing, S.B. (2001). Feasibility of bioremediation by white-rot fungi. *Appl. Microbiol. Biotechnol.* **57**, 20–33.

Pollard, S.J.T., Hrudey, S.E., and Fedorak, P.M. (1994). Bioremediation of petroleum- and creosote-contaminated soils—a review of constraints. *Waste Manag. Res.* **12**, 173–194.

Polnisch, E., Kneifel, F., Franzke, H., and Hofmann, K.H. (1992). Degradation and dehalogenation of monochlorophenols by the phenol-assimilating yeast *Candida maltosa. Biodegradation* **2**, 193–199.

Pommer, E.H. and Lorentz, G. (1982). Resistance of *Botrytis cinerea* to dicarboximide fungicides—a literature review. *Crop Protect.* **1**, 221–230.

Poutanen, K. (1997). Enzymes: An important tool in the improvement of the quality of cereal foods. *Trends Food Sci. Tech.* **8**, 300–306.

Prenafeta-Boldu, F.X., Luykx, D.M.A.M., Vervoort, J., and de Bont, J.A.M. (2001). Fungal metabolism of toluene: Monitoring of fluorinated analogs by F-19 nuclear magnetic resonance spectroscopy. *Appl. Environ. Microbiol.* **67**, 1030–1034.

Quan, C., Zhang, L., Wang, Y., and Ohta, Y. (2001). Production of phytase in a low phosphate medium by a novel yeast *Candida krusei. J. Biosci. Bioeng.* **92**, 154–160.

Rabinovich, M.L., Melnik, M.S., and Boloboba, A.V. (2002). Microbial cellulases. *Appl. Biochem. Microbiol.* **38**, 305–321.

Radwan, S.S. and Sorkhoh, N.A. (1993). Lipids of n-alkane-utilizing microorganisms and their application potential. *Adv. Appl. Microbiol.* **39**, 29–90.

Raghukumar, C., D'Souza, T.M., Thorn, R.G., and Reddy, C.A. (1999). Lignin-modifying enzymes of *Flavodon flavus*, a basidiomycete isolated from a coastal marine environment. *Appl. Environ. Microbiol.* **65**, 2103–2111.

Rahouti, M., Steiman, R., Seigle-Murandi, F., and Christov, L.P. (1999). Growth of 1044 strains and species of fungi on 7 phenolic lignin model compounds. *Chemosphere* **38**, 2549–2559.

Ralph, J.P., Graham, L.A., and Catcheside, D.E.A. (1996). Extracellular oxidases and the transformation of solubilised low rank coal by wood-rot fungi. *Appl. Microbiol. Biotechnol.* **46**, 226–232.

Ratledge, C. (1988). Hydrocarbons. Products of hydrocarbon microorganism interaction. In D.R. Houghton, R.N. Smith, and H.O.W. Eggins, (eds.) *Biodeterioration* **7**, Elsevier Appl. Sci., London, pp. 219–235.

Ravelet, C., Krivobok, S., Sage, L., and Steiman, R. (2000). Biodegradation of pyrene by sediment fungi. *Chemosphere* **40**, 557–563.

Read, G. and Vining, L.C. (1959). Telephoric acid. *Can. J. Chem.* **37**, 1442–1445.

Record, E., Punt, P.J., Chamkha, M., Labat, M., van den Hondel, C.A.M.J.J., and Asther, M. (2002). Expression of the *Pycnoporus cinnabarinus* laccase gene in *Aspergillus niger* and characterization of the recombinant enzyme. *Eur. J. Biochem.* **269**, 602–609.

Reddy, C.A. and D'Souza, T.M. (1994). Physiology and molecular biology of the lignin peroxidases of *Phanerochaete chrysosporium. FEMS Microbiol. Rev.* **13**, 137–152.

Reddy, G.V.B., Gelpke, M.D.S., and Gold, M.H. (1998). Degradation of 2,4,6-trichlorophenol by *Phanerochaete chrysosporium*: Involvement of reductive dechlorination. *J. Bacteriol.* **180**, 5159–5164.

Reddy, G.V.B., Joshi, D.K., and Gold, M.H. (1997). Degradation of chlorophenoxyacetic acids by the lignin-degrading fungus *Dichomitus squalens. Microbiology* **143**, 2353–2360.

Reyes, P., Pickard, M.A., and Vasquez-Duhalt, R. (1999). Hydroxybenzotriazole increases the range if textile dyes decolorized by immobilized laccase. *Biotechnol. Lett.* **21**, 875–880.

Richardson, A., Duncan, J., and McDougall, G.J. (2000). Oxidase activity in lignifying xylem of a taxonomically diverse range of trees: Identification of a conifer laccase. *Tree Physiol.* **20**, 1039–1047.

Richardson, A.E., Hadobas, P.A., and Hayes, J.E. (2001). Extracellular secretion of *Aspergillus* phytase from *Arabidopsis* roots enables plants to obtain phosphorous from hytate. *Plant J.* **25**, 641–649.

Ricotta, A., Unz, R.F., and Bollag, J.M. (1996). Role of a laccase in the degradation of pentachlorophenol. *Bull. Environ. Contam. Toxicol.* **57**, 560–567.

Rigling, D. and van Alfen, N.K. (1993). Extracellular and intracellular laccases of the chestnut blight fungus, *Cryphonectria parasitica. Appl. Environ. Microbiol.* **59**, 3634–3639.

Rigot, J. and Matsumura, F. (2002). Assessment of the rhizosphere competency and pentachlorophenol-metabolizing activity of a pesticide-degrading strain of *Trichoderma harzianum* introduced into the root zone of corn seedlings. *J. Environ. Sci. Health B* **37**, 202–210.

Roberts, S.A., Weichsel, A., Grass, G., Thakali, K., Hazzard, J.T., Tollin, G., Rensing, C., and Montfort, W.R. (2002). Crystal structure and electron transfer kinetics of CueO, a multicopper oxidase required for copper homeostasis in *Escherichia coli. Proc. Natl. Acad. Sci. U.S.A.* **99**, 2766–2771.

Robinson, T., Chandran, B., and Nigam, P. (2001). Studies on the production of enzymes by white-rot fungi for the decolourisation of textile dyes. *Enzyme Microb. Technol.* **29**, 575–579.

Rodriguez, E., Pickard, M.A., and Vazquez-Duhalt, R. (1999). Industrial dye decolorization by laccases from ligninolytic fungi. *Curr. Microbiol.* **38**, 27–32.

Rogalski, J., Lundell, T.K., Leonowicz, A., and Hatakka, A.I. (1991). Influence of aromatic compounds and lignin on production of ligninolytic enzymes by *Phlebia radiata. Phytochemistry* **30**, 2869–2872.

Romero, M.C., Hammer, E., Cazau, M.C., and Arambarri, A.M. (2001). Selection of autochthonous yeast strains able to degrade biphenyl. *World J. Microbiol. Biotechnol.* **17**, 591–594.

Rosenbrock, P., Martens, R., Buscot, F., Zadrazil, F., and Munch, J.C. (1997). Enhancing the mineralization of [U-^{14}C]dibenzo-*p*-dioxin in three different soils by addition of organic substrate or inoculation with white-rot fungi. *Appl. Microbiol. Biotechnol.* **48**, 665–670.

Rothemund, C., Amann, R., Klugbauer, S., Manz, W., Bieber, C., Schleifer, K.H., and Wilderer, P. (1996). Microflora of 2,4-dichlorophenoxyacetic acid degrading biofilms on gas permeable membranes. *Syst. Appl. Microbiol.* **19**, 608–615.

Rouvinen, J., Bergfors, T., Teeri, T., Knowles, J.K.C., and Jones, T.A. (1990). Three-dimensional structure of cellobiohydrolase II from *Trichoderma reesei. Science* **249**, 380–386.

Ruttimann-Johnson, C., Cullen, D., and Lamar, R.T. (1994). Manganese peroxidases of the white-rot fungus *Phanerochaete sordida. Appl. Environ. Microbiol.* **60**, 599–605.

Ruttimann-Johnson, C. and Lamar, R.T. (1996). Polymerization of pentachlorophenol and ferulic acid by fungal extracellular lignin-degrading enzymes. *Appl. Environ. Microbiol.* **62**, 3890–3893.

Ryazanova, L.P., Mikhaleva, N.I., Solveva, I.V., Boev, A.V., Okunev, O.N., and Kulaev, I.S. (1996). Pectolytic enzymes from *Aspergillus heteromorphus. Appl. Biochem. Microbiol.* **32**, 1–6.

Ryu, D.D.Y. and Nam, D.-H. (2000). Recent progress in biomolecular engineering. *Biotechnol. Progr.* **16**, 2–16.

Sack, U., Heinze, T.M., Deck, J., Cerniglia, C.E., Martens, R., Zadrazil, F., and Fritsche, W. (1997a). Comparison of phenanthrene and pyrene degradation by different wood-decaying fungi. *Appl. Environ. Microbiol.* **63**, 3919–3925.

Sack, U., Hofrichter, M., and Fritsche, W. (1997b). Degradation of polycyclic aromatic hydrocarbons by manganese peroxidase of *Nematoloma frowardii. FEMS Microbiol. Lett.* **152**, 227–234.

Safe, S., Ellis, B., and Hutzinger, O. (1976). The *in vitro* hydroxylation of 4'-chloro-4-biphenylol by a mushroom tyrosinase preparation. *Can. J. Microbiol.* **22**, 104–106.

Safe, S., Hutzinger, O., Ecobichon, O.J., and Grey, A.A. (1975a). The metabolism of 4'-chloro-4-biphenylol in the rat. *Can. J. Biochem.* **53**, 415–420.

Safe, S., Platonow, N., and Hutzinger, O. (1975b). Metabolism of chlorobiphenyls in the goat and cow. *J. Agric. Food Chem.* **23**, 259–261.

Sahasrabudhe, N.A. and Sankpal, N.V. (2001). Production of organic acids and metabolites of fungi for food industry. In G.G. Khachatourians and D.K. Arora (eds.), *Applied Mycology and Biotechnology. Vol. 1: Agriculture and Food Production* (pp. 387–425). Amsterdam: Elsevier.

Sahasrabudhe, S.R., Shailubhai, K., Vora, K.A., and Modi, V.V. (1987). Dehalogenation of chlorinated derivatives of phenoxyacetic acid by *Aspergillus niger. Microbios* **34**, 19–22.

Sanchez-Amat, A., Lucas-Elio, P., Fernandez, E., Garcia-Borron, J.C., and Solano, F. (2001). Molecular cloning and functional characterization of a unique multipotent polyphenol oxidase from *Marinomonas mediterranea. Biochim. Biophys. Acta—Protein Struct. Mol. Enzymol.* **1547**, 104–116.

Sanglard, D. and Loper, J.C. (1989). Characterization of the alkane-inducible cytochrome P-450 (*P-450 alk*) gene from the yeast *Candida tropicalis*. Identification of a new *P-450* gene family. *Gene* **76**, 121–136.

Sauer, J., Sigurskjold, B.W., Christensen, U., Frandsen, T.P., Mrgorodskaya, E., Harrison, A., Roepstorff, P., and Svensson, B. (2000). Glucoamylase: Structure/function relationships, and protein engineering. *Biochim. Biophys. Acta* **1543**, 275–293.

Saxena, R.K., Gupta, R., Saxena, S., and Gulati, R. (2001). Role of fungal enzymes in food processing. In G.G. Khachatourians and D.K. Arora (eds.) *Applied mycology and biotechnology: Agriculture and food production* (Vol. 1). Elsevier Science, Amsterdam, pp. 353–386.

Schäfer, A., Specht, M., Hetzheim, A., Francke, W., and Schauer, F. (2001). Synthesis of substituted imidazoles and dimerization products using cells and laccase from *Trametes versicolor. Tetrahedron* **57**, 7693–7699.

Schauer, F., Henning, K., Pscheidl, H., Wittich, R.M., Fortnagel, P., Wilkes, H., Sinnwell, V., and Francke, W. (1995). Biotransformation of diphenyl ether by the yeast *Trichosporon beigelii* SBUG 752. *Biodegradation* **6**, 173–180.

Schauer, F., Lindequist, U., Hammer, E., Jülich, W.D., Schäfer, A., and Jonas, U. (2001). Biotransformation von biologisch aktiven Verbindungen aus verschiedenen chemischen Stoffklassen mittels der Enzyme Laccase und Manganperoxidase. *Patentschrift, PCP/EP 01/07152.*

Scheibner, K., Hofrichter, M., and Fritsche, W. (1997). Mineralization of 2-amino-4,6-dinitrotoluene by manganese peroxidase of the white-rot fungus *Nematoloma frowardii. Biotechnol. Lett.* **19**, 835–839.

Scheller, U., Zimmer, T., Becher, D., Schauer, F., and Schunck, W.-H. (1998). Oxygenation cascade in conversion of n-alkanes to α,ω-dioic acids catalyzed by cytochrome P450 52A3. *J. Biol. Chem.* **273**, 32528–32534.

Scherer, M. and Fischer, R. (2001). Molecular characterization of a blue-copper laccase, TILA, of *Aspergillus nidulans. FEMS Microbiol. Lett.* **199**, 207–213.

Schlosser, D. and Hofer, C. (2002). Laccase-catalyzed oxidation of Mn^{2+} in the presence of natural Mn^{3+} chelators as a novel source of extracellular H_2O_2 production and its impact on manganese peroxidase. *Appl. Environ. Microbiol.* **68**, 3514–3521.

Schmidt, O., Dittberner, D., and Faix, O. (1991). On the reaction of some bacteria and fungi on coal-tar creosote. *Mater. Organismen* **26**, 13–30.

Scholz, F., Schädel, S., Schultz, A., and Schauer, F. (2000). Chronopotentiometric study of laccase-catalyzed oxidation of quinhydrone microcrystals immobilised on a gold electrode surface and of the oxidation of a phenol-derivatised graphite electrode surface. *J. Electroanalyt. Chem.* **480**, 241–248.

Schouten, A., Wagemakers, L., Stefanato, F.L., van der Kaaij, R.M., and van Kann, J.A.L. (2002). Resveratrol acts as a natural profungicide and induces self-intoxication by a specific laccase. *Mol. Microbiol.* **43**, 883–894.

Schultz, A., Jonas, U., Hammer, E., and Schauer, F. (2001). Dehalogenation of chlorinated hydroxybiphenyls by fungal laccase. *Appl. Environ. Microbiol.* **67**, 4377–4381.

Schultz, T.W., Kraut, D.H., Sayler, G.S., and Layton, A.C. (1998). Estrogenicity of selected biphenyls evaluated using a recombinant yeast assay. *Environ. Toxicol. Chem.* **17**, 1727–1729.

Schunck, W.-H., Kärgel, E., Grass, B., Wiedmann, B., Mauersberger, S., Köpke, K., Kiessling, U., Strauss, M. *et al.* (1989). Molecular cloning and characterization of the primary structure of the alkane hydroxylating cytochrome P-450 from the yeast *Candida maltosa. Biochem. Biophys. Res. Commun.* **161**, 843–850.

Schwartz, R.D., Williams, A.L., and Hutchinson, D.B. (1980). Microbial production of 4,4'-dihydroxybiphenyl: Biphenyl hydroxylation by fungi. *Appl. Environ. Microbiol.* **39**, 702–708.

Sealey, J., Ragauskas, A.J., and Elder, T.J. (1999). Investigations into laccase-mediator delignification of kraft pulps. *Holzforschung* **53**, 498–502.

Seghezzi, W., Meili, C., Ruffiner, R., Künzi, R., Sanglard, D., and Fiechter, A. (1992). Identification and characterization of additional members of the cytochrome P450 multigene family cyp 52 of *Candida tropicalis. DNA Cell Biol.* **11**, 767–780.

Seigle-Murandi, F.M., Krivobok, S.M.A., Steiman, R.L., Benoit-Guyod, J.-L.A., and Thiault, G.-A. (1991). Biphenyl oxide hydroxylation by *Cunninghamella echinulata. J. Agric. Food Chem.* **39**, 428–430.

Seigle-Murandi, F., Steiman, R., Benoit-Guyod, J.L., and Guiraud, P. (1993). Fungal degradation of pentachlorophenol by Micromycetes. *J. Biotechnol.* **30**, 27–35.

Seigle-Murandi, F., Toe, A., Benoit-Guyod, J.L., Steiman, R., and Kadri, M. (1995). Depletion of pentachlorophenol by deuteromycetes isolated from soil. *Chemosphere* **31**, 2677–2686.

Sephton, M.A., Looy, C.V., Veefkind, R.J., Visscher, H., Brinkhuis, H., and de Leeuw, J.W. (1999). Cyclic diaryl ethers in a Late P sediment. *Org. Geochem.* **30**, 267–273.

Shafiee, A., Harris, G., Motamedi, H., Rosenbach, M., Chen, T., Zink, D., and Heimbuch, B. (2001). Microbial hydroxylation of rustmicin (galbonolide A) and galbonolide B, two antifungal products produced by *Micromonospora* sp. *J. Mol. Catal. B—Enzym.*, **11**, 237–242.

Shailubhai, K., Sahasrabudhe, S.R., Vora, K.A., and Modi, V.V. (1983). Degradation of chlorinated derivatives of phenoxyacetic acid and benzoic acid by *Aspergillus niger. FEMS Microbiol. Lett.* **18**, 279–282.

Sheldon, R.A. (1994). *Metalloporphyrius in catalytic oxidation*. Marcel Dekker Inc., New York.

Sherry, J. (1994). Effects of 2,4-dichlorophenoxyacetic acid on fungal propagules in fresh-water ponds. *Environ. Toxicol. Water Qual.* **9**, 209–221.

Shimizu, M. (1993). Purification and characterization of phytase and acid phosphatase by *Aspergillus oryzae* K1. *Biosci. Biotechnol. Biochem.* **57**, 1364–1365.

Shuttleworth, K.L., Postie, L., and Bollag, J.M. (1986). Production of induced laccase by the fungus *Rhizoctonia praticola*. *Can. J. Microbiol.* **32**, 867–870.

Sietmann, R. (2002). Physiologische und biochemische Charakterisierung der Transformation umweltrelevanter Verbindungen mit Biarylstruktur durch Hefen der Gattung Trichosporon. Doctoral dissertation, University of Greifswald.

Sietmann, R., Hammer, E., and Schauer, F. (2002). Biotransformation of biarylic compounds by yeasts of the genus *Trichosporon*. *Syst. Appl. Microbiol.* **25**, 332–339.

Sietmann, R., Hammer, E., Moody, J., Cerniglia, C.E., and Schauer, F. (2000). Hydroxylation of biphenyl by the yeast *Trichosporon mucoides*. *Arch. Microbiol.* **174**, 353–361.

Sietmann, R., Hammer, E., Specht, N., Cerniglia, C.E., and Schauer, F. (2001). Novel ring cleavage products in the biotransformation of biphenyl by the yeast *Trichosporon mucoides*. *Appl. Environ. Microbiol.* **67**, 4158–4165.

Silva, C.M.M.D., de Melo, I.S., Maia, A.D.N., and Abakerli, R.B. (1999). Isolation of carbendazim degrading fungi. *Pesqui. Agropecu. Bras.* **34**, 1255–1264.

Sinnott, M.L. (1990). Catalytic mechanism of enzymatic glycosyl transfer. *Chem. Rev.* **90**, 1171–1202.

Skorobogatko, O.V., Stepanova, E.V., Gavrilova, V.P., and Yaropolov, A.I. (1996). Effects of inducer on the synthesis of extracellular laccase by *Coriolus hirsutus*, a basidial fungus. *Appl. Biochem. Microbiol.* **32**, 473–376.

Slomczynski, D., Nakas, J.P., and Tanenbaum, S.W. (1995). Production and characterization of laccase from *Botrytis cinerea*-61-34. *Appl. Environ. Microbiol.* **61**, 907–912.

Smirnov, S.A., Koroleva, O.V., Gavrilova, V.P., Belova, A.B., and Klyachko, N.L. (2001). Laccases from basidiomycetes: Physicochemical characteristics and substrate specificity towards methoxyphenolic compounds. *Biochemistry (Moscow)* **66**, 774–779.

Smith, J.G. and Christophers, A.J. (1992). Phenoxy herbicides and chlorophenols: A case control study on soft tissue sarcoma and malignant lymphoma. *Br. J. Cancer* **65**, 442–448.

Smith, R.V. and Rosazza, J.P. (1974). Microbial models of mammalian metabolism. *Arch. Biochem. Biophys.* **161**, 551–558.

Smith, R.V., Davis, P.J., Clark, A.M., and Glover-Milton, S. (1980). Hydroxylation of biphenyl by fungi. *J. Appl. Bacteriol.* **49**, 65–73.

Soares, G.M.B., Amorim, M.T.P., Hrdina, R., and Costa-Ferreira, M. (2002). Studies on the biotransformation of novel disazo dyes by laccase. *Process Biochem.* **37**, 581–587.

Soden, D.M. and Dobson, A.D.W. (2001). Differential regulation of laccase gene expression in *Pleurotus sajor-caju*. *Microbiology* **147**, 1755–1763.

Solano, F., Garcia, E., de Egea, E.P., and Sanchez-Amat, A. (1997). Isolation and characterization of strain MMB-1 (CECT 4803), a novel melanogenic marine bacterium. *Appl. Environ. Microbiol.* **63**, 3499–3506.

Solano, F., Lucas-Elio, P., Fernandez, E., and Sanchez-Amat, A. (2000). *Marinomonas mediterranea* MMB-1 transposon mutagenesis: Isolation of a multipotent polyphenol oxidase mutant. *J. Bacteriol.* **182**, 3754–3760.

Srebotnik, E. and Hammel, K.E. (2000). Degradation of nonphenolic lignin by the laccase/1-hydroxybenzotriazole system. *J. Biotechnol.* **81**, 179–188.

Steffens, J.J., Pell, E.J., and Tien, M. (1996). Mechanisms of fungicide resistance in phytopathogenic fungi. *Curr. Opin. Biotechnol.* **7**, 348–355.

Stemmer, W.P.C. (1994). Rapid evolution of a protein *in vitro* by DNA shuffling. *Nature* **370**, 389–391.

Stenroos, S. (1989). Taxonomic revision of the *Cladonia miniata* group. *Ann. Bot. Fenn.* **26**, 237–261.

Stephan, I., Leithoff, H., and Peek, R.D. (1996). Microbial conversion of wood treated with salt preservatives. *Mater. Organismen* **30**, 179–199.

Stope, M.B., Becher, D., Hammer, E., and Schauer, F. (2002). Cometabolic ring fission of dibenzofuran by Gram-negative and Gram-positive biphenyl-utilizing bacteria. *Appl. Microbiol. Biotechnol.* **59**, 62–67.

Sun, Y. and Cheng, J.Y. (2002). Hydrolysis of lignocellulosic materials for ethanol production: A review. *Biores. Technol.* **83**, 1–11.

Sunna, A. and Antranikian, G. (1997). Xylanolytic enzymes from fungi and bacteria. *Crit. Rev. Biotechnol.* **17**, 39–67.

Svenson, A., Kjeller, L.-O., and Rappe, C. (1989). Enzyme-mediated formation of 2,3,7,8-tetrasubstituted chlorinated dibenzodioxins and dibenzofurans. *Environ. Sci. Technol.* **23**, 900–902.

Takada, S., Nakamura, M., Matsueda, T., Kondo, R., and Sakai, K. (1996). Degradation of polychlorinated dibenzo-*p*-dioxins and polychlorinated dibenzofurans by the white rot fungus *Phanerochaete sordida* YK-624. *Appl. Environ. Microbiol.* **62**, 4323–4328.

Takagi, M., Ohkuma, M., Kobayashi, N., Watanabe, M., and Yano, K. (1989). Purification of cytochrome P450alk from normal-grown cells of *Candida maltosa*, and cloning and nucleotide sequencing of the encoding gene. *Agric. Biol. Chem.* **53**, 2217–2227.

Takahashi, A., Agatsuma, T., Matsuda, M., Ohta, T., Nunozawa, T., Endo, T., and Nozoe, S. (1992). Russuphelin A, a new cytotoxic substance from the mushroom *Russula subnigricans. Chem. Pharm. Bull.* **40**, 3185–3188.

Takahashi, A., Agatsuma, T., Ohta, T., Nunozawa, T., and Endo, T. (1993). Russuphelin-B, russuphelin-C, russuphelin-D, russuphelin-F, new cytotoxic substances from the mushroom *Russula subnigricans* Hongo. *Chem. Pharm. Bull.* **41**, 1726–1729.

Takamine, J. (1894). Process of making diastatic enzyme. *US Patent 525,823.*

Tanaka, C., Tajima, S., Furusawa, I., and Tsuda, M. (1992). The Pgr1 mutant of *Cochliobolus heterostrophus* lacks a *p*-diphenol oxidase involved in naphthalenediol melanin synthesis. *Mycol. Res.* **96**, 959–964.

Tanaka, T., Tonosaki, T., Nose, M., Tomidokoro, N., Kadomura, N., Fujii, T., and Taniguchi, M. (2001). Treatment of model soils contaminated with phenolic endocrine-disrupting chemicals with laccase from *Trametes* sp. in a rotating reactor. *J. Biosci. Bioeng.* **92**, 312–316.

Tekere, M., Ncube, I., Read, J.S., and Zvauya, R. (2002). Biodegradation of the organochlorine pesticide, lindane by a sub-tropical white rot fungus in batch and packed bed bioreactor systems. *Environ. Technol.* **23**, 199–206.

Temp, U., Meyrahn, H., and Eggert, C. (1999a). Extracellular phenol oxidase patterns during depolymerization of low-rank coal by three basidiomycetes. *Biotechnol. Lett.* **21**, 281–287.

Temp, U., Zierold, U., and Eggert, C. (1999b). Cloning and characterization of a second laccase gene from the lignin-degrading basidiomycete *Pycnoporus cinnabarinus. Gene* **236**, 169–177.

ten Have, R. and Teunissen, P.J.M. (2001). Oxidative mechanisms involved in lignin degradation by white-rot fungi. *Chem. Rev.* **101**, 3397–3413.

Thiele, S., Fernandes, E., and Bollag, J.M. (2002). Enzymatic transformation and binding of labeled 2,4,6-trinitrotoluene to humic substances during an anaerobic/aerobic incubation. *J. Environ. Qual.* **31**, 437–444.

Thomas, B.R., Yonekura, M., Morgan, T.D., Czapla, T.H., Hopkins, T.L., and Kramer, K.J. (1989). A trypsin-solubilized laccase from pharate pupal integument of the tobacco hornworm, *Manduca sexta. Insect Biochem.* **19**, 611–622.

Thomas, D.R., Carswell, K.S., and Georgiou, G. (1992). Mineralization of biphenyl and PCBs by the white rot fungus *Phanerochaete chrysosporium. Biotechnol. Bioeng.* **40**, 1395–1402.

Tomschy, A., Brugger, R., Lehmann, M., Svendsen, A., Vogel, K., Kostrewa, D., Lassen, S.F., Burger, D. *et al.* (2002). Engineering of phytase for improved activity at low pH. *Appl. Environ. Microbiol.* **68**, 1907–1913.

Tomschy, A., Tessier, M., Wyss, M., Brugger, R., Broger, C., Schnoebelen, L., van Loon, A.P.G.M, and Pasamontes, L. (2000). Optimization of the catalytic properties of *Aspergillus fumigatus* phytase based on the three dimensional structure. *Protein Sci.* **9**, 1304–1311.

Törrönen, A., Harkki, A., and Rouvinen, J. (1994). Three-dimensional structure of endo-1,4-beta-xylanase II from *Trichoderma reesei*: Two conformational states in the active site. *EMBO J.* **13**, 2493–2501.

Tsai, H.F., Wheeler, M.H., Chang, Y.C., and Kwon-Chung, K.J. (1999). A developmentally regulated gene cluster involved in conidial pigment biosynthesis in *Aspergillus fumigatus. J. Bacteriol.* **181**, 6469–6477.

Tsioulpas, A., Dimou, D., Iconomou, D., and Aggelis, G. (2002). Phenolic removal in olive oil mill wastewater by strains of *Pleurotus* spp. in respect to their phenol oxidase (laccase) activity. *Biores. Technol.* **84**, 251–257.

Tsujimoto, T., Uyama, H., and Kobayashi, S. (2001). Polymerization of vinyl monomers using oxidase catalysts. *Macromolec. Biosci.* **1**, 228–232.

Tuomela, M., Lyytikainen, M., Oivanen, P., and Hatakka, A. (1999). Mineralization and conservation of pentachlorophenol (PCP) in soil inoculated with the white rot fungus *Trametes versicolor. Soil Biol. Biochem.* **31**, 65–74.

Turner, W.B. and Aldridge, D.C. (1983). *Fungal metabolites II.* Academic Press, London.

Uchida, H., Fukuda, T., Miyamoto, H., Kawabata, T., Suzuki, M., and Uwajima, T. (2001). Polymerization of bisphenol A by purified laccase from *Trametes villosa. Biochem. Biophys. Res. Commun.* **287**, 355–358.

Ullah, M.A., Bedford, C.T., and Evans, C.S. (2000). Reactions of pentachlorophenol with laccase from *Coriolus versicolor. Appl. Microbiol. Biotechnol.* **53**, 230–234.

Uyama, H. (2001). Enzymatic synthesis and applications of new polymeric materials. *Kobunshi Ronbunshu* **58**, 382–296.

Valli, K. and Gold, M.H. (1991). Degradation of 2,4-dichlorophenol by the lignin-degrading fungus *Phanerochaete chrysosporium. J. Bacteriol.* **173**, 345–352.

van Aken, B. and Agathos, S.N. (2002). Implication of manganese (III), oxalate, and oxygen in the degradation of nitroaromatic compounds by manganese peroxidase (MnP). *Appl. Microbiol. Biotechnol.* **58**, 345–352.

van Aken, B., Hofrichter, M., Scheibner, K., Hatakka, A.I., Naveau, H., and Agathos, S.N. (1999). Transformation and mineralization of 2,4,6-trinitrotoluene (TNT) by manganese peroxidase from the white-rot basidiomycete *Phlebia radiata. Biodegradation* **10**, 83–91.

van den Brink, H.J.M., van Gorcom, R.F.M., van den Hondel, C.A.M.J.J., and Punt, P.J. (1998). Cytochrome P450 enzyme systems in fungi. *Fungal Genet. Biol.* **23**, 1–17.

Vares, T., Kalsi, M., and Hatakka, A. (1995). Lignin peroxidases, manganese peroxidases, and other ligninolytic enzymes produced by *Phlebia radiata* during solid-state fermentation of wheat-straw. *Appl. Environ. Microbiol.* **61**, 3515–3520.

Vares, T., Lundell, T.K., and Hatakka, A.I. (1993). Production of multiple lignin peroxidases by the white-rot fungus *Phlebia ochraceofulva. Enzyme Microb. Technol.* **15**, 664–669.

Vasconcelos, A.F.D., Barbosa, A.M., Dekker, R.F.H., Scarminio, I.S., and Rezende, M.I. (2000). Optimization of laccase production by *Botrysphaeria* sp. in the presence of veratryl alcohol by the response-surface method. *Process Biochem.* **35**, 1131–1138.

Vazquezduhalt, R., Westlake, D.W.S., and Fedorak, P.M. (1994). Lignin peroxidase oxidation of aromatic-compounds in systems containing organic-solvents. *Appl. Environ. Microbiol.* **60**, 459–466.

Venkov, P., Topashka-Ancheva, M., Georgieva, M., Alexieva, V., and Karanov, E. (2000). Genotoxic effect of substituted phenoxyacetic acids. *Arch. Toxicol.* **74**, 560–566.

Viney, I. and Bewly, R.J.F. (1990). Preliminary studies on the development of a microbiological treatment for polychlorinated biphenyls. *Appl. Environ. Contam. Toxicol.* **19**, 789–796.

Vroumsia, T., Steiman, R., Seigle-Murandi, F., and Benoit-Guyod, J.L. (1999). Effects of culture parameters on the degradation of 2,4-dichlorophenoxyacetic acid (2,4-D) and 2,4-dichlorophenol (2,4-DCP) by selected fungi. *Chemosphere* **39**, 1397–1405.

Vyas, B.R.M., Sasek, V., Matucha, M., and Bubner, M. (1994a). Degradation of 3,3',4,4'-tetrachlorobiphenyl by selected white-rot fungi. *Chemosphere* **28**, 1127–1134.

Vyas, B.R.M., Volc, J., and Sasek, V. (1994b). Ligninolytic enzymes of selected white-rot fungi cultivated on wheat-straw. *Folia Microbiol.* **39**, 235–240.

Wachtmeister, C.A. (1956). Studies on the chemistry of lichens. *Acta Chem. Scand.* **10**, 1404–1413.

Wagner, H.-C., Schramm, K.-W., and Hutzinger, O. (1990). Biogenes polychloriertes Dioxin aus Trichlorphenol. *Z. Umweltchem. Ökotox.* **2**, 63–65.

Wahleithner, J.A., Xu, F., Brown, K.M., Brown, S.H., Golightly, E.J., Halkier, T., Kauppinen, S., Pederson, A. *et al.* (1996). The identification and characterization of four laccases from the plant pathogenic fungus *Rhizoctonia solani. Curr. Genet.* **29**, 395–403.

Waldo, G.S., Standish, B.M., Berendzen, J., and Terwilliger, T.C. (1999). Rapid protein folding assay using green fluorescent protein. *Nat. Biotechnol.* **17**, 691–695.

Wallnöfer, R.R., Engelhardt, G., Safe, O., and Hutzinger, O. (1973). Microbial hydroxylation of 4-chlorobiphenyl and 4,4'-chlorobiphenyl. *Chemosphere* **2**, 69–72.

Ward, M., Wilson, L.J., Kodama, K.H., Rey, M.W., and Berka, R.M. (1990). Improved production of chymosin in *Aspergillus* by expression as a glucoamylase chymosin fusion. *Bio-Technology* **8**, 435–440.

Wariishi, H., Dunford, H.B., MacDonald, I.D., and Gold, M.H. (1989). Manganese peroxidase from the lignin-degrading basidiomycete *Phanerochaete chrysosporium. J. Biol. Chem.* **264**, 3335–3340.

Wild, B.L. (1983). Double resistance by citrus green mould *Penicillium digitatum* to the fungicides guazatine and benomyl. *Ann. Appl. Biol.* **103**, 237–241.

Williamson, P.R. (1994). Biochemical and molecular characterization of the diphenol oxidase of *Cryptococcus neoformans*—identification as a laccase. *J. Bacteriol.* **176**, 656–664.

Williamson, P.R., Wakamatsu, K., and Ito, S. (1998). Melanin biosynthesis in *Cryptococcus neoformans. J. Bacteriol.* **180**, 1570–1572.

Willmann, G. and Fakoussa, R.M. (1997). Extracellular oxidative enzymes of coal-attacking fungi. *Fuel Proc. Technol.* **52**, 27–41.

Wiseman, A., Gondal, J.A., and Sims, P. (1975). 4'-Hydroxylation of biphenyl by yeast containing cytochrome P450: Radiation and thermal stability, comparisons with liver enzyme (oxidized and reduced forms). *Biochem. Soc. Trans.* **3**, 278–281.

Wittich, R.-M. (1998a). *Biodegradation of dioxins and furans.* Springer-Verlag, Berlin.

Wittich, R.-M. (1998b). Degradation of dioxin-like compounds by microorganisms. *Appl. Microbiol. Biotechnol.* **49**, 489–499.

Wittich, R.-M., Wilkes, H., Sinnwell, V., Francke, W., and Fortnagel, P. (1992). Metabolism of dibenzo-*p*-dioxin by *Sphingomonas* sp. strain RW1. *Appl. Environ. Microbiol.* **58**, 1005–1010.

Wodzinski, R.J. and Ullah, A.H.J. (1996). Phytase. *Adv. Appl. Microbiol.* **42**, 263–302.

Wolfenden, R., Lu, X., and Young, G. (1998). Spontaneous hydrolysis of glycosides. *J. Am. Chem. Soc.* **120**, 6814–6815.

Wolter, M., Zadrazil., F., Martens, R., and Bahadir, M. (1997). Degradation of eight highly condensed polycyclic aromatic hydrocarbons by *Pleurotus* sp. Florida in solid wheat straw substrate. *Appl. Microbiol. Biotechnol.* **48**, 398–404.

Wong, K.K.Y. and Mansfield, S.D. (1999). Enzymatic processing for pulp and paper manufacture—a review. *Appita J.* **52**, 409–418.

Xia, G., Jin C., Zhou, J., Yang, S., Zhang, S., and Jin, C. (2001). A novel chitinase having a unique mode of action from *Aspergillus fumgatus* YJ-407. *Eur. J. Biochem.* **268**, 4079–4085.

Xu, F. (1999). Laccase. In M.C. Flickinger and S.W. Drew (eds). *Encyclopedia of bioprocess technology: Fermentation, biocatalysis and bioseparation.* Wiley, New York, pp. 1545–1554.

Yadav, J.S., Quensen, J.F., Tiedje, J.M., and Reddy, C.A. (1995). Degradation of polychlorinated biphenyl mixtures (Aroclor-1242, Aroclor-1254 and Aroclor-1260) by the white-rot fungus *Phanerochaete chrysosporium* as evidence by congener-specific analysis. *Appl. Environ. Microbiol.* **61**, 2560–2565.

Yaropolov, A.I., Kharybin, A.N., Emneus, J., Markovarga, G., and Gorton, L. (1995). Flow-injection analysis of phenols at a graphite electrode modified with co-immobilized laccase and tyrosinase. *Anal. Chim. Acta* **308**, 137–144.

Yaropolov, A.I., Skorobogatko, O.V., Vartanov, S.S., and Varfolomeyev, S.D. (1994). Laccase-properties, catalytic mechanism, and applicability. *Appl. Biochem. Biotechnol.* **49**, 257–280.

Yaver, D.S. and Golightly, E.J. (1996). Cloning and characterization of three laccase genes from the white-rot basidiomycete *Trametes villosa*: Genomic organization of the laccase gene family. *Gene* **181**, 95–102.

Yaver, D.S., Overjero, M.D., Xu, F., Nelson, B.A., Brown, K.M., Halkier, T., Bernauer, S., Brown, S.H. *et al.* (1999). Molecular characterization of laccase gene from the basidiomycete Coprinus cinereus and heterologous expression of the laccase Lcc1. *Appl. Environ. Microbiol.* **65**, 4943–4948.

Zeddel, A., Majcherczyk, A., and Hüttermann, A. (1993). Degradation of polychlorinated biphenyls by the white-rot fungi *Pleurotus ostreatus* and *Trametes versicolor* in a solid state system. *Toxicol. Environ. Chem.* **40**, 255–266.

Zeddel, A., Majcherczyk, A., and Hüttermann, A. (1994). Degradation and mineralization of polychlorinated biphenyls by white rot fungi in solid-phase and soil incubation experiments. In R.E. Hinchee (ed.) *Bioremediation of chlorinated and polychlorinated aromatic hydrocarbon compounds.* Lewis Publ. Boca Raton, pp. 436–440.

Zhang, S. and Williamson, P.R. (2001). Laccase gene expression in response to glucose starvation and temperature via Hsp70 and Hsf regulation. *Mol. Biol. Cell* **12** (Suppl. S), 1224.

Zhang, Z.B., Kornegay, E.T., Radcliffe, J.S., Denbow, D.M., Veit, H.P., and Larsen, C.T. (2000a). Comparison of genetically engineered microbial and plant phytases for young broilers. *Poultry Sci.* **79**, 709–717.

Zhang, Z.B., Kornegay, E.T., Radcliffe, J.S., Wilson, J.H., and Veit, H.P. (2000b). Comparison of phytase from genetically engineered *Aspergillus* and canola in weanling pig diets. *J. Anim. Sci.* **78**, 2868–2878.

Zhao, J. and Kwan, H.S. (1999). Characterization, molecular cloning, and differential expression analysis of laccase genes from the edible mushroom *Lentinula edodes. Appl. Environ. Microbiol.* **65**, 4908–4913.

Zhu, X.D., Gibbons, J., Garcia-Rivera, J., Casadevall, A., and Williamson, P.R. (2001). Laccase of *Cryptococcus neoformans* is a cell wall-associated virulence factor. *Infect. Immun.* **69**, 5589–5596.

Zilly, A., Souza, C.G.M., Barbosa-Tessmann, I.P., and Peralta, R.M. (2002). Decolorization of industrial dyes by a Brazilian strain of *Pleurotus pulmonarius* producing laccase as the sole phenol-oxidizing enzyme. *Folia Microbiol.* **47**, 273–277.

Zimmermann, R. (1958). Über phenolverwertende Hefen. *Naturwiss enschaften* **45**, 165–166.

Zouari, N., Romette, J.L., and Thomas, D. (1994). Laccase electrode for the continuous-flow determination of phenolic-compounds. *Biotechnol. Tech.* **8**, 503–508.

12

Organic Acid Production by Filamentous Fungi

Jon K. Magnuson and Linda L. Lasure

1. Introduction

Many of the commercial production processes for organic acids are excellent examples of fungal biotechnology. However, unlike penicillin, the organic acids have had a less visible impact on human well-being. Indeed, organic acid fermentations are often not even identified as fungal bioprocesses, having been overshadowed by the successful deployment of the β-lactam processes. Yet, in terms of productivity, fungal organic acid processes may be the best examples of all. For example, commercial processes using *Aspergillus niger* in aerated stirred-tank-reactors can convert glucose to citric acid with greater than 80% efficiency and at final concentrations in hundreds of grams per liter. Surprisingly, this phenomenal productivity has been the object of relatively few research programs. Perhaps a greater understanding of this extraordinary capacity of filamentous fungi to produce organic acids in high concentrations will allow greater exploitation of these organisms via application of new knowledge in this era of genomics-based biotechnology.

In this chapter, we will explore the biochemistry and modern genetic aspects of the current and potential commercial processes for making organic acids. The organisms involved, with a few exceptions, are filamentous fungi, and this review is limited to that group. Although yeasts including *Saccharomyces cerevisiae*, species of *Rhodotorula, Pichia*, and *Hansenula* are important organisms in fungal biotechnology, they have not been significant for commercial organic acid production, with one exception. The yeast, *Yarrowia lipolytica*, and related yeast species, may be in use commercially to produce citric acid (Lopez-Garcia, 2002). Furthermore, in the near future engineered yeasts may provide new commercial processes to make lactic acid (Porro *et al.* 2002).

This chapter is divided into two parts. The first contains a review of the commercial aspects of current and potential large-scale processes for fungal organic acid production. The second presents a detailed review of current knowledge of the biochemistry and genetic regulation of organic acid biosynthesis. The organic acids considered are limited

---

Jon K. Magnuson and Linda L. Lasure • Pacific Northwest National Laboratory, 902 Battelle Blvd., P.O. Box 999, MSIN: K2-12, Richland, WA 99352.

Advances in Fungal Biotechnology for Industry, Agriculture, and Medicine. Edited by Jan S. Tkacz and Lene Lange, Kluwer Academic/Plenum Publishers, 2004.

to polyfunctional acids containing one or more carboxyl groups, hydroxyl groups, or both, that are closely tied to central metabolic pathways. A major objective of the review is to link the biochemistry of organic acid production to the available genomic data.

2. Commercial Successes: Organic Acids from Filamentous Fungi

Although many organic acids are made by living cells, few are produced commercially. Citric, gluconic, itaconic, and lactic acids are manufactured via large-scale bioprocesses. Oxalic, fumaric, and malic acids can be made through fungal bioprocesses, but the market demand is small, since competing chemical conversion routes are currently more economical. A few other organic acids have been explored for the development of novel processes. To date, the largest commercial quantities of fungal organic acids are citric acid and gluconic acid, both of which are prepared by fermentation of glucose or sucrose by *A. niger*. Another *Aspergillus* species, *A. terreus*, is used to make itaconic acid. A significant commercial source of lactic acid at the time of this writing is a bioprocess employing the Zygomycete fungus *Rhizopus oryzae*.

These three species of fungi were initially chosen for process development because they exhibited the ability to produce large amounts of a particular organic acid. This prompts us to ask why these fungi produce seemingly ridiculous quantities of organic acids. One could reasonably argue that in their natural habitats (mostly soils) these fungi would not encounter high concentrations of free sugars very frequently and, consequently, may not have evolved tight regulation of acid production. Thus, when placed in an artificial medium with high carbohydrate, they may engage in profligate acid production, ultimately resulting in their own demise. Alternately, one might argue that the ability to acidify their environment confers a competitive advantage on these fungi. First, the chelating properties of citric acid in conjunction with the increased solubility of most metal compounds at acidic pH may allow *A. niger* to grow in environments where metals are present at very low concentrations or in an insoluble state. Second, acidification would inhibit the growth of competitors, as a majority of rapidly growing bacterial species and many fungi cannot grow below pH 3.

The three fungal species produce a variety of organic acids, which may reflect different strategies to compete with other microorganisms. Many strains of *A. niger* can lower the pH of their environment by oxidizing glucose outside the cell wall, converting it to gluconic acid via the action of the enzyme glucose oxidase. The ability to catabolize gluconic acid is more unusual than the catabolism of glucose, and gluconic acid is also an effective chelator and acidulant. Other strains of *A. niger* produce citric acid intracellularly and export the acid, perhaps as a chelator and acidifier that can also be reabsorbed for use as a carbon source. *A. terreus* acidifies the environment by producing itaconic acid. Itaconic acid is not a primary metabolite, so both the anabolism and catabolism of this acid are relatively rare metabolic attributes. Once again, acidification of the environment with itaconic acid will inhibit the growth of many microorganisms. Subsequently, the relatively unusual nature of itaconic acid would permit *A. terreus* and only a few other species to catabolize the acid. It is interesting to note that the aspergilli, and all the other filamentous fungi of the phylum Ascomycota fail to produce lactic acid. The ability to produce large amounts of lactic acid appears to be restricted to the phylum Zygomycota. Perhaps fungi classified as Zygomycetes, including *R. oryzae*, have developed a different strategy for acidifying the environment by producing lactic acid to compete with fungi unable to metabolize lactic acid. These fungi often produce both ethanol and lactic acid, a combination that would discourage many competitors.

The four commercial organic acids produced by fungi are employed in high-volume, low-value applications. For example, they are used in industrial metal cleaning or other metal treatments and in the food and feed industry as flavor enhancers, acidifiers, stabilizers, or preservatives. The commercial success of fungal bioprocesses is ultimately based on rapid and economical conversion of sugars to acid, but that alone does not explain the commercial situation for each of these acids. An understanding of the economic and business parameters that have contributed to the success of these four products may be useful in development and commercialization of new organic acid products from filamentous fungi.

2.1. Citric Acid

One of the most significant parameters explaining the commercial success of citric acid is the huge market size. In 1998, the worldwide production of citric acid was 879,000 metric tons (Lopez-Garcia, 2002), a number that far exceeds the production of any other organic acid made by fermentation. About 70% of the marketed citric acid is used in diverse food and beverage products, with carbonated beverages accounting for nearly 50% of the total production in the 1990s. The remainder of the market is mostly pharmaceutical formulations, though the metal cleaning and detergent markets are expected to increase. The market size continues to grow, largely because of expanding food and beverage markets in developing countries. The selling price of citric acid has continuously decreased over the last two decades as the market shifted from pharmaceutical to food applications, and this change in the economic climate is reflected in the ownership history of citric acid manufacturing plants. From about 1950 to 1980, citric acid was used primarily in pharmaceutical or health-related consumer products. In the early 1980s, the two largest manufacturers controlling the majority of the market were Pfizer and Miles/Bayer, suppliers of prescription drugs and over-the-counter remedies. By the early 2000s, with the expansion of the market into carbonated beverages and prepared foods, nearly all the citric acid manufacture worldwide was integrated into the corn wet milling industry either by acquisition (Archer Daniels Midland bought the Pfizer business and Tate and Lyle bought the Miles/Bayer business) or new process development (Cargill). Citric acid was a logical product line addition for the corn wet milling industry since the glucose syrups prepared from corn could be used to make the two major ingredients of carbonated beverages: citric acid by fermentation and high-fructose corn syrup by an enzymatic route (immobilized glucose isomerase). Today citric acid is considered a commodity chemical and is available as dry crystals in the anhydrous or monohydrate form. There are also several grades of 50% (w/w) solutions (saturated) made either from crystalline citric acid or from citric acid process recovery streams.

In addition to fermentation productivity and yield, two other attributes of organic acid manufacture are important economic factors: recovery and formulation of the final product. Two recovery processes are approved for food-grade citric acid in the United States: lime/sulfuric acid precipitation and liquid extraction (Title 21CFR173.280, 1984; Title 21CFR184.1033, 1994). The first step in either process is the separation of the fermentation liquor from the biomass by filtration or centrifugation (the by-product biomass can be sold as a supplement for animal feed). In the lime/sulfuric acid process the fermentation liquor is then treated with calcium hydroxide to precipitate calcium citrate, which is filtered from the slurry, washed free of impurities, and dissolved with sulfuric acid. The insoluble calcium sulfate generated is separated from the citric acid solution, and the solution is deionized and concentrated for crystallization to form either anhydrous or monohydrated citric acid. The major disadvantage of this process is the calcium sulfate by-product, which constitutes a significant disposal problem. The other approved process is counter-current liquid extraction, which employs a mixture of tri-laurylamine, n-octanol, and decane or undecane to extract citric acid from the fermentation broth, followed by extraction of the citric acid back into water at higher temperature (Baniel *et al.*, 1981). Subsequent purification steps include solvent washes, passage through activated carbon, concentration by evaporation, crystallization, and drying. In an alternative patented process, citric acid is recovered from aqueous solution via anion-exchange with a tertiary amine resin followed by thermal desorption (McQuigg *et al.*, 2000).

2.2. Gluconic Acid

Gluconic acid can be prepared in a bioprocess employing *A. niger*. Unlike the citric acid process where glucose is taken up by the organism, converted to citric acid, and exported, gluconic acid is produced extracellularly. Glucose in the medium is oxidized in a two-step reaction to gluconic acid through the action of glucose oxidase. The process can also be conducted in the absence of cells with glucose oxidase and catalase derived from *A. niger*, and under the appropriate conditions, nearly 100% of the glucose is converted to gluconic acid. The enzymatic process has the added advantage that no product purification steps are required. Both processes are approved by the FDA (Title 21CFR184.1318, 1986). In addition to many food applications, gluconic acid is used as an additive to improve cement hardening.

The economic parameters involved with the manufacture and marketing of this acid are complex. Gluconic acid can be a by-product of the glucose oxidase production processes, or it can be made very efficiently from glucose syrups by enzymatic conversion (Lantero and Shetty, 2001). Another common food additive, glucono-δ-lactone, can be prepared by crystallization from solutions of gluconic acid. Gluconic acid production has been an important example of a fungal bioprocess in the past, but it appears that the process may contribute a decreasing proportion of the commercial production volume as an enzyme-based process becomes cost-effective.

2.3. Itaconic Acid

A. terreus is employed for itaconic acid production in a process similar to that for citric acid. Both processes were invented about the same time (Kane 1945; Nubel and Ratajak, 1962; Batti and Schweiger, 1963), and both can be conducted in the same manufacturing facility. At this writing, the sole producer of itaconic acid in the USA also produces citric acid. Although the process is similar to that for citric acid, as would be expected for a by-product of the citric acid cycle, there is a significant difference: the sensitivity of the organism to the acid that necessitates neutralization to obtain yields above 80 g/L (Nubel and

Ratajak, 1962). Recovery of itaconic acid can be accomplished with the technology used for citric acid, and the final product can be prepared as a dry crystalline powder.

In contrast with citric, gluconic, and lactic acids, itaconic acid is used exclusively in non-food applications. Its primary application is in the polymer industry where it is employed as a co-monomer at a level of 1–5% for certain products. Itaconic acid is also important as an ingredient for the manufacture of synthetic fibers, coatings, adhesives, thickeners, and binders. The market volume has been estimated to be about 15,000 metric tons per year and is expected to grow if the selling price (estimated to be about US$4 per kg) can be reduced (Willke and Vorlop, 2001). To date very little research has been directed at the improvement of itaconic acid production. In contrast, there has been a larger research effort directed at lactic acid production to feed the market for biodegradable plastic.

2.4. L-Lactic Acid

There are several methods to prepare lactic acid. Among the biological routes is a process employing *R. oryzae*, a fungus unrelated to the aspergilli used for the other organic acids. Taxonomically *R. oryzae* belongs to a completely different phylum in the Fungi, and its strategy for acidifying the environment also appears to be distinct. The organism imports glucose and exports lactate, an acid that is not a component or by-product of the citric acid cycle. Lactate is produced by the organism aerobically, and the commercial process requires agitation and aeration just as the other fungal organic acid processes do.

The fungal lactic acid process faces several economics challenges. One dissimilarity with *A. niger* processes is that the growth and metabolic function of *R. oryzae* is inhibited below pH 4.5, and continuous neutralization of the fermentation is required to achieve the currently maximal yield of ~80 g/L. The pH sensitivity of the organism and the tendency for filamentous growth, which further complicates the process, increase manufacturing costs. Recent research efforts have been focused on discovering alternate producers or engineering lactic acid production in a more suitable microorganism (Porro *et al.*, 2002). The substrate for the *R. oryzae* process is glucose, and the manufacturers are corn-processing companies with readily available low-cost glucose. Process improvement research has included cloning the lactate dehydrogenase genes and metabolic engineering (Skory, 2000, 2001).

Lactic acid is recovered by the technologies used for the other organic acids, including precipitation from an alcoholic extract. In aqueous solution, lactic acid dimerizes to form lactide, an intermediate for the biodegradable plastic, polylactic acid.

Until recently, lactic acid was used primarily in the food industry as a preservative, flavor enhancer, and acidulant. The global market has been estimated to be about 100,000 tons/year. This is expected to grow rapidly when facilities for polylactic acid manufacture become operational: a single plant scheduled for start-up in 2002 will expand the market by 30%. Another non-food application for lactic acid is the manufacture of the biodegradable solvent, ethyl lactate.

2.5. Market Prospects

Initially, it seems surprising that so few fungi are used for organic acid manufacture, given their efficiency at producing high concentrations of various acids. However, the relevant economic driver is the continuing availability of inexpensive petroleum-derived carbon backbone molecules for chemical synthesis of organic acids. To date, the food applications market has demanded low cost organic acids. Similarly, the biodegradable polymer market demands low cost monomers. The increased demand for racemically pure lactic acid to feed the burgeoning polylactide manufacturing capability may lead to more research on lactic acid biosynthesis. Positive market pressure from the upward trend in the cost of finite petroleum resources and increased interest in the manufacture of biodegradable plastics may encourage further research into both the discovery and development of new organic acid processes as well as the refinement of known processes.

3. Biochemistry and Genetics of Organic Acid Production by Filamentous Fungi

A survey of current knowledge regarding the metabolic pathways, biochemistry, and genetics of organic acid production by filamentous fungi, and opportunities for improvement of these organisms are the objectives of the second part of this chapter.

3.1. *Aspergillus* and Organic Acid Production

The genus *Aspergillus* contains the workhorses of the fungal fermentation industry. Aspergilli have found application in the production of foods, enzymes, pharmaceuticals, and organic acids. *A. niger* is the source of three organic acids, gluconate, citrate, and oxalate, and *A. terreus* is the source of itaconate. The biochemistry and physiology of these organic acid fermentations will be treated as a group, since they share similar physiology and each product is no more than one enzymatic step from the primary pathway of D-glucose and D-fructose metabolism.

3.1.1. Citric Acid

The production of citric acid is the oldest and most thoroughly studied filamentous fungal fermentation (Currie, 1917). Many of the parameters important for a productive submerged citric acid fermentation process were determined by Shu and Johnson (1947, 1948a, b). The metabolic pathway is known, as are the fermentation conditions that result in high yields (approximately 200 g/L of citric acid from 240 g/L of glucose or sucrose) in submerged culture. The critical parameters for citric acid production by *A. niger* were defined empirically and include: high carbohydrate concentration, low but finite manganese concentrations (~10 ppb), maintenance of high dissolved oxygen, constant agitation, and low pH (Schreferl 1986; Zhang and Röhr, 2002a, b). These physical and chemical conditions are important for the adoption and maintenance of a pelleted morphology, which is also critical for citric acid production. Knowledge of these factors has enabled the development of highly efficient submerged fermentations for citric acid production. Research in the last 60 years has revealed some of the answers to why these parameters are important, but many questions about the physiological and biochemical mechanisms underlying these empirically derived fermentation conditions remain unanswered.

The production of citrate from glucose or sucrose involves a large number of enzymatic steps occurring in two different membrane-bound cellular compartments, namely, the cytosol and the mitochondrion. Glucose is taken into the cell and converted to the three-carbon acid, pyruvate, via the glycolytic pathway in the cytosol. One molecule of pyruvate is decarboxylated with the formation of acetyl-CoA by the mitochondrial pyruvate dehydrogenase complex and another is carboxylated to oxaloacetate in the cytosol by pyruvate carboxylase. Oxaloacetate must be transported into the mitochondrion (via malate) and condensed with acetyl-CoA to form citrate. The product is transported out of the mitochondrion and finally out of the cell. The mechanisms have been exhaustively reviewed (Kubicek and Röhr, 1986; Mattey, 1992; Ruijter *et al.*, 2002). The accumulated evidence on the biochemistry of citric acid production in *A. niger* will be presented with attention to possible targets for improvement.

The metabolic pathway leading to citric acid accumulation was elucidated through radioisotopic studies carried out in the 1950s (Martin 1950; Lewis and Weinhouse, 1951; Bomstein and Johnson, 1952; Cleland and Johnson, 1954). The earliest studies, performed with $^{14}CO_2$ and ^{14}C-labeled acetate, suggested a mechanism of citrate formation involving the condensation of "active acetate" (acetyl-CoA) and oxaloacetate that were derived by the decarboxylation of pyruvate and the carboxylation of pyruvate, respectively. The study by Cleland and Johnson utilized doubly labeled D-[3,4-^{14}C]glucose under citric acid production conditions (pH 2, high oxygen), which suppressed scrambling of the label by minimizing the formation and utilization of polyol by-products. Their study verified the C_2 plus C_4 mechanism, and they also demonstrated that the citric acid cycle was essentially shut down under citric acid production conditions as "very little shifting of labels in C_4 (oxaloacetate) due to equilibration with symmetrical fumarate" occurred. Subsequent studies demonstrated the fixation of carbon dioxide (Woronick and Johnson, 1960) and the enzyme involved, pyruvate carboxylase (Bloom and Johnson, 1962). Thus, the high yields observed in the citric acid production process are possible because all six carbons of the substrate, glucose or fructose, are conserved in the six-carbon product citric acid, through the glycolytic pathway and the actions of two additional enzymes, pyruvate carboxylase and citrate synthase (Figure 12.1).

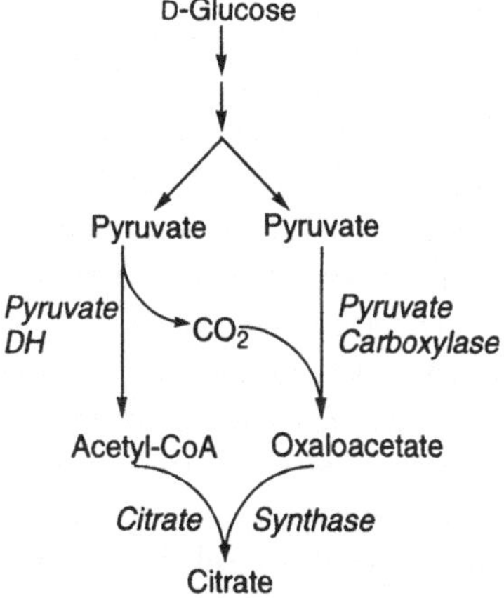

Figure 12.1. Simplified scheme of citrate synthesis in *Aspergillus niger*.

Recently, methods designed to view the citric acid production system as a whole have been applied to *A. niger*. These approaches fall under the freshly coined moniker of *systems biology*, and include metabolic modeling studies (discussed later) and NMR spectroscopy of intact cells. NMR has been used to study actively metabolizing cells of *A. niger* under various physiological conditions. Specifically, ^{13}C-NMR has been used to follow the intracellular metabolism of D-[1-^{13}C]glucose under conditions that are favorable or unfavorable for citric acid production (10% and 2% glucose, respectively) (Peksel *et al.*, 2002). The labeling patterns observed were consistent with the classical ^{14}C tracer studies showing the condensation of C_2 (acetyl-CoA) and C_4 (oxaloacetate) compounds to form citrate, thus confirming the role of pyruvate carboxylase in the formation of citrate.

Many industrial citric acid fermentations use molasses as a feedstock, which is principally sucrose (α-D-glucopyranosyl-β-D-fructofuranoside), therefore, the function of invertase is important for these processes. *A. niger* hydrolyzes sucrose through the action of invertase (β-D-fructofuranosidase), about 60% of which it exports extracellularly and 40% it retains in the cytosol and/or the periplasmic space (Vainstein and Peberdy, 1991). The gene encoding this invertase, *suc1*, has been cloned and sequenced (Boddy *et al.*, 1993). While low levels of invertase activity are constitutively expressed, the substrates sucrose and raffinose [α-galactosyl-(1→6)-α-D-glucopyranosyl-β-D-fructofuranoside] are strong inducers of expression. Glucose, fructose, and xylose are strong repressors of invertase expression in the absence of sucrose but only weak repressors in the presence of sucrose (Vainstein and Peberdy, 1991). The pH optimum of the invertase encoded by *suc1* is 5.5 whereas that of the *suc2* product, a second invertase with inulinase activity is 5.0 (Boddy *et al.*, 1993; Wallis *et al.*, 1997). Citric acid fermentations employing *A. niger* start at about pH 3 and drop to less than pH 2, suggesting that the majority of sucrose must be imported and hydrolyzed by invertase within the cytosol. Although sucrose–proton symporters are common in the yeast genus, *Kluyveromyces* (Kilian *et al.*, 1991), to our knowledge, they have not been studied in filamentous fungi.

Glucose and fructose must be transported across the cell membrane. A specific fructose transporter has not been identified in *A. niger*, but they are known in yeasts and *A. nidulans* (Mark, and Romano, 1971; Boles and Hollenberg, 1997; Heiland *et al.*, 2000). Steady state kinetic analyses of intact mycelia displayed biphasic uptake of radiolabeled glucose, indicating the presence of two glucose transporters in *A. niger*; a constitutively expressed high affinity glucose transporter (K_m = 0.3 mM) and a low affinity glucose transporter (K_m = 3.7 mM), which is induced at high glucose concentrations (15%) (Torres *et al.*, 1996a). Both transporters are susceptible to inhibition by citrate, although the low affinity transporter is less sensitive (Torres *et al.*, 1996a). Models have been developed for both active transport and passive diffusion mechanisms based on a number of assumptions, but neither model fits the experimental data precisely (Wayman and Mattey, 2000). Nevertheless, incorporating the citrate inhibition parameters into the model suggested that the two known glucose transporters would be inadequate for the observed glucose uptake rate at the high citric acid concentrations observed during the course of an industrial fermentation. Perhaps a combination of active transport and passive diffusion occur, or an additional high affinity glucose transporter is present with the same K_m but lower sensitivity to citrate. Additional experimental studies and model refinements will be required to resolve this issue. The *A. niger* genome sequence should greatly facilitate the identification and study of additional hexose and disaccharide transporters.

Upon entry into the cytosol, glucose and fructose are phosphorylated to glucose-6-phosphate and fructose-6-phosphate, which are key branch-points for glycolysis, synthesis of intracellular storage compounds, synthesis of cell wall components, and the pentose phosphate pathway. Fructose is phosphorylated by hexokinase and glucose is phosphorylated by either glucokinase or hexokinase. The genes for both of these enzymes have been cloned and characterized from *A. niger* (Panneman *et al.*, 1996, 1998). Based on the properties of the two enzymes shown in Table 12.1, it was proposed that glucokinase would account for most of the glucose phosphorylation at pH 7.5, whereas hexokinase would account for more of the phosphorylation at pH 6.5, at glucose concentrations exceeding 0.5 mM (Panneman *et al.*, 1998).

^{31}P-NMR has been used to examine intact cells of *A. niger*, resulting in valuable insights about the connections between oxygen concentration, intracellular pH, and glycolytic function. The acidophilic nature of *A. niger* is demonstrated by the finding that even under the extremely low ambient pH conditions of citric acid fermentation, the cytosolic and vacuolar (mitochondrial) compartments retain pH values of 7.5 and 6.1, respectively (Hesse *et al.*, 2000, 2002). Additional investigations demonstrated that intracellular pH declines as dissolved oxygen concentration falls (Legiša and Grdadolnik, 2002). Based on the kinetic parameters of the enzymes (Table 12.1), glucokinase would be expected to be responsible for a large portion of the flux from glucose to glucose-6-phosphate at pH 7.5. Nevertheless, flux of fructose or glucose through hexokinase appears to be significant during growth on sucrose greater than 5%, as disruption of the trehalose-6-phosphate synthase gene (*tpsA*) decreases the intracellular levels of the hexokinase inhibitor, trehalose-6-phosphate, and results in an increased citric acid production rate (Arisan-Atac *et al.*, 1996; Wolschek and Kubicek, 1997). The investigators note that this effect would probably not be observed in a glucose fermentation; in other words, the observed effect of the *tpsA* disruption on sucrose metabolism in *A. niger* could be decreased inhibition of hexokinase leading to increasing flux from fructose to fructose-6-phosphate. The dependence of intracellular pH on oxygen concentration and the decreased specific activities of critical enzymes at acidic pH explain one facet of the physiological requirement for vigorous aeration to support high flux from glucose to citric acid.

Trehalose and the polyols, glycerol, mannitol, and erythritol, are the primary small storage molecules produced from sugars or glycolytic intermediates in *A. niger*. The accumulation and utilization of these compounds vary with culture conditions (sugar concentration) and age (Röhr *et al.*, 1987; Witteveen and Visser, 1995; Peksel *et al.*, 2002). The proposed physiological roles of these compounds include osmotic balance, carbon storage,

Table 12.1. Properties of Hexose Phosphorylating Enzymes in *Aspergillus niger*

Enzyme	Hexokinase	Glucokinase
EC #	2.7.1.1	2.7.1.2
pH Optimum	7.5	7.5
Activity at pH 6.5	50 %	17 %
K_m for Glucose (mM)	0.350	0.063
K_m for Fructose (mM)	2	120
K_i for Trehalose-6-P (mM)	0.01	not inhibited

redox balance, and transport through hyphae (Witteveen and Visser, 1995). During growth on 2% sucrose, glycerol and erythritol accumulate early while mannitol becomes predominant as the culture ages (Witteveen and Visser, 1995). ^{13}C-NMR studies of citric acid producing cultures growing on 10% D-[1-^{13}C]glucose indicated that trehalose and mannitol accumulated during the growth phase, trehalose declined and glycerol accumulated during the production phase, while erythritol maintained a minimal concentration throughout (Peksel *et al.*, 2002). The observed labeling patterns in the polyols and the accumulation of glycerol suggest that glycolytic control is displaced from fructose-6-phosphate dehydrogenase to glyceraldehyde-3-phosphate dehydrogenase under citric acid production conditions (10% glucose media). Thus, the authors propose that glyceraldehyde-3-phosphate dehydrogenase would be a promising target for metabolic engineering.

Over the last decade, a number of glycolytic enzymes of *A. niger* have been purified, characterized, and cloned (see Table 12.2) (Habison *et al.*, 1983; Meixner-Monori *et al.*, 1984; Arts *et al.*, 1987; Panneman *et al.*, 1996; Panneman *et al.*, 1998; Ruijter and Visser, 1999). Subsequently, some of these genes have been over-expressed in an effort to increase flux through this critical pathway leading to citric acid production. *A. niger* 6-phosphofructokinase (PFK), like the PFK from other organisms, is very sensitive to activation by fructose-2,6-bisphosphate (half-maximal stimulation at less than 0.2 μM) (Arts *et al.*, 1987). Citrate is a strong inhibitor of PFK and the simultaneous presence of the activators AMP (0.1 mM), NH_4^+ ions (20 mM) and fructose-2,6-bisphosphate (0.2 μM) are required to overcome inhibition by citrate (5 mM) (Arts *et al.*, 1987). The apparent requirement for significant intracellular NH_4^+ concentrations to relieve this inhibition may be of practical interest in the control of commercial citrate processes where NH_3 introduction is carefully metered to provide sufficient nitrogen to maintain citric acid metabolism without promoting accumulation of biomass or increasing the pH of the fermentation. Sufficient intracellular NH_4^+ may be generated due to the decreased protein synthesis and concomitant amino acid accumulation observed under the low Mn^{2+} conditions of the citric acid production process (Kubicek *et al.*, 1979; Ma *et al.*, 1985).

The *A. niger pfkA* gene was cloned and used to increase the expression of its product 3 to 5 times that of the wild-type strain (Ruijter *et al.*, 1997). In the same study, pyruvate kinase (*pki*) was cloned and over-expressed alone or with PFK. However, in all three transformants (*pki, pfk, pki + pfk*) the citric acid production rate and the pools of glycolytic intermediates remained at the same levels observed for the wild-type strain, indicating no increase in flux through the glycolytic pathway. In the *pfkA* transformants, the concentration of fructose-2,6-bisphosphate was decreased by 40%. However, at the intracellular pH of 7.5, the levels of this PFK activator are still sufficient to keep PFK almost fully activated. Thus, some other mechanism for the control of PFK activity must be operable. One possibility is cAMP dependent kinase activation and phosphatase inactivation (Legiša and Benčina, 1994; Gradišnik-Grapulin and Legiša, 1997; Benčina, 1997). There were no significant differences between the PFK over-producers and the wild-type strain with regard to the activity levels of the other glycolytic enzymes and enzymes involved in removing glycolytic intermediates. Whatever the mechanism, it is apparent that *A. niger* carefully controls glycolytic flux even in the presence of substantial perturbations of PFK and pyruvate kinase activities.

At the terminus of glycolysis, a potential futile cycle exists involving the triad of phosphoenolpyruvate, pyruvate, and oxaloacetate catalyzed by cytosolic pyruvate kinase, pyruvate carboxylase, and PEP carboxykinase (Osmani and Scrutton, 1983; Bercovitz *et al.*,

Table 12.2. Proteins and Genes Relevant to Organic Acid Production in *Aspergillus* sp.

No.[a]	Protein	EC no.	Gene	Accession no.[b]	Organism
1	Glc[c] transporter, high aff.	n.a.	—	—	—
2	Glc transporter, low aff.	n.a.	—	—	—
3	Hexokinase	2.7.1.1	*hxkA*	AJ009973	*A. niger*
4	Glucokinase	2.7.1.2	*glkA*	X99626	*A. niger*
5	Glucose-6-P isomerase	5.3.1.9	*pgiA*	AB032269	*A. oryzae*
6	6-Phosphofructokinase	2.7.1.11	*pfkA*	Z79690	*A. niger*
7	Fructose-bisP aldolase	4.1.2.13	*fbaA*	AB032272	*A. oryzae*
8	Triosephosphate isomerase	5.3.1.1	*tpiA*	AB032273	*A. oryzae*
9	Glyceraldehyde-3-P DH	1.2.1.12	*gpdA*	Q12552	*A. niger*
10	Phosphoglycerate kinase	2.7.2.3	*pgkA*	D28484	*A. oryzae*
11	Phosphoglycerate mutase	5.4.2.1	*gpm*	X58789	yeast
12	PEP hydratase	4.2.1.11	*enoA*	D63941	*A. oryzae*
13	Pyruvate kinase	2.7.1.40	*pkiA*	S38698	*A. niger*
14	Pyruvate DH complex	1.2.4.1	—	—	—
15	Citrate synthase	4.1.3.7	*cit1*	D63376	*A. niger*
16	Citrate/malate antiporter	n.a.	—	—	—
17	Pyruvate carboxylase	6.4.1.1	*pyc*	AJ009972	*A. niger*
18	PEP carboxykinase	4.1.1.49	*acuF*	AY049067	*A. nidulans*
19	Malate dehydrogenase	1.1.1.37	—	—	*A. fumigatus*
20	Oxaloacetase	3.7.1.1	*oah*	AAA50372[d]	*A. niger*
21	Glucose oxidase	1.1.3.4	*gox*	X16061	*A. niger*
21	Glucose oxidase	1.1.3.4	*ggox*	AJ294936	*A. niger*
22	Gluconolactonase	3.1.1.17		—	—
23	Trehalose P synthase	2.4.1.15	*tspA*	U07184	*A. niger*
23	Trehalose P synthase	2.4.1.15	*tspB*	U63416	*A. niger*
24	Trehalose phosphatase	3.1.3.12	—	—	—
25	6-Phosphofructo-2-kinase	2.7.1.105	—	—	—
26	Citrate transporter, export	n.a.	—	—	—
27	Citrate transporter, uptake	n.a.	—	—	—
28	Oxalate transporter	n.a.	—	—	—
29	Aconitate hydratase	4.2.1.3	*aco*	AF093142	*A. terreus*
30	Aconitate decarboxylase	4.1.1.6	—		*A. terreus*
31	β-Fructofuranosidase	3.2.1.26	*suc1*	L06844	*A. niger*
31	β-Fructofuranosidase	3.2.1.26	*suc2*	—	*A. niger*
32	Sucrose transporter	n.a.	—	—	—
33	Fructose transporter	n.a.	—	—	—
	Mannitol-1-P DH	1.1.1.17	*mpdA*	AY081178	*A. niger*
	Isocitrate DH (NADP)	1.1.1.42	*icdA*	AB000261	*A. niger*
	Oxoglutarate DH	1.2.4.2	—	—	—
	Succinate DH (NADP)	1.3.5.1	—	—	—
	Fumarate hydratase	4.2.1.2	—	—	—
	Alternative oxidase	n.a.	*aox1*	AB046619	*A. niger*

[a]This number refers to the numbering in Figure 12.2.

[b]Accession numbers are for the GenBank/EMBL sequence databases, except [d]which is for the Derwent GENESEQ patent database.

[c]Abbreviations. Glc: D-Glucose; aff: affinity; PEP: phosphoenolpyruvate; P: phosphate; DH: dehydrogenase; n.a.: not applicable.

1990; Jaklitsch *et al.*, 1991). Each turn of this cycle consumes one molecule of ATP. ^{13}C-NMR studies indicated that during the growth phase, flux from pyruvate to oxaloacetate exceeded flux from oxaloacetate to citrate (Peksel *et al.*, 2002). Consistent with this observation, significant futile cycling was observed. This suggests two targets for minimizing flux through this futile cycle, decreased expression of PEP carboxykinase or increased expression of citrate synthase. Recently, a strain of *A. niger* expressing eleven times the wild-type levels of citrate synthase was successfully prepared (Ruijter *et al.*, 2000). However, this strain exhibited the same rate of citrate accumulation as the wild-type strain. The other strategy, cloning and over-expression of PEP carboxykinase, has not been reported in *A. niger*. The cumulative data also suggest that increasing pyruvate carboxylase activity would simply increase futile cycling, at least in the growth phase of the fermentation.

Citrate synthase is the terminal enzyme in the citric acid biosynthetic pathway. The enzyme from *A. niger* has been characterized and cloned (Kubicek and Röhr, 1980; Kirimura *et al.*, 1999a). The enzyme is inhibited by Mg^{2+} and ATP (K_i = 1 mM), but importantly, citrate is not an inhibitor of the enzyme. The results cited above showed that increasing citrate synthase expression was ineffective in increasing the citrate production rate (Ruijter *et al.*, 2000). Other enzymes of the citric acid cycle have been proposed to be significant. Aconitase inhibition would theoretically lead to an accumulation of citrate by blocking subsequent flux through the citric acid cycle. However, there were no significant differences in aconitase or isocitrate dehydrogenase activity levels in the *A. niger* parent strain and in mutant strains producing higher yields of citric acid (La Nauze, 1966). Furthermore, aconitase has been shown to be active even under citric acid production conditions (Kubicek and Röhr, 1985).

Isocitrate dehydrogenase could have a deleterious effect on citric acid production by decarboxylating isocitrate, which is in equilibrium with citrate via aconitase. The $NADP^+$-specific isocitrate dehydrogenase has been purified from *A. niger* and found to be present in both the cytosol and mitochondrion (Meixner-Monori *et al.*, 1986). These workers found that isocitrate dehydrogenase was inhibited by ATP and citrate via chelation of enzymatic Mg^{2+}. However, they concluded that at intracellular Mg^{2+} concentrations this would be of little significance. Recently, the *icdA* gene encoding the $NADP^+$-specific isocitrate dehydrogenase was cloned (Kirimura *et al.*, 2002). A single genomic copy of this gene produces two alternate mRNA transcripts. Although the predicted amino acid sequence of one transcript encodes a mitochondrial targeting sequence, peroxisome targeting sequences are encoded in both transcripts. Decreasing the expression of this gene is a possible strategy for decreasing the loss of citrate via TCA cycling. However, the previous data suggests this approach is unlikely to have a positive effect on citric acid accumulation and may have unintended negative effects on cell viability during the growth phase.

For citric acid to accumulate extracellularly at a final concentration of about 1.0 M, an active export system must exist to remove citrate from the cytosol where the concentration is only 2–30 mM (Kubicek and Röhr, 1985; Legiša and Kidric, 1989; Prömper *et al.*, 1993). Citrate export requires low Mn^{2+} concentrations in the range known to be required for efficient citric acid production (Netik *et al.*, 1997). On the other hand, citrate import required Mn^{2+} both for induction of expression of the citrate importer as well as for its function (Netik *et al.*, 1997). These results provide an additional explanation of the multiple effects of Mn^{2+} on the physiology of *A. niger* under citric acid production conditions.

The alternative oxidase (AOX) is an inducible component of the alternative respiratory pathway in fungi. In *A. niger*, active AOX is necessary for citric acid production and is another example of a component of *A. niger* that is sensitive to the presence of manganese ions (Kubicek *et al.*, 1980; Zehentgruber *et al.*, 1980; Kirimura *et al.*, 2000). In addition, AOX is apparently inactivated at low dissolved oxygen concentrations (Zehentgruber *et al.*, 1980). This is one component of the observed physiological requirement for undisrupted oxygen supply to maintain citric acid production. Although the protein appears to be constitutively expressed in *A. niger*, expression is increased on transition to the citric acid production phase (Kirimura *et al.*, 1987). The AOX has the desirable effect of regenerating the intracellular pool of NAD^+ by oxidizing the NADH generated in the glycolytic pathway, and it does so without excessive synthesis of ATP, since it transports only 40% as many protons as the standard respiratory chain (Joseph-Horne *et al.*, 2001). Hence, only 40% as much ATP is generated by this system, which is probably sufficient under the production phase of the citric acid fermentation where little cell growth is occurring. The cDNA corresponding to the *aox1* gene of *A. niger* has been cloned, sequenced, and functionally expressed in *E. coli* (Kirimura *et al.*, 1999b). Understanding the mechanism of the reportedly irreversible inactivation of AOX by low oxygen concentration could facilitate the design of a more robust *A. niger* production strain resistant to transient drops in oxygen concentration.

The net result of 60 years of work examining the biochemistry and molecular biology of citric acid production by *A. niger* has been to emphasize the exquisitely delicate control this organism exerts over its metabolic processes. Much has been learned about the pathways and the corresponding enzymes responsible for metabolism of glucose to citric acid, and the accumulated data on metabolite pool concentrations and steady state kinetic parameters of the relevant enzymes and transporters has been tabulated in support of model development (Alvarez-Vasquez *et al.*, 2000). Recent modeling studies in *A. niger* and other systems have provided insights on potentially successful strategies for manipulating metabolic flux (Torres, 1994; Cornish-Bowden *et al.*, 1995; Torres *et al.*, 1996b; Ruijter *et al.*, 1998; Alvarez-Vasquez *et al.*, 2000; Guebel and Torres Darias, 2001; Peksel *et al.*, 2002). The strategy of increasing the expression of an enzyme predicted to catalyze the rate-limiting step in a pathway has been calculated to be ineffective (Cornish-Bowden *et al.*, 1995). This is consistent with the experimental results from genetic manipulation of the levels of expression of enzymes alone or in tandem, which have failed to increase the citrate production rate (Ruijter *et al.*, 1997, 2000). It is predicted that simultaneously adjusting the expression of a large number of the enzymes in a pathway may lead to increases in metabolic flux, but this obviously raises considerable technical challenges (Cornish-Bowden *et al.*, 1995; Torres *et al.*, 1996b; Alvarez-Vasquez *et al.*, 2000). Adjusting the step that removes the desired product is predicted by models to have the desired effect and constitutes the most promising metabolic engineering strategy due to its relative simplicity (Cornish-Bowden *et al.*, 1995). This would suggest that manipulation of the transporters involved in citrate uptake and export would be a desirable strategy for increasing the rate of citrate production.

3.1.2. Oxalic Acid

Oxalic acid is produced by a wide variety of fungi, including brown-rots, white-rots, mycorrhizae, plant pathogens, and *A. niger*. In contrast to citric acid, the physiological roles of oxalic acid are well known and have been reviewed (Dutton and Evans, 1996). In *A. niger*, oxalate biosynthesis is due exclusively to the action of oxaloacetase, which catalyzes the hydrolysis of oxaloacetate to oxalate and acetate (Hayaishi *et al.*, 1956; Mueller, 1975; Lenz *et al.*, 1976; Kubicek *et al.*, 1988). The enzyme is located in the cytosol, and expression is induced at pH values greater than 4 and by carbonate

(Kubicek *et al.*, 1988; Pedersen *et al.*, 2000b). The enzyme requires Mn^{2+} for activity (K_m = 21 μM), which has obvious implications for the citric acid process, and is specific for oxaloacetate (K_m = 220 μM) (Hayaishi *et al.*, 1956; Lenz *et al.*, 1976). The known presence of pyruvate carboxylase in the cytosol of *A. niger* (Osmani and Scrutton, 1983; Bercovitz *et al.*, 1990; Jaklitsch *et al.*, 1991) together with the insensitivity of oxalate production to the TCA cycle inhibitor fluorocitrate (Kubicek *et al.*, 1988) indicates that oxalate is produced by a branch from the glycolytic pathway, as shown in Figure 12.2.

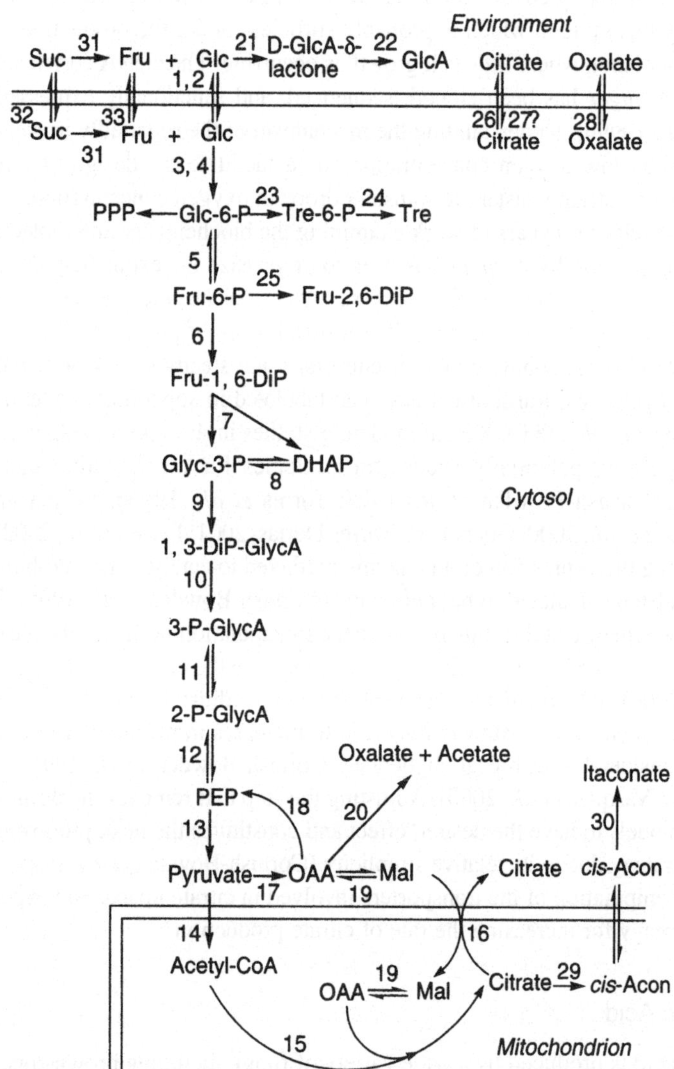

Figure 12.2. Critical pathways for organic acid synthesis in *Aspergillus* spp. Numbers refer to the proteins and genes listed in Table 12.2. Abbreviations: PPP, pentose phosphate pathway; Glc, glucose; Fru, fructose; Suc, sucrose; Tre, trehalose; GlcA, gluconic acid; P, phosphate; DiP, diphosphate; Glyc, glyceraldehydes; DHAP, dihydroxyacetone phosphate; GlycA, glyceric acid; PEP, phosphoenolpyruvate; OAA, oxaloacetic acid; Mal, malic acid; Acon, aconitic acid.

The *oah* gene has been cloned and sequenced from *A. niger* (Pedersen *et al.*, 2000b), and this gene is considered necessary and probably sufficient for expression of functional oxaloacetase based on the following evidence. Disruption of the *oah* gene resulted in an *A. niger* mutant defective in oxalic acid production, demonstrating that the *oah* gene is necessary for oxaloacetase activity (Pedersen *et al.*, 2000a). The relatively labile enzyme has not been purified to homogeneity, but four of the five bands observed on a denaturing polyacrylamide gel have been sequenced and demonstrated to be (proteolytic or glycosylation) variants of the polypeptide encoded by *oah*. The oxaloacetase is a large multimeric enzyme with a molecular weight of about 430 kDa, likely composed of 10–12 copies of a single polypeptide subunit of 37 kDa (Lenz *et al.*, 1976; Pedersen *et al.*, 2000b). Unequivocal demonstration of the sufficiency of the *oah* gene for oxaloacetase function awaits expression of the enzyme in a heterologous host. The 5′ untranslated region of the *oah* gene contains a putative FacB binding site (Pedersen *et al.*, 2000b), which in *A. nidulans* is involved in control of the expression of acetate utilization genes (Todd *et al.*, 1998). This is consistent with the observed maximal expression of *oah* mRNA and oxaloacetase activity during growth on acetate (Pedersen *et al.*, 2000b).

The *oah* gene belongs to the isocitrate lyase family containing a conserved active site motif of K(K/R)CGH(M/L)(A/E)GK. A TBLASTN search of the *Aspergillus fumigatus* unfinished genome revealed a gene with 84% amino acid identity (91% similarity) to the translated product of the *oah* gene containing the conserved active site motif (TIGR, 2002). The putative *orf* also contained two introns in the same positions as the *A. niger oah* gene. A TBLASTN search of the *Phanerochaete chrysosporium* unfinished genome identified a putative *oah* gene bearing 48% identity (64% similarity) to the *A. niger oah* gene over the C-terminal 270 amino acids (DOE Joint Genome Institute, 2002). This Basidiomycete is known to produce and utilize oxalic acid for lignin degradation (Wariishi *et al.*, 1992). These findings, though based on the limited publicly available fungal genome databases, indicate that oxaloacetase may be widely distributed in fungi, which is consistent with the widespread capability to produce oxalic acid in these organisms. However, the presence of a putative oxaloacetase does not preclude the function of the glyoxylate bypass cycle as a source of oxalate in *A. fumigatus* or *P. chrysosporium*.

Under the current economic conditions (relatively inexpensive petroleum and energy) there is little interest in producing oxalic acid by fungal fermentation but considerable interest in eliminating it as a contaminant in *A. niger* fermentations where citric acid or enzymes are the desired products. This was the goal of the study cited above where the *oah* gene was disrupted by homologous recombination in an industrial glucoamylase production strain, resulting in a transformant lacking the ability to produce oxalic acid (Pedersen *et al.*, 2000a). A second study dealing with mutants derived from a citric acid producing strain of *A. niger* demonstrated that a mutant lacking oxaloacetase activity (*prtF28*) did not produce oxalic acid (Ruijter *et al.*, 1999). A mutant derived from *prtF28*, which also lacked glucose oxidase activity, was able to produce citric acid at pH 5 in the presence of manganese. However, under standard citric acid production conditions (low pH) the mutant retained a sensitivity to manganese, emphasizing the multifactor nature of the manganese effect.

If the economic conditions for oxalic acid production by fungal fermentation become favorable in the future, then the accumulated information about oxalate production in *A. niger* will be useful for developing a production strain. The evidence is convincing that oxalate production in *A. niger* is mediated exclusively by cleavage of oxaloacetate to oxalate and acetate by oxaloacetase, though the sufficiency of the *oah* gene for oxaloacetase expression remains to be demonstrated. As mentioned, the expression of the *oah* gene is induced at pH values greater than 4 and by carbonate and is additionally controlled by the acetate utilization transcription factor (FacB). Genetic engineering of an *A. niger* strain to remove these controls would be helpful in a production strain. Removal of the toxic product, oxalic acid, from the cytosol could become problematic in a production strain; as with strategies proposed for citric acid production, oxalic acid export systems may have to be up-regulated in an oxalic acid production strain.

3.1.3. Gluconic Acid

Gluconic acid production by fermentation of glucose using *A. niger* is another mature bioprocess with literature reporting highly efficient processes dating back to 1940 (Moyer *et al.*, 1940; Blom *et al.*, 1952). We have noted that the gluconic acid process is unique among the organic acid fermentations as it occurs entirely outside of the cytoplasmic membrane. The first step is catalyzed by glucose oxidase, which oxidizes β-D-glucopyranose to D-glucono-1,5-lactone. The hydrolysis of the lactone to form gluconic acid occurs spontaneously in aqueous solutions, but the rate is six orders of magnitude greater with the enzyme gluconolactonase (Jermyn, 1960). A partial purification of the gluconolactonase from *A. niger* has been reported (Ogawa *et al.*, 2002). However, this second step is relatively unimportant from a practical standpoint, as gluconic acid, D-glucono-1,5-lactone, and D-glucono-1,4-lactone will rapidly reach equilibrium upon storage in aqueous solutions. More important is the action of catalase, which catalyzes the disproportionation of the cytotoxic hydrogen peroxide, formed by the action of glucose oxidase, into water and molecular oxygen.

The critical enzyme in this fermentation is glucose oxidase, which was first identified in *Penicillium* spp. (Coulthard *et al.*, 1945). Glucose oxidase has subsequently been purified and characterized from a variety of fungi, including a variety of *Penicillium* spp. (Coulthard *et al.*, 1945; Kusai *et al.*, 1960), *A. niger* (Pazur, 1966), and the Basidiomycete, *P. chrysosporium* (Kelley and Reddy, 1986, 1988). The enzyme from *A. niger* is the most thoroughly studied fungal glucose oxidase, and it has been cloned, sequenced, and expressed in yeast (Frederick *et al.*, 1990). It exists as a dimer of identical subunits containing one FAD per subunit. The FAD is reduced in the course of oxidizing glucose to gluconic acid, and the subsequent oxidation of the reduced FAD by molecular oxygen generates hydrogen peroxide (Gibson *et al.*, 1964). Both glucose oxidase and gluconolactonase are located outside the plasma membrane and by activity staining and immunocytochemical staining glucose oxidase appears to be associated specifically with the cell wall in *A. niger* N400 (Witteveen *et al.*, 1992). This is consistent with the observation that glucose oxidase from *A. niger* is glycosylated (Swoboda and Massey, 1965; O'Malley and Weaver, 1972), a general characteristic of fungal extracellular enzymes. The hydrogen peroxide generated by glucose oxidase inactivates the enzyme, probably through the oxidation of methionine residues (Kleppe, 1966). This emphasizes the need for catalase or some other mechanism of removing the hydrogen peroxide. In *A. niger*, there are two constitutive catalases and two catalases induced by dissolved oxygen concentrations of 30% or greater (Witteveen *et al.*, 1992). Both the induced and constitutive pairs of catalases consist of one intracellular and one extracellular enzyme. The importance of catalase is reflected in the design of a patented process for gluconic acid production by enzymes derived from *A. niger* where the catalase to glucose oxidase ratio is 200 or greater based on activity (Vroemen and Beverini, 1999). The identification of a large number of mutants affecting glucose oxidase production indicates that the expression of this enzyme, and gluconic acid production as a whole, is subject to complex regulation as are other organic acid production processes (Swart *et al.*, 1990; Witteveen *et al.*, 1990, 1993). Although gluconic acid production by fermentation of glucose with *A. niger* is in current practice, in the future it is likely that an enzymatic process will be utilized if glucose oxidase can be produced economically and in a form stable to the process conditions.

3.1.4. Itaconic Acid

The first reported biological source of itaconic acid was the descriptively named *Aspergillus itaconicus* (Kinoshita, 1931). Shortly thereafter, it was discovered that *A. terreus* produced itaconic acid (Calam *et al.*, 1939). Lockwood and Reeves (1945) screened over 300 strains of *A. terreus* and found eleven that were efficient producers of itaconic acid from glucose (45% yield). Most of the work on the fermentation parameters and biochemistry of itaconic acid production has been performed with *A. terreus*, largely with strain NRRL 1960. Itaconic acid is also produced by Basidiomycetes of the genus *Ustilago* (Haskins *et al.*, 1955; Guevarra and Tabuchi, 1990; Tabuchi, 1991), and a comprehensive review of itaconic acid production was recently published (Willke and Vorlop, 2001).

An efficient process for the fermentation of sucrose in molasses to itaconic acid using *A. terreus* was patented in 1962 (Nubel and Rabajak, 1962). The reported yield is 70%. In general, the parameters that are important for itaconic acid production by *A. terreus* include an incubation temperature of 37–40 °C, continuous aeration, a low starting pH (3–5), a lower operating pH (2.2–3.8), high glucose concentrations (10–20%), sufficient nitrogen, high magnesium sulfate concentration (0.5%), low phosphate to limit mycelial growth, and adequate levels of the trace metals, zinc, copper, and iron (Lockwood and Reeves, 1945; Nelson *et al.*, 1952; Pfeifer *et al.*, 1952; Larsen and Eimhjellen, 1955; Nubel and Ratajak, 1962; Gyamerah, 1995a, b; Willke and Vorlop, 2001). Since itaconic acid production by *A. terreus* shares many of the characteristics of citric acid production by *A. niger* it would be informative to perform a systematic survey of the effects of various trace metals in a highly defined synthetic media on fungal morphology and the production rate and yield of itaconic acid. *A. terreus* is able to grow well on a variety of monosaccharides, disaccharides, and polysaccharides but converts relatively few of these substrates to itaconic acid (Eimhjellen and Larsen, 1955). In this survey, sucrose and glucose were reported to give 57% and 52% yields on a weight basis, respectively, whereas the yields with the pentoses D-xylose and L-arabinose were only 31% and 18%, respectively (Eimhjellen and Larsen, 1955). A 45% yield of itaconic acid from D-xylose has been obtained with an immobilized *A. terreus* system growing in a 6.7% xylose medium (Kautola *et al.*, 1985), but this is considerably lower than the 70% yield obtained on sucrose (Nubel and Ratajak, 1962).

The biochemical pathway resulting in itaconic acid production has been determined but the unique enzyme in this pathway, *cis*-aconitate decarboxylase, has not been purified to homogeneity (Bentley and Thiessen, 1955). Bentley and Thiessen (1957a, b, c) performed studies with ^{14}C-labelled substrates to demonstrate that the pathway for itaconic acid production in *A. terreus* paralleled that of citric acid production in *A. niger* with two additional steps: namely, the dehydration of citrate by aconitase to form *cis*-aconitate and a decarboxylation by *cis*-aconitate decarboxylase to form itaconic acid. Aconitate decarboxylase activity was demonstrated in cell-free extracts, but its instability has been an obstacle to purification (Bentley and Thiessen, 1955, 1957c). These cell-free enzyme preparations, which contained both aconitase and *cis*-aconitate decarboxylase activities, were used to resolve the position of decarboxylation of *cis*-aconitate (Figure 12.3). Incubation with [1,6-^{14}C]citric acid resulted in the release of unlabeled CO_2, whereas incubation with D-[5,6-^{14}C]isocitric acid released labeled CO_2 (Bentley and Thiessen, 1957c). This indicated that *cis*-aconitic acid was decarboxylated at C-5. Both ^{13}C-NMR and radioisotope tracer studies with a variety of ^{13}C and ^{14}C labeled substrates have confirmed the mechanism of action of *cis*-aconitate decarboxylase and the similarity of the pathway to that of citric acid production in *A. niger* (Winskill, 1983; Bonnarme *et al.*, 1995).

Cis-aconitate decarboxylase resides exclusively in the cytosol in *A. terreus* (Jaklitsch *et al.*, 1991), but aconitase and citrate synthase are probably located only in the mitochondria. Therefore, *cis*-aconitate must be transported out of the mitochondrial compartment to the cytosol for decarboxylation to itaconic acid. It is not known whether *A. terreus* transports *cis*-aconitate via a specific *cis*-aconitate/malate antiporter or by the citrate/malate

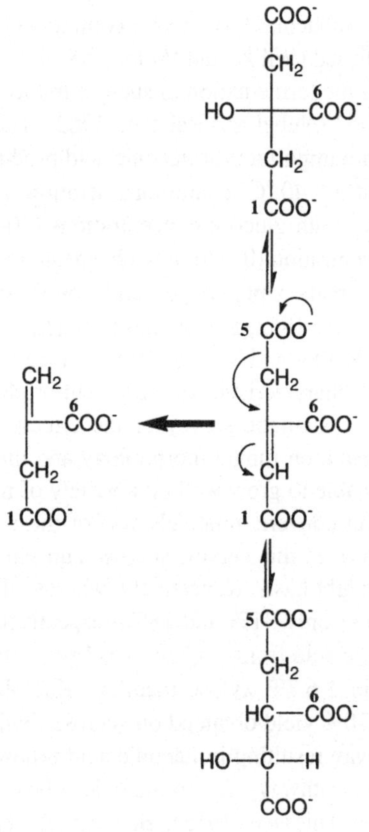

Figure 12.3. Conversion of citrate to itaconate. Labeling of carbon atoms follows that of Bentley and Thiessen (1957c).

antiporter. Work with an Egyptian isolate of *A. terreus* indicated that not only itaconic acid and citric acid but also *cis*-aconitic acid accumulated in the media (Shimi and Nour El Dein, 1962). These results indicate a need to determine the specificity of the dicarboxylic acid and tricarboxylic acid transporters of *Aspergillus* spp. Expression of *cis*-aconitate decarboxylase is induced only under itaconic acid production conditions, which also cause aconitase activity to increase 2.5– to 3.0-fold (Jaklitsch *et al.*, 1991; Bonnarme *et al.*, 1995).

The purification and cloning of the gene for *cis*-aconitate decarboxylase would have considerable scientific and biotechnological impact. The role of this enzyme in itaconic acid biosynthesis could be confirmed by transforming a citric acid producing strain of *A. niger*. In addition, a more efficient itaconic acid process might be devised by using a highly optimized citric acid production strain of *A. niger* as the recipient of the gene. However, the need for a specific *cis*-aconitate/malate antiporter for translocation of *cis*-aconitate from the mitochondrion to the cytosol could confound this approach. Nevertheless, such a negative result would suggest that this unique transporter is present in *A. terreus* (a somewhat circular argument). There is no theoretical reason to prevent the achievement of an itaconic acid production process that is as efficient as the citric acid production process. Genetic engineering of either *A. terreus* or *A. niger*, further refinement of fermentation conditions to obtain the optimal morphology of *A. terreus*, or the use of less expensive substrates may result in a more economical process for the production of itaconic acid.

3.2. Rhizopus and Organic Acid Production

Rhizopus spp. and related Zygomycetes (principally *R. oryzae*) are capable of producing significant amounts of L(+) lactic acid, fumaric acid, and potentially, L-malic acid. Although L-malic acid is also produced by *Aspergillus* spp. the production of these three acids will be discussed sequentially in the following section focusing on *Rhizopus* as a production organism. Generally, *Rhizopus* spp. have the desirable characteristics of growing on simple chemically-defined media, and utilizing complex carbohydrates, hexoses, and pentoses. Their main disadvantage is a tendency to produce more than one metabolic endproduct in significant yields.

3.2.1. L-Lactic Acid

The fungal production of L-lactic acid by a surface culture of *Rhizopus* spp. was reported early in the last century (Ehrlich *et al.*, 1911). However, the first report of an efficient submerged fermentation for the fungal production of L-lactic acid was in 1936 (Lockwood *et al.*, 1936; Ward *et al.*, 1938). This was the era in which the efficiencies of submerged fungal fermentations first became widely recognized. Ward *et al.* (1938) described a fermentation process utilizing the Zygomycete genera, *Rhizopus* and *Actinomucor* in general, and *R. oryzae* (syn. *arrhizus*) specifically, which resulted in 63–69% yields of L-lactic acid from chemically defined media containing 15% glucose. They also delineated the advantages of the fungal process over the bacterial process that remain true today: the use of a chemically defined medium, including inorganic nitrogen sources, which simplifies product purification; the ability to metabolize high concentrations of glucose, thus obtaining high product concentrations; and the production of enantiomerically pure L-lactic acid, necessary for food applications and preferred for PLA (poly-lactic acid, or polylactide) manufacture. The principal disadvantages of the *R. oryzae* process is the diversion of carbon away from the desired product into the byproducts ethanol and fumaric acid (see the review by Litchfield, 1996, for further details of the bacterial process, purification of lactic acid from fermentation broths, and uses of the product). Improvement in the L-lactic acid yield and product purification characteristics of the *R. oryzae* fermentation as described by Snell and Lowery (1964) consisted primarily of introducing calcium carbonate and increasing the temperature late in the production phase. These adjustments resulted in lactate yields of 72–79% and avoidance of calcium lactate crystallization during the fermentation, which simplified product purification.

From the late 1980s to the present, process optimization strategies have centered on the issue of morphology of *R. oryzae*, a universally critical parameter in fungal fermentations. These studies have taken two basic approaches, immobilization of cells (Hang *et al.*, 1989; Tamada *et al.*, 1992; Hamamci and Ryu, 1994; Dong *et al.*, 1996; Xuemei *et al.*, 1999), and promotion of mycelial or pellet morphology (Yang *et al.*, 1995; Kosakai *et al.*, 1997; Du *et al.*, 1998; Park *et al.*, 1998; Yin *et al.*, 1998; Zhou *et al.*, 2000). The term *pellet morphology* can be the source of some confusion in the discussion of optimal fungal morphology, since pellets of less than about one millimeter are associated with high production rates and yields, whereas larger pellets are not. Presumably, this is due to mass transfer limitations with regard to oxygen, substrates, and products. Process parameter optimization leading to consistent production of small pellets would probably be the most economical means of obtaining a high-yielding strain. A number of studies have reported yields (w/w) of 85% to 88% (Kosakai *et al.*, 1997; Longacre *et al.*, 1997; Yin *et al.*, 1997; Du *et al.*, 1998; Zhou *et al.*, 1999). These yields are comparable to the yields routinely obtained with the bacterial process, and if consistently obtained, would contribute greatly to the economic competitiveness of the fungal process.

Considerable progress has been made in understanding the physiology and biochemistry of acid production by *R. oryzae*. Early studies demonstrated that *R. oryzae* produced L-lactate via glycolysis with the concomitant production of ethanol and carbon dioxide (Waksman and Foster, 1938; Gibbs and Gastel, 1953; Margulies and Vishniac, 1961). These studies also showed that lactic acid yield was increased and ethanol formation decreased under aerobic conditions, while the opposite was true under low oxygen conditions. Wright and coworkers have developed a computational metabolic model for *R. oryzae* based on elegant radioisotope studies of intracellular and extracellular metabolite pools (Wright *et al.*, 1996; Longacre *et al.*, 1997). This model and the accompanying radiolabeling experiments with cultures of *R. oryzae*, have provided important insights for improving the yield of lactic acid (Longacre *et al.*, 1997; Wright *et al.*, 1996). A simplified scheme of metabolism in *R. oryzae* shows the critical reactions in the formation of organic acids and ethanol (Figure 12.4). The biosynthesis of L-lactic acid, L-malic acid, fumaric acid, and ethanol occur in the cytosol with pyruvate at the crossroads leading to production of each compound. The addition of carbonate has the desirable effect of decreasing ethanol production, presumably through the inhibition of pyruvate decarboxylase, but the undesirable effect of increasing malate and fumarate production through the stimulation of pyruvate carboxylase (Lockwood *et al.*, 1936; Waksman and Foster, 1938; Foster and Waksman, 1939). Through radiolabeling studies, this effect was quantified at four different sodium carbonate concentrations from 0 to 30 mM (Longacre *et al.*, 1997). These studies showed that ethanol did not reach a minimum until after fumarate and malate

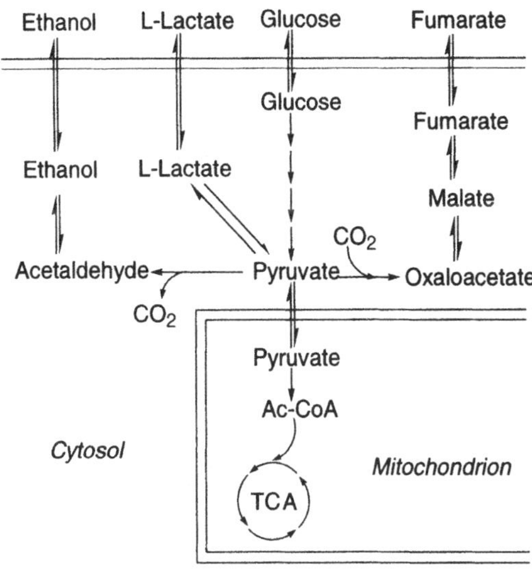

Figure 12.4. Critical pathways for organic acid synthesis in *Rhizopus oryzae*.

had begun to increase, indicating that minimizing the formation of these side products would not be possible through adjustment of carbonate concentrations alone. One strategy that has been pursued is to isolate mutants with decreased ethanol production, which is normally maximal under anaerobic conditions, so that the organism can be grown anaerobically in the production phase, which eliminates metabolic flux to malate and fumarate. Longacre *et al.* reported the isolation of such a mutant attaining 86% yield of lactic acid production. Similarly, Skory (1998) obtained a mutant with decreased alcohol dehydrogenase activity that produced relatively high concentrations of lactic acid under anaerobic conditions. However, even with the greatly decreased alcohol dehydrogenase activity levels, the mutant strain still produced substantial amounts of ethanol under anaerobic conditions. These results suggest that the complete elimination of ethanol production would be desirable and the logical genetic target would be the first committed step in ethanol biosynthesis. The cloning of two pyruvate decarboxylase genes from *R. oryzae* suggests this strategy is already being pursued (GenBank accession numbers AF282846 and AF282847).

The enzymology and genetic control of the key step in lactate synthesis by *R. oryzae* is now well understood. *R. oryzae* possesses three L-lactate dehydrogenases (LDH), including one NAD-independent LDH (Pritchard, 1971) and two NAD-dependent LDH isozymes (Pritchard, 1973; Yu and Hang, 1991; Skory, 2000, 2001). The D-(−)−lactate dehydrogenases present in bacteria, chytridiomycetes, and oomycetes have not been reported in the phyla Zygomycota, Ascomycota, and Basidiomycota (Gleason *et al.*, 1966; LéJohn, 1971; Wang and LéJohn, 1974). The cDNAs for two NAD-dependent LDH isozymes, *ldhA* (AF226154) and *ldhB* (AF226155), have been isolated and sequenced (Skory, 2000). PCR studies with gene-specific primers indicated that the two NAD-dependent LDH genes are expressed differentially; *ldhA* is expressed in the presence of glucose, xylose, or trehalose, whereas *ldhB* is expressed only on the non-fermentable carbon sources, ethanol, glycerol, and lactate (Skory, 2000). The cumulative results of this study suggested that *ldhA* encodes an LDH biased toward the reductive reaction (pyruvate to lactate) and *ldhB* encodes an LDH biased toward the oxidative activity (lactate to pyruvate). Unequivocal demonstration of these catalytic characteristics await independent expression of the genes in a heterologous host and analysis of the pure isozymes. If *ldhB* encodes an L-lactate oxidizing isozyme then inactivation of this gene might minimize degradation of product at later stages of L-lactate production processes.

3.2.2. Fumaric Acid

In addition to producing L-lactate, *Rhizopus* spp. are also the best of the identified fungal sources for fumarate production. Ehrlich (1911) first identified fungal fumaric acid production in a strain of *Rhizopus nigricans*. A later survey of 41 strains from eight genera of Mucorales identified *Rhizopus, Mucor, Cunninghamella*, and *Circinella* spp. as producers of fumarate, though this property occurred with the greatest frequency in *Rhizopus* spp. (Foster and Waksman, 1939). The nutritional and physical requirements of *R. oryzae* leading to maximum yields of fumarate have been examined (Rhodes *et al.*, 1959). Like other fungal fermentations accumulating high concentrations of organic acids, high carbohydrate concentrations, and high carbon to nitrogen ratios are conducive to high fumaric acid yields with minimal biomass accumulation. Conversion of 60% to 70% of the sugar to fumaric acid (w/w) was achieved in vigorously agitated submerged cultures containing 10–12% glucose and C : N ratios ranging from 120 : 1 to 150 : 1. Standard minerals and calcium carbonate were also added after 3–8 days, and the cultures were incubated at 33°C (Rhodes *et al.*, 1959). Recently, pH and metal (magnesium, zinc, iron, and manganese) concentrations were varied with the result that consistent pellet morphology (about 1 mm pellets) and relatively high fumarate output was obtained (Zhou *et al.*, 2000). Unfortunately, the strain of *R. oryzae* (ATCC 20344) used in these experiments produced high concentrations of ethanol under these conditions, leading to relatively low weight yields (39–46%) of fumarate.

The carbonate in fumaric acid production media is required to neutralize and precipitate the fumaric acid. In addition, carbonate is necessary for the formation of oxaloacetate by pyruvate carboxylase. The metabolic model developed for an L-lactate synthesizing strain of *R. oryzae* also has bearing on fumarate synthesis. The results of those modeling studies indicated that increasing carbonate concentrations raised fumarate and malate yields at the expense of lactate yields. This is consistent with the requirement for high concentrations of carbonate in fumarate production strains (Rhodes *et al.*, 1959). An increase in pyruvate carboxylase activity was observed to correlate with glucose utilization and fumarate production in *R. oryzae* (Overman and Romano, 1969). Pyruvate carboxylase is located in the cytosol of *R. oryzae* (Osmani and Scrutton, 1985), as is one of two fumarate hydratase isozymes (Peleg *et al.*, 1989b). A fumarate hydratase gene (*fumR*) has been cloned from *R. oryzae*, and found to encode a potential mitochondrial targeting sequence (Friedberg *et al.*, 1995). The cytosolic fumarate hydratase activity increases during the production stage of the fumarate fermentation (Peleg *et al.*, 1989b). However, it appears that *R. oryzae* contains a single *fumR* gene and a single mRNA transcript from this gene, so the mechanism of the observed increase in fumarate hydratase activity during fumarate production conditions remains unclear. *R. oryzae* contains both cytosolic and mitochondrial isozymes of NADP-malate dehydrogenase (decarboxylating) and NAD-malate dehydrogenase

(Osmani and Scrutton, 1985; Peleg *et al.*, 1989b). The decarboxylating malate dehydrogenases have the potential to create a futile cycle between pyruvate and malate. Thus, eliminating the gene encoding the cytosolic isozyme of NADP-malate (decarboxylating) might have a beneficial effect on yield. The cytosolic location of pyruvate carboxylase, malate dehydrogenase, and fumarate hydratase indicates that fumarate synthesis can occur by a reductive pathway located exclusively in the cytosol (Osmani and Scrutton, 1985; Peleg *et al.*, 1989b). However, this does not exclude a contribution from the TCA cycle (Kenealy *et al.*, 1986).

Strategies to increase fumarate production include obtaining consistent and favorable morphology as discussed above (Zhou *et al.*, 2000). It appears that the strain employed in the cited study would benefit from disabling the ethanol production pathway by eliminating the pyruvate decarboxylase or alcohol dehydrogenase genes. Kenealy *et al.* (1986) showed that fumarate hydratase specific activity is greater for the reverse reaction, L-malate to fumarate (at least *in vitro*), suggesting that the rate of fumarate production might be increased by removing the fumarate from the cytosol more rapidly. In analogy to the suggestion for the citric acid process, increasing the expression of the membrane transporter of fumarate (dicarboxylic acid transporter) may elicit the desired increase in rate and yield of fumarate production.

3.2.3. L-Malic Acid

L-Malic acid production has been observed in *R. oryzae* (Longacre *et al.*, 1997) and in *Aspergillus* spp. (Abe *et al.*, 1962; Peleg *et al.*, 1988, 1989a; Bercovitz *et al.*, 1990; Battat *et al.*, 1991). Generally, L-malate accumulation in *R. oryzae* is minor compared to L-lactate or fumarate. Clearly, the mechanism leading to malate production in *R. oryzae* is the same as the pathway leading to fumarate, abbreviated by one step. If the cytosolic isozyme of fumarate hydratase could be decreased, the transformation of a fumarate producing strain of *R. oryzae* into an L-malate producing strain would be possible. However, the ambiguity regarding the mechanism leading to increased cytosolic fumarate hydratase activity in *R. oryzae* renders manipulation of the organism problematic.

A variety of *Aspergillus* spp. have been found to produce L-malic acid (Abe *et al.*, 1962; Bercovitz *et al.*, 1990). Bercovitz *et al.* (1990) tested 13 strains, representing nine species, of *Aspergillus* for L-malic acid production and found yields of 1%–4% (w/v). An *A. flavus* strain (ATCC 13697), was found to be the best producer of L-malic acid, confirming the earlier results of Abe *et al.* (1962). However, this strain had some of the lowest levels of cytosolic pyruvate carboxylase and NAD-malate dehydrogenase activities of the strains tested, suggesting that flux through this portion of the metabolic pathway is not rate-limiting (Bercovitz *et al.*, 1990). Through manipulation of standard fermentation parameters (agitation, aeration, glucose, nitrogen, phosphate, and metals), an efficient process for production of L-malate was developed and yields up to 128 mole percent (95 weight percent) were reported (Battat *et al.*, 1991). Interestingly, the addition of 50 ppb of Mn^{2+} led to a precipitous decline in acid production, which is consistent with the effect of this metal on citric acid production by *A. niger*. Unfortunately, as the authors note, the use of *A. flavus* has undesirable implications for the production of food grade L-malate, viz., the possibility of aflatoxin contamination. Another characteristic of this organism was the production of significant quantities of succinic acid and, to a lesser extent, fumaric acid. *Aspergillus sojae* (ATCC 46250, a soy sauce producing strain) actually appeared to be the best candidate with regard to minimal contaminant, acid production, and food safety. Perhaps a broad survey of non-aflatoxigenic *Aspergillus* spp. would prove useful for identifying an organism producing high yield of L-malic acid and low levels of byproduct acids.

3.2.4. Succinic Acid

Most of the emphasis on biological succinic acid production has been on bacterial fermentations, as relatively efficient processes have been developed for anaerobic bacteria (see Zeikus *et al.*, 1999, for a recent review). However, *Fusarium* spp. (Foster, 1949), *Aspergillus* spp. (Bercovitz *et al.*, 1990), and *Penicillium simplicissimum* (Gallmetzer *et al.*, 2002) are known to produce and secrete the acid. The L-malic acid producing *Aspergillus* spp. secreted succinic acid as a secondary product at lower concentrations, the highest titer cited was only 1.3%, representing 25% of the total organic acid production (Bercovitz *et al.*, 1990). The formation of succinate from glucose by *P. simplicissimum* was investigated under aerobic and anaerobic conditions. This fungus secreted low levels of succinate; the highest rate was 0.063 grams succinic acid per gram dry weight mycelium per hour (final concentrations were not reported) with respiration inhibited by 5 mM sodium azide. Nevertheless, the results are interesting for three reasons: succinic acid was the predominant acid produced under anaerobic conditions, pellet formation (diameter not reported) was shown to be important in obtaining maximum succinate production rates, and it raised the possibility of fumarate respiration as a biochemical mechanism for succinate production under anaerobic conditions.

There are three possible metabolic mechanisms for production of succinate: the oxidative portion of the TCA cycle, the reductive portion of the TCA cycle, or the glyoxylate bypass (Figure 12.5). Metabolism by either the oxidative portion of the TCA cycle or the glyoxylate bypass pathway conserves only four of the six carbons from glucose in the four-carbon succinic acid product. On the other hand, the reductive portion of the TCA cycle produces two four-carbon acids for every glucose molecule metabolized via glycolysis operating in conjunction with pyruvate carboxylase. Thus, anaerobic metabolism is preferred for succinic acid producing microorganisms. Gallmetzer *et al.* (2002) suggested that succinate production in *P. simplicissimum* may occur via fumarate respiration but this has not been demonstrated. The physiology of succinate production by filamentous fungi is an emerging field but the current knowledge suggests it may be a promising field. A screening strategy might identify more promising fungal strains that combine the features of high succinate rates and final titers, low side-product concentrations, and tolerance to low pH.

3.2.5. (−)−*trans*-2,3-Epoxysuccinic Acid and *meso*-Tartaric Acid

(−)−*trans*-2,3-Epoxysuccinic acid (ESA) was first isolated from two species of fungi in 1939 (Sakaguchi *et al.*, 1939). These fungi are currently designated *Paecilomyces variotii* (NRRL 1282) and *Talaromyces flavus* (*Penicillium vermiculatum*, NRRL 1009) (Martin and Foster, 1955). A third ESA producing fungus identified as an *A. fumigatus* strain (NRRL 1986, no longer available) was identified in 1945 (Birkinshaw *et al.*, 1945). Optimization of fermentation parameters for *P. variotii* (NRRL 1123) on 12% (w/v) glucose, resulted in a 41% weight yield of ESA representing 61% of theoretical (one ESA molecule per glucose molecule) (Ling *et al.*, 1978).

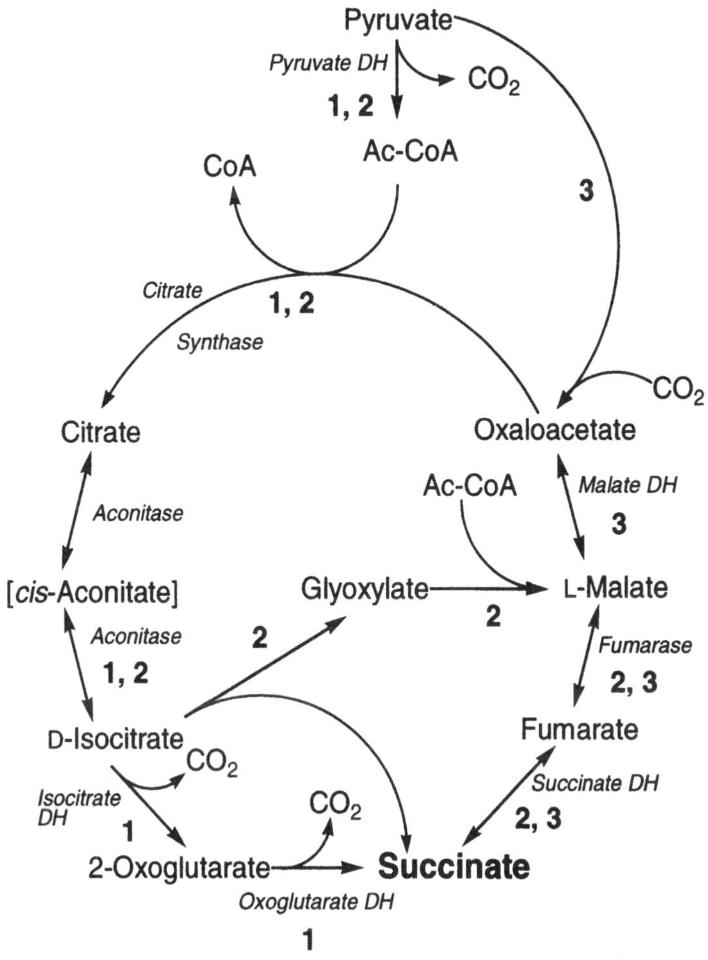

Figure 12.5. Three potential succinate biosynthetic pathways. 1. Oxidative TCA pathway. 2. Glyoxylate bypass pathway. 3. Reductive TCA pathway.

Like oxalate and itaconate biosynthesis, ESA production is only one step removed from the citric acid cycle. ESA appears to be formed directly from fumarate. Studies with $^{18}O_2$ and $H_2{}^{18}O$ showed that ^{18}O was incorporated into ESA from the former but not the latter (Aida and Foster, 1962; Wilkoff and Martin, 1963). Feeding [1,4-^{14}C]fumarate resulted in ESA labeled exclusively in the carboxyl carbons, whereas incorporation of [2,3-^{14}C] fumarate resulted in ESA labeled only in the epoxide carbons (Wilkoff and Martin, 1963). Thus, ESA appears to be derived via the addition of oxygen to the double bond in fumaric acid by an unidentified enzyme utilizing molecular oxygen (Figure 12.6). The hydration of ESA to *meso*-tartaric acid is catalyzed by fumarase from pig heart and is likely to be the general biological mechanism by which this reaction occurs (Albright and Schroepfer, 1971). Based on three-dimensional structural energy minimization *meso*-tartaric acid appears to have a rather linear structure (the carboxyl groups are "anti"), possibly suitable

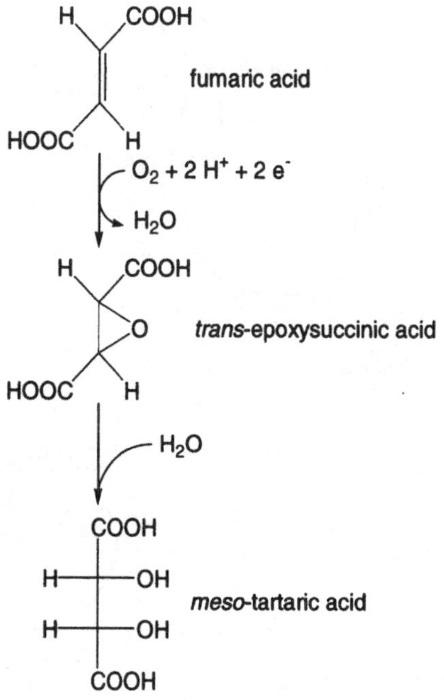

H COOH

fumaric acid

HOOC H

$O_2 + 2\,H^+ + 2\,e^-$

H_2O

H COOH

O *trans*-epoxysuccinic acid

HOOC H

H_2O

COOH

H—OH

meso-tartaric acid

H—OH

COOH

Figure 12.6. Presumed biosynthetic pathway for *trans*-epoxysuccinic acid and *meso*-tartaric acid.

for copolymer formation with diols, whereas the grape-derived tartaric acid of commerce L(+)-tartaric acid has the carboxyl groups in a "syn" conformation consistent with its use as a chelator.

4. Final Perspective

We are entering an era of accelerating development of novel fungal fermentations due to the explosion of information and tools to exploit this information. An increasing number of fungal genomes are being sequenced. The information from fungal genome sequences will accelerate and simplify the identification of metabolic pathways, the repertoire of catabolic enzymes available to an organism, the uptake and export mechanisms, and potential promoters, regulatory genes, etc. Microarray analysis and proteomics can be used to assess the expression and translation of functional gene products under different physiological (fermentation) conditions. Increasingly sophisticated models can be used to predict rational targets for metabolic engineering. Critical genetic loci implicated by traditional mutagenesis and screening programs may be functionally identified with the new information and tools available. An increasing appreciation of the diversity of fungi and the under-explored nature of this Kingdom is arising. High throughput culturing and screening tools are available to assess the potential utility of known and novel fungi. These developments in information, tools, and attitudes have the potential to accelerate the development of novel, efficient, economically feasible, and environmentally responsible fermentations.

One can now envision a fermentation development process where a desired product is chosen and the following sequence of questions is asked to identify or create a fungus with the requisite properties:

- Are there characterized fungi known to synthesize the product?
- Does a rapid screen of uncharacterized fungi reveal a producer?
- Are the biosynthetic pathways to the desired product known?
- Can a fungus with desirable fermentation behavior (e.g., wide substrate utilization range, and defined media requirements leading to simpler product purification) be genetically engineered to produce it?

Certain companies are already following this systematic approach to developing new processes and they will likely be the strongest competitors in the future.

References

Abe, S., Furuyu, A., Saito, T., and Takayama, K.I. (1962). Method of producing L-malic acid by fermentation. *US Patent 3,063,910.*

Aida, K. and Foster, J.W. (1962). Incorporation of molecular oxygen into *trans*-L-epoxysuccinic acid by *Aspergillus fumigatus. Nature* **196**, 672.

Albright, F. and Schroepfer, G.J.Jr. (1971). *l-trans*-2,3-Epoxy succinic acid a new substrate for fumarase. *J. Biol. Chem.* **246**, 1350–1357.

Alvarez-Vasquez, F., González-Alcón, C., and Torres, N.V. (2000). Metabolism of citric acid production by *Aspergillus niger*: Model definition, steady-state analysis and constrained optimization of citric acid production rate. *Biotechnol. Bioeng.* **70**, 82–108.

Arisan-Atac, I., Wolschek, M.F., and Kubicek, C.P. (1996). Trehalose-6-phosphate synthase A affects citrate accumulation by *Aspergillus niger* under conditions of high glycolytic flux. *FEMS Microbiol. Lett.* **140**, 77–83.

Arts, E., Kubicek, C.P., and Röhr, M. (1987). Regulation of phosphofructokinase from *Aspergillus niger*: Effect of fructose 2,6-bisphosphate on the action of citrate, ammonium ions and AMP. *J. Gen. Microbiol.* **133**, 1195–1200.

Baniel, A.M., Blumberg, R., and Hajdu, K. (1981). Recovery of acids from aqueous solutions. *US Patent, 4,275,234.*

Battat, E., Peleg, Y., Bercovitz, A., Rokem, J.S., and Goldberg, I. (1991). Optimization of L-malic acid production by *Aspergillus flavus* in a stirred fermenter. *Biotechnol. Bioeng.* **37**, 1108–1116.

Batti, M. and Schweiger, L.B. (1963). Process for the production of itaconic Acid. *US Patent 3,078,217.*

Bencina, M., Panneman, H., Ruijter, G.J.G., Legiša, M., and Visser, J. (1997). Characterization and overexpression of the *Aspergillus niger* gene encoding the cAMP-dependent protein kinase catalytic subunit. *Microbiology* **143**, 1211–1220.

Bentley, R. and Thiessen, C.P. (1957a). Biosynthesis of itaconic acid in *Aspergillus terreus*. I. Tracer studies with ^{14}C-labeled substrates. *J. Biol. Chem.* **226**, 673–687.

Bentley, R. and Thiessen, C.P. (1957b). Biosynthesis of itaconic acid in *Aspergillus terreus*. II. Early stages in glucose dissimilation and the role of citrate. *J. Biol. Chem.* **226**, 689–701.

Bentley, R. and Thiessen, C.P. (1957c). Biosynthesis of itaconic acid in *Aspergillus terreus*. III. The properties and reaction mechanism of *cis*-aconitic acid decarboxylase. *J. Biol. Chem.* **226**, 703–720.

Bentley, R. and Thiessen, C.P. (1955). *cis*-Aconitic decarboxylase. *Science* **122**, 330.

Bercovitz, A., Peleg, Y., Battat, E., Rokem, J.S., and Goldberg, I. (1990). Localization of pyruvate carboxylase in organic acid producing *Aspergillus* strains. *Appl. Environ. Microbiol.* **56**, 1594–1597.

Birkinshaw, J.H., Bracken, A., and Raistrick, H. (1945). Metabolic products of *Aspergillus fumigatus* Fresenius. *Biochem. J.* **39**, 70–72.

Blom, R.H., Pfeifer, V.F., Moyer, A.J., Traufler, D.H., Conway, H.F., Crocker, C.K., Farison, R.E., and Hannibal, D.V. (1952). Sodium gluconate production: Fermentation with *Aspergillus niger. Ind. Eng. Chem.* **44**, 435–440.

Bloom, S.J. and Johnson, M.J. (1962). The pyruvate carboxylase of *Aspergillus niger. J. Biol. Chem.* **237**, 2718–2720.

Boddy, L.M., Berges, T., Barreau, C., Vainstein, M.H., Dobson, M.J., Balance, D.J., and Peberdy, J.F. (1993). Purification and characterization of an *Aspergillus niger* invertase and its DNA sequence. *Curr. Genet.* **24**, 60–66.

Boles, E. and Hollenberg, C.P. (1997). The molecular genetics of hexose transport in yeasts. *FEMS Microbiol. Rev.* **21**, 85–111.

Bomstein, R.A. and Johnson, M.J. (1952). The mechanism of formation of citrate and oxalate by *Aspergillus niger. J. Biol. Chem.* **198**, 143–153.

Bonnarme, P., Gillet, B., Sepulchre, A.M., Role, C., Beloeil, J.C., and Ducrocq, C. (1995). Itaconate biosynthesis in *Aspergillus terreus. J. Bacteriol.* **177**, 3573–3578.

Calam, C.T., Oxford, A.E., and Raistrick, H. (1939). CLXXXIII. Studies in the biochemistry of micro-organisms. LXIII. Itaconic acid, a metabolic product of a strain of *Aspergillus terreus* Thom. *Biochem. J.* **33**, 1488–1495.

Cleland, W.W., and Johnson, M.J. (1954). Tracer experiments on the mechanism of citric acid formation by *Aspergillus niger. J. Biol. Chem.* **208**, 679–689.

Cornish-Bowden, A., Hofmeyr, J.-H.S., and Cardenas, M.L. (1995). Strategies for manipulating metabolic fluxes in biotechnology. *Bioorg. Chem.* **23**, 439–449.

Coulthard, C.E., Michaelis, R., Short, W.F., Sykes, G., Skrimshire, G.E.H., Standfast, A.F.B., Birkinshaw, J.H., and Raistrick, H. (1945). Notatin: An anti-bacterial glucose-aerodehydrogenase from *Penicillium notatum* Westling and *Penicillium resticulosum* sp. nov. *Biochem. J.* **39**, 24–36.

Currie, J.N. (1917). Citric acid fermentation. *J. Biol. Chem.* **31**, 15–37.

DOE Joint Genome Institute (2002). *JGI Programs: White Rot Genome Project.*; http://www.jgi.doe.gov/programs/whiterot.htm.

Dong, X.Y., Bai, S., and Sun, Y. (1996). Production of L(+)-lactic acid with *Rhizopus oryzae* immobilized in polyurethane foam cubes. *Biotechnol. Lett.* **18**, 225–228.

Du, J.X., Cao, N.J., Gong, C.S., and Tsao, G.T. (1998). Production of L-lactic acid by *Rhizopus oryzae* in a bubble column fermenter. *Appl. Biochem. Biotechnol.* **70**, 323–329.

Dutton, M.V. and Evans, C.S. (1996). Oxalate production by fungi: Its role in pathogenicity and ecology in the soil environment. *Can. J. Microbiol.* **42**, 881–895.

Ehrlich, F. (1911). Formation of fumaric acid by means of molds. *Ber. Dtsch. Chem. Ges.* **44**, 3737–3742.

Eimhjellen, K.E. and Larsen, H. (1955). The mechanism of itaconic acid formation by *Aspergillus terreus*. 2. The effect of substrates and inhibitors. *Biochem. J.* **60**, 139–147.

Foster, J.W. (1949). *Chemical activities of fungi.* Academic Press, New York.

Foster, J.W. and Waksman, S.A. (1939). The production of fumaric acid by molds belonging to the genus *Rhizopus. J. Am. Chem. Soc.* **61**, 127–135.

Frederick, K.R., Tung, J., Emerick, R.S., Masiarz, F.R., Chamberlain, S.H., Vasavada, A., Rosenberg, S., Chakraborty, S. *et al.*, (1990). Glucose oxidase from *Aspergillus niger*—cloning, gene sequence, secretion from *Saccharomyces cerevisiae* and kinetic analysis of a yeast-derived enzyme. *J. Biol. Chem.* **265**, 3793–3802.

Friedberg, D., Peleg, Y., Monsonego, A., Maissi, S., Battat, E., Rokem, J.S., and Goldberg, I. (1995). The *fumR* gene encoding fumarase in the filamentous fungus *Rhizopus oryzae*—cloning, structure and expression. *Gene* **163**, 139–144.

Gallmetzer, M., Meraner, J., and Burgstaller, W. (2002). Succinate synthesis and excretion by *Penicillium simplicissimum* under aerobic and anaerobic conditions. *FEMS Microbiol. Lett.* **210**, 221–225.

Gibbs, M. and Gastel, R. (1953). Glucose dissimilation by *Rhizopus. Arch. Biochem. Biophys.* **43**, 33–38.

Gibson, Q.H., Swoboda, B.E.P., and Massey, V. (1964). Kinetics and mechanism of action of glucose oxidase. *J. Biol. Chem.* **239**, 3927–3934.

Gleason, F.H., Nolan, R.A., Wilson, A.C., and Emerson, R. (1966). D(-)-Lactate dehydrogenase in lower fungi. *Science* **152**, 1272–1273.

Gradišnik-Grapulin, M. and Legiša, M. (1997). A spontaneous change in the intracellular cyclic AMP level in *Aspergillus niger* is influenced by the sucrose concentration in the medium and by light. *Appl. Environ. Microbiol.* **63**, 2844–2849.

Guebel, D.V. and Torres Darias, N.V. (2001). Optimization of the citric acid production by *Aspergillus niger* through a metabolic flux balance model. *Electron. J. Biotechnol.* **4**, 1–14.

Guevarra, E.D. and Tabuchi, T. (1990). Accumulation of itaconic, 2-hydroxyparaconic, itatartaric, and malic-acids by strains of the genus *Ustilago. Agric. Biol. Chem.* **54**, 2353–2358.

Gyamerah, M. (1995a). Factors affecting the growth form of *Aspergillus terreus* NRRL 1960 in relation to itaconic acid fermentation. *Appl. Microbiol. Biotechnol.* **44**, 356–361.

Gyamerah, M.H. (1995b). Oxygen requirement and energy relations of itaconic acid fermentation by *Aspergillus terreus* NRRL 1960. *Appl. Microbiol. Biotechnol.* **44**, 20–26.

Habison, A., Kubicek, C.P., and Röhr, M. (1983). Partial purification and regulatory properties of phosphofructokinase from *Aspergillus niger. Biochem. J.* **209**, 669–676.

Hamamci, H. and Ryu, D.D.Y. (1994). Production of L(+)-lactic acid using immobilized *Rhizopus oryzae*—Reactor performance based on kinetic model and simulation. *Appl. Biochem. Biotechnol.* **44**, 125–133.

Hang, Y.D., Hamamci, H., and Woodams, E.E. (1989). Production of L(+)-lactic acid by *Rhizopus oryzae* immobilized in calcium alginate gels. *Biotechnol. Lett.* **11**, 119–120.

Haskins, R.H., Thorn, J.A., and Boothroyd, B. (1955). Biochemistry of the Ustilaginales. XI. Metabolic products of *Ustilago zeae* in submerged culture. *Can. J. Microbiol.* **1**, 749–756.

Hayaishi, O., Shimazono, H., Katagiri, M., and Saito, Y. (1956). Enzymatic formation of oxalate and acetate from oxaloacetate. *J. Am. Chem. Soc.* **78**, 5126–5127.

Heiland, S., Radovanovic, N., Höfer, M., Winderickx, J., and Lichtenberg, H. (2000). Multiple hexose transporters of *Schizosaccharomyces pombe. J. Bacteriol.* **182**, 2153–2162.

Hesse, S.J.A., Ruijter, G.J.G., Dijkema, C., and Visser, J. (2000). Measurement of intracellular (compartmental) pH by ^{31}P NMR in *Aspergillus niger. J. Biotechnol.* **77**, 5–15.

Hesse, S.J.A., Ruijter, G.J.G., Dijkema, C.O.R., and Visser, J. (2002). Intracellular pH homeostasis in the filamentous fungus *Aspergillus niger. Eur. J. Biochem.* **269**, 3485–3494.

Jaklitsch, W.M., Kubicek, C.P., and Scrutton, M.C. (1991). Intracellular location of enzymes involved in citrate production by *Aspergillus niger. Can. J. Microbiol.* **37**, 823–827.

Jermyn, M.A. (1960). Studies on the glucono-delta-lactonase of *Pseudomonas fluorescens. Biochim. Biophys. Acta* **37**, 78–92.

Joseph-Horne, T.I.M., Hollomon, D.W., and Wood, P.M. (2001). Fungal respiration: A fusion of standard and alternative components. *Biochim. Biophys. Acta* **1504**, 179–195.

Kane, J., Finlay, A., and Amann, P. (1945). Production of itaconic acid. *US Patent 2,385,283.*

Kautola, H., Vahvaselka, M., Linko, Y.Y., and Linko, P. (1985). Itaconic acid production by immobilized *Aspergillus terreus* from xylose and glucose. *Biotechnol. Lett.* **7**, 167–172.

Kelley, R.L. and Reddy, C.A. (1986). Purification and characterization of glucose oxidase from ligninolytic cultures of *Phanerochaete chrysosporium. J. Bacteriol.* **166**, 269–274.

Kelley, R.L. and Reddy, C.A. (1988). Glucose oxidase of *Phanerochaete chrysosporium. Meth. Enzymol.* **161**, 307–316.

Kenealy, W., Zaady, E., du Preez, J.C., Stieglitz, B., and Goldberg, I. (1986). Biochemical aspects of fumaric acid accumulation by *Rhizopus arrhizus. Appl. Environ. Microbiol.* **52**, 128–133.

Kilian, S.G., van Deemter, A., Kock, J.L.F., and du Preez, J.C. (1991). Occurrence and taxonomic aspects of proton movements coupled to sugar transport in the yeast genus *Kluyveromyces. Antonie Van Leeuwenhoek* **59**, 199–206.

Kinoshita, K. (1931). Production of itaconic acid and mannitol by a new mold, *Aspergillus itaconicus. Acta Phytochim.* **5**, 271–287.

Kirimura, K., Hirowatari, Y., and Usami, S. (1987). Alterations of respiratory systems in *Aspergillus niger* under the conditions of citric acid fermentation. *Agric. Biol. Chem.* **51**, 1299–1304.

Kirimura, K., Yoda, M., Ko, I., Oshida, Y., Miyake, K., and Usami, S. (1999a). Cloning and sequencing of the chromosomal DNA and cDNA encoding the mitochondrial citrate synthase of *Aspergillus niger* WU-2223L. *J. Biosci. Bioeng.* **88**, 237–243.

Kirimura, K., Yoda, M., Kumatani, M., Ishii, Y., Kino, K., and Usami, S. (2002). Cloning and expression of *Aspergillus niger icdA* gene encoding mitochondrial NADP$^+$-specific isocitrate dehydrogenase. *J. Biosci. Bioeng.* **93**, 136–144.

Kirimura, K., Yoda, M., Shimizu, H., Sugano, S., Mizuno, M., Kino, K., and Usami, S. (2000). Contribution of cyanide-insensitive respiratory pathway, catalyzed by the alternative oxidase, to citric acid production in *Aspergillus niger. Biosci. Biotechnol. Biochem.* **64**, 2034–2039.

Kirimura, K., Yoda, M., and Usami, S. (1999b). Cloning and expression of the cDNA encoding an alternative oxidase gene from *Aspergillus niger* WU-2223L. *Curr. Genet.* **34**, 472–477.

Kleppe, K. (1966). The effect of hydrogen peroxide on glucose oxidase from *Aspergillus niger. Biochemistry* **5**, 139–143.

Kosakai, Y., Park, Y.S., and Okabe, M. (1997). Enhancement of L(+)-lactic acid production using mycelial flocs of *Rhizopus oryzae. Biotechnol. Bioeng.* **55**, 461–470.

Kubicek, C.P., Hampel, W., and Röhr, M. (1979). Manganese deficiency leads to elevated amino-acid pools in citric acid accumulating *Aspergillus niger. Arch. Microbiol.* **123**, 73–80.

Kubicek, C.P. and Röhr, M. (1980). Regulation of citrate synthase from the citric acid accumulating fungus, *Aspergillus niger. Biochim. Biophys. Acta* **615**, 449–457.

Kubicek, C.P. and Röhr, M. (1985). Aconitase and citric acid fermentation by *Aspergillus niger. Appl. Environ. Microbiol.* **50**, 1336–1338.

Kubicek, C.P. and Röhr, M. (1986). Citric acid fermentation. *Crit. Rev. Biotechnol.* **3**, 331–374.

Kubicek, C.P., Schreferl-Kunar, G., Wöhrer, W., and Röhr,M. (1988). Evidence for a cytoplasmic pathway of oxalate biosynthesis in *Aspergillus niger. Appl. Environ. Microbiol.* **54**, 633–637.

Kubicek, C.P., Zehentgruber, O., El-Kalak, H., and Röhr, M. (1980). Regulation of citric acid production by oxygen: Effect of dissolved oxygen tension on adenylate levels and respiration in *Aspergillus niger. Eur. J. Appl. Microbiol. Biotechnol.* **9**, 101–115.

Kusai, K., Sekuzu, I., Hagihara, B., Okunuki, K., Yamauchi, S., and Nakai, M. (1960). Crystallization of glucose oxidase from *Penicillium amagasakiense. Biochim. Biophys. Acta* **40**, 555–557.

La Nauze, J.M. (1966). Aconitase and isocitric dehydrogenases of *Aspergillus niger* in relation to citric acid production. *J. Gen. Microbiol.* **44**, 73–81.

Lantero, O.J. and Shetty, J.K. (2001). Process for the preparation of gluconic acid and gluconic acid produced thereby. *US Patent 20020119583.*

Larsen, H. and Eimhjellen, K.E. (1955). The mechanism of itaconic acid formation by *Aspergillus terreus*. 1. The effect of acidity. *Biochem. J.* **60**, 135–139.

Legiša, M. and Bencina, M. (1994). Evidence for the activation of 6-phosphofructo-1-kinase by cAMP-dependent protein kinase in *Aspergillus niger. FEMS Microbiol. Lett.* **118**, 327–333.

Legiša, M. and Grdadolnik, S.G. (2002). Influence of dissolved oxygen concentration on intracellular pH and consequently on growth rate of *Aspergillus niger. Food Technol. Biotechnol.* **40**, 27–32.

Legiša, M. and Kidric, J. (1989). Initiation of citric acid accumulation in the early stages of *Aspergillus niger* growth. *Appl. Microbiol. Biotechnol.* **31**, 453–457.

LéJohn, H.B. (1971). D(-)-Lactate dehydrogenases in fungi: Kinetics and allosteric inhibition by guanosine triphosphate. *J. Biol. Chem.* **246**, 2116–2126.

Lenz, H., Wunderwald, P., and Eggerer, H. (1976). Partial purification and some properties of oxalacetase from *Aspergillus niger. Euro. J. Biochem.* **65**, 225–236.

Lewis, K.F. and Weinhouse, S. (1951). Studies on the mechanism of citric acid production in *Aspergillus niger. J. Am. Chem. Soc.* **73**, 2500–2503.

Ling, E.T.M., Dibble, J.T., Houston, M.R., Lockwood, L.B., and Elliott, L.P. (1978). Accumulation of *l-trans*-2,3-epoxysuccinic acid and succinic acid by *Paecilomyces variota. Appl. Environ. Microbiol.* **35**, 1213–1215.

Litchfield, J.H. (1996). *Microbiological production of lactic acid* Academic Press, New York, (pp. 45–95).

Lockwood, L.B. and Reeves, M.D. (1945). Some factors affecting the production of itaconic acid by *Aspergillus terreus. Arch. Biochem.* **6**, 455–469.

Lockwood, L.B., Ward, G.E., and May, O.E. (1936). The physiology of *Rhizopus oryzae. J. Agric. Res.* **53**, 849–857.

Longacre, A., Reimers, J.M., Gannon, J.E., and Wright, B.E. (1997). Flux analysis of glucose metabolism in *Rhizopus oryzae* for the purpose of increasing lactate yields. *Fungal Genet. Biol.* **21**, 30–39.

Lopez-Garcia, R. (2002). Citric acid. In Kirk-Othmer (ed) *Kirk-Othmer encyclopedia of chemical technology.* John Wiley & Sons, Inc., New York, USA.

Ma, H., Kubicek, C.P., and Röhr, M. (1985). Metabolic effects of manganese deficiency in *Aspergillus niger*: Evidence for increased protein degradation. *Arch. Microbiol.* **141**, 266–268.

Margulies, M. and Vishniac, W. (1961). Dissimilation of glucose by the MX strain of *Rhizopus. J. Bacteriol.* **81**, 1–9.

Mark, C.G. and Romano, A.H. (1971). Properties of the hexose transport systems of *Aspergillus nidulans. Biochim. Biophys. Acta* **249**, 216–226.

Martin, S.M., Wilson, P.W., and Burris, R.H. (1950). Citric acid formation from $^{14}CO_2$ by *Aspergillus niger. Arch. Biochem.* **26**, 103–111.

Martin, W.R., and Foster, J.W. (1955). Production of *trans*-L-epoxysuccinic acid by fungi and its microbiological conversion to *meso*-tartaric acid. *J. Bacteriol.* **70**, 405–414.

Mattey, M. (1992). The production of organic acids. *Crit. Rev. Biotechnol.* **12**, 87–132.

McQuigg, D.W., Marston, C., Fitzpatrick, G., Crowe, E., and Vorhies, S. (2000). Processes for recovering citric acid. *US Patent 6,137,004*.

Meixner-Monori, B., Kubicek, C.P., Harrer, W., Schreferl, G., and Röhr, M. (1986). NADP-specific isocitrate dehydrogenase from the citric acid accumulating fungus *Aspergillus niger. Biochem. J.* **236**, 549–558.

Meixner-Monori, B., Kubicek, C.P., and Röhr, M. (1984). Pyruvate kinase from *Aspergillus niger* a regulatory enzyme in glycolysis. *Can. J. Microbiol.* **30**, 16–22.

Moyer, A.J., Umberger, E.J., and Stubbs, J.J. (1940). Fermentation of concentrated solutions of glucose to gluconic acid: Improved process. *Ind. Eng. Chem.* **32**, 1379–1383.

Mueller, H.-M. (1975). Oxalate accumulation from citrate by *Aspergillus niger*. I. Biosynthesis of oxalate from its ultimate precursor. *Arch. Microbiol.* **103**, 185–190.

Nelson, G.E.N., Traufler, D.H., Kelley, S.E., and Lockwood, L.B. (1952). Production of itaconic acid by *Aspergillus terreus* in 20-liter fermentors. *Ind. Eng. Chem.* **44**, 1166–1168.

Netik, A., Torres, N.V., Riol, J.-M., and Kubicek, C.P. (1997). Uptake and export of citric acid by *Aspergillus niger* is reciprocally regulated by manganese ions. *Biochim. Biophy. Acta* **1326**, 287–294.

Nubel, R.C. and Ratajak, E.J. (1962). Process for producing itaconic acid. *US Patent 3,044,941*.

O'Malley, J.J. and Weaver, J.L. (1972). Subunit structure of glucose oxidase from *Aspergillus niger. Biochemistry* **11**, 3527–3532.

Ogawa, K., Nakajima-Kambe, T., Nakahara, T., and Kokufuta, E. (2002). Coimmobilization of gluconolactonase with glucose oxidase for improvement in kinetic property of enzymatically induced volume collapse in ionic gels. *Biomacromolecules* **3**, 625–631.

Osmani, S. and Scrutton, M.C. (1983). The sub cellular localization of pyruvate carboxylase and of some other enzymes in *Aspergillus nidulans. Eur. J. Biochem.* **133**, 551–560.

Osmani, S. and Scrutton, M.C. (1985). The subcellular localization and regulatory properties of pyruvate carboxylase from *Rhizopus arrhizus. Eur. J. Biochem.* **147**, 119–128.

Overman, S.A. and Romano, A.H. (1969). Role of pyruvate carboxylase in fumaric acid accumulation by *Rhizopus nigricans. Bacteriol. Proc.* **69**, 128.

Panneman, H., Ruijter, G.J.G., van den Broeck, H.C., Driever, E.T.M., and Visser, J. (1996). Cloning and biochemical characterisation of an *Aspergillus niger* glucokinase. Evidence for the presence of separate glucokinase and hexokinase enzymes. *Eur. J. Biochem.* **240**, 518–525.

Panneman, H., Ruijter, G.J.G., van den Broeck, H.C., and Visser, J. (1998). Cloning and biochemical characterisation of *Aspergillus niger* hexokinase. The enzyme is strongly inhibited by physiological concentrations of trehalose-6-phosphate. *Eur. J. Biochem.* **258**, 223–232.

Park, E.Y., Kosakai, Y., and Okabe, M. (1998). Efficient production of L-(+)-lactic acid using mycelial cotton-like flocs of *Rhizopus oryzae* in an air-lift bioreactor. *Biotechnol. Prog.* **14**, 699–704.

Pazur, J.H. (1966). Glucose oxidase from *Aspergillus niger. Methods Enzymol.* **9**, 82–87.

Pedersen, H., Christensen, B., Hjort, C., and Nielsen, J. (2000a). Construction and characterization of an oxalic acid nonproducing strain of *Aspergillus niger. Metab. Eng.* **2**, 34–41.

Pedersen, H., Hjort, C., and Nielsen, J. (2000b). Cloning and characterization of *oah*, the gene encoding oxaloacetate hydrolase in *Aspergillus niger. Mol. Gen. Genet.* **263**, 281–286.

Peksel, A., Torres, N.V., Liu, J., Juneau, G., and Kubicek, C.P. (2002). ^{13}C-NMR Analysis of glucose metabolism during citric acid production by *Aspergillus niger. Appl. Microbiol. Biotechnol.* **58**, 157–163.

Peleg, Y., Barak, A., Scrutton, M.C., and Goldberg, I. (1989a). Malic acid accumulation by *Aspergillus flavus*. 3. ^{13}C-NMR and isoenzyme analyses. *Appl. Microbiol. Biotechnol.* **30**, 176–183.

Peleg, Y., Battat, E., Scrutton, M.C., and Goldberg, I. (1989b). Isoenzyme pattern and subcellular localization of enzymes involved in fumaric acid accumulation by *Rhizopus oryzae. Appl. Microbiol. Biotechnol.* **32**, 334–339.

Peleg, Y., Stieglitz, B., and Goldberg, I. (1988). Malic acid accumulation by *Aspergillus flavus*. 1. Biochemical aspects of acid biosynthesis. *Appl. Microbiol. Biotechnol.* **28**, 69–75.

Pfeifer, V.F., Vojnovich, C., and Heger, E.N. (1952). Itaconic acid by fermentation with *Aspergillus terreus. Ind. Eng. Chem.* **44**, 2975–2980.

Porro, D., Bianchi, M., Ranzi, B.M., Frontali, L., Vai, M., Winkler, A.A., and Alberghina, L. (2002). Yeast strains for the production of lactic acid transformed with a gene coding for lactic acid dehydrogenase. *US Patent 6,429,006*.

Pritchard, G.G. (1973). Factors affecting the activity and synthesis of NAD-dependent lactate dehydrogenase in *Rhizopus oryzae. J. Gen. Microbiol.* **78**, 125–137.

Pritchard, G.G. (1971). An NAD-independent L-lactate dehydrogenase from *Rhizopus oryzae. Biochim. Biophys. Acta* **250**, 25–34.

Prömper, C., Schneider, R., and Weiss, H. (1993). The role of the proton-pumping and alternative respiratory chain NADH: Ubiquinone oxidoreductases in overflow catabolism of *Aspergillus niger. Eur. J. Biochem.* **216**, 223–230.

Rhodes, R.A., Moyer, A.J., Smith, M.L., and Kelley, S.E. (1959). Production of fumaric acid by *Rhizopus arrhizus. Appl. Microbiol.* **7**, 74–80.

Röhr, M., Kubicek, C.P., Zehentgruber, O., and Orthofer, R. (1987). Accumulation and partial reconsumption of polyols during citric acid fermentation by *Aspergillus niger. Appl. Microbiol. Biotechnol.* **27**, 235–239.

Ruijter, G.J.G., Kubicek, C.P., and Visser, J. (2002). Production of organic acids by fungi. In H. D. Osiewacz (ed) *The mycota: A comprehensive treatise on fungi as experimental systems for basic and applied research. Industrial Applications* Springer-Verlag, Berlin, Germany, pp. 213–230.

Ruijter, G.J.G, Panneman, H., and Visser, J. (1998). Metabolic engineering of the glycolytic pathway in *Aspergillus niger. Food Technol. Biotechnol.* **36**, 185–188.

Ruijter, G.J.G., Panneman, H., and Visser, J. (1997). Overexpression of phosphofructokinase and pyruvate kinase in citric acid-producing *Aspergillus niger. Biochim. Biophys. Acta* **1334**, 317–326.

Ruijter, G.J.G., Panneman, H., Xu, D.-B., and Visser, J. (2000). Properties of *Aspergillus niger* citrate synthase and effects of *citA* overexpression on citric acid production. *FEMS Microbiol. Lett.* **184**, 35–40.

Ruijter, G.J.G., van de Vondervoort, P.J.I., and Visser, J. (1999). Oxalic acid production by *Aspergillus niger*: An oxalate-non- producing mutant produces citric acid at pH 5 and in the presence of manganese. *Microbiology* **145**, 2569–2576.

Ruijter, G.J.G. and Visser, J. (1999). Characterization of *Aspergillus niger* phosphoglucose isomerase. use for quantitative determination of erythrose-4-phosphate. *Biochimie* **81**, 267–272.

Sakaguchi, K., Inoue, T., and Tada, S. (1939). On the production of aethyleneoxide-alpha-beta-dicarboxylic acid by moulds. *Zentr. Bakteriol. Parasitenk. Abt. II* **100**, 302–307.

Schreferl, G., Kubicek, C.P., and Röhr, M. (1986). Inhibition of citric acid accumulation by manganese ions in *Aspergillus niger* mutants with reduced citrate control of phosphofructokinase. *J. Bacteriol.* **165**, 1019–1022.

Shimi, I.R. and Nour El Dein, M.S. (1962). Biosynthesis of itaconic acid by *Aspergillus terreus. Archive fur Mikrobiologie* **44**, 181–188.

Shu, P. and Johnson, M.J. (1947). Effect of the composition of the sporulation medium on citric acid production by *Aspergillus niger* in submerged culture. *J. Bacteriol.* **54**, 161–167.

Shu, P. and Johnson, M.J. (1948a). Citric acid production by submerged fermentation with *Aspergillus niger. Ind. Eng. Chem.* **40**, 1202–1205.

Shu, P. and Johnson, M.J. (1948b). The interdependence of medium constituents in citric acid production by submerged fermentation. *J. Bacteriol.* **56**, 577–585.

Skory, C.D. (2001). Fungal lactate dehydrogenase gene and constructs for the expression thereof. *US Patent 6,268,189.*

Skory, C.D. (2000). Isolation and expression of lactate dehydrogenase genes from *Rhizopus oryzae. Appl. Environ. Microbiol.* **66**, 2343–2348.

Skory, C.D., Freer, S.N., and Bothast, R.J. (1998). Production of L-lactic acid by *Rhizopus oryzae* under oxygen limiting conditions. *Biotechnol. Lett.* **20**, 191–194.

Snell, R.L. and Lowery, C.E. (1964). Calcium L (+) lactate and L (+) lactic acid production. *US Patent 3,125,494.*

Swart, K., van de Vondervoort, P.J.I., Witteveen, C.F.B., and Visser, J. (1990). Genetic localization of a series of genes affecting glucose oxidase levels in *Aspergillus niger. Curr. Genet.* **18**, 435–440.

Swoboda, B.E.P., and Massey, V. (1965). Purification and properties of the glucose oxidase from *Aspergillus niger. J. Biol. Chem.* **240**, 2209–2215.

Tabuchi, T. (1991). Manufacture of itaconic acid with *Ustilago. Japan Patent 3,035,785.*

Tamada, M., Begum, A.A., and Sadi, S. (1992). Production of L(+)-lactic acid by immobilized cells of *Rhizopus oryzae* with polymer supports prepared by gamma-ray induced polymerization. *J. Ferment. Bioeng.* **74**, 379–383.

TIGR (2002). The *Aspergillus fumigatus* genome database; *http://www.tigr.org/tdb/e2k1/afu1/.*

Title 21CFR173.280. (1984). Food and Drugs; Part 173-Secondary direct food additives permitted in food for human consumption; Subpart C—Solvents, lubricants, release agents and related substances; 173.280-solvent extraction process for citric acid.

Title 21CFR184.1318. (1986). Food and Drugs; Part 184-Direct food substances affirmed as generally recognized as safe; Subpart B—Listing of specific substances affirmed as Gras; 184.1318-glucono delta lactone.

Title 21CFR184.1033. (1994). Food and Drugs; Part 184-Direct food substances affirmed as generally recognized as safe; Subpart B-Listing of specific substances affirmed as Gras; 184.1033-citric acid.

Todd, R.B., Andrianopoulos, A., Davis, M.A., and Hynes, M.J. (1998). FacB, the *Aspergillus nidulans* activator of acetate utilization genes, binds dissimilar DNA sequences. *EMBO J.* **17**, 2042–2054.

Torres, N.V. (1994). Modeling approach to control of carbohydrate metabolism during citric acid accumulation by *Aspergillus niger*: I. Model definition and stability of the steady state. *Biotechnol. Bioeng.* **44**, 104–111.

Torres, N.V., Riol-Cimas, J.M., Wolschek, M., and Kubicek, C.P. (1996a). Glucose transport by *Aspergillus niger*: The low-affinity carrier is only formed during growth on high glucose concentrations. *Appl. Microbiol. Biotechnol.* **44**, 790–794.

Torres, N.V., Voit, E.O., and González-Alcón, C. (1996b). Optimization of nonlinear biotechnological process with linear programming: Application to citric acid production by *Aspergillus niger. Biotechnol. Bioeng.* **49**, 247–258.

Vainstein, M.H. and Peberdy, J.F. (1991). Regulation of invertase in *Aspergillus nidulans*: Effect of different carbon sources. *J. Gen. Microbiol.* **137**, 315–322.

Vroemen, A.J. and Beverini, M. (1999). Enzymatic production of gluconic acid or its salts. *US Patent 5,897,995*.

Waksman, S.A. and Foster, J.W. (1938). Respiration and lactic acid production by a fungus of the genus *Rhizopus. J. Agric. Res.* **57**, 873–899.

Wallis, G.L.F., Hemming, F.W., and Peberdy, J.F. (1997). Secretion of two beta-fructofuranosidases by *Aspergillus niger* growing in sucrose. *Arch. Biochem. Biophy.* **345**, 214–222.

Wang, H.S. and LéJohn, H.B. (1974). Analogy and homology of the dehydrogenases of oomycetes. Part 2. Regulation by GTP of D-*levo*-lactic dehydrogenases and isozyme patterns. *Can. J. Microbiol.* **20**, 575–580.

Ward, G.E., Lockwood, L.B., and May, O.E. (1938). Fermentation process for the manufacture of dextro-lactic acid. *US Patent 2,132,712*.

Wariishi, H., Valli, K., and Gold, M.H. (1992). Manganese(II) oxidation by manganese peroxidase from the basidiomycete *Phanerochaete chrysosporium*: Kinetic mechanism and role of chelators. *J. Biol. Chem.* **267**, 23688–23695.

Wayman, F.M. and Mattey, M. (2000). Simple diffusion is the primary mechanism for glucose uptake during the production phase of the *Aspergillus niger* citric acid process. *Biotechnol. Bioeng.* **67**, 451–456.

Wilkoff, L.J. and Martin, W.R. (1963). Studies on the biosynthesis of *trans-l*-epoxysuccinic acid by *Aspergillus fumigatus. J. Biol. Chem.* **238**, 843–846.

Willke, T. and Vorlop, K.D. (2001). Biotechnological production of itaconic acid. *Appl. Microbiol. Biotechnol.* **56**, 289–295.

Winskill, N. (1983). Tricarboxylic-acid cycle activity in relation to itaconic acid biosynthesis by *Aspergillus terreus. J. Gen. Microbiol.* **129**, 2877–2883.

Witteveen, C.F.B., van de Vondervoort, P., Swart, K., and Visser, J. (1990). Glucose oxidase overproducing and negative mutants of *Aspergillus niger. Appl. Microbiol. Biotechnol.* **33**, 683–686.

Witteveen, C.F.B., van de Vondervoort, P.J.I., van den Broeck, H.C., van Engelenburg, F.A.C., de Graaff, L.H., Hillebrand, M.H.B.C., Schaap, P.J., and Visser, J. (1993). Induction of glucose oxidase, catalase, and lactonase in *Aspergillus niger. Curr. Genet.* **24**, 408–416.

Witteveen, C.F.B., Veenhuis, M., and Visser, J. (1992). Localization of glucose oxidase and catalase activities in *Aspergillus niger. Appl. Environ. Microbiol.* **58**, 1190–1194.

Witteveen, C.F.B. and Visser, J. (1995). Polyol pools in *Aspergillus niger. FEMS Microbiol. Lett.* **134**, 57–62.

Wolschek, M.F. and Kubicek, C.P. (1997). The filamentous fungus *Aspergillus niger* contains two "differentially regulated" trehalose-6-phosphate synthase encoding genes, *tpsA* and *tpsB. J. Biol. Chem.* **272**, 2729–2735.

Woronick, C.L. and Johnson, M.J. (1960). Carbon dioxide fixation by cell-free extracts of *Aspergillus niger. J. Biol. Chem.* **235**, 9–15.

Wright B.E., Longacre A., and Reimers J. (1996). Models of metabolism in *Rhizopus oryzae. J. Theor. Biol.* **182**, 453–457.

Xuemei, L., Jianping, L., Mo'e, L., and Peilin, C. (1999). L-Lactic acid production using immobilized *Rhizopus oryzae* in a three-phase fluidized-bed with simultaneous product separation by electrodialysis. *Bioprocess Eng.* **20**, 231–237.

Yang, C.W., Lu, Z.J., and Tsao, G.T. (1995). Lactic acid production by pellet-form *Rhizopus oryzae* in a submerged system. *Appl. Biochem. Biotechnol.* **51**, 57–71.

Yin, P.M., Nishina, N., Kosakai, Y., Yahiro, K., Park, Y., and Okabe, M. (1997). Enhanced production of L(+)-lactic acid from corn starch in a culture of *Rhizopus oryzae* using an air-lift bioreactor. *J. Ferment. Bioeng.* **84**, 249–253.

Yin, P.M., Yahiro, K., Ishigaki, T., Park, Y., and Okabe, M. (1998). L(+)-Lactic acid production by repeated batch culture of *Rhizopus oryzae* in air-lift bioreactor. *J. Ferment. Bioeng.* **85**, 96–100.

Yu, R.C. and Hang, Y.D. (1991). Purification and characterization of NAD-dependent lactate dehydrogenase from *Rhizopus oryzae*. *Food Chem.* **41**, 219–225.

Zehentgruber, O., Kubicek, C.P., and Röhr, M. (1980). Alternative respiration of *Aspergillus niger. FEMS Microbiol. Lett.* **8**, 71–74.

Zeikus, J.G., Jain, M.K., and Elankovan, P. (1999). Biotechnology of succinic acid production and markets for derived industrial products. *Appl. Microbiol. Biotechnol.* **51**, 545–552.

Zhang, A. and Röhr, M. (2002a). Citric acid fermentation and heavy metal ions: II. The action of elevated manganese ion concentrations. *Acta Biotechnol.* **22**, 375–382.

Zhang, A. and Röhr, M. (2002b). Effects of varied phosphorus concentrations on citric acid fermentation by *Aspergillus niger. Acta Biotechnol.* **22**, 383–390.

Zhou, Y., Dominguez, J.M., Cao, N.J., Du, J.X., and Tsao, G.T. (1999). Optimization of L-lactic acid production from glucose by *Rhizopus oryzae* ATCC 52311. *Appl. Biochem. Biotechnol.* **77**, 401–407.

Zhou, Y., Du, J.X., and Tsao, G.T. (2000). Mycelial pellet formation by *Rhizopus oryzae* ATCC 20344. *Appl. Biochem. Biotechnol.* **84**, 779–789.

13

Flavors and Fragrances

Ralf G. Berger and Holger Zorn

1. Introduction

The awareness that fungi produce enticing flavors is not new. Presumably it was their pleasant aroma that prompted the notorious Roman emperor Nero to name mushrooms "*cibus deorum*," food of the gods. Not many years ago fungi were accessible for consumption if one knew how to identify edible species and had access to sites where they grew. Modern cultivation techniques have made several edible fungi much more widely available. For field mushroom, shiitake, and oyster mushroom, the three most important edible fungi worldwide, the annual production exceeds 3,400 million tons (Hobbythek, 1999). In one sense, this is an enormous agrobiotechnological production of fungal flavors. The aroma of the field mushroom (*Agaricus bisporus*) and of the oyster mushroom (*Pleurotus* spp.) originates mainly from the enzymatic oxidative degradation of linoleic acid, with 1-octen-3-ol (Figure 13.1) being the most prominent flavor compound (Venkateshwarlu *et al.*, 1999; Husson *et al.*, 2001). Shiitake (*Lentinus edodes*), which is widely consumed in China and Japan, has a very intensive aroma due to 1,2,3,5,6-pentathiepane (lenthionine; Figure 13.1) as well as sulfur containing degradation products of S-alkyl cysteine sulfoxide (lentinic acid) (Belitz and Grosch, 1987). Beyond these well known representatives, the spectrum of fungi producing flavor compounds is immense. In many cases their capability is reflected in genus or species names that point to aroma characteristics: *butyrace-* (butter-like), *delicat-* (delicious), *odor-/osm-* (fragrant), *olid-* (ambrosial), *suav-* (sweet), *nidoros-* (pungent) and *foetens-/foetid-* (fetid). Furthermore, trivial names in many different languages often give hints to floral, spicy, or aromatic flavor impressions. Currently, elucidation of flavor profiles either focuses on edible wild mushrooms or is employed as a tool in chemotaxonomic approaches. The volatile compounds emitted by the fruiting bodies of basidiomycetes are usually analyzed by coupled gas chromatography–mass spectrometry (GC–MS) or GC–olfactometry. Olfactometry requires dynamic headspace concentration through cryotrapping, solvent extraction, simultaneous distillation and solvent extraction (SDE), or solid phase microextraction (SPME). In an investigation of 26 basidiomycetes, more than 140 volatiles were detected, among which were mono- and sesquiterpenes,

---

Ralf G. Berger and Holger Zorn • Zentrum Angewandte Chemie, Universität Hannover, Institut für Lebensmittelchemie, Wunstorfer Straße 14, D-30453 Hannover, Germany.

Advances in Fungal Biotechnology for Industry, Agriculture, and Medicine. Edited by Jan S. Tkacz and Lene Lange, Kluwer Academic/Plenum Publishers, 2004.

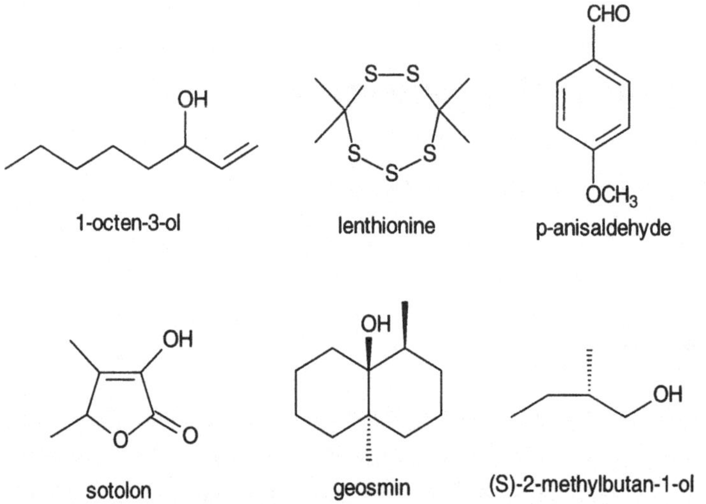

1-octen-3-ol lenthionine p-anisaldehyde

sotolon geosmin (S)-2-methylbutan-1-ol

Figure 13.1. Typical flavor compounds formed *de novo* by fruiting bodies of higher fungi.

aromatics, ketones, esters, alcohols, and sulfur-containing compounds (Talou *et al.*, 2000). In a series of 80 wild mushroom species, 34 strains synthesized a total of 28 different monoterpenes (Breheret *et al.*, 1997). Altogether 16 aromatic compounds, which were likely to be biogenically derived from lignin of the host tree, were identified in hot water extracts of the white-rot fungus *Gloeophyllum odoratum* (Rösecke and König, 2000).

The anise-like odor of *Clitocybe odora* can be ascribed to *p*-anisaldehyde (Rapior *et al.*, 2002), and the fenugreek odor of *Lactarius helvus* results mainly from sotolon (Figure 13.1) (Rapior *et al.*, 2000a). The musty-earthy smelling geosmin (Figure 13.1) is formed by *Cortinarius herculeus, Cystoderma amianthinum*, and *Cystoderma carcharias* (Breheret *et al.*, 1999). A complex almond odor with an anise note was reported for *Gyrophragmium dunalii* (Rapior *et al.*, 2000b), and (*S*)-2-methylbutan-1-ol (Figure 13.1) contributed to the flavor of the black Perigord truffle (*Tuber melanosporum* Vitt.) (Doumenc-Faure *et al.*, 2000). In most of these examples the overall odor is determined by traces of a single constituent with a low or ultra-low sensory threshold; these powerful compounds are usually termed "character impact components" (Berger, 1995).

The fungal kingdom has an enormous biochemical potential to serve as a biological aroma factory given that the fragrant fungal metabolites already known encompass a wide structural spectrum of acyclic, cyclic, heterocyclic, and aromatic products of varying oxidation states and also that these compounds are often produced at high levels. Moreover as many fungi propagate in nature by decomposing organic material, it is almost self-evident that the aroma compounds they produce will have originated either by de novo synthesis or by transformation of substrates they encounter in decomposing plants and other organic materials. Attempts to harness the potential of fungi for the flavor production and for the transformation of readily available substrates into valuable flavor compounds started in the early 1950s and have been pursued constantly since then. In recent years, phytopathogenic fungi (e.g., *Botrytis cinerea, Rhizoctonia solani, Glomerella cingulata*, and *Ceratocystis fimbriata*) have been found capable of converting substrates that are cytotoxic to most other microorganisms. A number of excellent up-to-date reviews and book chapters dealing inter alia with "fungal flavors" are available (Krings and Berger, 1998; Atta-ur-Rahman *et al.*, 1999; Lomascolo *et al.*, 1999b). Comprehensive summaries of previous studies on the transformation of terpenes and terpenoids by fungi have been published (Demyttenaere, 2001; Demyttenaere and de Kimpe, 2001; Schrader and Berger, 2001). To avoid redundancies, only publications since 1998 have been considered for this chapter. Further, the chapter focuses on the deliberate generation of volatile flavors by basidiomycetous and ascomycetous fungi; yeasts are not covered.

2. Biotransformation of Terpenoids by Fungi

Terpenoids (also known as isoprenoids) constitute the largest group of natural products. Many important flavor and fragrance compounds, contributing a wide range of pleasant scents, are biosynthetically derived from isoprene units.

In one study, more then 60 fungi, grown as sporulated surface cultures, were screened for the ability to modify (*R*)-(+)-limonene, the most widely distributed terpene in nature after α-pinene. Various species, notably *Penicillium digitatum* and *Corynespora cassiicola*, metabolized (*R*)-(+)-limonene mainly to (*R*)-(+)-α-terpineol and γ-terpinene (Figure 13.2). In submerged liquid cultures, product yields up to 45.7% were achieved for (*R*)-(+)-α-terpineol with *P. digitatum*, and *Corynespora cassiicola* gave a similar yield (46.7%) for

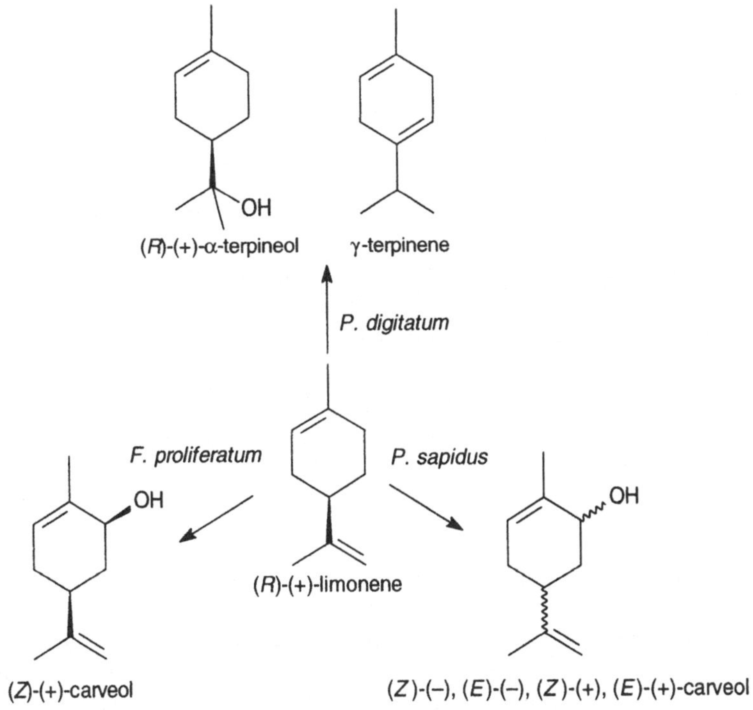

Figure 13.2. Some fungal oxidation products of (*R*)-(+)-limonene.

(1*S*,2*S*,4*R*)-limonene-1,2-diol (Demyttenaere *et al.*, 2001b). Further, the conversion of limonene to (*R*)-(+)-α-terpineol by *P. digitatum* was enantioselective and enantiospecific. (*R*)-(+)-Limonene was converted only to (*R*)-(+)-α-terpineol, no (*S*)-(−)-α-terpineol was detected. (*S*)-(−)-Limonene was not converted by the fungus (Tan *et al.*, 1998). The enzyme responsible for these conversions appeared to be inducible because bioconversion rates were enhanced significantly in cultures that were grown in the presence of low levels of substrate. Optimizing the composition of the growth medium, aeration rate, temperature, substrate concentration, and pH led to α-terpineol yields of up to 3.2 g L^{-1}. The process can be conducted with cells immobilized in calcium alginate beads, a stable form of the biocatalyst, and its efficiency can be enhanced by the addition of organic co-solvents, such as dioctylphthalate or methanol to solubilize the substrate (Tan and Day, 1998a, b).

An entirely different product spectrum was observed, when (*R*)-(+)-limonene was transformed by the basidiomycete *Pleurotus sapidus*. Allylic hydroxylation either alone or followed by oxidation resulted in (*Z*)/(*E*)-carveol and carvone, respectively (Onken and Berger, 1999a). The transformation efficiency could be markedly improved with production reaching more than 100 mg L^{-1} when the inoculum was grown in the presence of trace amounts of limonene and the substrate was fed continuously to submerged production cultures via the gas phase. Regardless of the limonene isomer fed, all four carveol stereoisomers were formed, namely (*Z*)-(−)-, (*E*)-(−)-, (*Z*)-(+)-, and (*E*)-(+)-carveol (Figure 13.2), and an enantiomeric distribution of 67% (*R*)-(−)- to 33% (S)-(+)-carvone was observed. To explain these findings, a two-step mechanism was envisaged: a monooxygenase catalyzed allylic hydroxylation to produce an intermediate carbo-cation or radical that rearranges along the double bond. The mechanism accounts for the loss of stereoselectivity. Another route to enantio-pure carvones was reported by Kaspera *et al.* (2002). With the ascomycete *Fusarium proliferatum* as biocatalyst, (*R*)-(+)-limonene is transformed exclusively to (*Z*)-(+)-carveol (enantiomeric excess (ee) > 99) (Figure 13.2), whereas (*S*)-(−)-limonene is converted to (*E*)-(−)-carveol (ee > 76). This points to an enzymatic mechanism different from that of *P. sapidus*. Following the oxidation that generates the enantio-pure terpene alcohols, further oxidation provides routes to (*S*)-(+)-carvone, the caraway essence, and to (*R*)-(−)-carvone, a spearmint-like essence.

As a major constituent of the strawberry, raspberry, and spearmint flavor complexes, verbenone is an attractive target compound for fungal terpene biotransformation. Verbenone is accessible through hydroxylation of the abundantly available substrate α-pinene to produce verbenol that can be further oxidized to the desired product. Strains of *Aspergillus* spp., *Penicillium* spp. (Agrawal *et al.*, 1999; Agrawal and Joseph, 2000), and the plant pathogen *Botrytis cinerea* catalyze these reactions (Figure 13.3; Farooq *et al.*, 2002). *B. cinerea* is also capable of converting α-pinene to 3β-hydroxy-(−)-β-pinene, 9-hydroxy-(−)-α-pinene, and 4β-hydroxy-(−)-α-pinen-6-one (Figure 13.3). Mutation of the *Aspergillus* and *Penicillium* strains, induced either chemically (ethyl methanesulphonate and colchicine) or by ultraviolet (UV)-irradiation provided strains that were able to produce verbenol in concentrations of up to 46 mg L^{-1} but that did not exhibit a parallel increase in verbenone yields (Agrawal *et al.*, 1999).

Diverse strains of *Aspergillus niger* were also capable of metabolizing a number of oxo-functionalized terpenoids. Biotransformation of (+/−)-linalool in surface and submerged cultures yielded mixtures of (*Z*)- and (*E*)-furanoid linalool oxide and of (*Z*)- and (*E*)-pyranoid linalool oxide. When (*R*)-(−)-linalool was fed as substrate, (*E*)-furanoid and

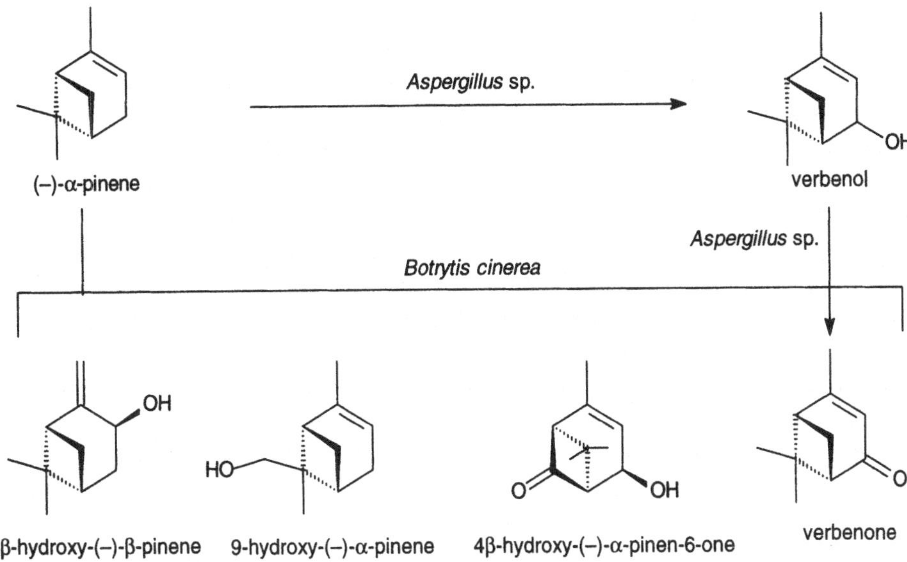

Figure 13.3. Products of the oxidative transformation of (−)α-pinene.

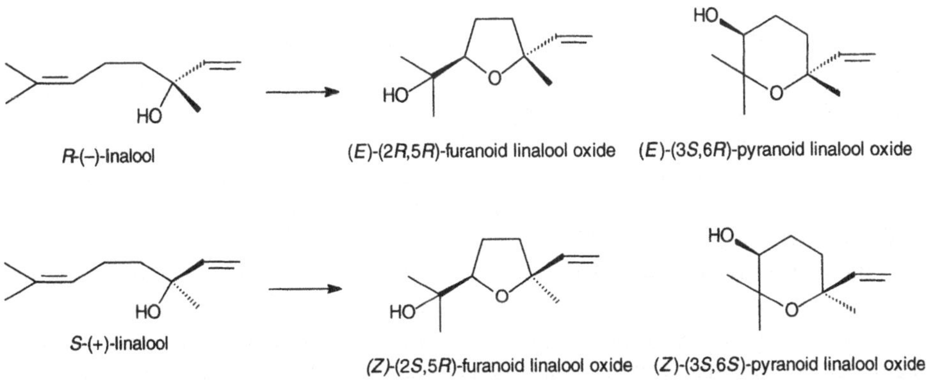

R-(–)-linalool

(E)-(2R,5R)-furanoid linalool oxide

(E)-(3S,6R)-pyranoid linalool oxide

S-(+)-linalool

(Z)-(2S,5R)-furanoid linalool oxide

(Z)-(3S,6S)-pyranoid linalool oxide

Figure 13.4. Biotransformation of (+)- and (−)-linalool by strains of *Aspergillus niger* (modified after Demyttenaere and Willemen, 1998; Demyttenaere *et al.*, 2001a).

(*E*)-pyranoid linalool oxides (ee > 95) were obtained almost exclusively (Figure 13.4). Evidence for the enantioselectivity of the reaction was derived from chiral GC analyses. Only one (*Z*)- and one (*E*)-form was present for each linalool oxide (Demyttenaere and Willemen, 1998). On the other hand, bioconversion of (*S*)-(+)-linalool, which took place with much higher yields, led to (*Z*)-(2*S*,5*R*)-furanoid and (*Z*)-(3*S*,6*S*)-pyranoid linalool oxide as the main products (Figure 13.4). When different solvents were evaluated as vehicles to dissolve the water-insoluble substrate for addition to the cultures, acetone provided the best results (Demyttenaere *et al.*, 2001a). *A. niger* spores entrapped in calcium alginate beads are able to transform the norisoprenoid compound β-ionone in an aerated two-phase liquid system. A dynamic model was established that accounted for consumption of the substrate by biocatalysis in the aqueous phase versus loss of substrate to the gas phase by aeration-related stripping. The data obtained demonstrated that the β-ionone taken up by the fungus was fully converted into metabolites, meaning that β-ionone was not incorporated into the anabolic metabolism of the fungus (Grivel *et al.*, 1999). Optimized conditions involving a fed-batch strategy for both the precursor and the energy source (sucrose) allowed the accumulation of products at combined levels of 3.5 g L^{-1} in 400 hr (Grivel and Larroche, 2001). Submerged liquid cultures, sporulated surface cultures, and spore suspensions of *A. niger* have been compared for their utility in converting nerol, geraniol, and citral. Linalool, α-terpineol, 6-methyl-5-hepten-2-one, and limonene were the main products obtained from nerol and citral with sporulated surface cultures, whereas geraniol was converted predominantly to linalool. Shaken liquid cultures with submerged *A. niger* cells gave only poor yields, but transformation yields of approximately 20% were obtained with the sporulated surface cultures. *A. niger* spores were able to convert nerol to α-terpineol and linalool. In this study, SPME proved to be a valuable tool for screening the biotransformation reactions. Being a fast, sensitive, and solvent-free technique, chromatograms comparable to those after dynamic headspace adsorption on Tenax® were obtained (Demyttenaere *et al.*, 2000).

Spores of *P. digitatum* converted geraniol, nerol, and citral, into 6-methyl-5-hepten-2-one and acetaldehyde. In the first step, geraniol and nerol were substrates for an NAD^+-dependent oxidiation to citral, which was subsequently deacetylated to 6-methyl-5-hepten-2-one without the further need for cofactor. The citral lyase responsible for the deacetylation was partially purified by anion exchange chromatography (Wolken and van der Werf, 2001). The first microbial formation of rose oxide was achieved by adding citronellol to submerged cultures of the basidiomycete *Cy. carcharias*. 3,7-Dimethyl-1,6,7-octanetriol was the main product of this biotransformation, and the following biosyntheic intermediates leading to rose oxide were identified: 3,7-dimethyl-6,7-epoxy-1-octanol, and the allylic diols 2, 6-dimethyl-2-octene-1,8-diol, 3,7-dimethyl-5-octene-1,7-diol, and 3,7-dimethyl-7-octene-1,6-diol (Onken and Berger, 1999b).

Initial investigations with the ascomycetous plant pathogen *G. cingulata* showed that it had an outstanding potential to transform various acyclic terpenoids. A recent extension of these studies explored (+/−)-lavandulol transformation in submerged cultures. Regioselective epoxidation at the double bond distal to the hydroxyl group and subsequent cyclization led to (2*S*,4*S*)-(−)-1,5-epoxy-5-methyl-2-(1-methylethenyl)-4-hexanol and to (*Z*)- and (*E*)-1,4-epoxy-5-methyl-2-(1-methylethenyl)-5-hexanol, respectively. A second reaction pathway involving allylic hydroxylation of a terminal methyl group produced 6-hydroxylavandulol (Figure 13.5; Nankai *et al.*, 1998, and references therein). *G. cingulata* was also capable of transforming (−)-dihydromyrcenyl acetate and dihydromyrcenol. The major metabolite in each case was identified as 3,7-dimethyloctane-1,2,7-triol. Time course experiments and structural studies of further metabolites led to the proposal of the metabolic pathway in Figure 13.6 for the biotransfomation of (−)-dihydromyrcenyl acetate by *G. cingulata* (Miyazawa *et al.*, 2000).

Another plant pathogen, *R. solani*, can hydroxylate (1*R*,2*S*,5*R*)-(−)-menthol at the C-4 position to give (1*R*,2*S*,4*S*,5*S*)-(−)-4-hydroxymenthol as the main product (Miyazawa *et al.*, 2001). The reaction proceeded stereoselectively, and almost all of the substrate was utilized within 3 days. The transformation of (−)-menthol by *Cephalosporium aphidicola*

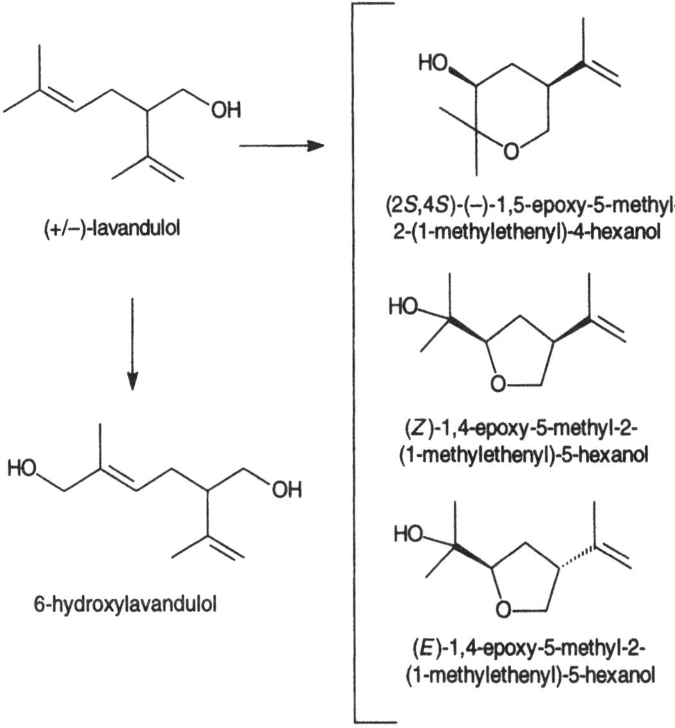

Figure 13.5. Biotransformation of (+/−)-lavandulol by *Glomerella cingulata* (adapted from Nankai *et al.*, 1998).

Figure 13.6. Biotransformation of (−)-dihydromyrcenyl acetate by *Glomerella cingulata* (modified after Miyazawa *et al.*, 2000).

provided six oxidized products, four of which had not been described before (Atta-ur-Rahman *et al.*, 1998).

Due to their low volatility, flavor compounds derived from diterpenes are not often studied. A prominent exception is (−)-Ambrox® with its moist, soft, persistent, warm, and animal-like odor. When Ambrox® is metabolized by the plant pathogen *B. cinerea* the major products are 1β-hydroxy-8-epiambrox, sclareolide (a cytotoxic lactone), and 3β-hydroxysclareolide (Farooq and Tahara, 2000a). Transformation of the diterpenoid compound sclareol, a potential Ambrox precursor, by the same fungus yielded epoxysclareol and 8-deoxy-14,15-dihydro-15-chloro-14-hydroxy-8,9-dehydrosclareol (Figure 13.7; Farooq and Tahara, 2000b). The formation of the latter compound is a good example of regioselective microbial halogenation.

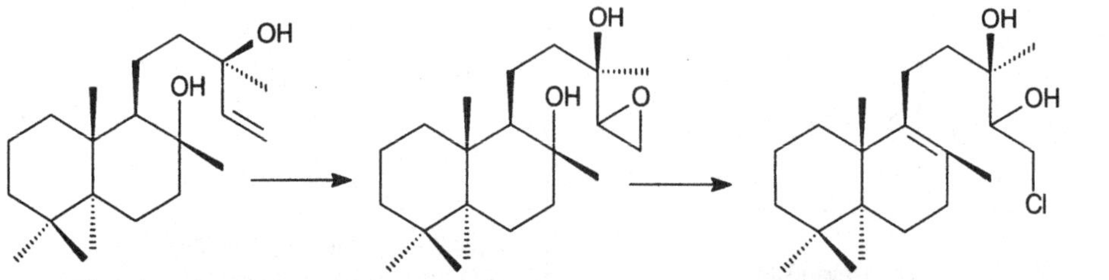

sclareol → epoxysclareol → 8-deoxy-14,15-dihydro-15-chloro-14-hydroxy-8,9-dehydrosclareol

Figure 13.7. Biotransformation of sclareol by *Botrytis cinerea* (adapted from Farooq and Tahara, 2000b).

3. Biosynthesis of Terpenyl Esters

Esters of terpene alcohols are often features of pleasant flavors, and consequently their synthesis by lipase catalysis has gained much attention during the past 20 years. Purified fungal lipases as well as lyophilized whole mycelia have been applied successfully. While the early attempts were concentrated mainly on aqueous, microaqueous, or microemulsion systems, more recent developments include esterification in organic solvents or approaches that are entirely solvent-free (reviewed by Schrader and Berger, 2001). Dry mycelium of *Rhizopus delemar* MIM catalyzed the formation of geranyl acetate from geraniol and acetic acid in heptane at 55°C, and with semicontinuous addition of one substrate, 10-day yields up to 75 g L^{-1} could be achieved. For an industrial process, it is far more economical to use whole cells as an insoluble source of enzymes than it is to extract the enzyme from cells and immobilize it (Molinari *et al.*, 1998). Direct esterification of geraniol and citronellol with short-chain fatty acids (acetic, propanoic, butanoic, and hexanoic acid) was catalyzed by free lipase from *Rhizomucor miehei* in *n*-hexane. To enhance yields, optimum enzyme and substrate concentrations were determined, and yields of terpenyl esters in a 2-L bioreactor were more than 100 g per day (Laboret and Perraud, 1999). *Trichosporon fermentans* lipase adsorbed on Celite esterified geraniol with fatty acids longer than eight carbons in yields higher than 90%. Shorter fatty acids, for example, butanoic or hexanoic acid, gave yields of only 10%, and acetic acid was not accepted as acyl donor at all. Short chain fatty acid esters of geraniol and citronellol became accessible via transesterification with vinyl esters as acylating reagents. The vinyl alcohol liberated was immediately tautomerized to acetaldehyde, eliminating the effect it would otherwise exert upon the equilibrium of the reaction (Nakagawa *et al.*, 1998). A promising alternative to these methods of producing geranyl and citronellyl esters was suggested by Chatterjee and Bhattacharyya (1998). They used a lipase of *R. miehei* (Lipozyme IM) immobilized on macroporous anionic resin added directly to the substrates without solvents. The optimum temperature was 55–60°C, and the transformation rates ranged from 96% to 99% in 6 hr. The immobilized lipase could be used repeatedly without significant loss of its catalytic activity. In flavor biotechnology, the area of terpenyl ester technology is quite mature and serves as a prime example of the application of fungal enzymes for the production of fine chemicals on an industrial scale.

To prepare lipase in a form that will be active in organic solvents, a procedure has been designed whereby the enzyme is "trapped in the presence of an amphiphile interface" (TPI). In aqueous environments the conformation of many lipases is such that a portion of the protein acts as the lid blocking access to the active site. Upon binding to an amphiphilic interface, the conformation changes and the lid moves sufficiently to expose the active site. This "open" conformation is trapped by rapid freeze-drying and is retained in the lyophilized powder. An enhancement of enzymatic activity as large as 90 fold is observed in TPI lipases from *R. miehei* and *R. delemar*, and the more open conformation induced by the interface broadens the fatty acid specificity allowing longer chain acids to be accepted as substrates (Gonzalez-Navarro and Braco, 1998).

4. Generation of Aromatic Flavor Compounds

Vanillin, the world's most important aromatic flavor compound, is used extensively in food, beverages, and cosmetics. Driven by the considerable difference in the price of natural vanillin and its chemically synthesized counterpart, intensive attempts have been made to find a biotechnological route to vanillin starting from abundant natural precursors. Plant tissue and plant cell cultures, bacteria, yeasts, and fungi have served as biocatalysts to perform the desired transformations. Although processes using fungal cells suffer from generally low yields, impressive vanillin concentrations of more than 10 g L^{-1} were recently obtained by feeding ferulic acid to the Gram positive bacteria, *Amycolatopsis* spp. or *Streptomyces setonii* (Rabenhorst and Hopp, 1997; Müller *et al.*, 1998). Detailed reviews summarizing the biotechnological production of vanillin have been published (Priefert *et al.*, 2001; Ramachandra *et al.*, 2000).

A study with [5-^{2}H]-ferulic acid showed that the white-rot basidiomycete, *Pycnoporus cinnabarinus*, degrades the phenyl propenoic side chain of ferulic acid by β-oxidation. The loss of a C_2 unit from ferulic acid yielded vanillic acid and its corresponding methyl ester (Krings *et al.*, 2001). Ferulic acid derived from sugar beet pulp is metabolized to vanillic acid analogously by *A. niger*. In a second step, vanillic acid can be reduced to vanillin by *P. cinnabarinus* (Lesage-Meessen *et al.*, 1999). *Phanerochaete chrysosporium* can likewise be employed for the reduction step, but an unwanted further reduction of vanillin to vanillyl alcohol is also catalyzed. Addition of an adsorption resin (XAD-2®) to the transformation medium to trap the target compound protects it from reduction (Stentelaire *et al.*, 1998). As an alternative, the aldehyde/alcohol ratio might be shifted toward the desired aldehyde by using the enzyme vanillyl alcohol oxidase (VAO, EC 1.1.3.38). This enzyme was cloned and sequenced from a *Penicillium simplicissimum* CBS 170.90 cDNA library. The open reading frame of 1,680 bp predicts a protein with 560 amino acids and a mass of ~63 kDa. Interestingly, the flavoprotein VAO acts on a wide range of phenolic compounds, converting both creosol and vanillylamine to vanillin in high yields; this fungal enzyme provides a new route to "natural" vanillin. Capsaicin, the pungent principle of red pepper, became a cheap feedstock for the production of vanillylamine, when a hydrolase was found that cleaved its amide bond (Figure 13.8; van den Heuvel *et al.*, 2001a). Moreover, the enzyme also stereospecifically hydroxylated 4-alkylphenols to optical pure aromatic alcohols. The efficiency of substrate hydroxylation is affected by the availability of water in the catalytic center and could be changed by site-directed mutagenesis that resulted in a single amino acid substitution in the enzyme. The stereoselectivity of the mutant VAO was inverted in comparison with the wild-type enzyme, demonstrating that protein engineering is a powerful tool to introduce new catalytic characteristics (Figure 13.9) (van den Heuvel *et al.*, 2000, 2001).

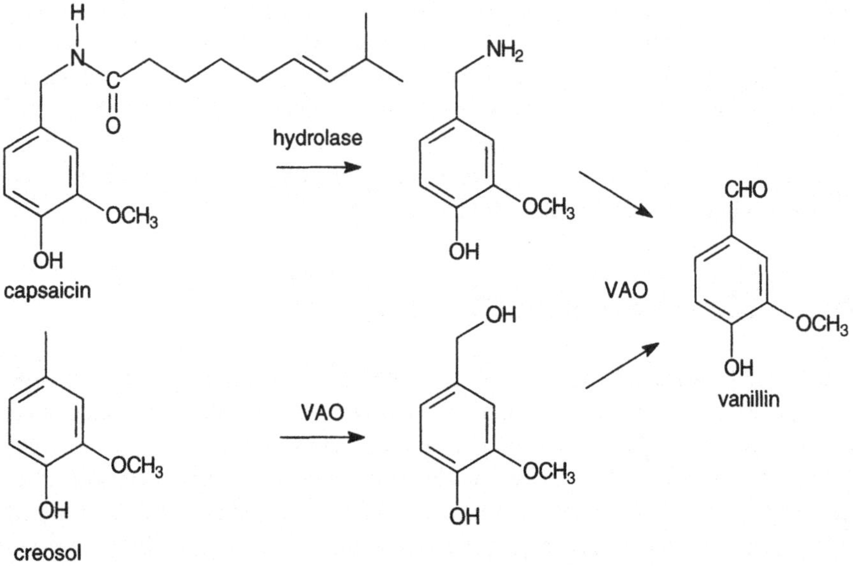

H
N
C
O
OCH3
OH
capsaicin
hydrolase
NH2
OCH3
OH
CHO
VAO
OCH3
OH
vanillin
OH
OCH3
OH
creosol
VAO
OCH3
OH

Figure 13.8. Biotechnological pathways to natural vanillin starting from capsaicin and creosol (van den Heuvel *et al.*, 2001a).

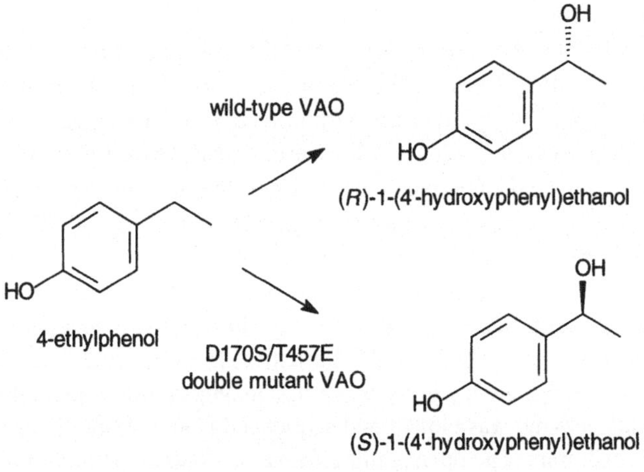

OH
wild-type VAO
HO
(R)-1-(4'-hydroxyphenyl)ethanol
HO
4-ethylphenol
OH
D170S/T457E
double mutant VAO
HO
(S)-1-(4'-hydroxyphenyl)ethanol

Figure 13.9. Inversion of stereospecificity of vanillyl alcohol oxidase (VAO) by exchange of two amino acid moieties through site-directed mutagenesis (modified after van den Heuvel *et al.*, 2000).

With a worldwide consumption of approximately 7,000 tons per year, the bitter almond aroma, benzaldehyde, is the second most important flavor after vanillin (Lomascolo *et al.*, 2001). The white-rot basidiomycete *Bjerkandera adusta* produces substantial amounts of benzaldehyde along with benzyl alcohol and benzoic acid when it is cultivated in a liquid medium supplemented with L-phenylalanine. The pathway leading to these aryl metabolites in *B. adusta* was studied with L-[U-^{14}C]-phenylalanine and ring-labeled L-[^{13}C]phenylalanine. (*E*)-Cinnamic acid proved to be a key metabolic intermediate that was hydroxylated to give either α- or β-hydroxyphenylpropanoic acid. β-Hydroxyphenylpropanoate was converted via β-oxidation to benzoic acid, as indicated by the presence of labeled acetophenone as a degradation product of β-oxophenyl propanoic acid. The reduction of benzoic acid to the target product, benzaldehyde, was catalyzed by an aryl-aldehyde dehydrogenase (Figure 13.10). This pathway is formally analogous to the one discussed for the vanillin production from ferulic acid by *P. cinnabarinus* (Krings *et al.*, 2001). Hydroxylation of (*E*)-cinnamic acid in α-position via phenylpyruvic acid, phenylacetaldehyde, phenylacetic acid, mandelic acid, and benzoylformic acid also leads to benzaldehyde and benzyl alcohol (Lapadatescu *et al.*, 2000). The intermediate phenylpyruvic acid may be formed directly from phenylalanine by action of a transaminase as the first biotransformation step. Benzaldehyde concentrations of more than 400 mg L^{-1} were obtained after 8 days when the cultures were supplemented with 10 g L^{-1} lecithin. In the absence of lecithin or with only low concentrations, benzoic acid was the major aryl metabolite synthesized. As aryl alcohol oxidase and lignin peroxidase were the only activities detected in the high lecithin condition and aryl alcohol oxidase activity and benzaldehyde production was maximal in this condition, the involvement of these enzymes in benzaldehyde formation by *B. adusta* seems likely (Lapadatescu *et al.*, 1999). HP20® adsorption resin, a styrene divinylbenzene copolymer with high selectivity for aromatic compounds, added to the *Trametes suaveolens* and *P. cinnabarinus* cultures shifted the accumulation of L-phenylalanine products to the target compound, benzaldehyde. Maximal yields reached with *T. suaveolens* and *P. cinnabarinus* were 710 and 790 mg L^{-1}, respectively. The resin increased the total yield of aromatic compounds by protecting benzaldehyde from further catabolism and facilitating its recovery (Lomascolo *et al.*, 1999, 2001).

Grown as a static culture in a liquid medium, *Pleurotus ostreatus* produced 4-methoxybenzaldehyde (*p*-anisaldehyde). Production was improved by supplementing the growth medium with L-tyrosine (5 mM) but not L-phenylalanine or other potential aromatic precursors. The activities of the lignin degrading enzymes, aryl alcohol oxidase, and manganese peroxidase rose in parallel with the formation of *p*-anisaldehyde, indicating

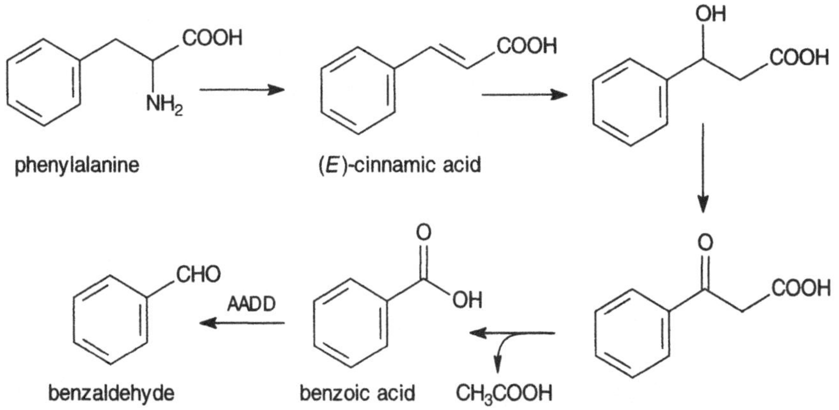

Figure 13.10. β-Oxidative degradation of phenylalanine to benzoic acid and benzaldehyde (AADD = aryl-aldehyde dehydrogenase).

again that the biosynthesis of aryl flavor compounds by a white-rot fungi is linked to the lignin degrading system (Okamoto *et al.*, 2002).

When grown on agar media, the basidiomycete, *Nidula niveo-tomentosa*, synthesizes *de novo* one of the characteristic impact compounds of raspberries, 4-(4-hydroxyphenyl)butan-2-one (raspberry ketone), together with the corresponding alcohol (betuligenol). If this highly sought flavor compound was extracted from raspberries, it would cost several million dollars per kilogram, based on the price of the fruit alone. Deriving it from the betuloside of birch bark, it would cost up to \$10,000 per kg. In light of these economic factors, a systematic attempt was made to improve raspberry ketone production by *N. niveo-tomentosa*. Optimizing the composition of the nutrient medium, the type and amount of precursor feed, and the time of precursor addition increased the yield 50 fold (~200 mg L^{-1}) (Böker *et al.*, 2001; Fischer *et al.*, 2001). Gas chromatography with atomic emission detection (GC–AED) or mass spectrometry (GC–MS) was used to elucidate the pathway to raspberry ketone from ^{2}H- or ^{13}C-labeled L-phenylalanine and from [1-^{13}C]glucose. Phenylalanine was converted to hydroxy benzoyl CoA, which was then elongated to the target compound through a poly-β-keto scheme (Figure 13.11). This fungal pathway differs from the route established for plant tissues (Zorn *et al.*, 2003).

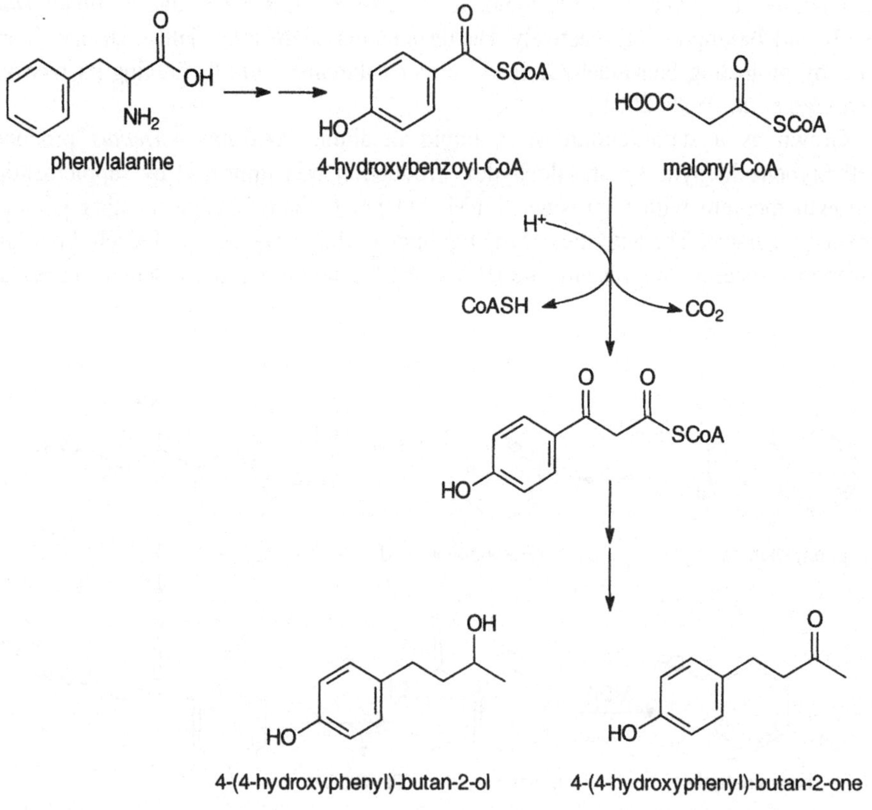

Figure 13.11. Biotransformation of L-phenylalanine to raspberry ketone and betuligenol by *Nidula niveo-tomentosa*.

5. Flavor Compounds from Other Chemical Classes

C. fimbriata is a fungal pathogen that affects a wide range of host plants, such as *Coffea arabica, Prunus* spp., and various citrus trees. Employing steam-treated coffee husk as the basal substrate for growth and flavor production and supplementing this substrate with glucose (20–35%) results in the formation of a strong pineapple aroma. Additional supplementation with leucine provided maximum yields of total volatiles of 8.3 mmol L^{-1} per g of dry mass. The headspace composition was dominated by esters, but traces of acetaldehyde, alcohols, and ketones were also identified (Soares *et al.*, 2000a, b). Methyl ketones, accounting for the characteristic "blue cheese odor," were generated by cultivating *Penicillium roqueforti* spores on copra (coconut) oil: the methyl ketones detected were 2-undecanone, the predominant product, followed by 2-nonanone and 2-heptanone. As these methyl ketones were produced essentially from the free fatty acids present in copra oil, a large increase of productivity followed lipolysis by an exogenous *Candida cylindracea* lipase (Chalier and Crouzet, 1998). Microencapsulated spores of *P. roqueforti* readily transformed free short to medium chain fatty acids and their methyl esters to the corresponding methyl ketones. Starting with methyl octanoate at a 1 M concentration in decane, the yield of 2-heptanone after 10 days was more than 11 g L^{-1}. In some cases, the spores were able to catalyze product formation even in the absence of the solvent decane (Park *et al.*, 2000).

In 2000, Yamashita and coworkers succeeded in characterizing and cloning an isoamyl alcohol oxidase (IAAOD) from *Aspergillus oryzae*. IAAOD catalyzes the oxidation of isopentyl alcohol to isopentanal, an odorous compound often associated with an off-flavor of sake. With a PCR-amplified DNA fragment corresponding to the partial amino acid sequences of the purified protein as a probe, the genomic DNA sequence encoding IAAOD was cloned and overexpressed under the control of the *amyB* promoter in *A. oryzae*. The isovaleraldehyde-producing activity of one transformant was over 800 times as high as that in transformants with the control vector (Yamashita *et al.*, 2000). This example provides a glimpse of what might be possible in the near future as deeper insight is gained into the metabolic pathways leading to volatile compounds. An approach such as this could help avoid off-flavors in certain foods or tailor flavor profiles for increased consumer acceptance.

6. Bioprocess Technology

Several promising examples have been presented for whole-cell processes that produce flavor compounds; however, none of these has been developed beyond laboratory scale at present. Typically, product yields of 1–20 g L^{-1} would be the prerequisite for an economically viable bioprocesses (Münch and Müller, 2000). However, many flavor compounds and/or their precursors inhibit fungal metabolism at concentrations far below the economic threshold. When inhibition is due to the precursor and not the product, sophisticated feeding strategies may help reduce the toxic effect. For example, feeding (*R*)-(+)-limonene continuously via the gas phase to *P. sapidus*, rather than simply including it in the growth medium, provides a 3-fold improvement in product yield (Onken and Berger, 1999a). Despite this, the overall yield of ~100 mg L^{-1} is still below that needed in a commercial process. Basing the time and amount of precursor to add on biosensor measurements that continuously monitor the physiological state of the biocatalyst can lead to maximal product yields and can prolong the viability of the fungal culture (Schäfer *et al.*, 2003).

Apart from toxicity, precursors may be poorly soluble in water, and this often limits their bioavailability and restricts transformation rates. To address this drawback, organic co-solvents are widely employed to give either single-phase systems, two-phase systems, or microemulsions. Two-phase systems have a continuous and a discontinuous phase formed by two immiscible liquids. While the aqueous phase usually harbors the biocatalyst, the organic solvent includes the substrate/product. Microemulsions, sometimes termed reversed micelles systems, are thermodynamically stable, single-phase systems, where the addition of an appropriate surfactant permits the single-phase coexistence of otherwise insoluble aqueous and organic media. Studying the bioconversion of (*R*)-(+)-limonene to (*R*)-(+)-α-terpineol by *P. digitatum*, Tan and Day (1998a) examined the influence of 22 co-solvents. The most striking effects were observed with dioctyl-phthalate and ethyl decanoate (1.5% v/v), which increased the relative activity of the biocatalysis 2.4 and 2.2 fold, respectively. Decane was a suitable and nontoxic organic solvent in a two-phase system developed for the production of flavor ketones from fatty acid esters by free and microencapsulated spores of *P. roqueforti* (Park *et al.*, 2000).

If only the biotransformation product, and not the starting material, inhibits the microbial or enzymatic activity, *in situ* product recovery may reduce the toxic effects significantly. *In situ* extraction into an immiscible liquid phase or trapping on an adsorbent (solid phase extraction) may have the added benefit of protecting the target compound from further metabolism or chemical degradation. This was illustrated earlier in the chapter in the discussion of vanillin. Solid phase adsorption of vanillin and benzaldehyde from the culture medium directed the process to the target compound (Stentelaire *et al.*, 1998; Lomascolo *et al.*, 2001). A systematic approach to introduce selectivity into an adsorptive separation of flavor compounds using reversed-phase polystyrene adsorbents has been presented by Gehrke *et al.* (2000). Thermodynamically and kinetically controlled adsorption capacities of various derivatized polystyrenes were compared to a commercial polystyrene resin (Amberlite XAD-16), a strong cationic exchanger (Amberlite 200), and a silica gel (Silikagel 100) in binary model systems: (−)-limonene/(−)-carvone and (−)-α-pinene/(−)-borneol. A sufficient separation of terpenoid products from their terpene precursors was achieved with sulfonated Amberlite XAD-16 in *n*-hexane.

Moving beyond adsorption, *in situ* product recovery can also be addressed by diverse membrane technologies. An integrated bioprocess for the production and recovery of aroma compounds synthesized *de novo* was designed by interlinking a pervaporation membrane module with a bioreactor. Removal of the flavor compounds via the pervaporation membrane as they were produced by *Ceratocystis moniliformis* permitted increased microbial growth rates that translated into higher total yields of flavor compounds (Bluemke and Schrader, 2001).

In case of oxidative bioprocesses, upgrading from shake flask to reactor requires careful consideration of oxygen supply. This is especially critical for pellet-forming fungi and highly viscous growth media. Continuous circulation of the cultures through a hollow-fiber aeration module is a successful strategy in bacterial or yeast-based transformation processes, but it is not applicable to a culture of *Cy. carcharias*. For the fungal process, a hydrophobic microporous aeration membrane was introduced directly into the reactor, and aeration via this membrane not only provided high gas exchange compared to direct aeration but also decreased losses of volatile substrates that were the consequence of large volumetric air flow rates (Onken and Berger, 1999b).

Immobilization of microorganisms is a common method to increase the operational stability of a bioprocess, to facilitate product separation, and to ensure reusability of the biocatalyst in repeated batch or continuous bioreactor transformations. However, immobilization is far less common for filamentous fungi than for bacteria and yeasts because mycelial morphology complicates the initial immobilization process and subsequent growth of the fungus may be impeded inside a rigid support structure. The bioconversion of (*R*)-(+)-limonene to (*R*)-(+)-α-terpineol was, however, feasible with *P. digitatum* mycelium immobilized in calcium alginate beads. Compared to free fungal mycelia, immobilized fungi proved to be somewhat protected at high concentrations of the toxic substrate. On the other hand, repeated batch bioconversions showed dramatic yield decreases in the second and third cycles, and regeneration of the beads in growth medium for 3 days was inadequate for restoring the specific activity (Tan and Day, 1998b). Spores of *P. roqueforti* remained active after entrapment within permeable polyamide microcapsules. At high substrate concentrations, no difference in the overall bioconversion rates were observed for free and microencapsulated spores (Park *et al.*, 2000).

7. Conclusion

In summary, higher fungi possess a significant potential for the generation of volatile flavors through *de novo* synthesis or by transformation of exogenous substrates. Because of the technical challenges involved in the development of economic whole-cell processes, research interest is shifting more and more from the products themselves, to the enzyme catalysts and the genes that encode them. Only through a multidisciplinary approach combining the tools of molecular biology, enzymology, analytical chemistry, and fermentation science will the potential, and perhaps the still hidden secrets, of fungi that produce flavors and fragrances be harnessed in commercial processes.

References

Agrawal, R., Deepika, N.-U.-A., and Joseph, R. (1999). Strain improvement of *Aspergillus* sp. and *Penicillium* sp. by induced mutation for biotransformation of α-pinene to verbenol. *Biotechnol. Bioeng.* **63**, 249–252.

Agrawal, R. and Joseph, R. (2000). Bioconversion of α-pinene to verbenone by resting cells of *Aspergillus niger. Appl. Microbiol. Biotechnol.* **53**, 335–337.

Atta-ur-Rahman, M. Y., Farooq, A., Anjum, S., Asif, F., and Choudhary, M. I. (1998). Fungal transformation of (1*R*,2*S*,5*R*)-(−)-menthol by *Cephalosporium aphidicola. J. Nat. Prod.* **61**, 1340–1342.

Atta-ur-Rahman, M. Y., Farooq, A., Anjum, S., and Choudhary, M. I. (1999). Microbial transformation of cytotoxic natural products. *Curr. Org. Chem.* **3**, 309–326.

Belitz, H.-D. and Grosch, W. (1987). *Food chemistry*. Springer-Verlag, Berlin.

Berger, R. G. (1995). *Aroma biotechnology*. Springer-Verlag, Berlin.

Bluemke, W. and Schrader, J. (2001). Integrated bioprocess for enhanced production of natural flavors and fragrances by *Ceratocystis moniliformis. Biomol. Eng.* **17**, 137–142.

Böker, A., Fischer, M., and Berger, R. G. (2001). Raspberry ketone from submerged cultured cells of the basidiomycete *Nidula niveo-tomentosa. Biotechnol. Progr.* **17**, 568–572.

Breheret, S., Talou, T., Rapior, S., and Bessiere, J. M. (1997). Mushrooms (Basidiomycetes): Novel source of monoterpenes bioproduction for aromatic industry? *Fr. Riv. Ital. EPPOS*, 592–602.

Breheret, S., Talou, T., Rapior, S., and Bessiere, J. M. (1999). Geosmin, a sesquiterpenoid compound responsible for the musty-earthy odor of *Cortinarius herculeus, Cystoderma amianthinum*, and *Cy. carcharias. Mycologia* **91**, 117–120.

Chalier, P. and Crouzet, J. (1998). Methyl ketone production from copra oil by *Penicillium roqueforti* spores. *Food Chem.* **63**, 447–451.

Chatterjee, T. and Bhattacharyya, D. K. (1998). Synthesis of terpene esters by an immobilized lipase in a solvent-free system. *Biotechnol. Lett.* **20**, 865–868.

Demyttenaere, J. C. R. (2001). Biotransformation of terpenoids by microorganisms. In H. E. J. Atta-ur-Rhaman (ed.) *Studies in natural products chemistry* (Part F, Vol. 25). Elsevier Science Ltd., New York, pp. 125–178.

Demyttenaere, J. C. R. and de Kimpe, N. (2001). Biotransformation of terpenes by fungi. Study of the pathways involved. *J. Mol. Catal. B: Enzym.* **11**, 265–270.

Demyttenaere, J. C. R. and Willemen, H. M. (1998). Biotransformation of linalool to furanoid and pyranoid linalool oxides by *Aspergillus niger. Phytochemistry*, **47**, 1029–1036.

Demyttenaere, J. C. R., Herrera, M. D. C., and de Kimpe, N. (2000). Biotransformation of geraniol, nerol, and citral by sporulated surface cultures of *Aspergillus niger* and *Penicillium* sp. *Phytochemistry* **55**, 363–373.

Demyttenaere, J. C. R., Adams, A., Vanoverschelde, J., and de Kimpe, N. (2001a). Biotransformation of (*S*)-(+)-linalool by *Aspergillus niger*: An investigation of the culture conditions. *J. Agric. Food Chem.* **49**, 5895–5901.

Demyttenaere, J. C. R., van Belleghem, K., and de Kimpe, N. (2001b). Biotransformation of (*R*)-(+)- and (*S*)-(−)-limonene by fungi and the use of solid phase microextraction for screening. *Phytochemistry* **57**, 199–208.

Doumenc-Faure, M., Giacinti-Marinie, G., and Talou, T. (2000). Dynamic headspace concentration versus solid phase microextraction (SPME) coupled with monodimensional chiral gas chromatography. In P. Schieberle and K.-H. Engel (eds.) *Frontiers of flavour science.* Deutsche Forschungsanstalt für Lebensmittelchemie, Garching, pp. 117–120.

Farooq, A. and Tahara, S. (2000a). Oxidative metabolism of ambrox and sclareolide by *Botrytis cinerea. Z. Naturforsch.* **55C**, 341–346.

Farooq, A. and Tahara, S. (2000b). Biotransformation of two cytotoxic terpenes, α-santonin and sclareol by *Botrytis cinerea. Z. Naturforsch.* **55C**, 713–717.

Farooq, A., Tahara, S., Choudhary, M. I., Atta-ur-Rahman, M. Y., Ahmed, Z., Can Baser, K. H. C., and Demirci, F. (2002). Biotransformation of (−)-α-pinene by *Botrytis cinerea. Z. Naturforsch.* **57C**, 303–306.

Fischer, M., Böker, A., and Berger, R. G. (2001). Fungal formation of raspberry ketone differs from the pathway in plant cell culture. *Food Biotechnol.* **15**, 147–155.

Gehrke, M., Krings, U., and Berger, R. G. (2000). Selective recovery of volatile flavour compounds using reversed-phase polystyrene adsorbents. *Flavour Fragrance J.* **15**, 108–114.

Gonzalez-Navarro, H. and Braco, L. (1998). Lipase-enhanced activity in flavour ester reactions by trapping enzyme conformers in the presence of interfaces. *Biotechnol. Bioeng.* **59**, 122–127.

Grivel, F. and Larroche, C. (2001). Phase transfer and biocatalyst behaviour during biotransformation of β-ionone in a two-phase liquid system by immobilized *Aspergillus niger. Biochem. Eng. J.* **7**, 27–34.

Grivel, F., Larroche, C., and Gros, J. B. (1999). Determination of the reaction yield during biotransformation of the volatile and chemically unstable compound β-ionone by *Aspergillus niger. Biotechnol. Progr.* **15**, 697–705.

Hobbythek/Westdeutscher Rundfunk (1999). Köln (November 11, 1999); http://www.hobbythek.de/archiv/291/grafik_1.html.

Husson, F., Bompas, D., Kermasha, S., and Belin, J. M. (2001). Biogeneration of 1-octen-3-ol by lipoxygenase and hydroperoxide lyase activities of *Agaricus bisporus. Process Biochem.* **37**, 177–182.

Kaspera, R., Krings, U., Onken, J., and Berger, R. G. (2003). Stereospecific allylic oxidation of limonene: A route to pure (*S*)-(+)-carvone. In J. L. Le Quéré and P. X. Étiévant (eds.) *Flavour research at the dawn of the twenty-first century.* Intercept publishers, Paris, pp. 397–400.

Krings, U. and Berger, R. G. (1998). Biotechnological production of flavours and fragrances. *Appl. Microbiol. Biotechnol.* **49**, 1–8.

Krings, U., Pilawa, S., Theobald, C., and Berger, R. G. (2001). Phenyl propenoic side chain degradation of ferulic acid by *Pycnoporus cinnabarinus*—elucidation of metabolic pathways using [5-^{2}H]-ferulic acid. *J. Biotechnol.* **85**, 305–314.

Laboret, F. and Perraud, R. (1999). Lipase-catalyzed production of short-chain acids terpenyl esters of interest to the food industry. *Appl. Biochem. Biotechnol.* **82**, 185–198.

Lapadatescu, C., Ginies, C., Djian, A., Spinnler, H.-E., Le Quere, J.-L., and Bonnarme, P. (1999). Regulation of the synthesis of aryl metabolites by phospholipid sources in the white-rot fungus *Bjerkandera adusta. Arch. Microbiol.* **171**, 151–158.

Lapadatescu, C., Ginies, C., Le Quere, J.-L., and Bonnarme, P. (2000). Novel scheme for biosynthesis of aryl metabolites from L-phenylalanine in the fungus *Bjerkandera adusta. Appl. Environ. Microbiol.* **66**, 1517–1522.

Lesage-Meessen, L., Stentelaire, C., Lomascolo, A., Couteau, D., Asther, M., Moukha, S., Record, E., Sigoillot, J.-C. *et al.* (1999). Fungal transformation of ferulic acid from sugar beet pulp to natural vanillin. *J. Sci. Food Agric.* **79**, 487–490.

Lomascolo, A., Lesage-Meessen, L., Labat, M., Navarro, D., Delattre, M., and Asther, M. (1999a). Enhanced benzaldehyde formation by a monokaryotic strain of *Pycnoporus cinnabarinus* using a selective solid adsorbent in the culture medium. *Can. J. Microbiol.* **45**, 653–657.

Lomascolo, A., Stentelaire, C., Asther, M., and Lesage-Meessen, L. (1999b). Basidiomycetes as new biotechnological tools to generate natural aromatic flavours for the food industry. *Trends in Biotechnology.* **17**, 282–289.

Lomascolo, A., Asther, M., Navarro, D., Antona, C., Delattre, M., and Lesage-Meessen, L. (2001). Shifting the biotransformation pathways of L-phenylalanine into benzaldehyde by *Trametes suaveolens* CBS 334.85 using HP20 resin. *Lett. Appl. Microbiol.* **32**, 262–267.

Miyazawa, M., Akazawa, S., Sakai, H., and Nankai, H. (2000). Biotransformation of (−)-dihydromyrcenal acetate using the plant parasitic fungus *Glomerella cingulata* as a biocatalyst. *J. Agric. Food Chem.* **48**, 4826–4829.

Miyazawa, M., Kawazoe, H., and Hyakumachi, M. (2001). Biotransformation of l-menthol by soil-borne plant pathogenic fungi (*Rhizoctonia solani*). *J. Chem. Technol. Biotechnol.* **77**, 21–24.

Molinari, F., Villa, R., and Aragozzini, F. (1998). Production of geranyl acetate and other acetates by direct esterification catalyzed by mycelium of *Rhizopus delemar* in organic solvent. *Biotechnol. Lett.* **20**, 41–44.

Müller, B., Münch, T., Muheim, A., and Wetli, M. (1998). Process for the production of vanillin. *European Patent EP0885968.*

Münch, T. and Müller, B. (2000). Bioengineering challenges of natural flavour production. In P. Schieberle and K.-H. Engel (eds.) *Frontiers of flavour science*. Deutsche Forschungsanstalt für Lebensmittelchemie, Garching, pp. 343–347.

Nakagawa, H., Watanabe, S., Shimura, S., Kirimura, K., and Usami, S. (1998). Enzymatic synthesis of terpenyl esters by transesterification with fatty acid vinyl esters as acyl donors by *Trichosporon fermentans* lipase. *World J. Microbiol. Biotechnol.* **14**, 219–222.

Nankai, H., Miyazawa, M., Akazawa, S., and Kameoka, H. (1998). Biotransformation of (+/−)-lavandulol by the plant pathogenic fungus *Glomerella cingulata. J. Agric. Food Chem.* **46**, 3858–3862.

Okamoto, K., Narayama, S., Katsuo, A., Shigematsu, I., and Yanase, H. (2002). Biosynthesis of *p*-anisaldehyde by white-rot basidiomycete *Pleurotus ostreatus. J. Biosci. Bioeng.* **93**, 207–210.

Onken, J. and Berger, R. G. (1999a). Effects of (*R*)-(+)-limonene on submerged cultures of the terpene transforming basidiomycete *Pleurotus sapidus. J. Biotechnol.* **69**, 163–168.

Onken, J. and Berger, R. G. (1999b). Biotransformation of citronellol by the basidiomycete *Cystoderma carcharias* in an aerated-membrane bioreactor. *Appl. Microbiol. Biotechnol.* **51**, 158–163.

Park, O.-J., Holland, H. L., Khan, J. A., and Vulfson, E. N. (2000). Production of flavour ketones in aqueous-organic two-phase systems by using free and microencapsulated fungal spores as biocatalysts. *Enzyme Microb. Technol.* **26**, 235–242.

Priefert, H., Rabenhorst, J., and Steinbüchel, A. (2001). Biotechnological production of vanillin. *Appl. Microbiol. Biotechnol.* **56**, 296–314.

Rabenhorst, J. and Hopp, R. (1997). Process for the preparation of vanillin and suitable microorganisms. *European Patent EP0761817.*

Ramachandra Rao, S. and Ravishankar, G. A. (2000). Vanilla flavour: Production by conventional and biotechnical routes. *J. Sci. Food Agric.* **80**, 289–304.

Rapior, S., Fons, F., and Bessiere, J.-M. (2000a). The fenugreek odor of *Lactarius helvus. Mycologia* **92**, 305–308.

Rapior, S., Mauruc, M.-J., Guinberteau, J., Masson, C.-L., and Bessiere, J.-M. (2000b). Volatile composition of *Gyrophragmium dunalii. Mycologia* **92**, 1043–1046.

Rapior, S., Breheret, S., Talou, T., Pelissier, Y., and Bessiere, J. M. (2002). The anise-like odor of *Clitocybe odora, Lentinellus cochleatus*, and *Agaricus essettei. Mycologia* **94**, 373–376.

Rösecke, J. and König, W. A. (2000). Odorous compounds from the fungus *Gloephyllum odoratum. Flavour Fragrance J.* **15**, 315–319.

Schäfer, S., Kaspera, R., Krings, U., Schrader, J., Sell, D., and Berger, R. G. (2003). Sensorgestützte Biokonversion zur Gewinnung natürlicher terpenoider Aromastoffe. *Lebensmittelchemie* **57**, 15.

Schrader, J. and Berger, R. G. (2001). Biotechnological production of terpenoid flavor and fragrance compounds. In H.-J. Rehm and G. Reed (eds.) *Biotechnology* (2nd edn., Vol. 10). Wiley-VCH, Weinheim, pp. 374–422.

Soares, M., Christen, P., Pandey, A., Raimbault, M., and Soccol, C. R. (2000a). A novel approach for the production of natural aroma compounds using agro-industrial residue. *Bioprocess Bioeng.* **23**, 695–699.

Soares, M., Christen, P., Pandey, A., and Soccol, C. R. (2000b). Fruity flavour production by *Ceratocystis fimbriata* grown on coffee husk in solid-state fermentation. *Process Biochem.* **35**, 857–861.

Stentelaire, C., Lesage-Meessen, L., Delattre, M., Haon, M., Sigoillot, J. C., Colonna Ceccaldi, B., and Asther, M. (1998). By-passing of unwanted vanillyl alcohol formation using selective adsorbents to improve vanillin production with *Phanerochaete chrysosporium. World J. Microbiol. Biotechnol.* **14**, 285–287.

Talou, T., Breheret/Hulin-Bertaud, S., and Gaset, A. (2000). Identification of the major key flavour compounds in odorous wild mushrooms. In P. Schieberle and K.-H. Engel (eds.) *Frontiers of flavour science*. Deutsche Forschungsanstalt für Lebensmittelchemie, Garching, pp. 46–50.

Tan, Q. and Day, D. F. (1998a). Organic co-solvent effects on the bioconversion of (*R*)-(+)-limonene to (*R*)-(+)-α-terpineol. *Process Biochem.* **33**, 755–761.

Tan, Q. and Day, D. F. (1998b). Bioconversion of limonene to α-terpineol by immobilized *Penicillium digitatum. Appl. Microbiol. Biotechnol.* **49**, 96–101.

Tan, Q., Day, D. F., and Cadwallader, K. R. (1998). Bioconversion of (*R*)-(+)-limonene by *P. digitatum. Process Biochem.* **33**, 29–37.

van den Heuvel, R. H. H., Fraaije, M. W., Ferrer, M., Mattevi, A., and van Berkel, W. J. H. (2000). Inversion of stereospecificity of vanillyl-alcohol oxidase. *Proc. Natl. Acad. Sci. U.S.A.* **97**, 9455–9460.

van den Heuvel, R. H., Fraaije, M. W., Laane, C., and van Berkel, W. J. (2001a). Enzymatic synthesis of vanillin. *J. Agric. Food Chem.* **49**, 2954–2958.

van den Heuvel, R. H. H., Partridge, J., Laane, C., Halling, P. J., and van Berkel, W. J. H. (2001b). Tuning the product spectrum of vanillyl-alcohol oxidase by medium engineering. *FEBS Lett.* **503**, 213–216.

Venkateshwarlu, G., Chandravadana, M. V., and Tewari, R. P. (1999). Volatile flavour components of some edible mushrooms (basidiomycetes). *Flavour Fragrance J.* **14**, 191–194.

Wolken, W. A. M. and van der Werf, M. J. (2001). Geraniol biotransformation-pathway in spores of *Penicillium digitatum. Appl. Microbiol. Biotechnol.* **57**, 731–737.

Yamashita, N., Motoyoshi, T., and Nishimura, A. (2000). Molecular cloning of the isoamyl alcohol oxidase-encoding gene (*mreA*) from *Aspergillus oryzae. J. Biosci. Bioeng.* **89**, 522–527.

Zorn, H., Fischer-Zorn, M., and Berger, R. G. (2003). A labeling study to elucidate the biosynthesis of 4-(4-hydroxyphenyl)-butan-2-one (raspberry ketone) by *Nidula niveo-tomentosa. Appl. Environ. Microbiol.* **69**, 367–372.

IV

Host–Fungal Interactions

14

Human Mycoses: The Role of Molecular Biology

Donald C. Sheppard, Ashraf S. Ibrahim, and John E. Edwards Jr.

1. Introduction

Since the early 1980s a dramatic rise in the incidence of invasive disease due to the filamentous fungi (molds) has occurred (Latgé, 1999). Because these fungi are primarily opportunistic pathogens, in large part this trend can be attributed to the increase in the number of patients undergoing highly immunosuppressive therapies including chemotherapy and solid organ and bone marrow transplantation. While the list of organisms infecting immunocompromised hosts is long, the vast majority of these infections are due to *Aspergillus* species. Among these species, *Aspergillus fumigatus* accounts for almost 90% of cases (Latgé, 1999). For those infections *not* due to *Aspergillus* sp., zygomycetes, *Pseudallescheria boydii*, and *Fusarium* sp., are also common pathogens. In this chapter, we will review advances in the study of these organisms as pathogens, particularly the progress made possible by the application of the techniques of molecular biology. We will focus our attention primarily on *Aspergillus* sp. and the zygomycetes as the most important pathogens in this group. While there has been substantial research on *Fusarium* sp., it has been predominately on the role of this organism in plant pathogenicity and is beyond the scope of this chapter.

Over the past decade, targeted gene disruption techniques, coupled with large-scale sequencing projects, facilitated the identification of dozens of putative virulence genes in the pathogenic yeasts such as *Candida albicans* and *Cryptococcus neoformans*. Progress in the study of the pathogenic filamentous fungi has been hindered by a lack of basic molecular tools, but now, fortunately, many of the molecular methods that were successful in other fungi have been adapted for use in filamentous fungi, and in particular for *A. fumigatus*.

Donald C. Sheppard, Ashraf S. Ibrahim, and John E. Edwards Jr. • Division of Infectious Diseases, Harbor-UCLA Research and Education Institute, 1000 W. Carson St., Torrance CA 90502.

Advances in Fungal Biotechnology for Industry, Agriculture, and Medicine. Edited by Jan S. Tkacz and Lene Lange, Kluwer Academic/Plenum Publishers, 2004.

2. Goals in the Study of Pathogenic Filamentous Fungi

2.1. Identification of Virulence Factors

Much of the research regarding bacterial pathogenesis has focused on the identification of well-defined virulence factors that play a clear role in the development of infection in the normal host. Secreted toxins, immune evasion factors, and invasins are classic examples of these virulence factors. When these factors are absent, many pathogenic bacteria become incapable of causing human disease. In contrast, the identification of virulence factors in opportunistic pathogens is much more complex (Tomee and Kauffman, 2000). Indeed, since most filamentous fungi are unable to cause infection in the normal host, one could argue that they lack "classic" virulence factors. However, there is clearly a biological distinction in the relative pathogenicity of these opportunistic pathogens. For example, while *A. fumigatus* is recovered as the most common cause of invasive disease due to filamentous fungi, it represents only a fraction of a percentage of the total fungal spores recovered in air sampling studies (Mullins and Seaton, 1978). Further, several studies have demonstrated a marked difference in virulence between clinical and environmental isolates of this fungus (Kothary *et al.*, 1984; Aufauvre-Brown *et al.*, 1998). Thus, there must be some intrinsic properties of certain strains of *A. fumigatus* that enhance its ability to infect the immunocompromised host. Much of the research interest in pathogenic filamentous fungi has therefore focused on the identification of these putative virulence factors. The ultimate goal of such research is to identify targets for the development of novel therapeutic strategies, both pharmacologic and immune based. The primary approach used in the search for putative virulence factors of filamentous fungi has been gene disruption, either via targeted or random mutagenesis.

2.2. Identification of Other Drug Targets

An alternate strategy for finding therapeutic targets is the identification of genes and gene products that are essential for the growth of the target organism *in vivo*. Essential genes are attractive antifungal targets for several reasons. Since many of these genes are conserved across fungal species, strategies to block their action may be applicable to more than one pathogenic organism. Additionally, interfering with the function of essential gene products has a high likelihood of resulting in a fungicidal rather than fungistatic therapy. Genome-wide searches for essential genes are currently underway in other fungi, such as *C. albicans*, with the goal of development of pharmacologic inhibitory strategies (Michel *et al.*, 2002; Willins *et al.*, 2002a). Such strategies are now beginning to be applied to *A. fumigatus* and other pathogenic filamentous fungi.

3. The Genus *Aspergillus*

Aspergilli comprise a group of more than 180 species of filamentous fungi. Of these, only a handful cause human disease, with *A. fumigatus* accounting for the vast majority of infections (Latgé, 1999; Denning, 2000). The natural reservoir of this thermophilic, saprophytic fungus is soil, where it plays an important role in recycling decaying organic matter

(Mullins *et al.*, 1976). *A. fumigatus* is ubiquitous and reproduces asexually by the production of abundant conidia which are rapidly dispersed by air currents. The small size (2–3 μm) and hydrophobicity of these conidia allow them to remain airborne for long periods. These physical characteristics also favor their direct delivery to the alveolar airspaces, allowing them to avoid being deposited in the airways where they could be removed by mucociliary action (Latgé, 1999; Denning, 2000).

3.1. Aspergillosis: Spectrum of Disease

Human disease due to *Aspergillus* encompasses a diverse group of conditions including allergic (i.e., asthma, allergic bronchopulmonary aspergillosis (ABPA), sinusitis, and extrinsic alveolitis), indolent infections (i.e., aspergilloma), and invasive disease (Latgé, 1999; Denning, 2000). Allergic diseases are beyond the scope of this chapter and will not be discussed further; we will focus on the epidemiology, pathogenesis, clinical manifestations, and treatment of indolent and invasive aspergillosis (IA).

3.2. Aspergilloma

Aspergilloma, commonly known as a fungus ball, is a distinctive clinical syndrome that develops as a complication of prior cavitary lung disease such as tuberculosis (Denning, 2000). In this condition, *A. fumigatus* colonizes the cavity, and grows as a large spherical mass that can expand to fill the entire cavity. Unlike most other forms of aspergillosis, conidiation can occur at the periphery of the mass. Aspergillomas are usually asymptomatic and are most commonly detected incidentally by pulmonary imaging studies. Dissemination does not generally occur, and the main complication of this condition is hemoptysis (expectoration of blood) due to erosion of blood vessels in the cavity wall (Latgé, 1999; Denning, 2000). The role of antifungal therapy, either systemic (Lebeau *et al.*, 1994) or by instillation into the cavity (Giron *et al.*, 1998), is unclear (Stevens *et al.*, 2000). Surgical excision remains the cornerstone of therapy for symptomatic disease (Denning, 2000; Stevens *et al.*, 2000).

3.3. Invasive Aspergillosis (IA)

3.3.1. Epidemiology and Significance

The overall incidence of invasive fungal disease has risen dramatically over the past two decades. Increased rates of invasive aspergillosis are a major part of this trend. Estimates are that the incidence of IA has increased over 4-fold during the 1990s, and develops in 10% to 25% of leukemic patients (Bodey *et al.*, 1992). Despite advances in therapeutics, the mortality of IA approaches 70–90% in persistently neutropenic patients (Denning, 1996).

3.3.2. Pathophysiology

A. fumigatus spores are ubiquitous, and it has been estimated that the average person inhales several hundred conidia of *A. fumigatus* daily (Hospenthal *et al.*, 1998; Latgé, 1999). The precise immune mechanisms involved in the response to this daily challenge have not been completely elucidated. It is believed that the mainstay of defense against infection is a combination of the elimination of conidia via pulmonary ciliary action and phagocytosis by pulmonary macrophages (Schaffner *et al.*, 1982; Schaffner, 1994; Latgé, 1999; Denning, 2000). The role of neutrophils in conidial killing remains undefined; however neutrophils are believed to be the most important element of the immune response against hyphae (Schaffner *et al.*, 1982). The contribution of humoral components to *A. fumigatus* immunity is also poorly understood; however, complement clearly plays an important role in host defense since C5 deficient mice (DBA/2N) are highly susceptible to experimental infection (Hector *et al.*, 1990). Acquired immunity, either cellular or humoral, contributes to host defense against these fungi to a lesser degree relative to phagocytic cells, although study of this aspect of immunity in IA is still in its infancy (Latgé, 1999).

Disease develops in susceptible patients in the following way (Latgé, 1999; Denning, 2000). Conidia germinate in the pulmonary alveoli and begin to form hyphae. These hyphae rapidly extend, traversing natural anatomic barriers, forming an area of focal pneumonitis. Blood vessel invasion is common. This angioinvasion is believed to form the basis for dissemination via hematogenous dispersion of hyphal fragments to other organs. Extensive tissue damage and eventual organ dysfunction are the inevitable result of unchecked infection.

3.3.3. Virulence Factors of *A. fumigatus*

As discussed, the identification of virulence factors in an opportunistic pathogen is a challenging task. Multiple *A. fumigatus* genes have been disrupted in the search for virulence factors. Surprisingly, many of these gene disruptions have had no detectable effect on virulence. A summary of the gene disruptions which have been evaluated for virulence is presented in Table 14.1. Excluding nutritional genes, the genes with the most convincing role in virulence are those governing the morphogenesis of conidia and hyphae. In contrast, none of the secreted factors, including ribotoxin and proteases, have been shown to contribute to virulence.

3.3.4. Clinical Presentation of IA

The clinical spectrum of IA encompasses several conditions including: rhinosinusitis, tracheobronchitis, pulmonary disease, and disseminated disease (Latgé, 1999; Denning, 2000). Acute invasive pulmonary disease accounts for the majority of cases with up to 40% progressing to dissemination. As dissemination occurs by hematogenous spread, multiple organs can be involved, such as brain, eye, heart, and kidneys. Fever is the most common initial presentation, followed by symptoms referable to the specific organ system involved. In the absence of immune reconstitution, and often despite therapy, the disease is rapidly progressive and often fatal.

3.3.5. Therapy of IA

Amphotericin B (AmB) has been the mainstay of therapy for decades (Stevens *et al.*, 2000). This agent binds to ergosterol in fungal cell membranes and kills cells by permeabilization. Unfortunately, AmB has some affinity for cholesterol in mammalian cell membranes

Table 14.1. *Aspergillus fumigatus* Genes that have been Disrupted to Evaluate their Contribution to Virulence

Gene	Gene product	Impact on virulence	Reference
afa1	Alkaline protease	No	Ikegami *et al.*, 1998
alb1	Naphthopyrone synthase	Yes	Tsai *et al.*, 1998
areA	Transcription factor	Yes	Hensel *et al.*, 1998
cat1	Catalase	No	Wysong *et al.*, 1998
chsD	Chitin synthase	No	Mellado *et al.*, 1996b
chsE	Chitin synthase	No	Aufauvre-Brown *et al.*, 1997
chsC	Chitin synthase	Yes[a]	Mellado *et al.*, 1996a
chsG	Chitin synthase	Yes[a]	Mellado *et al.*, 1996a
fos1	Two component histidine kinase	Yes	Clemons *et al.*, 2002
mepB	Metalloprotease	No	Ibrahim-Granet and d'Enfert, 1997
pabaA	PABA synthetase	Yes[b]	Brown *et al.*, 2000
pyrG	Orotodine-5′-phosphate decarboxylase	Yes[b]	d'Enfert *et al.*, 1996
ret	Ribotoxin restrictocin	No	Ikegami *et al.*, 1998
rodA	Rodlet protein RodA	No	Thau *et al.*, 1994

[a]In combination with *chsG* disruption.
[b]Auxotrophy.

and can therefore bind host cells as well. Thus toxicity, particularly nephrotoxicity, has been a major problem with this agent. Furthermore, efficacy is suboptimal, with up to 80% of patients progressing despite AmB therapy (Denning, 1996). New lipid-associated formulations of AmB are available. These formulations cause less toxicity than conventional AmB; however, no clear improvement in efficacy has yet been documented (Stevens *et al.*, 2000).

Newer antifungal agents, including echinocandins and novel azole drugs, are promising alternatives to AmB based regimens. Caspofungin, a novel echinocandin, inhibits β (1,3)-D-glucan synthesis, resulting in disruption of fungal cell walls. Caspofungin is well tolerated, but its efficacy remains unclear since no studies are available directly comparing caspofungin to AmB in human subjects. In a salvage therapy trial of patients with IA who were refractory to standard therapy (including AmB), a 50% response rate to caspofungin was observed (*http://www.cancidas.com/cancidas/shared/documents/english/protocol_019.pdf*). As a result, caspofungin is currently licensed for use only in patients who are refractory to standard therapy for IA.

Voriconazole is the newest member of the azole class of antifungals to be licensed for therapy of IA. Voriconazole, like all azoles, acts by inhibiting ergosterol biosynthesis, and subsequent fungal membrane synthesis. Voriconazole is extremely well tolerated and may be more effective than AmB in the treatment of IA. A recent study comparing voriconazole to AmB for IA showed a 52.8% response rate for voriconazole compared with 31.6% for AmB (Herbrecht *et al.*, 2002). In contrast to caspofungin, voriconazole is licensed for primary therapy of IA.

Combination therapy, especially with caspofungin and voriconazole, is currently under intensive investigation and may provide a promising alternative to standard single agent therapy.

3.4. Molecular Techniques for the Study of *Aspergillus* sp.

Molecular genetic approaches for the study of virulence involve manipulation of the fungal genome to test the contribution of specific genes to pathogenesis. A robust transformation system allowing the disruption and transfer of genes is the foundation for such experiments. Several transformation systems have been developed for *Aspergillus* species, and a subset of these has been investigated in *A. fumigatus*.

3.4.1. Selection Markers for *A. fumigatus*

All transformation systems require selection markers to identify organisms that have successfully incorporated the foreign DNA of interest. Such markers can be broadly divided into auxotrophic markers and dominant (usually drug resistance) markers. The use of an auxotrophic marker requires a parent strain that contains a mutation in a nutritional gene rendering it unable to grow on unsupplemented media. The defective mutation is then complemented with the wild-type version of the nutritional gene, rendering the transformants prototrophic. Markers which can also be selected against (counterselected) by compounds toxic to prototrophic but not auxotrophic strains are of particular utility. These compounds allow for the initial selection of auxotrophic mutants that can serve as host strains for transformation. Additionally, selection/counterselection provides the basis for a "recyclable" marker which can be used to perform sequential rounds of transformation (see *pyrG*-blaster below). Three such auxotrophic markers amenable to both selection and counterselection, *pyrG, niaA*, and *sC*, have been used for the molecular manipulation of *A. fumigatus* (Weidner *et al.*, 1998; de Lucas *et al.*, 2001; Firon *et al.*, 2002).

The *Aspergillus pyrG* gene encoding orotidine-5′-phosphate (OMP) decarboxylase has been used successfully as an auxotrophic marker for transformation (d'Enfert, 1996; Weidner *et al.*, 1998; Langfelder *et al.*, 2001, 2002). Mutations in this gene necessitate the addition of uracil and uridine to the medium for growth of the mutant. These mutants are able to grow on uridine/uracil supplemented medium containing 5-fluoroorotic acid (5-FOA), which is converted to a toxic product by the action of OMP-decarboxylase, the *pyrG* gene product. In contrast, wild-type strains with functional OMP decarboxylase can metabolize 5-FOA to its toxic intermediate and are poisoned. Thus, with media either deficient in uridine/uracil or supplemented with uridine/uracil and 5-FOA, both *pyrG* positive (ura$^+$/5FOA sensitive) and *pyrG* negative strains (ura$^-$/ 5-FOA resistant) can be selected.

This selectability/counterselectability has been exploited to develop a recyclable selection marker called the *pyrG*-blaster (Figure 14.1) (d'Enfert, 1996). This strategy is based on the Ura-blaster technique frequently used to disrupt *C. albicans* genes. The *pyrG*-blaster strategy works by flanking the *pyrG* gene with direct repeats of a sequence derived from the neomycin phosphotransferase gene (*neo*). This construct is used for specific gene disruption as follows: upstream and downstream sequences homologous to the gene of interest are subcloned to surround the *neo-pyrG-neo* construct. These homologous sequences serve as the basis for homologous recombination during transformation. A ura$^-$ strain of *A. fumigatus* is then transformed with the *pyrG*-blaster cassette. Those ura$^+$ transformants recovered by selection on unsupplemented minimal medium that contain the *pyrG*-blaster cassette interrupting the target gene as a consequence of double crossover events are identified. The presence of the flanking *neo* repeats in these strains allows for

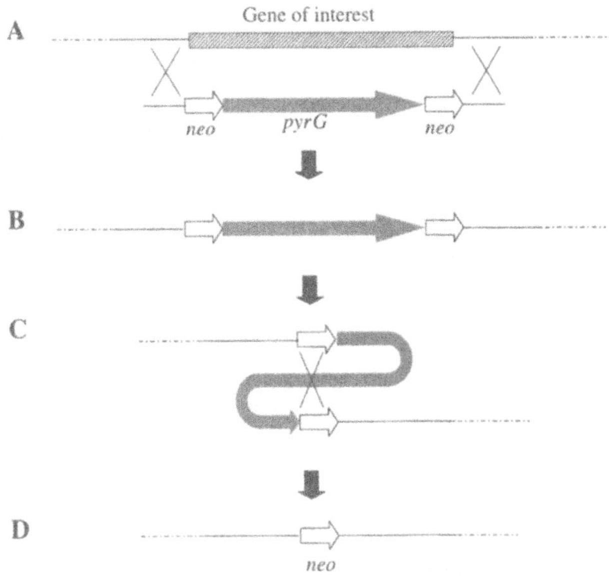

Figure 14.1. The *Pyr*-blaster strategy for gene disruption using a recyclable marker. (A) A disruption cassette consisting of *neo-pyrG-neo* is flanked by sequences homologous to the target gene. An ura$^-$ strain of *Aspergillus fumigatus* is transformed with this cassette. Homologous integration of the cassette occurs via a double crossover event. (B) Some transformants selected by plating on a uridine/uracil deficient medium contain the gene of interest disrupted by the insertion of the *neo-pyrG-neo* construct. (C) A disruptant is then plated on 5-FOA to select for clones which undergo an intrachromosomal recombination between the neo repeats (counterselection). This results in excision of the *pyrG* gene which renders the clone ura$^-$, and hence 5-FOA resistant. (D) The final result of selection and counterselection is disruption of the gene of interest with a single neo repeat, and regeneration of the ura$^-$ auxotrophy to permit transformation based on *pyrG* to be repeated.

an intrachromosomal recombination that results in excision of the *pyrG* gene at a frequency close to 10^{-4} and that leaves a single copy of *neo* behind, within the target gene. Strains undergoing this excision event can be selected by 5-FOA; only strains that have lost *pyrG* and are ura$^-$ will be recovered. The end result of this selection followed by counter-selection is the generation of a disruptant strain that retains uridine/uracil auxotrophy and that can be subjected to further rounds of transformation based upon *pyrG*. Even though this strategy was originally described for gene disruption, it is amenable to any transformation-based experiment that requires multiple rounds of transformation.

The *sC* gene of *A. fumigatus* encodes ATP-sulfurylase, which allows the organism to utilize sulfate as a sulfur source. Strains deficient in this gene require a source of reduced sulfur, such as methionine, for growth. Prototrophic strains can be selected against with selenate, which is reduced to a toxic metabolite by wild-type cells. Thus both *sC*$^-$ (sulfate$^-$/selenate resistant) and *sC*$^+$ (sulfate$^+$/selenate sensitive) strains can be selected. De Lucas *et al.* (2001) used this strategy first to isolate a *sC*$^-$ auxotroph, and then to complement the mutation via protoplast transformation with the wild-type *sC* allele.

A third counterselectable auxotrophic marker, *niaA* which encodes nitrate reductase, has been used successfully in other *Aspergillus* sp. and has recently been shown to be functional in *A. fumigatus* (Firon *et al.*, 2002). Strains that are deficient in *niaA*, are unable to grow on media containing nitrate as a sole nitrogen source. These *niaA*$^-$ mutants are resistant to chlorate, unlike wild-type strains containing functional *niaA*.

Most other auxotrophic markers cannot be selected against easily. Many auxotrophic markers known in *Aspergillus nidulans* are not available in *A. fumigatus*. The creation of parent strains with mutations in these nutritional genes is likely to permit their use in transformation strategies in the near future. Indeed, *pabaA* negative strains have been isolated, although not yet used for transformation (Sandhu *et al.*, 1976; Brown *et al.*, 2000).

While auxotrophic markers have been invaluable in the study of many other fungi including *A. nidulans, S. cerevisiae*, and *C. albicans*, there are important limitations to their use for the study of fungal pathogenesis. First, reliance on an auxotrophic marker precludes the transformation of wild-type isolates, as only auxotrophic mutants can serve as host strains. Additionally, recent studies with *C. albicans* auxotrophs have shown that, even after the mutant strain has been complemented with the nutritional marker, it may not behave as the wild-type strain due to a positional effect resulting from integration of the marker in a non-native locus (Sundstrom *et al.*, 2002). These differences can translate into variability in growth rate, morphogenesis, and even virulence.

Dominant selectable markers allow the transformation of a susceptible host strain based upon a gene that provides resistance to an inhibitor. Drug resistance genes are the most common source of dominant markers for transformation. Hygromycin B and phleomycin are two selection markers that function in *A. fumigatus*, providing alternatives to auxotrophic markers.

Hygromycin B is an aminoglycoside antibiotic, produced by *Streptomyces hygroscopicus*, that inhibits the growth of a number of fungi, including *A. fumigatus* (Gonzalez *et al.*, 1978). Resistance to this antibiotic is conferred by the *Escherichia coli* gene encoding hygromycin B phosphotransferase (*hph*). For fungal work, the most commonly used *hph* construct is the pAN-7 plasmid which contains the *hph* coding sequence under control of the *A. nidulans* glyceraldehyde-3-phosphate dehydrogenase (*gpd*) promoter and *trpC* terminator elements (Punt *et al.*, 1987). Transformation of *A. fumigatus* with this marker typically yields transformants resistant to hygromycin at ~200 μg/ml. Importantly, this construct contains no native *A. fumigatus* sequences, and therefore, biased integration at homologous sites is precluded, making this an attractive marker for genomic manipulations.

If recyclable markers such as *pyrG*-blaster are inappropriate, the construction of strains with two gene disruptions requires two dominant markers. Further, the two markers permit reintroduction of the native gene into a disrupted host strain. Such a complemented revertant constitutes an essential control when testing the virulence effect of a given gene. Phleomycin is another agent that is toxic to *A. fumigatus*. It has been used both for double disruptions, and as a second selection marker when complementing hygromycin-resistant disruption mutants with a wild-type copy of the gene of interest (Punt and van den Hondel, 1992; Smith *et al.*, 1994; Tsai *et al.*, 1998). Resistance to phleomycin is conferred by the *ble* gene from *Streptoalloteichus hindustanus*, which codes for a phleomycin-binding protein (Gatignol *et al.*, 1987). The gene has been cloned into constructs containing *A. nidulans* regulatory sequences and has been shown to function in *A. fumigatus*. This marker system has been used less frequently than *hph*, in part due to the cost of phleomycin (approximately twice that of hygromycin). The value of genes that provide resistance to nourseothricin and to mycophenolic acid, which have proven useful in other fungi, remains to be explored in *A. fumigatus*.

3.4.2. Transformation Techniques

The ability to introduce homologous or foreign DNA into an organism in a form that can be expressed to transform the organism's phenotype is the cornerstone of molecular genetics. Various techniques are available for the introduction of DNA into fungal cells, including electroporation, protoplasting, biolistic, and *Agrobacterium tumefaciens*-mediated transformation.

3.4.2a. Electroporation. Transformation by electroporation has been well described for several species of aspergilli including *A. niger, A. oryzae*, and *A. nidulans* (Chakraborty *et al.*, 1991; Ozeki *et al.*, 1994; Sanchez *et al.*, 1998). This technique has now been adapted for use with *A. fumigatus* (Brown *et al.*, 1998; Weidner *et al.*, 1998). Briefly, conidia are harvested and allowed to swell and begin germination. When these "competent" conidia are subjected to electroporation, transformation efficiencies of up to 1,000 transformants per μg of input DNA are achieved (Brown *et al.*, 1998). Competent conidia suffer only a minor decrease in transformation efficiency upon storage at −80°C. Unfortunately, high rates of non-homologous recombination are associated with the transformation of *A. fumigatus* by electroporation (Brakhage and Langfelder, 2002). This disadvantage limits the utility of electroporation-based transformation for strategies requiring targeted integration of DNA.

On the other hand, the illegitimate recombination seen with electroporation has been exploited to generate libraries of insertional mutants of *A. fumigatus* (Brown *et al.*, 1998). Since *A. fumigatus* is haploid, integration in either an open reading frame or a promoter region results in a strain that is deficient in the function normally provided by that gene. Libraries of random mutants can be screened for a desired phenotype, and the sequence flanking the insertion site can then be characterized by inverted PCR or plasmid retrieval to identify the gene responsible for the observed phenotype (Brown *et al.*, 2000; Firon *et al.*, 2002). Screening techniques for examining mutant banks will be described further in the subsequent sections on parasexual genetics and signature-tagged mutagenesis.

Recently, there has been increasing concern regarding the potential for genomic rearrangements occurring as a consequence of electroporation-mediated insertional mutagenesis. In a search for essential genes of *A. fumigatus*, Firon and colleagues (2002) used electroporation of a diploid strain of *A. fumigatus* to generate a bank of insertional mutants which were subsequently analyzed for essentiality using a parasexual genetics approach. They noted a significant number of genomic and plasmid DNA rearrangements at the sites of integration of the insertional cassette. However, as these rearrangements were seen with the transformation of a diploid strain, it remains unclear if similar frequencies of rearrangement occur during transformation of haploid strains.

3.4.2b. Protoplast formation. Protoplast transformation is the most commonly utilized transformation method for *A. fumigatus* (Brakhage and Langfelder, 2002). Briefly, germinated conidia are digested with a variety of lytic enzymes to release protoplasts which are collected by step gradient centrifugation. DNA is introduced into these competent protoplasts with a polyethylene glycol/$CaCl_2$-based protocol, and the protoplasts are allowed to regenerate hyphae under selective conditions. Sorbitol is used throughout to osmotically stabilize the protoplasts. In contrast to electroporation, transformation efficiency is low with this technique, yielding only 1–100 transformants per μg of input DNA (Brown *et al.*, 1998). Despite this low efficiency, homologous integration occurs with high frequency. Although cumbersome, this technique is extremely useful in generating strains in which specific integrations (i.e., gene disruptions, deletions, etc.) are sought.

3.4.2c. Microprojectile bombardment. Transformation by biolistic methods has been investigated in several fungi, including *A. nidulans* (Fungaro *et al.*, 1995). Microparticles of tungsten or gold are coated with the DNA of interest. Host fungi, either hyphae or conidia, are bombarded with these particles and subjected to selection. This method is fast and simple, but requires an expensive biolistic apparatus. Transformation frequencies in *A. nidulans* have been reported as ~100 transformants per μg of DNA, and the majority of integration events occurred by homologous recombination (Fungaro *et al.*, 1995). To date, there are no published reports of the use of this technology with *A. fumigatus*.

3.4.2d. Agrobacterium tumefaciens-mediated transformation. The bacterium *Agrobacterium tumefaciens* was initially discovered as an agent that induced tumors in plants (Mullins and Kang, 2001). *Agrobacterium* naturally provides the plant cells with a DNA fragment containing oncogenic and other genes whose function in the plant fosters the growth of *Agrobacterium*. These genes originate in a large (>200 Kb) plasmid, called the Ti (tumor inducing) element, where they are flanked by direct repeats forming the T-DNA. Products of the *vir* genes mediate transfer of the T-DNA region to the plant. These *vir* genes are also found on the Ti plasmid, outside of the T-DNA region. Successful transformation results from the transfer of a single copy of the T-DNA element and its integration into the host genome. This natural transformation system can be modified by replacing the native T-DNA of the Ti plasmid with DNA sequences of interest. A detailed review of the biology of T-DNA transfer is found in Chapter 4. It has been highly successful in the transformation of plants, and has now been applied to fungi. de Groot and colleagues have successfully transformed a variety of fungi, including *Aspergillus awamori* and *A. niger*, using this technique (de Groot *et al.*, 1998). The rates of transformation were higher than those observed with the protoplast technique, but more significantly, the majority of transformants resulted from single copy integrations. In *Agrobacterium*-mediated transformation, homologous recombination is readily achieved. This technique has not yet been applied to *A. nidulans* or *A. fumigatus*.

3.4.3. Parasexual Genetics

Unlike the model organism *A. nidulans, A. fumigatus* has no known sexual cycle. While this precludes the use of classical genetic methods in the study of this organism, a parasexual cycle is available for *A. fumigatus* (Levadoux *et al.*, 1981). When two haploid strains of *A. fumigatus* are grown on the surface of a solid medium in close proximity, heterokaryons are formed. If the strains contain different selection markers, stable diploid strains can be selected and isolated. Although meiosis does not occur in these strains, haploid progeny form by mitotic chromosomal non-dysjunction under the influence of destabilizing agents such as fluorophenylalanine (Brookman and Denning, 2000).

This approach has been combined with insertional mutagenesis to identify essential genes of *A. fumigatus* (Figure 14.2) (Firon *et al.*, 2002). In this study, diploid ura$^-$ organisms were subjected to insertional mutagenesis by electroporation, using the *pyrG* gene as a selection marker. This generated a bank of heterozygous ura$^+$ insertion mutants, containing one intact allele and one disrupted allele. These strains were individually subjected to haploidization, and the progeny were analyzed. If the *pyrG* insertion occurred in a nonessential gene, then half the progeny should contain the wild-type allele, and half would contain the *pyrG* inserted allele. Thus, ura$^+$ haploid progeny should be recoverable.

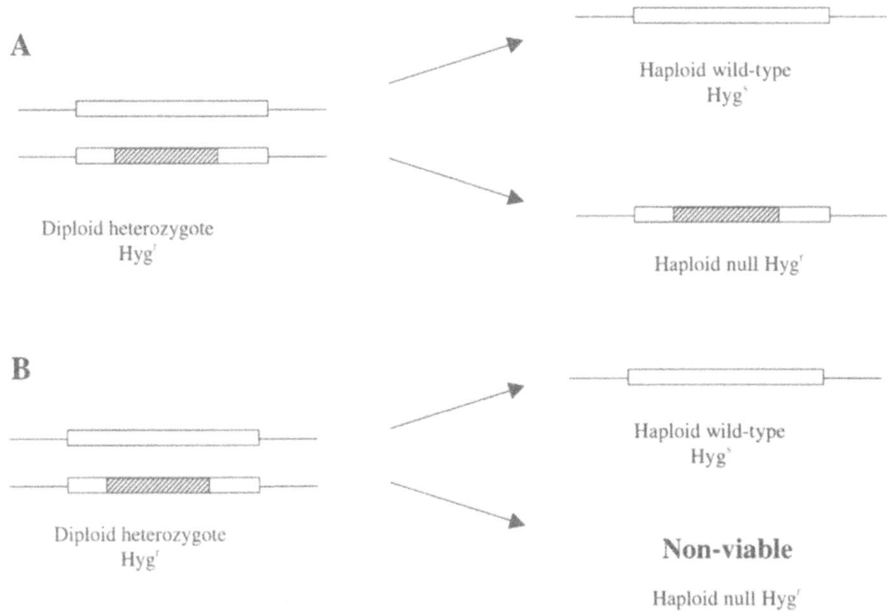

Figure 14.2. Discovery of essential genes through parasexual genetics. A diploid strain of *Aspergillus fumigatus* is subjected to insertional mutagenesis using a hygromycin resistance cassette. This generates a bank of strains which are heterozygous at the insertion site, and marked with the hygromycin resistance gene at the site of the insertion. These strains are then induced to undergo haploidization and the progeny are analyzed as follows. (A) Insertion of the resistance cassette into a non-essential gene. If the hygromycin cassette has integrated into a locus that does not encode an essential gene, then both the haploid progeny containing the insertion and the progeny with the wild-type allele will be viable. Hygromycin resistant progeny (containing the insertion) will be recoverable by hygromycin selection. (B) Insertion of the resistance cassette into an essential gene. In contrast, if the insertion event disrupts an essential gene, then only haploid progeny with the wild-type allele will be viable. Since these strains will have lost the hygromycin resistance cassette, there will be no hygromycin resistant haploids recovered upon selection with hygromycin.

In contrast, if integration of the *pyrG* marker had disrupted an essential gene in the diploid, only progeny containing the wild-type allele would be recovered upon haploidization. Since these haploid organisms would have lost the chromosome bearing the *pyrG* insertion, all progeny would be ura^{-}. In this way mutants containing insertions at essential loci can be identified.

3.4.4. Signature-Tagged Mutagenesis

Signature-tagged mutagenesis is a valuable tool for the detection of putative virulence factors. In this strategy, pools of mutants are created, each with a unique signature tag that marks the insertion site in the genome. Pools of mutants can be screened in a competitive assay either *in vitro* or *in vivo* and the output pool analysed for dropouts which have lost the phenotype of interest (i.e., virulence). Holden and colleagues have used this approach to search for *A. fumigatus* virulence genes (Figure 14.3) (Brown *et al.*, 2000). *A. fumigatus* insertional mutants were generated by electroporating a construct that

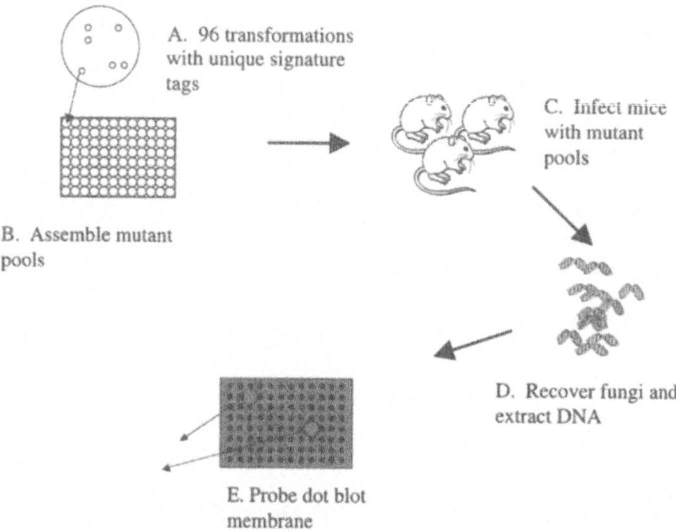

Figure 14.3. A simplified version of signature-tagged mutagenesis for the isolation of virulence genes. (A) Insertional mutagenesis with unique signature-tagged disruption cassettes is performed to generate a large bank of mutants. (B) Mutants are assembled into pools such that a single signature tag is represented only once per pool. This is designated the input pool. (C) Mice are infected with the input pools and, after a period, are sacrificed. (D) DNA is prepared from the fungi recovered from the mouse organs (output pool). (E) This output pool DNA is then hybridized with the signature tags to identify strains that are under-represented or absent after passage through the animals (dropout mutants). The transformants bearing the signature tags of the dropout mutants are analyzed to determine the genes that has been disrupted during the insertional mutagenesis.

incorporated the *hph* resistance marker as well as a unique signature tag. These transformants were divided into pools of 96 uniquely marked mutants. Mice were infected with pools of these strains ("input pool"), and the fungi that propagated in the animal were recovered as the "output" pool. DNA extracted from the output pool was hybridized on membranes containing the signature tags. Dropout strains which were present in the input pool but are absent or underrepresented in the output pool were examined for the presence of insertions in putative virulence genes. With this approach, the *pabaA* gene, encoding *para*-aminobenzoic acid synthetase, was identified as essential for *A. fumigatus* virulence. This was confirmed by demonstrating that the targeted *pabaA* deletion mutant was unable to cause lethal infection in mice.

3.4.5. Reporter Gene Systems

The detection of gene expression *in vitro* and *in vivo* is a valuable technique in the study of putative virulence genes. Two reporter genes have been modified for use in *A. fumigatus*: β-galactosidase (Smith *et al.*, 1994) and green fluorescent protein (Langfelder *et al.*, 2001; Wasylnka and Moore, 2002).

The *E. coli lacZ* gene, encoding β-galactosidase, has been used as a reporter gene in many microorganisms, including *S. cerevisiae, Fusarium oxysporum*, and *A. nidulans* (Weidner *et al.*, 2001; Notz *et al.*, 2002; Rupp, 2002). *lacZ* activity is detected by the cleavage of a chromogenic galactoside to a colored product (blue in the case of X-gal). Activity can be detected by quantitative liquid assay, *in situ* colorimetry, and colorimetric assays of colonies replicated onto filter paper. Smith *et al.* (1994) demonstrated that this reporter system is functional in *A. fumigatus* by driving the expression of the *E.coli lacZ* gene with the *A. fumigatus* alkaline protease promoter. β-galactosidase activity could be detected *in vitro* and *in vivo*.

The green fluorescent protein (GFP) from the jellyfish *Aequorea victorea* has proven to be an ideal reporter in several respects. It requires no cofactor for fluorescence and is active at the single cell level, permitting evaluation of intracellular localization of tagged or fusion proteins (Langfelder *et al.*, 2001). GFP-based expression analysis has been performed in several fungi, including *C. albicans* (Cormack *et al.*, 1997), *C. neoformans* (del Poeta *et al.*, 1999), and *Ustilago maydis* (Spellig *et al.*, 1996). Langfelder *et al.* (2001) have shown that GFP can also be used for the study of gene expression in *A. fumigatus*. Fusing GFP with the promoters of either the *pyrG* or *pksP/alb1* genes and introducing the fusions into *A. fumigatus* allowed the expression of GFP in cells growing in culture media or in animals. The importance of this technique is highlighted by the finding that the *pksP/alb1* construct was expressed in germlings during spore germination in animals but not during germination in artificial media.

3.4.6. Transposable Elements in Aspergilli

Transposon-mediated mutagenesis is effective in *Candida glabrata* (Willins *et al.*, 2002b), *Magnoporthe grisea* (Villalba *et al.*, 2001), and *S. cerevisiae* (Kontoyiannis, 1999), and it would be highly desirable for *Aspergillus*, particularly in view of the concerns raised regarding genomic rearrangement during electroporation-mediated mutagenesis. To date the only *Aspergillus* known to possess functional endogenous transposons is *A. niger* (Glayzer *et al.*, 1995; Amutan *et al.*, 1996; Nyyssonen *et al.*, 1996). In *A. nidulans*, a *Fot1*-like element is present, is transcribed, and is polymorphic among strains. Nevertheless, repeated efforts to demonstrate transpositions by this element were unsuccessful (Li Destri Nicosia *et al.*, 2001). Heterologous transposons (*Fot1* and *impala* from *F. oxysporum*) could migrate in *A. nidulans* (Li Destri Nicosia *et al.*, 2001), raising the possibility that a transposon-based method for gene tagging and mutagenesis could be developed in *Aspergillus*.

No functional transposons have yet been found in *A. fumigatus*, although the *A. fumigatus* sequencing project has identified several open reading frames that are highly homologous to transposases in other fungal organisms. Work in our laboratory has demonstrated that one such element (tentatively designated *taf1*) is present in multiple copies and is polymorphic among strains of *A. fumigatus*. It is not yet known if this element is transcribed or if transposition occurs naturally in *A. fumigatus*. Two defunct retrotransposons *Afut1*, and *Afut2*, have also been described in *A. fumigatus* (Neuveglise *et al.*, 1996; Paris and Latgé, 2001). As of the time of writing, there have not been any published reports of heterologous transposition in *A. fumigatus*, although this remains a subject of great interest.

3.4.7. Complementation and Heterologous Expression in Aspergilli

Complementation can be used to isolate and characterize fungal genes. Classically, an organism is subjected to mutagenesis, and mutants are screened or selected to find ones displaying the desired phenotype. DNA, either as a specific gene of interest or as a library, is introduced in the hope of complementing the defective mutation and transforming the phenotype back to that of the wild-type organism. This approach was used by Kwon-Chung and colleagues to study the genes that govern conidial pigment and morphology (Tsai *et al.*, 1998, 1999). Chemically-induced mutants of *A. fumigatus* were isolated with defective conidial pigmentation. A cosmid library of *A. fumigatus* genomic DNA was screened for DNA that restored normal conidial pigmentation. With this approach, the *arp1* and *alb1* genes were identified which are involved in conidial pigment biosynthesis and morphogenesis.

In an alternate approach, termed heterologous complementation, the DNA of one fungus is used to complement a mutation in a related but different fungus. This technique can exploit well-defined mutants that may be available in model organism. For example, numerous genes of *C. albicans* were found by complementing defined mutations in *S. cerevisiae*. The extensive collection of defined mutants of *A. nidulans* offers the possibility for similar studies with *A. fumigatus* gene libraries. Heterologous complementation was used by Parta *et al.* (1994) to characterize *hyp1/rodA*, an *A. fumigatus* hydrophobin gene. The *hyp1/rodA* gene was first retrieved from an *A. fumigatus* library using a probe generated by PCR with degenerate primers. The ability of the cloned gene to mediate the synthesis of a functional hydrophobin was then verified by complementing a *ΔrodA* strain of *A. nidulans* and demonstrating restoration of the normal hydrophobic rodlet layer on the conidial surface.

Heterologous expression can also provide a means for the detailed study of gene function, in a manner that is more tractable than is possible in the homologous organism. A classic example is the study of *C. albicans* adhesins (Fu *et al.*, 1998; Gaur *et al.*, 1999). Since *C. albicans* has many redundant adhesins, studying the role of any single adhesin is very difficult. In contrast, function can be easily detected when a single gene is expressed in *S. cerevisiae*, since this organism does not adhere to host substrates except as a consequence of the expression of the *C. albicans* adhesin. Watanabe *et al.* (2000) used heterologous expression to clarify the identity of the product made by the *A. fumigatus alb1* polyketide synthase. By overexpressing the *alb1* gene in *A. oryzae* and analyzing the polyketide that accumulated as a result, they were able to demonstrate that Alb1p is a naphthopyrone heptaketide synthase rather than a tetrahydroxynaphthalene pentaketide synthase as had been previously assumed (see Chapter 6).

3.4.8. Genome Sequencing

The availability of genome sequence data has revolutionized the study of several fungi, including *S. cerevisiae, C. albicans*, and *A. nidulans*. In addition to providing a wealth of information about individual genes and gene families, genome sequencing projects constitute the basis for large-scale functional genomics. Microarrays to examine gene expression globally have been a logical extension of genome sequencing; indeed whole genome arrays for *S. cerevisiae* and *C. albicans* are available and are in common use. A genome sequence project is currently underway for *A. fumigatus* as a cooperative effort between TIGR, the Sanger Center, University of Manchester, the Institut Pasteur, University of Salamanca, and Nagasaki University. The strain being sequenced, AF293, is a clinical strain isolated from the lung of a leukopenic patient who died with invasive pulmonary aspergillosis. The project's completion is anticipated in 2003 (Denning *et al.*, 2002). A searchable version of the most recent assembly is available through TIGR at *http://tigrblast.tigr.org/ufmg/*. Annotation of the *A. fumigatus* genome will facilitate the development of microarrays for genome-wide expression analysis.

4. The Agents of Mucormycosis

Mucormycoses are life-threatening infections caused by several fungi belonging to the class *Zygomycetes*, order Mucorales. In this group, *Rhizopus oryzae* (*Rhizopus arrhizus*) is the most common cause of infection followed by *Rhizopus microsporus* var. *rhizopodiformis* (Richardson and Shankland, 1999; Ribes *et al.*, 2000). Other, less frequently isolated pathogenic species of the Mucoraceae family include *Absidia corymbifera, Apophysomyces elegans, Mucor* species, *Rhizomucor pusillus, Cunninghamella bertholletiae*, and *Saksenaea* species (Sugar, 2000). Like the pathogenic *Aspergillus* species, these organisms are thermotolerant, ubiquitous organisms that grow on decaying organic matter. They produce abundant spores that are easily become air-borne and serve as the infectious agents via the respiratory tree.

Members of Mucorales cause both localized and disseminated infections in immunocompromised patients. Only rare case reports of invasive mucormycosis in normal hosts have been described (Al-Asiri *et al.*, 1996; del Valle-Zapico *et al.*, 1996). As with Aspergilli, allergic pulmonary disease does occur in immunocompetent hosts, but this is beyond the scope of the present chapter (Wimander and Belin, 1980; O'Connell *et al.*, 1995).

The major risk factors for mucormycosis include uncontrolled diabetes mellitus and other forms of metabolic acidosis, treatment with corticosteroid drugs in organ or bone marrow transplantation, trauma and burns, malignant hematological disorders, and deferoxamine therapy to chelate aluminum or iron in dialysis patients. Mucormycosis is relatively uncommon in neutropenic patients as compared with aspergillosis, but an increasing incidence of mucormycosis in this population has been observed. As the number of iatrogenically immunocompromised patients continues to rise, it is likely that the incidence of mucormycosis will also increase (Ribes *et al.*, 2000). The underlying risk factors influence the clinical manifestations of the disease. For example, diabetics in ketoacidosis usually develop rhinocerebral mucormycosis, whereas patients with malignant hematological disease, severe neutropenia, or a history of deferoxamine therapy usually suffer from pulmonary mucormycosis. Both forms of disease are acquired through inhalation (Lueg *et al.*, 1996; Ribes *et al.*, 2000). Other routes of infection include direct skin contact, that results in cutaneous mucormycosis, or ingestion of contaminated food, which leads to gastrointestinal mucormycosis. Cutaneous mucormycosis can occur in neonates and trauma and burn patients. Gastrointestinal mucormycosis is nearly always limited to malnourished infants.

The clinical hallmark of mucormycosis is the rapid onset of tissue necrosis, with or without fever. This necrosis is the result of invasion of the blood vessels and subsequent thrombosis. In most cases, the infection is relentlessly progressive and results in death unless treatment is promptly initiated. Treatment of mucormycosis requires a combination of medical and surgical approach. First, wherever possible, risk factors such as ketoacidosis or iron chelation therapy should be corrected. In addition, urgent surgical debridement of the infected and necrotic tissues can help prevent dissemination and disease progression. Finally, an appropriate antifungal agent should be administered. While amphotericin B deoxycholate remains the only antifungal agent approved for the treatment of this infection (Ribes *et al.*, 2000; Sugar, 2000), there are anecdotal reports of success using high doses of the lipid formulations of amphotericin B (Ericsson *et al.*, 1993; Weng *et al.*, 1998; Cagatay *et al.*, 2001). Unfortunately there are no controlled studies directly comparing lipid preparations with conventional amphotericin B in this fungal disease.

4.1. Molecular Techniques for the Study of Mucormycosis

Genetic manipulation of Mucorales is still in an early stage of development when compared to other filamentous fungi such as *Aspergillus*. This is due in part to the absence of information on the fate of transforming DNA in these species. Additionally, there are several unique challenges in transforming these organisms that make it even harder to establish a system whereby genes can be disrupted and their role in the pathogenesis evaluated. For example, unlike *Aspergillus* species, Mucorales are inherently resistant to most antifungals, and the utility of drug resistance markers is limited. Hygromycin B, phleomycin (which belongs to the bleomycin family of glycopeptide antibiotics), as well as nourseothricin (an amino-glycoside secreted by *Streptomyces noursei*; McDade and Cox, 2001) have all proven ineffective in our hands (unpublished). To date, only the neomycin and blasticidin S resistance genes have been used with success in Mucorales, specifically for the transformation of *Absidia glauca* (Wostemeyer *et al.*, 1987) and *Rhizopus niveus* (Yanai *et al.*, 1991), respectively. Drug selection markers are essential for transformation in this group because of the multi-nucleated nature of Mucorales spores (Takaya *et al.*, 1996). Following transformation, selection pressure must be maintained through repeated cycles of spore formation and single spore isolation to obtain homokaryotic transformants.

Nevertheless, transformation of Mucorales has usually been achieved on the basis of the complementation of auxotrophic mutations. A Leu^- mutant of *R. niveus* was generated by UV irradiation of sporangiospores (Liou *et al.*, 1992), and the *pyrG*$^-$ mutants of *R. oryzae* (Skory, 2002), and *R. pusilius* (Yamazaki *et al.*, 1999) were obtained by chemical mutagenesis and UV irradiation, respectively. A *Mucor circinelloides leuA*$^-$ strain was recovered through an enrichment strategy that makes use of the polyene antibiotic N-glycosylpolyfungin (Roncero, 1984). Having resulted from chemical or UV mutagenesis, these auxotrophic strains might harbor other undetected mutations that could complicate the delineation of the role of a specific gene in the pathogenesis of mucormycosis.

Transformation of Mucorales is often transient, presumably because the DNA that is introduced is not efficiently integrated into the host's genome (Skory, 2002). Rather it appears to replicate autonomously, even though it is devoid of sequences known to behave as initiation sites for replication. It is not known if Mucorales species can introduce ARS or telomeric sequences that would allow the transformed DNA to replicate autonomously. In the absence of integration, selection pressure must be maintained or the DNA introduced will be lost. A native ARS element from *M. circnelloides* has been characterized, but this element has not yet been used in transformation systems (Roncero *et al.*, 1989).

Gene disruption by homologous recombination has been claimed with members of the Mucorales (Liou *et al.*, 1992; Yamazaki *et al.*, 1999; Skory, 2002), but it is unclear if chromosomal integration actually occurred in these studies, since DNA introduced into

Mucorales tends to replicate in a concatenated high molecular weight form that can be mistaken for genomic DNA during electrophoresis. The absence of clear Southern hybridization data to support chromosomal integration raises the need for caution in evaluating these reports (Skory, 2002).

4.1.1. Transformation Techniques

Two techniques for transforming Mucorales have been described: a protoplast procedure and a biolistic method. Our own experience and that of others (C. Skory, personal communication) indicates that electroporation is not an appropriate method for these organisms. To our knowledge, *Agrobacterium*-mediated transformation has not been reported with Mucorales.

4.1.1a. Protoplast formation. Although the most widely used transformation method for both Aspergilli and Murorales is based on the generation of protoplasts, the members of the latter group have cell walls richer in chitosan (Suarez *et al.*, 1985). Therefore, Novozyme 234 must be combined with chitosanase and chitinase to be effective in generating protoplasts from germinated spores (Yamazaki *et al.*, 1999). Commercial chitosanase is expensive, but it can be generated as Streptozyme by growing *Sterptomyces* sp. no. 6 in media containing chitin and chitosan as sole carbon source (van Heeswijck *et al.*, 1998). Ungerminated sporangiospores are totally resistant to lysis by the action of this enzyme mixture, but they become susceptible upon germination and extension of the germling to a length of 35–50 μm (van Heeswijck, 1984; van Heeswijck *et al.*, 1998). We have found that if the hyphae are allowed to exceed this length, they tend to aggregate into balls, the inner portions of which resist lytic enzyme action, and protoplast yield decreases. Solutions of sugar alcohols such as sorbitol or mannitol (0.35–0.5 M) are necessary for osmotic stabilization of the protoplasts (van Heeswijck, 1984; van Heeswijck *et al.*, 1998). Other osmotic stabilizers that are useful with aspergilli, including KCl, $(NH_4)_2SO_4$ or $MgSO_4$, are inappropriate because they inhibit the activity of chitosanase (van Heeswijck, 1984; van Heeswijck *et al.*, 1998).

The transformation efficiency with protoplasts varies both with the genus and the nature of the gene being introduced (heterologous versus homologous). For example, when the *LeuA* of *M. circinelloides* is used to complement a *leu*$^-$ mutant of *R. niveus*, a transformation efficiency of 1–20 transformants/μg of DNA is obtained (Liou *et al.*, 1992). In contrast, when the same gene was used to complement a *leu*$^-$ *M. circinelloides* strain, the rate increases to 7,800 transformants/μg (Roncero *et al.*, 1989). Homologous complementation in *R. niveus* (Takaya *et al.*, 1996) and *Rhizomucor pusillus* (Yamazaki *et al.*, 1999) generated far fewer transformants (1.2 and 5 transformants/μg of DNA, respectively). Southern hybridization suggested that the DNA that was introduced either integrates into the chromosome in multiple, tandemly arrayed copies (Horiuchi *et al.*, 1995; Takaya *et al.*, 1996; Yamazaki *et al.*, 1999) or is maintained as a non-integrated plasmid, sometimes containing the selection marker in a multimeric form (Liou *et al.*, 1992). In some cases chromosomal integration of a single copy at a homologous locus seems to have occurred (Yamazaki *et al.*, 1999). Although linearization of the DNA used for transformation reduces the efficiency to 20% of that obtained with the same DNA in circular form, it seems to favor homologous recombination into the chromosome (Liou *et al.*, 1992).

4.1.1b. Biolistic delivery systems. The transformation of uracil auxotrophs of *R. oryzae* with wild-type *pyrG* gene by means of a microprojectile particle bombardment apparatus was recently described (Skory, 2002). Approximately 10–50 transformants appeared per μg of circular plasmid. Southern analysis of these transformants demonstrated that the plasmid almost always replicated autonomously in a high-molecular-weight form rather than integrating into genomic DNA, suggesting that the plasmid contained ARS-sequences. When integration did occur (1–5% of transformants), it resulted from Type-I homologous recombination (single crossover) into the *pyrG* locus. The integrants were mitotically stable even under non-selective conditions, unlike those harboring autonomously replicating elements. Introduction of a plasmid containing *pyrG* that had been linearized by restriction enzyme cleavege within the vector sequence resulted in transformants with autonomously replicating concatenated plasmids that formed by end-joining with restoration of the restriction site. When the same plasmid was digested with two different enzymes that cleaved within the vector sequence to release the *pyrG* containing fragment and the fragment was introduced into the cells, it integrated at the *pyrG* locus, replacing the defective *pyrG* gene. These results provide the strongest indication that targeted gene disruption can be achieved in Mucorales. Linearization of the plasmid with the *pyrG*, resulted in either Type-I homologous recombination at the *pyrG* locus, gene replacement of the *pyrG* gene, or concatenated plasmids that were replicating autonomously in high-molecular-weight form.

4.1.2. Sexual Cycle

Genera belonging to the class Mucorales are haploid organisms with vegetative and sexual life cycles. Mycelia can be one of two mating types, either (+) or (−) (Blakeslee, 1906). The sexual phase is initiated when nuclei of opposite mating types fuse in the zygospore to produce a germspore. These are resting spores, but when the conditions are appropriate, they grow into normal haploid vegetative mycelia. In general, the resting spores are dormant for several months. The length of this period and the difficulty in generating the zygospores in the laboratory represent a hurdle to the analysis of mutants and the generation of recombinants by classical genetics (van Heeswijck *et al.*, 1998).

4.1.3. Heterologous Expression

Genes originating in a member of the Mucorales that are devoid of introns can be expressed in *S. cerevisiae* (Kohno *et al.*, 1998). However if the gene has an intron, it must be removed (Horiuchi *et al.*, 1990) or modified by the addition of *S. cerevisiae* sequences believed to be involved in the splicing. Ashikari *et al.* (1990) found that the *R. niveus* aspartic proteinase gene provided a protein in *S. cerevisiae* only when the consensus sequence for a *S. cerevisiae* intron was introduced into the intron of the *R. niveus* gene (Ashikari *et al.*, 1990). A lipase gene of *R. niveus* that lacked introns was expressed in *S. cerevisiae* under the control of the phosphoglycerate kinase (PGK) promoter (Kohno *et al.*, 1998; Kohno *et al.*, 1999). Promoters of Mucorales can drive the expression of foreign genes. For example, the PGK promoter of *R. niveus* successfully drove the expression of *E. coli* β-glucuronidase gene in *R. niveus* (Takaya *et al.*, 1994, 1995).

Heterologous expression of genes among different genera of Mucorales has also been achieved. For example, the *LeuA* gene of *M. circinelloides* complemented a *R. niveus* *leu1*$^-$ mutant (Takaya *et al.*, 1996). Introns do not pose a barrier to gene expression in different species of the Mucorales genus, since the *pyrG* gene of *R. niveus* containing two putative introns was successfully expressed in *Rhizopus delemar* (Horiuchi *et al.*, 1995).

4.1.4. Summary

Mucormycoses are life-threatening infections that afflict the immunocompromised host. Molecular biology tools need to be developed for the manipulation of pathogenic species of Mucorales. No gene disruption system has yet been established with which the roles of specific fungal genes in the pathogenesis of mucormycosis could be delineated.

5. Other Pathogenic Filamentous Fungi

Several other species of filamentous fungi can cause invasive disease in immunocompromised patients, although these fungi are all less common than aspergilli or zygomycetes. For the most part, the molecular genetic tools discussed in this chapter have not been evaluated for these fungi. One exception is *P. boydii* (*Scedosporium apospermum*), the cause of an invasive syndrome clinically similar to invasive aspergillosis. The transformation to hygromycin resistance by electroporation with the *hph* gene has been described, though not in the pathogenic *Scedosporium* species (Ruiz-Diez and Martinez-Suarez, 1999). Inclusion of rDNA sequence in the DNA to be introduced, enhanced transformation efficiency and allowed homologous integration at the rDNA locus.

6. Future Directions

Molecular genetic analysis of medically important filamentous fungi has begun, as molecular tools have been adapted for the study of these organisms. Further, the wealth of information that will be generated by genome sequencing projects is likely to lead to the development of exciting new experimental methods for genome-wide analysis. These strategies, including whole genome expression analysis *in vivo*, and comparative genomics between pathogenic and non-pathogenic species (such as *A. fumigatus* versus *A. nidulans*), are highly likely to generate hypotheses regarding the molecular mechanisms governing pathogenicity. In turn, these hypotheses will require specific validation using the classic molecular methods that have been outlined in this chapter.

References

Al-Asiri, R.H., van Dijken, P.J., Mahmood, M.A., Al-Shahed, M.S., Rossi, M.L., and Osoba, A.O. (1996). Isolated hepatic mucormycosis in an immunocompetent child. *Am. J. Gastroenterol.* **91**, 606–607.

Amutan, M., Nyyssonen, E., Stubbs, J., Diaz-Torres, M.R., and Dunn-Coleman, N. (1996). Identification and cloning of a mobile transposon from *Aspergillus niger* var. *awamori. Curr. Genet.* **29**, 468–473.

Ashikari, T., Amachi, T., Yoshizumi, H., Horiuchi, H., Takagi, M., and Yano, K. (1990). Correct splicing of modified introns of a *Rhizopus* proteinase gene in *Saccharomyces cerevisiae. Mol. Gen. Genet.* **223**, 11–16.

Aufauvre-Brown, A., Brown, J.S., and Holden, D.W. (1998). Comparison of virulence between clinical and environmental isolates of *Aspergillus fumigatus. Eur. J. Clin. Microbiol. Infect. Dis.* **17**, 778–780.

Aufauvre-Brown, A., Mellado, E., Gow, N.A.R., and Holden, D.W. (1997). *Aspergillus fumigatus chsE*: A gene related to *CHS3* of *Saccharomyces cerevisiae* and important for hyphal growth and conidiophore development but not pathogenicity. *Fungal Genet. Biol.* **21**, 141–152.

Blakeslee, A.F. (1906). Zygospore germination in the Mucorineae. *Ann. Mycol.* **4**, 1–28.

Bodey, G., Bueltmann, B., Duguid, W., Gibbs, D., Hanak, H., Hotchi, M., Mall, G., Martino, P. *et al.* (1992). Fungal infections in cancer patients: An international autopsy survey. *Eur. J. Clin. Microbiol. Infect. Dis.* **11**, 99–109.

Brakhage, A.A. and Langfelder, K. (2002). Menacing mold: The molecular biology of *Aspergillus fumigatus*. *Annu. Rev. Microbiol.* **56**, 433–455.

Brookman, J.L. and Denning, D.W. (2000). Molecular genetics in *Aspergillus fumigatus*. *Curr. Opin. Microbiol.* **3**, 468–474.

Brown, J.S., Aufauvre-Brown, A., Brown, J., Jennings, J.M., Arst, H., Jr., and Holden, D.W. (2000). Signature-tagged and directed mutagenesis identify PABA synthetase as essential for *Aspergillus fumigatus* pathogenicity. *Mol. Microbiol.* **36**, 1371–1380.

Brown, J.S., Aufauvre-Brown, A., and Holden, D.W. (1998). Insertional mutagenesis of *Aspergillus fumigatus*. *Mol. Gen. Genet.* **259**, 327–335.

Çagatay, A.A., Öncü, S.S., Çalangu, S.S., Yildirmak, T.T., Özsüt, H.H., and Eraksoy, H.H. (2001). Rhinocerebral mucormycosis treated with 32 gram liposomal amphotericin B and incomplete surgery: A case report. *BMC Infect. Dis.* **1**, 22.

Chakraborty, B.N., Patterson, N.A., and Kapoor, M. (1991). An electroporation-based system for high-efficiency transformation of germinated conidia of filamentous fungi. *Can. J. Microbiol.* **37**, 858–863.

Clemons, K.V., Miller, T.K., Selitrennikoff, C.P., and Stevens, D.A. (2002). *fos-1*, a putative histidine kinase as a virulence factor for systemic aspergillosis. *Med. Mycol.* **40**, 259–262.

Cormack, B.P., Bertram, G., Egerton, M., Gow, N.A., Falkow, S., and Brown, A.J. (1997). Yeast-enhanced green fluorescent protein (yEGFP) a reporter of gene expression in *Candida albicans*. *Microbiology* **143**, 303–311.

de Groot, M.J., Bundock, P., Hooykaas, P.J., and Beijersbergen, A.G. (1998). *Agrobacterium tumefaciens*-mediated transformation of filamentous fungi. *Naure Biotechnol.* **16**, 839–842.

de Lucas, J.R., Dominguez, A.I., Higuero, Y., Martinez, O., Romero, B., Mendoza, A., Garcia-Bustos, J.F., and Laborda, F. (2001). Development of a homologous transformation system for the opportunistic human pathogen *Aspergillus fumigatus* based on the *sC* gene encoding ATP sulfurylase. *Arch. Microbiol.* **176**, 106–113.

del Poeta, M., Toffaletti, D.L., Rude, T.H., Sparks, S.D., Heitman, J., and Perfect, J.R. (1999). *Cryptococcus neoformans* differential gene expression detected *in vitro* and *in vivo* with green fluorescent protein. *Infect. Immun.* **67**, 1812–1820.

del Valle-Zapico, A., Rubio-Suarez, A., Mellado-Encinas, P., Morales-Angulo, C., and Cabrera-Pozuelo, E. (1996). Mucormycosis of the sphenoid sinus in an otherwise healthy patient. Case report and literature review. *J. Laryngol. Otol.* **110**, 471–473.

d'Enfert, C. (1996). Selection of multiple disruption events in *Aspergillus fumigatus* using the orotidine-5′-decarboxylase gene, *pyrG*, as a unique transformation marker. *Curr. Genet.* **30**, 76–82.

d'Enfert, C., Diaquin, M., Delit, A., Wuscher, N., Debeaupuis, J.P., Huerre, M., and Latgé, J.P. (1996). Attenuated virulence of uridine-uracil auxotrophs of *Aspergillus fumigatus*. *Infect. Immun.* **64**, 4401–4405.

Denning, D.W. (1996). Therapeutic outcome in invasive aspergillosis. *Clin. Infect. Dis.* **23**, 608–615.

Denning, D.W. (2000). *Aspergillus* species. In G.L. Mandell, J.E. Bennett and R. Dolin (eds.) *Mandell, Douglas, and Bennett's Principals and Practice of Infectious Diseases*, 5th edn. Churchill Livingston, Philadelphia, pp. 2674–2685.

Denning, D.W., Anderson, M.J., Turner, G., Latgé, J.P., and Bennett, J.W. (2002). Sequencing the *Aspergillus fumigatus* genome. *Lancet Infect. Dis.* **2**, 251–253.

Ericsson, M., Anniko, M., Gustafsson, H., Hjalt, C.A., Stenling, R., and Tarnvik, A. (1993). A case of chronic progressive rhinocerebral mucormycosis treated with liposomal amphotericin B and surgery. *Clin Infect Dis.* **16**, 585–586.

Firon, A., Beauvais, A., Latgé, J.P., Couve, E., Grosjean-Cournoyer, M.C., and d'Enfert, C. (2002). Characterization of essential genes by parasexual genetics in the human fungal pathogen *Aspergillus fumigatus*. Impact of genomic rearrangements associated with electroporation of DNA. *Genetics* **161**, 1077–1087.

Fu, Y., Rieg, G., Fonzi, W.A., Belanger, P.H., Edwards, J.E., Jr., and Filler, S.G. (1998). Expression of the *Candida albicans* gene *ALS1* in *Saccharomyces cerevisiae* induces adherence to endothelial and epithelial cells. *Infect. Immun.* 66, 1783–1786.

Fungaro, M.H., Rech, E., Muhlen, G.S., Vainstein, M.H., Pascon, R.C., de Queiroz, M.V., Pizzirani-Kleiner, A.A., and de Azevedo, J.L. (1995). Transformation of *Aspergillus nidulans* by microprojectile bombardment on intact conidia. *FEMS Microbiol. Lett.* **125**, 293–297.

Gatignol, A., Baron, M., and Tiraby, G. (1987). Phleomycin resistance encoded by the *ble* gene from transposon Tn 5 as a dominant selectable marker in *Saccharomyces cerevisiae. Mol. Gen. Genet.* **207**, 342–348.

Gaur, N.K., Klotz, S.A., and Henderson, R.L. (1999). Overexpression of the *Candida albicans ALA1* gene in *Saccharomyces cerevisiae* results in aggregation following attachment of yeast cells to extracellular matrix proteins, adherence properties similar to those of *Candida albicans. Infect. Immun.* **67**, 6040–6047.

Giron, J., Sans, N., Poey, C., Fajadet, P., Fourcade, D., Senac, J.P., and Railhac, J.J. (1998). CT-guided percutaneous treatment of inoperable pulmonary aspergilloma. Apropos of 42 cases. *J. Radiol.* **79**, 139–145.

Glayzer, D.C., Roberts, I.N., Archer, D.B., and Oliver, R.P. (1995). The isolation of *Ant1*, a transposable element from *Aspergillus niger. Mol. Gen. Genet.* **249**, 432–438.

Gonzalez, A., Jimenez, A., Vazquez, D., Davies, J.E., and Schindler, D. (1978). Studies on the mode of action of hygromycin B, an inhibitor of translocation in eukaryotes. *Biochim. Biophys. Acta* 521, 459–469.

Hector, R.F., Yee, E., and Collins, M.S. (1990). Use of DBA/2N mice in models of systemic candidiasis and pulmonary and systemic aspergillosis. *Infect. Immun.* **58**, 1476–1478.

Hensel, M., Arst, H.N., Jr., Aufauvre-Brown, A., and Holden, D.W. (1998). The role of the *Aspergillus fumigatus areA* gene in invasive pulmonary aspergillosis. *Mol. Gen. Genet.* **258**, 553–557.

Herbrecht, R., Denning, D.W., Patterson, T.F., Bennett, J.E., Greene, R.E., Oestmann, J.-W., Kern, W.V., Marr, K.A. *et al.* (2002). Voriconazole versus amphotericin B for primary therapy of invasive aspergillosis. *New Engl. J. Med.* **347**, 408–415.

Horiuchi, H., Ashikari, T., Amachi, T., Yoshizumi, H., Takagi, M., and Yano, K. (1990). High-level secretion of a *Rhizopus niveus* aspartic proteinase in *Saccharomyces cerevisiae. Agric. Biol. Chem.* **54**, 1771–1779.

Horiuchi, H., Takaya, N., Yanai, K., Nakamura, M., Ohta, A., and Takagi, M. (1995). Cloning of the *Rhizopus niveus pyr4* gene and its use for the transformation of *Rhizopus delemar. Curr. Genet.* **27**, 472–478.

Hospenthal, D.R., Kwon-Chung, K.J., and Bennett, J.E. (1998). Concentrations of airborne *Aspergillus* compared to the incidence of invasive aspergillosis: Lack of correlation. *Med. Mycol.* **36**, 165–168.

Ibrahim-Granet, O. and d'Enfert, C. (1997). The *Aspergillus fumigatus mepB* gene encodes an 82 kDa intracellular metalloproteinase structurally related to mammalian thimet oligopeptidases. *Microbiology* **143**, 2247–2253.

Ikegami, Y., Amitani, R., Murayama, T., Nawada, R., Lee, W.J., Kawanami, R., and Kuze, F. (1998). Effects of alkaline protease or restrictocin deficient mutants of *Aspergillus fumigatus* on human polymorphonuclear leukocytes. *Eur. Respir. J.* **12**, 607–611.

Kohno, M., Enatsu, M., and Kugimiya, W. (1998). Cloning of genomic DNA of *Rhizopus niveus* lipase and expression in the yeast *Saccharomyces cerevisiae. Biosci. Biotechnol. Biochem.* **62**, 2425–2427.

Kohno, M., Enatsu, M., Yoshiizumi, M., and Kugimiya, W. (1999). High-level expression of *Rhizopus niveus* lipase in the yeast *Saccharomyces cerevisiae* and structural properties of the expressed enzyme. *Protein Expr. Purif.* **15**, 327–335.

Kontoyiannis, D.P. (1999). Genetic analysis of azole resistance by transposon mutagenesis in *Saccharomyces cerevisiae. Antimicrob. Agents Chemother.* **43**, 2731–2735.

Kothary, M.H., Chase, T., Jr., and Macmillan, J.D. (1984). Correlation of elastase production by some strains of *Aspergillus fumigatus* with ability to cause pulmonary invasive aspergillosis in mice. *Infect. Immun.* **43**, 320–325.

Langfelder, K., Gattung, S., and Brakhage, A.A. (2002). A novel method used to delete a new *Aspergillus fumigatus* ABC transporter-encoding gene. *Curr. Genet.* **41**, 268–274.

Langfelder, K., Philippe, B., Jahn, B., Latgé, J.P., and Brakhage, A.A. (2001). Differential expression of the *Aspergillus fumigatus pksP* gene detected *in vitro* and *in vivo* with green fluorescent protein. *Infect. Immun.* **69**, 6411–6418.

Latgé, J.P. (1999). *Aspergillus fumigatus* and aspergillosis. *Clin. Microbiol. Rev.* **12**, 310–350.

Lebeau, B., Pelloux, H., Pinel, C., Michallet, M., Gout, J.P., Pison, C., Delormas, P., Bru, J.P., *et al.* (1994). Itraconazole in the treatment of aspergillosis: A study of 16 cases. *Mycoses* **37**, 171–179.

Levadoux, W.L., Gregory, K.F., and Taylor, A. (1981). Sequential cold-sensitive mutations in *Aspergillus fumigatus*. II. Analysis by the parasexual cycle. *Can. J. Microbiol.* **27**, 295–303.

Li Destri Nicosia, M.G., Brocard-Masson, C., Demais, S., Hua-Van, A., Daboussi, M.J., and Scazzocchio, C. (2001). Heterologous transposition in *Aspergillus nidulans. Mol. Microbiol.* **39**, 1330–1344.

Liou, C.M., Yanai, K., Horiuchi, H., and Takagi, M. (1992). Transformation of a Leu^- mutant of *Rhizopus niveus* with the *leuA* gene of *Mucor circinelloides. Biosci. Biotechnol. Biochem.* **56**, 1503–1504.

Lueg, E.A., Ballagh, R.H., and Forte, V. (1996). Analysis of the recent cluster of invasive fungal sinusitis at the Toronto Hospital for Sick Children. *J. Otolaryngol.* **25**, 366–370.

McDade, H.C. and Cox, G.M. (2001). A new dominant selectable marker for use in *Cryptococcus neoformans*. *Med. Mycol.* **39**, 151–154.

Mellado, E., Aufauvre-Brown, A., Gow, N.A., and Holden, D.W. (1996a). The *Aspergillus fumigatus chsC* and *chsG* genes encode class III chitin synthases with different functions. *Mol. Microbiol.* **20**, 667–679.

Mellado, E., Specht, C.A., Robbins, P.W., and Holden, D.W. (1996b). Cloning and characterization of *chsD*, a chitin synthase-like gene of *Aspergillus fumigatus*. *FEMS Microbiol. Lett.* **143**, 69–76.

Michel, S., Ushinsky, S., Klebl, B., Leberer, E., Thomas, D., Whiteway, M., and Morschhäuser, J. (2002). Generation of conditional lethal *Candida albicans* mutants by inducible deletion of essential genes. *Mol. Microbiol.* **46**, 269–280.

Mullins, E.D. and Kang, S. (2001). Transformation: A tool for studying fungal pathogens of plants. *Cell. Mol. Life Sci.* **58**, 2043–2052.

Mullins, J., Harvey, R., and Seaton, A. (1976). Sources and incidence of airborne *Aspergillus fumigatus* (Fres). *Clin. Allergy* **6**, 209–217.

Mullins, J. and Seaton, A. (1978). Fungal spores in lung and sputum. *Clin. Allergy* **8**, 525–533.

Neuveglise, C., Sarfati, J., Latgé, J.P., and Paris, S. (1996). *Afut1*, a retrotransposon-like element from *Aspergillus fumigatus*. *Nucleic Acids Res.* **24**, 1428–1434.

Notz, R., Maurhofer, M., Dubach, H., Haas, D., and Defago, G. (2002). Fusaric acid-producing strains of *Fusarium oxysporum* alter 2,4-diacetylphloroglucinol biosynthetic gene expression in *Pseudomonas fluorescens* CHA0 *in vitro* and in the rhizosphere of wheat. *Appl. Environ. Microbiol.* **68**, 2229–2235.

Nyyssonen, E., Amutan, M., Enfield, L., Stubbs, J., and Dunn-Coleman, N.S. (1996). The transposable element *Tan1* of *Aspergillus niger* var. *awamori*, a new member of the *Fot1* family. *Mol. Gen. Genet.* **253**, 50–56.

O'Connell, M.A., Pluss, J.L., Schkade, P., Henry, A.R., and Goodman, D.L. (1995). *Rhizopus*-induced hypersensitivity pneumonitis in a tractor driver. *J. Allergy Clin. Immunol.* **95**, 779–780.

Ozeki, K., Kyoya, F., Hizume, K., Kanda, A., Hamachi, M., and Nunokawa, Y. (1994). Transformation of intact *Aspergillus niger* by electroporation. *Biosci. Biotechnol. Biochem.* **58**, 2224–2227.

Paris, S. and Latgé, J.P. (2001). *Afut2*, a new family of degenerate gypsy-like retrotransposon from *Aspergillus fumigatus*. *Med. Mycol.* **39**, 195–198.

Parta, M., Chang, Y., Rülong, S., Pinto-DaSilva, P., and Kwon-Chung, K.J. (1994). *HYP1*, a hydrophobin gene from *Aspergillus fumigatus*, complements the rodletless phenotype in *Aspergillus nidulans*. *Infect. Immun.* **62**, 4389–4395.

Punt, P.J., Oliver, R.P., Dingemanse, M.A., Pouwels, P.H., and van den Hondel, C.A. (1987). Transformation of *Aspergillus* based on the hygromycin B resistance marker from *Escherichia coli*. *Gene* **56**, 117–124.

Punt, P.J. and van den Hondel, C.A. (1992). Transformation of filamentous fungi based on hygromycin B and phleomycin resistance markers. *Methods Enzymol.* **216**, 447–457.

Ribes, J.A., Vanover-Sams, C.L., and Baker, D.J. (2000). Zygomycetes in human disease. *Clin. Microbiol. Rev.* **13**, 236–301.

Richardson, M.D. and Shankland, G.S. (1999). *Rhisopus, Rhizomucor, Absidia*, and other agents of systemic and subcutaneous zygomycoses. In P.R. Murray, E.J. Baron, M.A. Pfaller, F.C. Tenover, and R.H. Yolken (eds) *Manual of Clinical Microbiology*, 5th edn. ASM Press, Washington DC, pp. 1242–1258.

Roncero, M.I. (1984). Enrichment method for the isolation of auxotrophic mutants of *Mucor* using the polyene antibiotic N-glycosylpolyfungin. *Carlsberg Res. Commun.* **49**, 658–690.

Roncero, M.I., Jepsen, L.P., Stroman, P., and van Heeswijck, R. (1989). Characterization of a *leuA* gene and an ARS element from *Mucor circinelloides*. *Gene* **84**, 335–343.

Ruiz-Diez, B. and Martinez-Suarez, J.V. (1999). Electrotransformation of the human pathogenic fungus *Scedosporium prolificans* mediated by repetitive rDNA sequences. *FEMS Immunol. Med. Microbiol.* **25**, 275–282.

Rupp, S. (2002). *lacZ* assays in yeast. *Methods Enzymol.* **350**, 112–131.

Sanchez, O., Navarro, R.E., and Aguirre, J. (1998). Increased transformation frequency and tagging of developmental genes in *Aspergillus nidulans* by restriction enzyme-mediated integration (REMI). *Mol. Gen. Genet.* **258**, 89–94.

Sandhu, D.K., Sandhu, R.S., Khan, Z.U., and Damodaran, V.N. (1976). Conditional virulence of a p-aminobenzoic acid-requiring mutant of *Aspergillus fumigatus*. *Infect. Immun.* **13**, 527–532.

Schaffner, A. (1994). Macrophage-*Aspergillus* interactions. *Immunol. Ser.* **60**, 545–552.

Schaffner, A., Douglas, H., and Braude, A. (1982). Selective protection against conidia by mononuclear and against mycelia by polymorphonuclear phagocytes in resistance to *Aspergillus*. Observations on these two lines of defense *in vivo* and *in vitro* with human and mouse phagocytes. *J. Clin. Invest.* **69**, 617–631.

Skory, C.D. (2002). Homologous recombination and double-strand break repair in the transformation of *Rhizopus oryzae*. *Mol. Genet. Genomics* **268**, 397–406.

Smith, J.M., Tang, C.M., van Noorden, S., and Holden, D.W. (1994). Virulence of *Aspergillus fumigatus* double mutants lacking restriction and an alkaline protease in a low-dose model of invasive pulmonary aspergillosis. *Infect. Immun.* **62**, 5247–5254.

Spellig, T., Bottin, A., and Kahmann, R. (1996). Green fluorescent protein (GFP) as a new vital marker in the phytopathogenic fungus *Ustilago maydis*. *Mol. Gen. Genet.* **252**, 503–509.

Stevens, D.A., Kan, V.L., Judson, M.A., Morrison, V.A., Dummer, S., Denning, D.W., Bennett, J.E., Walsh, T.J. *et al.* (2000). Practice guidelines for diseases caused by *Aspergillus*. Infectious Diseases Society of America. *Clin. Infect. Dis.* **30**, 696–709.

Suarez, T., Orejas, M., and Eslava, A.P. (1985). Isolation, regeneration, and fusion of *Phycomyces blakesleeanus* spheroplasts. *Exp. Mycol.* **9**, 203–211.

Sugar, A.M. (2000). Agents of mucormycosis and related species. In G.L. Mandell, J.E. Bennett, and R. Dolin (eds) *Mandell, Douglas, and Bennett's Principals and Practice of Infectious Diseases*, 5th edn. Churchill Livingston, Philadelphia, pp. 2685–2695.

Sundstrom, P., Cutler, J.E., and Staab, J.F. (2002). Reevaluation of the role of *HWP1* in systemic candidiasis by use of *Candida albicans* strains with selectable marker *URA3* targeted to the *ENO1* locus. *Infect. Immun.* **70**, 3281–3283.

Takaya, N., Yanai, K., Horiuchi, H., Ohta, A., and Takagi, M. (1994). Cloning and characterization of two 3-phosphoglycerate kinase genes of *Rhizopus niveus* and heterologous gene expression using their promoters. *Curr. Genet.* **25**, 524–530.

Takaya, N., Yanai, K., Horiuchi, H., Ohta, A., and Takagi, M. (1995). Analysis of the 3-phosphoglycerate kinase 2 promoter in *Rhizopus niveus*. *Gene* **152**, 121–125.

Takaya, N., Yanai, K., Horiuchi, H., Ohta, A., and Takagi, M. (1996). Cloning and characterization of the *Rhizopus niveus leu1* gene and its use for homologous transformation. *Biosci. Biotechnol. Biochem.* **60**, 448–452.

Thau, N., Monod, M., Crestani, B., Rolland, C., Tronchin, G., Latgé, J.P., and Paris, S. (1994). Rodletless mutants of *Aspergillus fumigatus*. *Infect. Immun.* **62**, 4380–4388.

Tomee, J.F. and Kauffman, H.F. (2000). Putative virulence factors of *Aspergillus fumigatus*. *Clin. Exp. Allergy* **30**, 476–484.

Tsai, H.F., Chang, Y.C., Washburn, R.G., Wheeler, M.H., and Kwon-Chung, K.J. (1998). The developmentally regulated *alb1* gene of *Aspergillus fumigatus*: Its role in modulation of conidial morphology and virulence. *J. Bacteriol.* **180**, 3031–3038.

Tsai, H.F., Wheeler, M.H., Chang, Y.C., and Kwon-Chung, K.J. (1999). A developmentally regulated gene cluster involved in conidial pigment biosynthesis in *Aspergillus fumigatus*. *J. Bacteriol.* **181**, 6469–6477.

van Heeswijck, R. (1984). The formation of protoplasts from *Mucor* species. *Carlberg Res. Commun.* **49**, 597–609.

van Heeswijck, R., Roncero, M.I., and Jepsen, L.P. (1998). Genetic analysis and manipulation of *Mucor* species by DNA-mediated transformation. In H.F. Linskens and J.F. Jackson (eds.) Modern Methods of Plant Analysis (New Series Vol. 7). Springer Verlag, pp. 207–220.

Villalba, F., Lebrun, M.H., Hua-Van, A., Daboussi, M.J., and Grosjean-Cournoyer, M.C. (2001). Transposon impala, a novel tool for gene tagging in the rice blast fungus *Magnaporthe grisea*. *Mol. Plant Microbe Interact.* **14**, 308–315.

Wasylnka, J.A. and Moore, M.M. (2002). Uptake of *Aspergillus fumigatus* conidia by phagocytic and non-phagocytic cells *in vitro*: Quantitation using strains expressing green fluorescent protein. *Infect. Immun.* **70**, 3156–3163.

Watanabe, A., Fujii, I., Tsai, H., Chang, Y.C., Kwon-Chung, K.J., and Ebizuka, Y. (2000). *Aspergillus fumigatus alb1* encodes naphthopyrone synthase when expressed in *Aspergillus oryzae*. *FEMS Microbiol. Lett.* **192**, 39–44.

Weidner, G., d'Enfert, C., Koch, A., Mol, P.C., and Brakhage, A.A. (1998). Development of a homologous transformation system for the human pathogenic fungus *Aspergillus fumigatus* based on the *pyrG* gene encoding orotidine 5′ monophosphate decarboxylase. *Curr. Genet.* **33**, 378–385.

Weidner, G., Steidl, S., and Brakhage, A.A. (2001). The *Aspergillus nidulans* homoaconitase gene *lysF* is negatively regulated by the multimeric CCAAT-binding complex AnCF and positively regulated by GATA sites. *Arch. Microbiol.* **175**, 122–132.

Weng, D.E., Wilson, W.H., Little, R., and Walsh, T.J. (1998). Successful medical management of isolated renal zygomycosis: Case report and review. *Clin. Infect. Dis.* **26**, 601–605.

Willins, D.A., Kessler, M., Walker, S.S., Reyes, G.R., and Cottarel, G. (2002a). Genomics strategies for antifungal drug discovery—from gene discovery to compound screening. *Curr. Pharm. Des.* **8**, 1137–1154.

Willins, D.A., Shimer, G.H., Jr., and Cottarel, G. (2002b). A system for deletion and complementation of *Candida glabrata* genes amenable to high-throughput application. *Gene* **292**, 141–149.

Wimander, K. and Belin, L. (1980). Recognition of allergic alveolitis in the trimming department of a Swedish saw mill. *Eur. J. Respir. Dis. Suppl*(107), 163–167.

Wostemeyer, J., Burmester, A., and Weigel, C. (1987). Neomycin resistance as a dominantly selectable marker for transformation of the zygomycete *Absidia glauca. Curr. Genet.* **12**, 625–627.

Wysong, D.R., Christin, L., Sugar, A.M., Robbins, P.W., and Diamond, R.D. (1998). Cloning and sequencing of a *Candida albicans* catalase gene and effects of disruption of this gene. *Infect. Immun.* **66**, 1953–1961.

Yamazaki, H., Ohnishi, Y., Takeuchi, K., Mori, N., Shiraishi, N., Sakata, Y., Suzuki, H., and Horinouchi, S. (1999). Genetic transformation of a *Rhizomucor pusillus* mutant defective in asparagine-linked glycosylation: Production of a milk-clotting enzyme in a less-glycosylated form. *Appl. Microbiol. Biotechnol.* **52**, 401–409.

Yanai, K., Horiuchi, H., Takagi, M., and Yano, K. (1991). Transformation of *Rhizopus niveus* using a bacterial blasticidin S resistance gene as a dominant selectable marker. *Curr. Genet.* **19**, 221–226.

15

Molecular Interactions of Phytopathogens and Hosts

Joanna M. Jenkinson and Nicholas J. Talbot

1. Introduction

Plant diseases are a significant constraint on agricultural production throughout the world. In developed countries the effects exerted by plant disease impacts on consumers by increasing the price of food, due to the cost of fungicide application, development of resistant cultivars, and refrigerated storage and transport (Talbot and Foster, 2001). For both individual farmers and whole countries, plant diseases can cause massive economic losses. In the United States, a country with access to a full range of pesticides and advanced agricultural practices, it is estimated that $9.1 billion worth of crops is lost to disease with a further $7.7 billion lost to insects and $6.2 billion to weeds (Agrios, 1997). However, in the developing world the costs of plant disease can be even more severe and can cause malnutrition and starvation.

The economic significance of plant diseases, the most important of which are caused by phytopathogenic fungi, has provided the impetus for a significant amount of research, both on host resistance mechanisms and the biology of fungal pathogens. The advent of molecular genetics provided the tools to carry out functional analysis of fungal pathogens for the first time, and in the 15 years since the first DNA-mediated transformation of a fungal pathogen (Oliver *et al.*, 1987; Parsons *et al.*, 1987), there has been rapid progress in identifying genes that condition the ability of a fungus to cause disease in plants. In spite of these advances, however, we are still some way from being able to define the molecular basis of fungal pathogenicity.

Pathogenicity is a difficult concept to define because it is such a complex phenotype that is likely to involve a wide variety of factors at various stages in the development of a fungus. Stages of obvious importance include attachment to the plant surface, spore germination, cuticle penetration, plant infection, tissue colonization, and then finally, and crucially, dissemination of the pathogen to new hosts. Pathogenicity also encompasses avoidance of the host defenses and use of the hosts' metabolism (Schäfer, 1994; Talbot,

---

Joanna M. Jenkinson and Nicholas J. Talbot • School of Biological Sciences, University of Exeter, Washington Singer Laboratories, Perry Road, Exeter, EX4 4QG, UK.

Advances in Fungal Biotechnology for Industry, Agriculture, and Medicine. Edited by Jan S. Tkacz and Lene Lange, Kluwer Academic/Plenum Publishers, 2004.

1995). Therefore we can be sure that diverse signaling pathways, alterations in fungal metabolism, and production of specific effector proteins are involved in pathogenicity. The emerging discipline of functional genomics offers a set of much more powerful tools than conventional molecular genetics and seems poised to lead to a far greater understanding of fungal pathogenicity. The ability to compare the genomes of saprotrophic fungi with pathogenic fungi provides a means of identifying genes that contribute to a pathogenic phenotype. When molecular genetics is carried out in tandem with whole genome analytical procedures such as transcript profiling and proteomics, there exists an unprecedented opportunity to define the molecular components of pathogenesis in a relatively rapid means and then to test their function in plant disease by high throughput mutagenesis.

In this chapter we will focus on reviewing the molecular genetics of two model foliar pathogens *Magnaporthe grisea* and *Ustilago maydis*. This will summarize what is currently known regarding the genetic regulation of plant infection, the clear parallels involved in infection-related signaling in both fungi, and the likely targets for future genomic analysis in these pathogenic species. The chapter will then go on to review further examples of genes and their products that are relevant to being a successful plant pathogen and look forward to the application of genomic approaches to investigating fungal pathogenicity.

1.1. The Life Cycles of *Magnaporthe grisea* and *Ustilago maydis*

The rice blast fungus *M. grisea* can parasitize more than 50 grass species, including economically significant crops such as rice, wheat, barley, and millet. Blast is a disease of great economic importance for most rice-growing regions of the world. *M. grisea* is a filamentous ascomycete, and it is the asexual conidial stage of the fungus that is important for spread of the disease in the field. Conidia are dispersed under conditions of high humidity; when a spore lands on a leaf it adheres tightly, due to release of mucilage from the spore tip, and quickly germinates (Hamer *et al.*, 1988). The germ tube is able to sense the topography and rigidity of a surface. Factors known to induce appressorium formation include hydrophobicity, cutin monomers, other plant wax components, and starvation stress. If the surface is conducive for appressorium formation then within 4 hr a form of cellular differentiation known as hooking occurs. The germ tube apex swells and bends, while flattening against the leaf surface (Bourett and Howard, 1990). After hooking, the tip of the germ tube swells to form the infection cell (appressorium) and a septum is formed. A thick cell wall containing melanin forms around the appressorium and the empty conidium and germ tube collapse. Within the appressorium, huge turgor pressure is generated (estimated to be as high as 8 MPa) (Talbot, 1995; Xu *et al.*, 1998). The source of appressorium turgor pressure is likely to be the accumulation of high concentrations of glycerol in the cell during appressorium maturation; concentrations of glycerol can reach as high as 3.2 M within the appressorium (de Jong *et al.*, 1997). Non-melanized mutants, which lack the glycerol-impermeable layer of the cell wall, are unable to generate cell turgor and are therefore non-pathogenic (Chumley and Valent, 1990). The generation of mechanical force allows a thin penetration peg to be produced at the base of the appressorium, and it is forced through the leaf cuticle and epidermal cell wall. Once within the epidermis the penetration peg widens into a primary infection hypha, which fills the first 3–4 cells encountered (Heath *et al.*, 1990). Runner hyphae then colonize the leaf tissue (Talbot and Foster, 2001). The necrotic leaf spots and associated chlorosis that result from successful infection cause a loss of photosynthetic activity, and therefore fewer resources are directed towards grain production. In older plants infection can spread to the neck of the plant, causing the panicle to fall away from the plant and loss of the entire crop can occur (Ou, 1985).

The corn smut fungus *U. maydis* is a foliar pathogen of maize, and the disease it causes is characterized by gall formation and neoplastic growth on the leaves and stems of the plant. *U. maydis* has an asexual yeast-like, or sporidial form that grows by budding, and a filamentous dikaryotic pathogenic form that invades plant tissue. The switch between these growth forms is fundamental to the ability of *U. maydis* to cause disease and is controlled by mating type recognition between two compatible haploid strains. The dikaryon is formed by cell fusion of conjugation tubes from two sporidia of different mating types. For conjugation and pathogenesis to occur the fungus must be heterozygous at two different multigenic, mating type alleles *a* and *b*. There are only two different allelic forms of *a* (*a1* and *a2*) but the *b* gene is multi-allelic with up to 33 alleles reported from nature (Banuett, 1992). The hyphae of the dikaryon infect the plant via an appressorium-like swollen hyphal tip, and proliferate rapidly in plant tissue, causing production of tumor-like structures. Tumors can form on leaves, stems, tassels, and ears and can grow to a large size. Tumors can become quite purple in color due to anthocyanin production. Karyogamy occurs to form basidia within tumors, and when the tumor bursts, large numbers of black, uninucleate, diploid teliospores are released. These teliospores subsequently undergo meiosis and germinate, each forming four haploid cells known as a promycelium. Basidiospores bud from this, germinate, and produce sporidia (Banuett, 1992; Lengeler *et al.*, 2000).

2. Pathogenicity Factors

2.1. Regulators of Infection

2.1.1. The cAMP Response Pathway

Cyclic AMP is a well-characterized secondary messenger in both prokaryotes and eukaryotes. Signaling via cAMP is known to be important in growth and morphogenesis of fungi but also for regulation of appressorium formation in pathogenic species (for review see Kronstad, 1997). Findings have suggested that increases in the intracellular concentration of cAMP may act as part of early signaling events involved in appressorium formation (Choi and Dean, 1997; Adachi and Hamer, 1998). In *M. grisea* cAMP is synthesized by the *MAC1*-encoded, membrane-associated, adenylate cyclase. The activity of the enzyme results in changes in the levels of cAMP and, therefore, the control of downstream cAMP-dependent proteins such as protein kinase A (PKA). Null *mac1* mutants are affected in vegetative growth, sporulation, mating, germination, and appressorium formation, and are therefore non-pathogenic, indicating that cAMP has a wide variety of roles. The *mac1* defect in appressorium formation can be overcome by exogenous application of cAMP, which confirms that the catalytic activity of Mac1 is required for pathogenesis (Choi and Dean, 1997; Kronstad, 1997; Adachi and Hamer, 1998).

In some strains of *M. grisea*, the Mac1 phenotype is unstable and subject to reversion due to bypass suppressor mutations. One identified suppressor mutation results from a single base change in the first cAMP-binding domain of the regulatory subunit of PKA. A mutant carrying this mutation, *mac1 sum1-99*, showed wild-type growth, morphology, and appressorium formation. However, the suppression of the *mac1* phenotype was not complete because the *sum1-99* strain is not fully pathogenic. Deletion of the *M. grisea CPKA* gene, which encodes a catalytic subunit of PKA, produced nonpathogenic mutants (Mitchell and Dean, 1995; Xu *et al.*, 1997). Although appressoria are produced by Δ*cpkA* mutants, compared to the isogenic wild-type their development is delayed, they can be variable in size, and they are unable to infect plants. *cpkA* mutants can colonize plants if inoculated through a wound, which suggests that the mutants display a turgor or penetration defect (Kronstad, 1997; Adachi and Hamer, 1998; Knogge, 1998; Kang *et al.*, 1999; Balhadère and Talbot, 2000). Taken together, the observations that MAC1 is required for appressorium development and that CPKA is essential for appressorium function, suggest that cAMP signaling is involved in the early stages of morphogenesis of infected cells and is also likely to regulate their function, perhaps acting at the level of metabolic regulation (described later). It is also apparent that not all of the components of these divergent cAMP signaling pathways have yet been identified. In *Saccharomyces cerevisiae*, for example, there are three PKA catalytic subunits, each with a distinct role (Kronstad, 1997; Hamer and Talbot, 1998; Lengeler *et al.*, 2000), and it is likely that the genomic sequence of *M. grisea* will reveal further genetic components of these pathways that can be empirically examined.

The upstream signaling components of the cAMP response pathway leading to appressorium formation may include heterotrimeric G-proteins which would most likely act in concert with a membrane-bound surface receptor (Figure 15.1). G-proteins

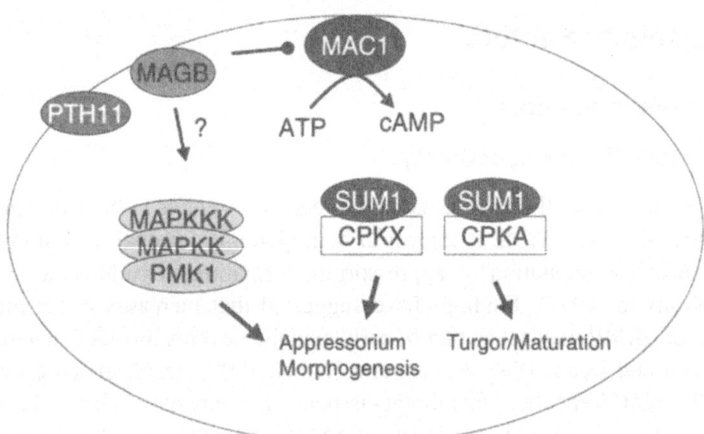

Figure 15.1. Model showing genetic control of appressorium development and maturation in the rice blast fungus *Magnaporthe grisea*. In this model external signals from the plant surface (hydrophobicity, hardness, and starvation) trigger a surface receptor PTH11 to activate a heterotrimeric G-protein, including the α subunit (magB). This, in turn, stimulates (or at least interacts with) a cyclic AMP response pathway involving adenylate cyclase (MAC1) that activates protein kinase A. cAMP binds to the regulatory sub-unit (SUM1), releasing a catalytic subunit that activates either appressorium development (an unknown catalytic subunit denoted CPKX in this model) or appressorium maturation and degradation of storage products for turgor generation (CPKA). In parallel a MAPK pathway is activated involving the MAPK PMK1, that is required for appressorium development and pathogenicity.

(guanine nucleotide-binding proteins) are important in transducing signals from cell surface receptors to intracellular effectors, such as protein kinases, and have roles in cell function, differentiation, pathogenicity, virulence, and mating in fungi (Bölker, 1998). Heterotrimeric G-proteins are composed of α-, β-, and γ-subunits. There are three subgroups of α-subunits, and one gene encoding a protein of each of these groups has been identified and characterized in *M. grisea* so far (Liu and Dean, 1997). *MAGB* encodes an inhibitory group I protein ($G\alpha_i$), and Δ*magB* mutants exhibit mating defects, and reduced vegetative growth, conidiation, and capacity to form appressoria. The reduced pathogenicity of Δ*magB* mutants appears partly due to the reduction in appressorium formation, but *MAGB* may also be important for invasive growth. The mutant phenotype of Δ*magB* mutants may result from repression (or de-repression) of adenylate cyclase activity, although no biochemical evidence has yet been presented regarding its effect in this regard. The appressorium defect of Δ*magB* mutants can, however, be reversed by the addition of cAMP, indicating that the gene product acts above or at the same level as the cAMP signal involved in appressorium formation (Liu and Dean, 1997; Deising *et al.*, 2000; Lengeler *et al.*, 2000; Tucker and Talbot, 2001).

In *U. maydis*, a cAMP response pathway provides one of the controls that regulates the dimorphic switch between haploid cells and dikaryotic filaments, probably acting during the pre-penetration phase of development in response to nutrient availability and/or plant signals. The human fungal pathogens *Candida albicans, Paracoccidioides brasiliensis, Cryptococcus neoformans*, and *Histoplasma capsulatum* also undergo a dimorphic switch that is controlled by a cAMP signaling pathway similar to that in *U. maydis*, and in each case the pathway is important for pathogenicity (Borges-Walmsley and Walmsley, 2000). In *U. maydis,* a G-protein containing the α-subunit encoded by *GPA3* appears to respond to starvation stress and is a positive regulator of adenylate cyclase. Haploid Gpa3 mutants are filamentous and show distorted morphology; these effects can be reversed by the addition of exogenous cAMP, returning the strain to a budding phenotype. Gpa3p, encoded by *UAC1* (equivalent to *MAC1* in *M. grisea*), therefore acts upstream of adenylate cyclase. Targeted *uac1* mutants are constitutively filamentous and exhibit reduced pathogenicity and pheromone expression. The filamentous phenotype of *uac1* mutants is suppressed by the addition of cAMP. One of the three catalytic subunits of PKA in *U. maydis* is encoded by *ADR1* (equivalent to *CPKA* in *M. grisea,* although not yet shown to be a direct functional homologue, experimentally), and the PKA regulatory subunit is encoded by the *UBC1* gene (equivalent to *M. grisea SUM1*). Mutations in these genes result in different phenotypes; *ubc1* mutants have constitutive PKA activity which mimics high levels of cAMP, reducing filamentation and eliminating tumor formation (Mayorga and Gold, 1998). In contrast *adr1* mutants are filamentous, in the same way as *uac1* mutants. A diagram showing the cAMP response pathway in *U. maydis* and the likely cross-talk with the MAP kinase pheromone response pathway is given in Figure 15.2.

2.1.2. *PMK1* and MAP Kinase Pathways in Fungal Pathogens

Protein phosphorylation is important for intracellular signal transduction in eukaryotic and prokaryotic cells and is catalyzed by protein kinases. These are broadly classified into three major groups based on their substrate specificity: serine/threonine kinases, tyrosine kinases, and histidine kinases (Urao *et al.*, 2000). Mitogen-activated protein kinases

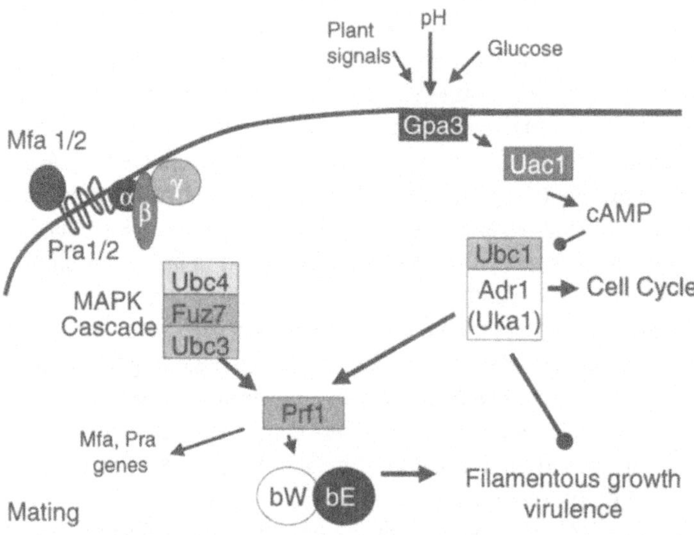

Figure 15.2. Model of pathogenic development by the corn smut fungus *Ustilago maydis*. A pheromone response pathway is triggered by binding of a lipopeptide pheromone to the receptor Pra1/2. This triggers a MAPK cascade involving Ubc4, Fuz7, and Ubc3. One of the targets of this pathway is the transcription factor Prf1, which is an HMG-class DNA-binding protein. The pathway stimulates filamentation. A cyclic AMP pathway is also triggered by external environmental signals and acts to repress hyphal growth. Cyclic AMP inhibits the catalytic subunit (Ubc1) releasing the catalytic subunits (Adr1 and Uka1). These repress hyphal growth (lines with round ends). There is also cross-talk in the pathways at the level of Prf1 affecting mating function.

(MAPKs) are serine/threonine kinases, which are particularly important components of signal transduction in a variety of developmental processes in all eukaryotes (for an introductory review see Herskowitz, 1995). The MAPK cascade is responsible for transduction of a signal from the cell surface to the nucleus, and the target of this cascade is normally one or more transcription factors, resulting in the increased transcription of a discrete set of target genes (Banuett, 1998; Lev *et al.*, 1999).

MAPK cascades have been well characterized in *S. cerevisiae*, which has six MAPKs, each involved in regulating specific developmental events or responses, with some cross-talk/redundancy occurring between pathways (Lev *et al.*, 1999). MAPKs are activated by MAPK/ERK kinases (MEK), which are activated in turn by MEK kinases. These have been shown to be activated by phosphorylation through upstream kinases, or raf or mos proteins (Hirt, 1997; Kim *et al.*, 2000).

The *PMK1* gene (Pathogenicity MAP-Kinase 1) from *M. grisea* is very similar to the yeast MAP kinase genes *FUS3* and *KSS1*, involved in yeast pheromone signaling and pseudohyphal growth regulation. *PMK1* can functionally substitute for both of these genes when expressed in *S. cerevisiae*. *M. grisea pmk1* mutants are unable to form appressoria and therefore cannot penetrate the cuticles of rice or barley plants. Staining cell walls and nuclei during germination showed that *pmk1* mutants germinated normally and seemed to initiate appressorium formation. However, they failed to form a melanized cell wall and discrete appressorium and instead underwent an unusual pattern of nuclear division and hyphal elongation (Xu and Hamer, 1996; Hamer and Talbot, 1998). Injecting *pmk1* mutants into rice leaves resulted in a small area of necrosis similar to a hypersensitive response, but no spreading lesion was formed, and the fungus could not be recovered from infected plants. These, and other experiments, have indicated that *PMK1* is necessary for invasive growth and fungal viability *in planta* (Xu and Hamer, 1996). The mating ability, vegetative growth rate, and conidiation of the *pmk1* strain was not affected however, indicating that this gene is only involved in specific virulence-associated processes and is important in transducing external signals to bring about infection-related development (Xu and Hamer, 1996; Deising *et al.*, 2000).

The morphogenetic effects of the *pmk1, cpkA*, and *mac1 sum1-99* mutations on appressorium development and maturation in *M. grisea* have led to a series of biochemical experiments to determine the roles of the proteins encoded by each of these genes in regulating metabolic pathways associated with appressorium function (Tucker and Talbot, 2001). Appressorium turgor in *M. grisea* appears to result from generation of large concentrations of glycerol which act to generate osmotic pressure. To study the likely origins of glycerol, the mobilization patterns of triacylglycerol and glycogen have been examined cytologically and biochemically in *M. grisea* during appressorium development. This showed that glycogen localizes to the appressorium during its formation but rapidly disappears at the onset of turgor generation. In the same way, lipid droplets localize to the hyphal apex during germ tube extension and hook formation, before coalescing and being taken up by vacuoles in developing appressoria. Rapid lipolysis then occurs and is associated with high triacylgycerol lipase activity (Thines *et al.*, 2000). In *pmk1* mutants glycogen and lipid mobilization to the hyphal apex and developing appressorium did not occur. In a *cpkA* mutant lipid and glycogen breakdown were both retarded, and conversely in a *mac1 sum1-99* mutant (with high non-cAMP dependent PKA activity), degradation was very rapid. These results indicate that the *PMK1* MAPK pathway is required for the mobilization of storage compounds to the appressorium and that their breakdown is reliant on the cAMP-dependent PKA encoded by *CPKA* (Thines *et al.*, 2000; Weber *et al.*, 2001).

The pheromone response MAPK pathway in *U. maydis* has been relatively well-characterized and appears to positively regulate filamentous growth in response to pheromone signaling. When haploid sporidia of opposite mating types meet, initial cell fusion is dependent on secretion and perception of lipopeptide pheromones encoded by the *Mfa1* and *Mfa2* genes (located within the *a1* and *a2* loci respectively) and perceived by the corresponding receptor proteins encoded by *Pra1* and *Pra2* (organized in the same loci). Pheromone perception triggers a signaling pathway that leads ultimately to activation of filamentous growth and *b* gene expression. Gpa3p is involved in the MAPK pathway as well as the cAMP pathway, and deletion of the *Gpa3* gene results in a mating defect. Pheromone binding triggers a MAP kinase pathway (Borges-Walmsley and Walmsley, 2000; Lengeler *et al.*, 2000); this involves a MEKK, encoded by *ubc4/kpp4*, a MEK encoded by *Fuz7* (*ubc5*) (Banuett and Herskowitz, 1994), and a MAPK encoded by *kpp2/ubc3* (Mayorga and Gold, 1999; Muller *et al.*, 1999). Deletion of the MEK gene *Fuz7* results in mutants that cannot respond to pheromone and fail to produce tumors or teliospores. This suggests that *FUZ7* is involved in the *a*-dependent pheromone response pathway, but also in regulating post-penetration functions in fungal pathogenesis (Banuett and Herskowitz, 1994). *Kpp2* (*Ubc3*) is a MAPK that is closely related to the *M. grisea PMK1* gene. Mutants lacking *Kpp2* are still able to cause disease but are deficient in pheromone production, conjugation tube production, and mating and also show defects in filamentation, tumor formation, and virulence. This phenotype is less dramatic than *fuz7* mutants, which suggests that there may be another partially redundant MAPK homologue that compensates for the *kpp2* mutation (Lengeler *et al.*, 2000; Tucker and Talbot, 2001).

In *U. maydis* the downstream targets of the MAPK pathway include the *Prf1*-encoded transcription factor, which belongs to the HMG-group of DNA-binding proteins. Prf1 is involved in activation of genes that are pheromone-responsive, and it is significant that Prf1 is activated post-transcriptionally by both the MAPK and cAMP response pathways, showing the interplay that exists between both signaling networks.

2.1.3. *PMK1*-Related MAP Kinases in Other Phytopathogenic Fungi

There are now nine known homologues of *PMK1* including *BMP1* from *Botrytis cinerea, CMK1* from *Colletotrichum lagenarium*, and *PTK1* from *Pyrenophora teres*. All of these genes could be proposed to have roles analogous to that of *PMK1* in regulating pathogenesis in *M. grisea* since their mutant phenotypes are very similar (Tucker and Talbot, 2001). However, despite the high homology between *PMK1* homologues, we know that the role of the gene products can be subtly different. For example, despite having 90% amino acid homology to Pmk1 and 59% to Fus3, *CHK1* from the ascomycete corn blight pathogen *Cochliobolus heterostrophus* shows important differences to *M. grisea* (Lev *et al.*, 1999). The *chk1* mutant phenotype is similar to *pmk1* mutants in that the *CHK1* gene product is required for leaf colonization, and mutants are unable to form conidia and appressoria and are therefore non-pathogenic. However, in *C. heterostrophus* appressoria are not essential for the penetration process which can occur passively via fungal invasion through stomata. *CHK1* differs from *PMK1* in other functions, including the apparent regulation of cell wall biosynthesis, and as such is more similar in function to *MPS1*, another *M. grisea* MAPK (discussed later). *CHK1* is important in sexual and asexual sporulation, cell wall integrity, and aerial hypha formation, suggesting it is also important in morphological transitions (Lev *et al.*, 1999; Tucker and Talbot, 2001).

A MEK gene identified from the avocado pathogen *Colletotrichum gloeosporioides, CgMEK1*, is involved in at least two processes, polarized cell division and appressorium differentiation. Disruption of *CgMEK1* resulted in loss of polarity preventing spore germination and hyphal development. This phenotype could be overcome by application of nutrients or the host signaling molecule ethylene. However, the mutant's inability to form appressoria and loss of virulence could not be suppressed by these exogenous compounds. Inoculation of the mutant strain onto wounded host tissue proved that the strain was still unable to grow which suggests that this MEK is also important for invasive growth (Kim *et al.*, 2000). The *PMK1*-related MAPK genes and related kinases identified in phytopathogenic fungi indicate that a common signaling pathway related to the yeast pheromone signaling pathway and pseudo-hyphal regulatory pathway may well be a common feature of phytopathogens regulating the ability to complete infection-related development and to colonize plant tissue (see Xu, 2000).

2.1.4. Alternative MAPK Pathways in *M. grisea*

In *M. grisea*, Mps 1 is a mitogen-activated protein kinase that is essential for production of a penetration peg. *mps1-1Δ* mutants are dramatically reduced in conidiation and completely non-pathogenic, and yet there is only minimal reduction in appressorium turgor. The *mps1-1*Δ pathogenicity defect is due to the lack of penetration peg production, and consistent with this, mutant strains are still able to grow invasively within the plant if inoculated through wounds (Hamer and Talbot, 1998; Xu *et al.*, 1998; Tucker and Talbot, 2001). Mps1 has high homology to Slt 2 from *S. cerevisiae* and is able to complement *slt2* null mutants. Slt 2 is a component of the protein kinase C pathway that controls cell wall growth in response to membrane stress (for review see Xu, 2000). Therefore *slt2* mutants show defective cell wall integrity and are unable to grow under stresses such as high temperature. The phenotype of *mps1-1*Δ mutants in *M. grisea* suggests that Mps1 is essential for maintaining cell wall integrity and that cuticle penetration may require remodeling of the appressorium wall through an Mps1-dependent signaling pathway (Xu *et al.*, 1998). It is likely that penetration peg production causes severe membrane stress and rapid cell wall biogenesis, and the mutant phenotype suggests that the MPS1 pathway may be involved in a number of morphogenetic processes in the fungus because sporulation, aerial hyphal growth, and female fertility were affected by the mutation (Xu *et al.*, 1998; Tucker and Talbot, 2001).

Another MAPK cascade in *M. grisea* is required for the maintenance of cellular turgor: *OSM1* is a component of this pathway and is related to the *S. cerevisiae HOG1* gene (Dixon *et al.*, 1999). In *M. grisea* arabitol is the major compatible solute accumulated under hyperosmotic stress. Solutes help to prevent water loss and maintain cellular turgor. An *osm1* mutant was sensitive to osmotic stress and showed a lower accumulation of arabitol compared to the wild type. However, importantly the mutant showed no reduction in appressorium turgor, which suggests that there is another independent pathway involved in turgor generation within the appressorium (Dixon *et al.*, 1999).

To summarize our knowledge regarding pathogenesis-related signaling in phytopathogenic fungi, it is already apparent that at least two conserved pathways operate in pathogenicity-related development and virulence. Although these pathways can have different roles in different organisms, depending on the type of pathogen and the disease symptoms it causes, there are likely to be common genetic components in the pathways. A conserved MAP kinase pathway clearly exists in a number of pathogens and is related either to morphogenetic transitions, or mating pheromone release, or to both. The cyclic cAMP pathway that operates in both *M. grisea* and *U. maydis* is somewhat related since in *U. maydis* low cAMP brings about the dimorphic switch from yeast-like growth to filamentous growth (in contrast to *S. cerevisiae* and *C. albicans* where filamentation involves high cAMP levels). In *M. grisea* it seems that a rise in cAMP is associated with a departure from filamentous growth and production of appressoria (a yeast-like growth stage?), perhaps showing the commonality between morphogenetic regulation. In both cases nutritional cues are likely to be important as the fungus is operating in the glucose-deficient environment of the plant surface.

2.1.5. Nutritional Regulatory Genes

In *S. cerevisiae SNF1* is known to be important for the expression of glucose-repressed genes, which are not expressed (de-repressed) in a *snf1* mutant strain even in the absence of glucose. The growth of *ccsnf1* mutants in the corn pathogen *Cochliobolus carbonum* is reduced on complex and simple sugars, and the activity of several cell wall-degrading enzymes are also reduced. As a result *ccsnf1* mutants are much less virulent on plants (Tonukari *et al.*, 2000). This suggests that the regulation of cell wall-degrading enzymes is important for the pathogenicity of *C. carbonum*. The *C. carbonum ccSNF1* gene is involved in the regulation of the extracellular enzymes, for example, pectinases and xylanases that are important in the degradation of the plant cell wall. These enzymes allow fungal penetration and their expression is inhibited by simple sugars (Deising *et al.*, 2000; Tonukari *et al.*, 2000; Tucker and Talbot, 2001). It will be intriguing to investigate the interplay between the glucose repression pathway and cAMP signaling in phytopathogens given the clear linkages between these processes in yeast.

2.2. Pathogen-Specific Molecules

2.2.1. Toxins and Host-Specific Toxins

Toxins are important in assisting fungal penetration and aiding colonization of the host plant, and may be responsible for disease symptoms (Schäfer, 1994; Punja, 2001; Panaccione *et al.*, 2002). Nonspecific toxins are toxic to a number of plants, whereas toxins produced by the fungus that are active only against susceptible hosts are termed host-specific toxins. The latter are intrinsically more interesting to a plant pathologist since they determine the host range of a fungus. The southern corn leaf blight pathogen *C. heterostrophus* can be classified into two pathotypes on the basis of production of the virulence factor T-toxin, partially reduced linear polyketols encoded by *Tox1* gene locus (see Chapter 6). Race T pathotypes are highly pathogenic on maize with Texas-type male sterile (tms) cytoplasm; race O are weakly pathogenic. Plant sensitivity to T-toxin is dependent on a small binding protein on the inner mitochondrial membrane called URF13, encoded by T-*urf13*, the presence of this novel gene is responsible for Texas-type cytoplasmic male sterility which is an agronomically desirable trait (Schäfer, 1994; Panaccione *et al.*, 2002).

The maize leaf spot pathogen *C. carbonum* race 1 produces a cyclic tetrapeptide host-selective HC-toxin, encoded by the *Tox2* locus, and causes large lesions on susceptible hosts. *Tox2* is a complex of multiple biosynthetic and regulatory genes, including the essential four-module nonribosomal peptide synthetase *HTS1* (see Chapter 7). The amount of toxin produced is determined by several genes. The toxin inhibits the host histone deacetylase and thereby prevents the activation of genes involved in defense responses. Therefore, this toxin is cytostatic rather than cytotoxic. Maize resistance to the pathogen is determined by the *HM1* allele, encoding a carbonyl HC-toxin reductase that is able to degrade the toxin (Schäfer, 1994; Deising *et al.*, 2000; Panaccione *et al.*, 2002).

There are many examples of toxins that have completely different effects on host plants. These include the cytotoxic Ptr ToxA produced by the causal agent of tan spot of wheat and other grasses, *Pyrenophora tritici-repentis*. This is another host-selective toxin, but Ptr ToxA acts by causing electrolyte leakage that results in host programmed cell death and necrosis. Host-specific PC toxins are produced by some isolates of *Periconia circinata* that cause milo disease of sorghum. PC toxins are pathogenicity determinants, as only toxin-producing strains are pathogenic. Plant resistance is conferred by a homozygous recessive condition at the *pc* locus, since this locus encodes either the toxin receptor or target (Panaccione *et al.*, 2002).

2.3. Plant Recognition Evasion

2.3.1. Saponin Detoxification

To colonize plant tissue effectively some phytopathogens have evolved the ability to evade recognition and to degrade pre-formed plant defense compounds or induced defense molecules, such as phytoalexins. Saponins are pre-formed plant antimicrobial compounds that cause pore formation in the membranes of fungal pathogens, resulting in loss of membrane integrity. Saponins normally consist of triterpenoid, steroid, or steroidal glycoalkaloid molecules bearing one or more sugar chains (Osbourn, 1996). Two monodesmosidic saponins that have been well characterized are avenacin A-1 from oats and α-tomatine from tomato. Bidesmosidic saponins have to be converted to a biologically active form by plant enzymes; for example, in oats the glycosyl hydrolase avenacosidase activates avenacosides A and B. Wounding of the plant triggers the de-compartmentalization of the enzymes and substrates, creating the active saponin which is then incorporated into the pathogen membrane (Osbourn, 1996).

Avenacin A-1 is found in the epidermis of oat roots and is important in determining resistance to the root pathogen *Gaeumannomyces graminis* that causes "take-all" in cereals and grasses. *G. graminis* var. *tritici* infects wheat, whereas *G. graminis* var. *avenae* is able to infect oats and wheat as it produces avenacinase, an extracellular β-glucosyl hydrolase that converts avenacin A-1 to less toxic products. Avenacinase is encoded by *AVN1*, and *avn1* mutants were unable to infect oats but were still pathogenic on wheat (reviewed by Osbourn, 1996). Tomatinases, which act on α-tomatine, have been isolated from the tomato pathogens *Septoria lycopersici, Verticillium albo-atrum, Botrytis cinerea, Fusarium oxysporum* f sp. *lycopersici*, and *Alternaria solani*. These pathogens have a variety of detoxification mechanisms that result from the production of extracellular enzymes (Osbourn, 1996; Bouarab *et al.*, 2002). *S. lycopersici* tomatinase mutants are able to infect tomato plants but are non-pathogenic in *Nicotiana benthamiana*. A recent study indicated that this was due to enhancement (de-repression) of plant defense responses when infections by the mutant and wild type were compared. This enhancement prevented the proliferation of the tomatinase mutant strain. This suggests that tomatinase and its resultant processed saponin product are important in suppression of plant defense responses (Bouarab *et al.*, 2002).

Detoxification is not the only strategy to avoid the effects of saponins. The biotrophic pathogen of tomato *Cladosporium fulvum* grows only in the intracellular spaces between cells where it does not come into contact with α-tomatine. *Pythium, Phytophthora*, and some isolates of *G. graminis* are also intrinsically resistant due to the lack of sterols in their membranes. However, saponins and their detoxification or avoidance are clearly important factors in determining the host range of a pathogen (Osbourn, 1996).

2.3.2. Phytoalexin Detoxification

Many plants produce low molecular weight antimicrobial compounds known as phytoalexins in response to infection. Successful fungal colonization of these plants requires either evasion or detoxification of these secondary metabolites. However, the ability to metabolize or tolerate a phytoalexin does not necessarily confer the ability to cause disease, as there does not appear to be a strict correlation between pathogenicity and phytoalexin metabolism. The rate of the detoxification reaction, the toxicity of the products produced from the reaction, and the rate of phytoalexin synthesis by the plant are also key factors (van Etten *et al.*, 1989; Delserone *et al.*, 1999; Punja, 2001).

The isoflavonoid pisatin in pea is a well-characterized phytoalexin. Most detoxification mechanisms involve 3-*O*-demethylation as a first step, although other mechanisms do exist and a pathogen may use more than one mechanism. Pea pathogens that are able to metabolize pisatin usually contain *PDA* genes, which encode pisatin demethylase enzymes. Nine *PDA* genes have been identified to date, and five have been shown to be important for maximum virulence on pea. Virulence on chickpea, tomato, and carrot is unrelated to pea pathogenicity. The genes have been shown to confer different phenotypes in terms of levels of activity and speed of induction. The pisatin demethylase from the ascomycete *Nectria haematococca* (anamorph: *Fusarium solani*), that all pathogenic strains produce, is a microsomal cytochrome P450 monooxygenase. In mammals as well, hepatic cytochrome P450s are important in detoxification. However, it has been proved that the *PDA* genes are not essential for the pathogenicity of *N. haematococca* (van Etten *et al.*, 1989; Schäfer, 1994; Delserone *et al.*, 1999; Funnell and van Etten, 2002). Delserone *et al.* (1999) have shown by sequencing that the P450s encoded by *PDA* genes are divergent and form a separate family of cytochrome P450s. Using this knowledge and Southern hybridization analysis it has been shown that the system of pisatin detoxification in other fungi differs from that of the most studied pea pathogen *N. haematococca*.

Medicarpin and maackiain are isoflavonoid phytoalexins found in leguminous plants, and their metabolism also involves a variety of reactions. *Ascochyta rabiei* performs four different initial reactions for detoxification: reductive cleavage of ring C, hydroxylation of ring A at position 1a and another unidentified position, and 9-*O*-demethylation (van Etten *et al.*, 1989). An isolate of *N. haematococca* performs three alternative reactions thought to be catalyzed by oxygenase, that are different from these four in *A. rabiei*. Three genes controlling maackiain degradation are known in *N. haematococca* *Makl, Mak2*, and *Mak3*, an isolate must contain at least one of these genes to be highly virulent (van Etten *et al.*, 1989). Consistent with this, over-expression of medicarpin genes resulted in a delayed development of disease in a number of plants (Punja, 2001).

The metabolic pathways involved in phytoalexin detoxification and the number of phytoalexins in a given plant can vary widely, but are clearly significant in a number of phytopathogenic species. However, pathogens may also circumvent the phytoalexin by avoiding its location or avoiding causing or actually repressing the induction of phytoalexins. Phytoalexins are clearly important defense compounds for a plant since their metabolism or tolerance can be prerequisites for pathogenicity and host range in a number of phytopathogenic fungi (van Etten *et al.*, 1989).

2.4. Proteins of Unknown Function

Restriction enzyme-mediated DNA integration (REMI) and related insertional mutagenesis strategies have been important tools for the identification of novel mutants (Sweigard *et al.*, 1998; Balhadère *et al.*, 1999). Of particular note in *M. grisea* are two genes recently identified by REMI mutant hunts, *PDE1*and *PTH11*.

PDE1 encodes a P-type ATPase, that is part of the DRS2 aminophospholipid translocase subfamily, and a functional homologue of the yeast *ATC8* gene. The ATPases are required for the maintenance of phospholipid asymmetry in membranes, which is important in cellular morphogenesis. The *pde1* mutant was identified from a REMI screen. The random insertion was located in the promoter region of a *M. grisea* 35-R-24 strain, a barley pathogen, and the resulting *pde1* strain was non-pathogenic on barley. Penetration is dramatically reduced because *pde1* mutants do not produce functional penetration pegs. Since the growth was non-polarized and did not extend beyond the first cell of the epidermis, the strain was non-pathogenic (Balhadère and Talbot, 2001). A *PDE1* gene replacement was made in a rice pathogen strain Guy-11, and the resulting *pde1* mutant was non-pathogenic on rice but pathogenic on barley. Therefore, the mutant phenotype is different in the two backgrounds, suggesting an important differential regulation of the genes in these two strains. In the *M. grisea* wild type a penetration peg emerges from a specialized area of the appressorium called the pore, where the plasma membrane directly contacts the plant cell wall (Bourett and Howard, 1990). The pore distends and is overlaid by a cell wall. There are several proposals regarding the role of *PDE1*, firstly that it might be required for the re-orientation of growth and formation of the peg. It may also be important in the extreme membrane stress that must be encountered during the formation of the peg. Alternatively, it may be important in membrane recycling in the same way as *DRS2* from yeast (Balhadère and Talbot, 2001).

The *PTH11* gene putatively encodes a membrane-bound sensor or receptor involved in appressorium development. The encoded transmembrane protein is located in vacuoles and the plasma membrane. The *pth11* mutant is non-pathogenic and reduced in appressorium differentiation: only 10%–15% of conidia form functional appressoria on inductive surfaces. This phenotype is similar to the phenotype of a wild type strain on a non-inductive surface such as glass, which suggests *PTH11* has a role in surface perception. Again, the role of that gene is different in different genetic strains of *M. grisea*. The gene product activates appressorium differentiation in two strains, represses it in another strain, and in Guy-11, activates morphogenesis on inductive surfaces and represses it on non-inductive surfaces. The *pth11* phenotype can be reversed by the addition of cAMP, which indicates that it acts upstream or convergent with a cAMP pathway. This suggests *PTH11* may act upstream of signaling pathways involved in appressorium differentiation, and since it performs different roles in closely related strains of *M. grisea*, this highlights the probable complexity of signaling pathways involved in appressorium differentiation (de Zwaan *et al.*, 1999; Lengeler *et al.*, 2000; Talbot and Foster, 2001; Tucker and Talbot, 2001).

2.5. Pathogen Associated Molecular Patterns

2.5.1. Plant Resistance Mechanisms

Plants need to defend themselves from a wide range of pathogens. They normally do this successfully, and in the majority of cases, the pathogen is recognized and controlled via a multi-component defense system (Richberg *et al.*, 1998; Martin, 1999). Pathogen recognition results in the activation of a variety of defense responses within the plant. These include a rapid localized form of programmed cell death known as the hypersensitive response, synthesis of a variety of pathogenesis-related (PR) proteins both locally and systemically, and an increased level of whole plant resistance. The hypersensitive response (HR) is correlated with disease resistance and is important in confining pathogen growth. HR is characterized by the rapid collapse of the challenged cell and the activation of defenses in the challenged and surrounding cells. An enhanced whole plant resistance which temporarily protects a plant against subsequent pathogen attack occurs within hours after infection. This is known as systemic acquired resistance (SAR), and (in most documented cases) requires the signal molecule salicylic acid (SA). It is now clear that other pathways also lead to broad-spectrum resistance; jasmonic acid (JA) and ethylene are important alternative signal molecules. Other more specific defense responses include a rapid oxidative burst, cell wall rigidification, and the synthesis of phytoalexins (Penninckx *et al.*, 1996; Cervone *et al.*, 1997; Shah *et al.*, 1997; Shirasu *et al.*, 1997; Dong, 1998; Innes, 1998; Maleck and Dietrich, 1999; Murphy *et al.*, 1999; Pieterse and van Loon, 1999). The generation of reactive oxygen species (ROS) is seen in many incompatible responses (Dinesh-Kumar *et al.*, 1995). ROS include hydrogen peroxide (H_2O_2), oxygen radicals, and hydroxyl radicals (Piffanelli *et al.*, 1999). ROS have an antimicrobial effect, and cross-link *O*-glycosylated glycoproteins in the plant cell wall, particularly those containing hydroxyproline, as well as other proteins, phenolics, and callose. This allows papillae formation. Papillae literally immobilize fungal hyphae that are in the process of penetrating the plant cell wall by sealing off the infection site (Bolwell, 1999). In humans ROS is also a critical part of the protective response to invasion by opportunistic fungi such as *Aspergillus fumigatus*. Hyphae germinating from conidia of this organism are normally killed by neutrophils through an oxidative mechanism that utilizes an NADPH oxidase (*phox*) to generate superoxide, the precursor of ROS species. Patients with a hereditary condition known as chronic granulomatous disease or mice with a targeted disruption of the $gp91^{phox}$ gene are deficient in the activity of this oxidase and are unusually susceptible to infection (Morgenstern *et al.*, 1997).

2.5.2. *R* Gene and *Avr* Gene Signaling

The more specific form of plant resistance is at the level of specific cultivars and particular physiological races of a pathogen. This form of resistance, known as gene-for-gene resistance occurs when plants carry dominant single resistance genes that encode products that interact directly or indirectly with pathogen-encoded molecules. This leads to a recognition event that triggers the deployment of plant resistance mechanisms described above (Somssich and Hahlbrock, 1998). Many crop plants have been bred to contain particular resistance genes to resist the most prevalent pathogens, although this form of resistance is often short-lived in the field (Punja, 2001). The pathogen-encoded molecules that are recognized by host plants are known as avirulence gene products because the genes that encode them will prevent disease from being established on hosts carrying corresponding resistance genes. Avirulence gene products appear to be variable in fungi and phytopathogenic bacteria. They probably fulfill roles in pathogenesis and are often secreted directly into plant cell by phytopathogenic bacteria. Plants have thus evolved a means of recognizing these pathogenicity factors and therefore perceiving pathogens at an early stage of the infection process.

Several avirulence gene products are known in phytopathogenic fungi including *Avr4, Avr9,* and *Ecp2* from the tomato leaf mold *Cladosporium fulvum, nip1* (*AvrRrs1*) from barley leaf scald *Rhynchosporium secalis*, and *Avr-Pita* from *M. grisea*. All of these genes encode extracellular proteins. *Ecp2* and *nip1* are known to encode important pathogenicity factors (Laugè and de Wit, 1998). The avirulence gene *AVR-Pita* (formerly *Avr2-YAMO*) from *M. grisea* encodes a protein homologue of a zinc-dependent metalloprotease.

AVR-Pita prevents the infection of rice carrying the complementary *Pita* gene (Talbot and Foster, 2001; Knogge, 2002). The products of *Pita* and *AVR-Pita* are thought to interact directly, and this suggests that the product of *AVR-Pita* is secreted into rice cells by *M. grisea*. The metalloprotease may be important, for example, in releasing nutrients from the rice plant or in degrading a particular plant product (Knogge, 2002). In *M. grisea* the species-specific *PWL* gene family determines host range on weeping lovegrass and effectively acts in the same way as cultivar-specific avirulence genes; both *PWL1* and *PWL2* prevent *M. grisea* isolates from infecting weeping lovegrass. *PWL3* is non-functional, and *PWL4* does not confer avirulence on weeping lovegrass, probably due to a defective promoter that prevents its expression (Laugè and de Wit, 1998; Knogge, 2002). The current challenge in studying avirulence gene products from fungal pathogens is to determine precisely which function they perform during normal compatible interactions, to define the targets of their activity in plant cells, and then to determine the molecular basis of their recognition by host resistance genes and the manner in which this transduces a signal to confer resistance. Such demanding challenges will be made somewhat more achievable by the recent development and application of functional genomics.

3. Genomics of Phytopathogens

The first draft sequence of a phytopathogenic fungus, *M. grisea*, has recently been generated and offers an unparalleled opportunity to investigate fungal pathogenicity (*http://www-genome.wi.mit.edu/annotation/fungi/magnaporthe/*). The fungal genome initiative at the Whitehead Institute is set to sequence 15 fungal genomes in the next years, and this will provide a wealth of information concerning fungal evolution and the relationships between pathogenic, mutualistic, and saprophytic species (for reviews see Tunlid and Talbot, 2002, and Chapter 2). Along with this information, however, will also come the technologies necessary to investigate the global regulation of gene expression in pathogens and the production of proteins at the level of the whole proteome. This form of analysis has already illuminated previously well-studied processes in yeast such as sporulation, pheromone responsiveness, and cell cycle regulation (for review see Brown and Botstein, 1999) and is likely to prove even more significant in less well-studied systems such as pathogenic fungi. Two key developments, however, are necessary in order to capitalize fully on genomic information. The first is a means of storing, interrogating, and accessing information easily and in the same format for several fungi. Such bioinformatic resources are in short supply, and it is imperative that such resources are developed in the forthcoming years. In phytopathogenic fungi the consortium for functional genomics of microbial eukaryotes has developed a relational database of EST information from seven phytopathogenic fungi and two oomycete species, which enables researchers to identify easily genes based on test searches or functional assignments (Soanes *et al.*, 2002).

The second necessary development is a means of rapidly generating targeted mutants. In the private sector this has been addressed by a group of researchers investigating *M. grisea* using an *in vitro* transposon mutagenesis approach to rapidly tag genes of interest utilizing the resulting clones as a template for genome sequence analysis and as vectors for gene disruption. This procedure, called TAG-KO, is useful in an industrial setting in providing a rapid scalable technology for functional gene analysis (Hamer *et al.*, 2001).

In academia, a similar approach may soon be necessary in order to utilize the wealth of genetic information in a timely and logical manner. Dissection of complex signaling pathways and metabolic processes will always require a multidisciplinary approach incorporating biochemical and cytological procedures, but if rapid mutagenesis was possible it would alleviate much of the labor-intensive aspects of this work allowing greater time for phenotypic analysis and the study of gene and protein interactions.

4. Future Prospects

In summary, it is likely that much will be learned in the next few years regarding the molecular basis of fungal pathogenicity and the genetic requirements necessary to be a pathogen. This holistic approach to pathogenesis will begin to replace the gene-by-gene analysis that has dominated the last 10 years. From such studies it is hoped that durable control strategies will also emerge, either through development of better chemical control methods or by deployment of genetically mediated resistance that is broad spectrum and long-lived. Only if such outcomes emerge will the efforts of so many laboratories be deemed by the public to have been successful, and our investment in genomics to have been worthwhile.

References

Adachi, K. and Hamer, J.E. (1998). Divergent cAMP signaling pathways regulate growth and pathogenesis in the rice blast fungus *Magnaporthe grisea. Plant Cell* **10**, 1361–1373.

Agrios, G.N. (1997). *Plant Pathology*, 4th edn. Academic Press, San Diego.

Balhadère, P.V., Foster, A.J., and Talbot, N.J. (1999). Identification of pathogenicity mutants of the rice blast fungus *Magnaporthe grisea* by insertional mutagenesis. *Mol. Plant–Microbe Interactions* **12**, 129–142.

Balhadère, P.V. and Talbot, N.J. (2000). Fungal pathogenicity—establishing infection. In M. Dickinson (ed.), *Molecular Plant Pathology, Annual Plant Reviews, Volume 4*, Sheffield Academic Press, Sheffield, pp. 1–25.

Balhadère, P.V. and Talbot, N.J. (2001). *PDE1* encodes a P-type ATPase involved in appressorium-mediated plant infection by the rice blast fungus *Magnaporthe grisea. The Plant Cell* **13**, 1987–2004.

Banuett, F. (1992). *Ustilago maydis*, the delightful blight. *Trends Genet.* **8**, 174–180.

Banuett, F. and Herskowitz, I. (1994). Identification of *Fuz7*, a *Ustilago maydis* MEK/MAPKK homologue required for a locus dependent and a locus independent steps in the fungal life-cycle. *Genes Dev.* **8**, 1367–1378.

Banuett, F. (1998). Signaling in the yeasts: An informational cascade with links to the filamentous fungi. *Microbiol. Mol. Biol. Rev.* **62**, 249–274.

Bölker, M. (1998). Sex and crime: Heterotrimeric G proteins in fungal mating and pathogenesis. *Fungal Genet. Biol.* **25**, 143–156.

Bolwell, G.P. (1999). Role of active oxygen species and NO in plant defence responses. *Curr. Opin. Plant Biol.* **2**, 287–294.

Borges-Walmsley, M.I. and Walmsley, A.R. (2000). cAMP signaling in pathogenic fungi: Control of dimorphic switching and pathogenicity. *Trends Microbiol.* **8**, 133–141.

Bouarab, K., Melton, R., Peart, J., Baulcombe, D., and Osbourn, A. (2002). A saponin-detoxifying enzyme mediates suppression of plant defences. *Nature* **418**, 889–892.

Bourett, T.M. and Howard, R.J. (1990). *In vitro* development of penetration structures in the rice blast fungus, *Magnaporthe grisea. Can. J. Bot.* **68**, 329–342.

Brown, P.O. and Botstein, D. (1999). Exploring the new world of the genome with DNA microarrays. *Nat. Genet.* **21**, 33–37.

Cervone, F., Castoria, R., Leckie, F., and de Lorenzo, G. (1997). Perception of fungal elicitors and signal transduction. In P. Aducci (ed.) *Signal Transduction in Plants, Molecular and Cell Biology Updates* Birkhauser Verlag, Basel, pp. 153–177.

Choi, W. and Dean, R.A. (1997). The adenylate cyclase gene *MAC1* of *Magnaporthe grisea* controls appressorium formation and other aspects of growth and development. *Plant Cell* **9**, 1973–1983.

Chumley, F.G. and Valent, B. (1990). Genetic analysis of melanin-deficient, nonpathogenic mutants of *Magnaporthe grisea. Mol. Plant–Microbe Interact.* **3**, 135–143.

Deising, H.B., Werner, S., and Wernitz, M. (2000). The role of fungal appressoria in plant infection. *Microbes Infect.* **2**, 1631–1641.

de Jong, J.C., McCormack, B.J., Smirnoff, N., and Talbot, N.J. (1997). Glycerol generates turgor in rice blast. *Nature* **389**, 244–245.

Delserone, L.M., McCluskey, K., Matthews, D.E., and van Etten, H.D. (1999). Pisatin demethylation by fungal pathogens and nonpathogens of pea: Association with piastin tolerance and virulence. *Physiol. Mol. Plant Pathol.* **55**, 317–326.

de Zwaan, T.M., Carroll, A.M., Valent, B., and Sweigard, J.A. (1999). *Magnaporthe grisea* Pth11p is a novel plasma membrane protein that mediates appressorium differentiation in response to inductive substrate cues. *Plant Cell*, **11**, 2013–2030.

Dinesh-Kumar, S.P., Whitham, S., Choi, D., Hehl, R., Corr, C., and Baker, B. (1995). Transposon tagging of tobacco mosaic virus resistance gene *N*: Its possible role in the TMV-*N* mediated signal transduction pathway. *Proc. Natl. Acad. Sci. USA* **92**, 4175–4180.

Dixon, K.P., Xu, J.-R., Smirnoff, N., and Talbot, N.J. (1999). Independent signaling pathways regulate cellular turgor during hyperosmotic stress and appressorium-mediated plant infection by *Magnaporthe grisea. Plant Cell* **11**, 2045–2058.

Dong, X. (1998). SA, JA, ethylene, and disease resistance in plants. *Curr. Opin. Plant Biol.* **1**, 316–323.

Funnell, D.L. and van Etten, H.D. (2002). Pisatin demethylase genes are on dispensable chromosomes while genes for pathogenicity on carrot and ripe tomato are on other chromosomes in *Nectria haematococca. Mol. Plant–Microbe Interact.* **15**, 840–846.

Hamer, J.E., Howard, R.J., Chumley, F.G., and Valent, B. (1988). A mechanism for surface attachment in spores of a plant pathogenic fungus. *Science* **239**, 288–290.

Hamer, J.E. and Talbot, N.J. (1998). Infection-related development in the rice blast fungus *Magnaporthe grisea. Curr. Opin. Microbiol.* **1**, 693–697.

Hamer, L., Adachi, K., Montenegro-Chamorro, M.V., Tanzer, M.M., Mahanty, S.K., Lo, C., Tarpey, R.W., Skalchunes, A.R. *et al.* (2001). Gene discovery and gene function assignment in filamentous fungi. *Proc. Natl. Acad. Sci. U.S.A.* **98**, 5110–5115.

Heath, M.C., Valent, B., Howard, R.J., and Chumley, F.G. (1990). Interactions of two strains of *Magnaporthe grisea* with rice, goosegrass, and weeping lovegrass. *Can. J. Bot.* **68**, 1627–1637.

Herskowitz, I. (1995). MAP kinases in yeast: For mating and more. *Cell* **80**, 187–197.

Hirt, H. (1997). Multiple roles of MAP kinases in plant signal transduction. *Trends Plant Sci.* **2**, 11–15.

Kang, S.H., Khang, C.H., and Lee, Y.-H. (1999). Regulation of cAMP-dependent protein kinase during appressorium formation in *Magnaporthe grisea. FEMS Microbiol. Lett.* **170**, 419–423.

Kim, Y.-K., Kawano, T., Li, D., and Kolattukudy, P.E. (2000). A mitogen-activated protein kinase kinase required for induction of cytokinesis and appressorium formation by host signals in the conidia of *Colletotrichum gloeosporioides. Plant Cell* **12**, 1331–1343.

Knogge, W. (1998). Fungal pathogenicity. *Curr. Opin. Plant Biol.* **1**, 324–328.

Knogge, W. (2002). Avirulence determinants and elicitors. In K. Esser and J.W. Bennett (eds.) *The Mycota, Volume XI, Agricultural Applications*. F. Kempken (Vol ed.) Springer-Verlag, Berlin Heidelberg, pp. 289–310.

Kronstad, J.W. (1997). Virulence and cAMP in smuts, blasts and blights. *Trends Plant Sci.* **2**, 193–199.

Innes, R.G. (1998). Genetic dissection of *R* gene signal transduction pathways. *Curr. Opin. Plant Biol.* **1**, 299–304.

Laugè, R. and de Wit, P.J.G.M. (1998). Fungal avirulence genes: Structure and possible functions. *Fungal Genet. Biol.* **24**, 285–297.

Lengeler, K.B., Davidson, R.C., D'Souza, C., Harashima, T., Shen, W.-C., Wang, P., Pan, X., Waugh, M. *et al.* (2000). Signal transduction cascades regulating fungal development and virulence. *Microbiol. Mol. Biol. Rev.* **64**, 786–820.

Lev, S., Sharon, A., Hadar, R., Ma, H., and Horwitz, B.A. (1999). A mitogen-activated protein kinase of the corn leaf pathogen *Cochliobolus heterostrophus* is involved in conidiation, appressorium formation, and pathogenicity: Diverse roles for mitogen-activated protein kinase homologs in foliar pathogens. *Proc. Natl. Acad. Sci. USA* **96**, 13542–13547.

Liu, S. and Dean, R.A. (1997). G protein α subunit genes control growth, development, and pathogenicity of *Magnaporthe grisea. Mol. Plant–Microbe Interact.* **10**, 1075–1086.

Maleck, K. and Dietrich, R.A. (1999). Defence on multiple fronts: How do plants cope with diverse enemies? *Trends Plant Sci.* **4**, 215–219.

Martin, G.B. (1999). Functional analysis of plant disease resistance genes and their downstream effectors. *Curr. Opin. Plant Biol.* **2**, 273–279.

Mayorga, M.E. and Gold, S.E. (1998). Characterization and molecular genetic complementation of mutants affecting dimorphism in the fungus *Ustilago maydis. Fungal Genet. Biol.* **24**, 364–376.

Mayorga, M.E. and Gold, S.E. (1999). A MAP kinase encoded by the *ubc3* gene of *Ustilago maydis* is required for filamentous growth and full virulence. *Mol. Microbiol.* **34**, 485–497.

Mitchell, T.K. and Dean, R.A. (1995). The cAMP-dependent protein kinase catalytic subunit is required for appressorium formation and pathogenesis by the rice blast pathogen *Magnaporthe grisea. Plant Cell* **7**, 1869–1878.

Morgenstern, D.E., Gifford, M.A.C., Li, L.-L., Doerschuk, C.M., and Dinauer, M.C. (1997). Absence of respiratory burst in X-linked chronic granulomatous disease in mice leads to abnormalities in both host defense and inflammatory response to *Aspergillus fumigatus. J. Exp. Med.* **185**, 207–218.

Muller, P., Aichinger, C., Feldbrugge, M., and Kahmann, R. (1999). The MAP kinase *kpp2* regulates mating and pathogenic development in *Ustilago maydis. Mol. Microbiol.* **34**, 1007–1017.

Murphy, A.M., Chivasa, S., Singh, D.P., and Carr, J.P. (1999). Salicylic acid-induced resistance to viruses and other pathogens: A parting of the ways? *Trends Plant Sci.* **4**, 155–160.

Oliver, R.P., Roberts, I.N., Harling, R., Kenyon, L., Punt, P.J., Dingemanse, M.A., and van den Hondel, C.A.M.J.J. (1987). Transformation of *Fulvia fulva*, a fungal pathogen of tomato, to hygromycin B resistance. *Curr. Genet.* **12**, 231–233.

Osbourn, A. (1996). Saponins and plant defence—a soap story. *Trends Plant Sci.* **1**, 4–9.

Ou, S.H. (1985). *Rice Diseases, 2nd Edn.* Commonwealth Mycological Institute, Kew, Surrey, UK, pp. 109–201.

Panaccione, D.G., Johnson, R.D., Rasmussen, J.B., and Friesen, T.L. (2002). Fungal Phytotoxins. In K. Esser and J.W. Bennett (eds.) *The Mycota, Volume XI, Agricultural applications*. F. Kempken (vol ed.) Springer-Verlag, Berlin, pp. 311–340.

Parsons, K.A., Chumley, F.G., and Valent, B. (1987). Genetic transformation of the fungal pathogen responsible for rice blast disease. *Proc. Natl. Acad. Sci. USA* **84**, 4161–4165.

Penninckx, I.A.M.A., Eggermont, K., Terras, F.R.G., Thomma, B.P.H.J., de Samblanx, G.W., Buchala, A., Métraux, J.-P., Manners, J.M. *et al.* (1996). Pathogen-induced systemic activation of a plant defensin gene in *Arabidopsis* follows a salicylic acid-independent pathway. *Plant Cell* **8**, 2309–2323.

Pieterse, C.M.J. and van Loon, L.C. (1999). Salicylic acid-independent plant defence pathways. *Trends Plant Sci.* **4**, 52–58.

Piffanelli, P., Devoto, A., and Schulze-Lefert, P. (1999). Defence signalling pathways in cereals. *Curr. Opin. Plant Biol.* **2**, 295–300.

Punja, Z.K. (2001). Genetic engineering of plants to enhance resistance to fungal pathogens—a review of progress and future prospects. *Can. J. Plant Pathol.* **23**, 216–235.

Richberg, M.H., Aviv, D.H., and Dangl, J.L. (1998). Dead cells do tell tales. *Curr. Opin. Plant Biol.* **1**, 480–485.

Schäfer, W. (1994). Molecular mechanisms of fungal pathogenicity to plants. *Ann. Rev. Phytopathol.* **32**, 461–77.

Shah, J., Tsui, F., and Klessig, D.F. (1997). Characterisation of a *s*alicylic *a*cid—*i*nsensitive mutant (*sai 1*) of *Arabidopsis thaliana*, identified in a selective screen utilising the SA-inducible expression of the *tms2* gene. *Mol. Plant–Microbe Interact.* **10**, 69–78.

Shirasu, K., Nakajima, H., Rajasekhar, V.K., and Dixon, R.A. (1997). Salicylic acid potentiates an agonist-dependent gain control that amplifies pathogen signals in the activation of defence mechanisms. *Plant Cell* **9**, 261–270.

Soanes, D.M., Skinner, W., Keon, J., Hargreaves, J., and Talbot, N.J. (2002). Genomics of phytopathogenic fungi and the development of bioinformatic resources. *Mol. Plant–Microbe Interact.* **15**, 421–427.

Somssich, I.E. and Hahlbrock, K. (1998). Pathogen defence in plants—a paradigm of biological complexity. *Trends Plant Sci.* **3**, 86–90.

Sweigard, J.A., Carroll, M.A., Farrall, L., Chumley, F.G., and Valent, B. (1998). *Magnaporthe grisea* pathogenicity genes obtained through insertional mutagenesis. *Mol. Plant–Microbe Interact.* **11**, 404–412.

Talbot, N.J. (1995). Having a blast: Exploring the pathogenicity of *Magnaporthe grisea. Trends Microbiol.* **3**, 9–16.

Talbot, N.J. and Foster, A.J. (2001). Genetics and genomics of the rice blast fungus *Magnaporthe grisea* developing an experimental model for understanding fungal diseases of cereals. *Adv. Bot. Res.* **34**, 263–287.

Thines, E., Weber, R.S., and Talbot, N.J. (2000). MAP kinases and protein kinase A—dependent mobilization of triacylglycerol and glycogen during appressorium turgor generation by *Magnaporthe grisea. Plant Cell* **12**, 1703–1718.

Tonukari, N.J., Scott-Craig, J.S., and Walton, J.D. (2000). The *Cochliobolus carbonum SNF1* gene is required for cell wall-degrading enzyme expression and virulence on maize. *Plant Cell* **12**, 237–247.

Tucker, S.L. and Talbot, N.J. (2001). Surface attachment and pre-penetration stage development by plant pathogenic fungi. *Annu. Rev. Phytopathol.* **39**, 385–417.

Tunlid, A. and Talbot, N.J. (2002). Genomics of parasitic and symbiotic fungi. *Curr. Opin. Microbiol.* **5**, 513–519.

Urao, T., Yamaguchi-Shinozaki, K., and Shinozaki, K. (2000). Two-component systems in plant signal transduction. *Trends Plant Sci.* **5**, 67–74.

van Etten, H.D., Matthews, D.E., and Matthews, P.S. (1989). Phytoalexin detoxification: Importance for pathogenicity and practical implications. *Annu. Rev. Phytopathol.* **27**, 143–64.

Weber, R.W.S., Wakley, G.E., Thines, E., and Talbot, N.J. (2001). The vacuole as central element of the lytic system and sink for lipid droplets in maturing appressoria of *Magnaporthe grisea. Protoplasma* **216**, 101–112.

Xu, J.-R.(2000). MAP kinases in fungal pathogens. *Fungal Genet. Biol.* **31**, 137–152.

Xu, J.-R. and Hamer, J.E. (1996). MAP kinase and cAMP signaling regulate infection structure formation and pathogenic growth in the rice blast fungus *Magnaporthe grisea. Genes Dev.* **10**, 2696–2706.

Xu, J.-R., Urban, M., Sweigard, J., and Hamer, J.E. (1997). The *CPKA* gene of *Magnaporthe grisea* is essential for appressorial penetration. *Mol. Plant–Microbe Interact.* **10**, 187–194.

Xu, J.-R., Staiger, C.J., and Hamer, J.E. (1998). Inactivation of the mitogen-activated protein kinase Mps1 from the rice blast fungus prevents penetration of host cells but allows activation of plant defence responses. *Proc. Natl. Acad. Sci. USA* **95**, 12713–12718.

16

Structural and Functional Genomics of Symbiotic Arbuscular Mycorrhizal Fungi

V. Gianinazzi-Pearson, C. Azcon-Aguilar, G. Bécard, P. Bonfante, N. Ferrol, P. Franken, A. Gollotte, L.A. Harrier, L. Lanfranco, and D. van Tuinen

1. Introduction

The absorbing organs (roots, rhizomes) of nearly all terrestrial plant families host an intimate symbiotic association, called a mycorrhiza, with specialized functional groups of soil fungi. The most common type of root symbiosis is the arbuscular mycorrhiza where soil fungi interact with a tremendous diversity of plant species, including many forest trees and agricultural, horticultural, and fruit crops (Gianinazzi *et al.*, 2002). The fungi involved are very ancient microorganisms compared to other true fungi. Fossil data and molecular phylogenetic analyses indicate that their origin dates back to the Ordovician–Devonian era some 460 to 400 million years ago (Remy *et al.*, 1994; Redecker *et al.*, 2000), coinciding with land colonization by early plants. Since then, arbuscular mycorrhizal (AM)

V. Gianinazzi-Pearson • UMR 1088 INRA/Université de Bourgogne BBCE-IPM, INRA-CMSE, BP 86510, 21065 Dijon cedex, France. **C. Azcon-Aguilar** • CSIC, Departmento de Microbiologia del Suelo y Sistemas Simbioticos, Estacion Experimental del Zaidin, Professor Albareda 1, 18008, Granada, Spain. **G. Bécard** • Université Paul Sabatier/UMR 5546 CNRS, Equipe de Mycologie Végétale, Pôle de Biotechnologie Végétale, 24 Chemin de Borde-Rouge, BP 17 Auzeville, 31326 Castanet-Tolosan, France. **P. Bonfante** • Università degli Studi di Torino, Dipartimento di Biologia Vegetale, Mycology IPP/CNR and Molecular Biology, Viale Mattioli 25, 10125 Torino, Italy. **N. Ferrol** • CSIC Departamento de Microbiologia del Suelo y Sistemas Simbioticos, Estacion Experimental del Zaidin, Profesor Albareda 1, 18008 Granada, Spain. **P. Franken** • Institute for Vegetable and Ornamental Plants, Theodor-Echtermeyer-Weg 1, D-14979 Grossbeeren, Germany. **A. Gollotte** • UMR 1088 INRA/Université de Bourgogne BBCE-IPM, INRA-CMSE, BP 86510, 21065 Dijon cedex, France. **L.A. Harrier** • The Scottish Agricultural College, Biotechnology Department, Plant Science Division, West Mains Road, EH9 3JG Edinburgh, United Kingdom. **L. Lanfranco** • Università degli Studi di Torino, Dipartimento di Biologia Vegetale, Mycology CSMT/CNR and Molecular Biology, Viale Mattioli 25, 10125 Torino, Italy. **D. van Tuinen** • UMR 1088 INRA/Université de Bourgogne BBCE-IPM, INRA-CMSE, BP 86510, 21065 Dijon cedex, France.

Advances in Fungal Biotechnology for Industry, Agriculture, and Medicine. Edited by Jan S. Tkacz and Lene Lange, Kluwer Academic/Plenum Publishers, 2004.

fungi have become an integral part and key components of most terrestrial ecosystems (Smith and Read, 1997). Their ability to enhance plant resistance to biotic and abiotic stresses makes them potentially powerful biotools for low-input agriculture (Gianinazzi *et al.*, 2002), and it is believed that their biodiversity can influence plant community structure in natural ecosystems (van der Heijden *et al.*, 1998). More than 100 species of AM fungi have been described, and many of them are held in international culture collections (IBG/BEG, 1993; INVAM, 1996). Their taxonomical status has been a matter of debate because of their asexual nature and the difficulty to affiliate them closely to existing fungal groups. Until recently, they were organized into six genera, distributed in four families, and grouped into a unique order, Glomales (Zygomycota) which comprises *Gigaspora* and *Scutellospora* (Gigasporaceae), belonging to the suborder Gigasporineae, and *Glomus*, *Sclerocystis* (Glomaceae), *Acaulospora*, and *Entrophospora* (Acaulosporaceae), clustered in the Glomineae (Morton and Benny, 1990). However, a revised classification is presently being considered which places them in a new phylum, the Glomeromycota with four new orders (Glomerales, Archeaosporales, Paraglomales, and Diversisiporales) (Schüßler *et al.*, 2001). For clarification, the generic term "AM fungi" will therefore be used above the species level throughout this chapter.

Notwithstanding their importance for plant health and survival in many ecosystems, relatively little is known about AM fungi at the genomic level (Gianinazzi-Pearson *et al.*, 2001). This is mainly because, despite (or due to) their long evolutionary history, they are obligate biotrophs, and they must establish a complex relationship with plants to ensure their survival. They are unable to produce extensive mycelia from germinating spores, and they will not complete their life cycle if contact is not made with a host root. Much of the spore's nutritional reserves remain intact during germination, suggesting that the fungus is unable to fully exploit its growth potential and that some important fungal activity may require regulatory elements of plant origin (Bécard and Piché, 1989; Bago and Bécard, 2002). In fact, molecules from host roots trigger developmental switches in mycelium, first seen by branching, then by appressorium formation at contact sites with plant cells (Giovannetti *et al.*, 1993; Buée *et al.*, 2000). This morphogenetic event is followed by root penetration and colonization which culminates in the development of intracellular haustoria (arbuscules), the sites of symbiotic exchanges in inner cortical cells, as well as the proliferation into surrounding soil or substrate of extraradical mycelium active in nutrient uptake and transfer to the host plant (Harrison, 1999). Structural and functional analyses of AM fungal genes is essential to understand (a) the molecular basis whereby these beneficial microorganisms have evolved to form intimate biotrophic associations with root cells, (b) how the fungi interact with plants, (c) how they reproduce, and (d) how they persist in the environment. Such information is critical for the effective use of AM fungi as biotools in low-input agriculture and for their development as new biological products for crop production systems which satisfy environmental and consumer concerns. Furthermore, some AM fungi host endobacteria within their cytoplasm (Bianciotto *et al.*, 1996), and the implication of the prokaryotic genome for the physiology or performance of the mycorrhizal symbiosis needs to be explored. It is with these aims in mind that researchers have joined forces within an international consortium (Genomyca, 2001). In this chapter, members present an overview of the current state of genomics of AM fungi and their endobacteria and discuss scientific avenues to exploit the parts of the genomes of AM fungi that determine symbiotic attributes.

2. Genome Structure and Organization

Considering that AM fungi have been interacting with plants for more than 460 million years and that they have adapted to a wide variety of ecosystems with very different soil types, the genomes of these fungi must harbor some interesting evolutionary features. AM fungi reproduce through asexual spores which, as individual cells, are unusual in that they carry an exceptionally high number of nuclei for the Mycota. This number has been determined to be about 720 for *Scutellospora castanea* (Hosny *et al.*, 1999a) and 2,400 to 2,600 for *Gigaspora* (Bécard and Pfeffer, 1993; Cooke *et al.*, 1987), although one study (Viera and Glenn, 1990) reported more than 20,000 nuclei per spore for members of this genus. These multinucleate spores are the sole source of DNA in quantities high enough for investigations of genome size, structure, and organization. The DNA content in AM fungal species from four different genera, measured in sporal nuclei by flow cytometry with chicken red blood cells as the calibration standard, has been reported to range from 0.13 pg of DNA/genome (1.3×10^8 bp) for *Scutellospora pellucida* (INVAM 337) to 1.08 pg of DNA/genome (1.06×10^9 bp) for *S. gregaria* (W1727) (Hosny *et al.*, 1998). Previous studies by Bianciotto and Bonfante (1992) had also determined the genome size of *Gigaspora margarita* to be of 0.65 pg (6.4×10^8 bp). Although this genome size range may seem broad, it is narrow when compared to the 167-fold range observed overall in the fungal kingdom. Nevertheless, the genome size of AM fungi is high when compared to most other true fungi, and it approaches the genome size of the Zygomycete *Entomophaga* with 8.1×10^9 bp, the largest reported in fungi (Murrin *et al.*, 1986).

Genome complexity has been explored in one AM fungus, *S. castanea*, by shearing genomic DNA ultrasonically, thermally denaturing it, and then measuring the rehybridization curve (Cot) as compared with that of *Escherichia coli*. The renaturation curve for *S. castanea* presents several slopes and contrasts with the curve for *E. coli* DNA that has a simple sigmoidal shape. This indicates a more complex genome in the AM fungus (Hosny, 1997). The curve for *S. castanea* suggests the existence of four sequence families. The first one, representing 7% of the genome, is generally believed to be made of palindromic sequences, and the three others represent three levels of repeated sequences. Although a complete rehybridization could not be achieved with the DNA of *S. castanea*, due to limited availability of DNA, it can be assumed from the renaturation experiments that repetitive DNA accounts for at least of 50% of the *S. castanea* genome. A number of repeated sequences have been isolated from genomic libraries of this and other AM fungi (Hosny *et al.*, 1999b). One of these sequences, SC1, is a tandemly repeated sequence representing about 0.24% of the *S. castanea* genome. Because this sequence does not hybridize with DNA of other AM fungi, it has been used to generate specific primers for the identification of this fungus in plant roots (Zézé *et al.*, 1996). Other repetitive DNA sequences that are presently being characterized in *Gigaspora rosea, G. margarita*, and *Glomus mosseae* appear mainly to be moderately repeated DNA elements. Most of these have no homology to known sequences and contain no open reading frame. Southern blot analyses and *in situ* fluorescence hybridization indicate that these sequences are mostly dispersed in the genome (Gollotte, van Tuinen, and Gianinazzi-Pearson, unpublished). One repeated sequence isolated from *S. castanea*, *Mycdire* (170 copies/genome), contains two perfect copies of the Centromeric DNA Element (CDEIs) found in *Saccharomyces cerevisiae* (Hieter *et al.*, 1985), and a third with a C to T transition which may have resulted from methylation of the cytosine (Zézé *et al.*, 1999). This *Mycdire* sequence also contains three 11-residue sequences identical in all but one residue to the autonomously replicating sequence (*ars*) of *S. cerevisieae* (Broach *et al.*, 1983). Furthermore, *G. rosea* and *G. caledonium* harbor a sequence with a similarity of 97% to *Mycdire*, while *Acaulospora laevis* has an element with a sequence similarity of only 65%. The occurrence of putative autonomously replicating elements in the genome of AM fungi strongly suggests the possible existence of transposable elements in these fungi. Although no active transposable element has been isolated from these fungi so far, recent work has confirmed the presence of sequences showing homology to gypsy and Non-Long Terminal Repeat retrotransposons in the genome of *G. margarita* (Gollotte *et al.*, unpublished). Whether such elements have played a role in the evolution of AM fungal genomes remains to be elucidated.

Typically the genomes of AM fungi are also characterized by their low GC content (Hosny *et al.*, 1998). Values determined for different species in the four families range from 29.9% to 35.25%, which is in the range of those reported for Zygomycetes (27.5%–59%) (Storck and Alexopoulos, 1970) but is at its lower end. Another unusual feature of genomic DNA of AM fungi is its high content of methylated cytosine relative to that of other fungi. The level of methylated cytosine determined for different isolates of nine species, ranges from 12.4% to 24.6% with the highest values observed for *A. laevis* (Hosny *et al.*, 1997). The content of methylated cytosine is generally low in fungi. For example in *Neurospora crassa* where methylated cytosine has been reported to be involved in the Repeated Induced Point mutation (RIP) mechanism associated with inactivation of duplicated genes (Singer *et al.*, 1995), only 1–2% of the cytosine residues are methylated (Russell *et al.*, 1987). Methylation of cytosine could favor cytosine to thymidine transitions and could account for a rapid divergence among repeated sequences. Furthermore methylated cytosine plays an important role in the silencing of transposable elements and in the regulation of gene expression (reviewed by Bird, 2002).

Data obtained from the analysis of ribosomal sequences of AM fungi (Sanders *et al.*, 1995; Lloyd-MacGilp *et al.*, 1996; Hijri *et al.*, 1998; Lanfranco *et al.*, 1999a) have indicated that different sequences are present in the same multinucleated spore. This finding is of great interest as it raises the question of a possible polymorphism at the nuclear level and, consequently, of the reproduction strategy of these fungi. Various studies have been undertaken to address this. Indirect evidence that nuclei from the same spore could be genetically different have been obtained by *in situ* hybridization of ribosomal DNA (Trouvelot *et al.*, 1999; Kuhn *et al.*, 2001), although clear data to confirm this hypothesis are not yet available. Likewise, contradictory data concerning clonal (Rosendahl and Taylor, 1997) versus non-clonal (Vandenkoornhuyse *et al.*, 2001) reproduction mode of AM fungi illustrate the difficulty of addressing such questions within the biological context of these symbiotic organisms.

3. Fungal Genes in the Symbiotic Context

3.1. Targeted Analyses of Gene Expression

In the last few years, there has been a considerable increase in our knowledge regarding AM fungal genes that are regulated in the asymbiotic life stage or during symbiotic

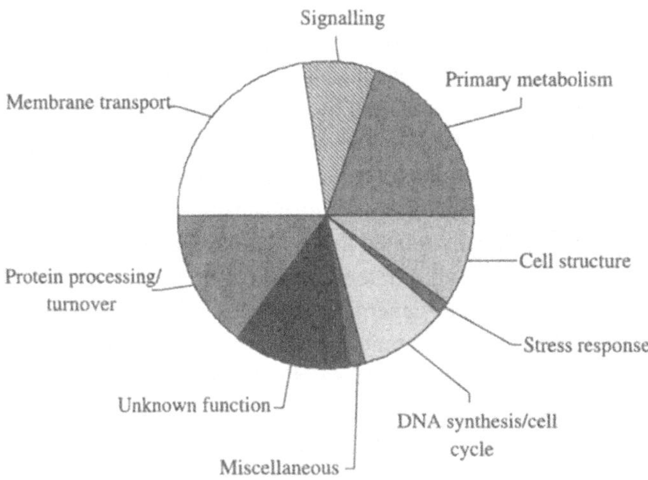

Figure 16.1. Distribution by functional categories of AM fungal genes showing differential expression during asymbiotic, presymbiotic, or symbiotic stages of the fungal life cycle.

interactions with host roots and the surrounding soil environment (Figure 16.1). Before Expressed Sequence Tag (EST) libraries were constructed (see following section), most AM fungal genes were isolated by targeted approaches based on the assumption that a gene or gene product plays a role in a certain developmental or metabolic process. Although relatively few AM fungal genes have been identified to date by such a strategy, a number of genes with important nutritional or morphogenetic functions have been isolated. Two main approaches have been successful: screening of cDNA libraries with heterologous probes and PCR amplification using degenerate primers.

The first AM fungal gene to be analyzed in detail was a plasma membrane phosphate transporter (Harrison and van Buuren, 1995). Phosphate uptake from the soil by AM fungi, the subsequent transport of this nutrient within the hyphae, and its release to the plant partner in the AM symbiosis are processes that have a significant impact on plant mineral nutrition and performance. Since phosphate uptake by all organisms is mediated by phosphate transporters and transporters from different species share a high degree of sequence similarity, Harrison and van Buuren (1995) used the *PHO84* gene from *S. cerevisiae* encoding a phosphate transporter to screen a cDNA library prepared from roots of *Medicago truncatula* colonized with the AM fungus *Glomus versiforme*. This strategy allowed isolation of a *G. versiforme* gene (*GvPT*) encoding a transmembrane, high-affinity phosphate transporter that operates by proton-coupled symport. The observation that *GvPT* is expressed only in the extraradical mycelium radiating out from mycorrhizal roots, but not in the fungal structures within the roots, indicates that it is involved in the initial uptake of phosphate from the soil by the AM fungus. Knowledge of the *GvPT* sequence allowed isolation of *GiPT*, the homolog gene from *Glomus intraradices*, by screening of a *M. truncatula*–*G. intraradices* cDNA library with a probe corresponding to the coding region of *GvPT* (Maldonado-Mendoza *et al.*, 2001). Expression analysis of *GiPT* in the extraradical mycelium of *G. intraradices* grown monoxenically in compartmented Petri dishes has considerably increased our knowledge about the mechanisms controlling phosphate uptake by AM fungi. These studies revealed that *GiPT* expression is induced in response to phosphate concentrations of 1–5 μM, typical of those found in the soil solution, and is modulated by the overall phosphate status of the mycorrhiza. Based on these data Maldonado-Mendoza *et al.* (2001) concluded that *G. intraradices* can perceive phosphate levels in the external environment and suggested the presence of an internal phosphate-sensing mechanism.

Other important nutritionally related genes that have been targeted in AM fungi are plasma membrane H^+-ATPases and nitrate reductase. The plasma membrane H^+-ATPase is an electrogenic pump which plays energetic and regulatory roles in fungi. The activity of this enzyme generates an electrochemical gradient of H^+ across the cell plasma membrane that drives a number of secondary transport systems, including those responsible for translocating cations, anions, amino acids, and sugars. With a nested PCR approach based on two sets of degenerate primers designed to match four highly conserved domains found in all H^+-ATPases, five partial genomic clones encoding putative H^+-ATPases (*GmHA1-5*) have been isolated from sporal DNA of *G. mosseae* and two cDNA clones from extraradical mycelium of *G. intraradices* (Ferrol *et al.*, 2000; Ferrol, unpublished). Expression analysis of the *G. mosseae* genes showed that *GmHA1* is expressed in the intraradical hyphae while *GmHA2* is expressed in the extraradical mycelium (Ferrol *et al.*, 2002). Expression of H^+-ATPase and phosphate transporter genes in the extraradical hyphae of AM fungi suggests that phosphate uptake is mediated by a high-affinity transporter, the function of which is directly dependent on the proton-motive force generated by a plasma membrane H^+-ATPase. In addition, Franken *et al.* (1997), using degenerate primers for P-type ATPases, isolated two partial cDNA clones from germinating spores of *G. rosea* encoding two putative P-type ATPases, one coding a transmembrane Ca^{2+}-ATPase and the other involved in transmembrane Li^+ and Na^+ effluxes.

AM fungi can also assist their host plant in nitrogen assimilation (Johansen *et al.*, 1992; Hawkins and George, 1999), and the basis of this activity has been addressed at the molecular level by cloning and analyzing fungal nitrate reductase genes. Part of a gene coding for a nitrate reductase apoprotein was isolated by PCR technology from different *Glomus* species (Kaldorf *et al.*, 1994). In situ RNA hybridization of the *G. intraradices* nitrate reductase gene in colonized maize roots revealed that the intracellular fungal haustoria (arbuscules) but not storage vesicles are preferential sites for the accumulation of nitrate reductase transcripts (Kaldorf *et al.*, 1998). While these data suggest that the AM fungus performs nitrate turnover, other metabolically-related genes need to be analyzed in order to envisage which nitrogen compounds may be exported from the fungus to the plant cell.

The developmental biology of AM fungi is still a largely unexplored domain, especially with relation to the molecular bases for control of their complex growth patterns. Attention has been focused, with the same target approaches, on genes involved in cell wall synthesis or genes coding for cytoskeleton elements. Complex spatial and temporal changes in cell wall morphology, in particular in chitin architecture of AM fungi, have been shown in different developmental stages of the symbiotic interaction (Bonfante, 2001). Chitin synthesis in yeast and other filamentous fungi involves multiple chitin synthase (*chs*) genes with transcriptional and/or post-transcriptional regulation. Molecular investigations have shown that AM fungi possess a multigene *chs* family, and according to the current classification, the genes presently known belong to classes I, II, and IV

(Lanfranco *et al.*, 1999b, c; Ubalijoro *et al.*, 2001). Screening a genomic library with a homologous PCR-derived probe led to identification of a complete class IV *chs* gene from *G. versiforme*; this was the first full sequence encoding a functional gene to be described in an AM fungus (Lanfranco *et al.*, 1999b). The differential *chs* gene expression pattern during the life cycle of two AM fungi suggests that regulation of chitin synthesis may contribute to the control of fungal morphogenesis during different steps of mycorrhiza formation (Lanfranco *et al.*, 1999c; Ubalijoro *et al.*, 2001).

Genes encoding β-tubulin have also been investigated to gain insight into the mechanisms controlling AM fungal development. The cytoskeleton is a cellular component that plays an important morphogenetic role in fungal growth, and there is evidence that microtubules of AM fungi are highly dynamic (Blancaflor *et al.*, 2001; Timonen *et al.*, 2001). β-Tubulin gene fragments obtained from a number of AM fungal isolates (Franken *et al.*, 1997; Butehorn *et al.*, 1999; Rhody *et al.*, 2003) can be grouped into two different sequence types with differences in GC content and codon usage suggesting a divergent origin. Interestingly, representatives of the two sequence types display different expression patterns in asymbiotic germinating spores compared to symbiotic extraradical hyphae, indicating that different isoforms of β-tubulin may be involved in specific cellular needs (Rhody *et al.*, 2002).

Heterologous probes and degenerate oligonucleotides have been fruitful in the identification of AM fungal genes, but they do not give direct information about the functions of the isolated sequences, meaning that these functions must be considered putative based solely on sequence similarity. An alternative and attractive approach would be the identification of genes starting from specific mutants. Unfortunately, the obligate nature of the AM fungi, coupled with the fact that they are asexual, multinucleate organisms, makes it impossible to perform classical genetic manipulations to create mutant strains. Nevertheless, AM fungal genes could be cloned using a cDNA library to complement corresponding mutants in a heterologous organism such as yeast (Rose and Broach, 1991; Frommer and Ninneman, 1995). However, heterologous expression systems are based on the assumption that protein expression and function are similar in all organisms, which excludes the possibility of identifying genes involved in specific morphogenic and differentiation processes or genes from organisms which have diverged widely during evolution. Although no AM fungal gene has been cloned to date by functional complementation of a yeast mutant, it is likely that this approach, which has been successful with mammalian and plant genes (Lee and Nurse, 1987; Minet *et al.*, 1992), will work equally well with AM fungi since it has been demonstrated that AM fungal genes are active in yeast. In fact, the *GvPT* cDNA successfully complemented the *pho84* yeast mutant strain that lacks the high affinity phosphate transporter; indeed this demonstration was key in biochemically characterizing the *GvPT* gene product (Harrison and van Buuren, 1995). Recently, a yeast complementation assay was also used to characterize the function of the cDNA for a gene encoding a metallothionein from *G. margarita* (*GmarMt1*) (Lanfranco *et al.*, 2002). Although the molecular basis for the ability of AM fungi to buffer heavy metal stress in host plants is unknown (Leyval *et al.*, 1997), metallothioneins play an essential role in managing the intracellular concentration of free thiophilic metal ions (copper, zinc, cadmium) in fungi and could represent a crucial component of the metal protection machinery in the AM symbiosis. Complementation of two yeast mutant strains highly sensitive to heavy metals with the *GmarMt1* cDNA unambiguously demonstrated that *GmarMT1* gene product can sequester metal ions conferring protection *in vivo* against heavy metals. These two examples clearly illustrate that yeast complementation assays offer promising experimental tools to clone AM fungal genes and characterize their function.

3.2. Transcriptome Profiling

A more global view of the functional genome of AM fungi is approached by monitoring transcriptional activity through the isolation and sequencing of cDNA fragments. Such ESTs can be used to analyze gene expression patterns, localize gene products, and isolate corresponding full-length genes. cDNA libraries have been established to identify ESTs from activated spores of *Gigaspora rosea* (Stommel *et al.*, 2001), germinated spores of *G. margarita* (Lanfranco *et al.*, 2002) and *G. intraradices* (Lammers *et al.*, 2001), and from extraradical hyphae of *G. intraradices* (Sawaki and Saito, 2001). Clone sequencing has revealed interesting similarities to known genes, which are consistent with postulated functions in AM fungi. The hypothesis that the glyoxylate cycle is involved in substantial carbon fluxes was, for example, reinforced by the identification and expression analysis of genes from a *G. intraradices* cDNA library which code for a malate synthase and an isocitrate lyase gene containing the motifs responsible for glyoxysomal targeting (Lammers *et al.*, 2001). An acyl-CoA-dehydrogenase gene was identified in the same library and expression analysis indicated similar mechanisms of lipid utilization in germinating spores and in the extraradical mycelium (Bago *et al.*, 2002).

Research aimed at characterizing AM fungal genes involved in interactions prior to contact with host root cells has focused on techniques which give direct access to differentially expressed gene information. Differential display RNA analyses has shown that a *Bacillus subtilis* strain, which promotes hyphal development of *G. mosseae* from spores, influences fungal gene expression (Requena *et al.*, 1999). Two differentially displayed cDNA fragments were identified, one corresponding to a gene encoding a fatty acid oxidase, *GmFox1*, that was down-regulated by the presence of the prokaryote (Requena *et al.*, 1999), and a second representing a gene, *GmTor2*, that was constitutively expressed. The sequence similarity of the putatively encoded protein GmTOR2 to a yeast protein controlling the cell cycle led to the hypothesis that the gene acts in development arrest, which occurs in AM fungi if hyphae do not come into contact with roots of a host plant (Requena *et al.*, 2000). Other studies of fungal genes expressed during hyphal elongation from spores in the absence of a host root have been based on a combination of suppressive subtractive hybridization (SSH) and reverse Northern blotting of the ESTs obtained by SSH. This strategy has been applied to *G. mosseae* to compare RNA accumulation patterns in germinating spores and extraradical hyphae growing out from mycorrhizal roots (Requena *et al.*, 2002). Numerous genes appear to be differentially expressed, and one, expressed only in germinating spores, encoded a protein with similarities to the metazoan hedgehog protein and might participate in signaling between the two partners prior to the establishment of symbiosis.

To define AM fungal genes which are modulated during the developmental switch from asymbiotic germinating spores to the presymbiotic stage of hyphal branching, which preempts fungal–root contact (Giovannetti and Sbrana, 1998), differential RNA display was used to screen for RNA accumulation in spores of *G. rosea* stimulated by a semipurified factor from root exudates of host plants that rapidly provokes hyphal ramification

(Buée *et al.*, 2000; Tamasloukht *et al.*, 2003). Changes in transcript patterns could be observed within 1 hr of exposure to the factor, although visible signs of induced branching were not detectable before 5 hr. Numerous *G. rosea* genes showing root exudate-induced RNA accumulation were subsequently identified by SSH and EST screening. The putative protein products of these exudate-activated genes were clustered into five functional groups: ATPases together with proteins involved in cell cycle regulation, signal transduction, regulation of gene expression, and mitochondrial function (Tamasloukht, Bécard, and Franken, unpublished). Further experiments proved that the root factor that elicits hyphal branching increases fungal respiratory activity and induces reorganization of the mitochondrial system, altering organelle number and shape before visible signs of branching (Tamasloukht *et al.*, 2003). Although the exact nature of the active host-exuded factor remains to be identified, these data show for the first time that compounds of plant origin can regulate a primary metabolic function in AM fungi and that gene activation and physiological changes are involved in the developmental switch from an asymbiotic to a presymbiotic state.

Identification of fungal genes regulated during subsequent interactions with host cells in AM is confounded by the presence of root tissues and the low levels of fungal RNA relative to plant RNA (Maldonado-Mendoza *et al.*, 2002). This problem was overcome by differential RNA display analyses of mycorrhizal and non-mycorrhizal tomato tissues, where a cDNA fragment with similarity to a phosphoglycerate kinase (PGK) gene from *G. mosseae* was identified (Harrier *et al.*, 1998). The corresponding protein accumulates in higher amounts in colonized roots compared to germinating spores. Promoter analysis of the whole gene isolated from a genomic library revealed the presence of several upstream activating sequences typical for carbon regulation in fungi (Harrier, 2001). Three differentially displayed cDNA fragments of *G. intraradices* which were obtained from barley mycorrhiza code for peptide sequences with interesting similarities to proteins involved in gene regulation (Delp *et al.*, 2000). Using this technique for comparing ESTs from roots of pea plants differentially affected in symbiotic development, an EST with no homologs was identified from *G. mosseae* (Lapopin *et al.*, 1999). The corresponding gene appears to be induced during arbuscule interactions in a wild type pea more strongly than in an arbuscule-defective plant mutant (Lapopin *et al.*, unpublished). Recently, a *G. intraradices* EST encoding a peptide with weak similarity to a glutamine synthetase was isolated by differential screening of cDNA libraries from mycorrhiza and non-mycorrhizal lettuce roots (Ruiz-Lozano *et al.*, 2002). The corresponding gene was expressed only in mycorrhizal roots and seemed to be up-regulated by nitrogen fertilization. Suppressive subtractive hybridization of mycorrhizal and non-mycorrhizal roots could be a more efficient way to isolate fungal genes related to symbiotic development and function. Six ESTs of *G. mosseae* were recently identified by this technique in *M. truncatula* mycorrhiza. Three of these were uniquely expressed in mycorrhiza interactions, two showing similarity to a thioredoxin homolog and to a peptidylprolyl *cis-trans* isomerase genes (Brechenmacher *et al.*, unpublished).

Screening of cDNA libraries, differential RNA display, and subtractive suppressive hybridization are all useful strategies to elucidate the molecular basis of AM fungal development and function. In the absence of genome sequencing projects for AM fungi, they can be used to answer specific questions without the need to sequence thousands of clones.

4. Manipulating the Symbiotic Genome

Demonstration of the functions of AM fungal genes identified through genome and EST-based sequencing projects necessitates an efficient tool for gene modification. Combination of genomic data with efficient gene manipulation should provide an excellent resource to understand and further appreciate the biology of AM fungi during different stages of their interactions with host plants. However, because of their recalcitrance to culture *in vitro* and the absence of an exploitable sexual recombination system for genetic analysis or manipulation, it is necessary to introduce recombinant DNA technology and develop a transformation system for such studies of AM fungi. Moreover, engineering of symbiotic traits in AM fungi not only will facilitate identification of genes essential to the asymbiotic, presymbiotic, and symbiotic stages of their life cycles but also will offer the potential of providing or enhancing specific attributes that would be useful in agriculture. To be successful a transformation system must deliver DNA efficiently into the fungal tissue in a format permitting the expression of the gene introduced (Harrier and Millam, 2001). Minimally the latter requires a functional promoter.

Transformation of fungi that are difficult to manipulate in the laboratory has been achieved by biolistic (or particle bombardment) technology, whereby microscopic particles coated with DNA are explosively accelerated into tissue of the organism to be transformed (Bailey *et al.*, 1993; Lorito *et al.*, 1993; Hilber *et al.*, 1994; Bills *et al.*, 1995, 1999; St. Leger *et al.*, 1995; Durand *et al.*, 1997; Forbes *et al.*, 1998; Yu and Cole, 1998; Chaure *et al.*, 2000). Physical biolistic parameters crucial for the delivery of the vector containing the transgene and/or reporter gene into cells have been optimized for a number of AM fungi by bombarding isolated spores (Table 16.1). An optimal transformation technique also requires enhanced transformation stability and integration of the transgene into the

Table 16.1. Optimized biolistic parameters for particle bombardment of *Gigaspora margarita, Gigaspora rosea*, and *Glomus mosseae*

Parameter	Optimized value
Amount of plasmid utilized per shot	5 μg
Final calcium chloride concentration	1 M
Final spermidine concentration	16 μM
Amount of gold particles	1.2 mg
Size of gold particles	0.6 micron
DNA load per bombardment	625 μg
Macrocarrier gold load per bombardment	150 μg
Acceleration pressure	1300 psi
Chamber vacuum pressure	27 inches Hg
Gap distance[a]	7 cm
Macrocarrier travel distance[b]	2 cm
Target distance[c]	12 cm
Stopping plate aperture	0.8 cm

[a]distance between rupture disk and macrocarrier.
[b]distance between macrocarrier and stopping screen.
[c]distance between stopping screen and target tissue.

fungal genome. To achieve this, plasmid constructs carrying transgenic DNA can be supplied to fungal tissue either without a fungal origin of replication, in which case the maintenance of transformation depends on integration of the DNA into the chromosomes, or ligated into a plasmid capable of autonomous replication. To date, successful biolistic transformation of AM fungi has only been reported for *G. rosea* where spores were bombarded with a plasmid vector containing a constitutive *gpd* promoter fused to the *gusA* reporter gene (Forbes *et al.*, 1998). Transient reporter gene expression was observed in *G. rosea* spores, and PCR with *gusA*-specific primers demonstrated that the gene was present within both colonized host roots and some second generation spores subsequently formed by extraradical hyphae (Harrier *et al.*, 2002). More recently, a gene encoding green fluorescent protein (GFP) was introduced into spores of this fungus, and gene expression within individual spores was evaluated. No selection pressure was exerted on the bombarded spores, and PCR analyses indicate that the proportion of spores carrying the GFP transgene declined from 60% at 8 days after biolistic transformation to only 20% by day 64 (Figure 16.2). To obtain more stable transformants of AM fungi, strategies to enhance genomic integration and identify autonomous replicating elements need to be developed. Some possibilities will now be discussed.

The integration of a transgene into a fungal genome can be facilitated by incorporating repetitive sequences of homologous DNA into the plasmid vector. Segments of ribosomal DNA genes, which are tandemly repeated multigene families, have been utilized in transformation vectors (Ruiz-Díez and Martínez-Suárez, 1999; Mackenzie *et al.*, 2000). The 18s rDNA region from *G. rosea* has been directionally cloned into a gGFP vector and is currently being studied for stability within biolistically transformed spores (Bergero *et al.*, unpublished). In addition, genetic elements such as transposable sequences can be used to enhance stability of transformants. For example, the genetic elements *seg1* (a single copy region that leads to high mitotic stability) and *rag1* (a highly repetitive interspersed DNA sequence that promotes plasmid integration) have been used to enhance the stability of transformants within the Zygomycete *Absidia glauca* (Schilde *et al.*, 2001). Transposable

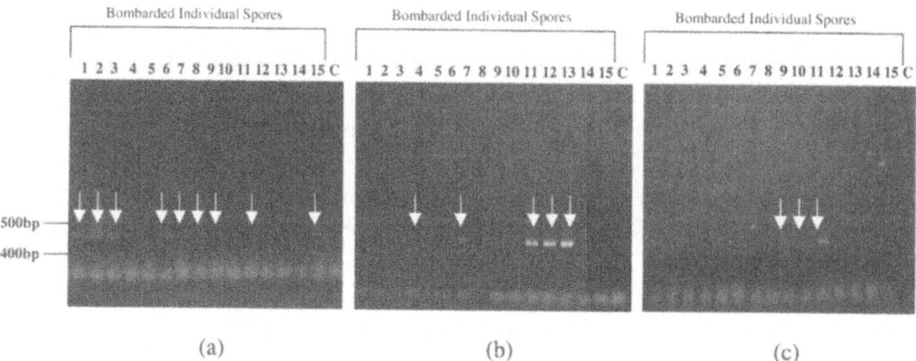

Figure 16.2. Detection of the GFP transgene by PCR using GFP gene specific oligonucleotide primers within gGFP biolistically transformed spores of *Gigaspora rosea* (BEG 9) (a) 8, (b) 24, and (c) 64 days after bombardment. (1–15—individual biolistically transformed spores; bp—base pairs; C—minus DNA control; arrows denote GFP gene amplicons) (Gollotte, Bergero, and Harrier, unpublished).

element-like sequences have recently been identified in the genome of AM fungi (Gollotte *et al.*, 2002), and these may provide an alternative source of repetitive DNA to be included into transformation vectors to improve transgene integration.

Zygomycota tend to replicate introduced plasmids autonomously (Revuelta and Jayaram, 1986; van Heeswijck, 1986; Yanai *et al.*, 1991), in contrast with Ascomycota or Basidiomycota where recombination events are more common and segments of the introduced plasmid are integrated into the fungal genome. Sequences responsible for autonomous replication of extrachromosomal plasmids can be derived from endogenous plasmids, from specific chromosomal fragments termed autonomous replication sequences (*ars*) and/or from telomeric sequences which have been isolated from many fungi (Stahl *et al.*, 1982; Grant *et al.*, 1984; van Heeswijck, 1986; Revuelta and Jayaram, 1986; Burmester and Wöstemeyer, 1987; Powell and Kistler, 1990; Kistler and Benny, 1992; Woods and Goldman, 1992, 1993; Aleksenko and Clutterbuck, 1997; Long *et al.*, 1998).

Incorporating an *ars* sequence into plasmids for the transformation of Zygomycota such as *Mucor* leads to improved transformation frequencies (van Heeswijck, 1986). Isolation of DNA fragments capable of acting as an *ars* sequence in AM fungi will be an important step toward preparing shuttle vectors that will provide high transformation frequencies and that can be recovered unmodified from the transformant. Furthermore, maintaining the transforming DNA as an extra-chromosomal element should also enable investigations of the expression and regulation of homologous and heterologous genes in AM fungi without the complications arising from single or multiple integration events within the genome.

An alternative to biolistic technology to generate stable transformants of AM fungi might be provided by *Agrobacterium tumefaciens*. This bacterium transfers a part of the T-DNA of its tumor inducing (Ti) plasmid to plant cells during crown gall induction, and the T-DNA integrates at random into the plant nuclear genome (Hooykaas *et al.*, 1979; see Chapter 4). This natural transformation system, which has been exploited in plant research for more than 25 years, has more recently been used to transform fungi (Bundock *et al.*, 1995, 2002; de Groot *et al.*, 1998; Gouka *et al.*, 1999; Abuodeh *et al.*, 2000; Covert *et al.*, 2001; Malonek and Meinhardt, 2001; Mikosch *et al.*, 2001; Mullins *et al.*, 2001; Rho *et al.*, 2001; Pardo *et al.*, 2002; Zhang *et al.*, 2003). One of the principal advantages of *A. tumefaciens*-mediated transformation (ATMT) over conventional transformation is the versatility in the type of starting fungal material that can be targeted for transformation. This technology avoids the isolation of protoplasts and can be used to transform conidia, protoplasts, hyphae, spores, and even blocks of mushroom mycelial tissue (de Groot *et al.*, 1998; Mullins *et al.*, 2001). ATMT can generate large libraries of fungal transformants, though the efficiency appears to be highly dependent on the fungal species transformed. In addition from 50% to 80% of transformants can possess a single T-DNA insert per genome (Mullins *et al.*, 2001). This is particularly important for rescuing the tagged gene from fungi lacking a sexual stage, such as *Fusarium oxysporum* or *Verticillium dahliae* (Mullins and Kang, 2001), and therefore ATMT could be a highly relevant approach to replace biolistic transformation of AM fungi.

Finally, modification of AM fungal traits may also be possible through the genetic engineering of the endosymbiotic bacteria present within cells of some species (see the following section) either by introducing genetically modified bacteria into AM fungal isolates that lack such bacteria and/or by transforming the bacterial cells within the AM fungi.

The latter may also be possible through biolistics as this technology has been applied to the transformation of structures within plant cells such as plastids and mitochondria (Sikdar *et al.*, 1998; Sidorov *et al.*, 1999; Bogorad, 2000; Daniell *et al.*, 2002).

5. Endobacterial Genes

Ecologists and evolutionary biologists are increasingly appreciating that many eukaryotic cells show some level of integration with bacteria (Moran and Wernergreen, 2000). In contrast, fungi, despite their huge diversity, do not provide many examples of symbiosis with bacteria. *Gigaspora* and *Scutellospora* species of AM fungi are unique in hosting bacteria in their cytoplasm as a rather common event (Hosny *et al.*, 1999b; Bianciotto *et al.*, 2000). To what extent such endocellular organisms contribute to the metabolic properties and physiology of the fungal host cell is an interesting question. Intracellular structures very similar to bacteria and termed BLOs (bacteria-like organisms) were first described in the 1970s (reviewed by Scannerini and Bonfante, 1991). Ultrastructural observations clearly revealed their presence in many field-collected fungal isolates. Further investigation of these BLOs, including the demonstration of their prokaryotic nature, was long hampered because of the inability to culture them. A combination of morphological observations (by electron and confocal microscopy) and molecular analyses has made it possible to identify BLOs as true bacteria and to start unravelling their symbiotic relationship with AM fungi (Bianciotto *et al.*,1996).

The AM fungus *G. margarita* (isolate BEG 34) contains a large number of endobacteria which can be easily detected by staining with fluorescent dyes specific for bacteria and capable of distinguishing between live and dead bacteria. The endocellular bacteria in *G. margarita* (BEG 34) were first identified as belonging to the genus *Burkholderia* on the basis of the 16S rDNA sequences (Bianciotto *et al.*, 1996). Amplification and sequencing of the 16S rDNA from isolates of *Scutellospora persica, S. castanea*, and *G. margarita* gave a phylogenetic branch which clustered with all endocellular bacteria so far sequenced in the Gigasporaceae (Bianciotto *et al.*, 2000), close to the genera *Burkholderia, Ralstonia*, and *Pandorea*. A new bacterial taxon was therefore proposed: *Candidatus* Glomeribacter gigasporarum (Bianciotto *et al.*, 2003). The extent to which these endobacteria, or other related microorganisms, are widespread in the Gigasporaceae remains an open question at present.

Despite many efforts to culture the endocellular *Candidatus* Glomeribacter gigasporarum (Jargeat and Becard, unpublished), the organism has proven recalcitrant, perhaps even unculturable. To overcome this experimental limitation, an "indirect" genetic approach was used to explore bacterial functions by constructing a genomic library rich in endobacterial sequences from DNA of *G. margarita* spores containing the endobacteria (van Buuren *et al.*, 1999). Random sequencing or screening with heterologous probes has led to the identification of some bacterial genes (Minerdi *et al.*, 2002a), the most interesting of which are those involved in nutrient uptake (a putative phosphate transporter operon, *pst*), in colonization events by bacterial cells (*vac*), and in chemotaxis (Ruiz-Lozano and Bonfante, 1999, 2000; Minerdi *et al.*, 2002b). A DNA region containing putative nitrogenase encoding genes (*nif* operon) was also found, which prompts the hypothesis that the endobacteria have nitrogen fixing activity (Minerdi *et al.*, 2001), a capability with the potential to enhance mycorrhiza effectiveness in agriculture. However, it is important to emphasize that presently there is no conclusive evidence demonstrating that these genes belong to the genome of *Candidatus* Glomeribacter gigasporum.

Recent data suggests that the endobacteria are stable cytoplasmic components of their fungal hosts. With mycorrhiza established *in vitro* in carrot root cultures, it has been demonstrated that bacteria are transmitted vertically through successive spore generations (Bianciotto and Jargeat, unpublished). A protocol to isolate the bacteria from fungal spores has recently been developed (Jargeat and Bécard, unpublished), which opens new possibilities to study the bacteria's genome size and structure, their free living capacities, and their ability to infect (horizontal transmission) other fungal isolates. Such investigations will help the understanding of their biological significance, whether they are obligate endocellular bacteria, and whether they contribute to the physiology and mycorrhizal performance of their fungal host. More work is required to understand more fully this fascinating new example of endosymbiotic life and to exploit the potential physiological traits of these bacteria for agronomic purposes.

6. Conclusions

Knowledge about genome complexity, diversity, and function in AM fungi is essential to understand their biology, the critical roles they play in agricultural systems, and their involvement in ecosystem conservation. With the advent of appropriate molecular techniques, research in the last few years has begun to uncover the identity of genes and gene products central to the biology of these fungi (Figure 16.1) and has advanced our appreciation of repetitive, non-coding DNA sequences that may be involved in genome dynamics and function. It has also become evident that AM fungi are unusual organisms. The biological issues posed by their multinucleate spores, large genomes, endobacterial symbionts, and asexuality are all challenges to be surmounted in the future engineering of symbiotic traits to provide specific attributes which can be exploited for plant production in sustainable agriculture and in fragile ecosystems.

Introduction of selected AM fungi into plant production systems and commercialization of inoculum are underway in several countries (Gianinazzi *et al.*, 2002). However, the impossibility of culturing these fungi alone necessitates time-consuming procedures for their isolation and maintenance on host plants, and their wide genetic diversity requires the development of well-defined molecular markers to monitor both their presence and activity in ecosystems. Moreover, biotic and abiotic ecofactors can trigger changes in isolate characteristics (Feldmann and Grotkass, 2002); thus, the ability to predict the success of inoculation to exert a positive effect in agriculture requires the definition of reliable genetic markers of fungal stability. This is likely to be achieved through the identification of that part of the AM fungal genome encoding key metabolic activities. This type of knowledge will also provide insights into genes or gene functions that are present or lacking in these organisms and will greatly assist in developing new inoculum production strategies like those targeting bioreactors or *in vitro* technologies. These goals will only be achieved by strengthening research efforts, through international initiatives, and by developing more ambitious strategies including broad genome sequencing, comparative genomics, and genetic transformation of AM fungi. Such investments will contribute information valuable in determining the reasons for obligate biotrophy in AM fungi. The information will also aid the development of tools to abrogate problems of inoculum production as well as tools to develop improved beneficial microbes for new biological products.

Acknowledgments

We express our sincere appreciation to Silvio Gianinazzi for comments on the manuscript. The authors' work is partly supported by the EU project Genomyca (QLK5-CT-2000-01319).

References

Abuodeh, R.O., Orbach, M.J., Mandel, M.A., Das, A., and Galgiani, J.N. (2000). Genetic transformation of *Coccodioides immitis* facilitated by *Agrobacterium tumefaciens. J. Infect. Dis.* **181**, 2106–2110.

Aleksenko, A. and Clutterbuck, A.J. (1997). Autonomous plasmid replication in *Aspergillus nidulans*: AMA1 and MATE Elements. *Fungal Genet. Biol.* **21**, 373–387.

Bago, B. and Bécard, G. (2002). Bases of obligate biotrophy of arbuscular mycorrhizal fungi. In S. Gianinazzi, H. Schüepp, J.M. Barea, and K. Haselwandter (eds.) *Mycorrhiza Technology in Agriculture: From Genes to Bioproducts*. Birkhaüser Verlag, Basel, pp. 33–48.

Bago, B., Zipfel, W., Williams, R.M., Jun, J., Arreola, R., Lammers, P.J., Pfeffer, P.E., and Shachar-Hill, Y. (2002). Translocation and utilization of fungal storage lipid in the arbuscular mycorrhizal symbiosis. *Plant Physiol.* **128**, 108–124.

Bailey, A.N., Mena, G.L., and Herrera-Estrella, L. (1993). Transfomation of four pathogenic *Phytophthora* sp. by microprojectile bombardment on intact mycelia. *Curr. Genet.* **23**, 42–46.

Bécard, G. and Piché, Y. (1989). New aspects on the acquisition of biotrophic status by a VAM fungus, *Gigaspora margarita. New Phytologist* **112**, 77–83.

Bianciotto, V. and Bonfante, P. (1992). Quantification of the nuclear DNA content of two arbuscular mycorrhizal fungi. *Mycol. Res.* **96**, 1071–1076.

Bianciotto, V., Bandi, C, Minerdi, D., Sironi, M., Tichy, H.V., and Bonfante, P. (1996). An obligately endosymbiotic fungus itself harbors obligately intracellular bacteria. *Appl. Environ. Microbiol.* **62**, 3005–3010.

Bianciotto, V., Lumini, E., Lanfranco, L., Minerdi, D., Bonfante, P., and Perotto, S. (2000). Detection and identification of bacterial endosymbionts in arbuscular mycorrhizal fungi belonging to Gigasporaceae. *Appl. Environ. Microbiol.* **66**, 4503–4509.

Bianciotto, V., Lumini, E., Bonfante, P., and Vandamme, P. (2003). "*Candidatus* Glomeribacter gigasporarum," an endosymbiont of arbuscular mycorrhizal fungi *Int. J. Systematics* **53**, 121–124.

Bills, S., Podila, G.K., and Hiremath, S. (1999). Genetic engineering of an ectomycorrhizal fungus *Laccaria bicolor* for use as a biological control agent. *Mycologia* **91**, 237–242.

Bills, S.N., Richter, D.L., and Horvath, B. (1995). Genetic transformation of the ectomycorrhizal fungus *Paxillus involutus* by particle bombardment. *Mycol. Res.* **99**, 557–561.

Bird, A. (2002). DNA methylation patterns and epigenetic memory. *Genes Dev.* **16**, 6–21.

Blancaflor, E.B., Zhao, L.M., and Harrison, M. (2001). Microtubule organization in root cells of *Medicago truncatula* during development of an arbuscular mycorrhizal symbiosis with *Glomus versiforme. Protoplasma* **217**, 154–165.

Bogorad, L. (2000). Engineering chloroplasts: An alternative site for foreign genes, proteins, reactions and products. *Trends Biotechnol.* **18**, 257–263.

Bonfante, P. (2001). At the interface between mycorrhiza fungi and plants: The structural organization of cell wall, plasma membrane and cytoskeleton. In B. Hock (ed) *The Mycota IX, Fungal Associations*. Springer-Verlag, Berlin, pp. 45–61.

Broach, J.R., Li, Y.Y., Feldman, J., Jayaram, M., Abraham, J., Nasmyth, K.A., and Hicks, J.B. (1983). Localization and sequence analysis of yeast origins of DNA replication. *Cold Spring Harb. Symp. Quant. Biol.* **47**, 1165–1173.

Buée, M., Rossignol, M., Jauneau, A., Ranjeva, R., and Bécard, G. (2000). The pre-symbiotic growth of arbuscular mycorrhizal fungi is induced by a branching factor partially purified from plant root exudates. *Mol. Plant–Microbe Interact.* **13**, 693–698.

Bundock, P., den Dulk-Ras, A., Beijersbergen, A., and Hooykaas, P.J.J. (1995). Tran-kingdom T-DNA transfer from *Agrobacterium tumefaciens* to *Saccharomyces cerevisiae. EMBO J.* **14**, 3206–3214.

Bundock, P., van Attikum, H., den Dulk-Ras, A., and Hooykaas, P.J.J. (2002). Insertional mutagenesis in yeasts using T-DNA from *Agrobacterium tumefaciens. Yeast* **19**, 529–536.

Burmester, A. and Wöstemeyer, J. (1987). DNA sequence and functional analysis of an ARS-element from the zygomycete *Absidia glauca. Curr. Genet.* **12**, 599–603.

Butehorn, B., Gianinazzi-Pearson, V., and Franken, P. (1999). Quantification of β-tubulin RNA expression during asymbiotic and symbiotic development of the arbuscular mycorrhizal fungus *Glomus mosseae. Mycol. Res.* **103**, 360–364.

Chaure, P., Gurr, S.J., and Spanu, P. (2000). Stable transformation of *Erysiphe graminis*, an obligate biotrophic pathogen of barley. *Nat. Biotechnol.* **18**, 205–207.

Cooke, J.C., Gemma, J.N., and Koske, R.E. (1987). Observations of nuclei in vesicular–arbuscular mycorrhizal fungi. *Mycologia* **79**, 331–333.

Covert, S.F., Kapoor, P., Lee, M., Briley, A., and Nairn, C.J. (2001). *Agrobacterium tumefaciens*-mediated transformation of *Fusarium circinatum. Mycol. Res.* **105**, 259–265.

Daniell, H., Khan, M.S., and Allison, L. (2002). Milestones in chloroplast genetic engineering: An environmentally friendly era in biotechnology. *Trends Plant Sci.* **7**, 84–91.

de Groot, M.J., Bundock, P., Hooykaas, P.J., and Beijersbergen, A.G. (1998). *Agrobacterium tumefaciens*-mediated transformation of filamentous fungi. *Nat. Biotechnol.* **16**, 839–842.

Delp, G., Smith, S.E., and Barker, S.J. (2000). Isolation by differential display of three partial cDNAs potentially coding for proteins from the VA mycorrhizal *Glomus intraradices. Mycol. Res.* **104**, 293–300.

Durand, R., Rascle, C., Fischer, M., and Fere, M. (1997). Transient expression of the β-glucuronidase gene after biolistic transformation of the anaerobic fungus *Neocallimastix frontalis. Curr. Genet.* **31**, 158–161.

Feldmann, F. and Grotkass, C. (2002). Directed inoculum production and predictable symbiotic efficiency. In S. Gianinazzi, H. Schüepp, J.M. Barea, and K. Haselwandter (eds.) *Mycorrhiza Technology in Agriculture: From Genes to Bioproducts* Birkhaüser Verlag, Basel, pp. 261–279.

Ferrol, N., Barea, J.M., and Azcón-Aguilar, C. (2000). The plasma membrane H^+-ATPase gene family in the arbuscular mycorrhizal fungus *Glomus mosseae. Curr. Genet.* **37**, 112–118.

Ferrol, N., Barea, J.M., and Azcón-Aguilar, C. (2002). Mechanisms of nutrient transport across interfaces in arbuscular mycorrhizas. *Plant Soil* **244**, 231–237.

Forbes, P.J., Millam, S., Hooker, J.E., and Harrier, L.A. (1998). Transformation of the arbuscular mycorrhizal fungus *Gigaspora rosea* Nicolson & Schenck using particle bombardment. *Mycol. Res.* **102**, 497–501.

Franken, P., Lapopin, L., Meyer-Gauen, G., and Gianinazzi-Pearson, V. (1997). RNA accumulation and genes expressed in spores of the arbuscular mycorrhizal fungus *Gigaspora rosea. Mycologia* **89**, 295–299.

Frommer, W.B. and Ninnemann, O. (1995). Heterologous expression genes in bacterial, fungal, animal, and plant cells. *Ann. Rev. Plant Physiol. Plant Mol. Biol.* **46**, 419–444.

Genomyca: An EU project on genes and genetic engineering for arbuscular mycorrhiza technology and applications in sustainable agriculture (QLK5-CT-2000-01319) (2001). Dijon (27 August, 2002); *http://www.dijon.inra.fr/bbceipm/genomyca/*

Gianinazzi, S., Schüepp, H., Barea, J.M., and Haselwandter, K. (2002). *Mycorrhiza Technology in Agriculture: From Genes to Bioproducts*. Birkhaüser Verlag, Basel.

Gianinazzi-Pearson, V., van Tuinen, D., Dumas-Gaudot, E., and Dulieu, H. (2001). Exploring the genome of glomalean fungi. In B. Hock (ed.) *The Mycota IX Fungal Associations.* Springer-Verlag, Berlin, pp. 3–17.

Giovannetti, M. and Sbrana, C. (1998). Meeting a non-host: The behaviour of AM fungi. Mycorrhiza **8**, 123–130.

Giovannetti, M., Sbrana, C., Avio, L., Citernesi, A.S., and Logi, C. (1993). Differential hyphal morphogenesis in arbuscular mycorrhizal fungi during pre-infection stages. *New Phytologist* **125**, 587–594.

Gollotte, A., Chatagnier, O., Arnould, C., van Tuinen, D., Gianinazzi, S., and Gianinazzi-Pearson, V. (2002). Identification of transposon-like sequences in the genome of fungi belonging to the Glomales. Proceedings of the 7th International Mycological Congress, 11–17 August, Oslo, Norway, p. 335.

Gouka, R.J., Gerk, C., Hooykaas, P.J.J., Bundock, P., Musters, W., Verrips, C.T., and de Groot, M.J.A. (1999). Transformation of *Aspergillus awamori* by *Agrobacterium tumefaciens*-mediated homologous recombination. *Nat. Biotechnol.* **17**, 598–601.

Grant, D.M., Lambowitz, A.M., Rambosek, J.A., and Kinsey, J.A. (1984). Transformation of *Neurospora crassa* with recombinant plasmids containing the cloned glutamate dehydrogenase (am) gene: Evidence for autonomous replication of the transforming plasmid. *Mol. Cell. Biol.* **4**, 2041–2051.

Harrier, L.A. (2001). Isolation and sequence analysis of the arbuscular mycorrhizal fungus *Glomus mosseae* (Nicol. & Gerd.) Gerdemann & Trappe. 3-phosphoglycerate kinase (PGK) gene promoter region. *DNA Seq.* **11**, 463–473.

Harrier, L.A. and Millam, S. (2001). Biolistic transformation of arbuscular mycorrhizal fungi—Progress and perspectives. *Mol. Biotechnol.* **18**, 25–33.

Harrier, L.A., Wright, F., and Hooker, J.E. (1998). Isolation of the 3-phosphoglycerate kinase gene of the arbuscular mycorrhizal fungus *Glomus mosseae* (Nicol. & Gerd.) Gerdemann & Trappe. *Curr. Genet.* **34**, 386–392.

Harrier, L.A., Millam, S., and Franken, P. (2002). Biolistic transformation of AM fungi: Advances and applications. In S. Gianinazzi, H. Schüepp, J.M. Barea, and K. Haselwandter (eds.) *Mycorrhiza Technology in Agriculture: From Genes to Bioproducts* Birkhaüser Verlag, Basel, pp. 59–71.

Harrison, M. (1999). Molecular and cellular aspects of the arbuscular mycorrhizal symbiosis. *Annu. Rev. Plant Physiol. Plant Mol. Biol.* **50**, 361–389.

Harrison, M.J. and van Buuren, M.L. (1995). A phosphate transporter from the mycorrhizal fungus *Glomus versiforme. Nature* **378**, 626–629.

Hawkins, H.J. and George, E. (1999). Effect of nitrogen status on the contribution of arbuscular mycorrhizal hyphae to plant nitrogen uptake. *Physiol. Plant.* **105**, 694–700.

Hieter, P., Pridmore, D., Hegemann, J.H., Thomas, M., Davis, R.W., and Philippsen, P. (1985). Functional selection and analysis of yeast centromeric DNA. *Cell* **42**, 913–921.

Hijri, M., Hosny, M., van Tuinen, D., and Dulieu, H. (1998). Intraspecific ITS polymorphism in *Scutellospora castanea* (Glomales, Zygomycetes) is structured within multinucleate spores. *Fungal Genet. Biol.* **26**, 141–151.

Hilber, U.W., Bodmer, M., Smith, F.D., and Köller, W. (1994). Biolistic transformation of conidia of *Botryotinia fuckeliana. Curr. Genet.* **25**, 124–127.

Hooykaas, P.J.J., Roobol, C., and Schillerpoort, R.A. (1979). Regulation of the transfer of Ti-plasmids of *Agrobacterium tumefaciens. J. Gen. Microbiol.* **110**, 99–109.

Hosny, M. (1997). Tailles et contenus en (G+C) des Glomales. Complexité du génome et polymorphisme des ADN ribosomiques chez une espèce-modèle. *Doctoral Dissertation.* University of Burgundy, France.

Hosny, M., de Barros, J.P.P., Gianinazzi-Pearson, V., and Dulieu, H. (1997). Base composition of DNA from glomalean fungi: High amounts of methylated cytosine. *Fungal Genet. Biol.* **22**, 103–111.

Hosny, M., Gianinazzi-Pearson, V., and Dulieu, H. (1998). Nuclear DNA content of 11 fungal species in Glomales. *Genome* **41**, 422–428.

Hosny, M., Hijri, M., Passerieux, E., and Dulieu, H. (1999a). rDNA units are highly polymorphic in *Scutellospora castanea* (Glomales, Zygomycetes). *Gene* **226**, 61–71.

Hosny, M., van Tuinen, D., Jacquin, F., Fuller, P., Zhao, B., Gianinazzi-Pearson ,V., and Franken, P. (1999b). Arbuscular mycorrhizal fungi and bacteria: How to construct prokaryotic DNA-free genomic libraries from the Glomales. *FEMS Microbiol. Lett.* **170**, 425–430.

IBG/BEG: International Bank of Glomales (1995). Kent, UK (January 31, 2002); *http://www.ukc.ac.uk/bio/beg/*

INVAM: International Culture Collection of Arbuscular and Vesicular–Arbuscular Mycorrhizal Fungi (1996). Morgantown, USA (June 2001); *http://invam.caf.wvu.edu/*

Johansen, A., Jacobsen, I., and Jensen, E.S. (1992). Hyphal transport of ^{15}N-labelled nitrogen by a vesicular-arbuscular mycorrhizal fungus and its effect on depletion of inorganic soil N. *New Phytologist* **122**, 281–288.

Kaldorf, M., Schmelzer, E., and Bothe, H. (1998). Expression of maize and fungal nitrate reductase genes in arbuscular mycorrhiza. *Mol. Plant–Microbe Interact.* **11**, 439–448.

Kaldorf, M., Zimer, W., and Bothe, H. (1994). Genetic evidence for the occurrence of assimilatory nitrate reductase in arbuscular mycorrhizal and other fungi. *Mycorrhiza* **5**, 23–28.

Kistler, H.C. and Benny, U. (1992). Autonomous replicating plasmids and chromosome rearrangement during transformation of *Nectria haematocca. Gene* **117**, 81–89.

Kuhn, G., Hijri, M., and Sanders, I.R. (2001). Evidence for the evolution of multiple genomes in arbuscular mycorrhizal fungi. *Nature* **414**, 745–748.

Lammers, P.J., Jun, J., Abubaker, J., Arreola, R., Gopalan, A., Bago, B., Hernandez-Sebastia, C., Allen, J.W. *et al.* (2001). The glyoxylate cycle in an arbuscular mycorrhizal fungus. Carbon flux and gene expression. *Plant Physiol.* **127**, 1287–1298.

Lanfranco, L., Delpero, M., and Bonfante, P. (1999a). Intrasporal variability of ribosomal sequences in the endomycorrhizal fungus *Gigaspora margarita*. *Mol. Ecol.* **8**, 37–45.

Lanfranco, L., Garnero, L., and Bonfante, P. (1999b). Chitin synthase genes in the arbuscular mycorrhizal fungus *Glomus versiforme*: Full sequence of a gene encoding a class IV chitin synthase. *FEMS Microbiol. Lett.* **170**, 59–67.

Lanfranco, L., Vallino, M., and Bonfante, P. (1999c). Expression of chitin synthase genes in the arbuscular mycorrhizal fungus *Gigaspora margarita*. *New Phytologist* **142**, 347–354.

Lanfranco, L., Bolchi, A., Cesale, R.E., Ottonello, S., and Bonfante, P. (2002). Differential expression of a metallothionein gene during the presymbiotic versus the symbiotic phase of an arbuscular mycorrhizal fungus. *Plant Physiol.* **130**, 58–67.

Lapopin, L., Gianinazzi-Pearson, V., and Franken, P. (1999). Comparative differential display analysis of arbuscular mycorrhiza in *Pisum sativum* and a mutant defective in late stage development. *Plant Mol. Biol.* **41**, 669–677.

Lee, M.G. and Nurse, P. (1987). Complementation used to clone a human homologue of the fission yeast cell cycle control gene *cdc2*. *Nature* **327**, 31–35.

Leyval, C., Turnau, K., and Haselwandter, K. (1997). Effects of heavy metal pollution on mycorrhizal colonization and function: Physiological, ecological and applied aspects. *Mycorrhiza* **7**, 139–153.

Long, D.M., Smidansky, E.D., Archer, A.J., and Strobel, G.A. (1998). *In vivo* addition of telomeric repeats to foreign DNA generates extrachromosomal DNAs in the taxol-producing fungus *Pestalotiopsis microspora*. *Fungal Genet. Biol.* **24**, 335–344.

Lorito, M., Hayes, C.K., di Pietro, A., and Harman, G.E. (1993). Biolistic transformation of *Trichoderma harzianum* and *Gliocadium virens* using plasmid and genomic DNA. *Curr. Genet.* **24**, 349–356.

Lloyd-MacGilp, S.A., Chambers, S.M., Dodd, J.C., Fitter, A.H., Walker, C., and Young, J.P.W. (1996). Diversity of the ribosomal internal transcribed spacers within and among isolates of *Glomus mosseae* and related mycorrhizal fungi. *New Phytologist* **133**, 103–111.

Mackenzie, D.A., Wongwathanarat, P., Carter, A.T., and Archer, D.B. (2000). Isolation and use of a homologous histone H4 promoter and a ribosomal DNA region in a transformation vector for the oil producing fungus *Mortierella alpina*. *Appl. Environ. Microbiol.* **66**, 4655–4661.

Maldonado-Mendoza, I.E., Dewbre, G.R., and Harrison, M.J. (2001). A phosphate transporter gene from the extra-radical mycelium of an arbuscular mycorrhizal fungus *Glomus intraradices* is regulated in response to phosphate in the environment. *Mol. Plant–Microbe Interact.* **14**, 1140–1148.

Maldonado-Mendoza, I.E., Dewbre, G.R., van Buuren, M.L., Versaw, W.K., and Harrison, M.J. (2002). Methods to estimate the proportion of plant and fungal RNA in an arbuscular mycorrhiza. *Mycorrhiza* **12**, 67–74.

Malonek, S. and Meinhardt, F. (2001). *Agrobacterium tumefaciens*-mediated genetic transformation of the phytopathogenic ascomycete *Calonectria morganii*. *Curr. Genet.* **40**, 152–155.

Mikosch, T.S.P., Lavrijssen, B., Sonnenberg, A.S.M., and van Griensven, L.J.L.D. (2001). Transformation of the cultivated mushroom *Agaricus bisporus* (Lange) using T-DNA from *Agrobacterium tumefaciens*. *Mycol. Res.* **39**, 35–39.

Minerdi, D., Bianciotto, V., and Bonfante, P. (2002a). Endosymbiotic bacteria in mycorrhizal fungi: From their morphology to genomic sequences. *Plant Soil* **244**, 211–219.

Minerdi, D., Fani, R., and Bonfante, P. (2002b). Identification and evolutionary analysis of putative cytoplasmic McpA-like protein in a bacterial strain living in symbiosis with a mycorrhizal fungus. *J. Mol. Evol.* **54**, 815–824.

Minerdi, D., Fani, R., Gallo, R., Boarino, A., and Bonfante, P. (2001). Nitrogen fixation genes in an endosymbiotic *Burkholderia* strain. *Appl. Environ. Microbiol.* **67**, 725–732.

Minet, M., Dufour, M.-E., and Lacroute, F. (1992). Complementation of *Saccharomyces cerevisiae* auxotrophic mutants by *Arabidopsis thaliana* cDNAs. *Plant J.* **2**, 417–422.

Moran, N.A. and Wernegreen, J.J. (2000). Lifestyle evolution in symbiotic bacteria: Insights from genomics. *Tree* **15**, 321–326.

Morton, J.B. and Benny, G.L. (1990). Revised classification of arbuscular mycorrhizal fungi (Zygomycetes): A new order, Glomales, two new suborders, Glomineae and Gigasporineae, and two new families, Acaulosporaceae and Gigasporaceae, with an emendation of Glomaceae. *Mycotaxon* **37**, 471–491.

Mullins, E.D. and Kang, S. (2001). Transformation: A tool for studying fungal pathogens of plants. *Cell. Mol. Life Sci.* **58**, 2043–2052.

Mullins, E.D., Chen, X., Romaine, P., Raina, R., Geiser, D.M., and Kang, S. (2001). *Agrobacterium*-mediated transformation of *Fusarium oxysporum*: An efficient tool for insertional mutagenesis and gene transfer. *Phytopathology* **91**, 173–180.

Murrin, F., Holtby, J., Nolan, R.A., and Davidson, W.S. (1986). The genome of the *Entomophaga aulicae* (Entomophthorales, Zygomycetes): Base composition and size. *Exp. Mycol.* **10**, 67–75.

Pardo, A.G., Hanif, M., Raudaskoski, M., and Gorfer, M. (2002). Genetic transformation of ectomycorrhizal fungi mediated by *Agrobacterium tumefaciens. Mycol. Res.* **106**, 132–137.

Powell, W.A. and Kistler, H.C. (1990). *In vivo* rearrangement of foreign DNA by *Fusarium oxysporum* produces linear self-replication plasmids. *J. Bacteriol.* **172**, 3163–3171.

Redecker, D., Kodner, R., and Graham, L.E. (2000). Glomalean fungi from the Ordovician. *Science* **289**, 1920–1921.

Remy, W., Taylor, T.N., Hass, H., and Kerp, H. (1994). Four hundred-million-year old vesicular arbuscular mycorrhizae. *Proc. Natl. Acad. Sci. USA* **91**, 11841–11843.

Requena, N., Füller, P., and Franken, P. (1999). Molecular characterisation of *GmFOX2*, an evolutionary highly conserved gene from the mycorrhizal fungus *Glomus mosseae*, down-regulated during interaction with rhizobacteria. *Mol. Plant–Microbe Interact.* **12**, 934–942.

Requena, N., Mann, P., and Franken, P. (2000). A homologue of the cell-cycle check-point TOR2 from *Saccharomyces cerevisiae* exists in the arbuscular mycorrhizal fungus *Glomus mosseae. Protoplasma* **212**, 89–98.

Requena, N., Mann, P., Hampp, R., and Franken, P. (2002). Early developmentally regulated genes in the arbuscular mycorrhizal fungus *Glomus mosseae*: Identification of *GmGIN*1, a novel gene with homology to the C-terminus of metazoan hedgehog proteins. *Plant Soil* **244**, 129–139.

Revuelta, J.L. and Jayaram, M. (1986). Transformation of *Phycomyces blakesleeanus* to G-418 resistance by an autonomously replicating plasmid. *Proc. Natl. Acad. Sci. USA* **83**, 7344–7347.

Rho, H.S., Kang, S., and Lee, Y.H. (2001). *Agrobacterium tumefaciens*-mediated transformation of the plant pathogenic fungus *Magnaporthe grisea. Mol. Cells* **12**, 407–411.

Rhody, D., Stommel, M., Roeder, C., Mann, P., and Franken, P. (2003). Differential RNA accumulation of two β-tubulin gene in arbuscular mycorrhizal fungi. *Mycorrhiza* **13**, 137–142.

Rose, M.D. and Broach, J.R. (1991). Cloning genes by complementation in yeast. *Methods Enzymol.* **194**, 195–320.

Rosendahl, S. and Taylor, J.W. (1997). Development of multiple genetic markers for studies of genetic variation in arbuscular mycorrhizal fungi using AFLPTM. *Mol. Ecol.* **6**, 821–829.

Ruiz-Díez, B. and Martínez-Suárez, J.V. (1999). Electrotransformation of the human pathogenic fungus *Scedosporium prolificans* mediated by repetitive rDNA sequences. *FEMS Immunonol. Med. Microbiol.* **25**, 275–282.

Ruiz-Lozano, J.M. and Bonfante, P. (1999). Identification of a putative P-transporter operon in the genome of a *Burkholderia* strain living inside the arbuscular mycorrhizal fungus *Gigaspora margarita. J. Bacteriol.* **181**, 4106–4109.

Ruiz-Lozano, J.M. and Bonfante, P. (2000). Intracellular *Burkholderia* of the arbuscular mycorrhizal fungus *Gigaspora margarita* possesses the *vacB* gene, which is involved in host cell colonization by bacteria. *Microb. Ecol.* **39**, 137–144.

Ruiz-Lozano, J.M., Collados, C., Porcel, R., Azcon, R., and Barea, J.M. (2002). Identification of a cDNA from the arbuscular mycorrhizal fungus *Glomus intraradices* that is expressed during mycorrhizal symbiosis and up-regulated by N fertilization. *Mol. Plant–Microbe Interact.* **15**, 360–367.

Russell, P.J., Rodland, K.D., Rachlin, E.M., and McCloskey, J.A. (1987). Differential DNA methylation during the vegetative cycle of *Neurospora crassa. J. Bacteriol.* **169**, 2902–2905.

Sanders, I.R., Alt, M., Groppe, K., Boller, T., and Wiemken, A. (1995). Identification of ribosomal DNA polymorphisms among and within spores of the Glomales: Application to studies on the genetic diversity of arbuscular mycorrhizal fungal communities. *New Phytologist* **130,** 419–427.

Sawaki, H. and Saito, M. (2001). Expressed genes in the extraradical hyphae of an arbuscular mycorrhizal fungus, *Glomus intraradices*, in the symbiotic phase. *FEMS Microbiol. Lett.* **195**, 109–113.

Scannerini, S. and Bonfante, P. (1991). Bacteria and bacteria like objects in endomycorrhizal fungi (Glomaceae). In L. Margulis and R. Fester (eds.) *Symbiosis as Source of Evolutionary Innovation: Speciation and Morfogenesis*. The MIT Press, Cambridge, MA, USA, pp. 273–287.

Schilde, C., Wöstemeyer, J., and Burmester, A. (2001). Green fluorescent protein as a reporter for gene expression in the mucoralean fungus *Absidia glauca. Arch Microbiol.* **175**, 1–7.

Schüßler, A., Schwarzott, D., and Walker, C. (2001). A new fungal phylum, the Glomeromycota: Phylogeny and evolution. *Mycol. Res.* **105**, 1413–1421.

Sikdar, S.R., Serino, G., Chaudhuri, S., and Malinga, P. (1998). Plastid transformation in *Arabidopsis thaliana. Plant Cell Rep.* **18**, 20–24.

Sidorov, V.A., Kasten, D., Pang, S.Z., Hajdukiewicz, P.T.J., Staub, J.M., and Nehra, N.S. (1999). Stable chloroplast transformation in potato: Use of green fluorescent protein as a plastid marker. *Plant J.* **19**, 209–216.

Singer, M.J., Marcotte, B.A., and Selker, E.U. (1995). DNA methylation associated with repeat-induced point mutation in *Neurospora crassa. Mol. Cell. Biol.* **15**, 5586–97.

Smith, S.E. and Read, D.J. (1997). *Mycorrhizal symbiosis*. Academic Press, San Diego.

Stahl, U., Tudzynski, P., Kuck, C., and Esser, K. (1982). Replication and expression of a bacterial-mitochondrial plasmid in the fungus *Podospora anserina. Proc. Natl. Acad. Sci. USA* **79**, 3641–3645.

Stommel, M., Mann, P., and Franken, P. (2001). Construction and analysis of an EST library using RNA from activated spores of the arbuscular mycorrhizal fungus *Gigaspora rosea. Mycorrhiza* **10**, 281–285.

Storck, R. and Alexopoulos, C.J. (1970). Deoxyribonucleic acid of fungi. *Bacteriol. Rev.* **34**, 126–154.

St. Leger, R.J., Shimizu, S., Joshi, L., Bidochka, M.J., and Roberts, D.W. (1995). Co-transformation of *Metarhizium anisopliae* by electroporation or using the gene gun to produce stable GUS transformants. *FEMS Microbiol. Lett.* **131**, 289–294.

Tamasloukht, M., Séjalon-Delmas, N., Kluever, A., Jauneau, A., Roux, C., Bécard, G., and Franken, P. (2003). Root factor induce mitochondrial-related-gene expression and fungal respiration during the developmental switch from asymbiosis to presymbiosis in the arbuscular mycorrhizal fungus *Gigaspora rosea. Plant Physiol.* **131**, 1468–1478.

Timonen, S., Smith, F.A., and Smith, S.E. (2001). Microtubules of the mycorrhizal fungus *Glomus intraradices* in symbiosis with tomato roots. *Can. J. Bot.* **79**, 307–313.

Trouvelot, S., van Tuinen, D., Hijri, M., and Gianinazzi-Pearson, V. (1999). Visualization of DNA loci in interphasic nuclei of glomalean fungi by fluorescence *in situ* hybridization. *Mycorrhiza* **8**, 203–208.

Ubalijoro, E., Hamel, C., McClung, C.R., and Smith, D.L. (2001). Detection of chitin synthase class I and II type sequences in six different arbuscular mycorrhizal fungi and gene expression in *Glomus intraradices. Mycol. Res.* **105**, 470–476.

van Buuren, M., Lanfranco, L., Longato, S., Minerdi, D., Harrison, M.J., and Bonfante, P. (1999). Construction and characterization of genomic libraries of two endomycorrhizal fungi: *Glomus versiforme* and *Gigaspora margarita. Mycol. Res.* **103**, 955–960.

van der Heijden, M.G.A., Boller, T., Wiemken, A., and Sanders, I.R. (1998). Different arbuscular mycorrhiza fungal species are potential determinants of plant community structure. *Ecology* **79**, 2082–2091.

van Heeswijck, R. (1986). Autonomous replication of plasmid in *Mucor* transformants. *Carlsberg Res. Commun.* **51**, 433–443.

Vandenkoornhuyse, P., Leyval, C., and Bonnin, I. (2001). High genetic diversity in arbuscular mycorrhizal fungi: Evidence for recombination events. *Heredity* **87**, 243–253.

Viera, A. and Glenn, M.G. (1990). DNA content of vesicular–arbuscular mycorrhizal fungal spores. *Mycologia* **82**, 263–267.

Woods, J.P. and Goldman, W.E. (1992). *In vivo* generation of linear plasmids with addition of telomeric sequences by *Histoplasma capsulatum. Mol. Microbiol.* **6**, 3603–3610.

Woods, J.P. and Goldman, W.E. (1993). Autonomous replication of foreign DNA in *Histoplasma capsulatum*: Role of native telomeric sequences. *J. Bacteriol.* **175**, 636–641.

Yanai, K., Horiuchi, H., Takagi, M., and Yano, K. (1991). Transformation of *Rhizopus niveus* using a bacterial blasticidin-S resistance gene as a dominant selectable marker. *Curr. Genet.* **19**, 221–226.

Yu, J. and Cole, G.T. (1998). Biolistic transformation of the human pathogenic fungus *Coccidioides immitis. J. Microbiol. Methods* **33**, 129–141.

Zézé, A., Hosny, M., Gianinazzi-Pearson, V., and Dulieu, H. (1996). Characterization of a highly repeated DNA sequence (SC1) from the arbuscular mycorrhizal fungus *Scutellospora castanea* and its detection *in planta. Appl. Environ. Microbiol.* **62**, 2443–2448.

Zézé, A., Hosny, M., van Tuinen, D., Gianinazzi-Pearson, V., and Dulieu, H. (1999). MYCDIRE, a dispersed repetitive DNA element in arbuscular mycorrhizal fungi. *Mycol. Res.* **103**, 572–576.

Zhang, A., Lu, P., Dahl-Roshak, A.M., Paress, P.S., Kennedy, S., Tkacz, J.S., and An, Z. (2003). Efficient disruption of a polyketide synthase gene (*pks*1) required for melanin synthesis through *Agrobacterium*-mediated transformation of *Glarea lozoyensis. Mol. Genet. Genomics* **268,** 645–655.

Index

O

P

The manufacturer's authorised representative in the EU is Springer Nature Customer Service Centre GmbH, Europaplatz 3, 69115 Heidelberg, Germany. If you have any concerns regarding our products, please contact ProductSafety@springernature.com

Printed and bound by CPI Group (UK) Ltd, Croydon, CR0 4YY

23/04/2026

02095594-0012